CRANIAL MAGNETIC RESONANCE IMAGING

CRANIAL MAGNETIC RESONANCE IMAGING

Allen D. Elster, M.D.

Assistant Professor
Department of Radiology
Bowman Gray School of Medicine of
Wake Forest University
Winston-Salem, North Carolina

CHURCHILL LIVINGSTONE
New York, Edinburgh, London, Melbourne 1988

Library of Congress Cataloging-in-Publication Data

Elster, Allen D.
Cranial magnetic resonance imaging.

Includes bibliographies and index.
1. Magnetic resonance imaging. 2. Brain—Diseases—Diagnosis. 3. Skull—Diseases—Diagnosis. I. Title.
[DNLM: 1. Brain—pathology. 2. Brain Diseases—diagnosis.
3. Nuclear Magnetic Resonance—diagnostic use.
WL 141 E49c]
RC386.6.M34E45 1988 617'.510757 87-22428
ISBN 0-443-08542-0

Distributed in the United Kingdom by Churchill Livingstone, Robert Stevenson House, 1–3 Baxter's Place, Leith Walk, Edinburgh EH1 3AF, and by associated companies, branches, and representatives throughout the world.

Accurate indications, adverse reactions, and dosage schedules for drugs are provided in this book, but it is possible that they may change. The reader is urged to review the package information data of the manufacturers of the medications mentioned.

Acquisitions Editor: *Robert A. Hurley*
Copy Editor: *Ozzievelt Owens*
Production Designer: *Angela Cirnigliaro*
Production Supervisor: *Sharon Tuder*

Printed in the United States of America

First published in 1988

To Jeanine and Allen II,
sine quibus non

Preface

When I began writing my first book and papers on magnetic resonance imaging (MRI), it was often necessary to provide lengthy explanations justifying the role and need for another expensive new technology. Since that time, worldwide experience has proved MRI's great value in the diagnosis of cranial disease. MRI has now been shown useful in so many situations that there is little need to justify its presence in the imaging armamentarium. MRI is clearly a proved modality, and one that will likely occupy an increasing share of the imaging market in the near future.

At the time of this writing there is no available textbook devoted exclusively to cranial MRI. So many journal articles have appeared in recent years, however, that it is difficult to distinguish established fact from opinion and uncorroborated data. There is a real need to try to consolidate what has been learned and substantiated during the first several years of cranial MRI. While this book will try to present only well accepted "facts," there is no doubt that facts will change as knowledge improves. Such is the nature of young science, new technology, and textbooks.

A book on cranial MRI is not merely a book on cranial computed tomography with different pictures. The MRI book must place considerably more emphasis on themes relating to biophysics and imaging strategies. Some topics, such as craniofacial fractures and anomalies of the middle ear, are not well suited for MRI study and will receive only abbreviated attention. On the other hand, disorders such as congenital malformation, posterior fossa lesions, brain stem anomalies, cerebral degenerations, and white matter diseases are very well evaluated by MRI and will be given extensive consideration. A number of very unusual disorders will also be presented, because patients referred for MRI evaluation are often diagnostic dilemmas, having either rare diseases or rare manifestations of common diseases. Be prepared, therefore, to encounter the bizarre. This is both the charm and the challenge of cranial MRI.

Allen D. Elster, M.D.

Acknowledgements

There are many people who have contributed either to this book or to my training as a neuroradiologist, to whom I am very grateful. First and foremost, I would like to thank Dr. Arnold Goldman and Dr. Stanley Handel of NMR-Associates in Houston. These two men gave me my start in MRI. They allowed me to study under them in the early days when the field was young and MR scanners were hard to find. They permitted me to use a number of their cases both in my first book and in this one. I hope that both books will serve as tributes to their pioneering work and service to the Houston medical community.

At Bowman Gray School of Medicine I must certainly thank the three fine men who trained me in the "classical" techniques of neuroradiology: Dr. Dixon Moody, Dr. Wayne Laster, and Dr. Marshall Ball. I am continually amazed by their wealth of knowledge in the neurologic diseases. Their teaching and comradery has been a tremendous resource to me.

I must also thank our Chairman, Dr. C. Douglas Maynard, for building such a fine department and having the wisdom to develop MRI in the early years. He has fostered a commitment to learning, research, and superb patient care, while maintaining a most pleasant atmosphere in which to work. This is a difficult combination of tasks to achieve. I know I reflect the opinion of many others on the staff when I say that there is probably no finer radiology chairman or person than Doug Maynard.

Finally, I must thank my wonderful wife Jeanine for helping and supporting me in so many ways during the writing of this book. Editing, typing, and organizing my notes was a labor of love; she did all this and more, maintaining our happy home and caring for little Allen. Such is her nature. Such is the essence of a true companion.

Contents

1

Instrumentation and Physical Principles

In the 1970s computed tomography (CT) revolutionized medical imaging, bringing new insights into the anatomic basis and natural history of many diseases. In the 1980s, magnetic resonance (MR) has carried the revolution a step further, and future ramifications of this new technology are not yet known. Exciting prospects for MR include the ability to perform spectroscopy, flow measurements, physiologic monitoring, and biochemical characterization of tissues. Even without these future advances, magnetic resonance imaging (MRI) has already proven itself to be a powerful and beneficial modality. This seems especially true in regard to the central nervous system, where MRI has had a long record of established success.

In the future, with more emphasis being placed on cost containment and efficiency in diagnosis, MRI must continue to prove its value relative to CT. CT is a mature technology, and a great deal is known concerning the CT appearance of various diseases. By contrast, MR is a new technology and relatively little is known about the MR appearances of normal and diseased tissue. It is the intent of this book to clarify the relative values of CT and MR in each disease and anatomic region. With further advances in knowledge, we may come to view CT and MR as complementary rather than competitive modalities.

RELATIVE VALUES OF CT AND MR IN CRANIAL IMAGING

The advantages and disadvantages of MRI relative to CT have been expounded upon in many texts.[1–6] At present, certain clinical situations seem to be better evaluated by one modality or the other (Table 1-1).[7] Future clinical experience and technical advancements will further define the precise roles of CT and MR in the imaging of cranial diseases.

A major advantage of MR is that MR images, unlike those of CT, do not contain beam-hardening or streak artifacts from metals or dense bone at the base of the skull. Accordingly, structures in the posterior fossa and brain stem are exquisitely well visualized by MR. MR has the ability to directly image the brain in multiple planes, including the coronal, sagittal, and oblique. Obtaining equivalent planes on CT images would require moving the patient into uncomfortable or impossible positions, or using crude and inaccurate reformatting techniques. This multiplanar capacity for MRI may aid in understanding anatomic relationships of certain tumors and thus help in the planning of surgical approach.

Perhaps the greatest advantage of MR relative to CT is that the MR signal depends upon a number

Table 1-1. Cranial Applications of CT and MR

Situations Where MRI is Usually Superior to CT
Multiple sclerosis and other demyelinating diseases
Subcortical arteriosclerotic encephalopathy
Posterior fossa lesions (tumors, infarcts)
Small extra-axial fluid collections
Patients with dense foreign materials (bullet fragments, wire sutures, dental fillings, etc.) that degrade CT image with streak artifacts
Certain metabolic and degenerative diseases
Patients with temporal lobe seizures
Situations Where CT is Usually Superior to MRI
Acute trauma, fractures
Acute hemorrhage, especially subarachnoid
Suspected meningioma
Uncooperative patients
Patients with cerebral aneurysm clips or cardiac pacemakers
Calcified lesions

of physical and chemical properties of the imaged tissue, including hydrogen density, flow, and molecular environment. Potentially it is possible to study each of these factors independently by MR. This makes MR very sensitive in the dection of certain diseases, particularly those producing demyelination or degeneration, where subtle biochemical changes preceed anatomic alterations. Conversely, CT is limited to the evaluation of a single tissue parameter, x-ray attenuation, and is relatively insensitive to early disease.

Despite the apparent advantages of MRI, CT remains the imaging modality of choice for a majority of cranial applications. First, CT is cheaper and faster than MRI, and in many situations supplies equivalent or superior information. CT has proven capabilities and well-defined limitations, while many aspects of MRI remain unclear. CT is much better than MR in detecting small areas of calcification within tissues, an important feature in diagnosing and differentiating certain tumors. Cortical bone is also better visualized by CT, making CT superior to MR in acute trauma. Additionally, acute subarachnoid and intraparenchymal hemorrhages may be impossible to detect by MR, again supporting the use of CT in most acute situations. Finally, until intravenous MR contrast material becomes widely available, contrast-enhanced CT will retain an advantage over MRI in delineation of the disrupted blood-brain barrier.

CONTRAINDICATIONS FOR MR SCANNING

Persons with cardiac pacemakers should be absolutely excluded from the MR imaging suite, preferably beyond a 10 G fringe field limit.[8] Several disasterous interactions between pacemakers and magnetic fields may potentially occur. First, the static component of an MR field is capable of closing the reed switch of certain pacemakers, thus converting their mode of operation from demand to asynchrony. Second, time varying components of the magnetic field may induce currents within the cardiac pacing wires. These currents can confuse or inhibit a demand pacemaker if the induced pulses are mistakenly interpreted as cardiac activity. Currents in pacing wires can potentially precipitate a fatal cardiac arrhythmia as well. Finally, any of the numerous small electronic components within a pacemaker can be caused to fail or malfunction when placed in powerful or rapdily changing electromagnetic fields. The dangers to pacemaker recipients apply to patients being scanned as well as to MR personnel.

The presence of a cerebral aneurysm clip has long been considered an absolute contraindication for MR scanning.[9] Some aneurysm clips contain ferromagnetic material (martensite) and may move significantly in strong magnetic fields.[10] This motion could potentially shear the clip from an aneurysm neck or cause the clip to press against a sensitive structure in the brain. However, recent research suggests that several commercially available clips are nonferromagnetic and will not move during MR scanning (Table 1-2).[9–11] It would appear, therefore, that if one were absolutely certain about the composition of a certain nonferromagnetic clip, a patient having such a clip could be safely scanned. Additionally, it should be noted that patients with tantalum hemostatic clips, tantalum mesh, craniotomy wires, stapedectomy prostheses, skin staples, and ventriculostomy devices may be safely imaged, and the presence of any of these should not contraindicate MR scanning.[9,12,12a] We routinely scan patients with cochlear implants and neurostimulator wires, though isolated cases of patient discomfort with these devices have been reported.

Table 1-2. Composition of Various Commercially Available Cerebral Surgical Clips[a]

Nonferromagnetic	*Ferromagnetic*
Heifetz (Eligiloy)	Downs Multipositionsl
Hemoclip	Drake (DR 14, DR 16, DR 24)
Sugita standard	Heifetz (17-7PH alloy)
Tantalum ligating	Housepain
Tantalum hemostatic	Kapp (curved and straight)
Vari-Angle McFadden	Mayfield
Yasargil	Pivot
	Schwartz
	Scoville
	Silver Olivecrona
	Smith
	Sundt-Kees Multi-Angle
	Vari-Angle

[a] Information compiled from References 9–12.

The presence of a prosthetic heart valve was originally considered an absolute contraindication for scanning.[13] However, most of the alloys used in prosthetic valves today are intrinsically nonmagnetic. Upon testing in fields up to 2.35 T most of these valves have been found to exhibit little deflection, torque, or heating.[14] The small deflections produced by the magnet seem particularly insignificant when compared to the torque exerted on a prosthesis by the beating heart itself. On the basis of in vivo testing, Soulen et al. have suggested that patients with present-day prosthetic heart valves may be safely imaged in present-day MR scanners.[14] A notable exception might be those with Pre 6000 series Starr-Edwards valves, which were found to undergo significant deflection in fields over 0.35 T.

Another contraindication to MR scanning would be in those patients having ferromagnetic foreign bodies implanted in critical locations within the body. In a single reported case, a sheet metal worker was permanently blinded during MR scanning when motion of an occult metallic foreign body in his eye induced gross vitreous hemorrhage.[15] It is thus conceivable that ferromagnetic shrapnel in other critical locations (such as the brain) could move with similar disasterous consequences.

MR INSTRUMENTATION

Three types of MR scanners are in use today, classified by mechanism of magnetic field production.[1] *Permanent magnet* systems (like the Fonar Beta 3000, Fig. 1-1) are composed of blocks of ferromagnetic material resulting in a giant horseshoelike magnet. Permanent magnets are cheap to operate because they require no power source or cooling system to maintain the magnetic field. However, they are very heavy, initially expensive, and are not capable of generating high strength magnetic fields. *Resistive magnet* systems (like the Picker 0.15 T Resistive unit, Fig. 1-2) produce their magnetic fields on the solenoid principle, that is, generation of a magnetic field by passing current through coils of wire wrapped around a ferromagnetic core. Although relatively cheap to build, these magnets have a high power consumption for daily operation and produce relatively weak magnetic fields. *Superconductive magnet* systems are the most popular in use today (exemplified by the Siemens Magnetom, Fig. 1-3). These machines generate their magnetic fields by passing current through coils

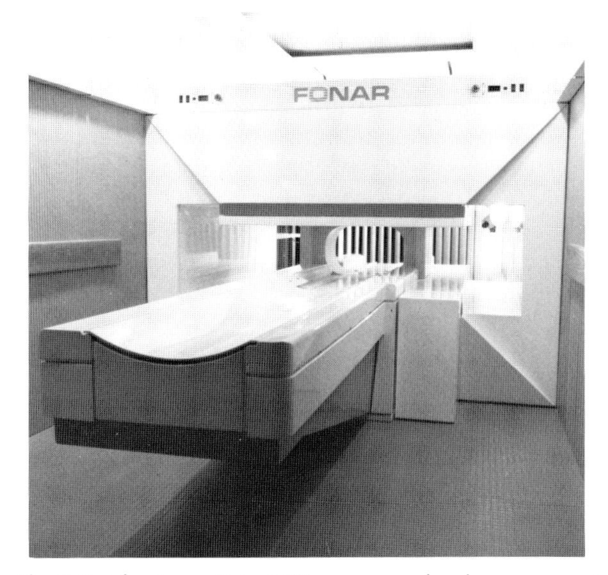

Fig. 1-1. The Fonar Beta 3000, an example of a permanent magnet MR imager that operates at 0.3 T. (Courtesy of Fonar Corporation, 110 Marcus Drive, Melville, NY.)

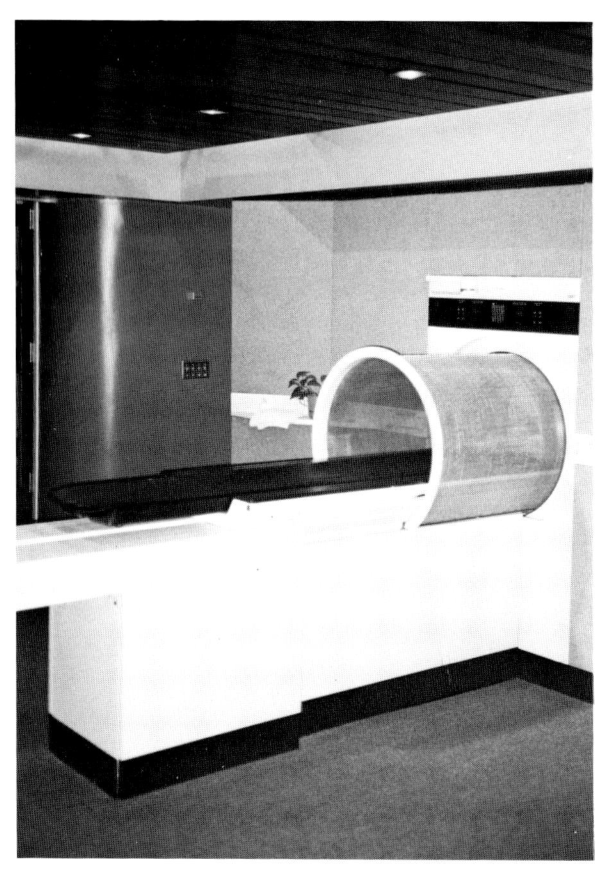

Fig. 1-2. A Picker 0.15 T resistive MR scanner that has operated faithfully at the author's institution since 1983.

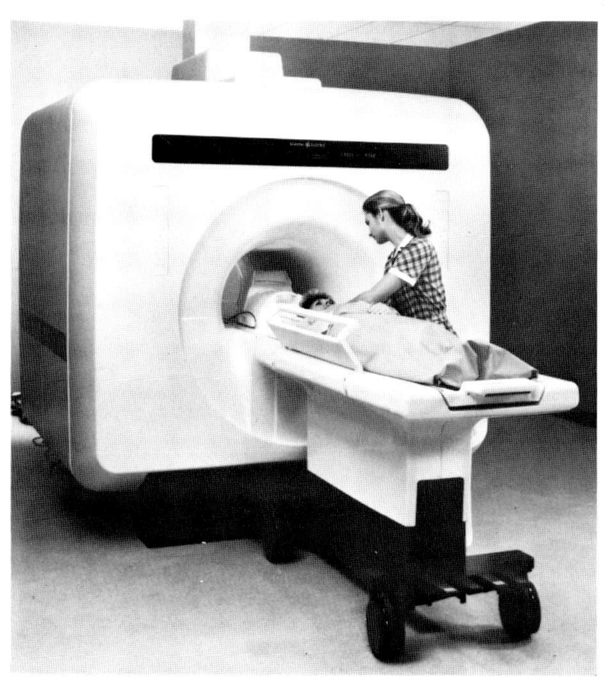

Fig. 1-3. A General Electric Signa superconductive scanner, operating at 1.5 T. (Courtesy of General Electric Medical Systems, Inc.)

of wire like the resistive magnets. However, liquid helium and nitrogen are used to cool the coils to very low temperatures where the physical phenomenon of superconduction occurs and high magnetic fields may be maintained with little power consumption. Superconductive magnet systems are the most expensive to build, but provide the strongest and most uniform magnetic fields currently available. Most of the images in this book were obtained on superconductive MR scanners.

The magnetic field of an MR scanner is measured in units called *Tesla* (T). An older unit of measurement, *Gauss* (G), is sometimes also used, with 1 T = 10,000 G. Most commercial MR scanners today operate in the range of 0.1 to 2.0 T.

A typical commercial superconductive MR scanner at the author's institution is shown in Figure 1-4.

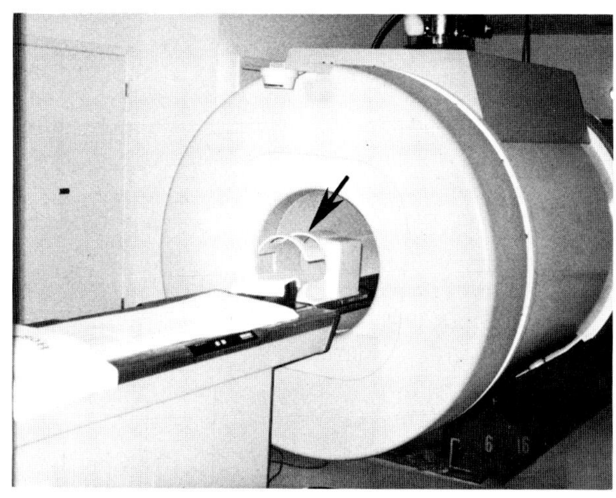

Fig. 1-4. A Diasonics superconductive MR scanner, operating at 0.35 T from NMR Associates-Houston. Arrow indicates radio frequency (head) coil.

Housed within the cylindrical body of the main structure are the coils that generate the powerful main magnetic field. This field is directed longitudinally along the gantry of the magnet and is denoted B_0.

Fig. 1-5. Radio frequency (RF) coil is placed over the patients head, and both are moved into the bore of the magnet for scanning. (Courtesy of Siemens, Inc.)

These main magnetic coils are insulated and bathed in liquid helium, which is replenished every month or so through pipes seen at the top of the structure. Also within the main magnet housing are smaller *shim coils* and *gradient coils*. The shim coils are small electromagnets that are activated to correct for irregularities in the main magnetic field. The gradient coils are small electromagnets that are switched on and off throughout the scan to change the phase and frequency of resonating nuclei within the patient. By varying the strength and direction of these gradient magnetic fields, it is possible to encode each volume element within the patient with a specific frequency and phase, and hence determine the location of a signal from within the patient.

In addition to the electromagnets incorporated into the main cylinder, another magnetic coil must be placed around the patient's head before scanning. This second device is called the *radio frequency (RF)-coil* (Fig. 1-5). The RF-coil generates a burst of electromagnetic radiation, called an *RF-pulse*, at high frequencies during scanning. As a result of this RF-pulse, hydrogen nuclei in the patient absorb energy and are said to undergo *nuclear magnetic resonance*. When the nuclei release the RF energy they have absorbed, an MR signal can be recorded. The RF-coil, which was initially used as a transmitter of the RF-pulse, is now used as an antenna to receive and record the MR signal. A computer processes the complex signals by a mathematical procedure known as *Fourier transformation*, and generates the MR image.

MR PHYSICS SIMPLIFIED

The reader may find comfort in knowing that a great many physicians involved in clinical MR imaging actually know very little about MR physics. Nevertheless, all have attained a sort of working familiarity with the basic principles, which allows them to be quite effective diagnosticians. The goal of this section is to provide a simple, practical, working knowledge of the physical phenomenon of nuclear magnetic resonance. The interested reader is referred to several more detailed descriptions of MR physics available elsewhere.[1,3–5,16,17] Applications of the physical principles to imaging is the subject of Chapter 2.

To begin, certain atomic nuclei, like the proton of hydrogen, possess a quantum mechanical property known as *spin*. Because the proton is positively charged as well as spinning, a magnetic field surrounds the nucleus. It is sometimes helpful to think of the atomic nucleus as a tiny spinning magnet. Because the proton is so small, the magnetic field around an individual nucleus cannot be detected. Each proton magnet does interact with small magnetic fields in neighboring molecules, however. These interactions at the atomic and molecular level form the basis for tissue contrast during MR imaging.

Although the magnetic field of an individual nucleus is too small to detect, when many of these tiny fields align with one another, a measurably large magnetic field is produced. This small but measurable field induced in a tissue is called the *magnetization*, (**M**) and is illustrated in Figure 1-6. In the absence of an external magnetic field, the individual proton "magnets" are randomly oriented, and no magnetization exisits. When the tissue is placed in a strong external magnetic field (such as in the bore of a MR imager) many of these proton magnetic moments will tend to align with the external field. As a result, the tissue will become weakly magnetized. The resultant small tissue magnetization is represented by the vector **M.**

When a patient is placed in the bore of a typical

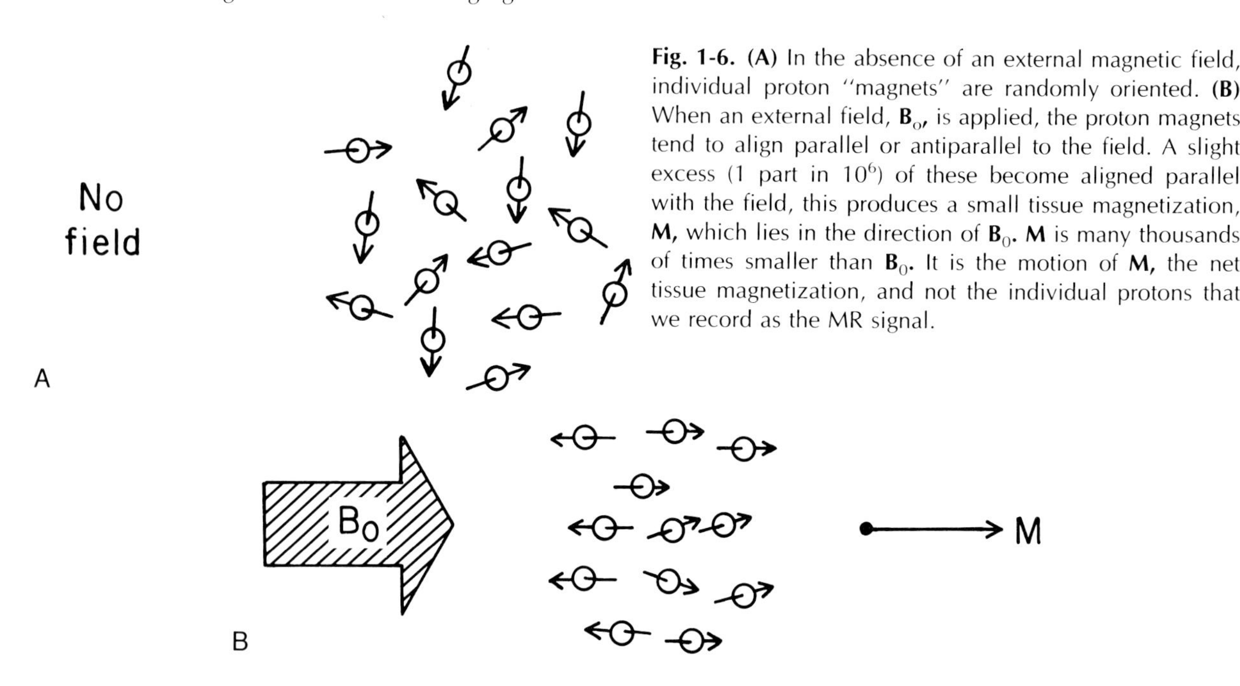

Fig. 1-6. **(A)** In the absence of an external magnetic field, individual proton "magnets" are randomly oriented. **(B)** When an external field, $\mathbf{B}_0$, is applied, the proton magnets tend to align parallel or antiparallel to the field. A slight excess (1 part in 10^6) of these become aligned parallel with the field, this produces a small tissue magnetization, **M**, which lies in the direction of $\mathbf{B}_0$. **M** is many thousands of times smaller than $\mathbf{B}_0$. It is the motion of **M,** the net tissue magnetization, and not the individual protons that we record as the MR signal.

MR scanner, the external static field, $\mathbf{B}_0$, is directed along the gantry (Fig. 1-7). Coordinate axes, x, y, and z, are commonly defined in relation to $\mathbf{B}_0$, as shown in the same figure. At equilibrium, the tissue magnetization, **M**, is directed along the z-axis (i.e., parallel to $\mathbf{B}_0$, Fig. 1-8A). An MR signal, however, can only be recorded by moving **M** out of alignment with $\mathbf{B}_0$. This is accomplished by stimulating the tissue with a second external magnetic field, $\mathbf{B}_1$, which is directed in the xy-plane (Fig. 1-8B).

The $\mathbf{B}_1$ field is produced by the small RF coil placed over the patient's head during scanning (Fig. 1-5). The $\mathbf{B}_1$ field is only turned on for a few dozen milliseconds at a time, in brief bursts of energy called *pulses.* The effect of these RF pulses is to tip the magnetization vector, **M,** away from its equilibrium alignment with $\mathbf{B}_0$.

In order to deflect the magnetization vector, $\mathbf{B}_1$ must rotate in the xy-plane rapidly at a specific frequency, known as the *Larmor resonance frequency* (ν_0). The Larmor frequency is directly related to the strength of the external magnetic field, $\mathbf{B}_0$, by the equation

$$\nu_0 = \gamma \mathbf{B}_0. \qquad (1\text{-}1)$$

where γ = the gyromagnetic ratio, which is a constant for each nuclear species. For the hydrogen nucleus, the gyromagnetic ratio is 42.58 MHz/T. As an exam-

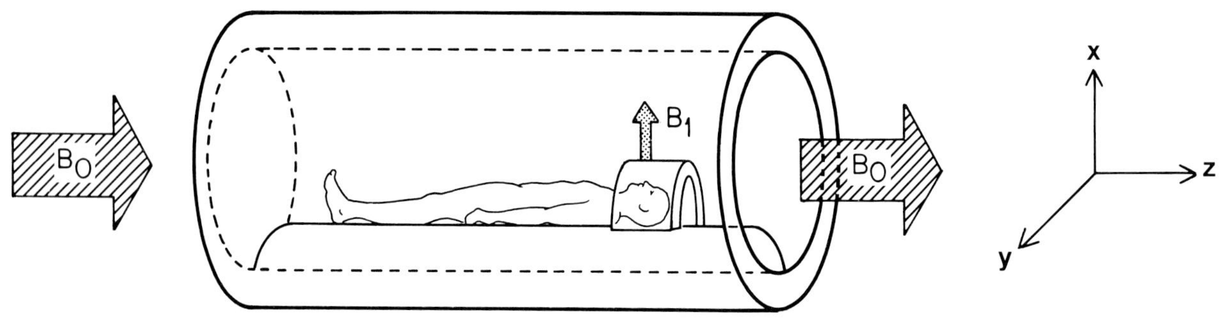

Fig. 1-7. In a typical superconductive scanner, $\mathbf{B}_0$ is directed along the gantry (z-axis). The RF-field, $\mathbf{B}_1$, is induced by the head coil and rotates in the xy-plane as shown.

Fig. 1-8. **(A)** At equilibrium, **M** is directed parallel to $\mathbf{B}_0$ (along the z-axis). **(B)** The RF field, $\mathbf{B}_1$, lies in the xy-plane. **(C)** During RF stimulation, $\mathbf{B}_1$ can be thought of as rotating in the xy-plane as shown.

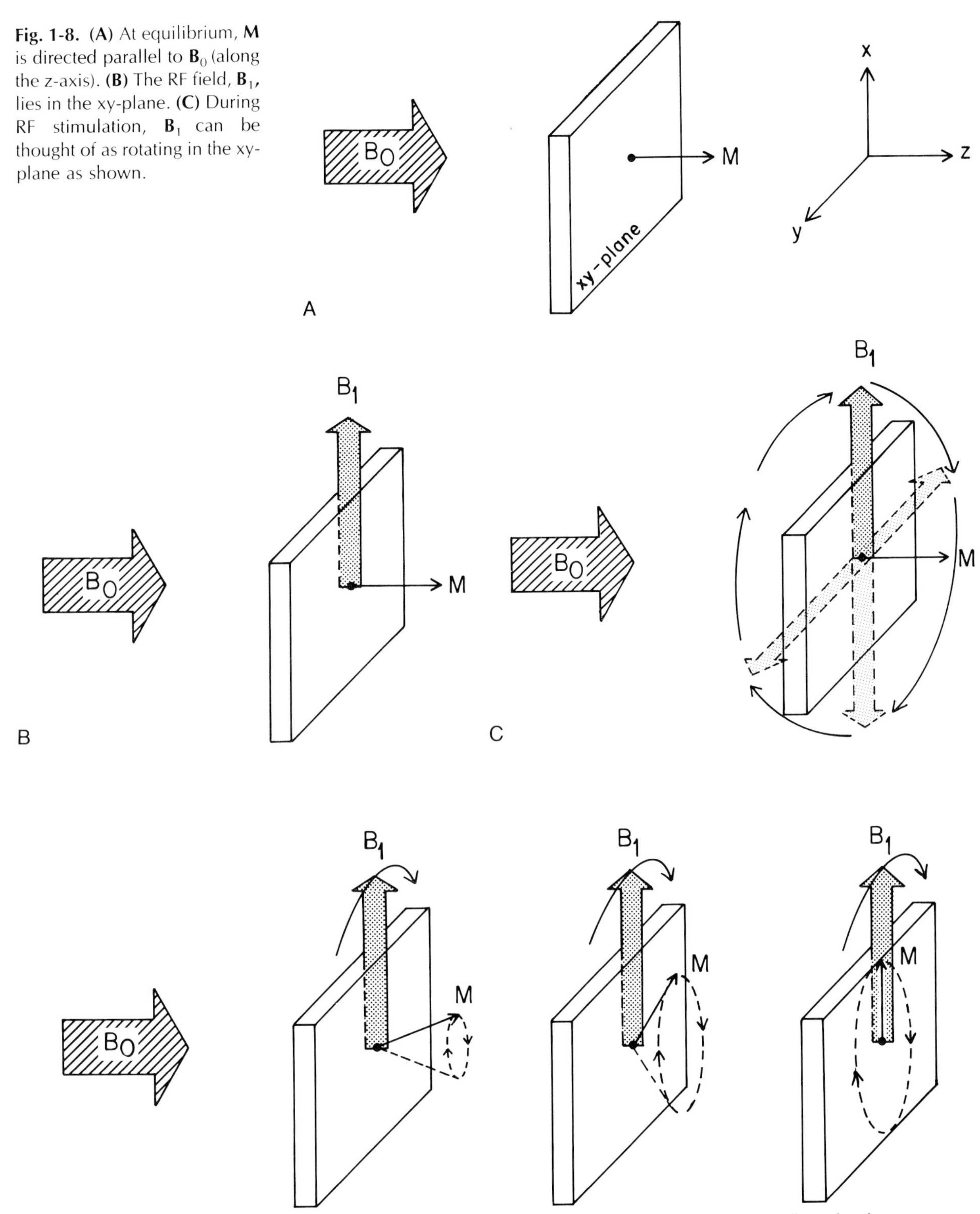

Fig. 1-9. The rotation of $\mathbf{B}_1$ causes **M** to rotate as well. The longer $\mathbf{B}_1$ is left on, the more deflected **M** becomes. After a few dozen milliseconds, **M** has been tipped into the xy-plane and rotates with $\mathbf{B}_1$.

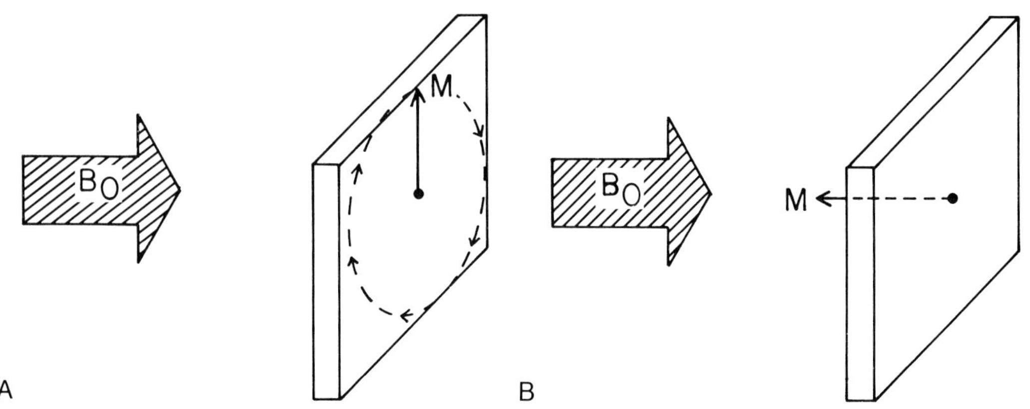

Fig. 1-10. **(A)** When **M** reaches the xy-plane, $\mathbf{B}_1$ is shut off. A 90° pulse is said to have been applied, because **M** has been rotated by 90°. **(B)** If $\mathbf{B}_1$ were left on for twice as long, **M** would continue to spiral and come to point in exactly the opposite direction to which it started. This is called a 180° pulse.

ple, protons imaged in an MR unit operating at 1.5 T would have a ν_0 of (42.58 MHz/T) × (1.5 T) = 63.87 MHz. The applied frequency from the RF coil in this scanner would have to be 63.87 MHz in order to deflect the magnetization vector **M** away from equilibrium alignment with $\mathbf{B}_0$.

The $\mathbf{B}_1$ field must therefore have two characteristics in order to induce an MR signal: (1) it must be directed in the xy-plane (Fig. 1-8B), and (2) it must rotate in this plane at ν_0 (Fig. 1-8C). When these two criteria are met, the $\mathbf{B}_1$ field causes **M** to rotate, tipping **M** progressively away from its equilibrium alignment along the z-axis (Fig. 1-9). After a few tens of msecs, **M** may be tipped totally into the xy-plane (Fig. 1-10A). The specific magnitude and duration of $\mathbf{B}_1$ required to induce this transition is called a *90°-pulse*. If the $\mathbf{B}_1$ field is twice as strong or left on for twice as long, **M** will continue to rotate so that it now points along the −z-axis (Fig. 1-10B). The $\mathbf{B}_1$ pulse that induces this transition is called a *180°-pulse*. Most imaging sequences used today (like spin echo and inversion recovery) are merely combinations of these 90°- and 180°-pulses.

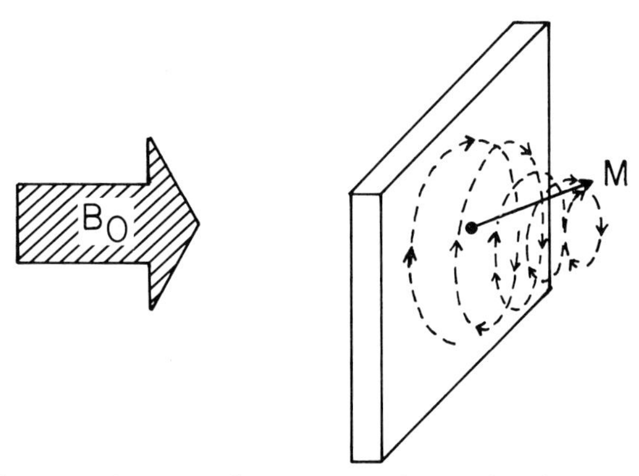

Fig. 1-11. Recovery from a 90° pulse. In the absence of $\mathbf{B}_1$, **M** will spiral back exponentially toward its equilibrium alignment with $\mathbf{B}_0$. This process is called recovery or relaxation.

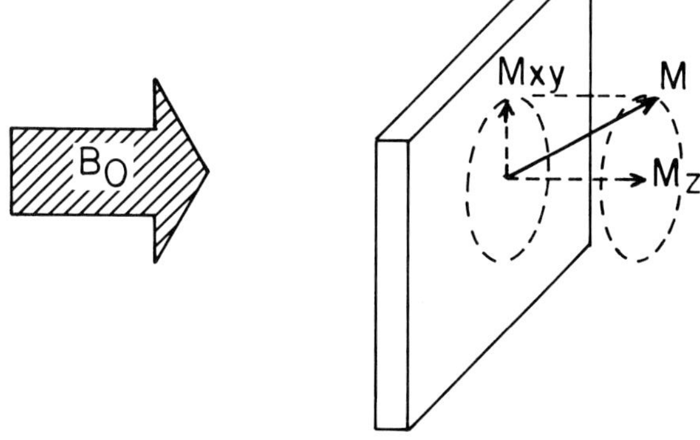

Fig. 1-12. When **M** is spiraling back toward equilibrium, it lies at an angle to the defined axis system. The **M** vector may be analyzed by two components. M_z is the component of **M** along the z axis, while M_{xy} is the component of **M** in the xy-plane.

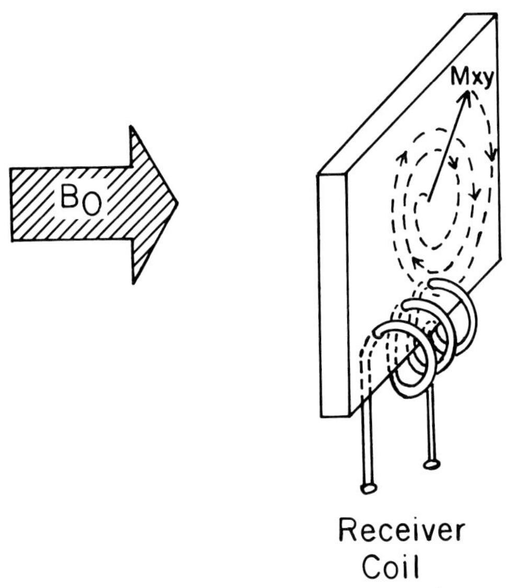

Fig. 1-13. Only the M_{xy} component can induce an MR signal. This signal is produced when the transverse magnetization (M_{xy}) induces a current in the RF (receiver)-coil.

After **M** has been rotated a desired amount, say 90°, the $\mathbf{B}_1$ field is shut off. In a process known as relaxation, the **M** vector slowly realigns itself toward its starting position parallel to $\mathbf{B}_0$ (Fig. 1-11). Note that **M** is still spinning in the xy-plane at ν_0. As it returns toward equilibrium following a 90°-pulse, it spirals toward the z-axis in an exponential fashion. The vector **M**, directed obliquely in space, may be decomposed into two components, M_{xy} and M_z (Fig. 1-12). These quantities represent the components of **M** transverse and parallel to $\mathbf{B}_0$, respectively.

Relaxation parallel to the direction of $\mathbf{B}_0$ (i.e., along the z-axis) is called *T1-relaxation* or *spin-lattice relaxation*. Relaxation transverse to $\mathbf{B}_0$ (i.e., in the xy-plane) is called *T2-relaxation* or *spin-spin relaxation*. The vector components M_z and M_{xy} can be used to measure these relaxation processes. After a 90°-pulse, M_z is zero while M_{xy} is a maximum (because **M** has been deflected totally into the xy-plane). With relaxation, M_z increases while M_{xy} decreases in an exponential fashion. With return to equilibrium, M_z is a maximum and M_{xy} is zero. Two time constants, *T1* and *T2*, may be defined for these exponential growth and decay processes. As will be described in more detail later, several physical properties contribute to T1 and T2, and these allow for tissue differentiation in MRI.

The M_z component is exceedingly small compared to $\mathbf{B}_0$ and lies in the same direction; furthermore, it is changing much too slowly to be detected by an MR imager. The M_{xy} component, however, is rapidly rotating in the transverse plane and can easily be detected. The same RF-coil used to transmit $\mathbf{B}_1$ or a second specialized receiver coil may be used to detect the field of M_{xy} as it sweeps by (Fig. 1-13). The resulting MR signal, called a *free induction decay*, is easily recorded and has a form similar to that shown in Figure 1-14.

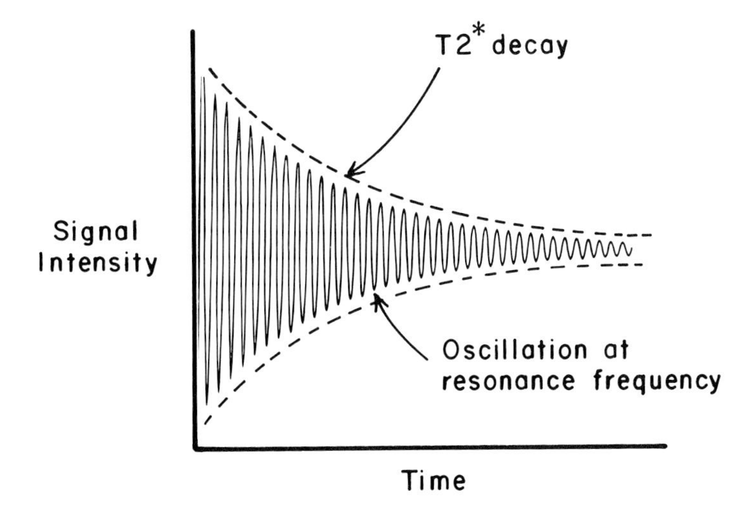

Fig. 1-14. The MR signal is called a free induction decay (FID).

TISSUE-SPECIFIC FACTORS WHICH INFLUENCE THE MR SIGNAL

The strength of the MR signal from a small volume of tissue depends on several factors (Table 1-3). Some of these factors are intrinsic to the tissue under study. Others relate to technical aspects of scanning and are under operator control. In this section, four tissue-specific factors that influence the MR signal are discussed: hydrogen density, flow, T1-relaxation, and T2-relaxation. In the next chapter those extrinsic factors that can be controlled or selected by the machine operator to affect the MR signal are discussed: technical aspects, pulse sequences, and pulse timing parameters. Choosing an appropriate pulse sequence and timing parameters allows *weighting* of the MR signal so that it reflects one or more of the tissue-specific factors. Image contrast is therefore partially under operator control, and the knowledgeable diagnostician can direct the performance of the scan to obtain maximal information from a given clinical scenario.

It should be recognized that MR signal intensity directly translates into how bright a dot appears on the final MR image. Tissues that produce a strong MR signal are displayed as white, intermediate signals are gray, and weak signals are black. Contrast within an image, therefore, relates to differential MR signal intensities between adjacent regions of tissue under study.

Hydrogen Density

Hydrogen density (spin density, ρ) reflects the relative number of hydrogen nuclei in a tissue that contribute to the MR signal. Tissues with a large number of available protons will cast a large signal; those with a paucity of hydrogen will produce little signal. Fortunately for MRI, hydrogen is abundant in most tissues of the body, being found in water, carbohydrates, and fats.

Table 1-3. Factors Determining the Appearance of an MR Image

Tissue-Specific (Intrinsic) Factors
Hydrogen (spin) density
Velocity (motion)
T1-relaxation
T2-relaxation
Operator-Dependent (Extrinsic) Factors
Magnetic field strength and homogeneity
Pulse sequence
Sequence timing parameters
Number of signal averages

Two substances in the body have very few hydrogen protons, cortical bone and air. Consequently, little MR signal can be generated and these substances appear dark on all MR sequences. The paucity of signal from cortical bone is one reason MR can so well image the posterior fossa and brain stem, which are surrounded by the dense skull base. By contrast, CT is significantly limited in these regions because of streak and beam-hardening artifacts from the same dense bone.

Do not get the mistaken impression that bone casts no MR signal, however. This statement is true only for cortical bone (and areas of dense tissue calcification). The medullary cavities of bone often contain blood, fat, or marrow. Accordingly, appreciable signal may be seen centrally, such as in the calvarial diploë (Fig. 1-15).

Table 1-4 lists relative hydrogen densities of several normal cerebral tissues. Note that gray matter has

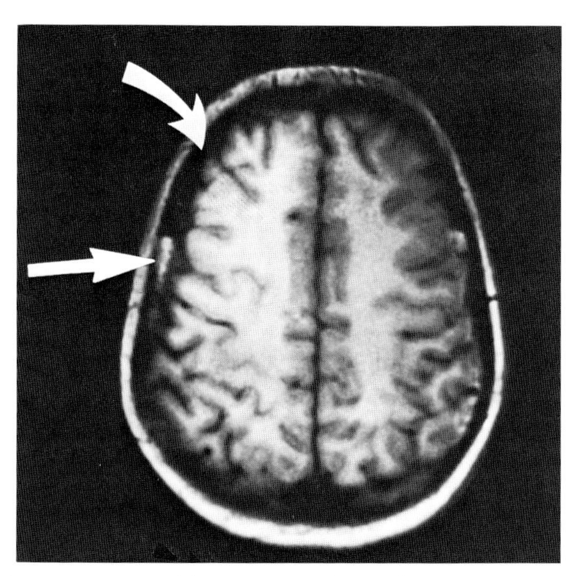

Fig. 1-15. Cortical bone (as in the skull) has few mobile hydrogen protons and causes little MR signal (curved arrow). Fat and marrow in the medullary space, however, casts a high signal (straight arrow).

Table 1-4. Relative Hydrogen Densities of Several Normal Brain Components[a]

Tissue	*Relative Hydrogen Density (%)*
Cerebrospinal fluid (CSF)	100
Cortical gray matter	95
Cerebellar gray matter	95
Caudate nucleus	94
Putamen	94
Thalamus	93
Subcortical white matter	77
Cerebellar white matter	77
Corpus callosum	76
Internal capsule	69

[a] Composite of data from References 18 and 30 normalized to CSF = 100%. Such values should be considered only approximate because sources are somewhat at variance and are methodology dependent.

an average hydrogen density about 20 percent higher than white matter. This difference largely explains gray matter–white matter contrasts on certain imaging sequences (Fig. 1-16). Additionally Kjos et al. have reported that the internal capsule has a hydrogen density less than other white matter.[18] This may account for the commonly observed slightly lower intensity of the internal capsule relative to gray and white matter (Fig. 1-16).

Although calculated hydrogen densities vary up to 25 percent throughout the brain, the absolute hydrogen concentrations vary by less than 0.1 percent.[19] In the brain parenchyma, for example, it is thought that the fraction of hydrogen most contributing to the MR signal is contained in water molecules in a bound hydration layer associated with macromolecules.[20] Chemical shift experiments have confirmed that the hydrogen in brain phospholipids contribute negligibly to the recorded MR signal.[21]

Flow Effects

Flow (velocity) effects may contribute significantly to MR signal intensity. Intracranially, flow-related signal changes are observed in blood vessels and in the cerebrospinal fluid (CSF) pathways. In general the effects of flow are difficult to predict and quantitate. Signal changes depend upon the composition of the flowing fluid, pulse sequence, magnetic field strength, angle of flow relative to plane of imaging, velocity, turbulence, and other factors. Several excellent reviews are available.[22–26]

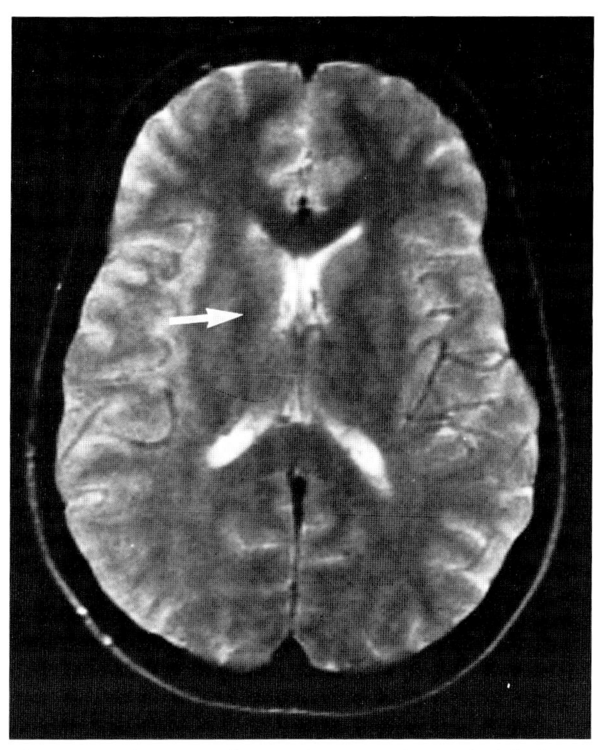

Fig. 1-16. The internal capsule (arrow) has a slightly lower hydrogen density than other cerebral structures, perhaps accounting for its lower intensity on most imaging sequences.

Despite this apparent complexity two general principles about flow may be stated: (1) Rapidly flowing blood or CSF usually appears dark. (2) Slowly flowing blood or CSF often appears bright. Increased signal from slowly moving fluids has been called paradoxical enhancement. Mechanisms responsible for signal loss or augmentation by flow are presented in Table 1-5.

Two factors are responsible for signal loss from flow-

Table 1-5. Flow Effects in MRI

Flow Effects that Decrease MR Signal
High velocity loss
Turbulence
Flow Effects that Increase MR Signal
Entry phenomenon
Even-echo rephasing
Diastolic pseudogating

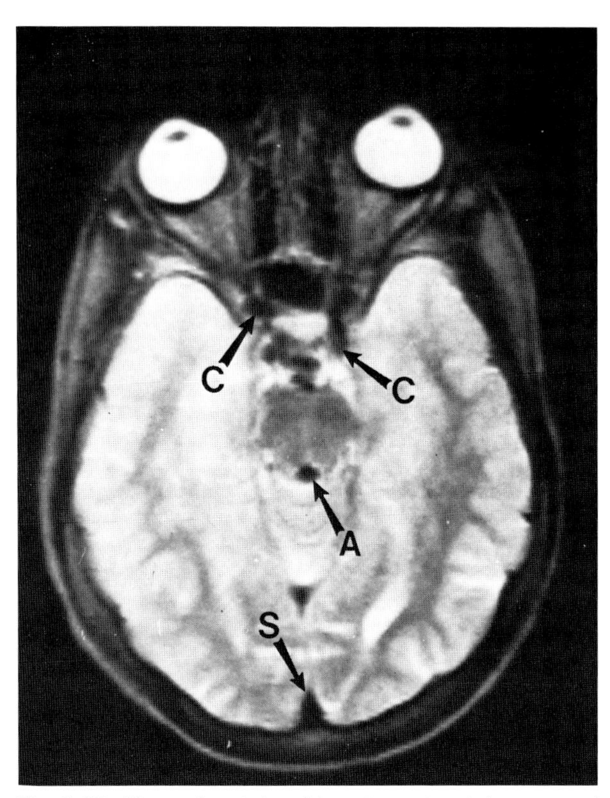

Fig. 1-17. Flow related signal loss. Low signal due to rapid flow is seen in the cerebral aqueduct (*A*), carotid arteries (*C*), and superior sagittal sinus (*S*).

ing fluids.[24] High velocity produces MR signal loss because rapidly flowing protons do not remain within a selected slice long enough to be stimulated with a full complement of RF-pulses. Turbulence produces decreased signal because the random motion of molecules leads to loss of phase coherence among nuclei in a given slice. Examples of flow induced signal loss in the carotid arteries and cerebral aqueduct are shown in Figure 1-17.

Three factors may account for paradoxical enhancement in moving fluids.[22,26] The first is an *entry phenomenon* that occurs when slowly flowing fluid enters the first section of a multisection volume. Partially saturated fluid protons remaining from the previous sequence are replaced by totally unsaturated protons flowing in. The unsaturated protons produce a stronger MR signal than saturated ones. Flow related enhancement from this entry phenomenon is most prominent in vessels with fluid velocities below 1 cm/sec and on pulse sequences performed with short repetition times. An example of this entry enhancement of blood is shown in Figure 1-18A.

Even-echo rephasing, the second factor, refers to paradoxical enhancement of slowly flowing fluids, which occurs only when the multiple spin-echo pulse sequence (Ch. 2) is used.[22] Even-numbered echoes in this sequence result in unusually high intensities from these slowly flowing fluids. The explanation for this phenomenon is somewhat complex, but involves the fact that flow is occurring across a gradient magnetic field (e.g., the readout gradient). Accordingly, dephasing of individual proton spins is not refocused or corrected completely into the usual spin-echo. Instead, phase changes build up nonlinearly and become abnormally collected following even-numbered pulses in this sequence. Even-echo rephasing is commonly noted both in large and small cerebral vessels (Fig. 1-18B).

A third mechanism that may produce paradoxical enhancement is called *diastolic pseudogating*. This occurs when the repetition time of a pulse sequence serendipitously coincides with the length of the cardiac cycle (or an integral multiple thereof). When this occurs, each section in a multisection acquisition is obtained at the same time in the cardiac cycle. Those sections containing blood vessels imaged in diastole contain more slowly flowing blood and may have increased intraluminal signal. Diastolic pseudogating is more commonly noted in the great vessels of the thorax than intracranially.

T1-Relaxation

After a 90° or 180° RF-pulse, the tissue magnetization **M** has been tipped away from its equilibrium alignment parallel to the static field $\mathbf{B}_0$ (Figure 1-10). Over the next few seconds M will slowly realign itself with $\mathbf{B}_0$. The rate of return of M to its equilibrium alignment is exponential in nature with characteristic time constant T1. This realignment of M along the $\mathbf{B}_0$ direction is called *T1-relaxation*.

Other names for T1-relaxation include *spin-lattice relaxation*, *longitudinal relaxation*, and *thermal relaxation*. These other names indicate some of the physical phenomena involved in the T1-relaxation process. In order for relaxation to occur, stimulated hydrogen nuclei must give up the energy they have absorbed

A 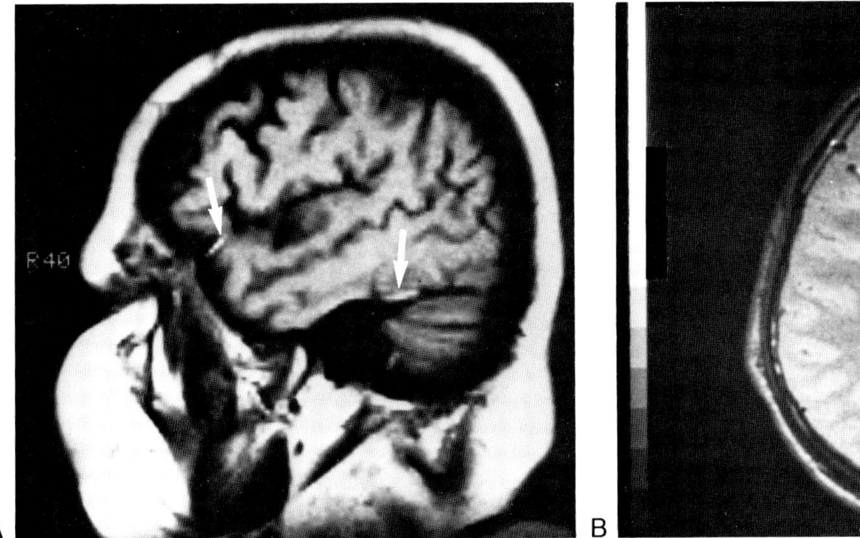

B 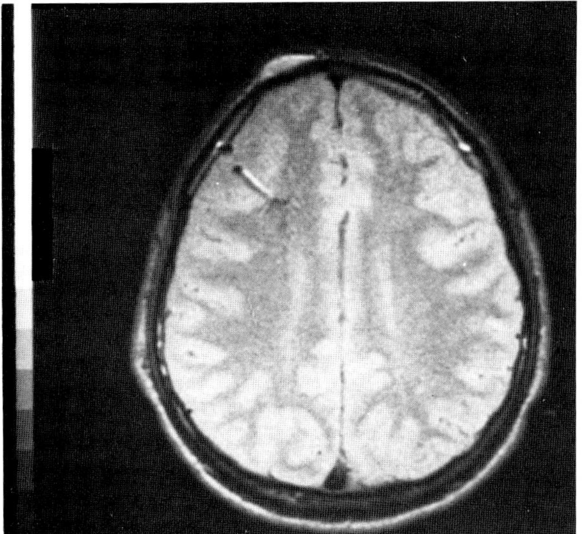

Fig. 1-18. Flow related signal enhancement. **(A)** Entry phenomenon may cause enhancement in slowly flowing blood seen in the first section of a multisection image acquisition. **(B)** Even echo rephasing is seen in a venous angioma. A spatial misregistration artifact is also seen. (The rephasing signal is not perfectly superimposed over the blood vessel.)

for the RF-pulse. This energy is transferred to neighboring atoms or molecules ("the lattice"), which dissipate the energy as heat. In biologic systems this dissipation of energy results in molecular rotation, translation, or vibration.[17] The transfer of energy may be either between nuclei in the same molecule (intramolecular) or between those in adjacent molecules (intermolecular).

A quantum feature of the relaxation process is that stimulated nuclei can only give up their energy by interacting with others at ν_0. That is, stimulated nuclei must come into contact with magnetic fields in neighboring atoms, which fluctuate near ν_0 for hydrogen. Accordingly, certain molecules and chemical environments are more efficient at inducing T1-relaxation than others.

Most liquids (like CSF) are very inefficient in allowing T1-relaxation to occur. The small free water molecules tumble and move very quickly in the unstructured liquid. Their motion is much more rapid than the resonance frequency and the transfer of energy to them is very inefficient. Consequently, T1 times for fluids like CSF are very long, on the order of 2000 msec.

In biologic tissues, however, a significant fraction of water molecules are not free as they are in pure liquids. Instead, these water molecules are relatively tightly bound in a hydration layer associated with macromolecules. The hydrogen nuclei in this hydration layer are more efficient at T1 interactions; they may transfer their energy to other bound H_2O molecules or to the macromolecules associated with them. Current evidence suggests that T1 in biologic tissues is determined predominantly by intermolecular (possibly rotational) interactions between macromolecules and this single bound hydration layer.[20] The elevated T1 seen in tumors and other pathologic processes may relate to an abnormal relative increase in the ratio of free to bound water.[28]

Table 1-6. Approximate T1 and T1 Values of Various Cranial Tissues Measured at 0.15–0.3 T[a]

Tissue	T_1 *(msec)*	T_2 *(msec)*
Gray matter	450–650	60–100
White matter	350–460	50–90
Brain stem	450–500	75–100
CSF	1200–2700	150–250
Fat	200–290	60–85
Skeletal muscle	330–380	40–50

[a] Approximate average values compiled from References 3, 18, 20, 28–31.

Table 1-7. Ratio of T1 at Fields Different from 0.15 T[a]

	Field Strength (T)				
Tissue	*0.15*	*0.3*	*0.5*	*1.0*	*1.5*
Gray matter	1	1.24	1.45	1.79	2.03
White matter	1	1.27	1.52	1.93	2.23
Muscle	1	1.34	1.66	2.22	2.63
Fat	1	1.13	1.24	1.39	1.50

[a] Compiled from Reference 20.

Fatty tissues are somewhat unique in their high efficiency of T1-relaxation. It is thought that the multiple hydrogen-laden side chains of fats are ideally suited for interacting with each other at frequencies near resonance. Accordingly, fatty tissues have very short values of T1, usually on the order of 200 to 300 msec. It should be noted, however, that lipids in gray and white matter exist primarily in the form of phospholipids.[29] This hydrogen in phospholipids does not participate significantly in the MR signal. Therefore, the MR signal from brain predominantly depends upon protons in tissue H_2O. By contrast, lipids in orbital, scalp, and marrow fat exist in the form of fatty acid triglycerides. The hydrogen in these lipids do contribute to the MR signal and these tissues all demonstrate very short T1 values.

As previously noted, T1-relaxation requires atomic interactions at the resonance frequency to occur. Since ν_0 depends upon the external field $\mathbf{B}_0$, it is not surprising to learn that T1-values are field dependent. As a rule, T1 values in biologic tissues increase with increasing field strength.[20,30] Table 1-7 shows the ratio of T1 values measured at various fields to those measured at 0.15 T. Note that as field strength is increased from 0.15 T to 1.5 T, T1 values approximately double. Fat is seen to have less field dependence than other tissues, possibly reflecting a different mechanism of spin-lattice relaxation for the $-CH_2$ protons.[30]

T2-Relaxation

T2-relaxation, also called *spin-spin* or *transverse relaxation*, is a process in which the transverse component of magnetization, M_{xy}, decays. The physical processes underlying this phenomenon are as complex and poorly understood as those of nuclear magnetic resonance itself. At a quantum mechanical level, T2-relaxation means that resonating hydrogen protons are losing phase coherence. In other words, the synchronous energy level transitions of these protons induced by the 90°-pulse are becoming increasingly inharmonious. This dispersion of magnetic spins occurs much more rapidly than T1-relaxation.

The loss of transverse magnetization, M_{xy}, can be viewed as an exponential decay process with time constant T2*. This time constant represents the envelope of the free induction decay signal shown in Figure 1-14. T2* decay can be thought of as arising from two independent processes: (1) decay due to inhomogeneities (irregularities) in the static field ($T2^i$), and (2) decay due to true tissue spin-spin interactions (T2). As an approximation

$$\frac{1}{T2^*} = \frac{1}{T2^i} + \frac{1}{T2} \qquad (1\text{-}2)$$

For all MR imaging systems, T2* is predominantly determined by field inhomogeneities and not true tissue T2. This is unfortunate, because we are not interested in measuring the imperfections of our magnet but rather the true T2 of the tissue under study. Fortunately, a technique known as *spin echo* has been devised, which largely corrects for magnet inhomogeneities. Using spin echo, it becomes possible to measure true tissue T2 values and construct images whose contrast depends upon relative T2's of various tissues. How the spin echo technique accomplishes this feat is the subject of the next chapter.

True tissue T2-relaxation (spin-spin relaxation) is a different phenomenon than T1-relaxation. T1-relaxation results in energy loss from resonating nuclei to the lattice of surrounding mostly nonresonating molecules. T2-relaxation results in changes of phase between resonating nuclei, without conversion of magnetic energy to thermal motion in the lattice.

The physical bases for T2-relaxation are the stable and slowly fluctuating local magnetic fields in a tissue of interest. Atoms, molecules, and ions, including those not involved in resonance, generate local magnetic fields in the vicinity of resonating nuclei. These local fields are not constant but vary throughout the tissue. Even if the external field of the MR magnet were perfectly uniform, these local (internal) fields would subject resonating nuclei to varying forces within the tissue. Those nuclei exposed to slightly higher local fields would resonate at higher frequencies than those exposed to lower fields. Differences in local resonance frequencies would cause harmoniously resonating nuclei in fall out of phase with one another. This loss of phase coherence results in loss of magnetization in the transverse plane.

A tissue's efficiency at T2 interactions is more closely related to tissue state (solid or liquid) than to any other single factor. The local magnetic fields in a tissue responsible for T2-relaxation are primarily those that are stable or only slowly fluctuating. If local magnetic fields fluctuate too quickly, the average local field experienced by an ensemble of protons is zero, so very little dephasing occurs. Accordingly, T2 will be long. This situation occurs in liquids, where mobile water protons move so quickly that the average of many rapidly fluctuating local fields is zero. The T2 value of liquids (like CSF, urine) are quite long, on the order of 150 to 300 msec.

By contrast, solid tissues contain many closely packed hydrogen nuclei that are efficient at spin-spin interactions. The nuclei in solids and large molecules, rigidly held, possess local fields that fluctuate very slowly compared to the period of nuclear precession ($1/\nu_0$). These local fields are nonzero when averaged on the time scale of MRI. Dephasing is accelerated, and T2 is shortened. Thus in "solid" biological tissues, T2 may range from 10 to 100 msec.

In some large macromolecules such as brain phospholipids, spin-spin relaxation is extremely efficient. In fact some protons in these molecules relax in microseconds, too rapidly to be observed in a conventional MR imaging experiment. This is why we say some protons in a tissue are "MR silent." Although they participate in resonance they relax so quickly that their signals are undetectable on routine imaging.

While T1 relaxation can be described as a simple exponential growth curve, T2 decay may have multiple components.[20] In muscle, for example, the major component (T2 ~ 40 msec) constitutes about 75 percent of the signal, and derives from tissue water. There also exist very long T2 ~ 140 msec) and very short (T2 ~ msec) T2 components from macromolecules each accounting for about 10 percent of the signal. A remaining short T2 component (T2 ~ 5 msec) accounts for about 7 percent of the signal and is derived from both water and macromolecules. For most tissues, therefore, the dominant T2 component is governed primarily by exchange diffusion of water between the bound layer and a free water phase.[20] A comparison of the T1 and T2 relaxation processes is presented in Table 1-8.

Table 1-8. A Comparison of T1 and T2 Relaxation Processes

	T1-Relaxation	*T2-Relaxation*
Synonyms	Spin-lattice relaxation Longitudinal relaxation Thermal relaxation	Spin-spin relaxation Transverse relaxation
Direction of Action	Exponential recovery (growth) of magnetization parallel to static field $\mathbf{B}_0$ (single component)	Exponential decay perpendicular to $\mathbf{B}_0$ (multiple components)
Typical Values	250–3000 msec	10–250 msec (Always shorter than T1)
Value Relates Primarily To	Fat content (short T1) Free water content (long T1)	Tissue state: solid (short T2) liquid (long T2)
Physical Mechanisms Responsible	Intermolecular dipole-dipole interactions near the Larmor frequency	Inter- and intramolecular interactions of resonating nuclei with static or slowly fluctuating local magnetic fields
Dependence On External Magnetic Field ($\mathbf{B}_0$)	Yes, roughly proportional to $\mathbf{B}_0^{1/3}$	No

REFERENCES

1. Elster AD: Magnetic Resonance Imaging: A Reference Guide and Atlas. JB Lippincott, Philadelphia, 1986
2. Health Technology Case Study 27: Nuclear Magnetic Resonance Imaging Technology: A Clinical, Industrial, and Policy Analysis. U.S. Congress Office of Technology Assessment, Washington, D.C., September, 1984
3. Newton TH, Potts DG (eds): Modern Neuroradiology. Vol. 2. Advanced Imaging Techniques. Clavadel Press, San Anselmo, CA, 1985
4. Margulis AR, Higgins CB, Kaufman L, Crooks LE (eds): Clinical Magnetic Resonance Imaging. CV Mosby, St. Louis, 1984
5. Young SW: Nuclear Magnetic Resonance Imaging: Basic Principles. Raven Press, New York, 1984
6. Partain CL (ed): Nuclear Magnetic Resonance and Correlative Imaging Modalities. Society of Nuclear Medicine, New York, 1984
7. Bradley WG Jr, Waluch V, Yadley RA et al: Comparison of CT and MR in 400 patients with suspected disease of the brain and cervical spinal cord. Radiology 152:695, 1984
8. Pavlicek WA, Weisinger M, Castle L et al: Effects of nuclear magnetic resonance on patients with cardiac pacemakers. Radiology 147:149, 1983
9. New PFJ, Rosen BR, Brady TJ et al: Potential hazards and artifacts of ferromagnetic and nonferromagnetic surgical and dental materials and devices in nuclear magnetic resonance imaging. Radiology, 147:139, 1983
10. Dujovny M, Kossovsky N, Kossovsky R et al: Aneurysm clip motion during magnetic resonance imaging: in vivo experimental study with metallurgical factor analysis. Neurosurgery 17:543, 1985
11. Finn EJ, Di Chiro G, Brooks RA et al: Ferromagnetic materials in patients: detection before MR imaging. Radiology 156:139, 1985
12. Laakman RW, Kaufman B, Han JS et al: MR imaging in patients with metallic implants. Radiology 157:711, 1985

12a. Applebaum EL, Valvassori GE: Effects of magnetic resonance imaging fields on stapedectomy prostheses. Arch Otolaryngol Head Neck Surg 111:820, 1985

13. Pavlicek W, Meaney TF: The special environmental needs of magnetic resonance. Appl Radiol 7:23, 1984
14. Soulen RL, Budinger TF, Higgins CB: Magnetic resonance imaging of prosthetic heart valves. Radiology 155:705, 1985
15. Kelly WM, Paglen PG, Pearson JA et al: Ferromagnetism of intraocular foreign body causes unilateral blindness after MR study. AJNR 7:243, 1986
16. Pavlicek W, Modic M, Weinstein M: Pulse sequence and significance. RadioGraphics 4:49, 1984
17. Fukushima E, Roeder SBW: Experimental Pulse NMR: A Nuts and Bolts Approach. Addison-Wesley, Reading, Massachusetts, 1981
18. Kjos BO, Ehman RL, Brant-Zawadski M et al: Reproducibility of relaxation times and spin density calculated from routine MR imaging sequences: clinical study of the CNS. AJR 144:1165, 1985
19. Brooks RA, Di Chiro G, Keller MR: Explanation of cerebral white-gray contrast in computed tomography. J Comput Assist Tomogr 4:489, 1980
20. Bottomley PA, Foster TH, Argersinger RE et al: A review of normal tissue hydrogen NMR relaxation times and relaxation mechanisms from 1–100 MHz: Dependence on tissue type, NMR frequency, temperature, species, excision, and age. Med Phys 11:425, 1984
21. Pykett IL, Rosen BR: NMR: In vivo proton chemical shift imaging (work in progress). Radiology 149:197, 1983
22. Waluch V, Bradley WG: NMR even echo rephasing in slow laminar flow. J Comput Assist Tomogr 8:594, 1984
23. Axel L: Blood flow effects in magnetic resonance imaging. AJR 143:1157, 1984
24. Bradley WG Jr, Waluch V, Lai K-S et al: The appearance of rapidly flowing blood on magnetic resonance images. AJR 143:1167, 1984
25. von Schulthess GK, Higgins CB: Blood flow imaging with MR: spin phase phenomena. Radiology 157:687, 1985
26. Bradley WG Jr, Waluch V: Blood flow: magnetic resonance imaging. Radiology 154:443, 1985
27. Mills CM, Brant-Zawadski M, Crooks LE et al: Nuclear magnetic resonance: principles of blood flow imaging. AJNR 4:1161, 1983
28. Mansfield P, Morris PG: NMR Imaging in Biomedicine. p. 29. Academic Press, New York, 1982
29. Wehrli FW, MacFall JR, Newton TH: Parameters determining the appearance of NMR images. p. 81. In Newton TH, Potts DG (eds): Modern Neuroradiology. Vol. 2. Advanced Imaging Techniques. Clavadel Press, San Anselmo, CA 1985
30. Wehrli FW: Basic principles of NMR. Data from General Electric Medical Systems Group presented at Neuroradiology at Vail Conference, Vail, CO, March, 1984
31. Johnson GA, Herfkens RJ, Brown MA: Tissue relaxation time: in vivo field dependence. Radiology 156:805, 1985

2

Pulse Sequences and Parameters

As discussed in the last chapter, nuclear magnetic resonance occurs when tissue protons in a static field ($\mathbf{B}_0$) absorb energy from an oscillating radio frequency (RF) field. This abosorption of energy perturbs the tissue magnetization vector **M**, causing it to rotate away from its initial equilibrium alignment with $\mathbf{B}_0$. During relaxation the resonating protons give up their absorbed energy to each other and to the surrounding tissue lattice. The magnetization **M** returns to its equilibrium alignment. As **M** sweeps by the receiver coil of the MR scanner, it induces a current in the coil, which is a bulk MR signal for the entire imaged volume.

The manner in which this bulk MR signal is decoded into an image lies beyond the scope of this book. Briefly, small additional magnetic fields, called *gradient fields*, are superimposed upon the static and RF fields during scanning. These gradient fields, induce subtle differences in the frequency and phase of resonating nuclei throughout the sample. The bulk MR signal can be decoded by Fourier transformation into smaller signals arising from each volume element (voxel) of tissue. The signal from each voxel is uniquely identified by a signature of frequency and phase, which were assigned to that voxel by the gradient fields during scanning. The intensity (I) of the MR signal from each voxel is directly proportional to brightness of the corresponding picture element (pixel) in the MR image.

The duration and strength of the RF field determines to what degree the tissue magnetization **M** is shifted from equilibrium. Although it is possible to shift this vector by any amount, in current scanning techniques, **M** is usually rotated either 90° or 180° by a given RF-pulse. An MR pulse sequence represents a series of closely spaced 90°- and 180°-pulses that generate an MR signal by perturbing **M.** The two pulse sequences in most common use today are spin-echo and inversion recovery, and will now be described in detail.

THE SPIN ECHO PULSE SEQUENCE

Spin echo (SE) is the most widely used pulse sequence in clinical MRI today. In the 1950s Hahn, Carr, Purcell, Meiboom, and Gill developed this technique to correct for effects produced by inhomogeneities in the fields of their NMR spectrometers.[1–3] As adopted in modern clinical imaging, SE is a powerful technique that produces MR signals that reflect to varying degrees three intrinsic physical characteristics of a tissue: ρ, T1 and T2. Furthermore, by adjusting certain operator-selectible timing parameters, the SE technique makes it possible to manipulate the MR signal to maximize contrast differences between tissues based upon each of these intrinsic characteristics.

In its simplest form the SE sequence is defined as a series of alternating 90°- and 180°-pulses (Fig. 2-1A). The interval between successive 90°-pulses is called the *repetition time* (TR). In clinical MRI, TR is usually chosen in the range 300–3000 msec. The MR signal, called the *echo*, is recorded after each pair of 90°- and 180°-pulses. The time from the center of the 90°-pulse to the center of the echo is called the *echo time* (TE). In clinical usage, TE is usually chosen in the range 20–120 msec. Note that the time between the 90°- and 180°-pulses is exactly ½ TE.

A multiple spin echo (MSE) technique is also often used. The MSE is a modification of the basic SE sequence where a second, third, or higher order 180°-pulses are applied after the first 180°-pulse (Fig. 2-1B). Second, third, and later echoes are then recorded following each 180°-pulse. In clinical practice usually only two to four echoes are recorded because signal strength falls off exponentially with increasing TE.

The spin echo pulse sequence is commonly abbreviated SE TR/TE, where TR and TE are the repetition and echo times respectively, expressed in milliseconds. For example, a spin echo technique with TR = 2500 msec and TE = 40 msec could be abbreviated SE 2500/40. A multiple spin echo sequence with TR = 2500 msec and echoes at 40, 80, and 120 msec could be denoted SE 2500/40, 80, 120.

The physical principles underlying the formation of spin echoes lie well beyond the scope of this book. Suffice it to say that by recording the MR signal as an echo following an 180°-pulse has several theoretical and practical advantages over recording the signal as a free induction decay (FID) following a 90°-pulse. The echo technique helps eliminate phase errors that

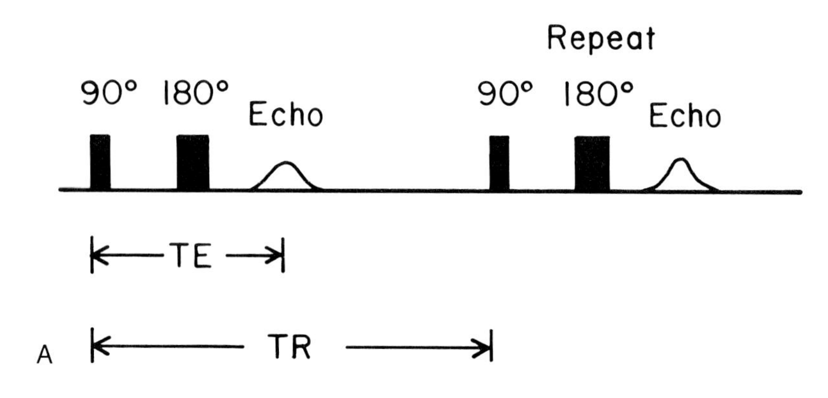

Fig. 2-1. (A) Spin echo pulse sequence consists of alternating 90°-and 180°-pulses. The echo signal is recorded at time TE after the 90°-pulse. The cycle is repeated at time TR milliseconds later. **(B)** Multiple spin echo technique. Successive 180°-pulses are applied before the cycle repeats. Several echoes can be recorded, although their strength falls off exponentially.

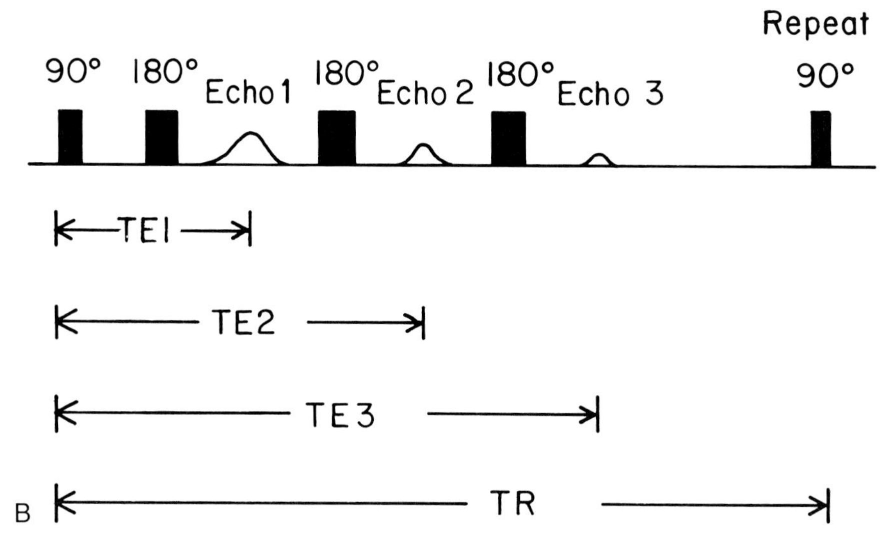

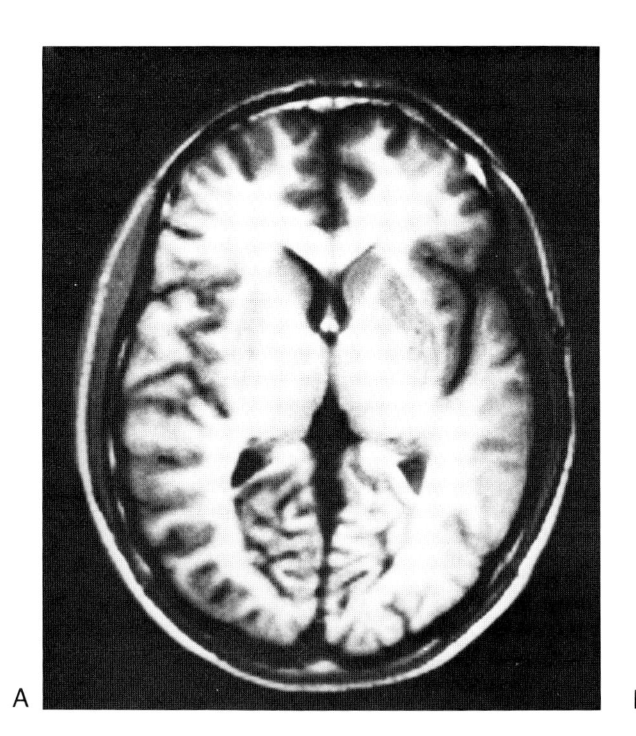

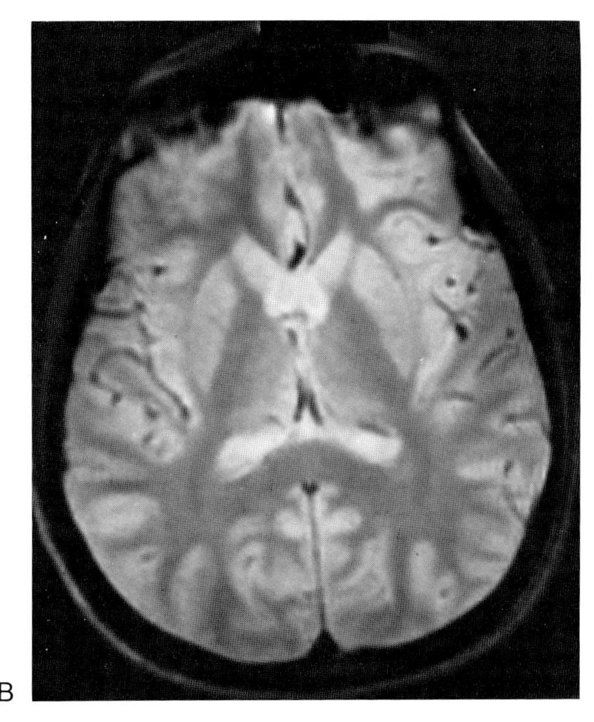

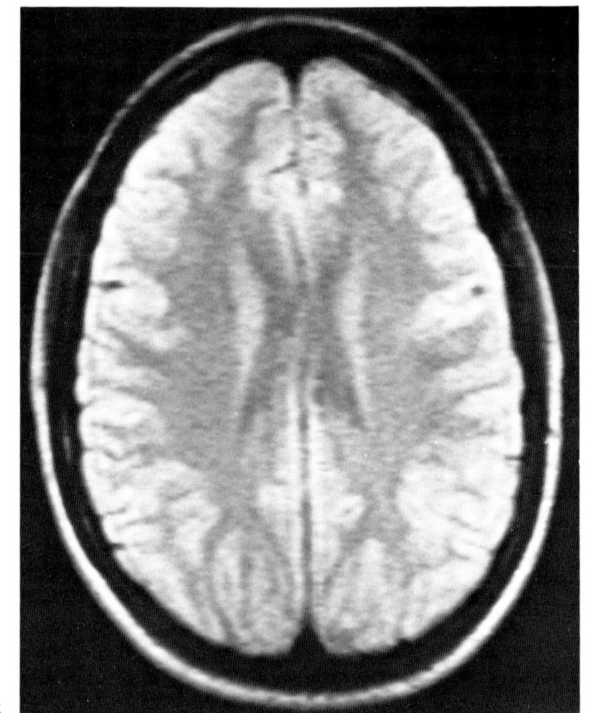

Fig. 2-2. (A) T1-weighted image (IR 1500/400/40). **(B)** T2-weighted image (SE 2500/70). **(C)** Hydrogen-weighted image (SE 2500/25).

occur in the signal secondary to irregularities (inhomogeneities) in the static magnetic field. Furthermore, the echo technique provides the MR signal with significant dependence on tissue T2. Because change in tissue T2 may be the earliest and most sensitive indicator of tissue pathology, this T2-dependence of the MR signal can be quite desirable.

The intensity of an MR signal (and hence how bright a dot appears on the MR image) recorded from a small volume of tissue excited with the SE sequence depends upon five parameters. Three of these (T1, T2, ρ) are tissue-specific, while the other two (TR, TE) are under operator control. By varying TR and TE it is possible to obtain MR images whose contrast may be made to vary based on differences in tissue T1, T2, or ρ (Fig. 2-2).

An equation has been developed to predict the MR signal intensity, I, from a volume of tissue stimulated by an SE pulse sequence:[4]

$$I = K \rho \, (1 - e^{-TR/T1}) \, e^{-TE/T2} \qquad (2\text{-}1)$$

where K is a scaling factor.

The weighting of this SE sequence depends upon the relative values of TR and TE with respect to T1 and T2. An overly simplified, but commonly used scheme for predicting MR parameter weighting is presented in Table 2-1.

Recall that most intracranial tissues have T1 values of 200 to 700 msec and T2 values of 40 to 100 msec (Table 1-6). For cranial imaging, therefore, "long TR" generally means TR >2500 msec while "short TR" means TR <500 msec. Similarly "long TE" means TE >75 msec and "short TE" means TE <25 msec. It should be recognized, however, that such numbers are entirely arbitrary. An oversimplified but important corollary follows consideration of Eq. 2-1: *Signal intensity varies directly with T2 but inversely with T1.* In other words, if two tissues have identical T1 values but differ in T2, that tissue with the higher T2 value will generally appear brighter. Conversely, if two tissues have identical T2 values, but different T1s, the tissue with the higher T1 value will appear darker. Such differences are accentuated when the pulse sequence is weighted in favor of T1 or T2 contrast.

A contrast reversal phenomenon may occur between two tissues as TR or TE is increased. For example, consider the contrast between brain and CSF on SE 500/30 (short SE, "T1-weighted") and SE 3000/80 (long SE, "T2-weighted") images (Fig. 2-3). Recall that the T1 and T2 values of CSF are much higher than brain, because loosely structured liquids are very inefficient at magnetic relaxation. On the "T1-

Table 2-1. Weighting of the MR Signal for Various Spin-Echo Sequences

TR	*TE*	*Weighting*	
Short	Short	T1	short spin-echo
Long	Long	T2	long spin-echo
Long	Short	ρ	

Short and long are arbitrary. Usually short TR means TR <500 msec; long TR means TR >2500 msec; short TE means TE <25 msec; long TE means TE >75 msec.

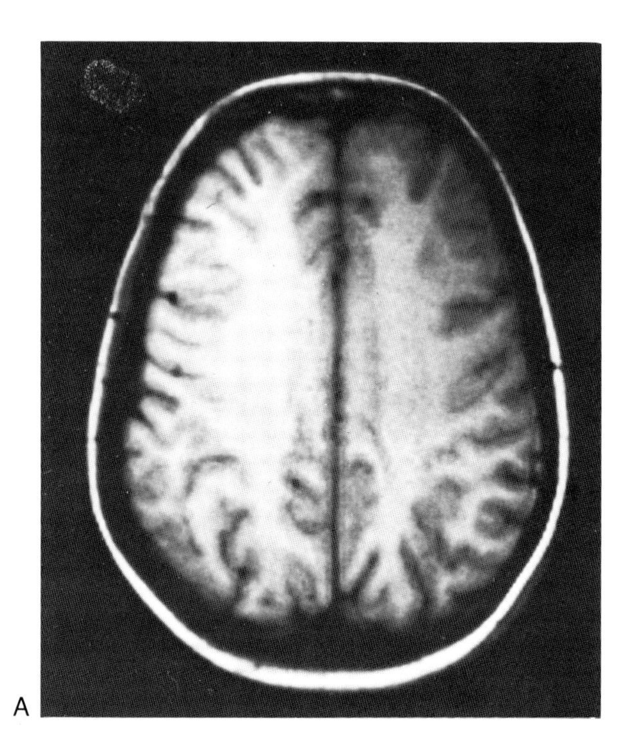

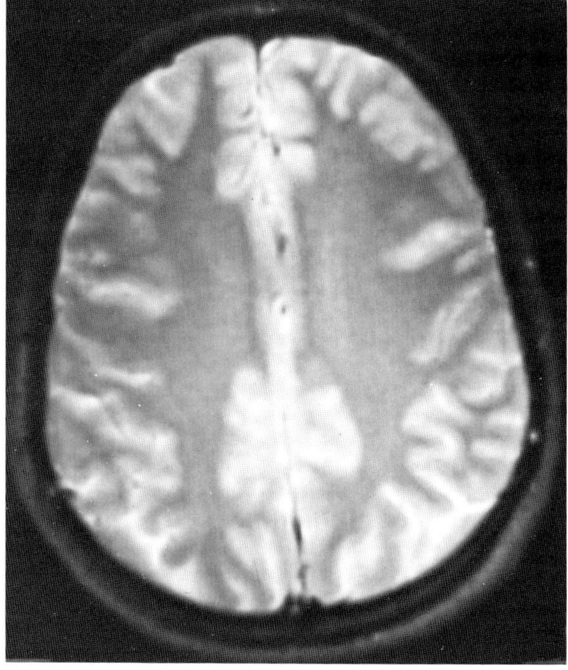

Fig. 2-3. Contrast reversal phenomenon. **(A)** T1-weighted SE 500/30 image. White matter signal > gray matter > CSF. **(B)** T2-weighted SE 2500/70 image. CSF signal > gray matter > white matter. A contrast reversal can take place between certain tissues as either TR or TE is increased.

weighted image" T1 contrast predominates; CSF with a long T1 appears darker than brain. On the T2-weighted image, however, the reverse holds true. When image contrast is weighted towards T2 contrast, the long T2 of CSF makes it appear brighter than that of brain. A "contrast reversal" is said to have taken place.

Unfortunately, the simplified scheme presented in Table 2-1 is not really correct. While widely adopted in the literature and popular among practicing radiologists, its arbitrary use may lead to logical paradoxes and errors in image interpretation. Several factors tend to make the scheme of Table 2-1 somewhat treacherous.

First, several important substances with high water content have T1 and T2 values that are much longer than those of other intracranial tissues. Such substances include the ocular vitreous, CSF, certain tumors, and parenchymal edema. With such very long T1 values (3000 to 5000 msec), significant T1-diminution of the MR signal may still occur even on heavily T2-weighted sequences. For example, consider the high intensity MS plaque seen on the T2-weighted SE 2000/80 image of Figure 2-4. The fact that this plaque has higher signal intensity than CSF does not necessarily mean its T2 is longer than that of CSF. It may merely mean that its T1 is shorter. Therefore, even with so-called heavily T2-weighted pulse sequences appreciable dependence on T1 and still exists for some tissues. Similar conclusions may be drawn concerning the so-called T1- and hydrogen density-weighted sequences, which always retain considerable T2 dependence.

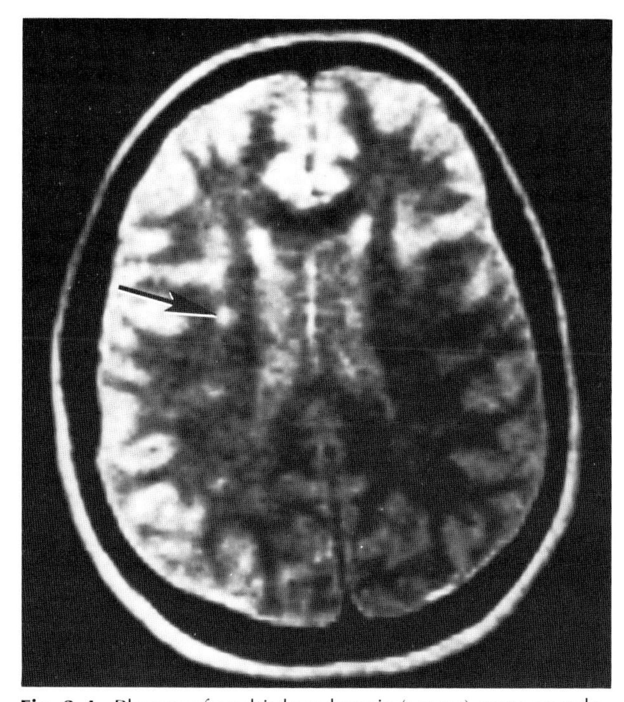

Fig. 2-4. Plaque of multiple sclerosis (arrow) seen on relatively T2-weighted SE 2000/56 image. Because the plaque is higher intensity than CSF it does not mean the plaque has a longer T2 than CSF. The signal from CSF is lower than that of the plaque because the T1 of CSF is much longer. Hence, even though SE 2000/56 is considered to be a "T2-weighted" sequence, one should always remember that appreciable T1 and ρ-dependence remains.

In an effort to quantitatively understand parameter weighting in cranial MRI, a set of weighting indices have been developed.[5] These indices quantify the fractional sensitivities of the SE signal to changes in T1, T2, and ρ. For example, an SE 1000/25 sequence when applied to brain parenchyma (with T1 $\approx$ 500 msec and T2 $\approx$ 50 msec) is approximately 55 percent ρ-weighted, 17 percent T1-weighted, and 28 percent T2-weighted. By comparison, an SE 2000/100 sequence is approximately 33 percent ρ-weighted, 65 percent T2-weighted, and only 2 percent T1-weighted.

These weighting indices give a measure of the fractional sensitivity of each pulse sequence to changes in ρ, T1, and T2. For example, consider the SE 2000/100 sequence, which was said to be 33 percent ρ-weighted and 65 percent T2-weighted. If tissue T2 increased by one unit (arbitrary), its MR signal would also increase. However, an identical increase in signal could be obtained by increasing tissue ρ by two units, leaving T2 constant. Thus although the SE 2000/100 sequence is considered "T2-weighted," it may be seen that considerable ρ-dependence still remains.

The relative weighting of an SE sequence will also depend upon field strength. As field strength increases from 0.15 T to 1.5 T, T1 values of most tissues approximately double.[6] Therefore, an SE 500/20 image at 0.15 T possesses about the same degree of T1-weighting as an SE 1000/20 image at 1.5 T. However, since ρ and T2 are largely independent of field strength an SE 3000/100 sequence at 0.15 T and 1.5 T are similar in T2-weighting.

In conclusion, it is necessary to know field strength,

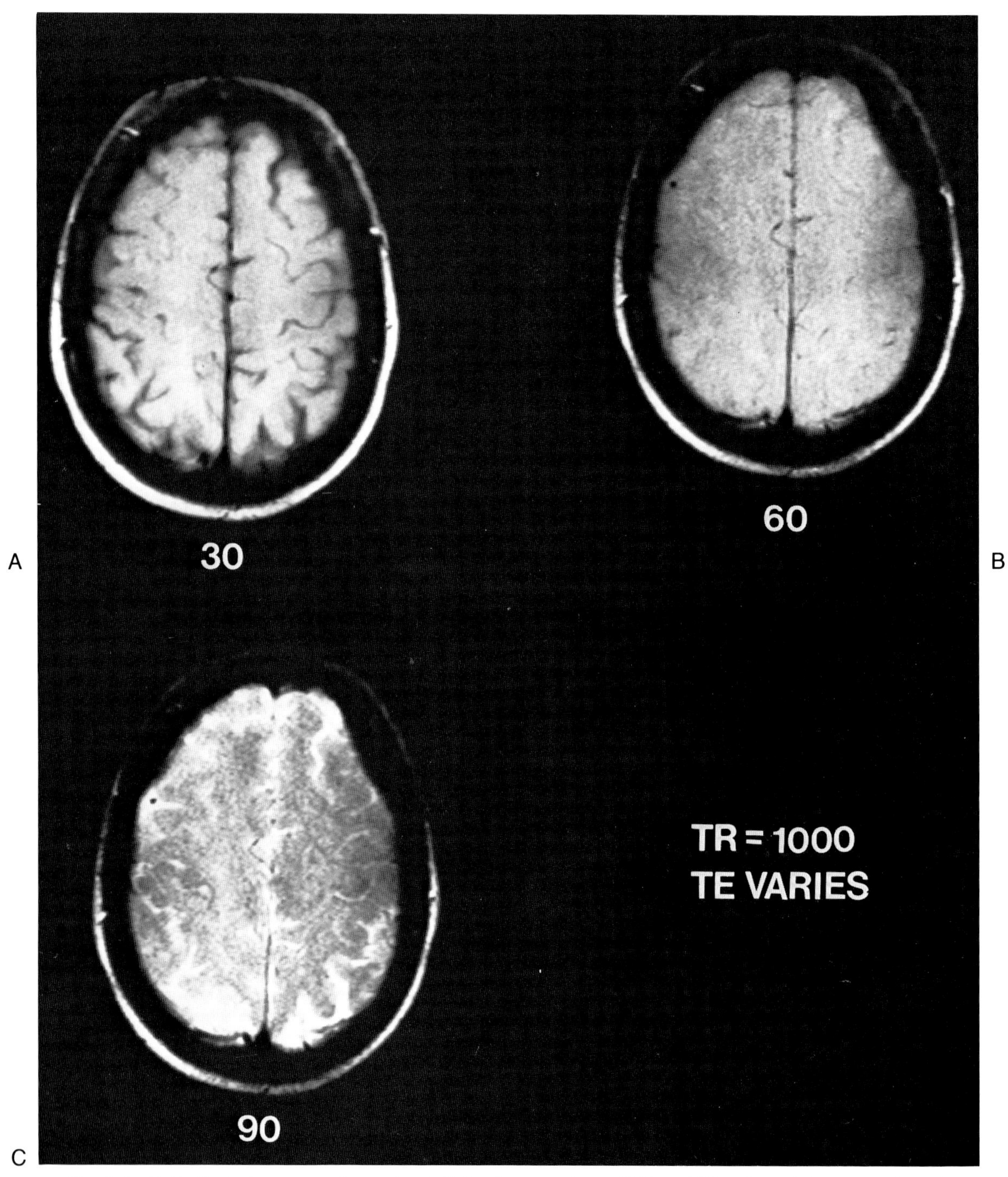

Fig. 2-5. (A-I). Brain images performed at 1.5 T using a variety of TR and TE values. (*Figure continues*).

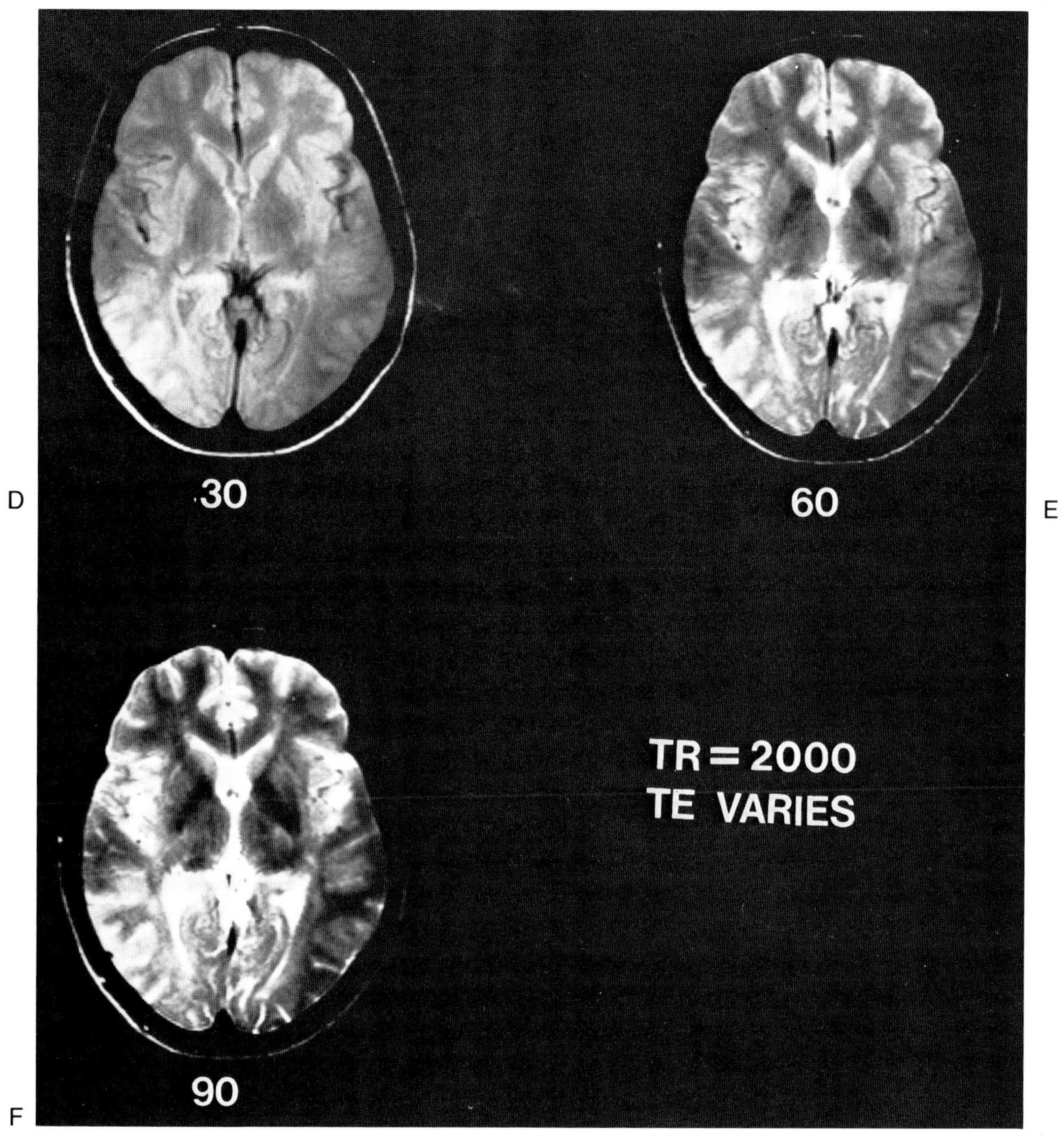

Fig. 2-5 *(Continued).*

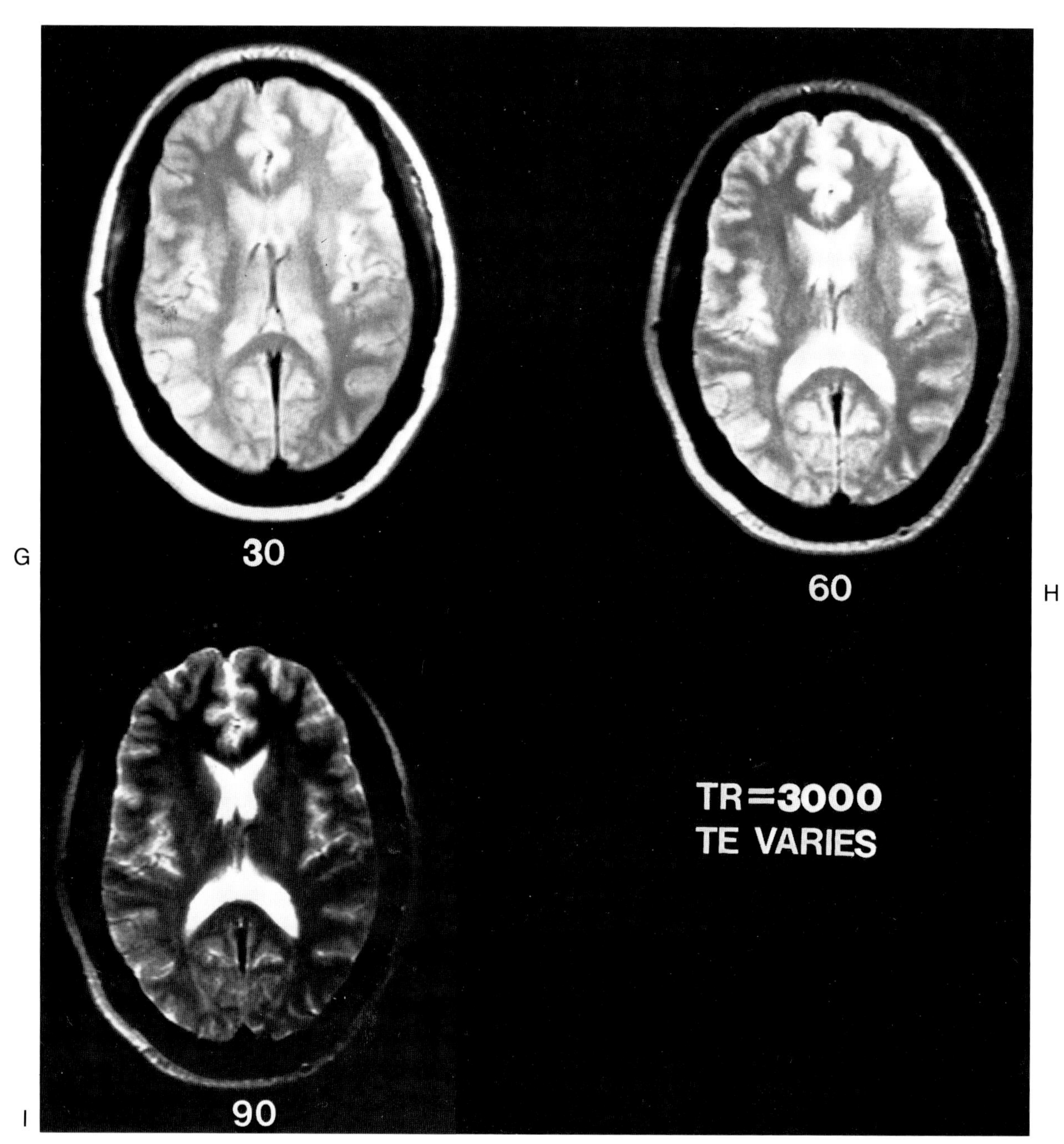

Fig. 2-5 *(Continued).*

pulse sequence, and intrinsic tissue parameters to predict how a given tissue will appear on an MR image. Such predictions may require considerable experience. It may be difficult to compare two published reports where different pulse sequences and field strengths were employed. What is "T1-weighted" to one person is not necessarily so to another.

For reference, a set of brain images performed on 0.5 T scanner with different values of TR and TE are presented in Figure 2-5.

THE INVERSION RECOVERY PULSE SEQUENCE

The inversion recovery (IR) sequence is less commonly used than SE, but remains an important technique for cranial imaging. While SE is basically a 90°-180°-signal sequence, IR is a 180°-90°-signal sequence (Fig. 2-6). A *repetition time* (TR) may be defined as that between successive 180°-pulses. An *inversion time* (TI) is defined as the time between the 180°- and 90°-pulses. In common clinical usage TR is chosen in the range 1000 to 2500 msec, while TI is varied between about 80 and 800 msec.

The IR sequence is difficult to understand because it is implemented on commercial scanners in at least four different ways.[7] These four implementations represent combinations of two modes of data collection (free induction decay, spin echo) and two modes of signal processing (magnitude reconstruction, phase-corrected reconstruction). This confusion has perhaps contributed to IR's loss of popularity in many imaging centers.

Data collection for an IR sequence may be obtained in two ways: (1) by recording a free induction decay (FID) or (2) by generating a spin echo. The FID technique is illustrated in Figure 2-6A. The FID technique results in a signal that has little T2-dependence (i.e., is strongly T1-weighted). However, the FID technique is noisy and susceptible to field inhomogeneities. Because of this, most commercial scanners modify the IR sequence by the addition of a second 180°-pulse (Fig. 2-6B). This 180°-pulse generates a spin echo at time TE following the 90°-pulse. The use of this echo technique imparts considerable T2 dependence to the IR sequence. For this reason the IR sequence with spin echo data collection is sometimes called the *inversion spin echo technique*.

A shorthand technique for denoting inversion re-

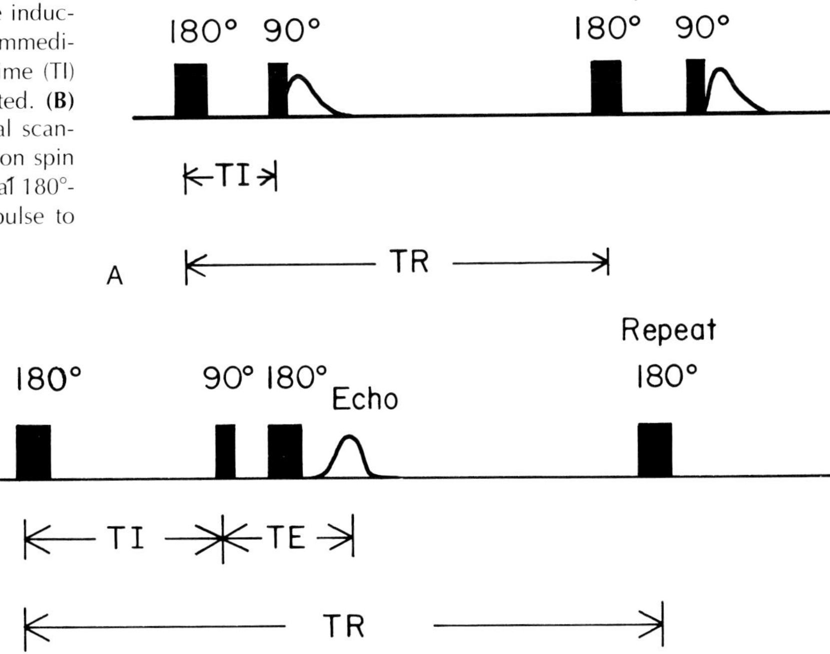

Fig. 2-6. Inversion recovery pulse sequence. **(A)** In its original development a free induction decay (FID) signal was sampled immediately after the 90°-pulse. Inversion time (TI) and repetition time (TR) are illustrated. **(B)** As implemented on most commercial scanners today, a modified IR or "inversion spin echo" technique is used. An additional 180°-pulse is applied following the 90°-pulse to create an echo signal at time TE.

covery sequences is in common use. A given sequence is denoted IR TR/TI/TE where TR, TI, and TE are in milliseconds. For example, an IR technique with TR = 1500, TI = 400, and TE = 25 msec would be denoted IR 1500/400/25.

Regardless of the method of data collection, two very different methods for processing the MR signal exist: (1) magnitude reconstruction and (2) phase-corrected magnitude reconstruction. The difference between these two techniques is based upon whether the detector can distinguish a tissue magnetization parallel to the external field from a vector of equal magnitude pointed opposite to the field. A phase-sensitive (quadrature) detector is needed to make this distinction.

The difference between these two modes of signal processing can be understood by following the recovery of the tissue magnetization vector **M** from the 180°- (inverting) pulse (Fig. 2-7). Immediately following the 180°-pulse, tissue magnetization is flipped to the −z direction. As longitudinal recovery occurs this magnetization decreases to zero then increases in the +z direction (in the direction of $\mathbf{B_0}$). The position of this recovery at time TI determines the strength of the MR signal.

In Figure 2-8 the longitudinal recovery for brain and CSF is plotted as a function of time following a 180°-pulse. At time TI = a, the magnetization vectors for brain and CSF both point in the −z direction (opposite the external field). Using simple magnitude reconstruction, CSF appears whiter than brain because its absolute magnitude is greater. Using phase-corrected magnitude, however, CSF will appear darker than brain because more negative magnetization is considered to be a smaller quantity on this scale.

At time TI = b, the signal from brain and CSF are of equal magnitude but lie in opposite directions. Using magnitude construction, both brain and CSF would be the same shade of gray, and there would be no visual contrast between them. Using phase-corrected reconstruction, however, the signal of brain would be appreciably different from that of CSF.

At time TI = c, CSF has no longitudinal magnetization while brain has appreciably recovered. While both magnitude and phase-corrected reconstruction could both differentiate the two substances, use of the simple magnitude method might be more advantageous here. Using magnitude reconstruction the CSF is imaged as purely black; its signal is said to have been *suppressed*. The IR technique with magnitude reconstruction may be useful in certain situations to suppress bothersome high signals from fat or edema in order to better delineate an abnormality.

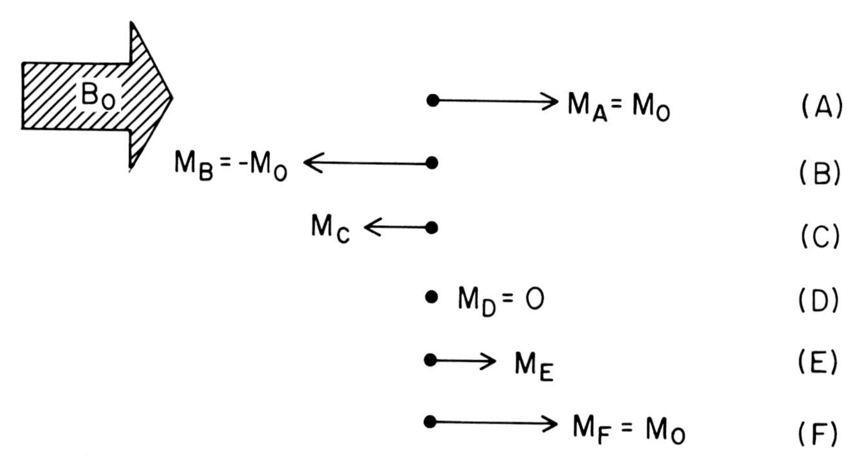

Fig. 2-7. Recovery of tissue magnetization vector, **M,** during the inversion recovery sequence. **(A)** At equilibrium **M** is at maximum value M_0, pointed in the +z direction. **(B)** Immediately following the 180°- (inverting)-pulse, **M** has been flipped to point opposite $\mathbf{B_0}$, i.e., in the −z direction. Its magnitude is the same, but now negative, $-M_0$. **(C–E)** During the recovery, **M** grows exponentially from negative to positive values. **(F)** After complete recover M now points in the +z direction with full initial value, M_0.

Fig. 2-8. Graphic representation of the longitudinal magnetization (M_z) as a function of time.

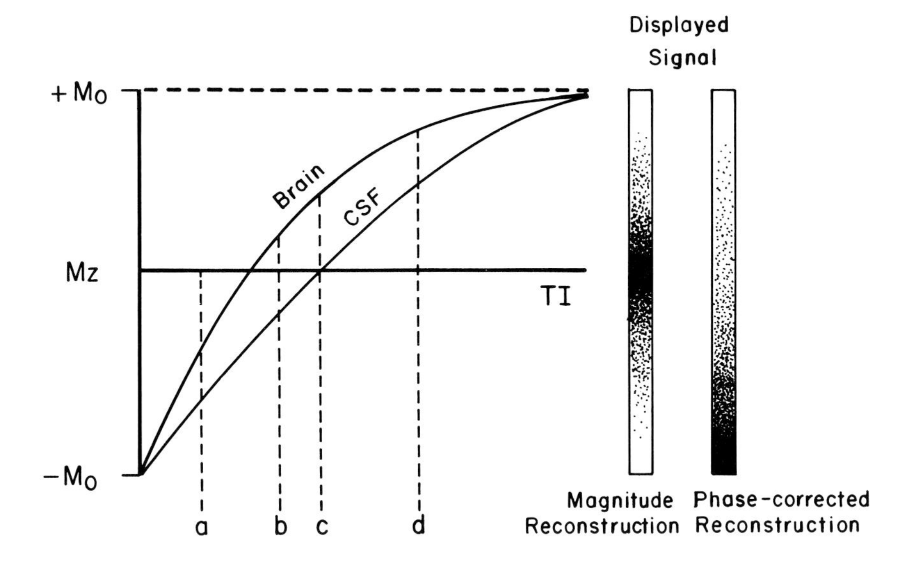

At time TI = d, both CSF and brain possess positive magnetizations. There is no particular advantage to one type of reconstruction over another. Examples of the brain imaged with short TI, medium TI, and long TI sequences with either magnitude or phase-corrected magnitude reconstruction are shown in Figures 2-9 and 2-10.

An equation similar to Eq. 2-1 has been developed

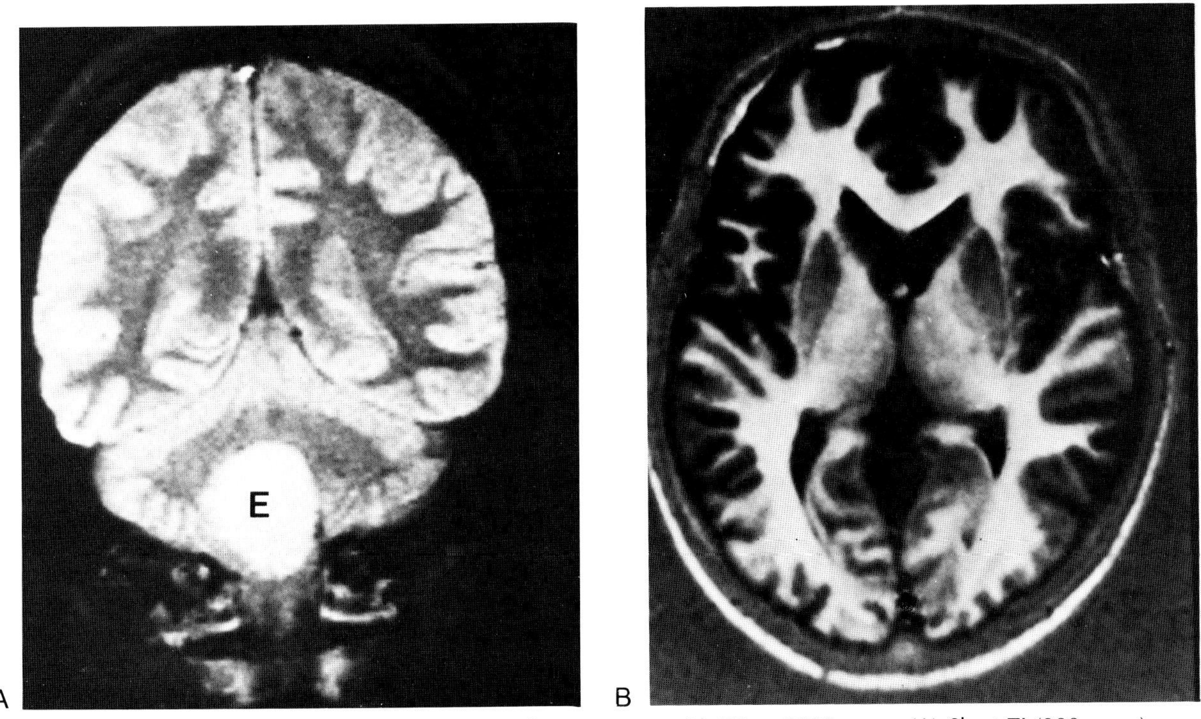

Fig. 2-9. Examples of the inversion recovery pulse sequence with TR = 2000 msec. **(A)** Short TI (200 msec). **(B,C)** Long TI (700 msec).

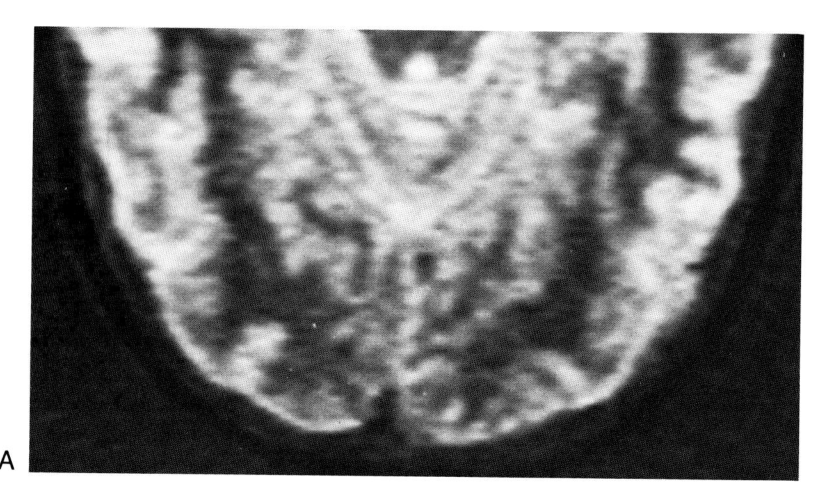

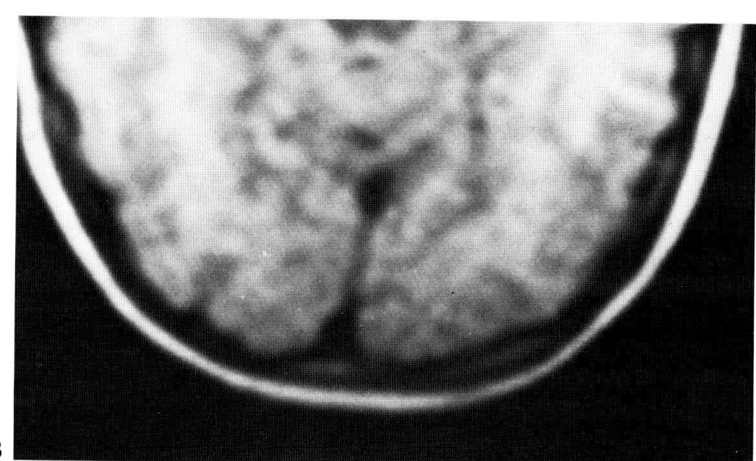

Fig. 2-10. Two modes for processing an IR signal (IR 750/400/40). **(A)** Magnitude reconstruction. **(B)** Phase-corrected magnitude reconstruction.

for the IR sequence.[4] The basic structure of this equation is

$$I = k\,\rho\,(1 - 2e^{-TI/T1} + e^{-TR/T1}) \qquad (2\text{-}2)$$

where I = MR signal intensity, k = constant, ρ = proton density, TI = inversion time, TR = recovery time, and T1 = transverse relaxation time for the imaged tissue. This equation is accurate for IR imaging using phase reconstruction and FID signal detection. If magnitude reconstruction is used, the absolute value of the term in parentheses must be taken. If spin echo signal detection is used an additional term $e^{-TE/T2}$ must appear as a factor in the equation.

As a general rule IR is a proton and T1 weighted sequence, with signal intensity directly related to ρ and 1/T1. However, many exceptions exist depending on the choice of TR, TI, type of signal collection and type of data reconstruction. A simplified scheme is presented in Table 2-2.

Table 2-2. Dependence of IR Signal Intensity Using Various IR Techniques

For most IR sequences, signal $\propto \frac{1}{T1}$
For two sequences, signal $\propto$ T1
Long TR/short TI with magnitude reconstruction
Short TR/short TI with phase-corrected reconstruction
Medium IR sequences using magnitude reconstruction have largely unpredictable contrast reversals.

Long TR means TR >2000 msec; short TR means TR <1000 msec; short TI means TI <250 msec; medium TI means 250< TI<750 msec.

OTHER IMAGING CONSIDERATIONS

In addition to selecting a pulse sequence, several other factors need to be considered in clinical MR imaging.

Matrix Size

The image acquisition matrix on most commercial scanners today ranges from 128 × 128 to 512 × 512, representing a product of the number of phase- and frequency-encoding steps. As with CT, increasing the number of matrix elements increases spatial resolution. However, the noise per pixel is also increased.[8] Additionally, imaging time is directly proportional to the number of phase-encoded steps. As a tradeoff, many clinicians choose a 128 × 256 matrix for routine head imaging today.

Signal Averaging

The single best way to improve the quality of an MR image is to increase the number of signal averages. Signal-to-noise ratio, in MRI increases with $\sqrt{N}$, where N is the number of excitations.[8] However, scanning time is directly proportional to N. Hence to obtain a twofold increase in signal-to-noise, scanning time must be prolonged by a factor of 4. Prolonged scanning time carries with it another complicating factor, patient motion. Subtle patient motion during prolonged scanning may obviate any potential benefits obtained in signal-to-noise.[9] For these reasons only 1 to 4 signal averages are performed for most applications in the CNS.

Field of View

A field of view option is available on nearly all commercial MR scanners. This allows setting of the gradient fields so that recording and processing of signal information occurs only over a certain limited region of space. The field of view is usually specified in terms of a sphere of a certain diameter, say 10 to 30 cm. The field of view should be chosen such that the imaged area or body part is nearly matched in size. For routine cranial imaging, a 25 cm field of view is usually best. If the body part is larger than the chosen field of view a *fold-over artifact* will be produced.

Selection of Frequency- and Phase-Encoded Gradients

Most scanners allow the operator to select which imaging axis (x or y) will be encoded by frequency and which by phase. In the majority of cases this choice is inconsequential. However, it may be important when the imaged slice contains some tissues that are in motion. For example, slices that include the orbit or brain stem frequently contain motion artifacts produced by eye motion or flowing blood. These "ghost" artifacts propagate across the entire MR image along the direction of the phase-encoded axis.[9] When imaging the brain in axial section, therefore it is useful to specify that the horizontal x-axis be encoded in phase so that ghost artifacts from the orbits not cross through the brain parenchyma.

Imaging Time Considerations

Image acquisition time can be easily calculated as the product of (number of phase-encoding steps) × (repetition time) × (number of excitations for signal averaging). Note that it is the number of phase-encoding steps, not frequency-encoding increments that determine imaging time. Furthermore, multiple slices may be acquired during this period, the precise number depending upon pulse sequence and scanner design. As a general rule, more simultaneous slices can be obtained when TR is long and only one echo is recorded.

Cardiac Gating and Motion Suppression Techniques

Pulsatile flow of blood and CSF can produce artifacts that may obscure or mimic pathology in the basal cisterns, temporal lobes, and brain stem. These artifacts are particularly prominent on T2-weighted images. Cardiac gating has been shown effective in removing some of these artifacts.[10] However, cardiac gating is inconvenient and can add considerably to set-up time. Elster et al. have recently shown the ease and utility of a gradient motion artifact suppression technique (MAST) and its superiority over cardiac gating in reducing such artifacts.[11]

REFERENCES

1. Hahn EL: Spin echoes. Phys Rev 80:580, 1950
2. Carr HY, Purcell EM: Effects of diffusion on free precession in nuclear magnetic resonance experiments. Phys Rev 94:630, 1954
3. Meiboom S, Gill D: Modified spin-echo method for measuring nuclear relaxation times. Rev Sci Instrum 29:688, 1958
4. Dixon RL, Ekstrand KE: The physics of proton NMR. Med Phys 9:807, 1982
5. Elster AD: An index system for comparative parameter weighting in MR imaging. Radiology 161(p):71, 1986
6. Bottomley PA, Foster TH, Argersinger RE, Pfeifer LM: A review of normal tissue hydrogen NMR relaxation times and relaxation mechanisms from 1–100 MHz: dependence on tissue type, NMR frequency, temperature, species, excision, and age. Med Phys 11:425, 1984
7. Bydder GM, Young IR: MR imaging: clinical use of the inversion recovery sequence. J Comput Assist Tomogr 9:659, 1985
8. Edelstein WA, Bottomley PA, Hart HR, Smith LS: Signal, noise, and contrast in nuclear magnetic resonance (NMR) imaging. J Comput Assist Tomogr 7:391, 1983
9. Elster AD: Magnetic Resonance Imaging: A Reference Guide and Atlas. JB Lippincott, Philadelphia, 1986
10. Enzmann DR, Rubin JB, O'Donohue J, et al: Use of cerebrospinal fluid gating to improve T2-weighted images. Radiology 162:768, 1987
11. Elster AD, Moody DM, Laster DW et al: A motion artifact suppression technique (MAST) for cerebral MR imaging. Presented at American Society of Neuroradiology annual meeting, New York, May, 1987

3

Normal Cranial Anatomy and Variants

Even for diagnosticians well versed in neuroanatomy and intracranial diseases, certain aspects of head MRI will still require some study. Some anatomic relationships, particularly those in the coronal and sagittal planes, may not be immediately apparent. The different normal signals cast by brain components and surrounding soft tissues may also be somewhat confusing. Finally, a number of artifacts and normal variants must be recognized to avoid mistaking them for pathologic processes. Once these few tricks are learned, the quality of the diagnostician's interpretive ability in cranial MRI will depend largely on a knowledge of neurologic diseases, not on a background in physics.

BONES AND SOFT TISSUES

The bones and soft tissues of the skull are generally best appreciated on T1-weighted images, which allow good contrast between fat and the denser structures (Fig. 3-1). The scalp and superficial connective tissue are the first layers encountered, followed by high intensity fat. The galea aponeurotica is not normally seen because it lies close to the outer table of the skull. Subgaleal masses, however, may be identified by their characteristic extrinsic elevation of the scalp fat layer (Fig. 3-2).

The first dark band encountered (Fig. 3-1) is the outer table of the skull, which casts little signal due to its paucity of hydrogen protons. The inner table may also be distinguished, separated from the outer table by high intensity diploic marrow or fat. Where the diploë are thin and at cranial sutures, no MR signal is identified (Fig. 3-1, curved arrow).

Several scalp lesions are frequently encountered. Redundant scalp fat may cause "pseudo" masses (Fig. 3-3A). Subcutaneous lipomas have similar intensity to other fat and are often separated from it by a capsule (Fig. 3-3B). Sebaceous cysts are frequently located in the suboccipital region and are diagnosed clinically, but may be seen on MR (Fig. 3-3C).

Distortions in calvarial shape may also be recognized by MRI. Other than trauma or surgery, the most frequent cause of skull deformity is a premature fusion of one or more of the cranial sutures, a process called *craniosynostosis*. Usually craniosynostosis is a selective process involving one or a pair of sutures, but various combinations may occur (Table 3-1). In most cases the cause is unknown, although over 40 genetically determined syndromes are associated with craniosynostosis and in other cases an infantile metabolic disorder (rickets, hypophosphatasia, hyperthyroidism) may be responsible.[1] Examples of several head shapes produced by the more common craniosynostoses are presented in Figure 3-4.

Several congenital and developmental lesions may be encountered in the bony calvarium. The *Davidoff-*

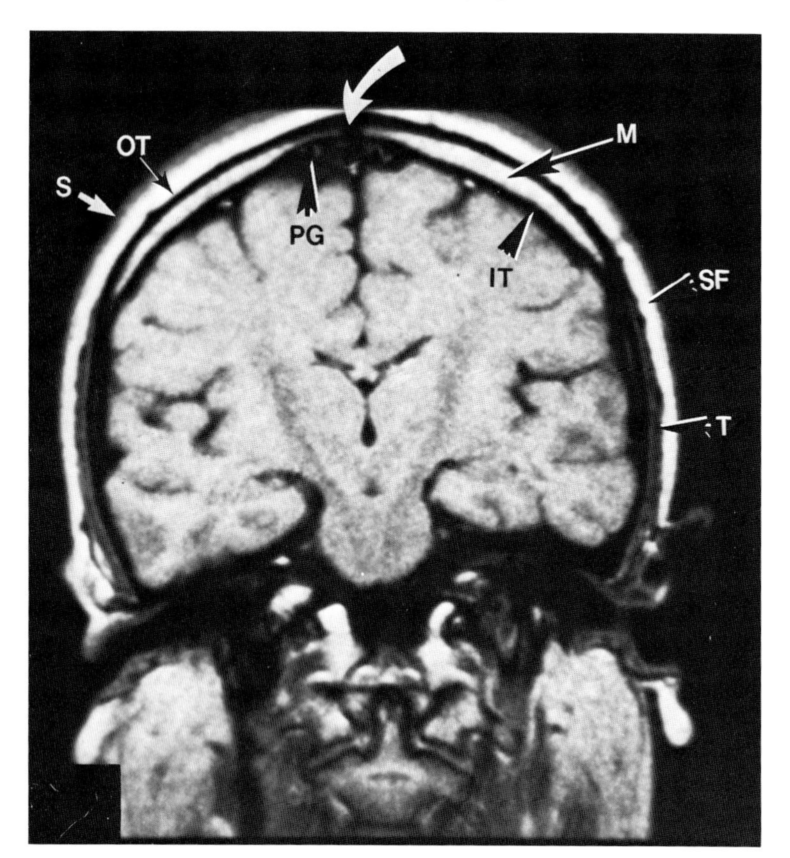

Fig. 3-1. Bones and soft tissues of the head seen on coronal SE 400/28 image. *S*, skin and subcutaneous connective tissue layer; *OT*, outer table of skull; *PG*, Pacchionian granulation; *M*, marrow of skull diploic space; *IT*, inner table of skull; *SF*, scalp fat; *T*, temporalis muscle; curved white arrow, sagittal suture.

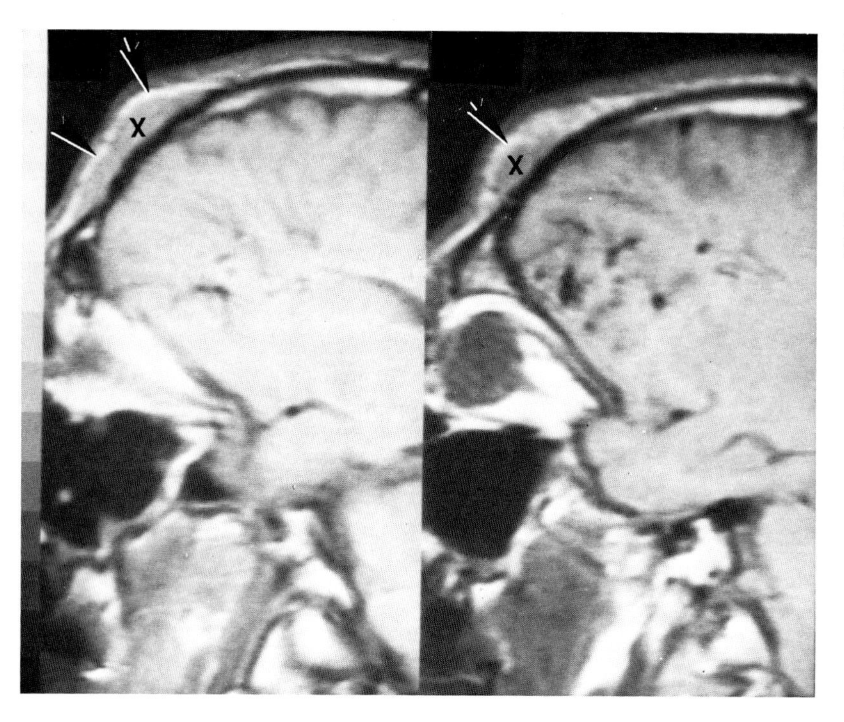

Fig. 3-2. Subgaleal mass (*X*) displaces scalp fat layer (arrowheads) away from outer table of the skull. The mass was part of a venous malformation that emptied into the superior sagittal sinus through a frontal bone defect, a sinus pericranii.

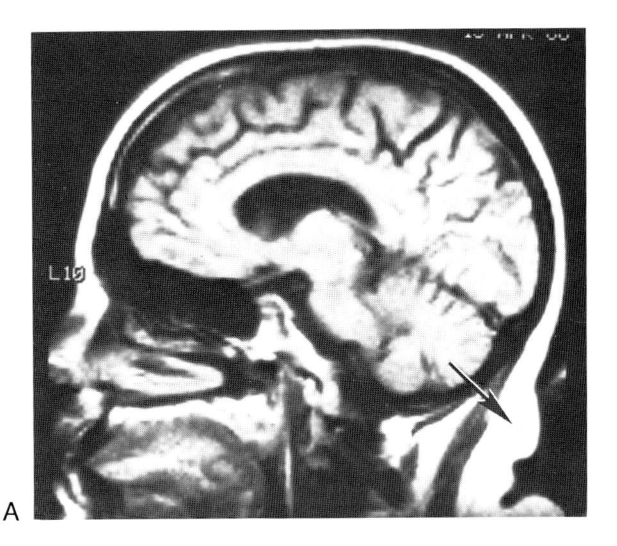

A

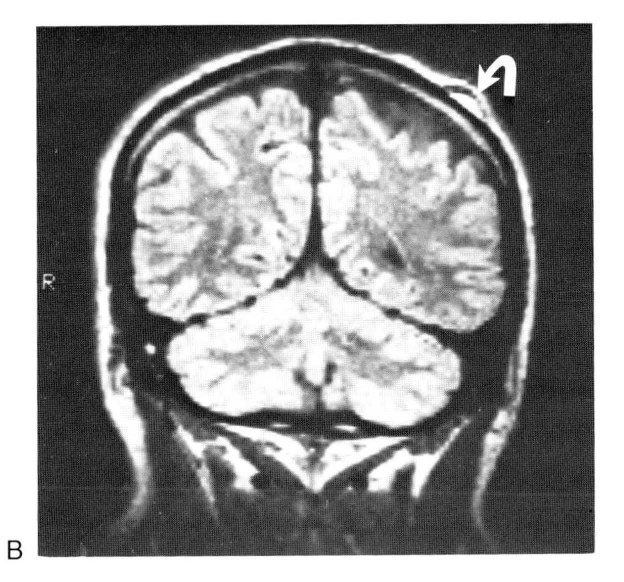

B

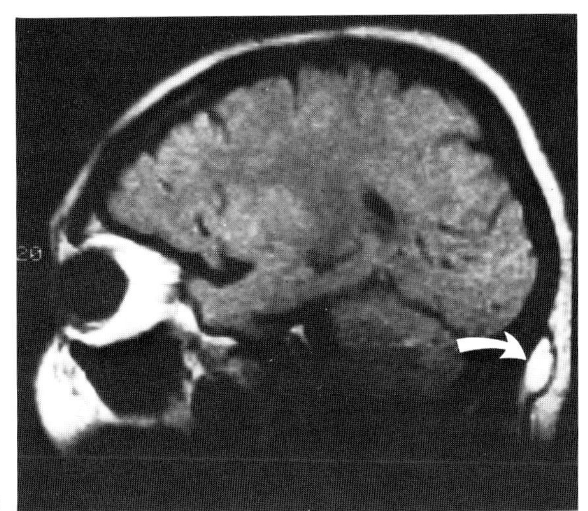

C

Fig. 3-3. Common scalp lesions. **(A)** "Pseudo" mass caused by redundant scalp fat. **(B)** Lipoma. **(C)** Sebaceous cyst.

Table 3-1. Frequencies and Types of Craniosynostoses

Suture Involved	*Skull Shape*	*Frequency* (%)
Sagittal alone	Dolicocephaly (scaphocephaly)	55
Coronal		24
Bilateral	Brachycephaly	
Unilateral	Plagiocephaly	
Metopic	Trigonencephaly	4
Lambdoid	Flat occiput	2
All three sutures	Oxycephaly (acrocephaly, turricephaly)	7
Other combinations		8

(Shellito J, Matson DD: Craniosynostosis. A review of 519 surgical patients. Pediatrics 41:829, 1968.)

Dyke syndrome refers to skull changes relating to hemiatrophy of the brain. These include thickening of the diploic space and enlargement of the frontal sinus on the involved side (Fig. 3-5A). Congenital inclusions of epidermal elements may result in an *epidermoid tumor* (Fig. 3-5B). *Meningoceles* and *encephaloceles* may also be recognized on MRI, though they are usually apparent clinically (Fig. 3-5C,D).

Hyperostosis frontalis interna is a benign common lesion seen in at least 15 percent of women older than 40.[2] It is exceedingly rare in men of all ages. Radiolographically, thickening occurs along the frontal bone inner table. The outer table is spared and usually some diploic space remains. Hyperostosis frontalis interna is nearly always bilateral and does

A
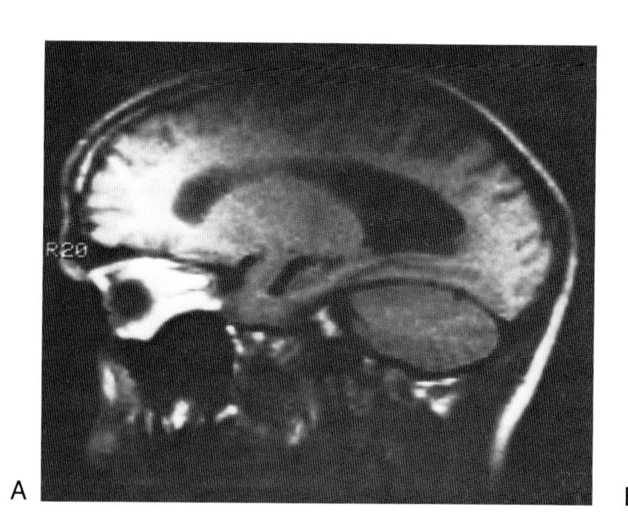

B
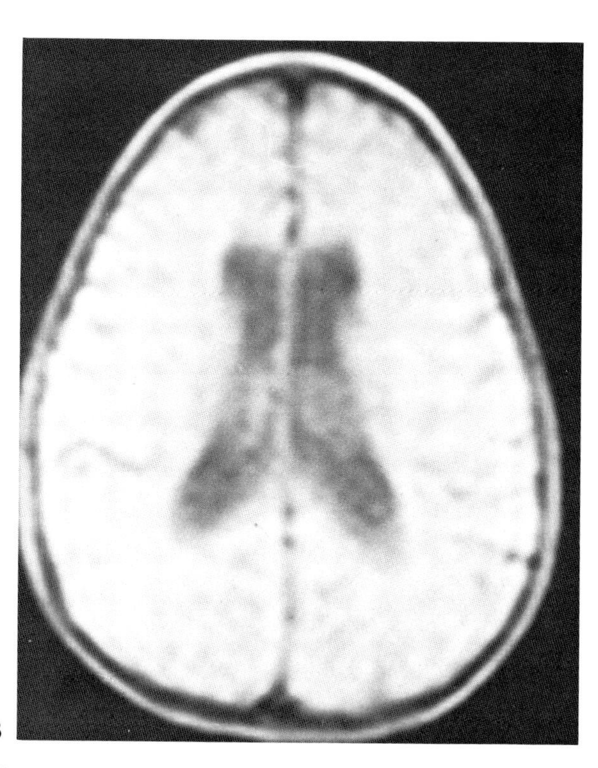

C
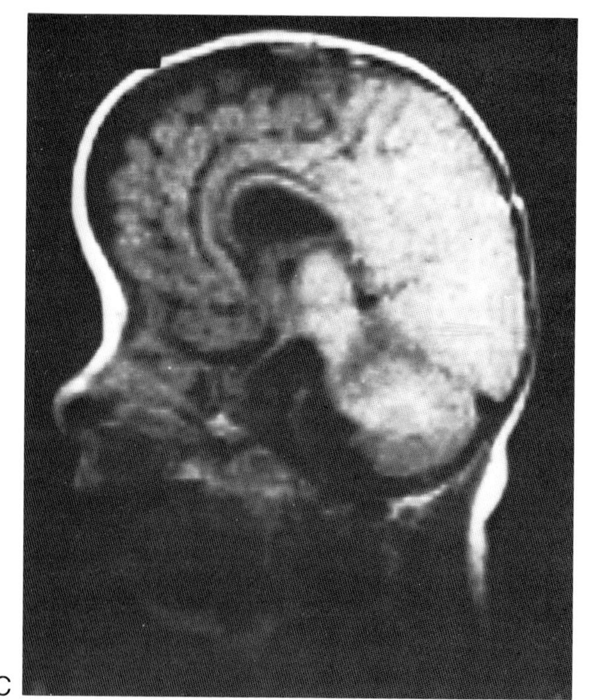

Fig. 3-4. Craniosynostoses. **(A)** Dolicocephaly, caused by premature fusion of the sagittal suture. **(B)** Trigonencephaly, caused by premature fusion of the metopic suture. **(C)** Oxycephaly due to fusion of coronal, lamboid, and sagittal sutures. In this case it was associated with fusion of the fingers, a condition known as Apert syndrome.

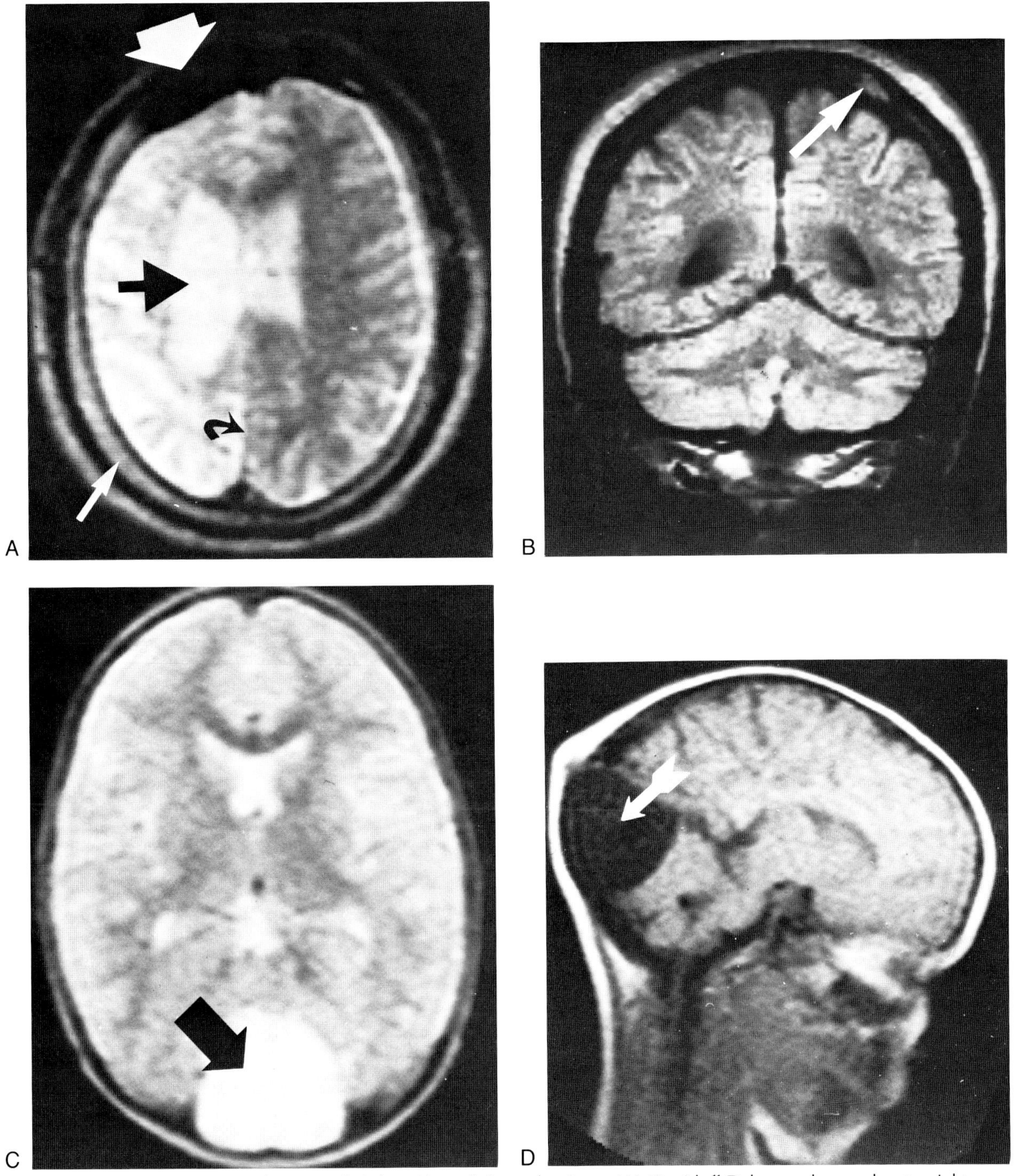

Fig. 3-5. Congenital/Developmental deformities of the calvarium. **(A)** Davidoff-Dyke syndrome due to right hemisphere hemiatrophy. The frontal sinus is enlarged (large white arrow) and the diploic space is thickened (small white arrow). The lateral ventricle is enlarged (large black arrow) and the falx is shifted to the right (curved black arrow). **(B)** Calvarial epidermoid (arrow). Nonspecific on MR, requires skull x-ray and/or bone scan to confirm. **(C)** Small meningocele in a 2-year old.

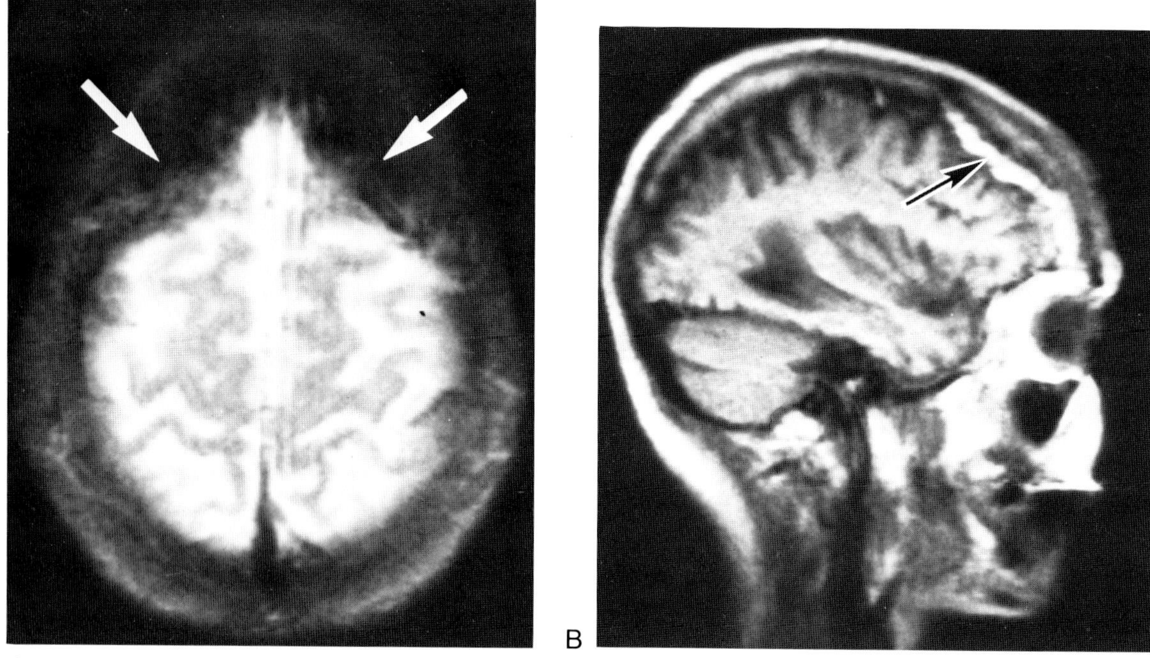

Fig. 3-6. Hyperostosis frontalis interna. **(A)** Usually the dense calcification has little MR signal. **(B)** On T1-weighted images, fat may occasionally be seen within the lesions as a normal variant.

not cross the midline, a feature serving to differentiate the disorder from bifrontal meningiomas. On MRI the hyperostosis characteristically has little signal due to its paucity of hydrogen protons (Fig. 3-6A). However, the hyperostosis will sometimes contain fatty marrow resulting in central high intensity on T1-weighted images[3] (Fig. 3-6B).

THE SKULL BASE

The skull base is a dense bony structure probably best visualized by high resolution CT using a bone algorithm. However, certain aspects of its anatomy are well demonstrated by MR and should be recognized. The detailed anatomy of the temporal bone and skull base will be presented more fully in Chapter 11. The section here can serve as an introduction to this region.

The term *clivus* is often used to refer to the bony complex consisting of the body of the sphenoid and basiocciput, though more properly it is defined to be only the dorsal surface of this bone. In adults the clivus frequently contains fatty marrow and casts a high signal on T1-weighted images. It is identified by its characteristic C-shape on axial images and triangular shape on sagittal images (Fig. 3-7A, B). Adjacent to the clivus on axial images, fat in the petrous pyramid apices may also be seen (Fig. 3-7C). In children, the spheno-occipital synchondrosis may be identified as a well-defined line on sagittal images (Fig. 3-7D). With aging, sclerosis and indistinctness may normally occur along this synchondrosis (Fig. 3-7E). Variable pneumatization of the sphenoid body may also be seen, ranging from none (Fig. 3-7E) to marked (Fig. 3-7F).

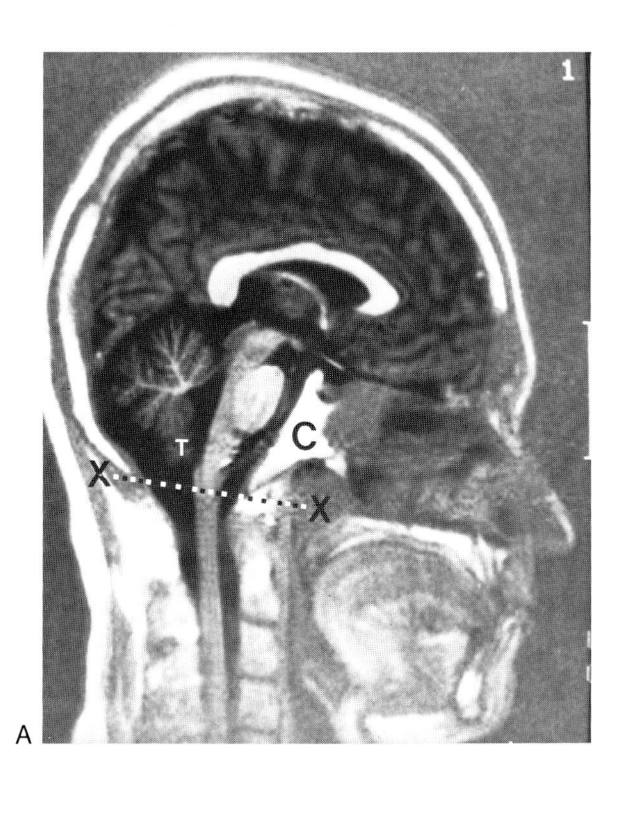

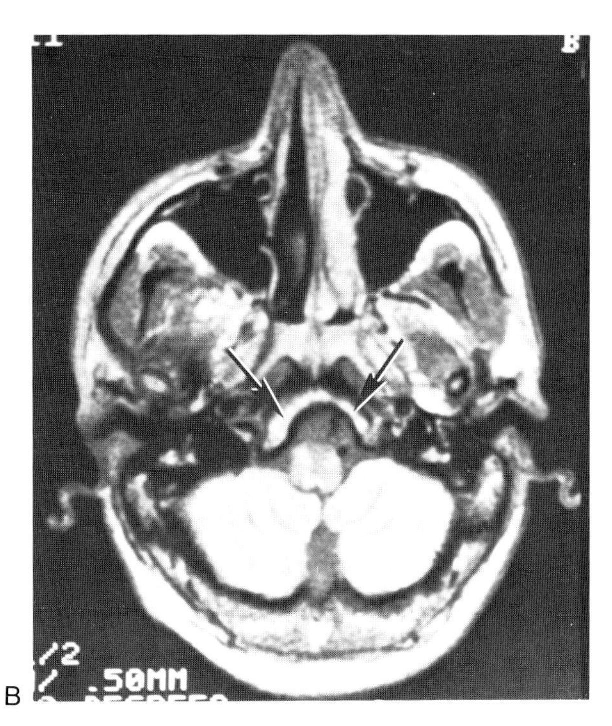

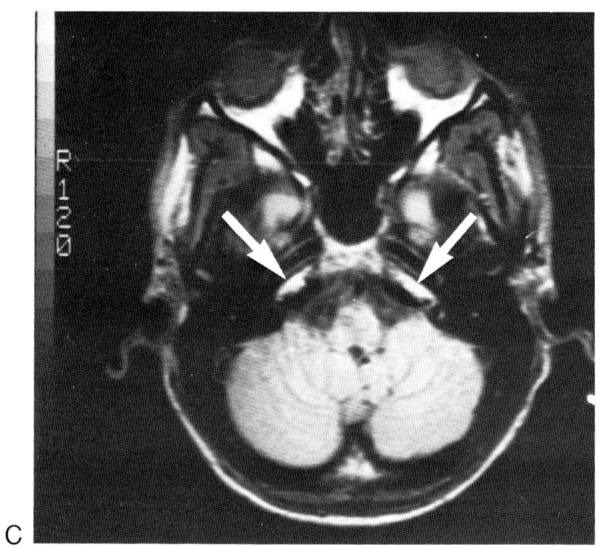

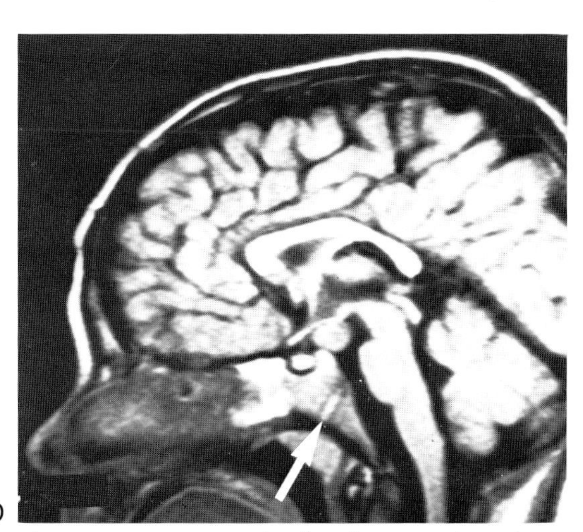

Fig. 3-7. MR anatomy of the skull base. **(A)** In the adult, the clivus (C) appears as a bright white triangle with sharp margins on T1-weighted images. The plance of the foramen magnum (dotted line, *X–X*) is defined by the tip of the clivus and occipital bone. The cerebellar tonsil (*T*) is also noted. **(B)** On axial images the clivus has a characteristic C-shape. **(C)** Fat in the petrous pyramid tups (arrows) may cause a confusing high signal on T1-weighted images if not recognized. **(D)** The spheno-occipital synchrondrosis is seen in younger patients (arrow). (*Figure continues*).

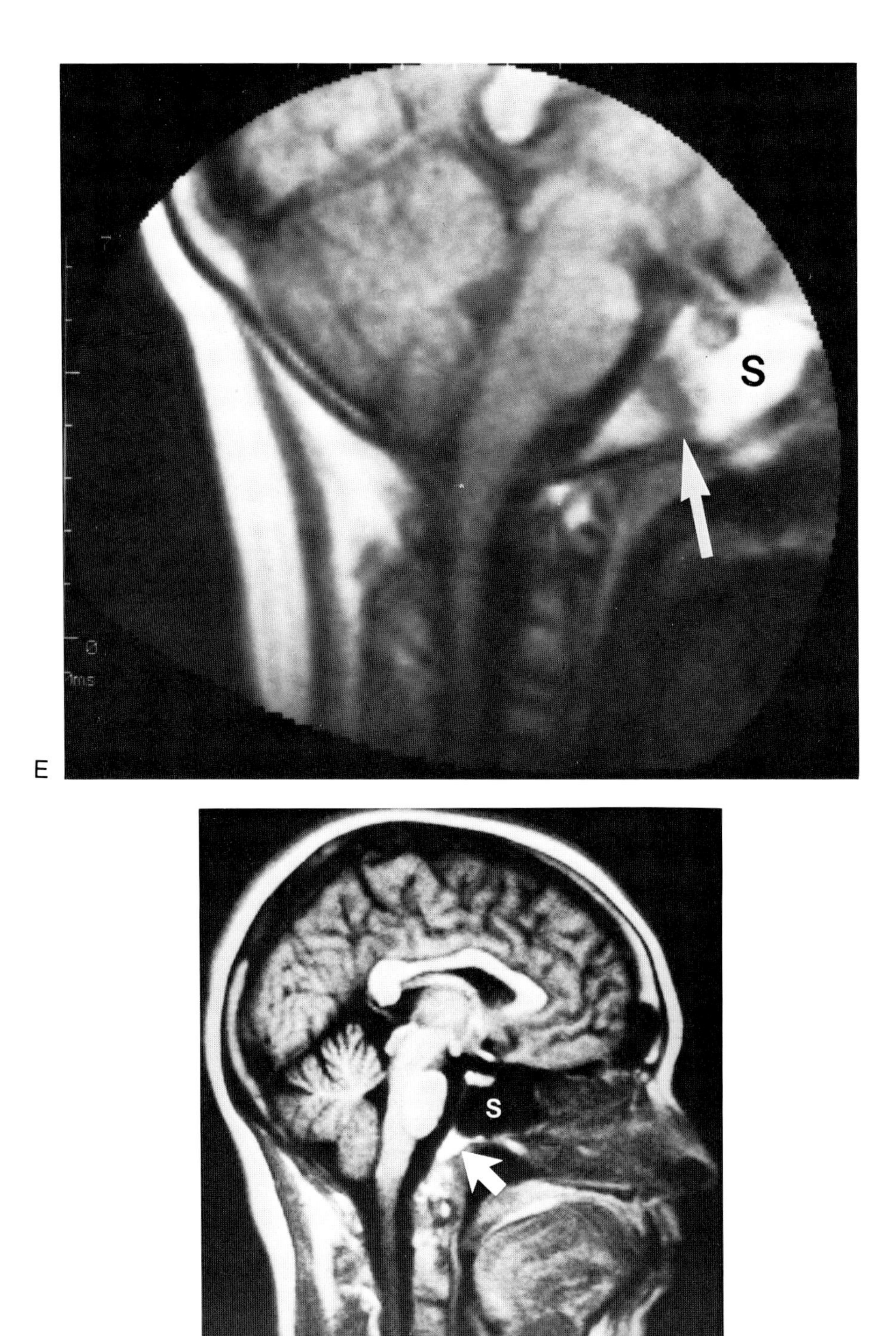

Fig. 3-7 (*Continued*). **(E)** With aging, the synchondrosis may become sclerotic and poorly defined. The body of the sphenoid bone (*S*) is not pneumatized here. **(F)** In this adult there is extensive aeration of the sphenoid sinus (*S*). Only a small triangular region of clivus marrow can be identified (arrow).

CEREBRAL VASCULATURE

Magnetic resonance imaging allows direct visualization of the cerebral vasculature. Relatively high velocity arterial and venous vessels appear dark due to turbulence and high velocity signal loss. More slowly flowing blood may appear bright. Some examples of normal arterial and venous anatomy on MRI are shown in Figures 3-8A,B.

CRANIAL NERVES

Each of the cranial nerves may be identified using high resolution MRI. The olfactory nerves may be seen in the subfrontal region on coronal or sagittal images (Fig. 3-9A). The optic nerves, chiasm, and tract are well visualized in multiple planes (see Fig. 3-9A and multiple illustrations, Ch. 12). The oculomotor nerve is best seen on parasagittal images that are

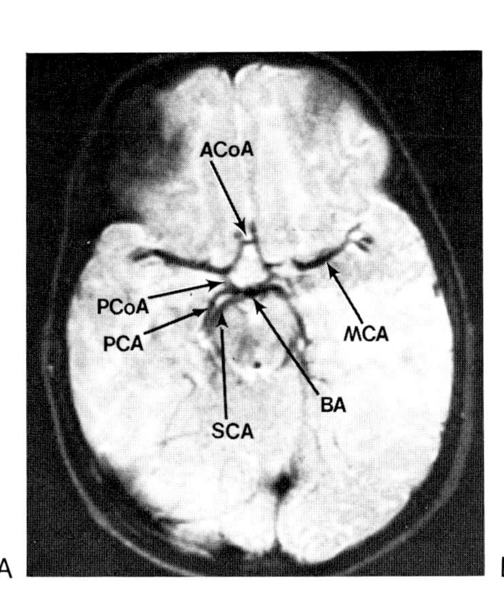

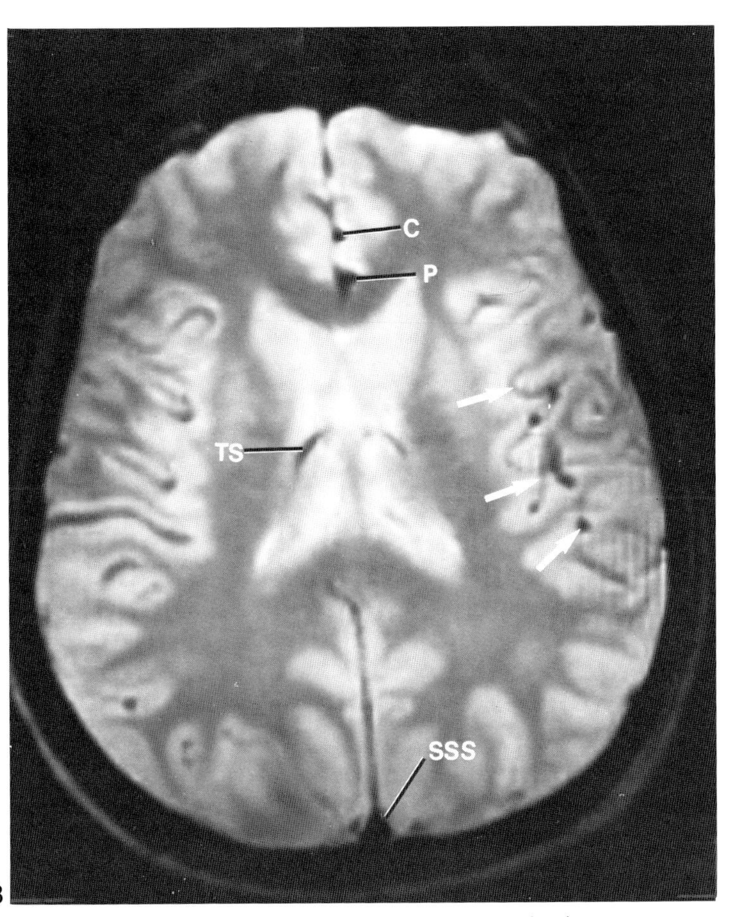

Fig. 3-8. Normal vascular anatomy on axial T2-weighted MR images. *AcoA*, anterior communicating artery; *BA*, basilar artery; *C*, callosal marginal artery; *MCA*, middle cerebral artery, M1 segment; *P*, pericallosal artery; *PCA*, posterior cerebral artery; *PCoA*, posterior communicating artery; *SCA*, superior cerebellar artery; *SSS*, superior sagittal sinus; *TS*, thalamostriate vein.

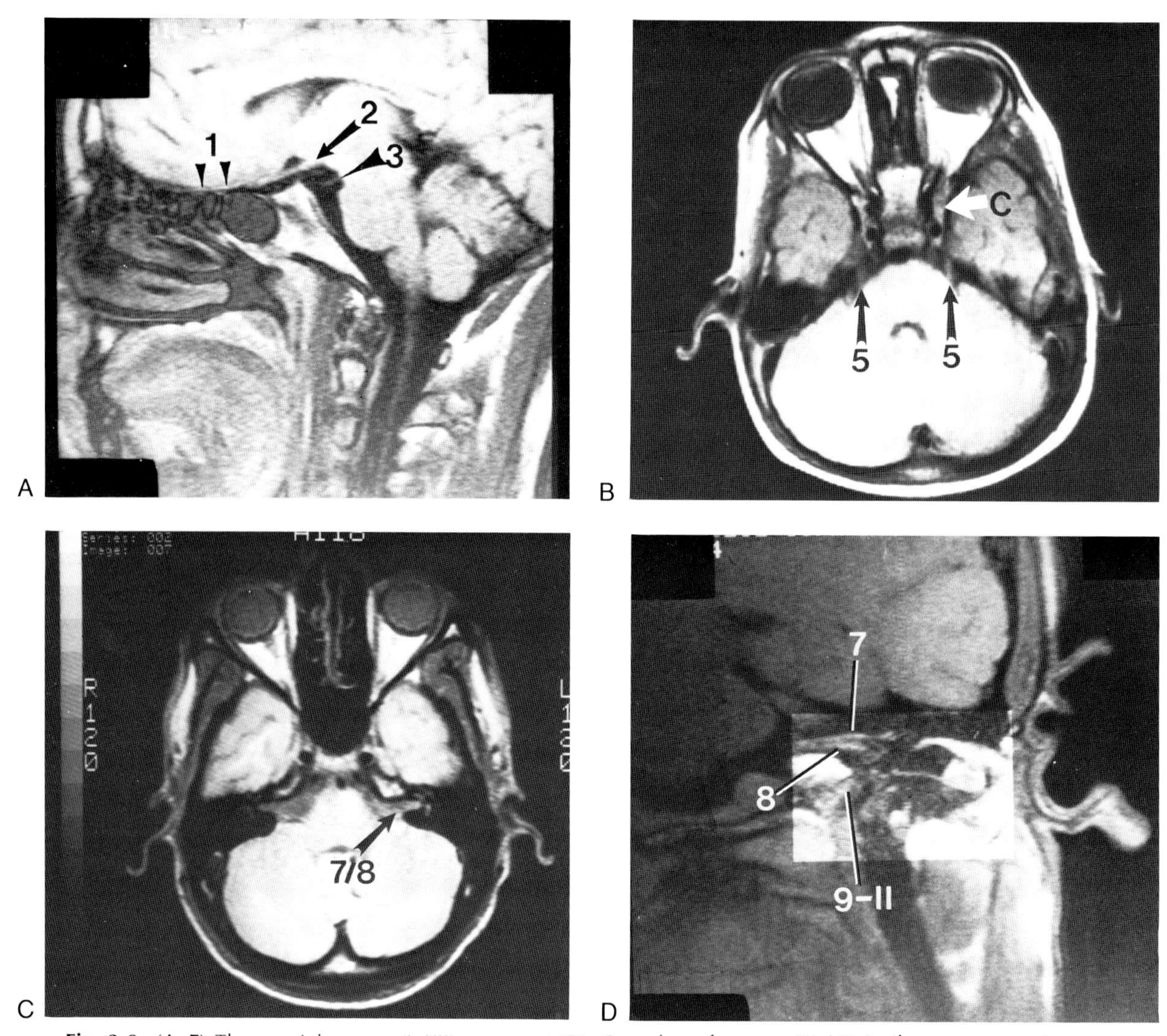

Fig. 3-9. (A–F) The cranial nerves (I–XII) seen on MRI. Complex of nerves (III–VI) in the cavernous sinus are labelled (C). Images of the cavernous sinus in more detail appear in Chapter 11. (*Figure continues*).

T1-weighted (Fig. 3-9A). The third, fourth, fifth, and sixth cranial nerves are seen in the cavernous sinus (Fig. 3-9B and multiple illustrations, Ch. 11). The facial and vestibulocochlear nerves are seen in the internal auditory canal and in the temporal bone (Figs. 3-8C, D, E). Nerves IX through XI in the jugular foramen may be seen (Fig. 3-8D, E). The hypoglossal nerve is well seen in its canal on axial MR images (Fig. 3-8F).

BRAIN IRON AND ITS EFFECTS ON THE MR IMAGE

It has been known for several decades that certain regions of the brain accumulate ferric iron as part of the normal aging process.[4,5] These areas, in order of decreasing concentration, include the globus pallidus, red nucleus, zona reticulata of the substantia nigra,

FIG. 3-9 (*Continued*).

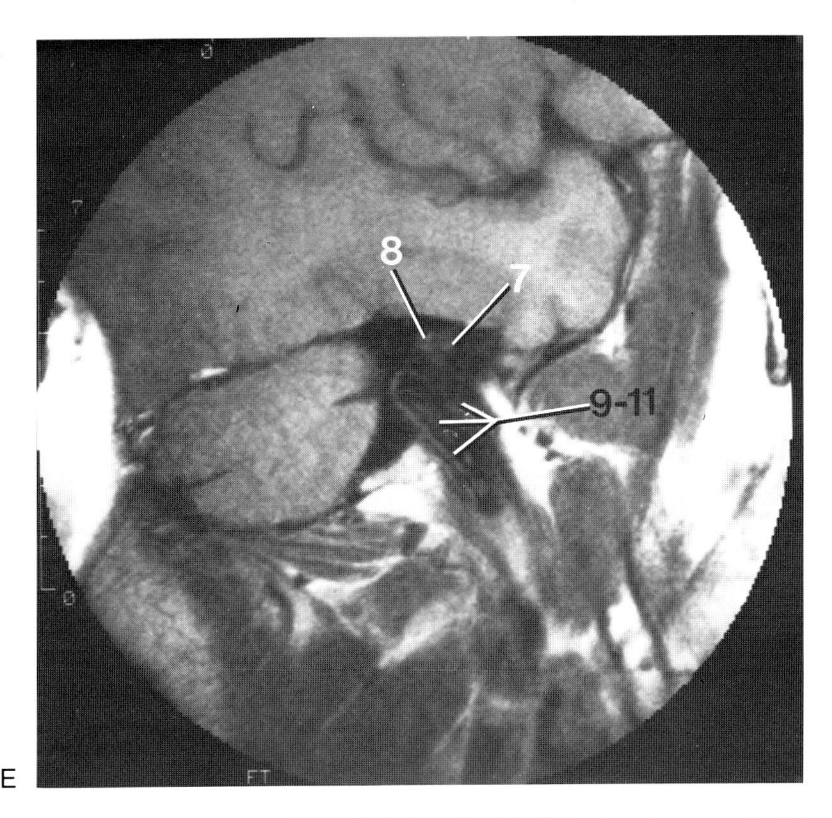

E

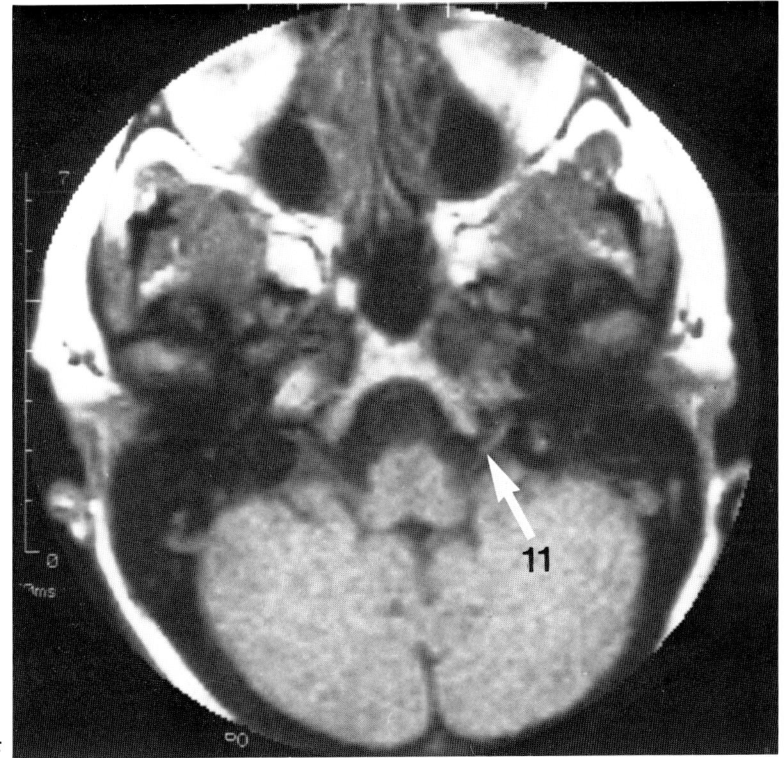

F

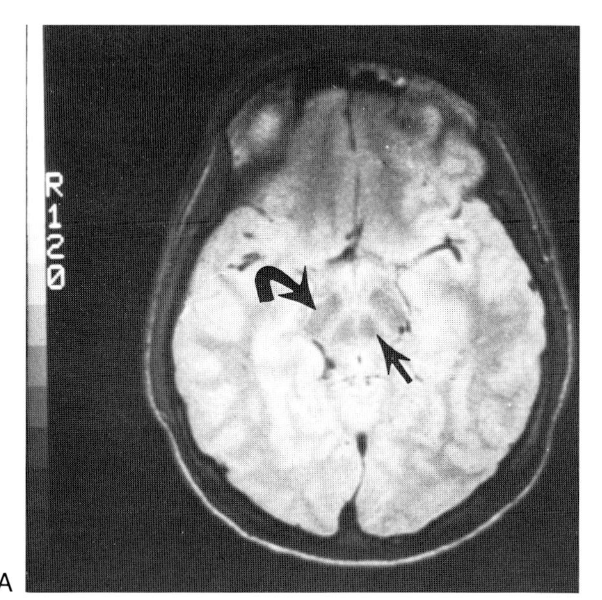

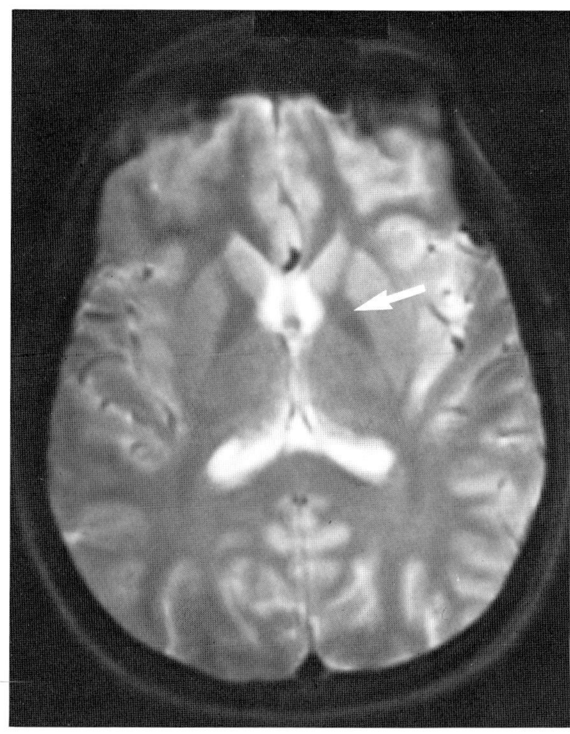

Fig. 3-10. How brain iron affects high field MR images. **(A)** Iron accumulation in the zona reticulata of the substantia nigra (curved arrow) and red nuclei (straight arrow) produce low signal on this SE 2500/75 MR image performed at 1.5 T. **(B)** The globus pallidus (arrow) has lower signal than the putamen. The frontal white matter is also seen to have lower signal than the occipital white matter as a consequence of preferential iron accumulation in the foramen.

subthalamic nucleus, dentate nucleus, and the putamen. The frontal white matter has greater iron distribution than occipital white matter. Subcortical "U" fibers have greater iron concentration than cerebral gray or white matter. There is almost no ferric iron in the posterior limb of the internal capsule or optic radiations.

Drayer and colleagues first applied this knowledge of the normal distribution of brain iron to explain certain aspects of cerebral MR imaging.[6] At higher field strengths (1.5 T) the presence of ferric iron produces relative shortening of T2 relaxation times. As a result, those regions of brain that accumulate iron may be seen to have low signal on T2-weighted images performed on high field MR scanners.

These features are illustrated in Figure 3-10. Note the low signal especially in the substantia nigra, red nuclei, and globus pallidi. The frontal white matter and subcortical "U" fibers have lower signal than the occipital white matter or posterior limb of the internal capsule.

Table 3-2. The Accumulation of Brain Iron in Various Pathologic States

Disease	*Site(s) of Iron Deposition*
Hallervorden-Spatz	Globus pallidus, reticular zone of substantia nigra
Huntington	Caudate, putamen
Parkinson and Parkinson-plus syndromes	Putamen, globus pallidus, substantia nigra compacta
Alzheimer	Cerebral cortex
Multiple sclerosis	Adjacent to plaques
Radiation effects	Vascular endothelium
Intracerebral hematoma	At periphery

(Drayer B, Burger P, Darwin R et al: Magnetic resonance imaging of brain iron. AJNR 7:373, 1986.)

The accumulation of brain iron has important implications in the study of various neurologic diseases. Excessive iron deposition has been shown to occur in a number of CNS disorders (Table 3-2). Additionally, the ability to map brain iron by MRI has important potential ramifications including the better understanding of aging, iron storage, and in the metabolism of certain neurotransmitters (e.g., GABA) that utilize iron-containing enzymes.

MRI OF NORMAL BRAIN MATURATION

The chemical composition of the infant brain undergoes a number of major changes, particularly during the first year of life. Such changes include decrease in water content and progressive myelination.[7] These rapid alterations in brain parenchyma result in different appearances of the infant brain on MRI at different ages.[8–13] Recognition of the normal appearance of the infant brain on MRI is necessary before one can evaluate disorders of brain development and myelination.

One of the most important changes the infant brain undergoes during the first year of life is a decrease in its water content.[7] The water content of the brain of a third trimester fetus is about 90 percent, falling to 72 percent in a child of 2 years. Simultaneously with this decrease in water, lipid content rises. The most rapid increase in brain lipids is seen during the first 6 months, with a more gradual increase through age 5 years.

Changes in water and lipid content in the maturing brain help explain its appearance on CT.[11,12] Similarly, changes in chemical composition alter the T1 and T2 relaxation times of gray and white matter on MRI (Table 3-3). In the neonate, T1 and T2 relaxation times for gray and white matter are roughly twice as long as in the adult. Nearly adult T1 values are reached by ages 1½ to 2 years, while adult T2 values are reached by age 3.[9]

The MR appearance of the developing brain is highly dependent upon the pulse sequence used, a fact not well appreciated in the current limited literature.[8–13] With MRI it is possible to study the process of myelination noninvasively. While no norms are yet well established, Figures 3-11 (A through L) show a typical sequence of normal brain maturation on MRI. It should be noted that the T1-weighted images show slightly different characteristics than the T2-weighted images. For example, no peripheral (cortical) gray-white differentiation is noted on the T2-weighted sequence of the 8-month-old, while it is clearly seen on the T1-weighted image. A second feature to note is that the process of myelination continues much later into childhood than is detectable with current MR strategies. Except for size, the MR appearance of a 5-year-old's brain is largely indistinguishable from that of a 15-year-old, although it is known that myelination continues pathologically into the third decade.

Table 3-3. Relaxation Times of Gray and White Matter by Age

	Gray Matter		*White Matter*	
Age (years)	*T1(msec)*	*T2(msec)*	*T1(msec)*	*T2(msec)*
Birth	1590	88	1615	91
½	1300	67	1150	64
1	890	68	580	57
1½	840	68	570	59
2	820	69	505	56
Adult	800	60	500	50

(Holland BA, Haas DK, Norman D et al: MRI of normal brain maturation. AJNR 7:201, 1986.)

Several characteristic patterns of myelination can be distinguished on MR images. As a general principle, myelination proceeds from deep to superficial and from posteriorly to anteriorly. In the first month of life (of term infants) myelination may be observed in the deep white matter of the internal capsule, brain stem, and thalami. This produces a relative shortening of T1 and T2 in these areas compared to the rest of the neonatal brain. As a result, the deep portions of the neonatal brain may have slightly low signal on T2-weighted images and slightly high signal on T1-weighted images (Fig. 3-11 A,B).[10,13]

During the next 3 months of life myelination becomes further established in these deep areas resulting in further decrease in T2. Peripherally the distinction between gray and white matter remains poor (Fig. 3-11 C,D).

In the second 6 months of life remarkable changes in the infant brain occur on MRI. The corpus callosum is seen to myelinate from posteriorly to anteriorly (Figs. 3-11 E,G). T1-weighted images now clearly show peripheral gray-white differentiation (Figs. 3-11 F,H).

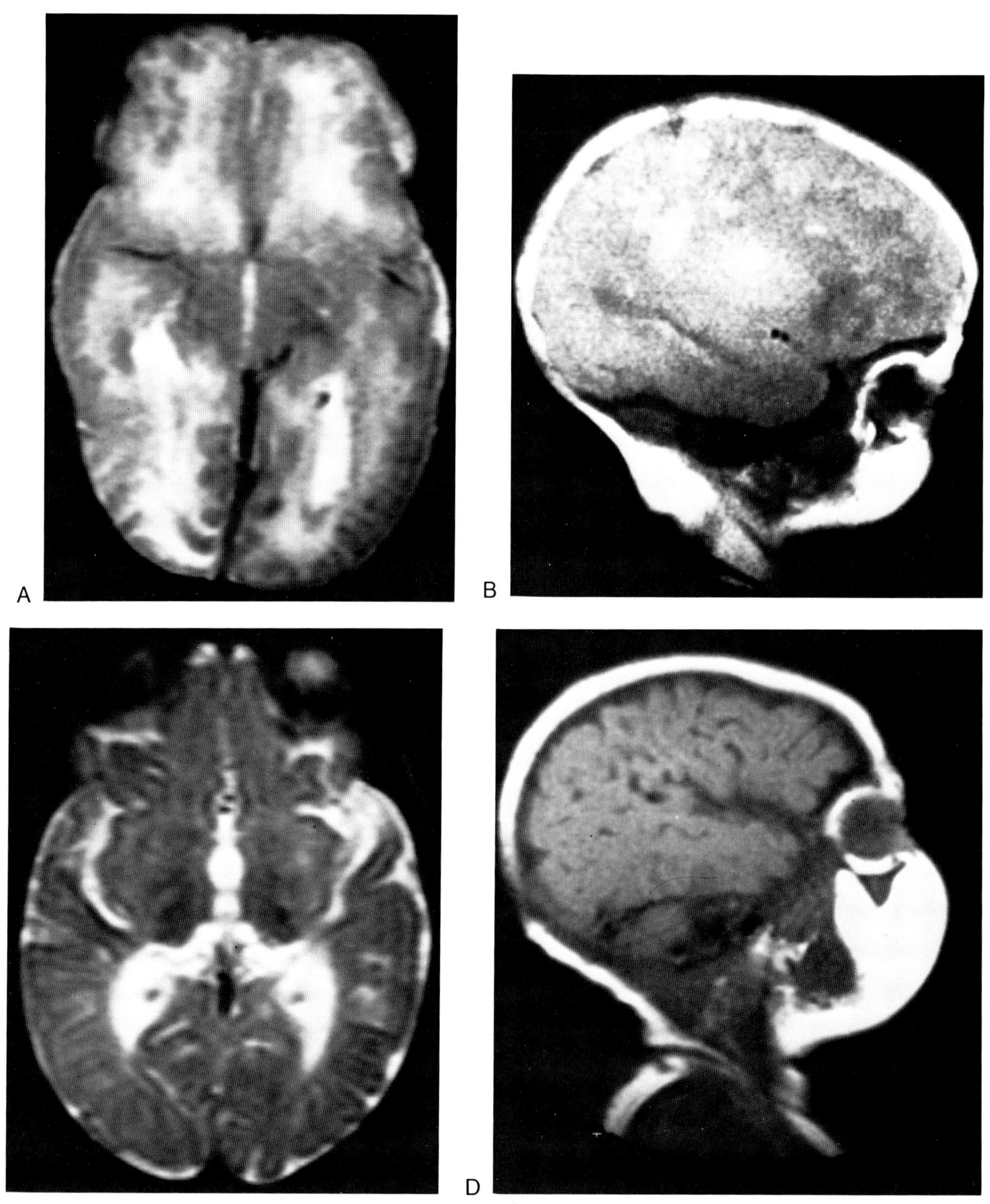

Fig. 3-11. MRI of normal brain maturation. T1-weighted (SE 450/26) and T2-weighted (SE 3000/100) images in six patients, aged 1 day to adulthood. **(A,B)** One-day-old. There is reversal of gray-white differentiation on either T1- or T2-weighted images. The thalami and internal capsules show decreased signal on T2-weighted images compatible with early myelination. **(C,D)** Two-month-old. Myelination is better established in the posterior limb of the internal capsule (arrow). There is marked reduction of differential signal between cortical gray and white matter also. (*Figure continues*).

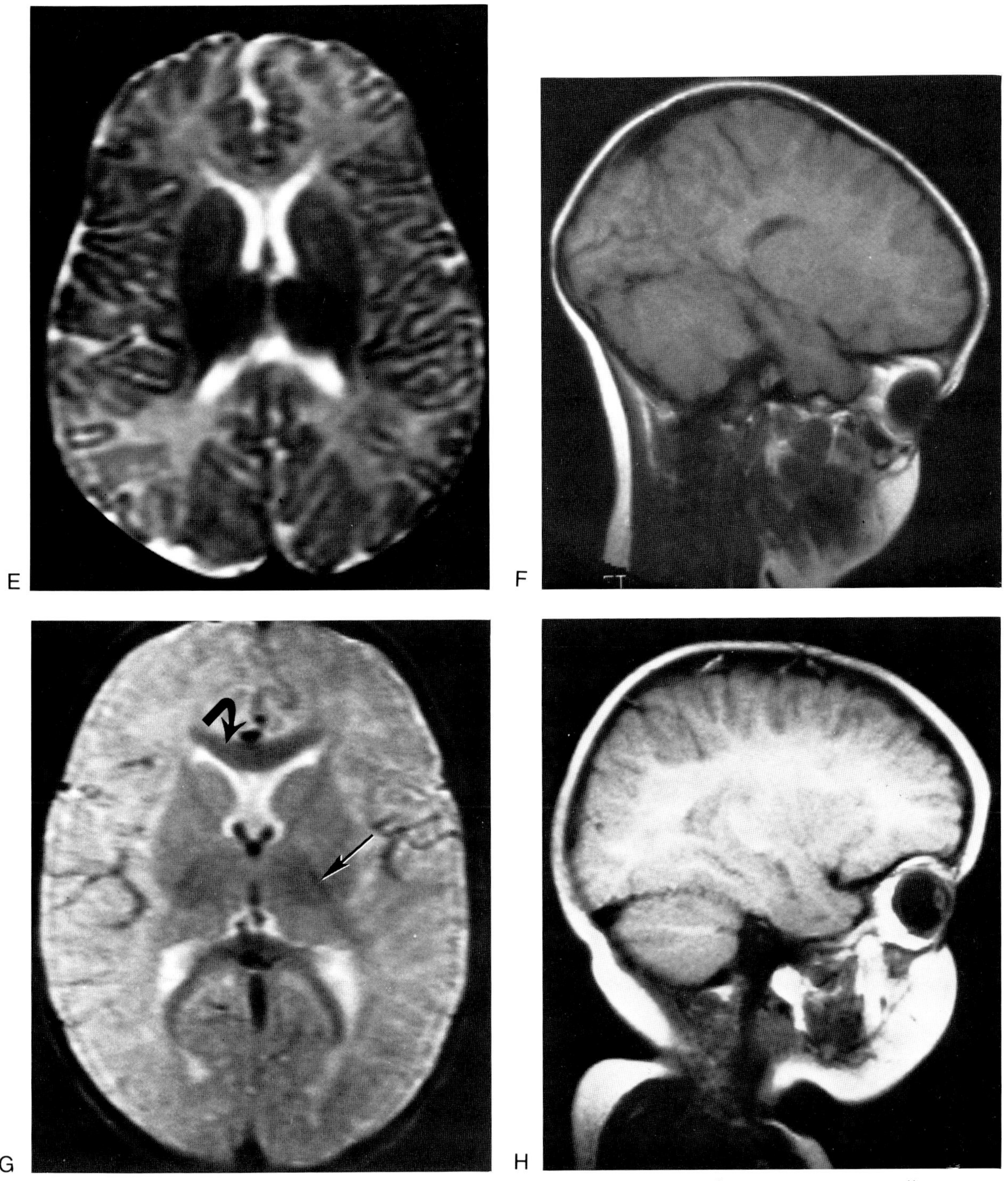

Fig. 3-11 (*Continued*). **(E,F)** Eight-month-old. Myelination is now apparent in the posterior corpus callosum (arrow) but not anteriorly. On the T1-weighted image, cortical gray and white matter may be clearly distinguished, though they could not be on the T2-weighted image. **(G,H)** One-year-old. The T2-weighted image shows progressive myelination of the corpus callosum so that its anterior portion now has how signal (curved arrow). The internal capsule is well visualized (straight arrow). Excellent cortical gray-white differentiation is available on the T1-weighted image.

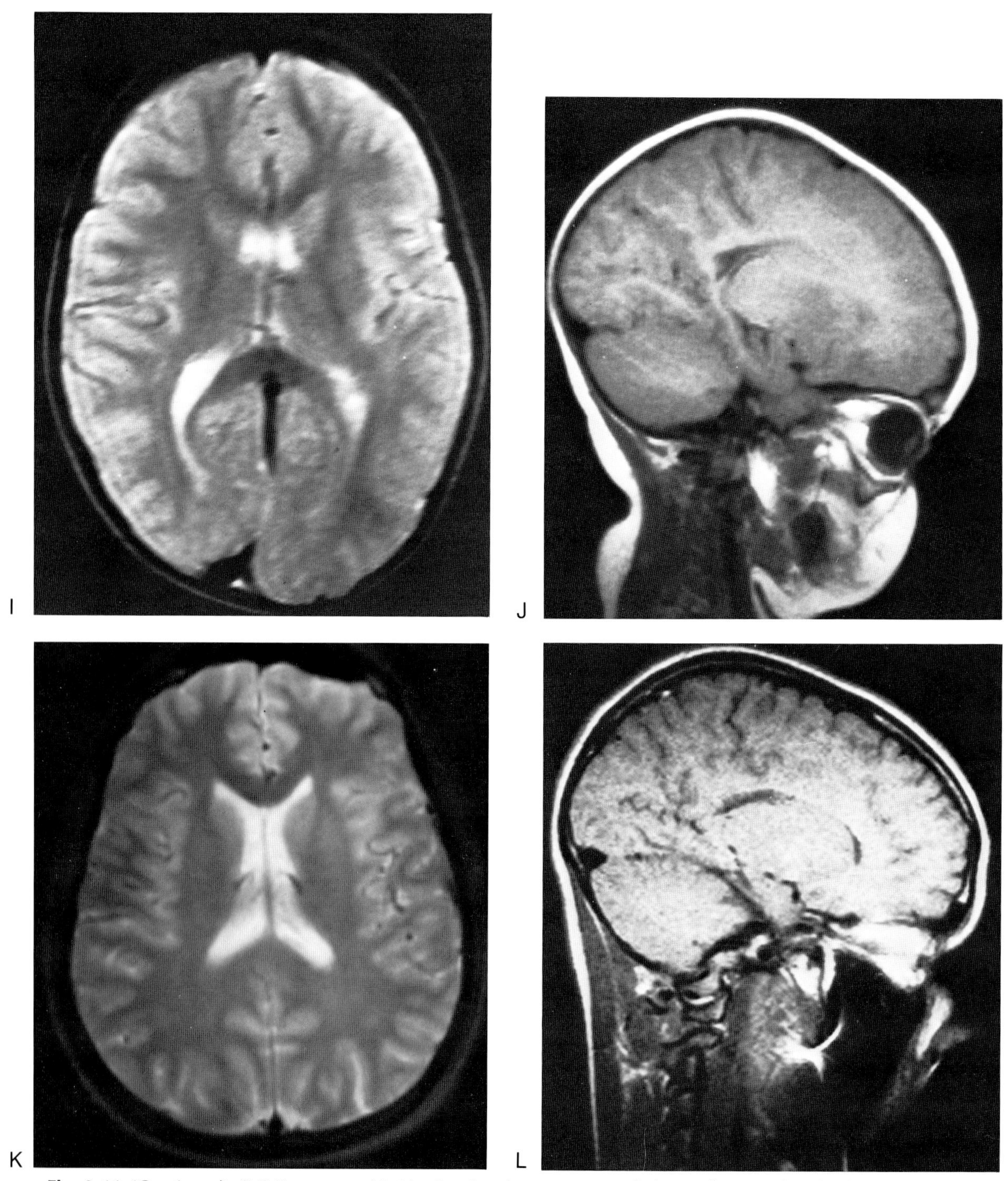

Fig. 3-11 (*Continued*). **(I,J)** Two-year-old. Myelination has now extended into the peripheral white matter so that it can now be recognized here on T2-weighted images. Excellent gray-white differentiation remains on the T1-weighted image. **(K,L)** Adult. Myelination is complete with extension to the subcortical "U" fibers recognized on T2-weighted images. Because of age-related changes in water content, the same T1-weighted sequence (SE 450/26), which provided excellent cortical gray-white differentiation in the 1- and 2-year-olds, provides poor contrast in this adult.

By age 2 years, the brain on MRI largely resembles that of the adult (Figs. 3-11 I,J). All that is lacking is further myelination of subcortical association fibers and some further minor changes in water content of gray matter relative to white. It should also be noted that adults tend to progressively accumulate iron in the basal ganglia, substantia nigra, and elsewhere which may be appreciated on T2-weighted images performed at high field strengths. This constitutes another important difference in aging not discussed here, but considered in the preceeding section. Clearly much further work is needed to elucidate the many aspects of brain maturation and aging detectible by MRI.

REFERENCES

1. Shellito J, Matson DD: Craniosynostosis. A review of 519 surgical patients. Pediatrics 41:829, 1968
2. Taveras JM, Wood EH: Diagnostic Neuroradiology. 2nd Ed. Williams & Wilkins, Baltimore, 1976
3. Elster AD, Wiggins TB: Indentification of fat within hyperostosis frontalis interna on MRI. J Comput Assist Tomogr, 1987 (in press)
4. Diezel PB: Iron in the brain: a chemical and histochemical examination, p. 145. In Diezel PB (ed): Biochemistry of the Developing Nervous System. Academic Press, New York, 1955
5. Hallgran B, Sourander PL: The effect of age on nonhaemin iron in the human brain. J Neurochem 3:41, 1958
6. Drayer B, Burger P, Darwin R et al: Magnetic resonance imaging of brain iron. AJNR 7:373, 1986
7. Dobbing J, Sands J: Quantitative growth and development of human brain. Arch Dis Child 48:757, 1973
8. Johnson MA, Pennock JM, Bydder GM et al: Clinical NMR imaging of the brain in children: normal and neurologic disease. AJR 141:1005, 1983
9. Holland BA, Haas DK, Norman D et al: MRI of normal brain maturation. AJNR 7:201, 1986
10. Lee BCP, Lipper E, Nass R et al: MRI of the central nervous system in neonates and young children. AJNR 7:605, 1986
11. Penn RD, Trenko B, Baldwin L: Brain maturation followed by computed tomography. J Comput Assist Tomogr 4:614, 1980
12. Brant-Zawadzki M, Enzmann DR: Using computed tomography of the brain to correlate low white matter attenuation with early gestational age in neonates. Radiology 139:105, 1981
13. McArdle CB, Richardson CJ, Nicholas DA et al: Developmental features of the neonatal brain: MR imaging. Radiology 162:223, 1987

4

Congenital and Developmental Anomalies

Brain development may be divided into several stages, each with a well-defined time of occurrence. A disturbance at a given point in time may affect that one stage or the several stages that follow it; the final result is a congenital malformation. In the majority (60 percent) of cases, no clear pathogenetic factor producing the malformation can be identified. About 20 percent of cases result from inheritance and about 10 percent represent spontaneous chromosomal mutation. The remaining 10 percent may be attributed to intrauterine environmental factors such as infection, ischemia, or toxins.[1]

CLASSIFICATION OF MALFORMATIONS

Several classifications for congenital craniocerebral malformations have been proposed, but that of DeMyer has gained greatest popularity (Table 4-1).[2–4] In DeMyer's scheme cerebral malformations are divided into two broad groups: *disorders of histogenesis* or *disorders of organogenesis*. Disorders of histogenesis form a small but important group of mostly hereditary diseases, including neurofibromatosis, tuberous sclerosis, and Sturge-Weber syndrome. Disorders of organogenesis is a large category containing all remaining craniocerebral malformations. Each disorder

Table 4-1. Classification of Cerebral Malformations

Disorders of Histogenesis
Tuberous sclerosis
Neurofibromatosis
Encephalotrigeminal angiomatosis
Congenital neoplasms
Disorders of Closure
Cranioschisis (amencephaly, encephalocele, menigocele)
Agenesis of corpus callosum
Aicardi syndrome
Lipoma of corpus callosum
Teratoma
Chiari malformation
Dandy-Walker syndrome
Disorders of Diverticulation
Aventricular cerebrum
Holoprosencephaly (alobar, semilobar, lobar)
Septo-optic dysplasia
Trisomies 13–15, 18
Disorders of Neuronal Proliferation
Micrencephaly
Megalencephaly
Disorders of Neuronal Migration
Lissencephaly (agyria)
Polymicrogyria
Schizencephaly
Heterotopias
Destructive Lesions
Hydrancephaly
Porencephaly
Hypoxia
Toxicoses
Inflammatory disease

(Modified from DeMyer W: Classification of cerebral malformations. Birth Defects 7:78, 1971.)

in this group is considered to have arisen from a disturbance at a specific stage of brain development, such as closure, diverticulation, or migration. It should be kept in mind that this is an oversimplified scheme, because some anomalies result from disruption in several stages of organogenesis, and the origin of other anomalies may be difficult to identify.

Disorders of Closure

Disorders of closure are thought to result primarily from defective formation or closure of the neural tube between the 3rd and 20th weeks of gestational life. Incomplete closure is called *cranioschisis* and ranges from severe (anencephaly) to mild (meningocele). Imperfect closure results in midline defects (e.g., dysgenesis of the corpus callosum) or incorporation of abnormal tissue deep within the brain (lipoma, teratoma). DeMyer considered the Chiari and Dandy-Walker malformations to be disorders of closure, but it is clear that their pathogenesis is more complex, possibly involving several stages and mechanisms.

Disorders of Diverticulation

Disorders of diverticulation occur when there is faulty cleavage of the prosencephalon (forebrain) between the 5th and 6th weeks of embryogenesis. During this period the primitive three-vesicle brain is being transformed to the five-vesicle stage, and separation of the ventricles and cerebral hemispheres is occurring. Interference with this process results in forebrain (holoprosencephalies) and facial malformations.

Disorders of Neuronal Proliferation

Disorders of neuronal proliferation may occur during a 3 month period from about 8 to 20 weeks. The production of too many cells results in megalencephaly and the production of too few results in micrencephaly. The term *megalencephaly* describes an abnormally large brain, while the term *megalocephaly* describes an abnormally large head. It should be noted that most cases of megalocephaly are due to hydrocephalus and not primarily megalencephaly, which is rare. Similarly, the terms *micrencephaly* describes an abnormally small brain while *microcephaly* means a small head. Most cases of micrencephaly result from intrauterine infection or ischemia and not from failed neuronal proliferation. The latter cause of micrencephaly is exceedingly rare and is a diagnosis of exclusion.

Disorders of Neuronal Migration

Disorders of neuronal migration occur during a slightly later time period than disorders of proliferation (i.e., between the 12th and 24th weeks). Migration anomalies may result in abnormally located brain tissue (heterotopias). The formation of sulci is also interfered with (lissencephaly, schizencephaly).

Destructive Lesions

Destructive lesions produce congenital anomalies on the basis of intrauterine insult. These insults usually occur late in gestation and destroy normally formed brain. Destructive processes include infections, toxins, and ischemia. Global (hydrancephaly) or localized (porencephaly) regions of brain destruction result.

MRI VERSUS CT IN BRAIN ANOMALY EVALUATION

Magnetic resonance imaging is capable of providing important information for the evaluation of patients with various congenital brain anomalies. It has been found particularly useful in evaluating those malformations of the posterior fossa and corpus callosum, and may be superior to CT. MR has also been helpful in the phakomatoses (disorders of histogenesis) where it may find or delineate lesions better than CT. Destructive lesions and disorders of sulcation are well seen by CT, and MR seems to have little advantage for these. Accordingly, this chapter will not attempt to discuss every congenital malformation, but will concentrate on those well-suited for evaluation by MRI.

ANOMALIES OF THE CORPUS CALLOSUM

Dysgenesis of the Corpus Callosum

The corpus callosum is a large collection of myelinated fibers connecting the two cerebral hemispheres and forming the roof of the lateral ventricles. It develops in an anterior to posterior direction between weeks 12 and 20 of fetal life.[5] Intrauterine insults such as infection or ischemia, which occur early in this period may result in total agenesis of the corpus callosum. Insults occurring late may affect only the posterior portion (splenium). Dysgenesis of the corpus callosum may also have a genetic basis, and is associated with a number of other anomalies (Table 4-2).[6,7]

The corpus callosum has a characteristic curvilinear shape best seen on midline sagittal MR images (Fig. 4-1A). The splenium is typically bulbous in shape.

Table 4-2. Anomalies Associated with Dysgenesis of the Corpus Callosum[a]

Aicardi syndrome
Aqueduct stenosis
Arachnoid cyst
Cerebral atrophy
Chiari malformation
Dandy-Walker syndrome
Encephalocele
Heterotopic gray matter
Holoprosencephaly
Interhemispheric cyst
Lipoma
Microcephaly
Polymicrogyria
Porencephaly
Septo-optic dysplasia
Trisomies 13–15, 18

[a] Compiled from References 6 and 7.

When this bulbous contour is missing, partial agenesis should be suspected. While callosal development occurs in a cephalocaudad direction, callosal myelination occurs in the reverse direction (i.e., from posterior to anterior) between the 4th and 20th months of postnatal life.[8] This process of myelination of the corpus callosum can be observed by MRI. The T1 value of the corpus callosum shortens progressively with myelin deposition during the first 2 years of infancy.[9]

Axial plane findings with dysgenesis of the corpus callosum are similar on both CT and MR. Wide separation of the frontal horns is noted. The third ventricle, often dilated superiorly, is interposed between the bodies of the lateral ventricles. This abnormal configuration has been called a "bat wing" or "devil's horns" pattern.[10] The occipital horns are often dilated while the frontal horns may be small or narrow.[11] The lateral ventricular margins may be irregular, secondary to associated heterotopias.[12]

Coronal MR images demonstrate well the upward extension of the third ventricle between the frontal horns. Also well seen in this plane is an abnormally acute lateral ventricular angle of the frontal horn.[11] Coronal MR may also demonstrate the relation of midline cysts and lipomas to the dysgenetic corpus callosum.

Sagittal MR seems to be the single most useful technique for demonstrating dysgenesis and other anomalies of the corpus callosum. Even minor structural abnormalities indiscernible on CT scans may be detected by sagittal MR.[13,13A] A corpus callosum-cerebrum (CCC) ratio may be defined as the ratio of the maximal anterioposterior diameter of the corpus callosum to the maximal fronto-occipital diameter of the cerebrum on a midline sagittal image. In normal patients the CCC ratio is 0.45 or greater while those with partial agenesis the ratio may be 0.30 or less.[13] Examples of agenesis of the corpus callosum are shown in Figure 4.2.

The *Aicardi syndrome* is a rare disorder that includes dysgenesis of the corpus callosum, seizures, mental retardation, and ocular abnormalities (chorioretinitis).[14,15] All patients are female. Other findings sometimes seen include Dandy-Walker cyst, vertebral anomalies, skull asymmetry, and heterotopic gray matter.[16] In one of two reported cases, MR was superior to CT in detecting partial agenesis of the corpus callosum in this condition (Fig. 4-3).[17]

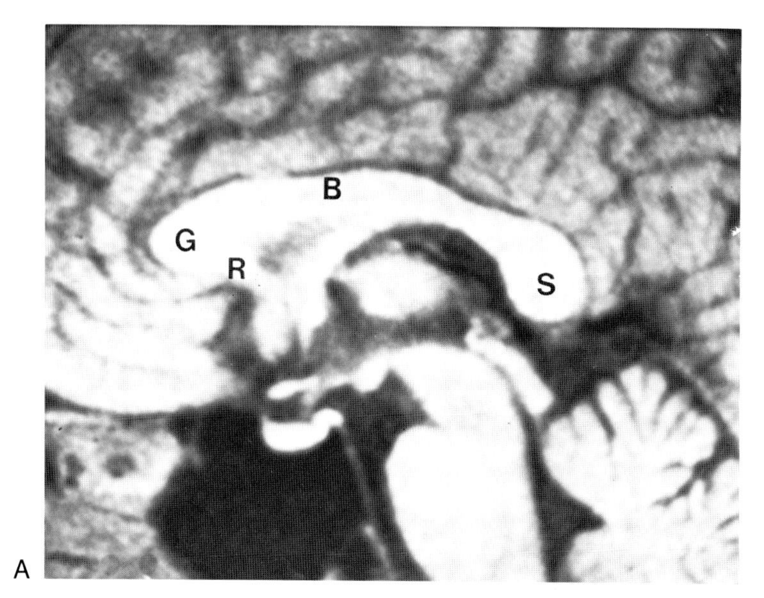

Fig. 4-1. **(A)** Normal corpus callosum on sagittal SE 500/30 image. *R,* rostrum; *G,* genu; *B,* body; *S,* splenium. **(B)** Normal variant of corpus callosum. Because the corpus callosum is composed primarily of transversely oriented fibers, focal lobulations and defects (arrows) are common. **(C)** Partial agenesis of corpus callosum. The corpus is thin and hypoplastic, lacking a bulbous splenium posteriorly.

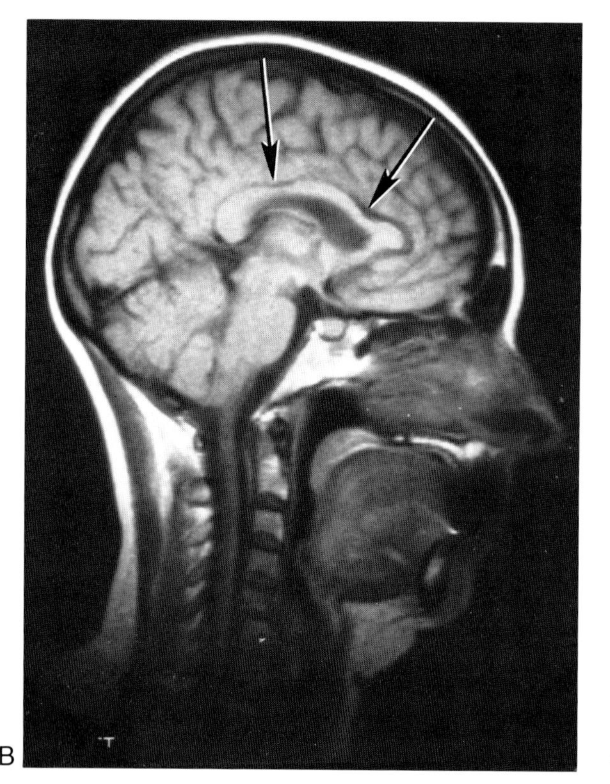

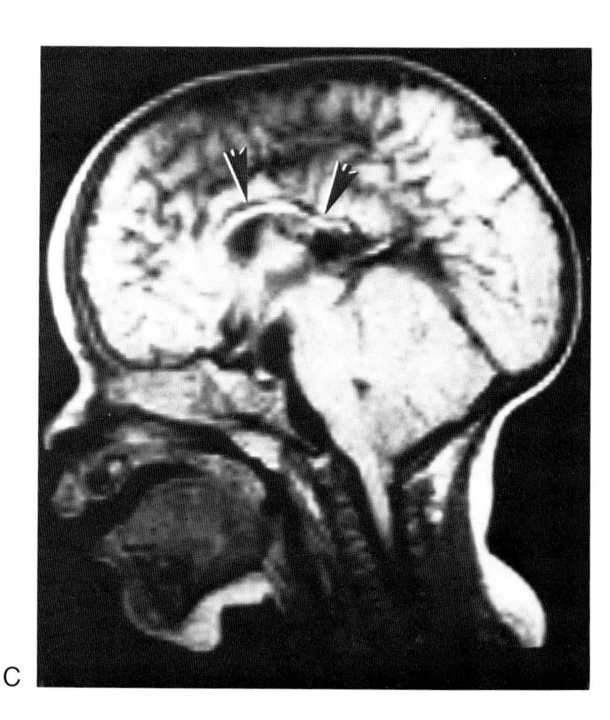

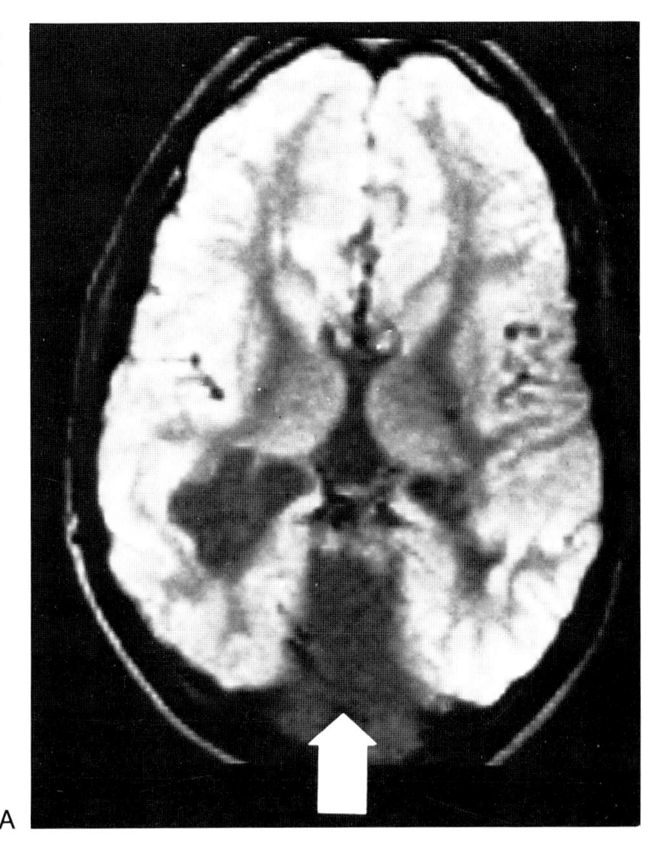

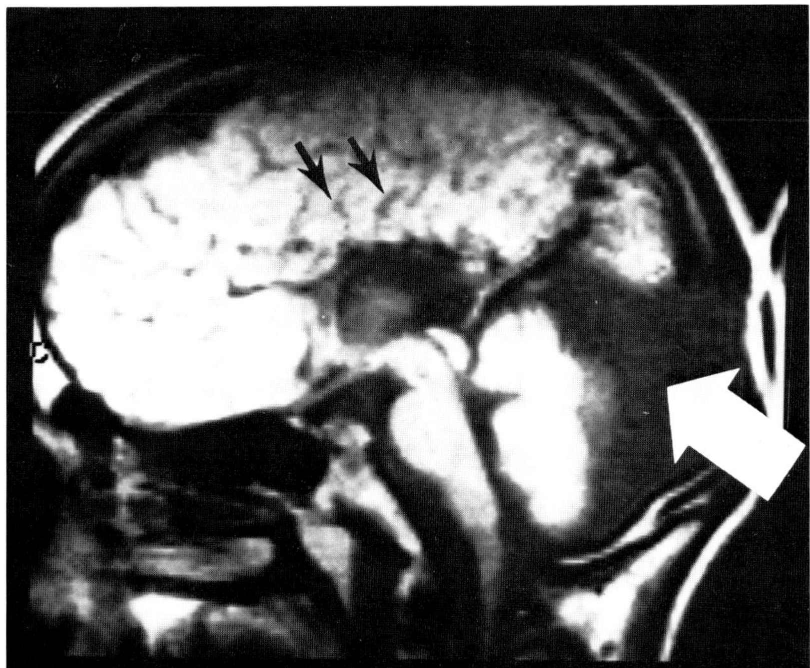

Fig. 4-2. Complete agenesis of corpus callosum. Note in **(A)** the cingulate gyrus is also disordered with abnormal vertical sulci (black arrows). A posterior fossa arachnoid cyst is also present (white arrow).

MR at present, therefore, seems to be the procedure of choice in the diagnosis of partial or complete agenesis of the corpus callosum.[13,17] CT and MR may be complementary in evaluating the spectrum of other associated abnormalities.

Lipoma of the Corpus Callosum

Abnormal incorporation of mesodermal fat into the neural tube during its closure may result in a midline intracranial lipoma.[18] Most frequently this lipoma lies within the anterior portion (genu) of the corpus callosum and is relatively small. Occasionally it may be large and involve the entire corpus callosum.[19] Intracranial lipomas may also occur in the quadrigeminal cistern, suprasellar region, crural cistern, and cerebellopontine angle.[19,20]

About 50 percent of patients with a midline lipoma will have dysgenesis or agenesis of the corpus callosum.[21] However, lipoma of the corpus callosum is much rarer than agenesis, and most patients with agenesis will not have an associated lipoma. Frontal bone defects are frequently associated with callosal lipomas,[22] and a second lipoma in the choroid plexus may be found in up to 25 percent of patients.[23]

Only about half of callosal lipomas become symptomatic, presenting with seizures, dementia, retardation, headaches, or hemiplegia.[24] Seizures are the most common symptom produced by either lipoma or agenesis of the corpus callosum.[25] Seizures are thought to result from infiltration of the adjacent brain by fibrous connective tissue around the lipoma.[26] It is most unusual for a lipoma to produce symptoms by growth or mass effect.[27] Rarely, obstructive hydrocephalus requiring shunting may occur.[28] Encasement of the anterior cerebral artery by tumor has been reported.[18]

Lipomas of the corpus callosum are easily diagnosed by CT based on their characteristic location and fat density.[26,27,29] Dense, curvilinear calcification in the walls of the lipoma are common in large callosal

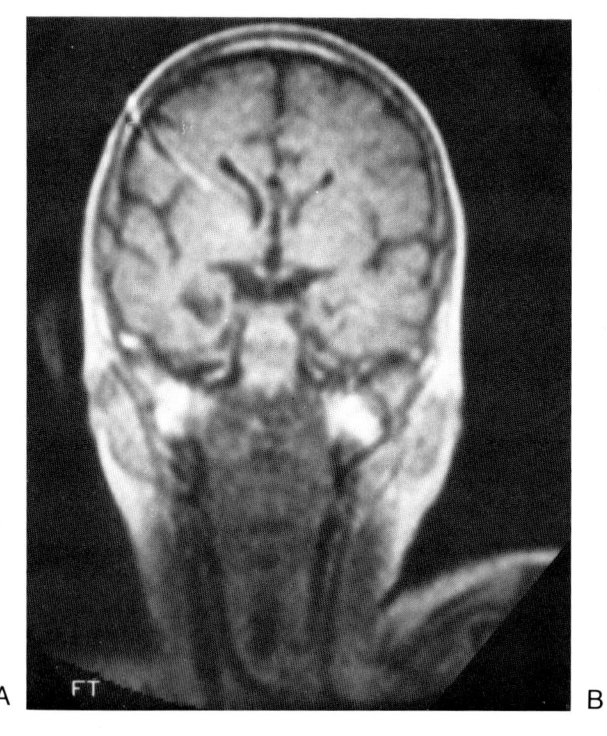

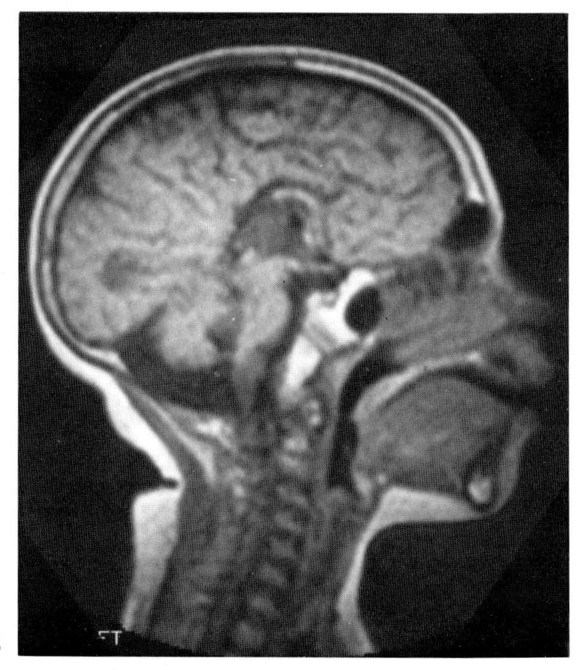

Fig. 4-3. Aicardi syndrome. The lateral ventricles are widely separated and display a bicornuate appearance due to infolding of cingulate gyri and the presence of longitudinal callosal bundles of Probst. The superior portion of the third ventricle is interposed between the lateral ventricles.

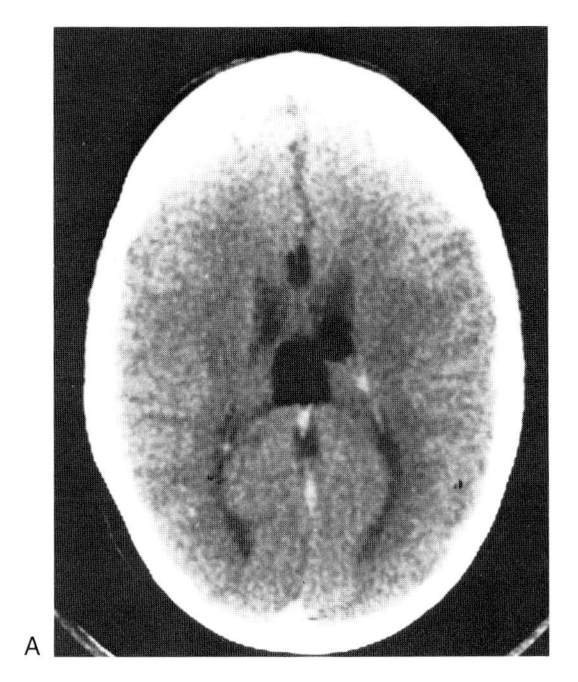

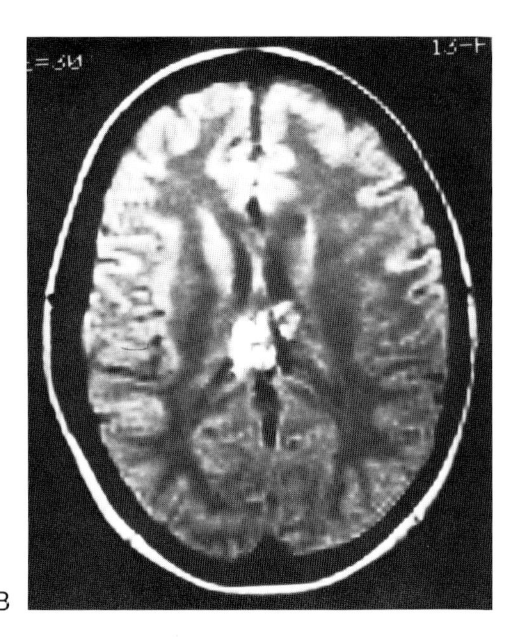

A B

Fig. 4-4. Lipoma of corpus callosum on CT **(A)** and MR **(B)**. Lipoma has high signal on T1-weighted image due to its short T1 value.

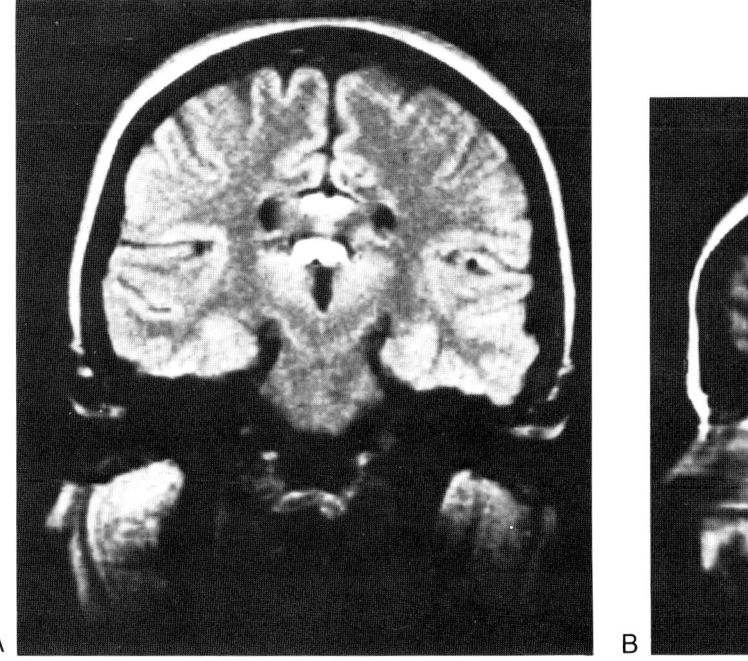

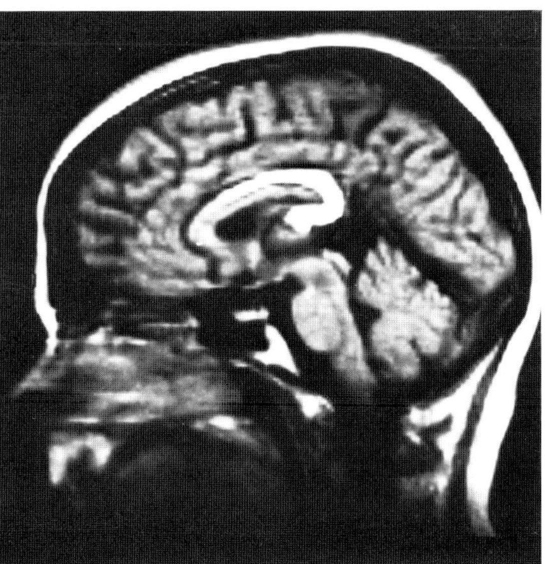

A B

Fig. 4-5. A curvilinear lipoma replacing the posterior portion of the corpus callosum.

tumors, but not in lipomas elsewhere.[19,29] Calvarial defects may also be appreciated.[21,22]

MRI can also easily diagnose intracranial lipomas by their unique signal characteristics.[30–32] Although the T2 values of fat and brain are similar, the T1 value of fat is much shorter than that of brain. Accordingly, a T1-weighted pulse sequence will result in high signal intensity from the intracranial lipoma.[31] The peripheral calcification may sometimes be seen as an area of signal drop-out due to low proton density. Whereas MR seems to offer no advantage over CT in the diagnosis of callosal lipomas per se, MR does seem to be better suited for evaluating the callosal dysgenesis frequently associated with the lipoma.[13,17,32] Examples of lipomas of the corpus callosum imaged by CT and MR are presented in Figures 4-4 and 4-5.

DANDY-WALKER SYNDROME

The Dandy-Walker syndrome refers to a group of congenital hindbrain malformations having two essential features: (1) a massively dilated fourth ventricle or posterior fossa cyst communicating with that ventricle, and (2) dysplasia of the vermis and fourth ventricular roof. While the original descriptions of the syndrome included hydrocephalus and atresia of fourth ventricular outlet foramina, these features are not invariably present.[33,34] Associated CNS malformations are present in more than 50 percent of cases.[35] These include midline dysplasias (of the corpus callosum, anterior commissure, cingulate gyrus, inferior olives, choroid plexus, or cerebral aqueduct), midline inclusion tumors (lipomas, hamartomas), heterotopias, and polymicrogyria. Of these, dysgenesis of the corpus callosum is the most common.[19] Coexisting skeletal anomalies (polydactyly, syndactyly, cleft palate, Klippel-Feil syndrome, occipital bone defects) occur in up to 25 percent.[35]

The etiology and pathogenesis of the disorder are still debated.[36,37] From the observed coexistence of polydactyly, the defect presumably occurs about the 6th to 7th week of embryonic life.[35] Primary vermian dysgenesis and impaired permeability of the fourth ventricular roof at present seem to provide the best embryologic explanation for the malformation.[38,39]

MR findings in the Dandy-Walker syndrome correspond to those observed with multiplanar CT.[38,40–42] The posterior fossa is enlarged, with elevation of the transverse sinus and torcula above the lambda. This inversion of the normal torcula-lambda relationship may aid in distinguishing the Dandy-Walker malformation from other posterior fossa cysts (which do not cause this inversion).[43] The tentorium is elevated, a finding best appreciated on coronal or sagittal images. A large cyst occupies much of the posterior fossa in the midline, and may be seen to communicate with an enlarged fourth ventricle (Fig. 4-6). Hydrocephalus of the lateral and third ventricles is nearly always noted. Kinking of the cerebral aqueduct as a cause of the hydrocephalus is sometimes appreciated in sagittal images. MRI is frequently better than CT for displaying associated abnormalities such as dysplasia of the corpus callosum or gray matter heterotopias (q.v.)

The *Dandy-Walker variant* (Fig. 4-7) is a milder form of hindbrain malformation, where the upper fourth ventricle and superior vermis are relatively normal.[44] A diverticular outpouching of variable size and shape arises from the inferior medullary velum. The vallecula is widened, and there is incomplete development of the inferior vermis. In contrast to the classic Dandy-Walker syndrome, which presents in infancy with symptoms related to hydrocephalus, the variant form may not be recognized until adulthood and hydrocephalus is not a prominent feature. MR imaging may demonstrate communicatio of the cyst with the fourth ventricle, a finding confirming suspicion of the Dandy-Walker malformation. MR also aids in the differential diagnosis of other posterior fossa lesions of similar appearance, including retrocerebellar arachnoid cyst, ependymal cyst, cystic midline tumor, trapped fourth ventricle, and megacisterna magna.

CHIARI MALFORMATIONS

The Chiari malformations are a complex set of cerebral anomalies that primarily involve the brain stem and cerebellum. Although first described by Cleland in 1883, the first comprehensive analysis and classification scheme was provided by Chiari in 1891.[45] An

Fig. 4-6. (A), (B) Dandy-Walker cyst (arrows). Note aplasia of inferior vermis and communication of cyst with the fourth ventricle, especially noted on sagittal view. This patient also has agenesis of the corpus callosum (arrowheads). **(C)** Another example, this one also with an interhemispheric arachnoid cyst.

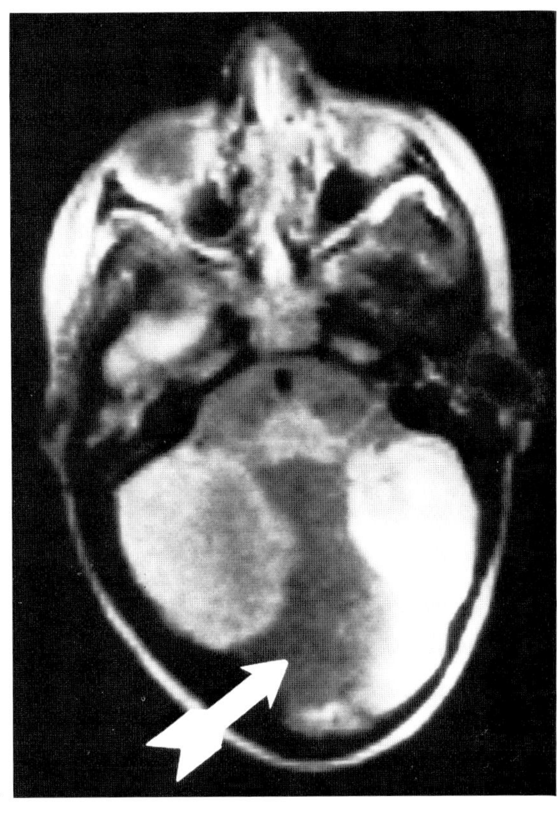

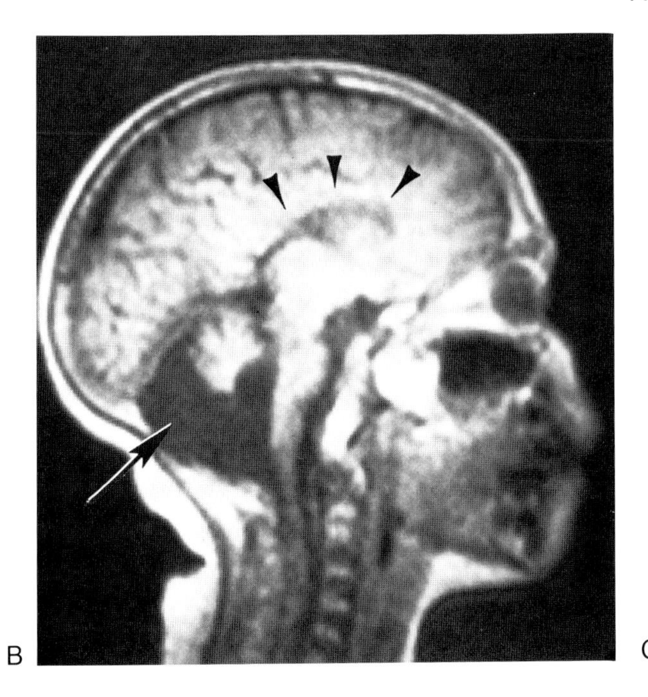

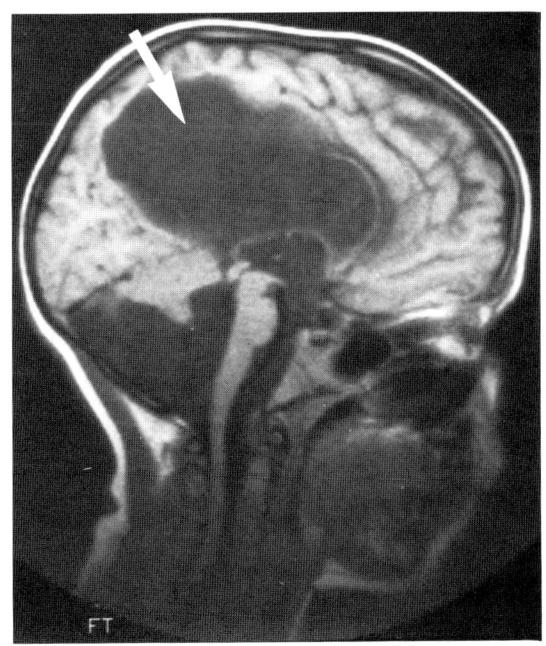

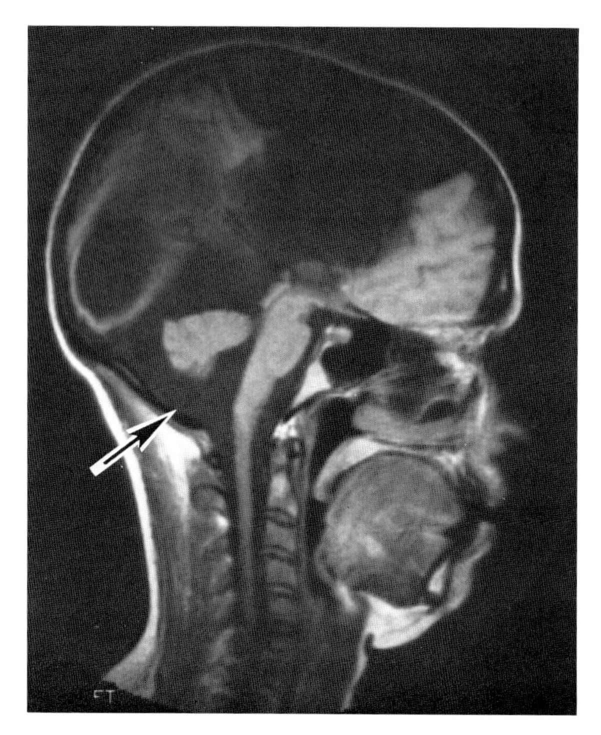

Fig. 4-7. Dandy-Walker variant. The inferior vermis is hypoplastic but not absent.

English language translation of Chiari's original paper is available.[46] Chiari described three types of hindbrain malformations in order of increasing severity:

Type I consists of downward displacement of the cerebellar tonsils and inferior cerebellum into the spinal canal. The medulla and fourth ventricle are in normal position. These patients present in late childhood or adulthood with lower cranial nerve and spinal nerve symptomatology. Syringomyelia is commonly associated with the Type I malformation.

Type II was also described by Arnold in 1894.[47] In a strict sense the term "Arnold-Chiari malformation" should be applied only to the Chiari Type II anomaly, but in common parlance all three Chiari types are sometimes called Arnold-Chiari malformations. Chiari Type II is similar to Type I in that there is downward displacement of the cerebellum into the spinal canal. Additionally, the medulla and fourth ventricle are elongated and caudally displaced. Patients present in infancy and nearly all have an associated myelomeningocele. Hydrocephalus and a large number of other cerebral and calvarial abnormalities may also be seen.

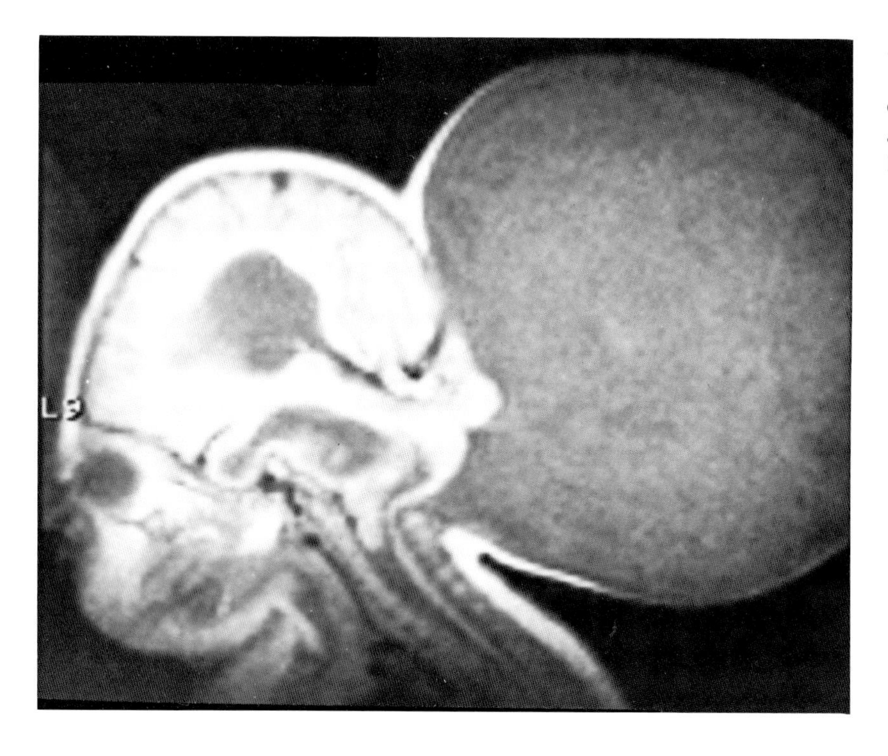

Fig. 4-8. Chiari III malformation. Note hindbrain herniation into large occipital meningocele. (Diagn Imaging 8(6):74, 1986, S. Karger AG, Basel)

Type III is extremely rare and involves displacement of the medulla, cerebellum, and fourth ventricle into an occipital and high cervical encephalocele. This type is clinically apparent in the newborn and uncommonly evaluated by MRI (Fig. 4-8).

A Type IV Chiari malformation has also been described that manifests extreme cerebellar hypoplasia but no downward displacement. This type appears to be a distinctly different anomaly, and is probably better classified as a form of cerebellar agenesis rather than a Chiari lesion.[48] As such, it will not be included in the discussion here.

The pathogenesis of the Chiari malformations is poorly understood, and none of the several available theories is entirely satisfactory.[49] The time of occurrence of the malformation is near the third month of fetal life.[50] Possible mechanisms include primary overgrowth of neural tissue, primary brain stem dysgenesis, and dysbalance of CSF dynamics between the ventricular system and subarachnoid spaces.

Chiari I Malformation

The Chiari I is the mildest form of the Chiari hindbrain malformations, characterized by downward herniation of the cerebellar tonsils through the foramen magnum. In contrast to the Chiari II malformation, the fourth ventricle remains in normal position, though it may be stretched. Symptoms of cranial nerve and spinal compression typically occur in late childhood and adulthood. Downbeat nystagmus is characteristic.[51] About half of these patients have syringomyelia of the cervical cord and present with a central cord syndrome.[52]

A number of other associated abnormalities may be seen (Table 4-3).[52–55] Between one-third and one-half of patients have craniovertebral junction anomalies including basilar invagination, atlantoaxial assimilation, high odontoid, short clivus, or Klippel-Feil syndrome.[52,53] At surgery, dural bands and arachnoidal adhesions may be observed.[54]

Table 4-3. Associations of Chiari I Malformation

Association
Syringomyelia (44–56%)
Craniovertebral junction anomalies (37%)
Arachnoidal adhesions (41%)
Dural bands (30%)
Cervicomedullary kinking (12–80%)
Hydrocephalus (3%)

The diagnosis of Chiari I malformation may be made by pneumoencephalography, vertebral angiography, myelography, or metrizamide-enhanced CT. All of these procedures are uncomfortable and potentially dangerous. MRI has become the imaging modality of choice to establish this diagnosis because it is noninvasive, delineates posterior fossa anatomy well, and allows direct visualization of syringomyelic cavities.[55]

The essential feature for MR diagnosis is demonstration of cerebellar tonsil herniation into the spinal canal. It should be noted that many normal patients have tonsils that lie below the plane of the foramen magnum. In a quantitative MR study it was found that extension of the tonsils below the foramen magnum was normal up to 3 mm, was borderline between 3 mm and 5 mm, and was clearly pathologic when exceeding 5 mm.[56] In the same study it was also noted that the sagittal diameter of the foramen magnum was frequently greater in patients with symptomatic Chiari I malformations than in the normal population (40 mm vs. 37 mm, respectively).

Tonsilar position and shape are best evaluated on T1-weighted sagittal images just off the midline (Fig. 4-9).[55] The herniated tonsils are frequently peglike in shape in contrast to their normal rounded inferior surfaces. Occasionally coronal MR images may reveal asymmetric tonsilar hernation not appreciated on sagittal images (Fig. 4-10). The tonsils are sometimes tightly compressed against the posterior margin of the spinal cord and may be confused with localized tumor expansion of the medulla or cervical cord.[57] Cervicomedullary kinking (Fig. 4-11) has been noted by MR in 74 percent of patients, a finding not easily detected by other modalities or even at surgery.[54,55]

A T1-weighted sequence is also optimal for diagnosing associated syringomyelia, which occurs in about half of symptomatic Chiari I patients.[58] Both sagittal and coronal images may be useful in demonstrating the internal architecture of the syrinx in regard to septations and skip areas (Fig. 4-12). Occasionally, small slitlike syrinxes may only be appreciated on axial images.[55] When the cord is small, however, de-

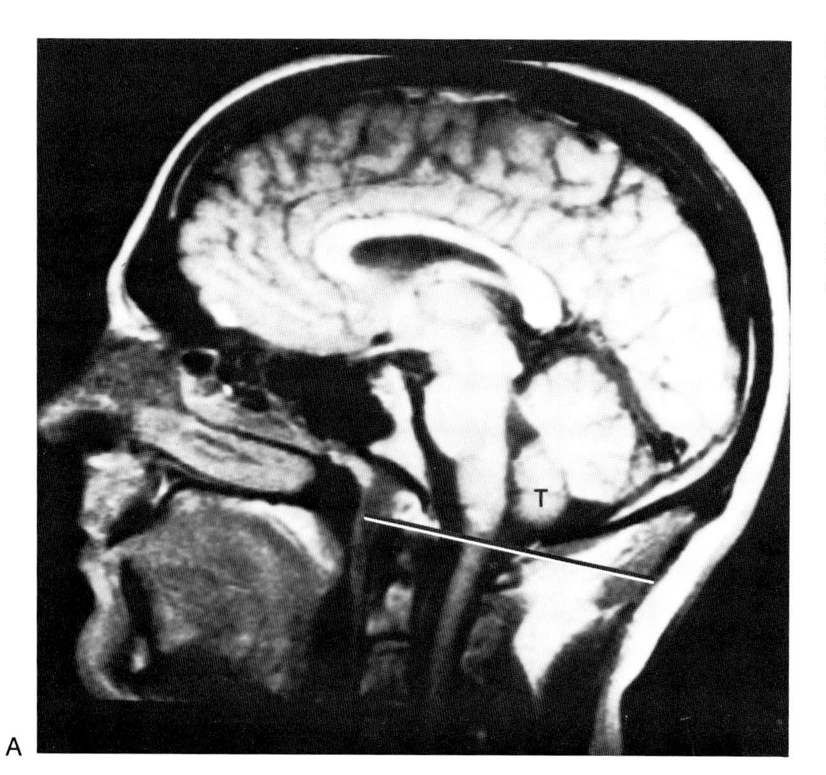

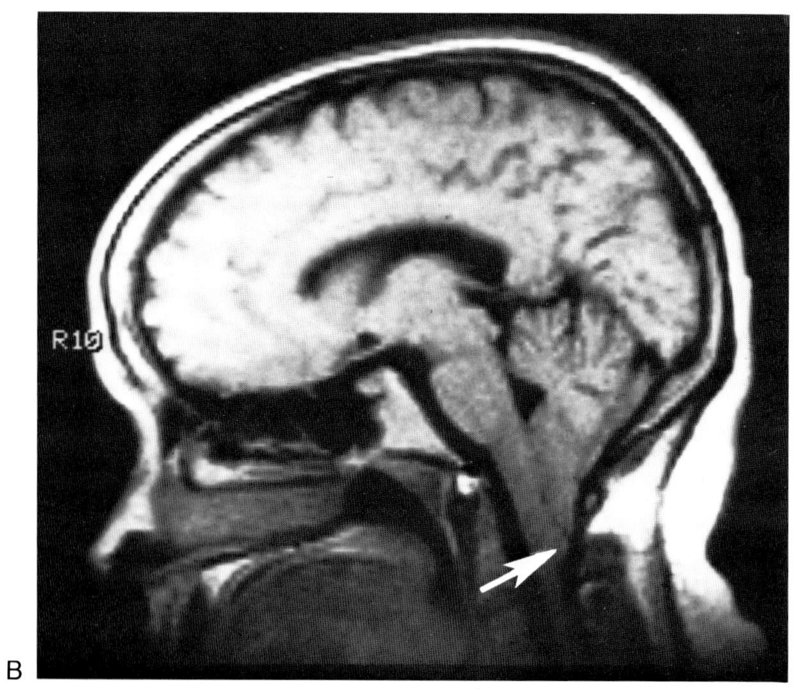

Fig. 4-9. (A) Normally the cerebellar tonsil (*T*) lies above the level of the foramen magnum (line). Up to 3 mm below this line is still considered within normal limits, provided the tonsils are not deformed. **(B)** Chiari I malformation. The tonsils are herniated well below the foramen magnum. (*Figure continues*).

Fig. 4-9 (*continued*). **(C)** Metrizamide CT demonstrates tonsilar herniation into the spinal canal.

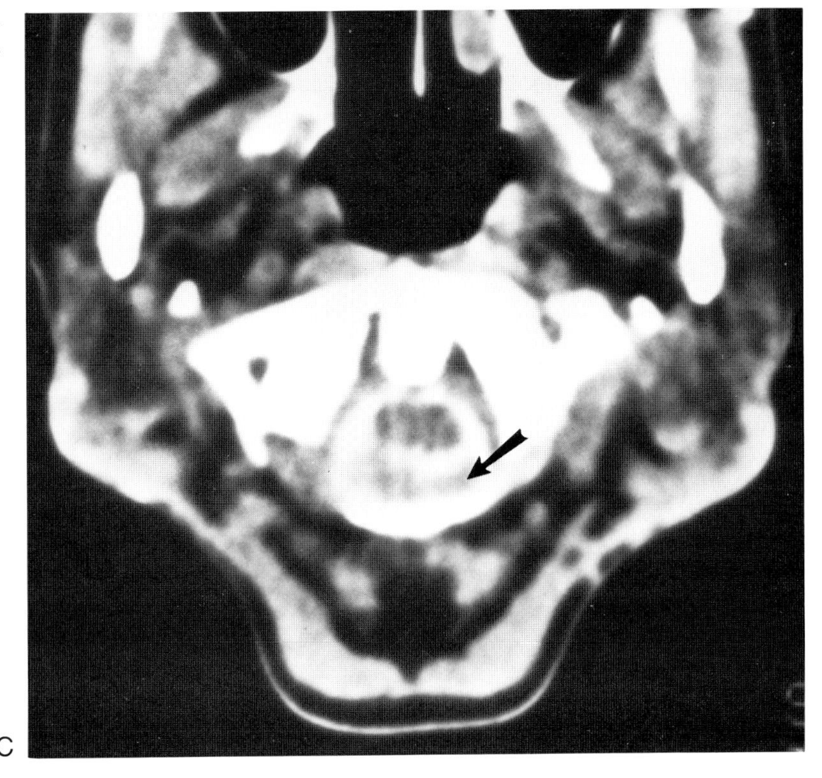

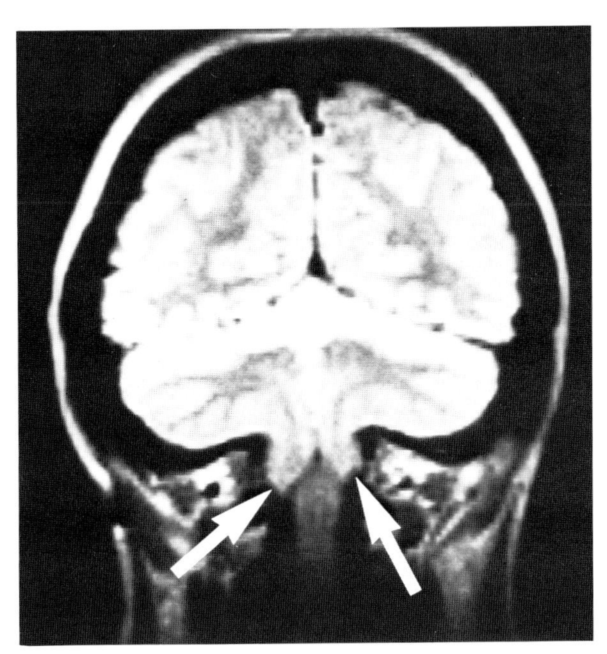

Fig. 4-10. Chiari I malformation seen on coronal SE 1500/28 image. The peglike tonsils (arrows) extend well below the foramen magnum. Asymmetric tonsilar herniation is best appreciated coronally.

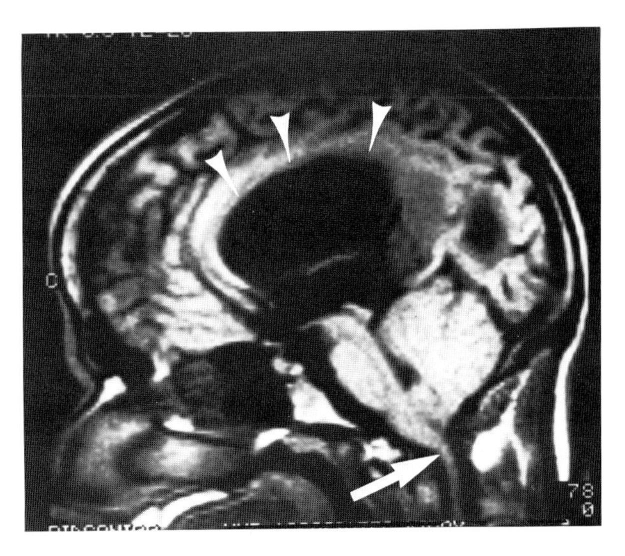

Fig. 4-11. Cervicomedullary kinking (arrow) and cord atrophy in Chiari I malformation. Also note hydrocephalus, with uplifting of corpus callosum (arrowheads).

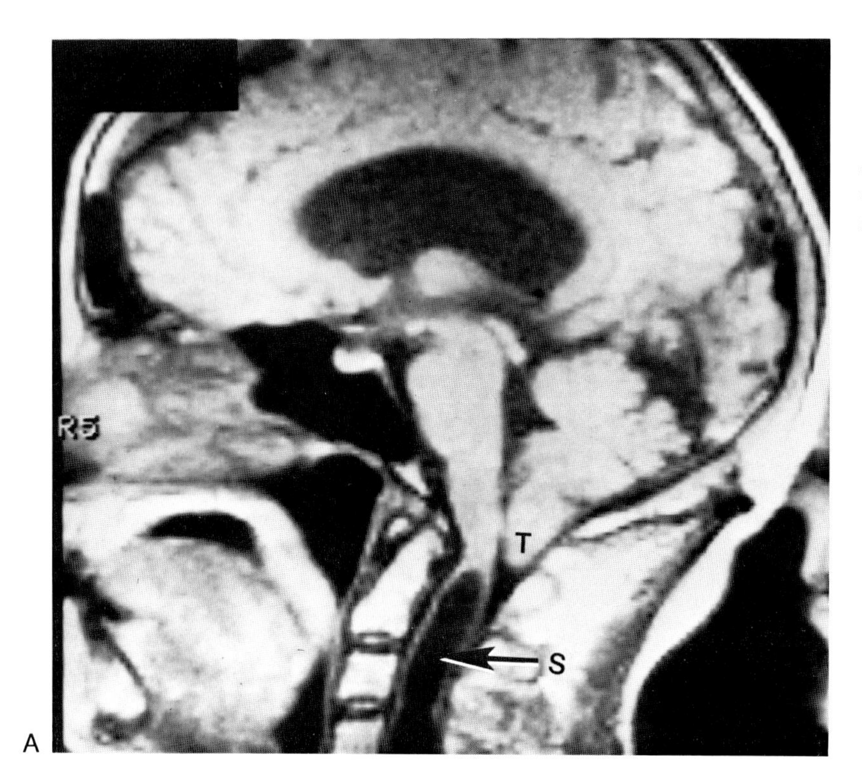

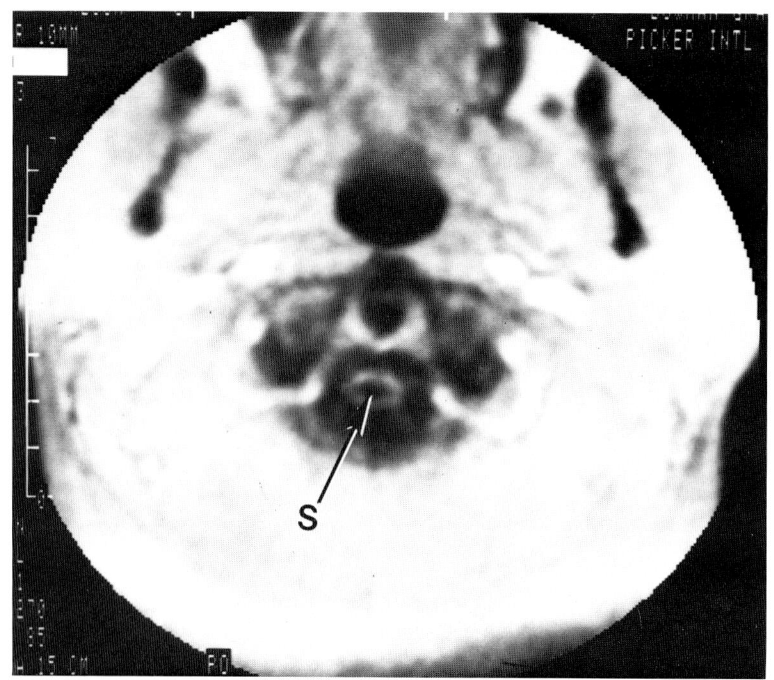

Fig. 4-12. Chiari I with syrinx. **(A)** Note large syrinx cavity (*S*) expanding the cord. Cerebellar tonsil (*T*) is well below foramen magnum. **(B)** A syrinx (*S*) is seen on axial surface coil image of the upper cervical spine.

layed metrizamide enhanced CT may be more sensitive than MRI in syrinx detection.[58]

Chiari II Malformation

The CT and pathologic findings of the Chiari II malformation are perhaps best described in a series of classic papers by Naidich and colleagues (1980).[59–61] The observed abnormalities may be divided into those of the skull and dura, midbrain and cerebellum, ventricles and cisterns, and others (Table 4-4). The imaging findings reflect primarily the following pathologic features: (1) hydrocephalus, (2) caudal displacement of the cerebellum and fourth ventricle, (3) crowding of the posterior fossa, and (4) hypoplasia of certain structures. While MR is inferior to CT in evaluating bony changes of the Chiari II malformation, for most other features it seems equal or superior to CT. Following the scheme of Naidich et al., the Chiari II complex will now be discussed.

Table 4-4. Characteristic CT and MR Findings in Chiari II Malformations

Skull and Dural Anomalies
Craniolacuniae (85%)
Clivus and petrous scalloping (90%)
Enlargement of foramen magnum and posterior fossa
Partial absence/fenestration of falx (>90%)
Hypoplasia of the tentorium (95%)
Midbrain and Cerebellar Anomalies
Tectal beaking (89%)
Towering cerebellum (43% of unshunted, 76% of postshunted)
Overlapping of cerebellum around brain stem (93%)
Pointing of anterior cerebellar margins (83%)
Anomalies of Ventricles and Cisterns
Elongation, downward displacement, and flattening of 4th ventricle (100%)
Large massa intermedia (75%)
Absence of septum pellucidum (50%)
Asymmetric dilatation of lateral ventricles, with beaking of frontal horns
Compression of posterior fossa cisterns (100%)
Other Associations
Myelomeningocele (>99%)
Syringomyelia
Diastematomyelia
Heterotopias
Cerebral microgyria
Aqueduct stenosis/forking
Meningeal gliosis
Fourth ventricular cyst
Absence of corpus callosum
Accessory anterior commissure

Anomalies of the Skull and Dura

Craniolacunia. Craniolacunia (lacunar skull, Lückenschädel) are pits, holes, or areas of calvarial thinning resulting from dysplasia of membranous bone. They are present in at least 85 percent of patients with Chiari II malformation and nearly always indicate the presence of a myelomeningocele.[62,63] Craniolacuniae are not visualized by MR because the newborn calvarium is usually thin and has low signal intensity like the lacunae. Craniolacuniae disappear spontaneously by 6 months of age and bear no relation to hydrocephalus or prognosis.[19,64]

Clivus and Petrous Scalloping. Pressure erosion by the tight posterior fossa contents causes clivus and petrous scalloping. A concavity in the clivus involving the basiocciput but not the basisphenoid is seen in 94 percent (Fig. 4-13).[65] A smooth erosion of the pos-

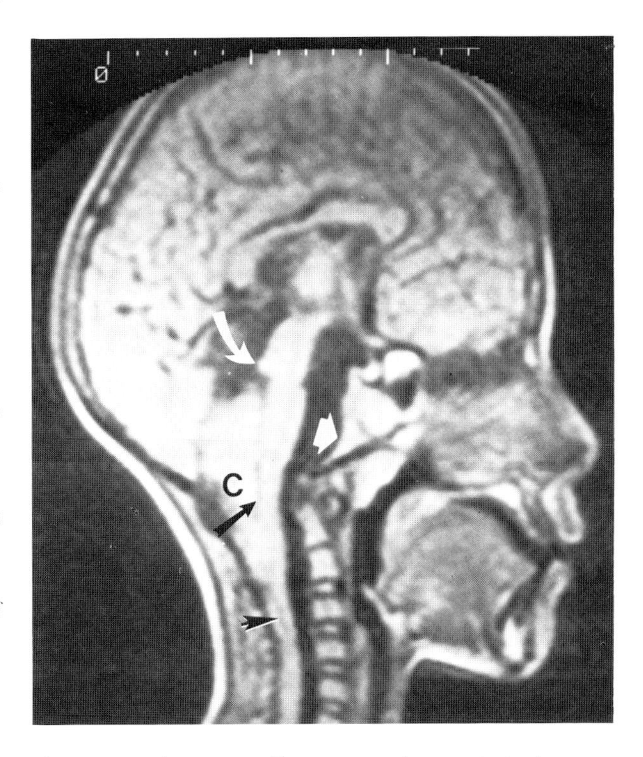

Fig. 4-13. Chiari II malformation, characteristic features. There is anterior concavity of the clivus (short white arrow). The cerebellum (C) is displaced downward through the foramen magnum and the fourth ventricle (black arrow) is elongated and flattened. A small cervical cord syrinx is seen (black arrowhead). Note beaking of the tectum (curved white arrow).

teromedial petrous pyramids is seen that typically spares both the petrous ridge and the jugular tubercle.[66] The internal auditory canals are shortened and appear directed more posteromedially than normal.[66]

Enlargement of the Foramen Magnum. This disorder occurs particularly in the sagittal direction.[59,66] It is more commonly seen in older children and in those with myelomeningoceles.[19,66] The enlargement may be best appreciated on sagittal MR images or axial images through the plane of the foramen magnum.[59]

Dysgenesis of the Falx. Some form of dysgenesis of the falx occurs in nearly 100 percent of patients with the Chiari II malformation.[63] The falx may be partially absent, hypoplastic, or fenestrated. The anterior and middle portions are most frequently involved. Interdigitations of the cerebral gyri seen on coronal or axial images provide MR evidence for falcine dysgenesis (Fig. 4-14). This is most commonly noted after shunting.[59]

Hypoplasia of the Tentorium. About 95 percent of cases have hypoplasia of the tentorium.[63] Posteriorly the tentorial attachment is very low, sometimes at or near the foramen magnum (Fig. 4-15A).[67] The straight sinus is foreshortened and directed upward. The free tentorial margin, instead of having a normal V-configuration, has a U-shape.[19] The edges of the tentorium are bowed laterally and the cerebellum projects upward through the wide incisura (Fig. 4-15B).

Midbrain and Cerebellar Anomalies

Tectal Beaking. The midbrain is abnormal in nearly 90 percent of patients with the Chiari II malformation.[60] In particular the tectum (quadrigeminal plate) becomes distorted into a beaklike configuration (Fig.4-15).[68] In its mild form there is distortion of the tectum with bulging of the inferior colliculi. In more severe cases the superior and inferior colliculi are fused into a conical mass that, together with the pineal, are elongated caudally. Tectal beaking is more easily appreciated in postshunted, older children.[60]

Cerebellar Anomalies. Cerebellar anomalies are seen in every case of the Chiari II malformation. First, the midline cerebellum becomes invaginated to receive the beaked tectum. Secondly, marked upward growth of the cerebellum throught the tentorial incisura results in a "towering cerebellum" (Fig. 4-16). This is seen in 43 percent of unshunted and 76 percent of postshunted patients.[60] The towering cerebellum may act as an extra-axial supratentorial mass that displaces the temporal and occipital lobes. Growth of the cerebellar hemispheres around the brain stem

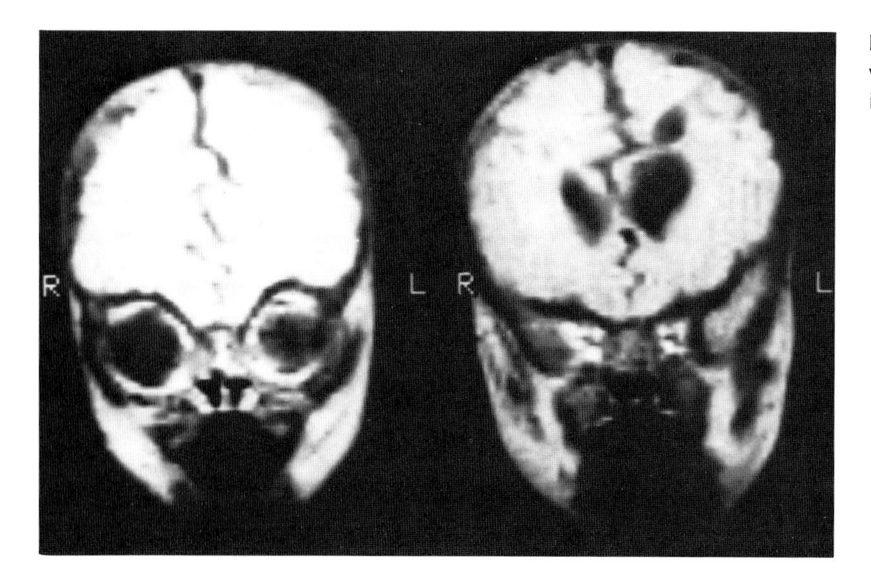

Fig. 4-14. Dysgenesis of the falx with Chiari II malformation results in interdigitation of the gryi.

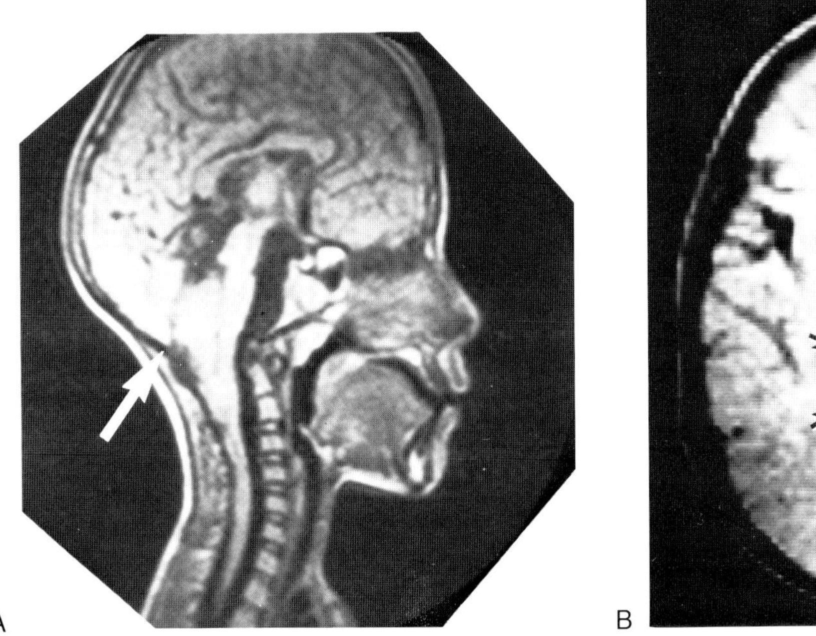

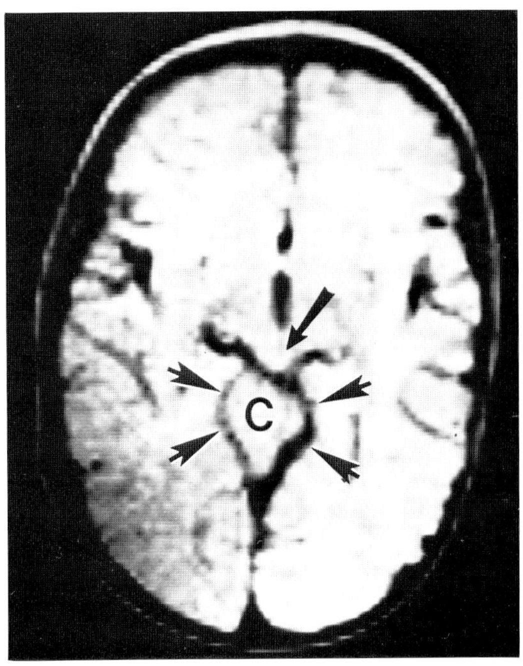

Fig. 4-15. Chiari II malformation. **(A)** Sagittal view shows low attachment of tentorium (arrow) with resulting small posterior fossa. **(B)** The edges of the tentorium are bowed laterally (arrowheads) and the cerebellum (C) projects upward through the wide incisura. Also note tectal beaking (arrow).

results in several abnormalities. As the cerebellum grows lateral to the midbrain it may become interposed between the cerebral peduncles and the hippocampus. As it grows lateral to the pons it may bulge into the cerebellopontine angle cisterns, presenting radiologically as bilateral cerebellopontine angle "masses" (Fig. 4-17). The advancing cerebellar margins become pointed anteriorly in over 80 percent.[60] The cerebellar margins may cover the cranial nerve origins so that the nerves emerge between the cerebellar folia.[67] In the extreme case, the cerebellar margins may meet in the midline anteriorly covering the basilar artery.[67]

Anomalies of the Ventricles and Cisterns

Fourth Ventricle. The essential diagnostic feature of the Chiari II malformation is downward displacement of the fourth ventricle (Fig. 4-18). In nearly all cases, the fourth ventricle is elongated and sagittally flattened with parallel walls and no obvious

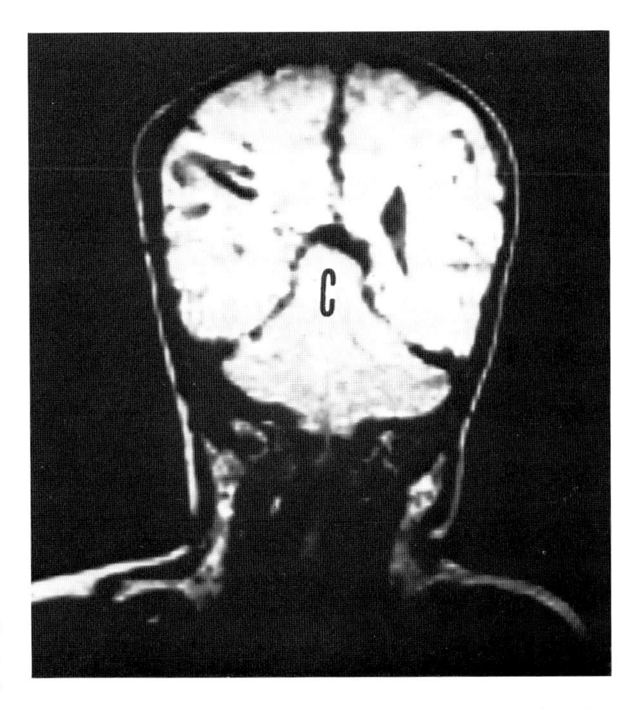

Fig. 4-16. "Towering" cerebellum (C) seen after shunting of Chiari II malformation.

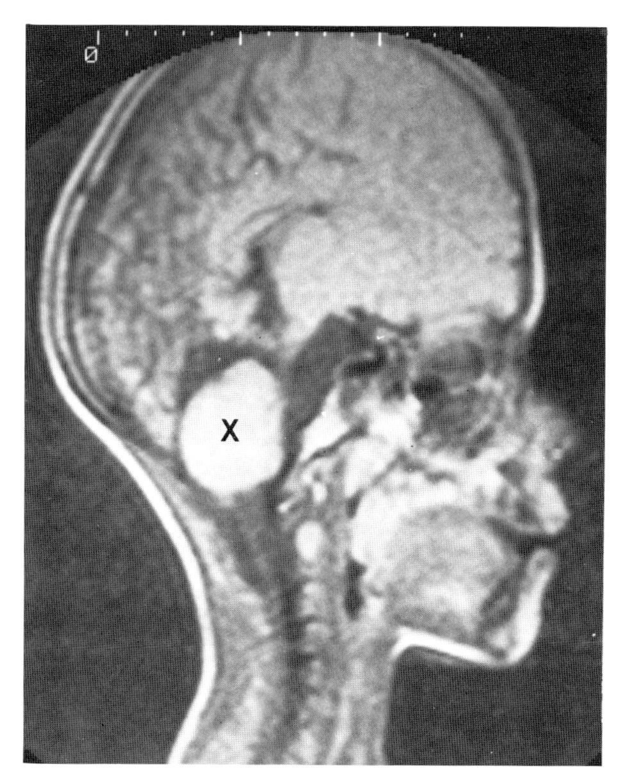

Fig. 4-17. Chiari II malformation. The cerebellum may wrap laterally around the brain stem, mimicking a cerebellopontine angle mass (*X*).

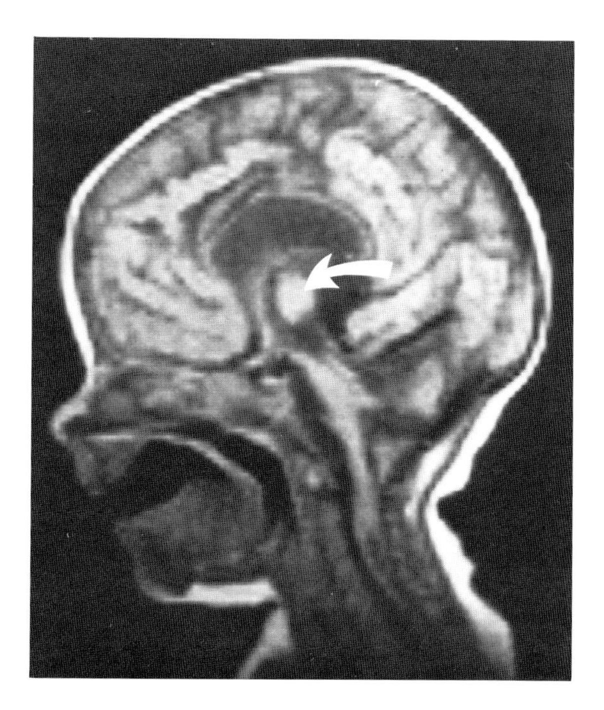

Fig. 4-18. Chiari II malformation. The massa intermedia (arrow) is enlarged and lies closer than usual to the foramina of Monro.

recesses.[67] It may be twice as long as a normal fourth ventricle. It may extend into the cervical spinal canal, with maximal sagittal diameter at or below the foramen magnum.[63,67] On axial images the normal fourth ventricle should be at or above the level of the upper third of the petrous pyramid. Failure to visualize the fourth ventricle at this level indicates pathologic displacement or malposition.[69]

Third Ventricle. The third ventricle is only mildly dilated in most patients with the Chiari II malformation.[61] This stands in contrast to simple aqueduct stenosis where the third ventricle is markedly enlarged.[70] The massa intermedia is enlarged and lies closer to the foramen of Monro in 75 to 90 percent of patients (Fig. 4-19).[61,71,72]

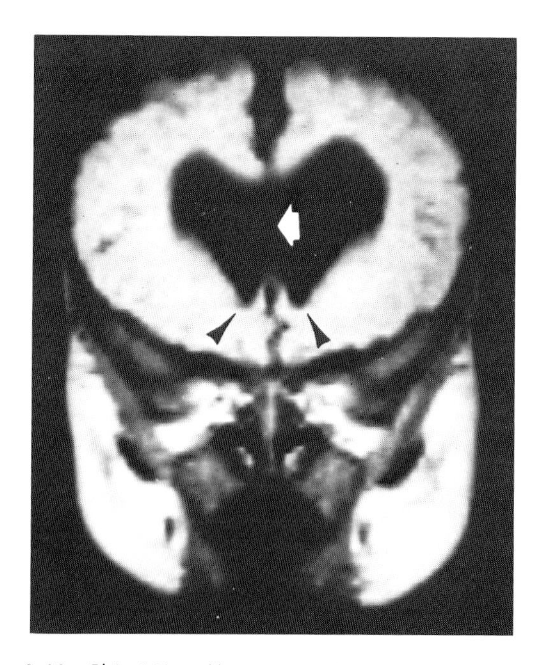

Fig. 4-19. Chiari II malformation. The septum pellucidum is absent (white arrow). There is characteristic pointing of the frontal horns inferiorly (arrowheads).

Lateral Ventricles. The lateral ventricles are usually asymmetrically dilated with the occipital horns larger than the frontal horns.[61] This finding may be explained on the basis of dysgenesis of the corpus callosum, thalamus, or telencephalon. On coronal images, inferior pointing of the frontal horns may be noted (Fig. 4-20). The septum pellucidum is absent or deficient in 50 percent of cases.[61] Heterotopic gray matter may occasionally be present and produce irregularities in the walls of the lateral ventricles.

Hydrocephalus. Over 90 percent of patients display hydrocephalus with the Chiari II malformation, but it cannot be ascribed to any single cause.[38] Aqueductal stenosis or forking is often present. The subarachnoid spaces may also be compressed at the foramen magnum or tentorial incisura. In some cases the fourth ventricular outlet foramina are obstructed. Increased venous pressure may also play a role. Finally, ventriculomegaly may sometimes be a manifestation of cerebral dysgenesis rather than of obstructive or functional hydrocephalus.

DISORDERS OF DIVERTICULATION: THE PROSENCEPHALIES

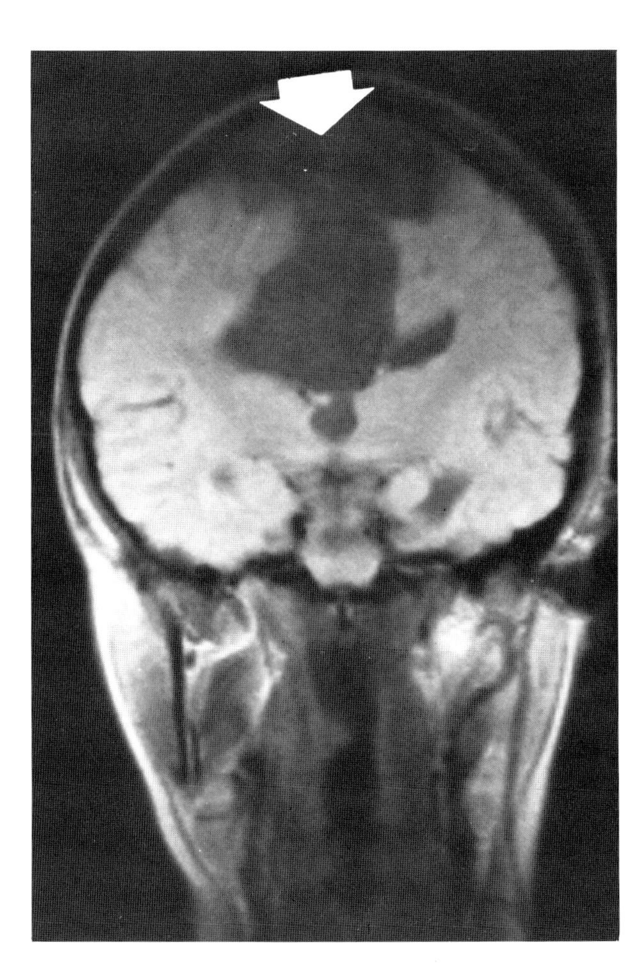

Fig. 4-20. Lobar holoprosencephaly. Midline defect in a patient who survived to adulthood.

The primitive forebrain (prosencephalon) normally undergoes cleavage and diverticulation between the 4th and 8th weeks of fetal life to form the telencephalon, diencephalon, and divided ventricular system.[38] Disturbances in this process result in a complex set of cranial and facial anomalies called the prosencephalies. When the forebrain remains largely undivided, severe malformation called *holoprosencephaly* (holo = "entire") results. Three major subtypes of holoprosencephaly have been defined, based on the degree of separation of the cerebral hemispheres and ventricles: the *alobar*, *semilobar*, and *lobar* types. *Septo-optic dysplasia* is a milder disorder of diverticulation that is usually distinguished from the more severe holoprosencephalies. However, it shares many common embryologic features and can be considered to be closely related to them.

Midline facial anomalies are frequently associated with the prosencephalies, including cyclopia, ethmocephaly, cebocephaly, and median cleft lip.[73] These facies have in common aplasia or hypoplasia of the medial facial skeleton with close-set eyes (hypotelorism). As a rule the clinician should search for holoprosencephaly in a patient with hypotelorism and midline facial anomaly.[74] Patients with the milder prosencephalies may have no recognizable facial defect, however. Other associations of prosencephaly include trigonocephaly,[38] dysgenesis of the corpus callosum,[75] encephalocele,[76] and Dandy-Walker malformation.[77] Familial occurrence and occurrence with chromosomal anomalies has also been reported.[78,79]

The prosencephalies are infrequently encountered in radiologic practice, because many cases are spontaneously aborted or die shortly after birth. Overall, however, they are among the most common malformations of the central nervous system.[80] While patients with alobar and semilobar holoprosencephaly frequently do not survive infancy, those with lobar holoprosencephaly and septo-optic dysplasia often survive to adulthood. These latter types are more likely to undergo investigation by MRI.

Alobar Holoprosencephaly

The alobar type is the most severe form of holoprosencephaly, nearly always associated with a clinically obvious midline craniofacial anomaly. Little or minimal cleavage of the primitive prosencephalon has taken place. There is a single large ventricle central representing the unseparated lateral and third ventricles. A thin mantle of cerebral cortex is often seen anteriorly, and a dorsal sac of CSF may be present posteriorly. The thalami are fused. There is no falx, corpus callosum, or septum pellucidum. The posterior fossa structures are normal. Virtually all affected infants die within the first year of life.[81,82]

In a single reported case of alobar holoprosencephaly, MR accurately delineated the monoventricle, dorsal sac, and callosal agenesis.[83] The single ventricle morphology was best appreciated on coronal views, while the dorsal sac and callosal agenesis were better delineated on sagittal images. While CT could only infer indirectly the absence of the corpus callosum, MR was able to document this feature of the malformation unequivocally.

Semilobar Holoprosencephaly

The semilobar type is of milder degree than the alobar variety, although the same fundamental pathology is present. The central monoventricle is not quite as large, and there is partial formation of the frontal or occipital horns. Again there is absence of the corpus callosum, falx, and septum pellucidum. The cerebral hemispheres and thalami remain fused, although partial formation of the posterior interhemispheric fissure and third ventricle may occasionally be noted. Facial anomalies are less severe, most commonly being cleft lip or palate. Olfactory bulbs and tracts are absent or hypoplastic, as they are in the alobar variety. While infants with semilobar holoprosencephaly may survive, they are profoundly retarded.[81,84]

Lobar Holoprosencephaly

The lobar type is much less severe than the alobar or semilobar varieties. Facial anomalies usually are not seen, although minor midline defects may be present. Distinct lateral ventricles are seen, and the thalami are separated by a well-formed third ventricle. The lateral ventricles are frequently dilated, and the roof of the frontal horn has a characteristic flat or squared-off appearance, best seen on coronal CT or MR. The septum pellucidum is usually absent, but the falx cerebri and corpus callosum are at least partially formed. Although the interhemispheric fissure is well defined, some residual lobar fusions may still often be noted, particularly anteriorly or along the cingulate gyri. Survival into adulthood is common, but mental retardation is the rule (Fig. 4-20).[38,84]

Septo-Optic Dysplasia

Also called deMorsier syndrome, septo-optic dysplasia represents a mild variant of the holoprosencephalies, characterized by absense of the septum pellucidum, enlarged third ventricular chiasmatic recess, and hypoplasia of the optic nerves, chiasm, and infundibulum. Clinical features include a 3:1 female preponderance, seizures, blindness, diabetes insipidus, and other hypothalamic dysfunction. On physical examination there is hypoplasia of one or both optic discs with visual loss and wandering nystagmus.

Magnetic resonance findings mirror those observed on CT. The septum pellucidum is absent. The lateral and third ventricles are moderately enlarged. The frontal horns are squared superiorly and pointed inferiorly (Fig. 4-21A). The optic nerves and chiasm are

Fig. 4-21. Septo-optic dysplasia. **(A)** Characteristic flattening of the roof of the lateral ventricles (arrows) with pointing of inferior aspects (arrowheads). The septum pellucidum is absent. **(B)** The corpus callosum is attenuated, lacking its normal bulbous splenium. The optic chiasm (black arrowhead) is more vertically oriented than normal. A primitive optic ventricle is present (white arrows) representing a dilated optic recess of the third ventricle. **(C)** Axial image confirms absence of septum pellucidum.

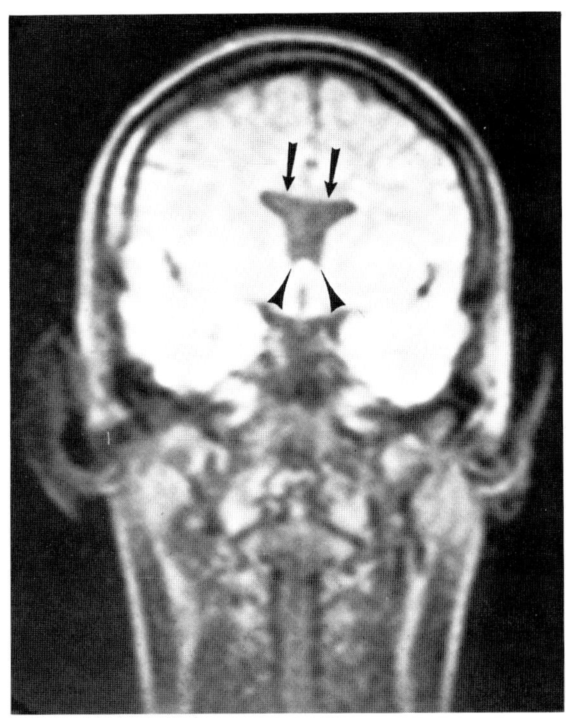

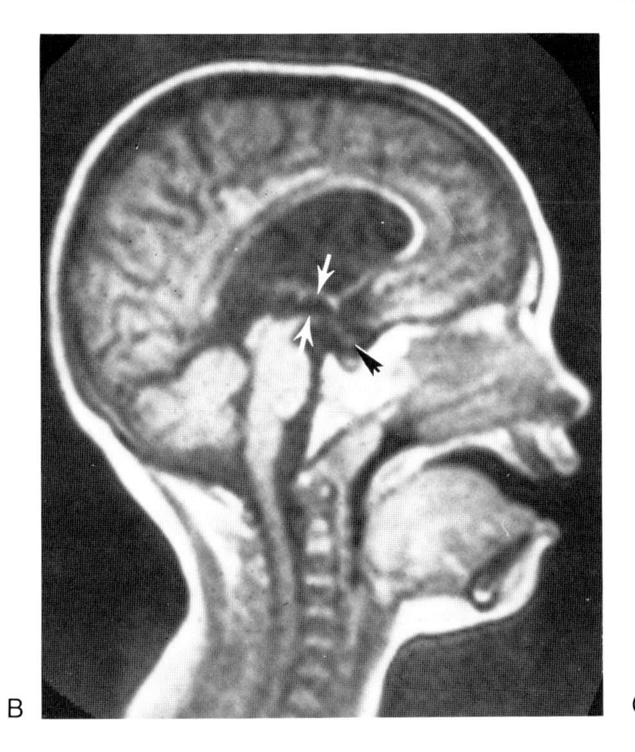

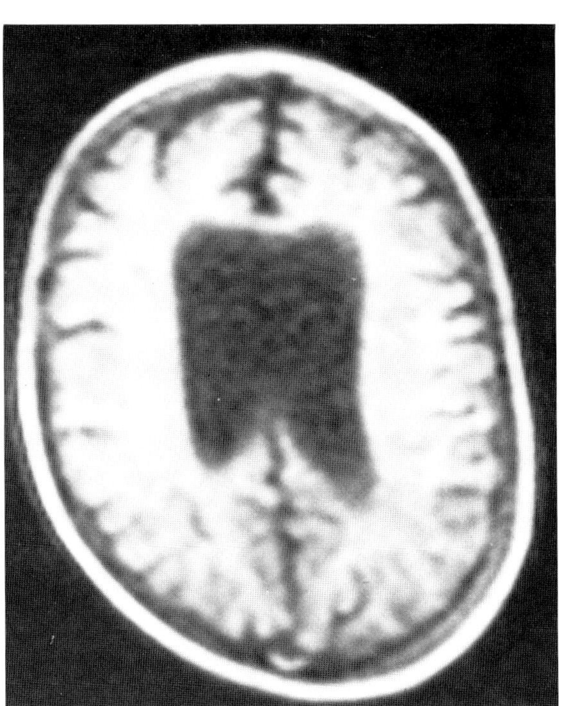

small, and there is enlargement of the chiasmatic recess of the third ventricle. The chiasm may be abnormally positioned and appear to divide in a vertical rather than a horizontal fashion (Fig. 4-21B). Sagittal and coronal images seem most useful for defining the pathology.[17] Partial agenesis of the corpus callosum posteriorly is appreciated by MR but not CT (Fig. 4-21C).[17,81,85–87]

HETEROTOPIC GRAY MATTER

Heterotopic gray matter occurs when neuroblasts in the developing embryo fail to properly migrate to the cortical surface. As a result, islands of abnormally placed gray matter persist deep within the brain parenchyma. This disruption of normal migration is thought to take place about the 12th week of fetal life.[88]

Small foci of heterotopic gray matter are usually asymptomatic, but may induce seizures that can be difficult to control with medication.[89] Typically, these small islands of gray matter are periventricular and may impinge upon the ependyma and protrude into the lateral ventricles (Fig. 4-22). These lesions were commonly recognized in the days of pneumoencephalography, requiring differentiation from subependymomas and the hamartomas of tuberous sclerosis.[90]

Larger masses of heterotopic gray matter are often associated with mental retardation, seizures, and abnormal brain development. Concomitant lesions include microcephaly, hypoplasia or agenesis of the corpus callosum, cerebellar anomalies, aqueduct stenosis, and cardiovascular and skeletal anomalies.[90] A rare neurocutaneous syndrome, Hypomelanosis of Ito (incontinentia pigmenti achromians) features heterotopic gray matter with hypopigmented skin and other disorders relating to both neuroblast and melanocyte migration.[91] An example of large masses of heterotopic gray matter in a severely retarded patient are presented in Figure 4-23.

Computed tomography has proved useful in confirming the diagnosis of heterotopic gray matter.[92] The tissue should not enhance with contrast administration and should have x-ray attenuation values in the range of other gray matter. However, at times heterotopic gray matter may be difficult to recognize or characterize by CT. The question may arise whether the mass of heterotopic gray matter is actually a neoplastic lesion.

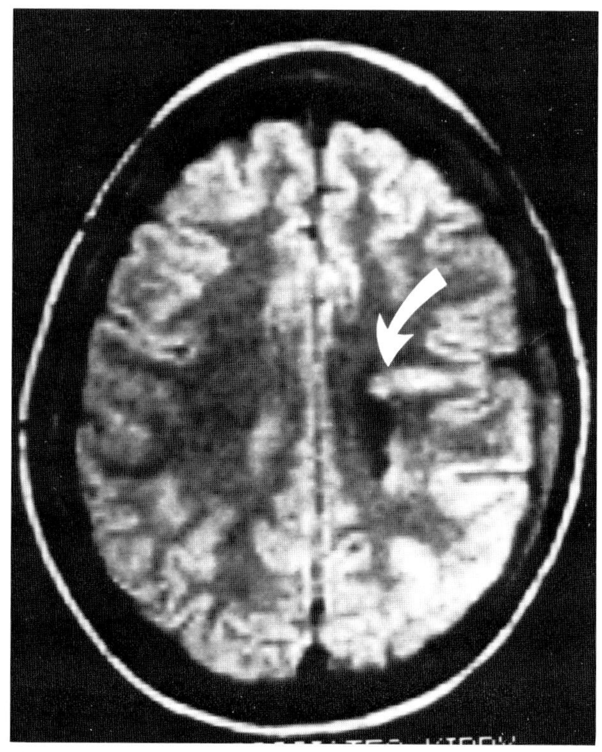

Fig. 4-22. Heterotopic gray matter (arrow) protrudes into lateral ventricle on this SE 2000/28 axial image. Note signal intensity is identical to gray matter elsewhere. This held up on multiple other pulse sequences, confirming the diagnosis. The patient was otherwise normal.

Magnetic resonance imaging can provide further support for the diagnosis of heterotopic gray matter, and is superior to CT for its recognition and evaluation.[93] Because heterotopic gray matter should have similar T1 and T2 values to cortical gray matter, the two tissues should be isointense on all pulse sequences.[94] Identification of heterotopic gray matter by MRI may be helpful in evaluation of seizure patients. Furthermore, MRI may potentially obviate unnecessary surgery in asymptomatic patients in whom the heterotopia is interpreted to be a neoplasm on CT.

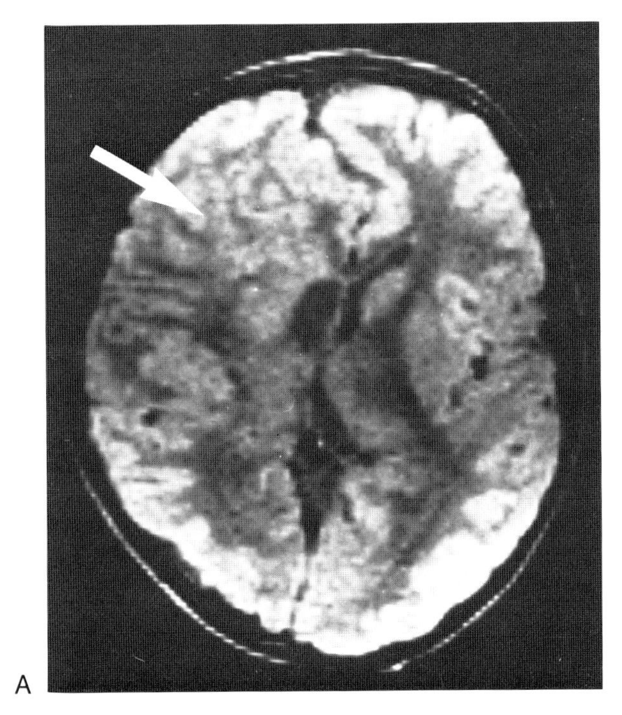

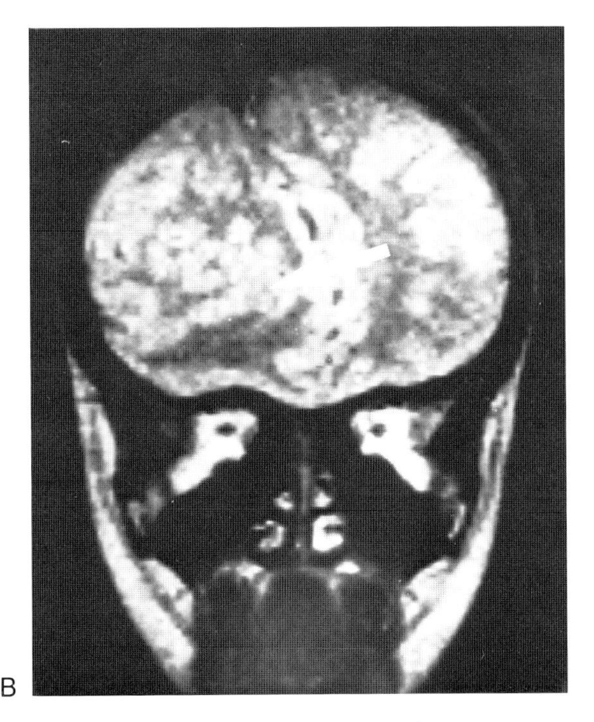

Fig. 4-23. Large mass of heterotopic gray matter in a severely retarded infant with seizures. Signal isointense with gray matter both on **(A)** axial SE 2000/28 and **(B)** coronal SE 1500/56 images.

NEUROFIBROMATOSIS (VON RECKLINGHAUSEN DISEASE)

Neurofibromatosis is an inherited disorder of neuroectodermal and mesodermal tissues, first described extensively in 1882 by von Recklinghausen.[95] The disorder is relatively common, being observed in about 1 in 3000 live births.[38] Transmission is autosomal dominant. It is characterized by multisystem organ involvement, although some of its more important manifestations involve the central nervous system.

The classic clinical presentation includes pigmented skin lesions (cafe-au-lait spots), cutaneous fibromas (fibroma molluscum), and neurofibromas of the cranial and peripheral nerves. Skeletal, vascular, and endocrine abnormalities may be noted as well. A wide spectrum of manifestations may be seen intracranially (Table 4-5).[96,97]

Table 4-5. Intracranial Abnormalities in patients with Neurofibromatosis

Skull
- Hemihypertrophy or hemiatrophy
- Macrocranium
- Lytic defects, especially near lambdoid suture
- Enlargement of skull foramina
- Dysplasia of sphenoid, especially absent greater wing
- Enlarged orbit

CNS Tumors
- Optic gliomas
- Acoustic schwannomas, frequently bilateral
- Meningiomas, including intraventricular and multiple meningiomas
- Plexiform neurofibromas of orbit
- Gliomas elsewhere in brain
- Other tumors (neuroblastomas, hamartomas, ependymoma)

Other Findings
- Arterial stenosis or occlusion
- Basal ganglia calcification
- Heavy calcification of choroid plexus
- Aqueduct stenosis
- Ventriculomegaly

A variety of findings are noted in the skull. The calvarium is frequently enlarged. A lytic defect near the lambdoid suture, more commonly on the left side, is virtually pathognomonic of the disease.[98] The sphenoid bone is frequently dysplastic, especially its greater wing. This is discussed more thoroughly in Chapter 12. The basilar skull foramina may be enlarged, either by neurofibromas or by dural ectasia.[99] In general the bony changes are more easily appreciated by CT or plain radiographs than by MR.[100]

A number of CNS tumors are associated with neurofibromatosis. Optic nerve gliomas are frequently noted, and may be imaged directly by MRI (Fig. 4-24). These gliomas frequently involve the posterior visual pathways, spreading to the optic tracts, geniculate bodies, and optic radiations. MRI may detect this posterior extension while it is unrecognized by CT.[101,102] Plexiform neurofibromas may also involve the orbit (see Chapter 13).

Gliomas in other parts of the brain are also increased in incidence in neurofibromatosis. These gliomas may be multiple (Fig. 4-25). Meningiomas, meningiomatosis, and multiple meningiomas are also commonly seen (Fig. 4-26). Acoustic schwannomas, frequently bilateral, are characteristic, and should suggest the diagnosis of neurofibromatosis. MR imag-

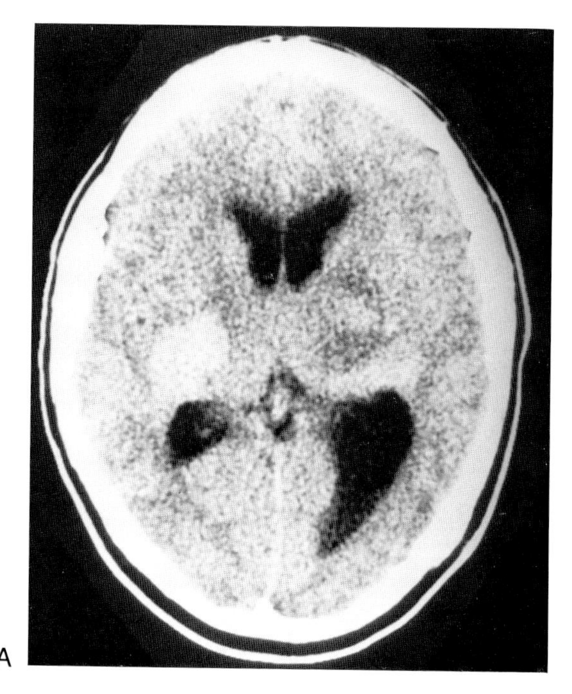

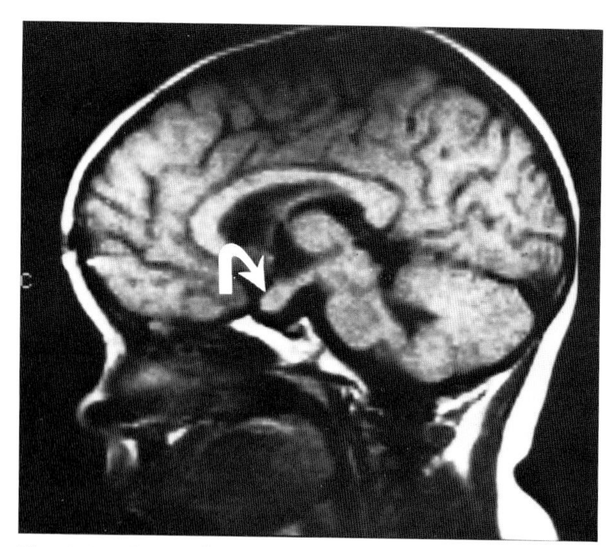

Fig. 4-24. Optic glioma (arrow) in a patient with neurofibromatosis.

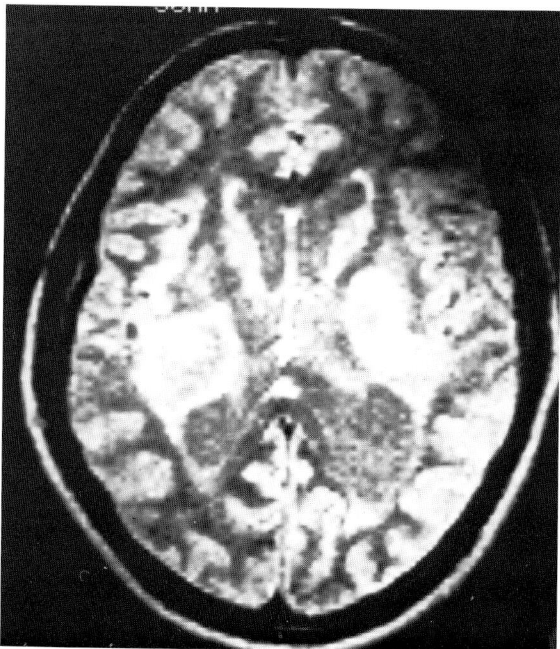

Fig. 4-25. Multiple gliomas in a patient with neurofibromatosis. **(A)** CT image. **(B)** Axial SE 2000/56. (*Figure continues*).

ing of acoustic tumors is discussed in detail in Chapter 12.

MRI seems generally superior to CT for evaluating the brains of patients with neurofibromatosis.[100,102,103] However, meningiomas and bony changes are probably better seen by CT than MR.[100] MRI is exquisitely sensitive for the detection of brain parenchymal tumors in neurofibromatosis, especially those along the optic pathways and in the brain stem. One group of investigators (as well as ourselves) have observed

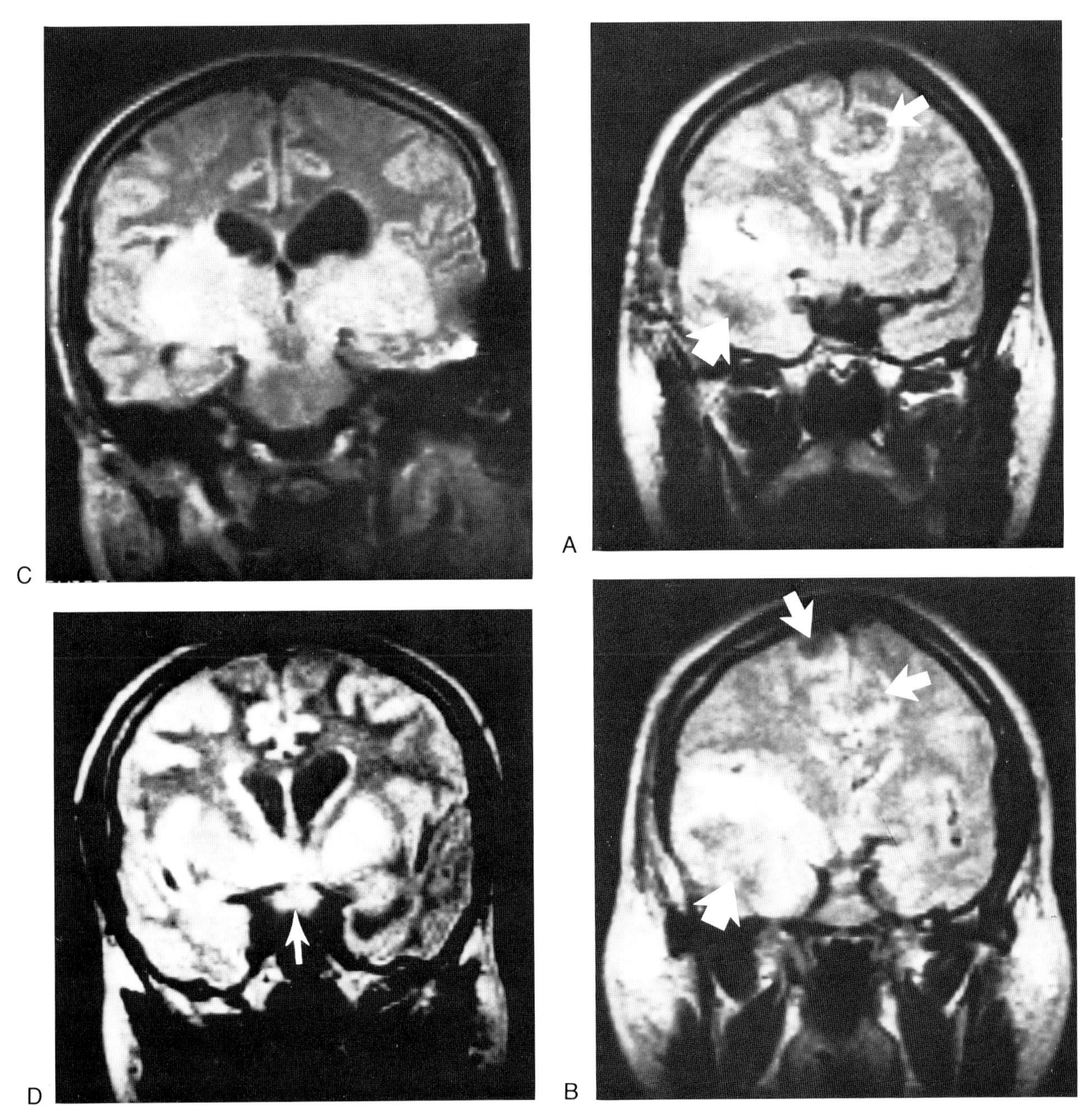

Fig. 4-25 (*continued*). **(C)** Coronal SE 1500/28. **(D)** Coronal SE 1500/56. In **(D)** note optic chiasm glioma (arrow).

Fig. 4-26. (**A** and **B**) Multiple meningiomas (arrows) in a patient with neurofibromatosis, SE 1500/56.

small areas of increased T2 scattered throughout the brain parenchyma of several patients with neurofibromatosis.[100] The nature of these lesions is at present unknown, with the possibility raised that these are small glial tumors.

TUBEROUS SCLEROSIS (BOURNEVILLE DISEASE)

Tuberous sclerosis is a rare, inherited neurocutaneous disorder characterized by the presence of hamartomas in multiple organ systems. These hamartomas (or "tubers") may be found in the skin, brain, kidney, heart, spleen, skeleton, lungs, and gastrointestinal tract.[104] Bourneville (1880) is credited with first identifying and naming the sclerotic brain lesions, although von Recklinghausen had provided an earlier description of the disease.[105] Today the disease is recognized in about 1 in 100,000 people, although it is much more common among inmates of mental institutions.[106] Transmission is autosomal dominant, with variable penetrance and skip generations.[106]

The classic clinical triad consists of skin lesions (adenoma sebaceum), seizures, and mental retardation. While facial adenoma are the most common, other characteristic skin lesions may be present including depigmented nevi, shagreen patches, cafe-au-lait spots, and fibromas.[107] Seizures are the most common presenting symptom, occurring in 80 to 100 percent of cases, and usually developing in infancy or early childhood.[107,108] While mental deficiency is the rule, an increasing number of patients with normal or above normal intelligence are being recognized.[107] This is probably because milder forms of the disease are now being recognized in patients outside of mental institutions.

Intracranially, hamartomas are seen predominantly supratentorially, though they may be found in the cerebellum or brain stem.[108] These hamartomas are composed of bizarrely shaped, very large cells that resemble gemistocytic astrocytes.[109] Most of these hamartomas occur in the subependymal regions of the lateral ventricles, often near the foramen of Monro.[104] These periventricular lesions are multiple in 75 percent and bilateral in 50 percent.[104] They typically calcify during the first 2 years of life. Cortical tubers are less common than the periventricular variety, and are composed primarily of fibrillary gliosis. The cortical lesions are frequently solitary and are much less likely to calcify than the periventricular ones.[110]

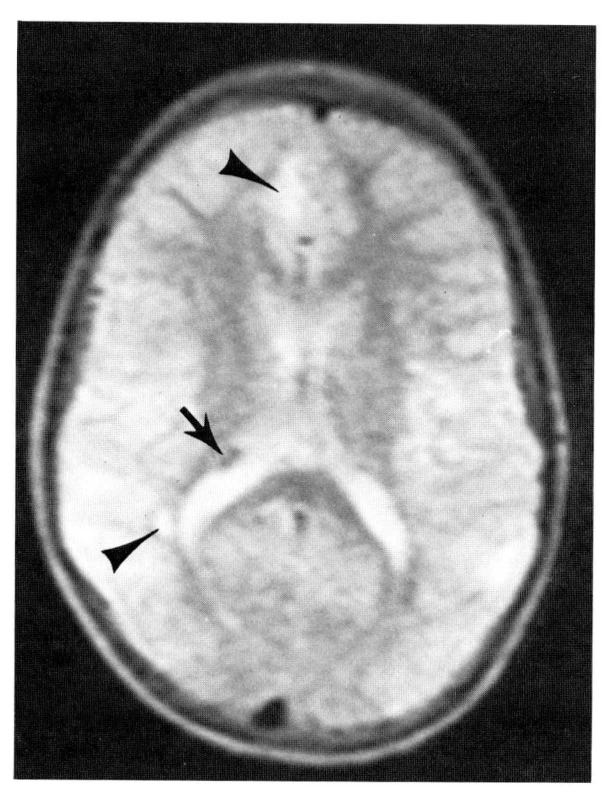

Fig. 4-27. Tuberous sclerosis on SE 3000/100 axial image. Note subependymal calcification (arrow) as area of decreased signal adjacent to right lateral ventricle. High intensity parenchymal tubers (arrowheads) are also seen.

Malignant transformation of the hamartomas occurs in about 10 to 15 percent of cases.[104,109] The resultant tumor is usually classified as a giant cell astrocytoma, but other gliomas sometimes occur.[111] Giant cell astrocytomas often develop in subependymal tubers near the foramen of Monro. They grow slowly and only rarely metastasize, but may cause massive hemorrhage or obstructive hydrocephalus.[104]

Subependymal calcifications are characteristically seen on CT scans, and these strongly suggest the diagnosis.[110] However, because many of the nodules may be small and noncalcified, they may be hard to see on conventional CT. Therefore, a negative CT examination, particularly in an infant, will not totally exclude the diagnosis.[112] Additionally, CT is insensi-

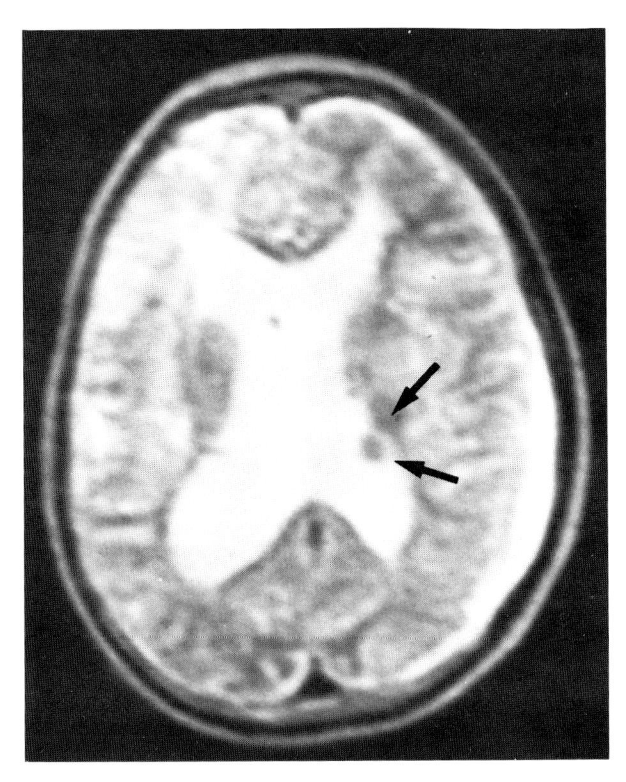

Fig. 4-28. Tuberous sclerosis with subependymal giant cell astrocytoma (arrows).

tive for the detection of small areas of focal demyelinization, which is also known to occur in the brains of these patients.[112]

While MRI cannot detect calcified hamartomas as readily as CT, many subependymal lesions can be identified, because they have signal characteristics different than CSF or normal brain (Fig. 4-27). Subependymal nodules that are undergoing malignant change have been found to have increased T2 values (Fig. 4-28).[113] Focal areas of demyelination can also be identified by MR.[112] Cortical hamartomas are depicted with greater size and number on MR compared to CT.[113] Parenchymal changes of tuberous sclerosis are detected at an earlier stage than by CT (Fig. 4-29).[103] This may have significant prognostic implications because the degree of clinical severity seems proportional to the number of cortical, not subependymal lesions.[114] We have also observed that while many tubers have similar appearance on CT, they may differ in their MR imaging characteristics considerably. This finding awaits confirmation and pathologic correlation.

REFERENCES

1. Ebaugh FG, Holt GW: Congenital malformations of the nervous system. Am J Med Sci 246:106, 1963
2. Yakovlev PI: Pathoarchitectonic studies of cerebral malformation. J Neuropathol Exp Neurol 18:22, 1959
3. Adams RD, Sidman RL: Introduction to Neuropathology. McGraw-Hill, New York, 1968
4. DeMyer W: Classification of cerebral malformations. Birth Defects. 7:78, 1971
5. Byrd SE, Harwood-Nash DC, Fitz CR: Absence of the corpus callosum: computed tomographic evaluation in infants and children. J Can Assoc Radiol 20:108, 1978
6. Parrish ML, Roessmann U, Levinsohn MW: Agenesis of the corpus callosum: a study of the frequency of associated malformations. Ann Neurol 6:349, 1979
7. Ludwin SK, Malmud N: Pathology of congenital anomalies of the brain. p.2979. In Newton TH, Potts DC (eds): Radiology of the Skull and Brain. CV Mosby, St. Louis, 1977
8. Yakolev PI, Lecours AR: The myelogenetic cycles of regional maturation in the brain. p.3. In Minkowski A (ed): Regional Development of the Brain in Early Life. Blackwell Scientific, Oxford, 1967
9. Johnson MA, Bydder GM: NMR imaging of the brain in children. Br Med Bull 40:175, 1983
10. Williams AL: Congenital anomalies. p.316. In Williams AL, Haughton VM (eds): Cranial Computed Tomography: A Comprehensive Text. CV Mosby, St. Louis, 1985
11. Davidoff LM, Dyke CG: Agenesis of the corpus callosum: diagnosis by encephalography. AJR 32:1, 1934
12. Kendall BE: Dysgenesis of the corpus callosum. Neuroradiology 25:239, 1983
13. Davidson HD, Abraham R, Steiner RE: Agenesis of the corpus callosum: magnetic resonance imaging. Radiology 155:371, 1985

13a. Atlas SW, Zimmerman RA, Bilaniuk LT et al: Corpus callosum and limbic system: neuroanatomic MR evaluation of developmental anomalies. Radiology 160:355, 1986

14. Aicardi J, Lefebvre J, Lerique-Koechlin A: A new syndrome: spasm in flexion, callosal agenesis, ocular abnormalities. Electroencephalogr Clin Neurophysiol 19:606, 1965
15. Erenberg G: Aicardi's syndrome: report of an autopsied case and review of the literature. Cleve Clin Q 50:341, 1983
16. Phillips HE, Carter AP, Kennedy JL et al: Aicardi's syndrome: radiologic manifestations. Radiology 127: 453, 1978

17. Curnes JT, Laster DW, Koubek TD et al: MRI of corpus callosal syndromes. AJNR 7:617, 1986
18. Zimmerman RA, Bilaniuk LT, Dolinskas C: Cranial computed tomography of epidermoid and congenital fatty tumors of maldevelopmental origin. Comput Tomogr 3:40, 1979
19. Rao KCVG, Harwood-Nash DC: Craniocerebral anomalies. p.115. In Lee SH, Rao KCVG (eds): Cranial Computed Tomography. McGraw-Hill, New York, 1983
20. Fukui M, Tanaka A, Kitmurak K et al: Lipoma of the cerebellopontine angle: case report. J Neurosurg 46:544, 1977
21. Zee CS, McComb JG, Segall HD et al: Lipomas of the corpus callosum associated with frontal dysraphism. J Comput Assist Tomogr 5:201, 1981
22. Kushnet MW, Goldman RL: Lipoma of the corpus callosum associated with a frontal bone defect. AJR 131:517, 1978
23. Yock DH: Choroid plexus lipomas associated with lipoma of the corpus callosum. J Comput Assist Tomogr 4:678, 1978
24. Wallace D: Lipoma of the corpus callosum. J Neurol Neurosurg Psychiatry 39:1179, 1976
25. Gastaut H, Regis JL, Gastaut E et al: Lipomas of the corpus callosum and epilepsy. Neurology 30:132, 1980
26. Zimmerman RA, Bilaniuk LT, Grossman RI: Computed tomography in migratory disorders of human brain development. Neuroradiology 25:257, 1983
27. Nabawi P, Dobben GD, Mafer M et al: Diagnosis of lipoma of the corpus callosum by CT in five cases. Neuroradiology 21:159, 1981
28. Kazner E, Stochdorp O, Wende S et al: Intracranial lipoma: diagnostic and therapeutic considerations. J Neurosurg 52:243, 1980
29. Faerber EN, Wolpert SM: The value of computed tomography in the diagnosis of intracranial lipoma. J Comput Tomogr 2:297, 1979
30. Kean DM, Smith MA, Douglas RHB et al: Two examples of CNS lipomas demonstrated by computed tomography and low field (0.08 T) MR imaging. J Comput Assist Tomogr 9:494, 1985
31. MacKay IM, Bydder GM, Young IR: MR imaging of central nervous system tumors that do not display increase in T1 or T2. J Comput Assist Tomogr 9:1055, 1985
32. Friedman RB, Segal R, Latchaw RE: Computerized tomographic and magnetic resonance imaging of intracranial lipoma. J Neurosurg 65:407, 1986
33. Dandy WE, Blackfan KD: Internal hydrocephalus: an experimental, clinical, and pathological study. Am J Dis Child 8:406, 1914
34. Taggart JK, Walker AE: Congenital atresia of the foramens of Luschka and Magendie. Arch Neurol Psychiatr 48:583, 1942
35. Hart MN, Malamud N, Ellis WG: The Dandy-Walker syndrome: a clinico-pathological study based on 28 cases. Neurology 22:771, 1972
36. Boulter TR: The dysraphic states. Surg Gynecol Obstet 124:1091, 1967
37. Wolpert MS: Vascular studies of congenital anomalies. p. 1071. In Newton TH, Potts DG (eds): Radiology of the Skull and Brain. Vol 2. Book 4. Angiography. CV Mosby, St. Louis, 1974
38. Sarwar M: Computed Tomography of Congenital Brain Malformations. Warren H. Green, St. Louis, 1985
39. Gardner WJ: The Dysraphic States, from Syringomyelia to Anencephaly. p.201. Excerpta Medica, Amsterdam, 1975
40. Masden JC, Dobben GD, Azar-Kia B: Dandy-Walker syndrome studied by computed tomography and pneumoencephalography. Radiology 147:109, 1983
41. Raybaud C: Cystic malformations of the posterior fossa: abnormalities associated with the development of the roof of the fourth ventricle. J Neuroradiol 9:103, 1982
42. Naidich TP, Radkowski MA, McLone DG, Leestma J: Chronic cerebral herniation in shunted Dandy-Walker malformation. Radiology 158:431, 1986
43. Harwood-Nash DC: Congenital cranio-cerebral abnormalities and computed tomography. Semin Roentgenol 12:39, 1977
44. Lipton HL, Preziosi TJ, Moses H: Adult onset of the Dandy-Walker syndrome. Arch Neurol 35:672, 1978
45. Chiari H: Uever Veranderungen des Kleinhirns infolge von Hydrocephalie des Grosshirns. Dtsch Med Wochenschr 17:1172, 1891
46. Wilkins RH, Brody IA: The Arnold-Chiari malformation. Arch Neurol 25:376, 1971
47. Arnold J: Myelocyste Transposition von Gewebskeimen und Sympode. Beitr Pathol Anat 16:1, 1894
48. Harwood-Nash DC, Fitz CR: Neuroradiology in Infants and Children. CV Mosby, St. Louis, 1976
49. Gardner E, O'Rahilly R, Prolo D: The Dandy-Walker and Arnold-Chiari malformations. Clinical developmental and teratological considerations. Arch Neurol 32:393, 1975
50. Cameron AH: The Arnold-Chiari and other neuro-anatomical malformations associated with spina bifida. J Path Bact 73:195, 1957
51. Brazis PW, Masdeu JC, Biller J: Localization in Clinical Neurology. Little, Brown, Boston, 1985
52. Logue V, Edwards MR: Syringomyelia and its surgical

treatment—an analysis of 75 patients. J Neurol Neurosurg Psychiatry 44: 273, 1981

53. DeBarros MC, Farias W, Ataide L et al: Basilar impression and Arnold-Chiari malformation: a study of 66 cases. J Neurol Neurosurg Psychiatry 31:596, 1968
54. Paul KS, Lye RH, Strang AF et al: Arnold-Chiari malformaiton: review of 71 cases. J Neurosurg 58:183, 1983
55. Spinos E, Laster DW, Moody DM et al: MR evaluation of Chiari I malformations at 0.15 T. AJR 144:1143, 1985
56. Aboulezzo AO, Sartor K, Geyer CA et al: Position of cerebellar tonsils in the normal population and in patients with Chiari malformation: a quantitative approach with MR imaging. J Comput Assist Tomogr 9:1033, 1985
57. Wickbom I, Hanafee W: Soft tissue anatomy within the spinal cord as seen on computed tomography. Radiology 134:649, 1980
58. Lee BCP, Zimmerman RD, Manning JJ et al: MR imaging of syringomyelia and hydromyelia. AJR 144:1149, 1985
59. Naidich TP, Pudlowski RM, Naidich JB et al: Computed tomographic signs of Chiari II malformation. Part I: skull and dural partitions. Radiology 134:65, 1980
60. Naidich TP, Pudlowski RM, Naidich JB: Computed tomographic signs of Chiari II malformation. Part II: midbrain and cerebellum. Radiology 134:391, 1980
61. Naidich TP, Pudlowski RM, Naidich JB: Computed tomographic signs of the Chiari II malformation. Part III: ventricles and cisterns. Radiology 134:657, 1980
62. Vogt EC, Wyatt GM: Craniolacunia (luckenschadel). Radiology 36:147, 1941
63. Peach B: Arnold-Chiari malformation. Anatomic features of 20 cases. Arch Neurol (Chicago) 12:613, 1965
64. McRae DL: Lacunar skull (leuckenschadel). pp. 648, 652. In Newton TH, Potts DG (eds): Radiology of the Skull and Brain. Vol. 1, CV Mosby, St. Louis, 1971
65. Yu HC, Deck MDF: The clivus deformity of the Arnold-Chiari malformation. Radiology 101:613, 1971
66. Kruyff E, Jeffs R: Skull abnormalities associated with the Arnold-Chiari malformation. Acta Radiol (Diag) 5:9, 1966
67. Daniel PM, Strich SJ: Some observations on the congenital deformity of the central nervous system known as the Arnold-Chiari malformation, Neuropathol Exp Neurol 17:255, 1958
68. Emery JL: Deformity of the aqueduct of Sylvius in children with hydrocephalus and myelo-meningocele. Dev Med Child Neurol 16 (Suppl 32):40, 1974
69. Zimmerman RD, Breckbill D, Dennis MW et al: Cranial CT findings in patients with meningomyelocele. AJR 132:623, 1979
70. McCoy WT, Simpson DA, Carter RF: Cerebral malformations complicating spina bifida. Radiologic studies. Clin Radiol 18:176, 1967
71. Davie JC, Baldwin M: Radiographic-anatomic study of the massa intermedia. J Neurosurg 26:483, 1967
72. Gooding CA, Carter A, Hoare RD: New ventriculographic aspects of the Arnold-Chiari malformation. Radiology 89:626, 1967
73. DeMyer W: Holoprosencephaly (Cyclopia-arhinencephaly). p.431. In Vinken PJ, Bruyn GW (eds). Handbook of Clinical Neurology. Congenital Malformations of the Brain and Skull, Part I. Vol 30. North-Holland Publishing, New York, 1977
74. DeMyer W: Median facial malformations and their implications for brain malformations. Birth Defects Orig Art Series 11:155, 1975
75. Jellinger K, Gross H, Kaltenback G: Holoprosencephaly and agenesis of the corpus callosum: frequency of associated malformations. Acta Neuropathol 55:1, 1981
76. Hutchison JW, Stovring J, Turner PT: Occipital encephalocele with holoprosencephaly and aqueduct stenosis. Surg Neurol 12:331, 1979
77. Hayashi T. Takasi S, Kuramoto S: A case of holoprosencephaly: with possible association of Dandy-Walker cyst. Brain Dev 3:97, 1981
78. Dellaire L, Fraser FC, Siglesworth FW: Familial holoprosencephaly. Birth Defects 7:136, 1971
79. Agbata IA, Kovi J, Parshad R et al: Holoprosencephaly and Trisomy 13 in cyclopia. JAMA 241:1109, 1979
80. Osaka K, Matsumoto S: Holoprosencephaly in neurosurgical practice. J Neurosurg 48:787, 1978
81. Fitz CR: Holoprosencephaly and related entities. Neuroradiology 25:225, 1983
82. Manelfe C, Sevely A: Neuroradiological study of holoprosencephalies. J Neuroradiol 9:15, 1982
83. Lee BCP, Lipper E, Nass R et al: MRI of the central nervous system in neonates and young children. AJNR 7:605, 1986
84. Byrd SE, Harwood-Nash DC, Fitz CR, Rogovitz DM: Computed tomography evaluation of holoprosencephaly in infants and children. J Comput Assist Tomogr 1:456, 1977
85. Wilson DM, Enzmann DR, Hintz RL, Rosenfeld G: Computed tomographic findings in septo-optic dysplasia: discordance between clinical and radiologic findings. Neuroradiology 26:279, 1984
86. O'Dwyer JA, Newton TH, Hoyt WF: Radiologic features of septo-optic dysplasia: de Morsier syndrome. AJNR 1:443, 1980

87. Manelfe C, Rochiccioli P: CT of septo-optic dysplasia. AJR 133:1157, 1979
88. Mueller CF: Heterotopic gray matter. Radiology 94:357, 1970
89. Layton DD: Heterotopic cerebral gray matter as an epileptogenic focus. J Neuropath 1:244, 1962
90. Bergeron RT: Pneumographic demonstration of subependymal heterotopic cortical gray matter in children. AJR 101:168, 1967
91. Ross DL, Liwnicz BH, Chun RWM et al: Hypomelanosis of Ito (incontinentia pigmenti achromians)—a clinicopathologic study: macrocephaly and gray matter heterotopias. Neurology 32:1013, 1982
92. Mikhael M, Mattar AG: Malformation of the cerebral cortex with heterotopia of the gray matter. J Comput Assist Tomogr 2:291, 1978
93. Dunn V, Mock T, Bell WE et al: Detection of heterotopic gray matter in children by magnetic resonance imaging. Magnetic Resonance Imaging 4:33, 1986
94. Deeb ZL, Rothfus WE, Maroon JC: MR imaging of heterotopic gray matter. J Comput Assist Tomogr 9:1140, 1985
95. von Recklinghausen F: Ueber die multiplen Fibrome der Haut und ihre Beziehung zu den multiplem Neuromen. Berlin. A Hirschwald, 1882
96. Holt JF: Neurofibromatosis in children. AJR 130:615, 1978
97. Klatte EC: The radiographic spectrum in neurofibromatosis. Semin Roentgenol 11:17, 1976
98. Joffe N: Calvarial bone defects involving the lambdoidal suture in neurofibromatosis. Br J Radiol 38:23, 1965
99. Jacoby CG, Go RT, Beren RA: Cranial CT of neurofibromatosis. AJR 135:553, 1980
100. Edwards MK, Lee TA, Klatte EC, Scott JA: Neurofibromatosis, MR and CT correlation (abstr). AJNR 7:556, 1986
101. Lourie GL, Kirks DR, Osborne DRS: Involvement of posterior visual pathways by optic nerve gliomas. Radiology 157:128, 1985
102. Pomeranz SM: MR imaging of the visual pathways in neurofibromatosis. Radiology 157(P):328, 1985
103. Bilaniuk LT, Zimmerman RA, Kemp S et al: MR imaging of phakamatoses. Radiology 157:212, 1985
104. Medley BE, McLeod RA, Houser OW: Tuberous sclerosis. Semin Roentgenol 11:35, 1976
105. Bourneville DM: Sclerose tubereuse des circonvolutions cerebrales: idiotie et epilepsie hemiplegique. Arch Int Neurol 1:81, 1880
106. Nevin NC, Pearce WG: Diagnostic and genetical aspects of tuberous sclerosis. J Med Genet 5:273, 1968
107. Lagos JC, Gomez MR: Tuberous sclerosis: reappraisal of a clinical entity. Mayo Clin Proc 42:26, 1967
108. Magib MG, Haines SJ, Erickson DL, Mastri AR: Tuberous sclerosis: a review for the neurosurgeon. Neurosurgery 14:93, 1984
109. Kapp JP, Pauson GW, Odom GL: Brain tumors with tuberous sclerosis. J Neurosrug 26:191, 1967
110. Gardeur D, Palmier A, Mashaly R: Cranial computed tomography in the phakomatoses. Neuroradiology 25:293, 1983
111. Fitz CR, Harwood-Nash DC, Thompson JR: Neuroradiology of tuberous sclerosis in children. Radiology 110:635, 1974
112. Fritts H, Katz D, Pribram H, Friedenberg R: MRI of the head in patients with tuberous sclerosis. Mag Reson Imag 4:160, 1986
113. McMurdo SK, Moore SG, Brant-Zawadzki M et al: MR of intracranial tuberous sclerosis, AJNR 7:542, 1986
114. Roach ES, Williams DP, Laster DW: Magnetic resonance imaging in tuberous sclerosis. Arch Neurol 44:301, 1987

5

The Ventricles and Subarachnoid Spaces

The ventricles and subarachnoid spaces contain cerebrospinal fluid (CSF), a dynamic medium with important protective and homeostatic functions. CSF is essentially an ultrafiltrate of plasma that maintains a constant external environment for neurons and glia. Mechanically it serves as a hydrostatic cushion to buffer impact of the brain with the bony calvarium. The buoyant action of the CSF reduces the weight of the brain from about 1400 g in air to only 50 g in situ.[1] The CSF may in some ways function as a lymphatic system for the brain, which lacks true lymphatics. It may be involved in the transport of hormones and neurotransmitters between remote parts of the brain. Changes in its pH affect cerebral circulation as well as pulmonary ventilation. The flow of CSF through the ventricular system and subarachnoid spaces has been called the "third circulation."

A variety of intracranial diseases may upset this delicately balanced "third circulation." Hydrocephalus, an abnormal enlargement of ventricular volume, is the most important manifestation of altered CSF dynamics. Tumors, cystic lesions, and developmental anomalies may also affect the ventricles and subarachnoid spaces. These are all subjects of this chapter.

APPLIED ANATOMY AND PHYSIOLOGY OF THE CSF PATHWAYS

Anatomic Considerations for MRI

The detailed anatomy of the ventricular system is well known to older radiologists and neuroscientists who trained during the heyday of pneumoencephalography. For an entire younger generation of physicians, raised in an era of axial CT, some aspects of ventricular anatomy, particularly those relationships in the sagittal and coronal planes, are not well appreciated. This section gives a brief overview of some of the multiplanar anatomy of the ventricular system as demonstrated by MRI.

Lateral Ventricles

Each lateral ventricle is divided into several parts: frontal (anterior) horn, body, trigone (isthmus, atrium), temporal (inferior) horn, and occipital (posterior) horn. The major portion of the lateral ventricle follows a C-shaped curve, lying between the gray matter of the caudate nucleus and thalamus inferiorly and the corpus collosum and other white matter tracts superiorly.

Frontal Horn

The frontal horn is a gently rounded anterior extension of the lateral ventricle (Fig. 5-1). Its anterior wall and roof are formed by the rostrum, genu, and forceps minor of the corpus callosum. The floor and lateral wall are indented by the head of the caudate nucleus. Medially, lies the most rostral portion of the septum pellucidum. In the floor of the frontal horn medially can be found the interventricular foramen (of Monro), which leads into the third ventricle.

Just posterior to the frontal horn lies the *body* of the lateral ventricle. Like the frontal horn, its roof is also formed by the corpus callosum. The floor is composed of the dorsum of the thalamus medially and the body of the caudate laterally, separated by a groove in which runs the thalamostriate vein. Just medial to this groove lies the choroid plexus. Laterally and superiorly the caudate and corpus callosum meet at a rather acute angle, called the lateral ventricular angle. The medial wall of the body of the lateral ventricle is composed of the septum pellucidum and fornix.

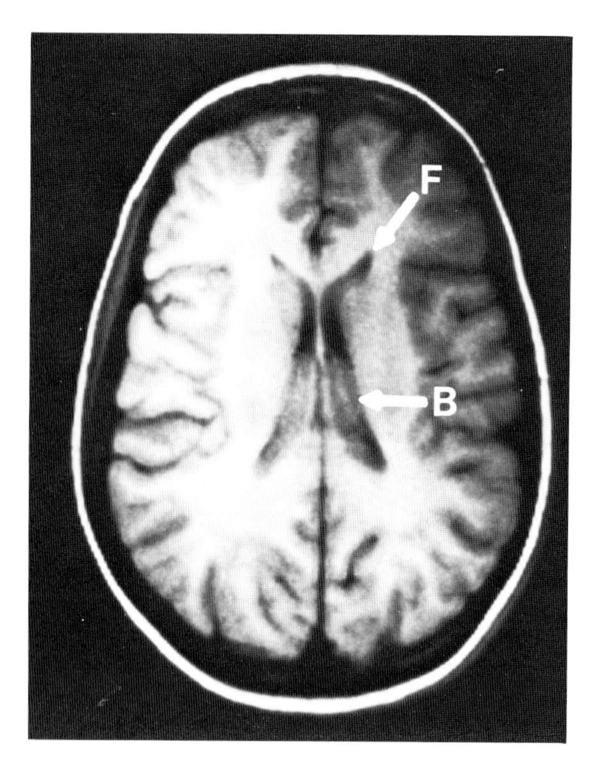

Fig. 5-1. The frontal horn (*F*) and body (*B*) of the lateral ventricle on axial MR image (SE 400/25).

Temporal Horn

The temporal horn runs posterior to the thalamus, downward, and forward into the temporal lobe (Fig. 5-2). Normally the temporal horn is rather slitlike by comparison to the other fuller parts of the lateral ventricle. In coronal section, this crescent-shaped horn has walls that meet at two acute angles, the medial and lateral clefts. The choroid plexus extends into the temporal horn within the medial cleft. Three indentations may be seen on the temporal horn. On its floor, an indentation called the collateral eminence is formed as a response to infolding of cortex from the collateral sulcus. On the medial wall, an indentation is formed by the pes hippocampi. Anteriorly and superiorly, the amygdaloid body forms an impression.

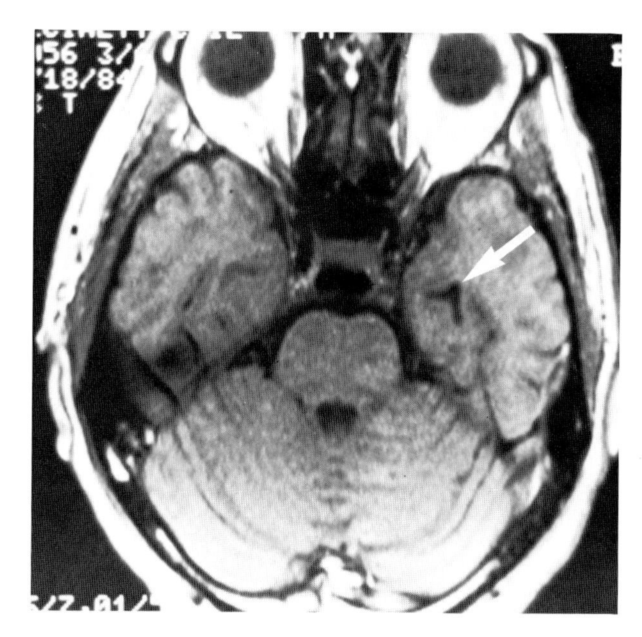

Fig. 5-2. Characteristic, "comma" shape of the temporal horn (arrow) on axial SE 500/28 MR image. (Courtesy Fonar Corporation, 110 Marcus Drive, Melville, NY.)

The roof of the temporal horn is composed of white matter tracts in the sublenticular portion of the internal capsule.

Occipital Horn

The occipital horn is the most variable in size and shape of any portion of the lateral ventricle (Fig. 5-3). It extends posteriorly from the splenium of the corpus callosum into the occipital lobe, tapering into a point. Its roof, lateral wall, and floor are formed by a sheet of white matter fibers (the tapetum) passing from the corpus callosum into the occipital lobe. The lateral wall fibers compose the optic radiation. Medially and inferiorly, two indentations may be seen. The upper and smaller of these is called the bulb and is formed by fibers of the forceps major. The lower and larger ridge is called the calcar avis and is produced by infolding of cortex along the calcarine sulcus, which dips deeply into the occipital lobe.

Trigone

The *trigone* of the lateral ventricle is that portion where the three major components (body, temporal horn, and occipital horn) merge to form a common cavity. It is indented by the collateral eminence inferiorly, the calcar avis medially, and the bulb superiorly. Anteriorly in the trigone, the glomus of the choroid plexus lies on the thalamus along with the posterior portion of the fornix.

Third Ventricle

The third ventricle is a narrow midline cavity lying between the thalami superiorly and hypothalami inferiorly (Fig. 5-4). A shallow groove, the hypothalamic

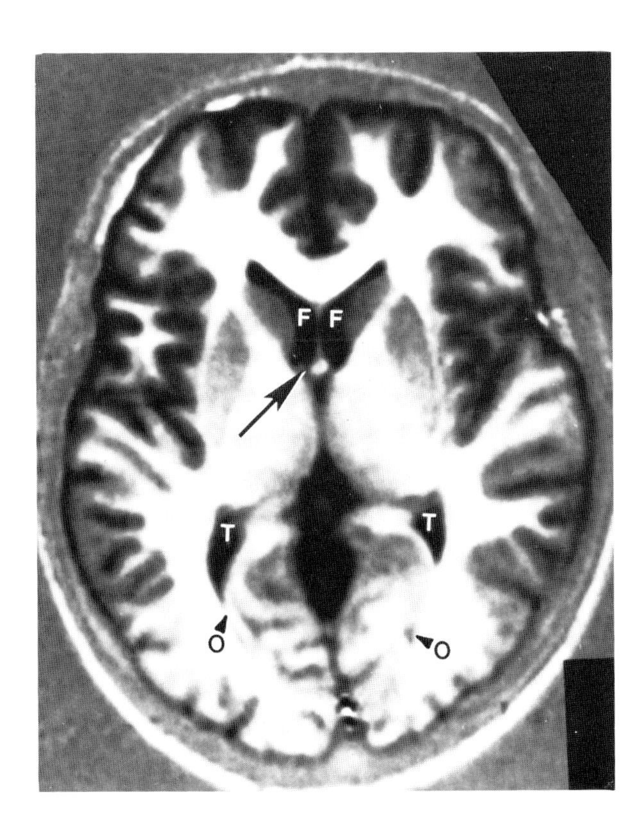

Fig. 5-3. The frontal horns (*F*), trigone (*T*), and occipital horns (*O*) of the lateral ventricle. The foramen of Monro is indicated by an arrow.

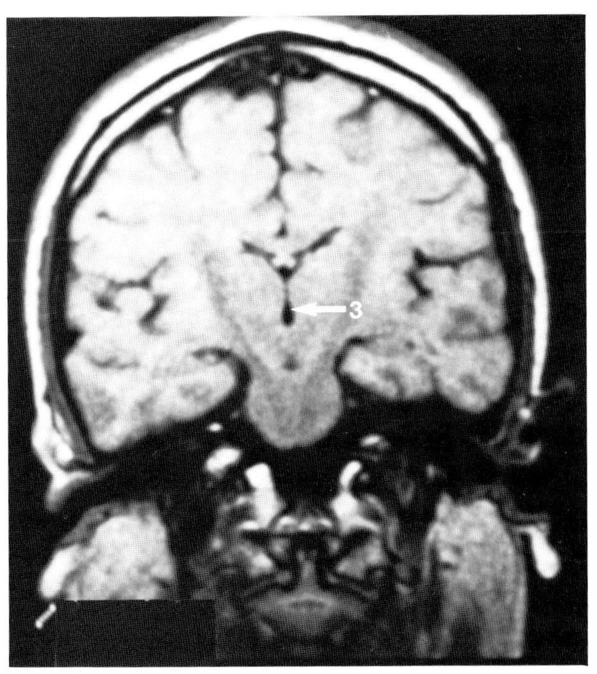

Fig. 5-4. The third ventricle (3), a narrow midline cavity between the thalami, best seen on coronal MR images. The walls of the third ventricle should be flat or concave, never convex.

sulcus, extends along the lateral wall from the foramen of Monro to the cerebral aqueduct and separates the thalamus and hypothalamus. Partial fusion between the lateral walls may occur during development. The resulting adhered area is called the massa intermedia. The anterior wall of the third ventricle is formed by the anterior commissure and lamina terminalis. Its floor is formed by the hypothalamus and subthalamus. Its roof is the velum interpositum, along which is found the choroid plexus. Sagittal view reveals the anteriorly and inferiorly placed suprachiasmatic and infudibular recesses. Posteriorly may be seen the pineal recess and iter (the point where the third ventricle and cerebral aqueduct join).

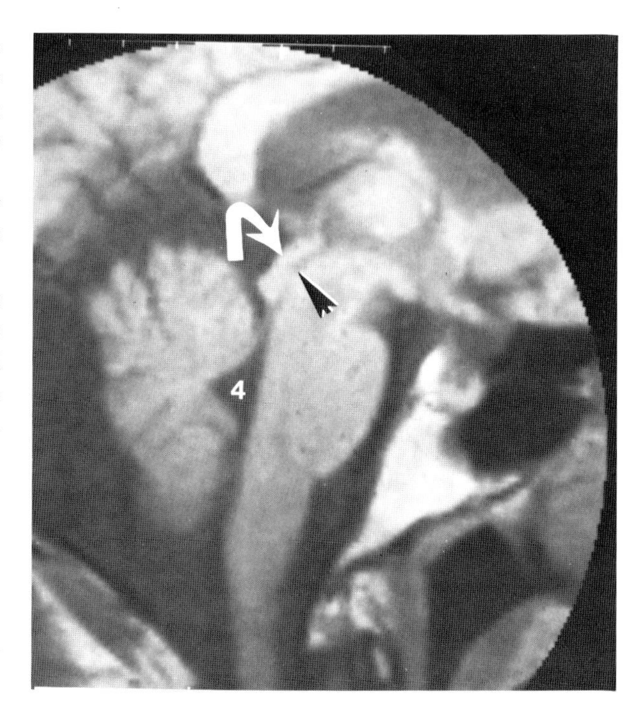

Fig. 5-5. The cerebral aqueduct (arrowhead) whose roof is the quadrigeminal plate (arrow) and floor is the midbrain tegmentum, well seen on this modline sagittal SE 500/25 image. The triangular fourth ventricle (4) is also well seen.

Cerebral Aqueduct

The cerebral aqueduct (of Sylvius) is a narrow passage connecting the third and fourth ventricles (Fig. 5-5). Its floor is the tegmentum of the midbrain and its roof is the tectum (quadrigeminal plate). The aqueduct has a gentle dorsal curvature in the sagittal plane. Sometimes a slight dilatation is seen in the aqueduct between the colliculi. This has been called the "ventricle of the midbrain."

Fourth Ventricle

The fourth ventricle lies between the pons and medulla anteriorly and the cerebellum posteriorly (Fig. 5-5). The floor of the fourth ventricle is called the tegmentum and contains the nuclei of most of the cranial nerves, as well as many ascending and descending fiber tracts. The roof of the fourth ventricle is formed by two thin sheets of tissue, the anterior and posterior medullary vela, which meet at an apex named the fastigium. The posterior medullary velum is invaginated by the choroid plexus as well as the nodule of the vermis. The caudal part of the fourth ventricular roof contains a midline aperture, the foramen of Magendie. The lateral recesses of the fourth ventricle sweep around the upper medulla on either side and contain openings (foramina of Luschka) that allow communication with the subarachnoid space (Fig 5-6).

Subarachnoid Space

The subarachnoid space receives CSF, which effluxes from the fourth ventricle. Liliequist has divided the subarachnoid space into 15 different cisterns, a convenient classification from both an anatomic and radiologic point of view (Table 5-1).[2,3] It should be noted, however, that this division of the subarachnoid spaces is somewhat artificial, because the various cisterns merge and blend into one another, and their boundaries are rather arbitrary.

The subarachnoid space provides several pathways for CSF flow after leaving the fourth ventricle. A small portion of the CSF will flow downward into the spinal subarachnoid space where it will later be reabsorbed. Most of the CSF, however, travels upward through the tentorial incisura. Supratentorially, the CSF may ascend laterally along the Sylvian fissures and cerebral convexities or medially around the perimesencephalic cistern and into the interhemispheric fissure. Finally CSF reaches the arachnoid

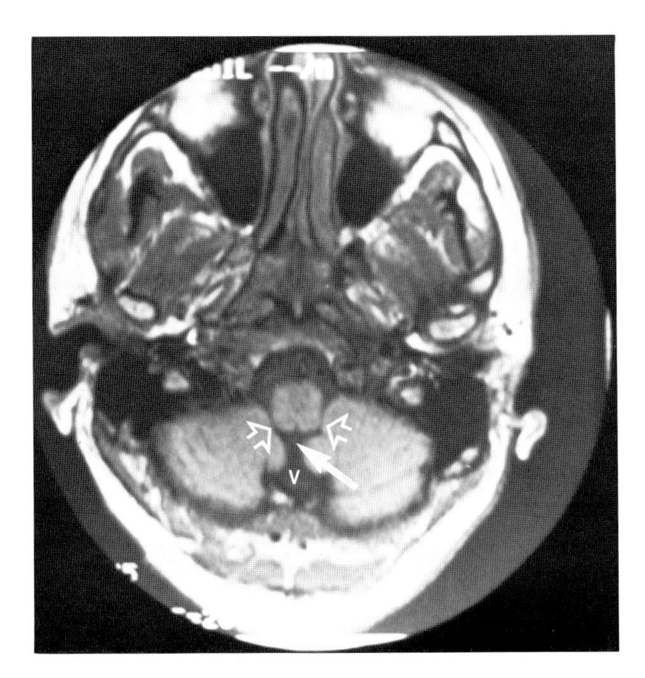

Fig. 5-6. Fourth ventricular outlet foramina of Luschka (open arrows) and Magendie (solid arrow). The foramen of Magendie empties into the vallecula (V). (Courtesy of Fonar Corporation, 110 Marcus Drive, Melville, NY.)

Table 5-1. The Subarachnoid Cisterns

Ventral Cisterns
Medial (unpaired)
Medullary cistern
Pontine cistern
Interpedundular cistern
Chiasmatic cistern
Paramedian (paired)
Cerebello-pontine-angle cisterns
Crural cisterns
Sylvian cisterns
Dorsal Cisterns
Medial (unpaired)
Superior cerebellar cistern
Quadrigeminal plate cistern (cistern of the Vein of Galen)
Pericallosal cistern
Paramedian (paired)
Wings of the ambient cisterns (retrothalamic cisterns)
Communicating Cisterns
Medial (unpaired)
Cistern of the lamina terminalis
Cisterna magna
Paramedian (paired)
Ambient cisterns (perimesencephalic cisterns)
General Subarachnoid Space Over the Cerebral Convexities

(Adapted from Wilson M: The Anatomic Foundation of Neuroradiology of the Brain. 2nd Ed. Little, Brown, Boston, 1972.)

villi, which are responsible for its reabsorption and transport into the superior sagittal sinus. It can thus be seen that obstruction to CSF flow could potentially occur at many locations throughout the ventricular and subarachnoid system.

Physiologic Considerations for MRI

Cerebrospinal fluid is produced by two mechanisms: (1) active secretion by the choroid plexus and ependyma, and (2) bulk diffusion of extracellular fluid elaborated by the cerebral vasculature. Although choroid plexus secretion has traditionally been described as the dominant mechanism, recent evidence suggests that bulk diffusion may account for up to 60 percent of CSF production.[4] Whatever the mechanism of its formation, CSF is produced at a rate of about 550 ml/d.[5,6] In a normal adult the total volume of cerebrospinal fluid is about 135 ml, of which 35 ml is intraventricular.[6] Comparing production to total capacity, it is seen that the complete CSF volume turns itself over three or four times each day. CSF is primarily reabsorbed by the parasagittal arachnoid villi. This resorption may be a passive, pressure-dependent process or an active process involving vesicular transport.[7,8]

Cerebrospinal fluid is thus a dynamic medium, flowing along several pathways that form both series and parallel circuits. Several factors are involved with the propulsion of CSF along this pathway: (1) the continuous formation of CSF, (2) arterial pulsations, (3) ciliary action of ependymal cells, and (4) the pressure gradient across the arachnoid villi.[3] Perhaps the most important are arterial pulsations, which propel CSF from the third ventricle as the thalami squeeze together in systole.[9]

In some places CSF flow may be surprisingly fast. For example, CSF flow velocity is greatly accelerated in the small diameter cerebral aqueduct in accordance with the Bernoulli principle. CSF velocity in the aqueduct and adjacent third and fourth ventricles may be as high as 7.8 cm/sec.[10] Such rapid flow may result in high velocity signal loss on MR images, a phenomenon known as the *CSF flow-void sign* (q.v.).[11] The pulsatile nature of CSF flow has also been demonstrated using cardiac gated MR.[12] Cycled cardiac gating may also be used to study dynamic changes in the ventricular morphology during arterial pulsations.[13] Quantitative measurements of CSF flow in various disease states may one day be possible.

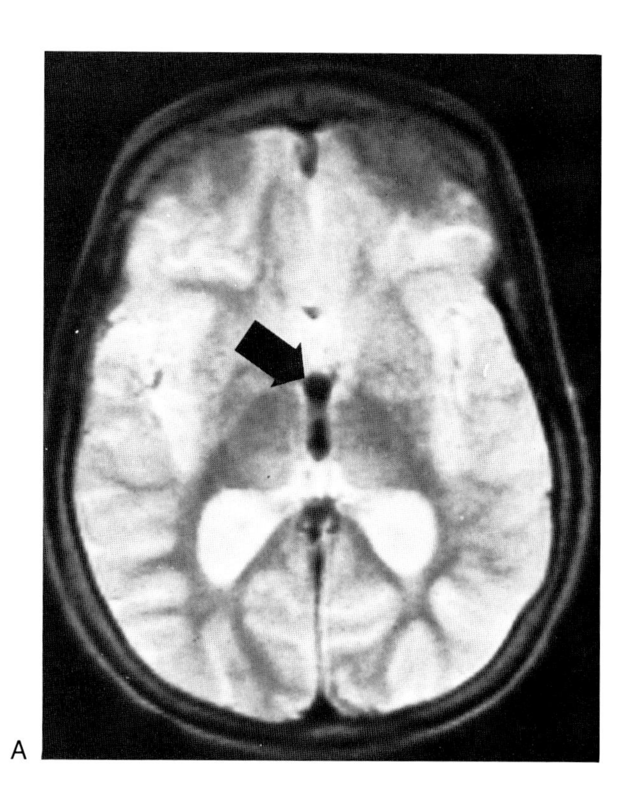

The CSF Flow-Void Sign

Turbulent or rapid flow of CSF in narrower parts of the ventricular system may result in high-velocity signal loss on MR images. In these areas CSF will appear to have an abnormally low or absent signal, a phenomenon called the "CSF flow-void sign."[11] This signal loss is probably caused by a combination of phase shift and time-of-flight effects.

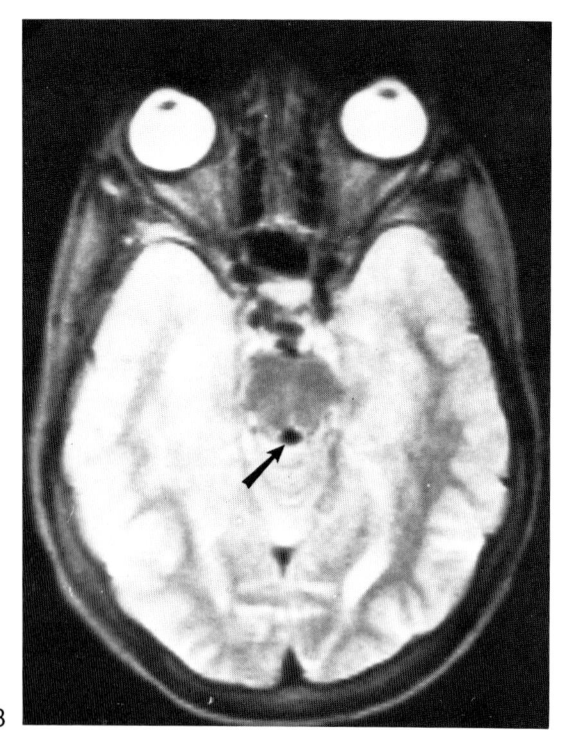

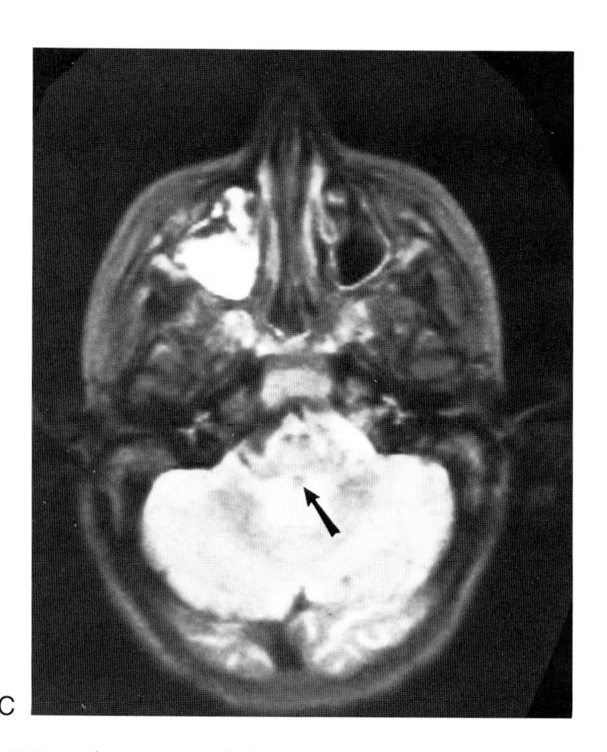

Fig. 5-7. The CSF flow void phenomenon. Rapid flow of CSF produces areas of signal dropout in the posterior third ventricle **(A)**, cerebral aqueduct **(B)**, and proximal fourth ventricle **(C)**.

The CSF flow-void sign can be observed in the cerebral aqueduct in about two-thirds of normal patients.[14] (Fig. 5-7.) Similar signal changes are also frequently encountered in the distal third ventricle, proximal fourth ventricle, and near the foramina of Monro, Magendie, and Luschka. The plane of imaging, type of scanner, pulse sequence, and timing parameters all affect the visualization of this sign.[15]

The CSF flow-void sign is best seen on T2-weighted images, because CSF signal is already low on T1-weighted images. We have observed the sign more frequently in children, perhaps because of their more hyperdynamic circulation. The sign is more frequently noted in patients with ventriculomegaly or communicating hydrocephalus.[14,16]

The CSF flow-void sign may be important in the diagnosis of hydrocephalus and other forms of CSF obstruction. *When seen the sign virtually assures patency of that portion of the CSF pathway.*[14] However, the converse of this statement is not necessarily true; failure to visualize the sign does not always imply obstruction to flow of CSF.

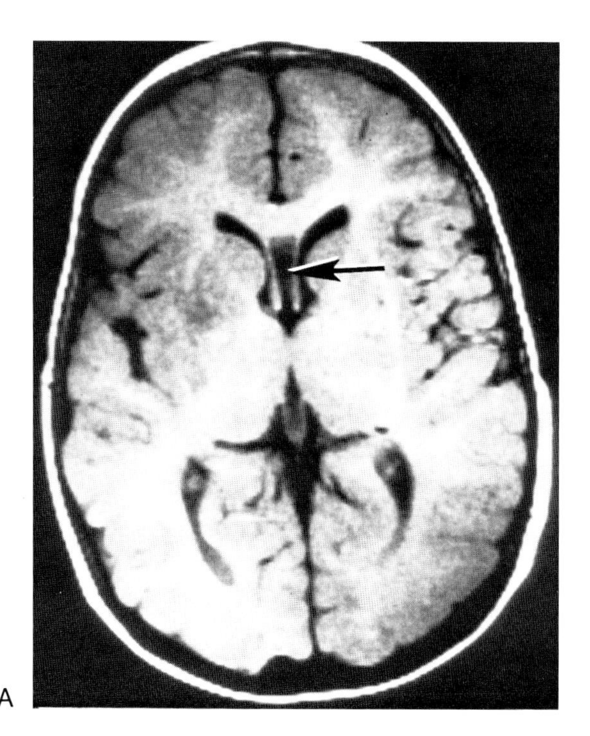

A

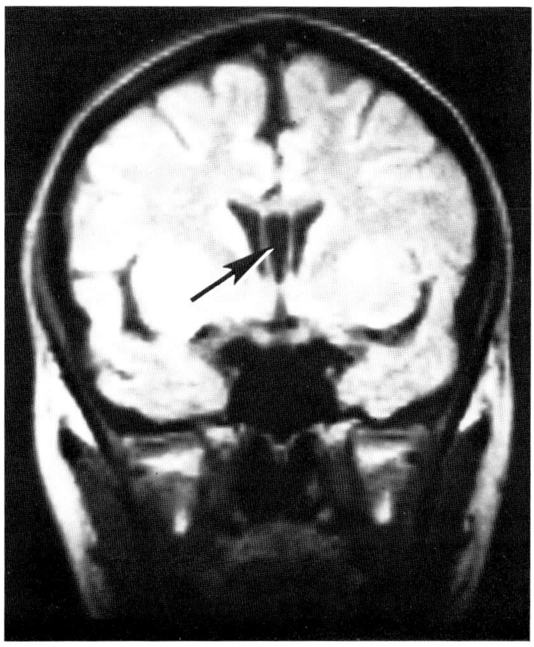

B

Fig. 5-8. Cavum septum pellucidi, a normal variant seen on axial **(A)** and coronal **(B)** MR images.

ANATOMIC VARIANTS OF THE CSF-CONTAINING SPACES

Cavum Septi Pellucidi

During the 4th month of fetal life a midline cleft forms in the primitive septum pellucidum, which then develops in separate leaflets. The space between these leaflets is called the cavity of the septum pellucidum or *cavum septi pellucidi*. This space is normally filled with cerebrospinal fluid and communicates with the ventricles at the foramina of Monro. A constant finding in fetal life, this cavity may persist and be seen in about 60 percent of premature and 40 percent of full-term neonates.[17] By about 2 months postnatally, most of these cava have been obliterated by fusion of the two leaflets. Persistence of the cavum septi pellucidi occurs in about 12 to 15 percent of adults.[18,19]

The cavum septi pellucidi is bounded anteriorly and superiorly by the corpus callosum, inferiorly and posteriorly by the fornix, and laterally by the leaflets of the septum pellucidum. It may be easily recognized on coronal or axial MR images (Fig. 5-8). The walls of the cavum are parallel and the cavity is filled with fluid isointense with CSF on all pulse sequences. High resolution images may reveal communication of the cavum with the ventricular system at the foramina of Monro.

Cyst of the Cavum Septi Pellucidi

Occasionally cystic dilatation of a cavum septi pellucidi may occur secondary to fluid production within the cavum. Although most are asymptomatic, some may become sufficiently large to cause ventricular obstruction at the foramina of Monro.[20] Symptoms may be similar to those of a colloid cyst, with intermittent headache, nausea, or neurologic findings.

A cyst of the cavum septi pellucidi frequently involves the cavum vergae (q.v.) as well. The leaves of the septum pellucidum are laterally displaced and outwardly convex, in contrast to their parallel alignment normally.[21] Distention of one or both lateral ventricles may be seen. Coronal MRI may better aid in differentiating this condition from a large colloid cyst, agenesis of the corpus callosum, or interhemispheric arachnoid cyst. An example of a symptomatic cyst of the cavum septi pellucidi is shown in Figure 5-9.

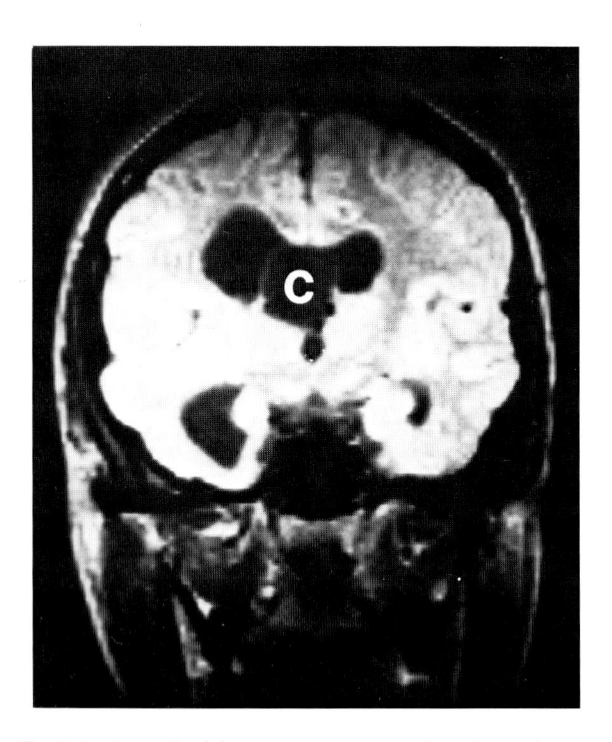

Fig. 5-9. Cyst (C) of the cavum septum pellucidi, producing somewhat asymmetric obstructive hydrocephalus.

Cavum Vergae

The cavum vergae (or *cavum fornicatum*) is a potential space conveniently visualized as a posterior extension of the cavum septi pellucidi. The cavum vergae is named after the Italian anatomist Andrea Verga who first described it in 1851. Cavum vergae is almost never seen without a coexisting cavum septi pellucidi. The converse is not true, however, because only about 10 percent of patients with cavum septi pellucidi will also have a cavum vergae.[22]

The cavum vergae is a quadrilateral space posterior to the cavum septi pellucidi and about one-third its size. The roof of the cavum vergae is formed by the body and splenium of the corpus callosum. Its floor is the hippocampal commissure (psalterium), which separates the cavum vergae from the cistern of the velum interpositum. Anteriorly and laterally the cavity is bounded by the body and posterior columns of the fornix.

The cavum septi pellucidi-cavum vergae complex typically forms a continuous elongated cavity inferior to the corpus callosum, and superior-medial to the fornix. Occasionally the complex has an hourglass configuration when the two cava are not so openly united.[23] The cava septi pellucidi and vergae usually communicate with each other and with the ventricular system at the foramina of Monro. In older literature they were sometimes referred to as the fifth and sixth ventricles, respectively. An example of cavum vergae on axial and coronal MR images appears in Figure 5-10.

Cavum Veli Interpositi

The cavum veli interpositi (CVI) or interventricular cavum is a dilatation of the normal cistern of the velum interpositum. It lies immediately above the roof of the third ventricle and communicates posteriorly

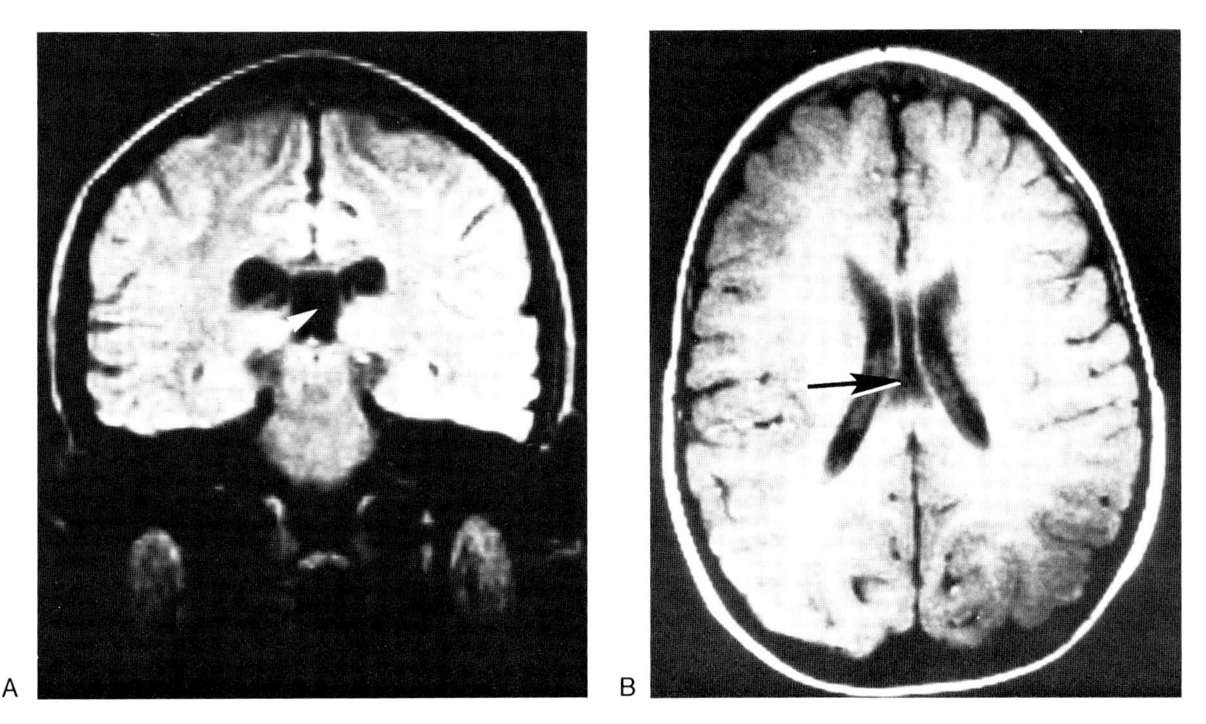

Fig. 5-10. Cavum vergae, a normal anatomic variant seen on axial **(A)** and coronal **(B)** MR images.

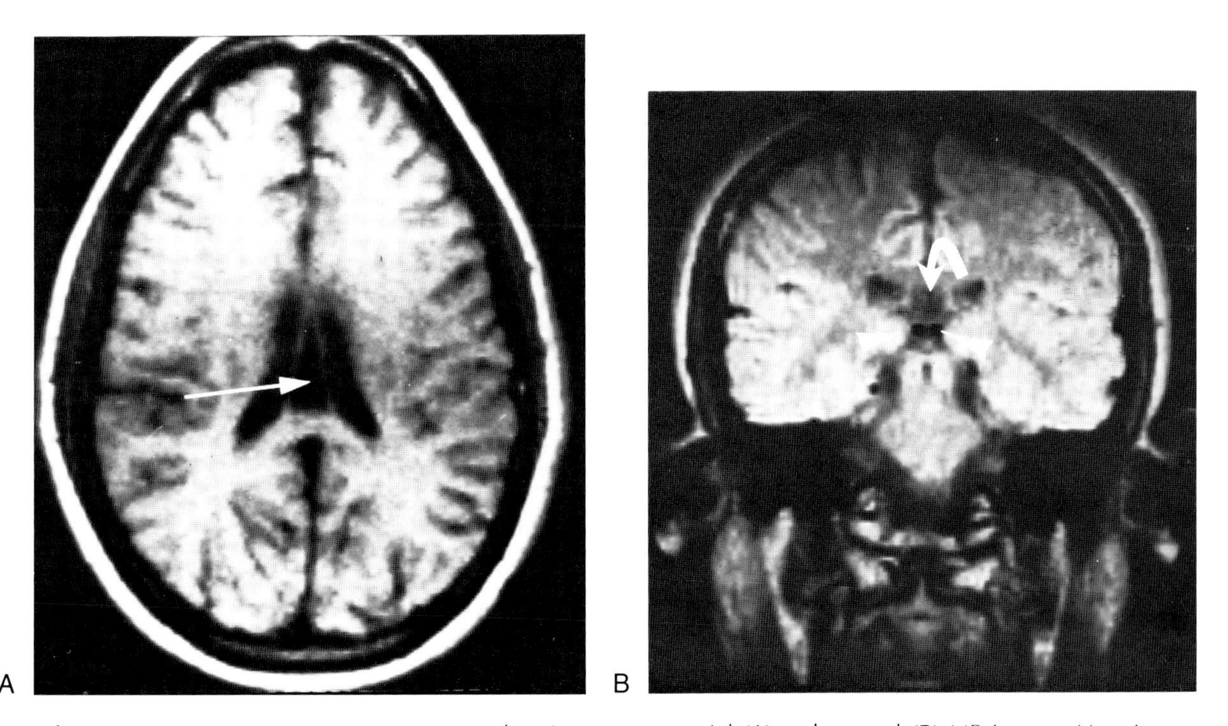

Fig. 5-11. Cavum velum interpositi, a normal variant seen on axial **(A)** and coronal **(B)** MR images. Note in **(B)** the internal cerebral veins (arrowheads), which lie below the cavum velum interpositum.

with the quadrigeminal cistern. The cavity develops because of an unusually wide divergence of the crura of the fornix. In contrast to the cavum vergae, which lies above the fornix and hippocampal commissure, the CVI lies beneath these structures. Within this pass the internal cerebral veins, superior thalamic veins, and posterior medial choroidal arteries.

A CVI has been observed in 60 percent of children under 1 year old and in 30 percent of children aged 1 to 10.[24] Its incidence in adults has not been tabulated but it is observed less frequently than cava septi pellucidi or vergae.[25] Like these latter two malformations, CVI carries with it no associated pathologic significance, and should be recognized merely as another anatomic variant. An example of a CVI shown on MRI is presented in Figure 5-11.

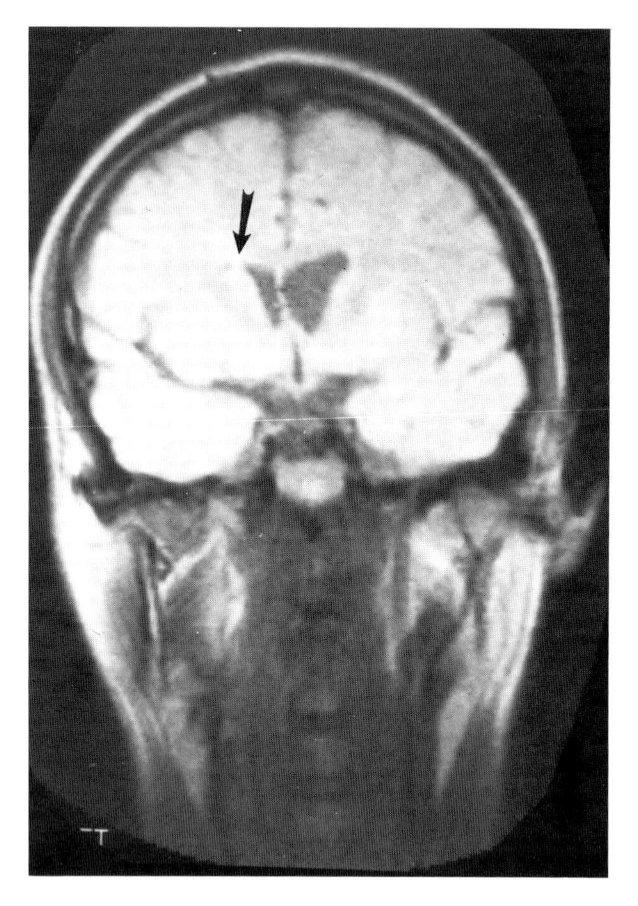

Fig. 5-12. Coaptation of the frontal horn, a normal variant. This coronal SE 1400/40 image demonstrates a synechia (arrow) with slight bowing of the septum pellucidum to the side of the smaller ventricle.

Coaptation of the Lateral Ventricle

One of the more frequent causes of ventricular asymmetry is partial fusion of ventricular surfaces of the lateral ventricles, a deformity known as coaptation (coarctation). This fusion probably occurs between the 4th and 6th fetal months when white matter is growing rapidly.[26] Synechiae form when adjacent ependymal surfaces are forced together by the enlarging white matter mass. A favorite site for this fusion is in the upper lateral aspect of the frontal horn and body where the caudate nucleus and corpus callosum form an acute angle. The condition may be unilateral or bilateral and is found to some degree in several percent of the normal population.[27] Occasionally small cysts may form in the line of fusion between the ventricular walls.

The recognition of ventricular coaptation by CT is usually straightforward. Occasionally, however, the question of a parenchymal mass causing the ventricular deformity may be raised. MRI can support the diagnosis of coaptation in two ways: (1) by demonstrating homogeneous signal in the frontal lobe to exclude a parenchymal mass, and (2) by demonstrating the synechiae within the ventricle more clearly. An example of ventricular coarctation imaged by MR is shown in Figure 5-12.

Megacisterna Magna and Retrocerebellar Arachnoid Pouch

Megacisterna magna (giant cisterna magna) is a normal anatomic variant found in about 0.4 percent of the population.[28] The size of the normal cisterna magna varies widely, often extending upward to the level of the internal occipital protuberance. Sometimes the cisterna magna communicates directly with the superior cerebellar cistern or may even rarely protrude through a defect in the tentorium.[29] Currently it is thought that many cases of megacisterna magna are the result of a retrocerebellar arachnoid pouch (Blake's pouch cyst).[29a] This benign entity is characterized by evagination of the tela choroidea of the fourth ventricle above and behind an intact vermis to form a pouch that freely communicates with the fourth ventricle and subarachnoid space (Fig. 5-13).

HYDROCEPHALUS

Hydrocephalus may be defined as an abnormal enlargement of ventricular volume not caused by brain dysgenesis, destruction, or atrophy. Many disease processes may result in hydrocephalus, including neoplasia, infection, and hemorrhage. The diagnosis and treatment of hydrocephalus forms a significant portion of neurosurgical practice.

Hydrocephalus may be divided into obstructive and functional types based on its pathophysiology. *Obstructive hydrocephalus* implies mechanical impedance to flow of CSF at some point between its site of production (choroid plexus of the lateral ventricles) and its site of absorption (arachnoid villi). If the point of obstruction is proximal to the fourth ventricular outlets, the hydrocephalus is called *noncommunicating* or *intraventricular obstructive hydrocephalus*. If the obstruction is distal to the fourth ventricular outlet foramina, the hydrocephalus is called *communicating* or *extraventricular obstructive hydrocephalus*. *Functional hydrocephalus* occurs when the rate of CSF production exceeds its rate of absorption. This may happen from overproduction or underabsorption, when no mechanical obstruction exists. A classification scheme and common causes for each type of hydrocephalus are presented in Table 5-2.

Table 5-2. Classification and Causes of Hydrocephalus

- Obstructive Hydrocephalus
  - Intraventricular (noncommunicating)
    - Tumor
    - Postinflammatory
    - Posthemmorhagic
    - Congenital (e.g., Dandy-Walker)
  - Extraventricular (communicating)
    - Posthemorrhagic (subarachnoid or subdural)
    - Postinflammatory
    - Acute obstruction of arachnoid villi by blood or protein
    - Congenital (e.g. Chiari II)
    - Meningeal carcinomatosis
    - Normal pressure hydrocephalus
- Functional Hydrocephalus (nonobstructive, communicating)
  - Overproduction of CSF
    - Choroid plexus papilloma
  - Underabsorption of CSF
    - Elevated venous sinus pressure (jugular vein thrombosis, superior vena cava obstruction, arteriovenous malformation, achondroplasia)
    - Congenital (agenesis of arachnoid granulations)

Obstructive Hydrocephalus

Mechanical obstruction to CSF flow accounts for the vast majority of cases of hydrocephalus observed in clinical practice. The point of obstruction is identified radiographically as the site of transition from dilated to normal CSF spaces. For example, dilatation of the lateral ventricles but not the third or fourth ventricles suggests bilateral obstruction at the foramen of Monro. Similarly, dilatation of the ventricles and basal cisterns, but not the Sylvian or high convexity cisterns implies obstruction at the tentorial notch.

Unforturately, this simplified scheme for identifying the site of obstruction has limited accuracy. The fourth ventricle will be normal or small in about 30 percent of patients with extraventricular obstructive hydrocephalus.[35] Furthermore, at least 20 percent of patients with communicating hydrocephalus have more than one site of obstruction.[31] Occasionally a fluid containing mass such as an arachnoid or cysticercus cyst may masquerade as an enlarged ventricle, whose nature escapes detection by either MR or CT.[32]

A number of congenital lesions may cause obstructive hydrocephalus. The most common form of congenital hydrocephalus is that associated with the Chiari II malformation (q.v.). In this anomaly the obstruction may occur at the aqueduct or tentorial notch.[33] The Dandy-Walker syndrome (q.v.) features developmental obstruction of the fourth ventricular outlets as a constant feature. Hypoplasia or aplasia of the vermis is associated with fourth ventricular cyst formation and obstructive hydrocephalus.[34] Congenital stenosis or atresia of the cerebral aqueduct may result in infantile hydrocephalus. This anomaly is often associated with Chiari II malformation, but may be an isolated finding, or rarely is hereditary.[35] Congenital webs or stenosis in the aqueduct may not become clinically apparent until late childhood or puberty when symptoms of acutely raised intracranial pressure may occur. Finally, an extremely rare congenital atresia of the foramen of Monro has been reported that can present in infancy with lateral ventricular enlargement.[36]

Hemorrhage and infection are extremely common causes of obstructive hydrocephalus, particularly the extraventricular type. Acutely, clots, pus, or necrotic debris my block the foramen of Monro, aqueduct, fourth ventricular outlets, or resorptive surfaces of arachnoid villi. Later fibrosis may ensue in these same areas. Parenchymal hematoma and abscess may cause obstruction to CSF flow by their mass effect. Subdural hematomas and empyemas also produce extraventricular obstructive hydrocephalus. Inflammatory septations may also occur obstructing flow.[37]

Tumors and other mass lesions commonly produce intraventricular obstructive hydrocephalus. Virtually any neoplasm that is large enough or strategically located may result in hydrocephalus. Particularly common offenders include metastases, colloid cysts, ependymomas, pineal region tumors, medulloblastomas, and gliomas. Congenital and developmental arachnoid cysts in the ventricles, posterior fossa and quadrigeminal cistern can produce obstruction. Vascular lesions like giant aneurysms and vein of Galen aneurysms may result in hydrocephalus.[38] Leptomeningeal spread of tumor may result in extraventricular obstructive hydrocephalus. This may be seen with metastases,[39] leukemia,[40] and primary leptomeningeal melanoma.[41]

Functional Hydrocephalus

Functional hydrocephalus occurs when the rate of CSF production exceeds its rate of absorption. No mechanical barrier to CSF flow exists and the ventricles freely communicate with the subarachnoid space. Functional hydrocephalus is therefore both communicating and nonobstructive. Either overproduction of CSF by the choroid plexus or underabsorption by the arachnoid villi may be responsible. Because the normal absorptive capacity of the arachnoid villi is at least three times the normal rate of CSF production, functional hydrocephalus can occur when this differential threshold is reached.[42]

Overproduction of CSF is commonly seen with certain ventricular neoplasms, namely choroid plexus papilloma and carcinoma.[43] Rarely a primary overproduction of CSF in the absence of tumor may occur.[44]

Functional underabsorption of CSF usually occurs when dural venous sinus pressure is elevated. CSF absorption is directly related to pressure differential across the arachnoid villi.[45] Increased venous sinus pressure may occur with jugular vein thrombosis, dural sinus thrombosis, superior vena cava obstruction, and high-flow arteriovenous malformations.[46] Elevated venous pressure caused by sigmoid sinus or jugular foramen stenosis may contribute to the hydrocephalus seen in certain diseases that deform the skull, such as Hurler syndrome, achondroplasia,[47] and craniometaphyseal dysplasis.[48] Very rarely, dysplasia[49] or agenesis[50] of parasagittal arachnoid granulations may result in functional hydrocephalus from underabsorption.

MR Imaging of Hydrocephalus

The MR diagnosis of hydrocephalus relies upon two types of evidence: (1) demonstration of increased ventricular size, based primarily on traditional CT and pneumoencephalographic criteria, and (2) showing transependymal resorption of CSF, which is manifest by periventricular hyperintensity on T2-weighted images.

Evaluation of ventricular size has been the subject of considerable study since the early days of pneumoencephalography and CT. Ventricular size changes significantly as part of the normal aging process.[51–58] In children the frontal horns, bodies of the lateral

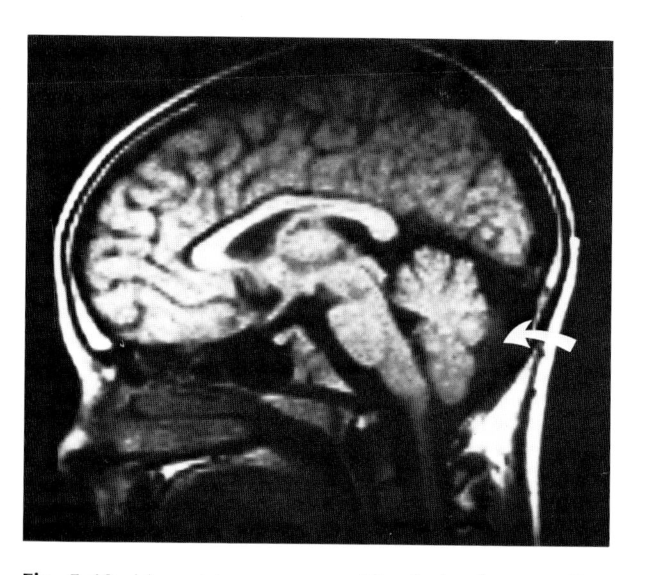

Fig. 5-13. Megacisterna magna. The lack of mass effect on the cerebellum, isointensity with CSF elsewhere, and continuity with the foramen magnum and superior cerebellar cistern speak against this being an arachnoid cyst, cystic tumor, or Dandy-Walker cyst.

ventricles, third ventricle, and subarachnoid cisterns are relatively larger than in the adult. With brain growth, adult proportions are reached by the mid-teens.[56] The ventricular system increases in size only minimally between the 2nd and 6th decades. From the 6th decade on, the sulci and ventricles significantly increase in size as part of the normal aging process.[58]

In addition to linear measurements of the ventricular system, volumetric methods using planimetry have also been employed.[59] A variety of indexing techniques utilize ratios of the width of the ventricles to standard landmarks on the skull. Of these, the frontal horn ratio,[53] Evan's ratio,[60] and ventricular size index[61] are the most popular. Less frequently used indices include the Huckman number[62] and cella media index.[55]

Gado and colleagues have proposed a scoring system that may be useful in distinguishing atrophy from hydrocephalus.[63] In this system lateral ventricular enlargement is scored as mild (+1), moderate (+2), or severe (+3); the third ventricle is either normal (0) or enlarged (−2); the sulci are either normal (0) or enlarged (−2). If the sum of these scores is 3 or more, communicating hydrocephalus is suggested.

Despite these many available indices and scoring methods, few practicing radiologists routinely employ them. With experience, ventricular size may be reliably estimated by visual inspection. Several such visual clues include:

1. Recognition of the disproportionate enlargement of the ventricles with respect to the size of the subarachnoid space (Fig 5-14A).
2. Bulging and rounding of the atria and temporal horns, perhaps best recognized in the coronal plane. (In atropy the ventricles enlarge but tend to maintain their shape.) (Fig. 5-14B.)
3. Ballooning of the third ventricle, with compression and inferior displacement of the thalami (Fig. 5-14C).
4. Elevation and upward stretching of the corpus callosum over the enlarged third ventricle, best appreciated on midline sagittal MR images (Fig. 5-14D).

An important pathologic feature seen in all forms of hydrocephalus (especially the acute obstructive types) is extracellular periventricular edema.[64] This fluid has been equated by most workers as CSF un-

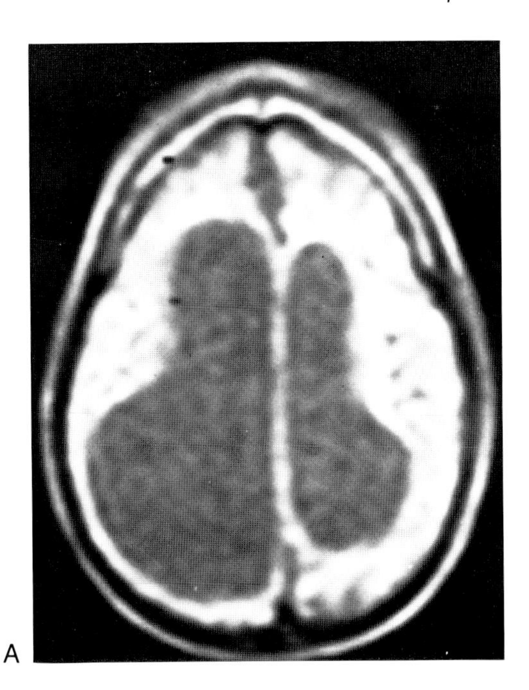

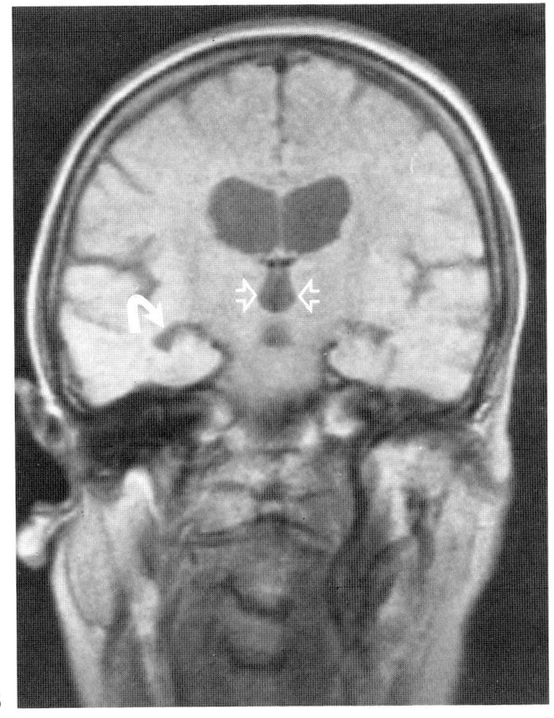

Fig. 5-14. Signs of hydrocephalus. **(A)** The lateral ventricles are disproportionately enlarged relative to the size of the subarachnoid spaces (congenital aqueduct stenosis). **(B)** The temporal horns are rounded and enlarged (curved arrow). Ballooning of the third ventricle (arrowheads). (*Figure continues.*)

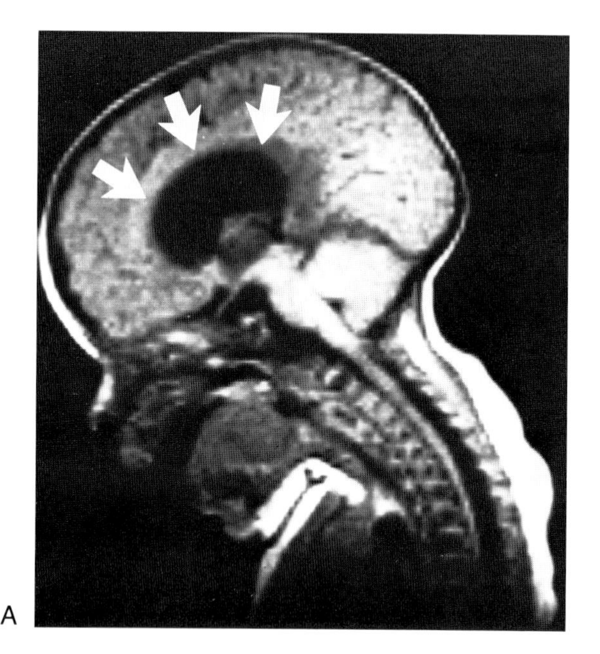

Fig. 5-14 (*continued*). **(C)** Upward stretching of the corpus callosum.

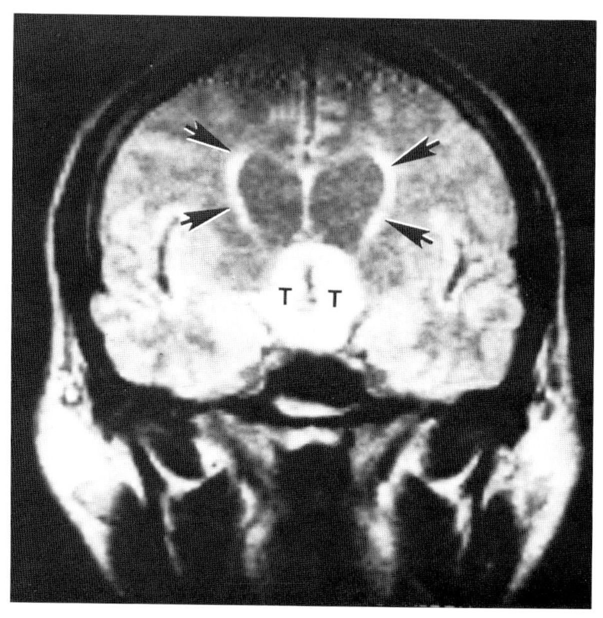

Fig. 5-15. Transependymal absorption of CSF in obstructive hydrocephalus seen on this SE 2000/28 image. Obstruction was caused by a pilocytic astrocytoma tumor (*T*) invading the fourth ventricle.

dergoing transependymal absorption and by some others as merely CSF extravasation.[65,66] Periventricular edema in hydrocephalus may be recognized on CT.[67,68] MR, however, seems exquisitely more sensitive in detecting this edema (Fig. 5-15).[69–74]

It should be recognized, however, that several disease processes other than hydrocephalus may cause periventricular edema and produce high intensity lesions on MR.[74,75] Foci of high signal in the white matter are commonly observed in normal patients adjacent to the anterior and lateral borders of the frontal horn (Fig. 5-16). This anatomic variant is thought to be due to a combination of three factors; (1) naturally low myelin content of the region, (2) convergent concentration of normal interstitial fluid, and (3) ependymitis granularis, a patchy loss of ependyma with astrocytic gliosis.[76] These lesions, however, are confined to the ventricular angles, primarily at the frontal horn. Conversely transependymal absorption of CSF is distributed around the whole ventricular system.

Multiple sclerosis and other demyelinating diseases may result in periventricular lesions.[77] Although extensive periventricular plaques involving the whole length of the lateral ventricle, have been described, skip areas are nearly always present, aiding in differ-

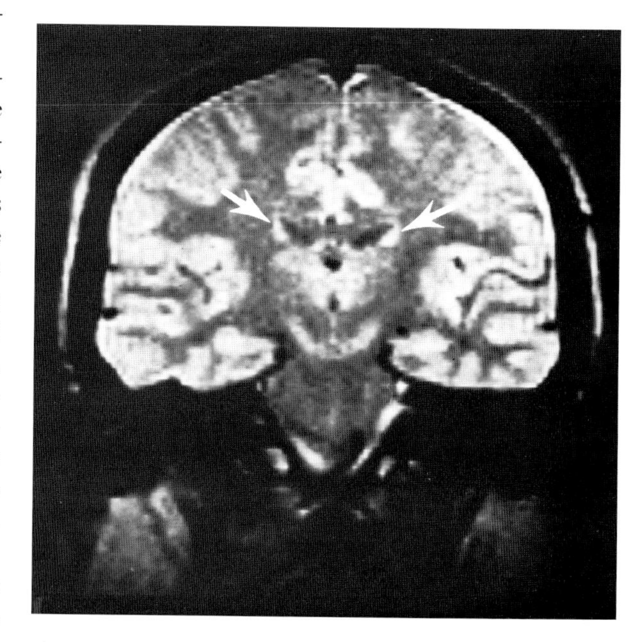

Fig. 5-16. High intensity foci adjacent to the lateral angle of the lateral ventricles, a normal variant. This is said to be due to several factors, including ependymitis granularis.

ential diagnosis. In neonates periventricular leukomalacia may produce a similar appearance, but can usually be distinguished from hydrocephalus with transependymal absorption of CSF.[78]

In conclusion, MRI is more sensitive than CT in defining ventricular enlargement and in detecting periventricular edema. However, the value of this increased sensitivity is offset somewhat by its decreased specificity. Even with MRI no infallible criteria to predict the success of shunting, prognosis, or etiology of hydrocephalus in most cases have yet been devised.

CYSTIC LESIONS CONTAINING CSF

Arachnoid Cysts

True arachnoid cysts are uncommon benign congenital masses thought to arise from splitting, duplication, or outpouching of the arachnoid membrane.[79] They should be distinguished from secondary arachnoid cystic fluid collections that may occur following infection, surgery, or hemorrhage. True arachnoid cysts are most commonly located near the tip of the temporal pole.[80] The posterior fossa is the next most common location. They may also arise over the cerebral convesities, in the interhemispheric fissure, in the quadrigeminal plate cistern, or in the suprasellar region.

Arachnoid cysts are filled with a clear or slightly anthochronic fluid. They frequently communicate with the subarachnoid space. They may be asymptomatic or cause clinical manifestations by way of mass effect. Symptoms are often mild or nonspecific. Erosion of the inner table of the skull or localized calvarial remodelling may be seen.

Uncomplicated arachnoid cysts contain fluid with long T1 and T2 values, usually identical to CSF.[81,82] (Figs. 5-17 to 5-20). Postinflammatory and complicated cysts may have high protein content in their fluid[80] and might be expected to have T1 and T2 values different than CSF. MRI is superior to CT in demonstrating the mass effects of an arachnoid cyst particularly in the temporal lobes.[83] The identity of signal between arachnoid cyst fluid and CSF on multi-

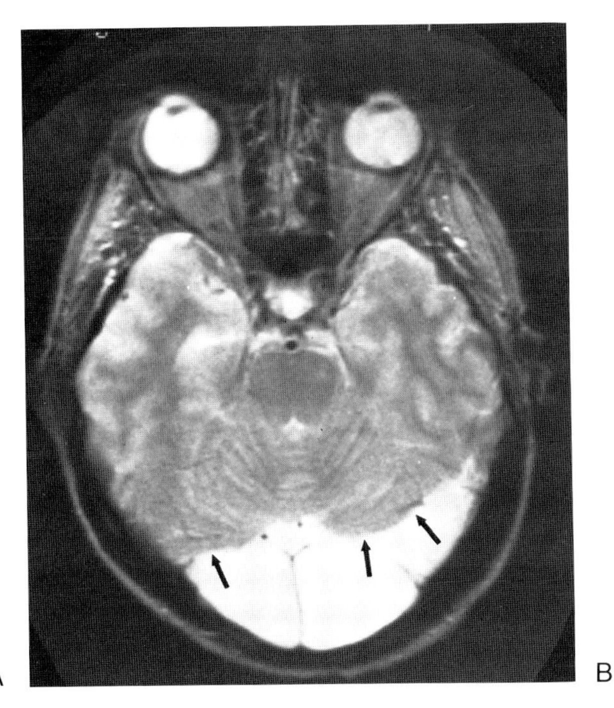
A

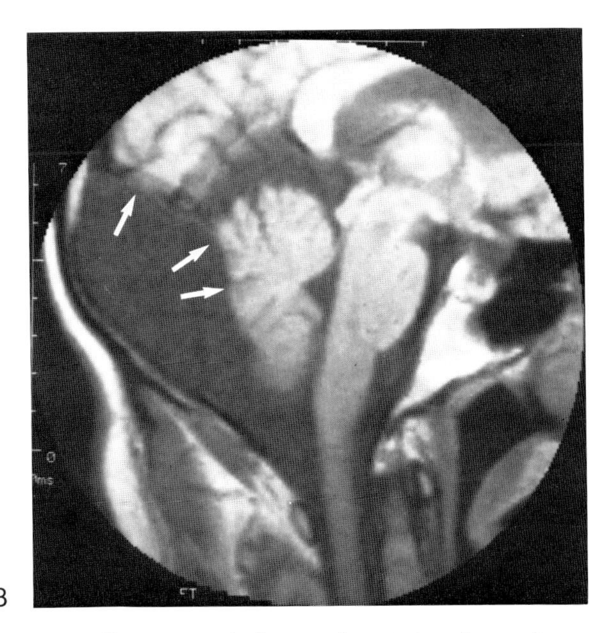
B

Fig. 5-17. (**A** and **B**) Posterior fossa arachnoid cyst. Note mass effect on cerebellum and tentorium (arrows) distinguishing this from a megacisterna magna.

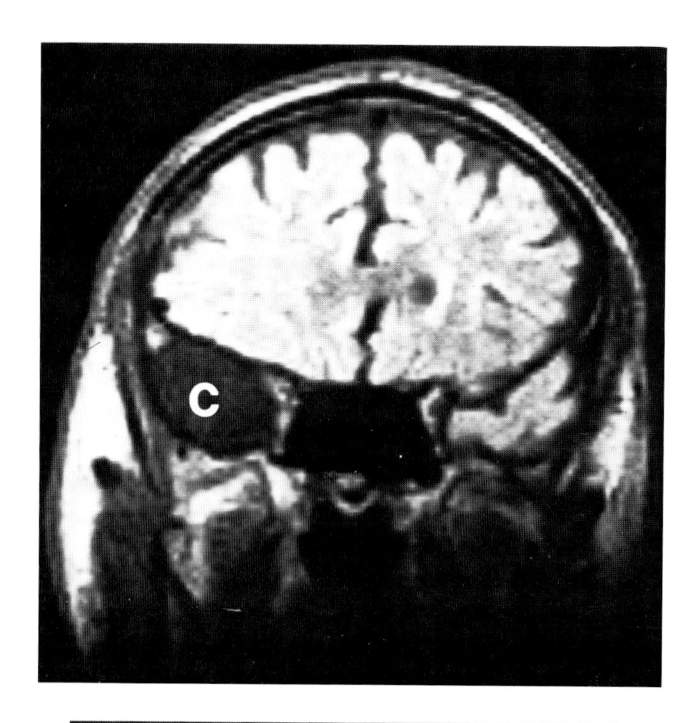

Fig. 5-18. The floor of the middle cranial fossa is the most common location for arachnoid cysts. They are well demonstrated on coronal MR images. C = cyst.

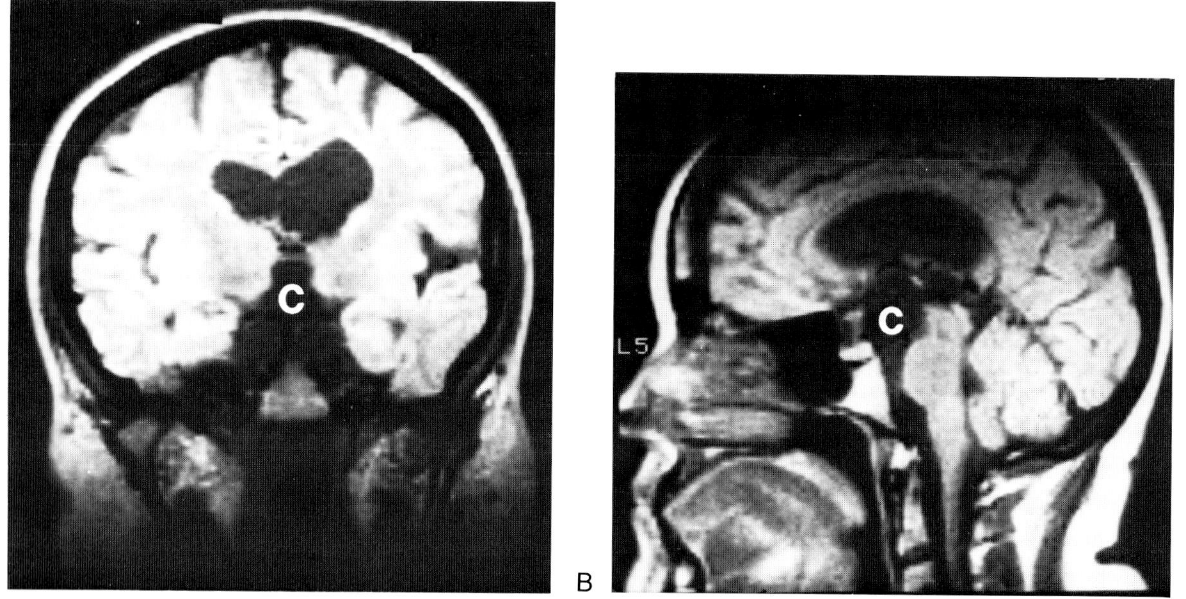

Fig. 5-19. (**A** and **B**) Suprasellar arachnoid cyst (C) causing obstructive hydrocephalus.

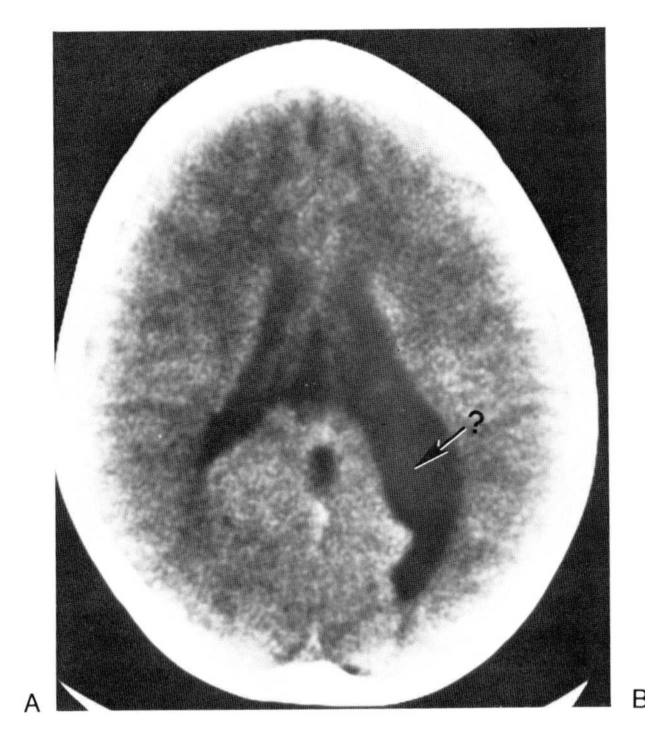

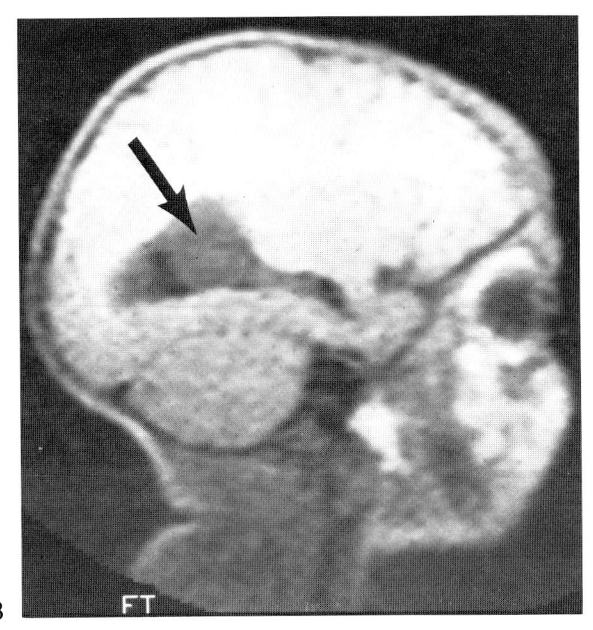

Fig. 5-20. Intraventricular arachnoid cyst, difficult to see on CT **(A)**. More easily appreciated on sagittal MR image.

ple imaging sequences usually allows a confident diagnosis to be made for uncomplicated cysts.[83] The CT diagnosis, however, is often less clear, and differential diagnosis of epidermoid, craniopharyngioma, hygroma, and cystic astrocytoma must often be entertained.

Ependymal Cysts

Ependymal cysts are much rarer than arachnoid cysts, and differ from them in that they are lined by cuboidal ependymal cells. These cysts frequently arise in relation to the third or fourth ventricles, and are thought to develop as a diverticular outpouching between the 8th and 16th weeks of gestational life.[84] The cysts do not usually communicate with the ventricular system, but may be intraventricular (Fig. 5-21). The most common location is in the quadrigeminal plate cistern. They cannot be distinguished from arachnoid cysts radiographically, only pathologically.

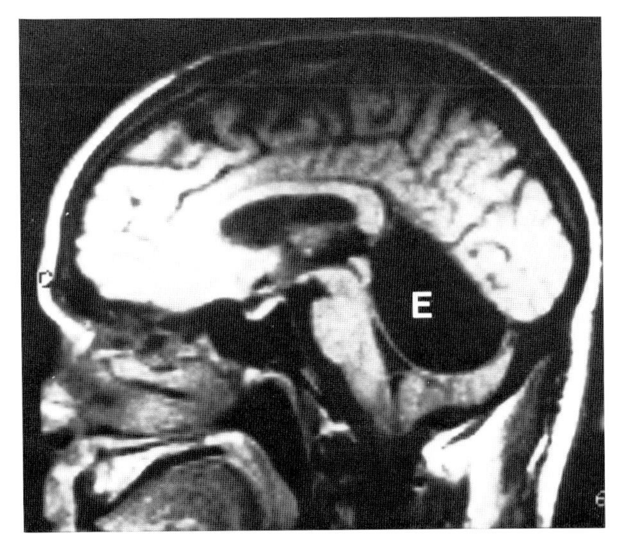

Fig. 5-21. Ependymal cyst (*E*) cannot be distinguished radiographically from more common arachnoid cysts.

The Encysted ("Trapped") Ventricle

Encystment of a ventricle or part of it is a well-described entity that most frequently involves the lateral or fourth ventricles. The disorder usually results from occlusion of ventricular CSF outlets following intraventricular hemorrhage or infection.[85] Occasionally tumors and intraventricular masses such as cysticercosis cysts may represent the cause of the obstruction.[86] When the fourth ventricle is isolated or encysted the term "double compartment hydrocephalus" has been applied.

The encysted fourth ventricle has received special attention because of its somewhat unique etiology and clinical presentation. Isolation of the fourth ventricle requires obstruction of both inlet (cerebral aqueduct) and outlet (Magendie and Luschka) foramina. Continued secretion by the choroid plexus results in distension of the fourth ventricle. The encysted fourth ventricle usually occurs in patients with aqueduct stenosis who have undergone shunting of the lateral ventricles.[87] Secondary hemorrhage or infection then obstructs the fourth ventricular outlets, thereby isolating the ventricle. Similarly, secondary aqueduct stenosis may occur in patients shunted for hydrocephalus caused by fourth ventricular outlet obstruction.

The clinical presentation of the encysted fourth ventricle may be like that of a posterior fossa mass, with ataxia, diplopia, and drowsiness.[87] However, the disorder may be asymptomatic or present insidiously. The patients have no signs of raised intracranial pressure. Selective shunting of the isolated ventricle may be required in symptomatic patients.

Computed tomography or MR diagnosis is usually straight forward. The isolated fourth ventricle tends to maintain its normal shape despite enlargement.[88] (Fig. 5-22.) Clinical history may be helpful particularly in regard to prior hemorrhage, shunting, or inflammation. (Fig. 5-23.) Differential diagnosis includes cystic astrocytoma, arachnoid cysts, and parasitic cysts. In the several cases of trapped ventricle we have seen on MR, the encysted fluid had T1 and T2 values similar to CSF, and the CSF flow-void sign was not present.

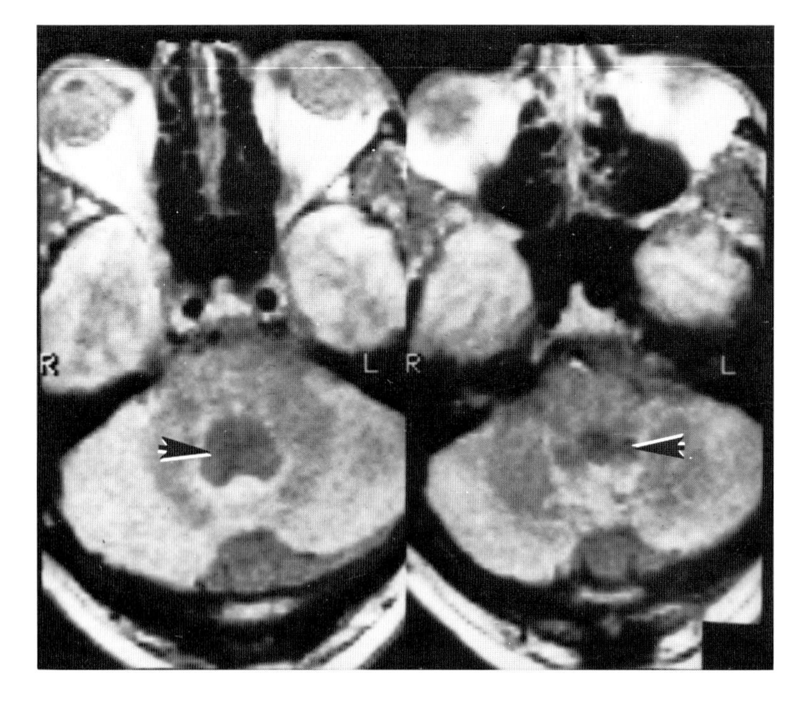

Fig. 5-22. Trapped fourth ventricle in a patient status postbrain abscess and shunt. Despite enlargement the ventricle retains its normal shape.

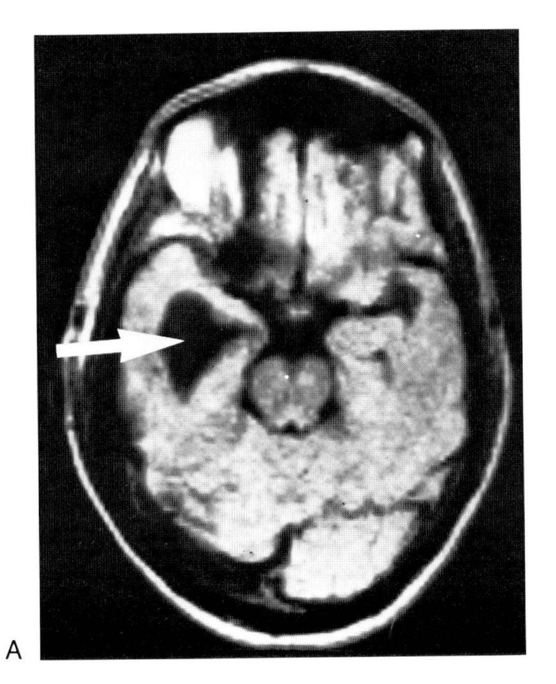

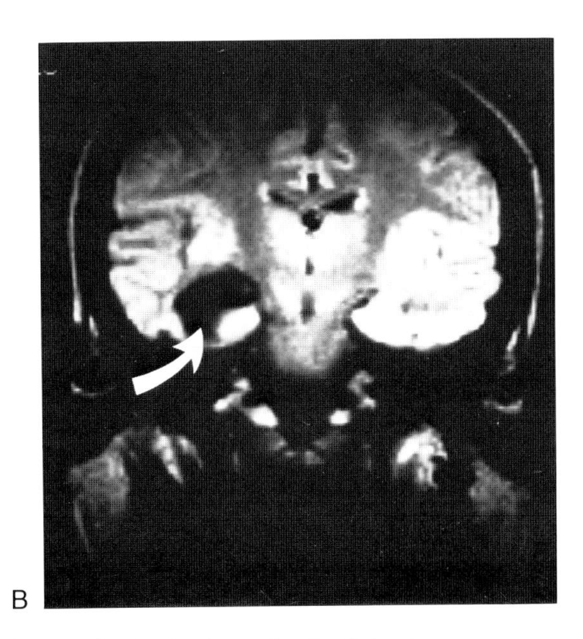

Fig. 5-23. (**A** and **B**) Trapped temporal horn, which maintains its shape despite distension.

VENTRICULAR AND EPENDYMAL NEOPLASMS

The mature ependyma is a thin layer of epithelial cells that line the ventricles of the brain and the central canal of the spinal cord. Embryologically, the ependyma originates from neuroectodermal cells of the primitive neural tube. Unlike other neuroectodermal cells, which migrate radially to become mature neurons and glia, the ependymal cells remain in a relatively central position. These ependymal cells then differentiate into a number of important structures that may give rise to tumors in the adult (Table 5-3).

Ependymal cell rests may also be found in several locations far from the ventricular lining. These rests typically occur near points where the ventricular system has become angled or twisted during embryologic development. These sites include a band posterior to the occipital horns, in white matter lateral to the ventricular trigone, near the ventral spur of the cerebral aqueduct, near the foramina of Luschka, along the tela choroidea, in the central canal of the spinal cord, and in the filum terminale.[89] This distribution of ectopic ependymal cell rests may explain the locations of ependymal neoplams, which develop remote from the ventricular system.

Table 5-3. Ependymal Derivatives and Potential Lesions

Ependymal Derivative	*Potential Neoplasm*
Mature ependyma	Ependymoma Neuronal and glial heterotopias
Choroid plexus	Choroid plexus papilloma Choroid plexus carcinoma Choroid plexus cyst
Pineal body	Pineal cell tumors Gliomas Germ cell tumors
Neurohypophysis	Glioma Germ cell tumors
Paraphysis	Colloid cyst
Subcommissural organ	Mesencephalic cyst
Area postrema	No known pathology associated

Ependymoma

Ependymomas are the most common of the ependymal-derived tumors accounting for up to 6 percent of gliomas in some series.[90] Ependymomas typically arise within or adjacent to the ventricular system. They occur most frequently in children and young adults with a 2:1 preference for males.[91] Pathologic subtypes of ependymoma include cellular, myxopapillary, and papillary histologies. More malignant varieties include anaplastic ependymoma and ependymoblastoma. Subependymoma is a variant of ependymoma in which the subependymal neural glial elements are present within the tumor. Radiologic characteristics of ependymomas depend upon their location and histology.

Infratentorial Ependymomas

Infratentorial ependymomas are the most common, seen primarily in infants and children. They arise from the walls of the fourth ventricle, filling and enlarging the ventricle while maintaining a midline position. Obstructive hydrocephalus is nearly always present. Occasionally, the dilated fourth ventricle may be seen to surround the tumor in a thin rimlike horseshoe, a sign highly suggestive of ependymoma. Fourth ventricular ependymomas may extend desmoplastically through the outlet foramina to wrap around the medulla and upper cors. Although occurring in only a minority of cases, this "melted wax" appearance of tumor extending through the lateral recesses and vallecula cerebelli is nearly diagnostic of ependymoma.[92]

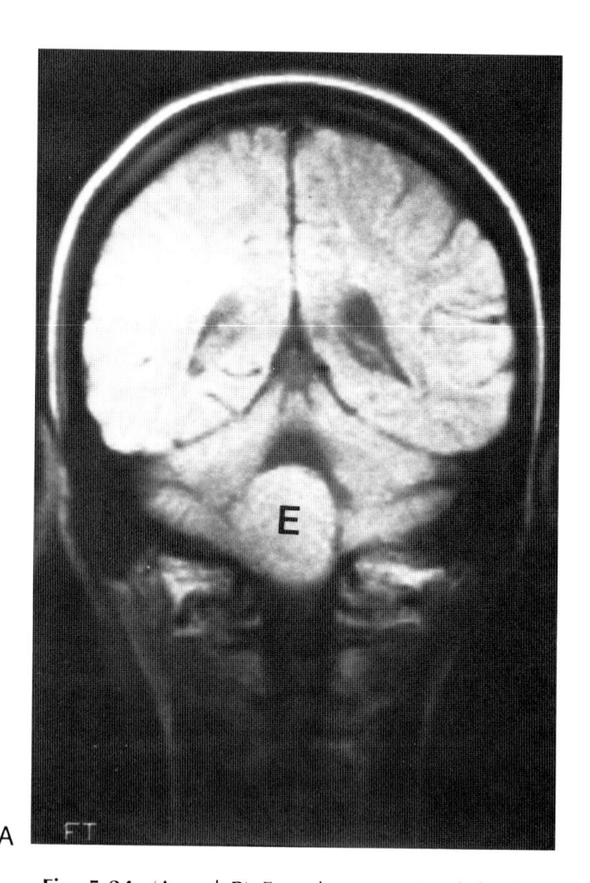

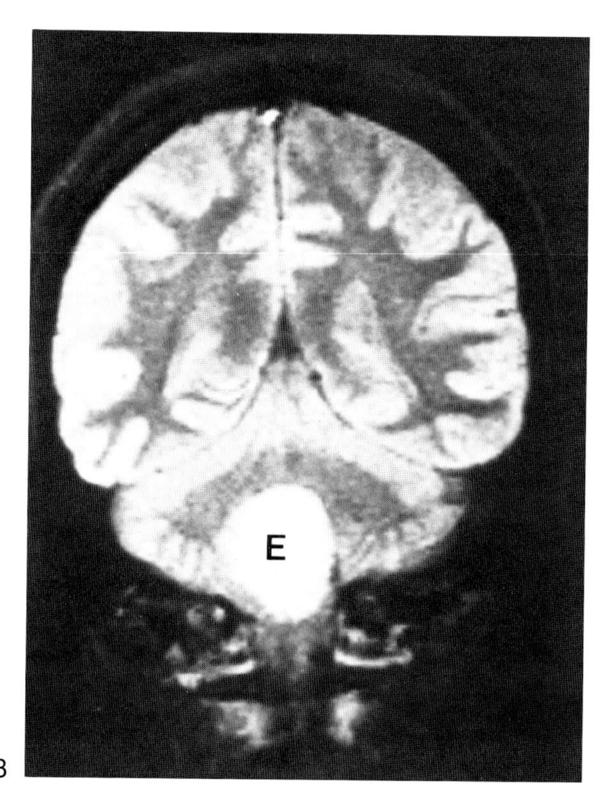

Fig. 5-24. (**A** and **B**) Ependymoma (*E*) of the fourth ventricle. Mass arises from floor of ventricle inferiorly. Note in (**A**) a characteristic rim of ventricle still surrounding the tumor. (*Figure continues*).

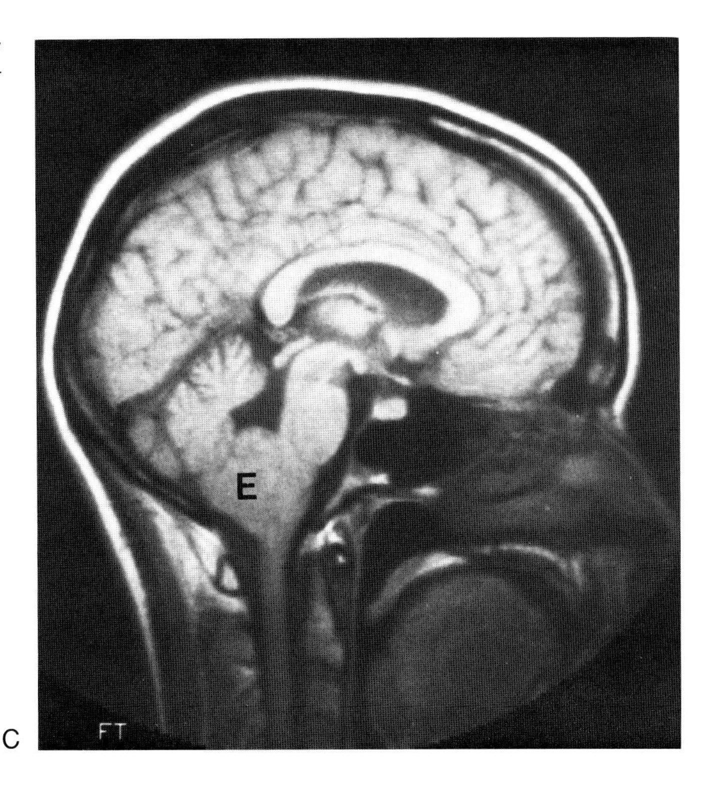

Fig. 5-24 (*continued*). Sagittal image **(C)** elegantly demonstrates that the mass arises from the floor of the ventricle, not the vermis.

Table 5-4. Relative Frequency of Various Intraventricular Mass Lesions

Lateral Ventricle
[Astrocytoma][a]
Meningioma
Ependymoma
Choroid plexus papilloma
Dermoid tumors
Arachnoid and ependymal cysts
Third Ventricle
Colloid cyst
[Astrocytoma][a]
[Craniopharyngioma][a]
Choroid plexus papilloma
Ependymoma
Fourth Ventricle
Ependymoma
[Medulloblastoma][a]
Epidermoid and dermoid tumors
Choroid plexus papilloma
Cysticercosis cyst

[a] Lesions in brackets [ ] originate outside of ventricular system, but commonly mimic primary intraventricular masses when they flow into the ventricles.

(Morrison G, Sobel DF, Kelley WM: Intraventricular mass lesions. Radiology 153:435, 1984.)

Infratentorial ependymomas are usually solid tumors, with cystic changes seen in only a few percent.[89] Dense, punctate calcifications occur in 25 to 50 percent and are occasionally demonstrated on MRI.[93] Peritumoral edema is seen in about half of patients. The tumors are vascular, but hemorrhage is rare.

Infratentorial ependymomas have been demonstrated on MRI to have a characteristic location in the midline caudal fourth ventricle with prolonged T1 and T2 values (Fig. 5-24).[93] Sagittal imaging assists in differentiating these tumors from the more common medulloblastomas, which usually lie rostrally within the ventricle near the inferior vermis. Dense calcifications are sometimes noted on MR.

Supratentorial Ependymomas

Supratentorial ependymomas are much less common than the infratentorial variety, and demonstrate several differences in appearance (Table 5-4). Supratentorial ependymomas are more frequent in adults than the infretentorial variety. They are usually intraparenchymal rather than intraventricular, often found lateral to the ventricular trigone.[91] Cystic changes are common. No specific MR characteristics are known that may distinguish supratentorial epedymomas from the more common glial tumors.

Like other glial neoplasms, ependymomas demonstrate varying grades of malignancy. Periventricular spread of tumor is much more common than subarachnoid spread, although both occur.[94] Rapid and extensive seeding of the entire ventricular system may take place almost forming a cast of the ventricle. Local recurrence following resection is the rule, and all ependymomas have a poor prognosis regardless of histology.[95] Distant metastases to lung, liver, lymph nodes, and bone have been reported.[96]

Ependymoblastoma

Ependymoblastoma is a particularly malignant form of ependymoma that grows rapidly with local infiltration and emningeal seeding. Most occur in the aupratentorial compartment, but may be found infratentorially as well.[97] Malignant ependymomas and ependymoblastomas may also contain malignant glial elements such as glioblastoma multiforme. Ependymoblastomas are usually large (> 4 cm) with cystic components seen in 50 percent.[91] In contrast to the cellular ependymomas, malignant ependymoblastomas demonstrate less distinct margination, a lower frequency of calcification, and more marked vascularity.

Subependymoma

Subependymoma is a variant of ependymoma in which subependymal glial elements are present within the tumor. Most of these tumors arise in relation to the ventricular system and are incidental findings at autopsy. The most common location is within the fourth ventricle. The second most common site is within the lateral ventricles, where the tumor may be attached to the septum pellucidum (Fig. 5-25). Calcification is often present.

Subependymomas rarely obstruct the ventricles or produce symptoms unless they are large. They may be noted incidentally in MR scans in elderly males;

Table 5-5. Comparison of Infratentorial and Supratentorial Ependymomas

Characteristic	*Infratentorial*	*Supratentorial*
Incidence	70%	30%
Age	Usually < 20 years Uncommon in adults	Usually < 20 years, but more common in adults
Location	Over 90% are midline within 4th ventricle	85% are lateral and intraparenchymal, especially parietal or frontal lobes
Size	Often < 4 cm	Much larger, nearly all > 4 cm
Cystic components	Unusual	Common (over 60%)
Associated hydrocephalus	Nearly always	Less than half. Degree of hydrocephalus is often slight to moderate

(Composite findings from several series summarized by Armington WG, Osborn AG, Cubberley DA et al: Supratentorial ependymoma: CT appearance. Radiology 157:367, 1985.)

Fig. 5-25. Subependymoma (*S*) attached to the septum pellucidum, seen on CT **(A),** T1-weighted MR **(B),** and T2-weighted MR **(C).** The patient also has Binswanger disease, evidenced by multiple matter infarcts (arrowheads).

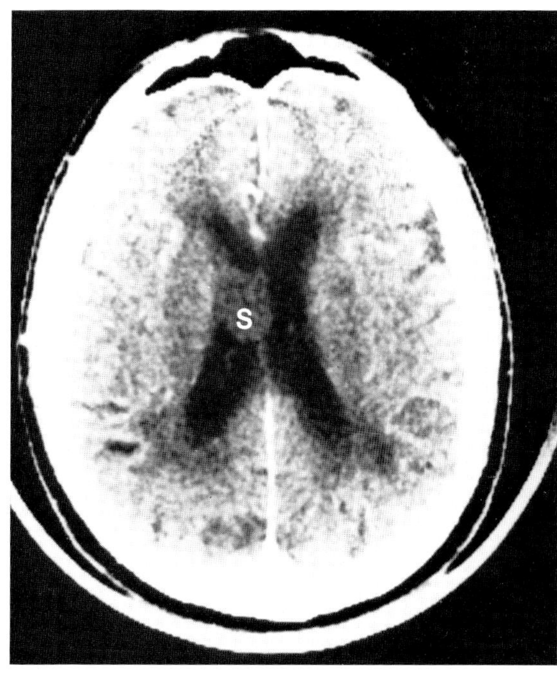

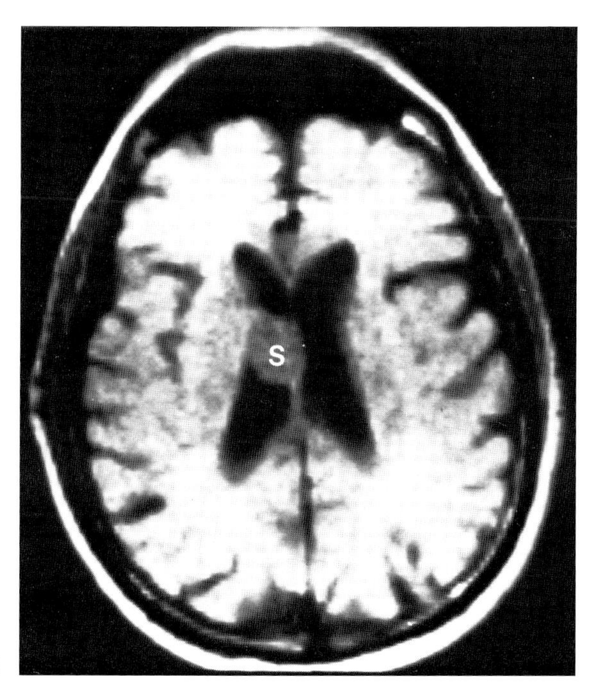

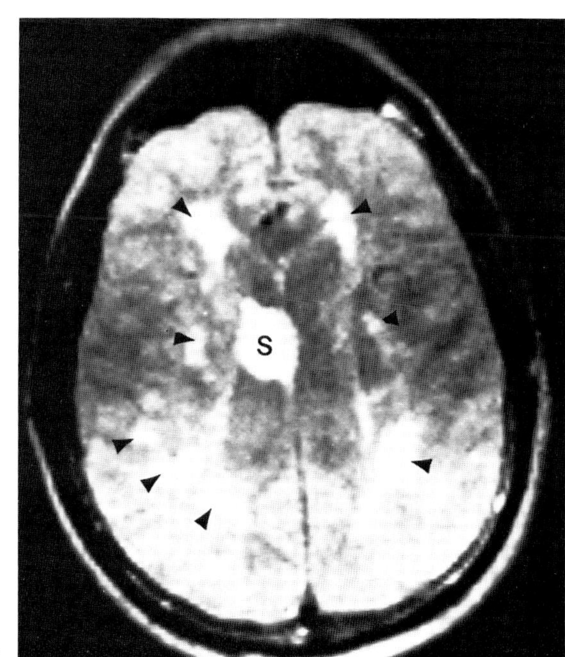

symptomatic patients tend to be younger. The prognosis is much less favorable when they arise on the floor of the fourth ventricle and contain elements of ependymoma.[98]

Choroid Plexus Papilloma

Choroid plexus papilloma is an uncommon intraventricular tumor occurring in children and young adults.[90] Most occur in children under 10 years old, but may be seen at any age. There is a slight male predilection.

Pathologically, the tumors are globular intraventricular masses measuring one to a few centimeters in diameter. They are highly vascular with a fibrous central stroma and frondlike (papillated) surface. Although usually sessile, the tumors may be pedunculated and move with changes in the patient's position. The tumors occur in the fourth, lateral, and third ventricles, in that order of frequency, but the location depends upon the age at presentation.[99] In adults and teenagers, the vast majority of choroid plexus papillomas occur in the fourth ventricle. In children under the age of 10, these tumors predominantly occur in the lateral or third ventricles. The atrium of the lateral ventricle is a particularly common location in the pediatric age group. Curiously, these tumors show a remarkable preference for the left lateral ventricle. Rarely, choroid plexus papillomas may arise in the lateral recess of the fourth ventricle and extend into the cerebellopontine angle cistern. As many as 4 percent may be multiple. Only a minority are calcified, but they may engulf normal choroid plexus glomus calcifications. Occasionally they may be heavily calcified and even undergo ossification. Because glomus calcification is unusual in children, its presence

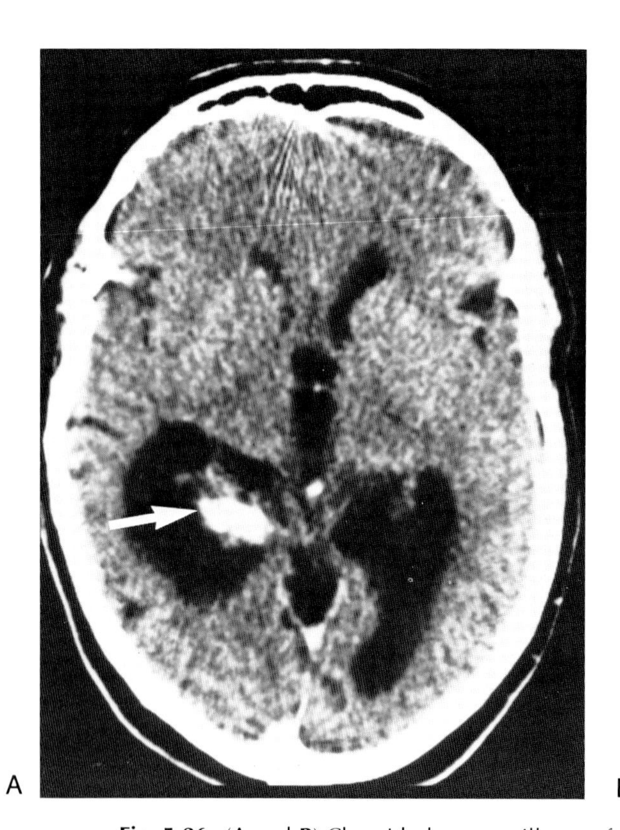

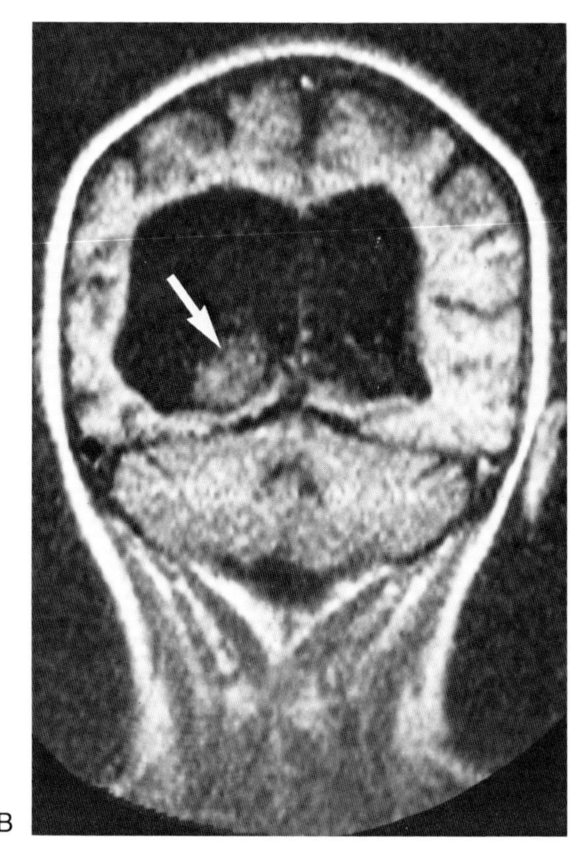

Fig. 5-26. (**A** and **B**) Choroid plexus papilloma of the lateral ventricle well seen on CT and MR.

together with an intraventricular mass should strongly suggest the diagnosis of choroid plexus papilloma.

Hydrocephalus is usually present in conjunction with a choroid plexus papilloma. The primary mechanism is thought to be an overproduction of cerebrospinal fluid by the tumor, which exceeds the capacity of the arachnoid granulations to reabsorb it.[99] Additional mechanisms that may play a role in the production of hydocephalus in some cases includes communicating hydrocephalus from prior tumor subarachnoid hemorrhage, and direct mechanical obstruction of CSF outlet foramina.

The MR and CT appearances of a choroid plexus papilloma are shown in Figure 5-26. This tumor, as well as those scattered examples in the literature, has prolonged T1 and T2 values. The relaxation times are nonspecific; the diagnosis was suggested by age, location, and clinical presentation.

Choroid Plexus Carcinoma

Choroid plexus carcinoma is a very rare malignant form of the choroid plexus papilloma. Aggressive behavior of the tumor is demonstrated by early subependymal and intraparenchymal invasion. Of the cases reported in the literature, the vast majority have been in the lateral ventricles in children.[100] Hydrocephalus is invariable present. The tumor mass has been described as a "ball of spaghetti" with multiple frondlike projections.[99] Areas of hemorrhage and necrosis may be seen within the tumor.

Colloid Cysts

Colloid cysts are uncommon benign tumors that arise exclusively in the anterior superior part of the third ventricle.[90] Most of these cysts occur in young adults, with very few being seen in infancy or childhood. Both sexes are affected equally.

Grossly, colloid cysts are smooth, white balls about the size of a cherry (2 to 4 cm) filled with a gelatinous material. They are loosely attached to the choroid plexus on the roof of the third ventricle between the foramina of Monro. Because of this location they may obstruct a foramen and produce hydrocephalus. Common presenting symptoms therefore relate to increased intracranial pressure and include headache, vomiting, and visual impairment. Intermittent obstruction may occur due to mobility of the cyst, and the patient's symptoms may sometimes be relieved or exacerbated by a change of head position.

The pathogenesis of these cysts has not been established, but the most accepted view is that they arise as remnants of the paraphysis or parapineal organ.[101] The paraphysis is an ependymal pouch that regresses in the human embryo at about 100 days gestation. In some primitive reptiles, the paraphysis develops into a third eye (parietal eye) on the dorsum of the head.

Computed tomography findings include a homogeneous high density (42 to 76 HU) on precontrast scans, possible due to calcium distributed evenly throughout the cyst mucin.[102] The anterior third ventricle is commonly obliterated, there is widening of the septum pellucidum, and separation of the frontal horns. With obstruction of the foramen of Monro, the third ventri-

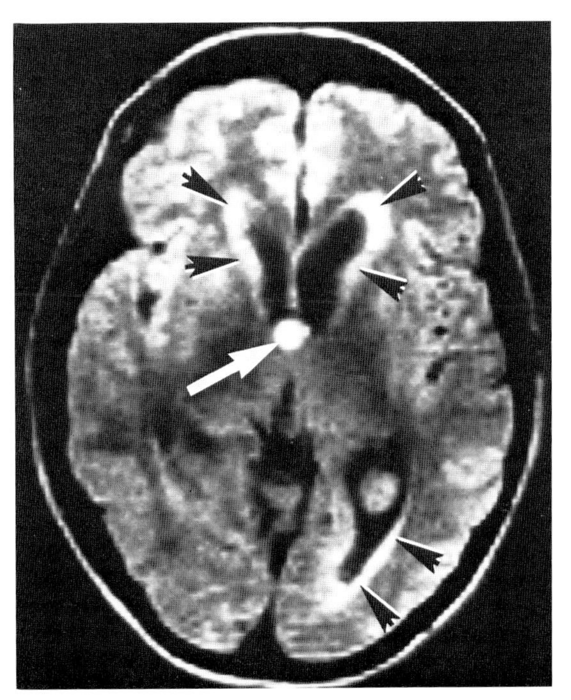

Fig. 5-27. Typical colloid cyst, containing short T1 and long T2 gelatinous material (arrow). Note obstructive hydrocephalus manifest by ventricular dilatation and transependymal absorption of CSF (arrowheads). Technique is SE 2000/28

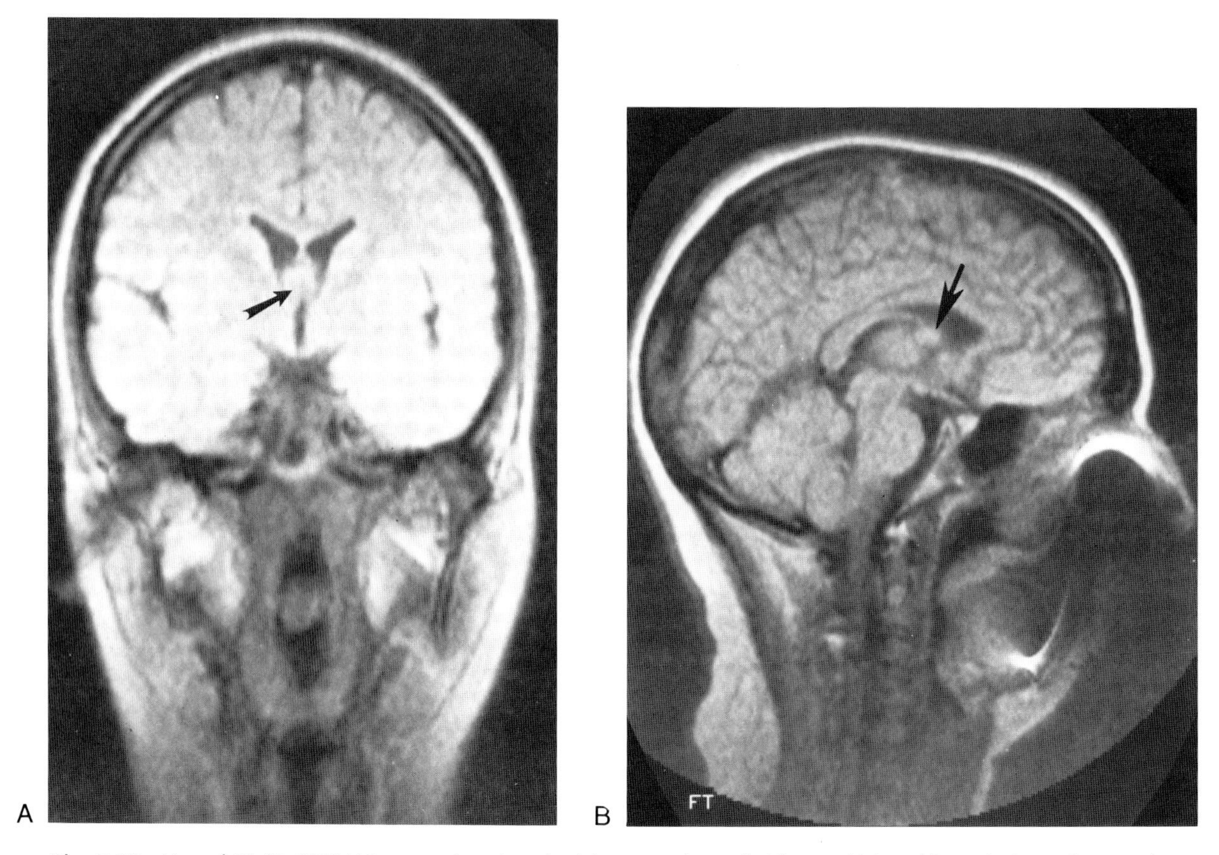

Fig. 5-28. (**A** and **B**) SE 1000/40 coronal and sagittal images of a colloid cyst. Using this technique the cyst is isointense with brain.

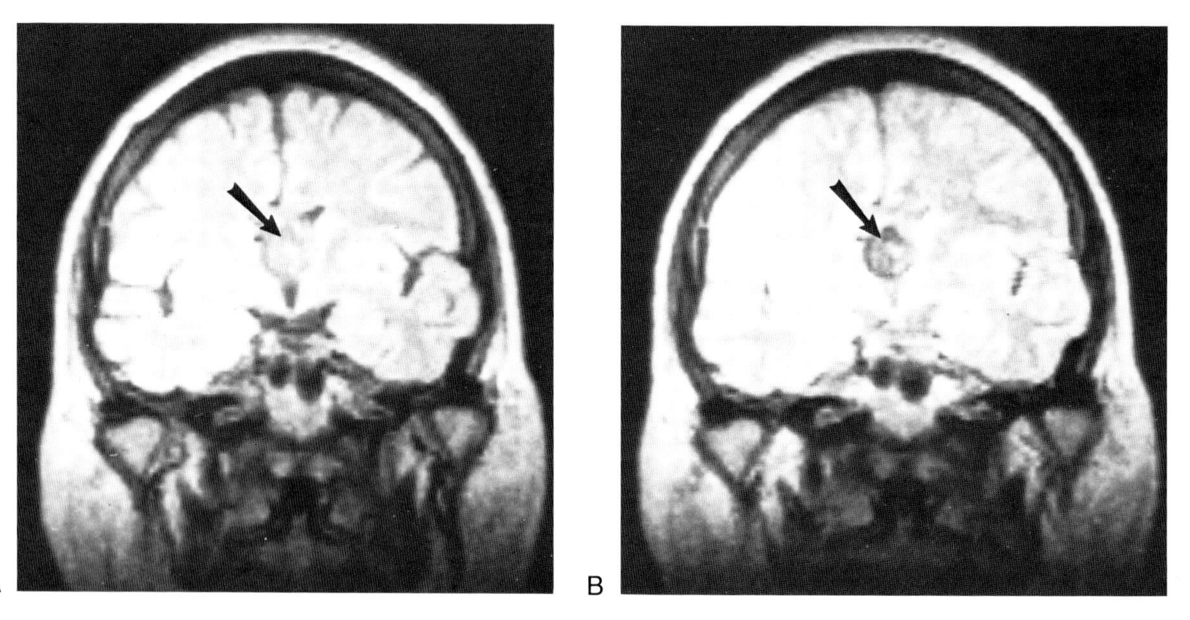

Fig. 5-29. A large, somewhat inhomogeneous colloid cyst seen on SE 1500/28 (**A**) and SE 1500/56 (**B**) coronal MR images.

cle may collapse posteriorly.[103] The cysts are not particularly vascular and enhance only slightly on CT with intravenous contrast.

Colloid cysts, because of their high protein content, have short T1 and long T2 values.[104] Accordingly, they appear hyperintense relative to brain on most T1 and T2 weighted pulse sequences (Figs. 5-27, 5-28). In a few colloid cysts we have observed an inhomogeneous signal pattern (Fig. 5-29). The cause of this inhomogeneity remains unclear, but is not due to CT detectable calcification.

REFERENCES

1. Rowland LP: Blood-brain barrier, cerebrospinal fluid, brain edema, and hydrocephalus. p.837. In Kandel ER, Schwartz JH (eds): Principles of Neural Science. 2nd Ed. Elsevier, New York, 1985
2. Liliequist B: The subarachnoid cisterns. Acta Radiol 185 (Suppl): 23, 1959
3. Wilson M: The Anatomic Foundation of Neuroradiology of the Brain. 2nd Ed. Little, Brown, Boston, 1972
4. Milhorat TH: The third circulation revisited. J Neurosurg 42:629, 1975
5. Cutler RWP, Page L, Galicich J et al: Formation and absorption of cerebrospinal fluid in man. Brain 91:707, 1968
6. Rubin RC, Henderson ES, Ommarja AK et al: The production of cerebrospinal fluid in man and its modification by acetazolamide. J Neurosurg 25:430, 1966
7. Bradley KC: Cerebrospinal fluid pressure. J Neurol Neurosurg Psychiatry 33:387, 1970
8. Fishman RA: Cerebrospinal Fluid in Diseases of the Nervous System. WB Saunders, Philadelphia, 1980
9. DuBoulay GH: Pulsatile movements in CSF pathways. Br J Radiol 39:255, 1966
10. Lane B, Kricheff IT: Cerebrospinal fluid pulsations at myelography: videodensitometric study. Radiology 110:579, 1974
11. Sherman JL, Citrin CM: Magnetic resonance demonstration of normal CSF flow. AJNR 7:3, 1986
12. Bergstrand G, Bergstrom M, Nordell B: Cardiac gated MR imaging of cerebrospinal fluid flow. J Comput Assist Tomogr 9:1003, 1985
13. Elster AD: Dynamic changes in the morphology of the cerebral ventricles with arterial pulsation: demonstration using cardiac cycled MRI (in press).
14. Sherman JL, Citrin CM, Bowen BJ, Gangarosa RE: MR demonstration of altered cerebrospinal fluid flow by obstructive lesions. AJNR 7:571, 1986
15. Elster AD: Technical considerations which affect visualization of the CSf flow-viod sign on MRI (in press).
16. Bradley WG Jr, Kortman KE, Burgoyne B: Flowing cerebrospinal fluid in normal and hydrocephalic states: appearance on MR images. Radiology 159:611, 1986
17. Farruggia S, Babcock DS: The cavum septi pellucidi: its appearance and incidence with cranial ultrasonography in infancy. Radiology 139:147, 1981
18. Nokano S, Hojo H, Kataoka K et al: Age related incidence of cavum septi pellucidi and cavum vergae on CT scans of pediatric patients. J Comput Assist Tomogr 5:348, 1981
19. Shaw CM, Alvord EC: Cava septi pellucidi et vergae: their normal and pathological states. Brain 92:213, 1969
20. Heiskanen O: Cyst of the septum pellucidum causing increased intracranial pressure and hydrocephalus. J Neurosurg 28:771, 1973
21. Cowley AR, Moody DM, Alexander E et al: Distinctive CT appearance of cyst of the cavum septi pellucidi. AJR 133:548, 1979
22. Schwidde JT: Incidence of cavum septi pellucidi and cavum vergae in 1,032 human brains. Arch Neurol Psychiat 67:625, 1952
23. Taveras JM, Wood EH: Diagnostic Neuroradiology. 2nd Ed. Williams & Wilkins, Baltimore, 1976
24. Strother CM, Harwood-Nash DC: Congenital malformations in radiology of the skull and brain p.3712. In Newton TH, Potts DG (eds): Radiology of the Skull and Brain: Ventricles and Cisterns. Vol.4. CV Mosby, St. Louis, 1978
25. Picard L. Leymarie F, Roland J et al: Cavum veli interpositi. Roentgen anatomy-pathology and physiology. Neuroradiology 10:215, 1976
26. Bates JI, Netsky MG: Developmental anomalies of the horns of the lateral ventricle. J Neuropath Exp Neurol 14:316, 1955
27. Kozlowski P, Dymecki J: Deformation of the lateral ventricles of the brain due to ependymal fusions. Acta Radiol (Diagn) 9:187, 1969
28. Adams RD: The mega cisterna magna. J Neurosurg 48:190, 1970
29. Liliequist B, Tovi D, Schisano G: Developmental defects of the tentorium and cisterna magna. Acta Psychiatr Neurol Scand 35:223, 1960

29a. Raybaud C: Cystic malformations of the posterior fossa: Abnormalities associated with the development of the roof of the fourth ventricle and adjacent meningeal structures. J Neuroradiol 9:103, 1982

30. Naidich TP, Schott LH, Baron RL: Computed tomography in evaluation of hydrocephalus. Radiol Clin North Am 20:143, 1982
31. Hoare RD, Strand RD; The third ventricle in hydrocephalus due to basal cistern block. Pediatr Radiol 2:15, 1974

32. Latchaw RE, Gold LHA, Moore JS et al: Nonspecificity of absorption coefficients in the differentiation of solid tumor and cystic lesions. Radiology 125:141, 1977
33. Fitz CR, Harwood-Nash DC: Computed tomography in hydrocephalus. Comput Tomogr 2:91, 1978
34. Hart MN, Malmud N, Ellis WG: The Dandy-Walker syndrome: a clinicopathological study based on 28 cases. Neurology 22:771, 1972
35. Edwards JH: The syndrome of sex-linked hydrocephalus. Arch Dis Child 36:486, 1961
36. Taboada D, Alonso A, Alvarez JA et al: Congenital atresia of the foramen of Monro. Neuroradiology. 17:161, 1979
37. Schultz P, Leeds NE: Intraventricular septations complicating neonatal meningitis. J Neurosurg 38:620, 1973
38. Spalline A: Computed tomography in aneurysm of the vein of Galen. J Comput Assist Tomogr 3:779, 1979
39. Enzmann DR, Kricorian J, Yorke C et al: Computed tomography in leptomeningeal spread of tumor. J Comput Assist Tomogr 2:448, 1978
40. Fitz CR: The ventricles and subarachnoid spaces in children. p.201. In Lee SH, Rao KCVG (eds): Cranial Computed Tomography. McGraw-Hill, New York, 1983
41. Flodmark O, Fitz CR, Harwood-Nash DC et al: Neuroradiological findings in a child with primary leptomeningeal melanoma. Neuroradiology 18:153, 1979
42. Cutler RWP, Page L, Galicich J et al: Formation and absorption of cerebrospinal fluid in man. Brain 91:707, 1968
43. Eisenberg HM, McComb JG, Lorenzo AV: Cerebrospinal fluid overproduction and hydrocephalus associated with choriod plexus papilloma. J Neurosurg 40:381, 1974
44. Cutler RWP, Murray JE, Moody RA: Overproduction of cerebrospinal fluid in communicating hydrocephalus. Neurology 23:1, 1973
45. Bradley KC: Cerebrospinal fluid pressure. J Neurol Neurosurg Psychiatry 33:387, 1970
46. Rosman NP, Shands KN: Hydrocephalus caused by increased intracranial venous pressure: a clinicopathological study. Ann Neurol 3:445, 1978
47. Friedman WA, Mickle JP: Hydrocephalus in achondroplasia: a possible mechanism. Neurosurgery 7:150, 1980
48. Allen HA, Haney P, Rao KCVG: Vascular involvement in cranial hyperostosis. AJNR 3:193, 1982
49. Gilles FH, Davidson RI: Communicating hydrocephalus associated with deficient dysplastic parasagittal arachnoid granulations. J Neurosurg 35:421, 1971
50. Gutierrez Y, Friede RL, Kaliney WJ: Agenesis of arachnoid granulations and its relationship to communicating hydrocephalus. J Neurosurg 43:553, 1975
51. Barron SA, Jacobs L, Kinkel WR: Changes in size of normal lateral ventricles during aging determined by computerized tomography. Neurology 26:1011, 1976
52. Gyldensted C: Measurements of the normal ventricular system and hemispheric sulci of 100 adults with computed tomography. Neuroradiology 14:183, 1977
53. Hahn FJY, Rim K: Frontal ventricular dimensions on normal computed tomography. AJR 126:593, 1976
54. Haug G: Age and sex dependence on the size of normal ventricles on computed tomography. Neuroradiology 14:201, 1977
55. Meese W, Kluge W, Grumme T et al: CT evaluation of the CSF spaces of healthy persons. Neuroradiology 19:131, 1980
56. Pedersen H, Gyldensted M, Gyldensted C: Measurement of the normal ventricular system and subarachnoid space in children with computed tomography. Neuroradiology 17:231, 1979
57. Wolpert S: The ventricular size on computed tomography. J Comput Assist Tomogr 1:222, 1977
58. Zatz LM, Jernigan TL, Ahumada AJ et al: Changes on computed tomography with aging: intracranial fluid volume. AJNR 3:1, 1982
59. Synek V, Reuben JR, Gawler J et al: Comparison of the measurements of the cerebral ventricles obtained by CT scanning and pneumoencephalography. Neuroradiology 17:149, 1976
60. Evans WA Jr: An encephalographic ratio for estimating ventricular enlargement and cerebral atrophy. Arch Neurol Psychiatry 47:931, 1942
61. Heinz ER, Ward A, Drayer BP et al: Distinction between obstructive and atrophic dilatation of ventricles in children. J Comput Assist Tomogr 4:320, 1980
62. Huckman MS, Fox J, Topel J: The validity of criteria for the evaluation of cerebral atrophy by computed tomography. Radiology 116:85, 1975
63. Gado MH, Coleman RE, Lee KS et al: Correlation between computerized transaxial tomography and radionuclide cisternography in dementia. Neurology 26:555, 1976
64. Moseley IF, Radu EW: Factors influencing the development of periventricular lucencies in patients with raised intracranial pressure. Neuroradiology 17:65, 1979
65. Fishman RA: Brain edema. N Engl J Med 293:706, 1975
66. Rubin RC, Hochwald G, Liwnicz B: The effect of severe hydrocephalus on size and number of brain cells. Dev Med Child Neurol 14:117, 1972

67. DiChiro G, Arimitsu T, Brooks RA et al: Computed tomography profiles of periventricular hypodensity in hydrocephalus and leukoencephalopathy. Radiology 130:661, 1979
68. Mori K, Handa T, Murata T, Nakano Y: Periventricular lucency in computed tomography of hydrocephalus and cerebral atrophy. J Comput Assist Tomogr 4:204, 1980
69. Johnson MA, Pennock JM, Bydder GM et al: Clinical NMR imaging of the brain in children: normal and neurologic disease. AJR 141:1005, 1983
70. Novetsky GL, Berlin L: Aqueductal stenosis: demonstration by MR imaging. J Comput Assist Tomogr 8:1170, 1984
71. Bydder GM, Steiner RE, Young IR et al: Clinical NMR imaging of the brain: 140 cases. AJNR 3:459, 1982
72. Brant-Zawadzki M, Norman D, Newton TH et al: Magnetic resonance of the brain: the optimal screening technique. Radiology 152:71, 1984
73. Bradley WG, Waluch V, Yadley RA, Wycoff RR: Comparison of CT and MR in 400 patients with disease of the brain and cervical spinal cord. Radiology 152:695, 1984
74. Zimmerman RD, Fleming CA, Lee BCP et al: Periventricular hyperintensity as seen by magnetic resonance: prevalence and significance. AJR 146:443, 1986
75. Bradley WG, Walluch V, Brant-Zawadzki M et al: Patchy, periventicular white matter lesions in the elderly: a common observation during NMR imaging. Noninvasive Med Imaging 1:35, 1984
76. Sze G, DeArmond SJ, Brant-Zawadzki M et al: Foci of MRI signal (pseudolesions) anterior to the frontal horns: histologic correlations of a normal finding. AJNR 7:381, 1986
77. Simon JH, Holtas SL, Schiffer RB et al: Corpus callosum and subcallosal-periventricular lesions in multiple sclerosis: detection with MR. Radiology 160:363, 1986
78. Wilson DA, Steiner RE: Periventricular leukomalacia: evaluation with MR imaging. Radiology 160:507, 1986
79. Lee BCP: Intracranial cysts. Radiology 130:667, 1979
80. Leo JS, Pinto RS, Hulvat GF et al: Computed tomography of arachnoid cysts. Radiology 130:675, 1979
81. Kjos BO, Brant-Zawadzki M, Kucharczyk W et al: Cystic intracranial lesions: magnetic resonance imaging. Radiology 155:363, 1985
82. Brant-Zawadzki M, Kelly W, Kjos B et al: Magnetic resonance imaging and characterization of normal and abnormal intracranial cerebrospinal fluid (CSF) spaces. Neuroradiology 27:3, 1985
83. Heier LA, Zimmerman RD, Russell EJ et al: MR imaging of arachnoid cysts. Radiol 157:287, 1985
84. Brocklehurst G: The development of the human cerebrospinal fluid pathway with particular reference to the roof of the fourth ventricle. J Anat 105:467, 1969
85. Harwood-Nash DC, Fitz CR: Neuroradiology in Infants and Children. CV Mosby, St. Louis, 1976
86. DeFeo D, Foltz EL, Hamilton AE: Double compartment hydrocephalus in a patient with cysticercosis meningitis. Surg Neurol 4:247, 1975
87. Scotti G, Musgrave MA, Fitz CR, Harwood-Nash DC: The isolated fourth ventricle in children: CT and clinical review of 16 cases. AJR 135:1233, 1980
88. Zimmerman RA, Bilaniuk LT, Gallo E: Computed tomography of the trapped fourth ventricle. AJR 130:503, 1978
89. Swartz JD, Zimmerman RA, Bilaniuk LT: Computed tomography of intracranial ependymomas. Radiology 143:97, 1982
90. Russell DS, Rubinstein LJ: Pathology of Tumors of the Nervous System. 4th Ed. Williams & Wilkins, Baltimore, 1977
91. Armington WG, Osborn AG, Cubberley DA et al: Supratentorial ependymoma: CT appearance. Radiology 157:367, 1985
92. Courville CB, Broussalian SL: Plastic ependymomas of the lateral recess. Report of eight verified cases. J Neurosurg 18:792, 1961
93. Han JS, Benson JE, Kaufman B et al: MR imaging of pediatric cerebral anomalies. J Comput Assist Tomogr 9:103, 1985
94. Enzmann DR, Norman D, Levin V et al: Computed tomography in the follow-up of medulloblastomas and ependymomas. Radiology 128:57, 1978
95. Kricheff II, Becker M, Schneck SA et al: Intracranial ependymomas: factors influencing progosis. J Neurosurg 21:7, 1964
96. Glasauer FE, Yuan RHP: Intracranial tumors with extracranial metastases. Case report and review of the literature. J Neurosurg 20:474, 1963
97. Dohrmann GJ, Farwell JR, Flannery JT: Ependymomas and ependymoblastomas in children. J Neurosurg 45:273, 1976
98. Scheithauer BW: Symptomatic subependymoma. Report of 21 cases with review of the literature. J Neurosurg 49:689, 1978
99. Zimmerman RA, Bilaniuk LT: Computed tomography of choroid plexus lesions. J Comput Assist Tomogr 3:93, 1979
100. Carpenter DB, Michelsen WJ, Hays AP: Carcinoma of the choroid plexus. J Neurosurg 56:722, 1982
101. Ariens-Kappers J: The development of the paraphysis cerebri in man with comments on its relationship to

the intercolumnar tubercle and its significance for the origin of cystic tumors in the third ventricle. J Comput Neurol 102:425, 1955
102. Donaldson JO, Simon RH: Radiodense ions within a third ventricular colloid cyst. Arch Neurol 37:246, 1980
103. Ganti SR, Antunes JL, Lousi KM et al: Computed tomography in the diagnosis of colloid cysts of the third ventricle. Radiology 138:385, 1981
104. Kjos BO, Brant-Zawadzki M, Kucharczyk W et al: Cystic intracranial lesions: magnetic resonance imaging. Radiology 155:363, 1985

6

Neoplasms

A wide variety of neoplasms, both primary and secondary, may affect the central nervous system. While the relative frequencies of these lesions vary from hospital to hospital, a typical distribution of tumors that might be encountered in a metropolitan medical center setting is presented in Table 6-1.[1] As a general trend intratentorial and ependymal tumors tend to occur at younger ages, while supratentorial gliomas, metastases, and meningiomas predominate in older age groups. In a community hospital a brain mass imaged by MR would more likely represent a metastasis. In a large university hospital or referral center where many neurosurgeons practice, statistics would favor such a lesion being a primary neoplasm.

Computed tomography has a long history of established success in the evaluation of cerebral neoplasms. MRI, while a relatively new technology, has already made significant contributions to the diagnosis and management of certain tumors.[2–5] However, limitations as well as advantages are being recognized in the use of MR to evaluate cerebral neoplasms. This chapter will first address the general roles of MR and CT in cerebral neoplasia, highlighting those tumors that might be better suited by one modality or the other. Next the discussion will turn to the use of paramagnetic contrast agents to enhance the ability of MR to evaluate lesions that disrupt the blood-brain barrier. Finally, brief analyses and examples of MRI of the more common brain tumors in children and adults will be presented. While most of the major tumors will be presented in this chapter, it should be noted that some regional tumors are discussed elsewhere (ependymal neoplasms, Chapter 5; pituitary and hypothalamic tumors, and tumors of the skull base and temporal bone, Chapter 11; and orbital tumors, Chapter 12).

Table 6-1. Distribution of Cerebral Neoplasms by Age and Location

Location	*0–20 Years*	*20–60 Years*	*Over 60 Years*
Supratentorial	Glioma (10–14%)	Glioblastoma (25%)	Glioblastoma (35%)
	Craniopharyngioma (5–13%)	Meningioma (14%)	Meningioma (20%)
	Ependymoma (3–5%)	Astrocytoma (13%)	Metastases (10%)
	Choroid plexus papilloma (1.5–3%)	Metastases (10%)	
	Optic glioma (1–3.5%)	Pituitary tumors (5%)	
Infratentorial	Cerebellar astrocytoma (15–20%)	Metastases (5%)	Acoustic (20%)
	Medulloblastoma (14–18%)	Acoustic (3%)	Metastases (5%)
	Brain stem glioma (9–12%)	Meningioma (1%)	Meningioma (5%)
	Ependymoma (4–8%)		

(Adapted from Youmans JR: Neurological Surgery. WB Saunders, Philadelphia, 1982, reprinted with permission from WB Saunders, Co.)

GENERAL PRINCIPLES OF TUMOR IMAGING BY MRI

Most Tumors Have Long T1 and T2 Values

While the precise mechanisms are far from clear, the ratio of free to bound water is thought to increase within most tumors. As a result, tissue relaxation times increase. This increase in T1 and T2 produces conflicting effects on MR signal intensity. On spin echo sequences, prolonged T1 tends to reduce signal while prolonged T2 increases signal. With the use of T2-weighted images (long TR and TE) the competing effects of T1 can be somewhat elminated. T2-weighted images in general have been found to be the most sensitive for tumor detection. Benign tumors often have T1 and T2 relaxation times that are shorter than more malignant ones. However, the overlap between benign and malignant values is too broad for its use in diagnosis. More immature tumors have greater free water content than those with more mature cell lines; again, however, the overlap is too great to be definitive.[3,6–10]

Several Important Tumors Have T1 and T2 Similar to Normal Brain

Most meningiomas, most hamartomas, and some acoustic neurinomas have T1 and T2 values similar to normal brain. Thus except for morphologic changes these tumors may not be readily visualized on T1- or T2-weighted MR images. Isolated examples of other tumors that are occasionally isointense with brain have been reported, and include prolactinomas, choriod plexus papillomas, metastases, gliomas, and certain tumors treated with radiotherapy.[11,12]

Fat-Containing Tumors Usually Have Short T1 Values

Lipomas and dermoids nearly always display high signal on T1-weighted images, by virtue of their short T1 values caused by fat. Epidermoids, teratomas, and craniopharyngiomas may often have short T1 values due to fat content. However, if they contain cholesterol crystals instead of typical fat, T1 values may be normal or elevated. Teratomas may also contain edematous mesenchymal tissue with prolonged T1 and T2 values.[11]

Hemorrhage, Necrosis, and Cystic Degeneration Affect the Appearance of Tumors

Tumoral hemorrhage results in the deposition of paramagnetic iron compounds that alter T1 and T2 characteristics. The precise MR signal changes depend upon the age of the hemorrhage and are discussed more fully in Chapter 10. In general, subacute hemorrhage has short T2 and normal T1 values while chronic hemorrhage often shows short T1 and normal or prolonged T2. Areas of tumor nectosis and cyst formation show prolonged T1 and T2 values and may be indistinguishable from adjacent tumor tissue on MRI.[2,3,11]

Tumor May Be Difficult to Distinguish from Surrounding Edema

Because of prolonged T1 and T2 values in both tissues, it may be difficult or impossible to distinguish tumor margin from edematous normal brain. While often applying a number of different pulse sequences will overcome this problem (Fig. 6-1), in at least 10 percent of cases tumor margins remain indeterminate. The use of intravenous paramagnetic MR contrast agents in the future may better define the margins of tumor from edema.[2,4,5,13]

Tumoral Calcification is Unreliably Detected by MRI

Calcification within cerebral neoplasms is an important characteristic that directs the differential diagnosis by CT. Unfortunately MR is unable to reliably detect such tumoral calcifications. Fine calcifications cannot be seen at all, while larger punctate and con-

Fig. 6-1. Difficulty separating tumor form edema on MRI. **(A)** On SE 2000/100 image metastatic adenocarcinoma and vasogenic edema are inseparable. **(B)** Increasing TR to 3000 msec allows separation to be made. **(C)** Inversion recovery (IR 1400/400/40) scan may also be useful.

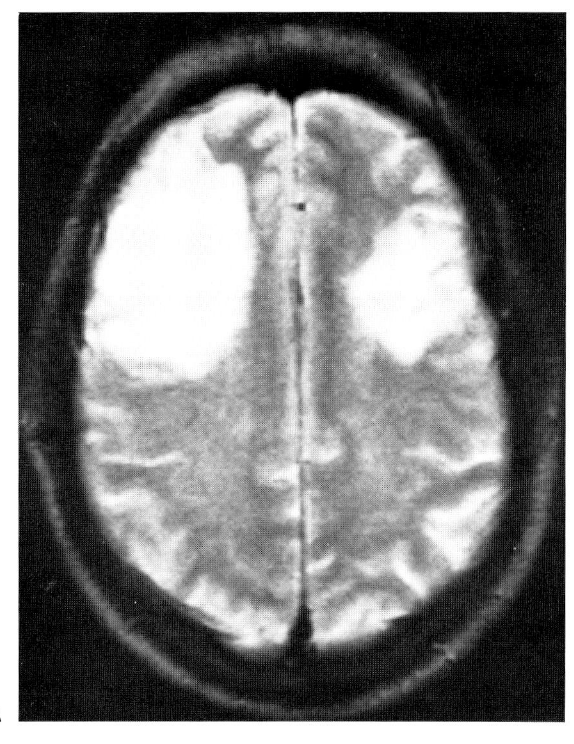

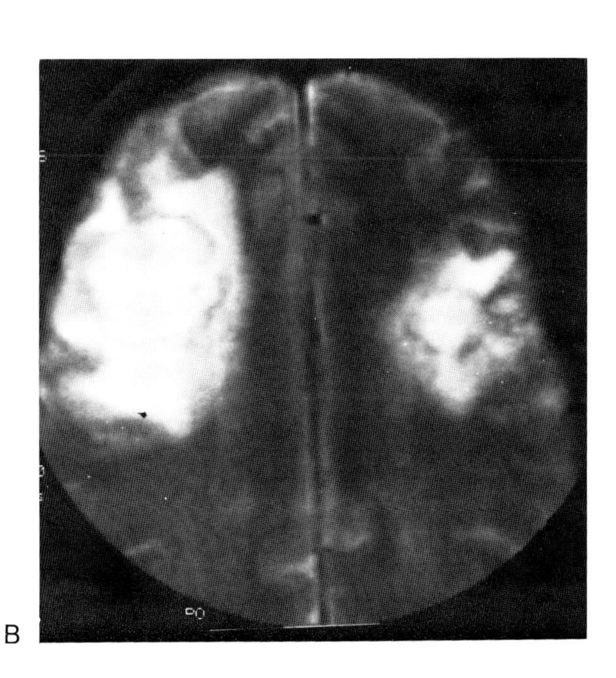

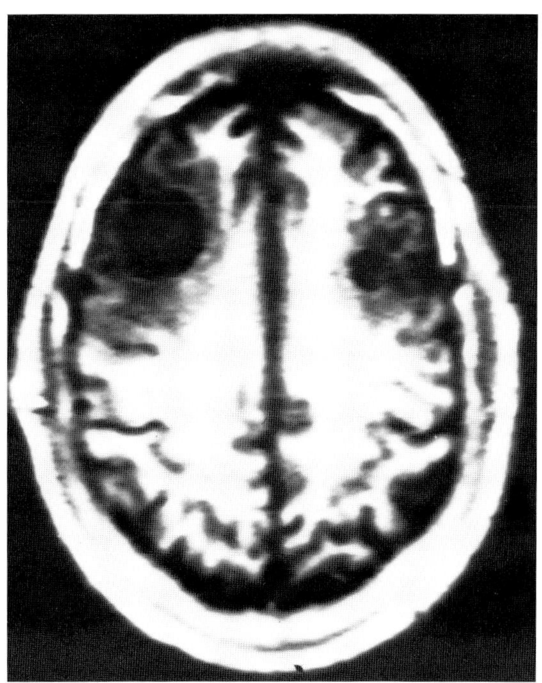

glomerate calcifications can be diagnosed in less than half of cases with assurance. When detected, calcification is best noted on T2-weighted images as foci of diminished or absent MR signal.[14,15]

THE USE OF PARAMAGNETIC CONTRAST AGENTS FOR MR IMAGING OF TUMORS

Physical Basis of Paramagnetic Contrast Enhancement

As discussed in Chapter 1, resonating protons escalated to higher energy levels must give up the energy they have absorbed from the RF field in order to undergo relaxation. This absorbed energy is transferred to the surrounding tissue lattice (T1 relaxation) or to other resonating nuclei (T2 relaxation). By changing the local magnetic environment experienced by a proton, T1 and T2 relaxation times may be altered.

Certain small molecules (NO, O_2), transition metals (Fe^{+3}, Mn^{+2}), and rare earth element ions (Gd^{+3}) have multiple unpaired outer shell electrons. When brought into the vicinity of a resonating hydrogen proton, these substances significantly alter the local magnetic field experienced by that proton. As a result, T1 and T2 relaxation times are shortened, a phenomenon called *relaxation enhancement*. These ions and small molecules with unbalanced electron spins that can induce proton relaxation enhancement are known as *paramagnetic substances*.

Gadolinium-DTPA: Properties and Kinetics

Gadolinium ion (Gd^{+3}) is among the more powerful paramagnetic substances, possessing seven unpaired electrons in its outer shell.[16] Chelation of the ion to diethylenetriaminepentaaceticacid (DTPA) neutralizes the toxicity of the ionic form. Gd-DTPA has undergone extensive clinical testing and will likely be approved by the FDA for general use in the USA in late 1987 or 1988.

Gd-DTPA is an extraordinarily stable complex with no adverse reactions noted in human subjects to date. The drug's LD_{50} is approximately 20 mmol/kg, while the usual dose for imaging is only 0.1 mmol/kg. It thus has a margin of safety of over 200. By comparison, iodinated contrast agents for CT have a safety margin of only 8 to 10.[16] Rapid administration of Gd-DTPA produces slight (approximately 10 percent) elevation blood pressure and cardiac output.[17] Higher doses can cause transient EKG changes in dogs.[18] Free gadolinium ion may interfere with blood formation and coagulation.[19] However, it should again be noted that more of these potential complications have yet been encountered in humans when slow infusion of lower doses is used.

For routine cranial MR imaging, approximately 0.1 mM/kg of Gd-DTPA is administered by slow bolus IV infusion. Within five minutes after infusion the drug reaches peak serum levels but is quickly redistributed into the extracellular spaces. Gd-DTPA does not enter intact cells and does not cross an intact blood-brain barrier.[20] The drug is not metabolized and undergoes rapid excretion by the kidneys through glomerular filtration. With normal renal function the half time of clearance from the circulation is about 35 minutes.

Optimal time for imaging is about 20 to 60 minutes postinjection.[21] This timing allows for good enhancement of normal cranial structures and for diffusion into tumors and disrupted portions of the blood-brain barrier. Maximal Gd-DTPA accumulation in these tissues results in shortening of both T1 and T2 by about 20 to 25 percent. This relaxation time enhancement is best appreciated on T1-weighted images as regions of increased signal.[22] Preinjection, both T1- and T2-weighted images should be obtained. Postinjection T1 images will usually suffice, although occasionally additional information is obtained on postcontrast T2-weighted images.[22]

Clinical Imaging with Gd-DTPA

Gd-DTPA enhancement of normal cranial structures on MR is not entirely the same as iodine-enhancement on CT.[23] Most of the brain shows no appreciable enhancement with either agent, because both Gd-DTPA and iodinated contrast are excluded by the blood-brain barrier. Enhancement of the pituitary, infundibulum, intracavernous cranial nerves, choroid plexus, and ocular choroid are well seen both on contrast MRI and contrast CT. However, enhancement of large blood vessels the flax, tentorium, and dura is not nearly so prominent on contrast MRI as on contrast CT. The sinus and nasopharyngeal mucosa are tremendously enhanced with Gd-DTPA. Inspection of the image for mucosal enhancement can be used to determine the efficacy of Gd-DTPA administration. A summary of the normal enhancement patterns of various intracranial structures following Gd-DTPA administration is presented in Table 6-2.

In pathologic states regions of Gd-DTPA enhancement are largely equivalent to those seen with iodinated contrast.[4,18,24–26] These sites of contrast enhancement correspond to areas of breakdown of the blood-brain barrier. Contrast enhanced MR may detect this breakdown earlier than contrast CT because MR can more sensitively detect small amounts of Gd-DTPA while relatively large amounts of iodinated contrast must be present to be detected by CT.[24]

Table 6-2. Enhancement of Normal Cranial Tissues on MRI Following Gd-DTPA Infusion

Marked Enhancement
Sinus mucosa[a]
Nasopharynx[a]
Cavernous sinus
Infundibulum
Choroid plexus
Periphery of petrous internal caroid artery[a]
Moderate Enhancement
Dura, falx, tentorium
Retinal choroid
Pituitary gland
Pineal
Distal cranial nerves[a]
Small cortical veins
Skeletal muscle
Little or No Enhancement
Bone
Large cerebral vessels[a]
CSF
Normal white matter or gray matter
Fat

[a] Signifies significantly different enhancement (more or less marked) than with contrast CT.

(Kilgore DP, Breger RK, Daniels DL et al: Cranial tissues: normal MR appearance after intravenous injection of Gd-DTPA. Radiology 160:757, 1986.)

A

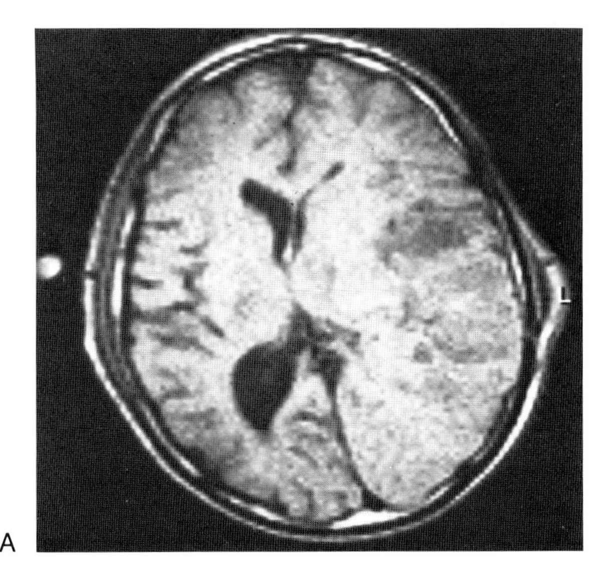

B

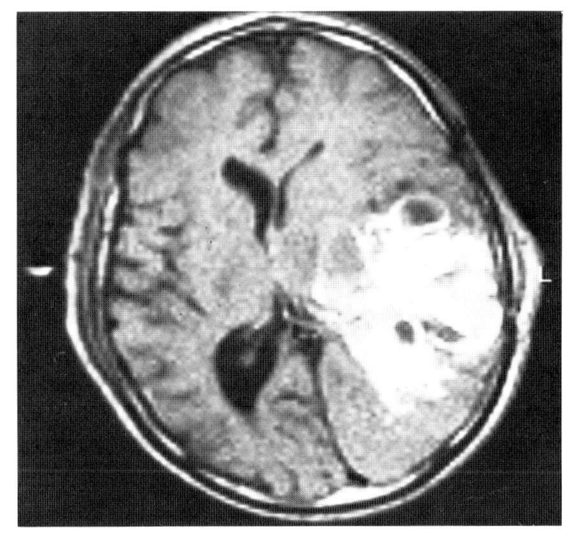

Fig. 6-2. Glioblastoma multiforme of the left hemisphere imaged using SE 500/28 technique. In **(A)** precontrast image shows poor delineation of tumor and normal brain. In **(B),** 30 minutes after administration of Gd-DTPA, there is excellent visualization of tumor. (Diagn Imaging 8(6):75, 1986.)

On T1-weighted images Gd-DTPA enhancement results in increased signal intensity, due to shortening of T1. In general the degree of T1 shortening is equivalent to the concentration of Gd-DTPA in the tissue. Accordingly, one might expect progressively higher signals with increasing accumulation of Gd-DTPA. However, this occurs only over a limited range. With increasing concentration of paramagnetic contrast T2 relaxation time also decreases. The decrease in T2 at higher concentrations may overcome the signal enhancing effects of decreased T1, thus diminishing the intensity of contrast enhancement.[28]

Worldwide clinical trials to date have reported favorable experience with Gd-DTPA enhanced MR images. The use of paramagnetic contrast improves visualization of small metastases and meningiomas.[27] The distinction between tumor and edema is remarkably improved, because edematous brain does not accumulate Gd-DTPA while tumor margin will.[4,5] It should be noted, however, that enhancement at the periphery of a tumor does not necessarily delineate tumor margin, either by CT or MR.[28] Furthermore, contrast-enhanced MR cannot replace precontrast T2-weighted sequences for detection of all types of brain pathology.[28] Furthermore, some small lesions will occasionally become isodense following Gd-DTPA administration.[28,29] This implies that a precontrast scan should always be performed in addition to the postcontrast one. As example of Gd-DTPA enhancement of a glioma is shown in Figure 6-2.

In summary, Gd-DTPA should probably be used liberally in MRI once the FDA approves it for general use. Contrast enhancement of a lesion provides important functional information such as degree of perfusion and breach of the blood-brain barrier. While Gd-DTPA may not improve sensitivity for the detection of intra-axial lesions, this functional information may be vital for proper diagnosis in many instances. Extra-axially, Gd-DTPA holds promise to remarkably improve detection of meningiomas and nerve sheath tumors, which are too frequently missed on noncontrast MR.

METASTATIC DISEASE

In most large series metastatic disease accounts for 10 to 30 percent of brain tumors detected by CT.[30–33] The high incidence of intracranial metastatic disease is further underscored by the fact that at least one in five patients who dies with cancer has a brain metastasis.[34] Because of different referral patterns and current usage of MRI, a much smaller percentage of lesions represent metastasis. Nevertheless, as the applications of MRI are broadened in the future, it is expected that metastatic disease will become a more common diagnosis.

Most parenchymal brain metastases are manifest by neurologic signs of mass effect: headache, nausea, vomiting, papilledema, or ataxia.[35] Sometimes an acute onset similar to infarct or hemorrhage may be seen. About 10 percent may be asymptomatic. Most patients are known to have a primary cancer at the time of neurologic presentation. However, as many as 14 percent of patients may present with neurologic symptoms as their first sign of cancer.[36]

In adults the most common sources of parenchymal brain metastases are lung, breast, kidney, melanoma, and colon. When calvarial and dural metastases are considered, prostate cancer and myeloma are also frequent. Most brain tumors in children are primary. When metastases are seen they are usually from a round cell tumor or sarcoma.

Recognition of the precise number of brain metastases present in a given patient is becoming important, because aggressive surgeons in some cancer centers will resect single or small numbers of brain metastases. The actual incidence of solitary metastases at the time of clinical presentation is unknown, but is probably about 25 percent. In autopsy series the incidence has been quoted from 14 to 40 percent.[35,37] Autopsy series are composed of patients who have often died with disseminated carcinomatosis. The actual incidence of solitary metastasis in patients at the time of clinical presentation may be higher than autopsy series would suggest. On the other hand, data from CT series tend to overestimate the percentage of solitary metastasis, because small metastases may be below the resolution of the scanner. With today's fourth generation CT scanners and high dose contrast about 38 percent of metastases are said to be solitary.[36] MRI is capable of demonstating metastases missed by CT, particularly those in the posterior fossa. The future use of MRI has obvious implications for evaluation of the resectability of presumed solitary metastases.

Several characteristics of parenchymal metastases may aid in their identification by MRI, although it should be kept in mind that these are only generalizations:

1. Metastases may appear in any location but they are characteristically found when small at the junc-

tion between cortex and white matter. When larger they may reach the brain surface and penetrate the meninges. Pure white matter lesions may be found but their presence alone should suggest other diagnoses such as a demyelinating disease or vasculitis. Deep lesions are also seen. A tumor mass in the cerebellum of an adult is highly likely to represent a metastasis.

2. Multiplicity of lesions is the hallmark of metastastic disease. Nevertheless, probably one-fourth of metastases are solitary at time of presentation. In patients with known primary carcinoma, a solitary brain lesion should be considered a metastasis until proven otherwise.
3. Hemorrhage within metastasis is most suggestive of lung, melanoma, choriocarcinoma, thyroid, or renal cell (Fig. 6-2). Up to 30 percent of melanoma metastases may demonstrate intratumoral hemorrhage, and melanoma is also the most likely metastasis to present with subarachnoid hemorrhage.[38] Lung carcinoma metastases only occasionally bleed. However, because lung carcinoma is much more common than melanoma, statistically it becomes a likely diagnosis when hemorrhage is seen.

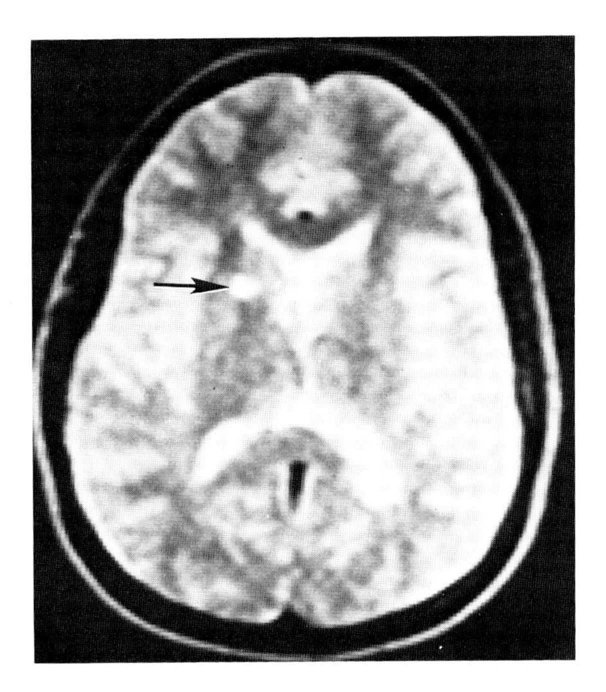

Fig. 6-4. Small cell carcinoma of the lung metastases sometimes demonstrate little or no vasogenic edema.

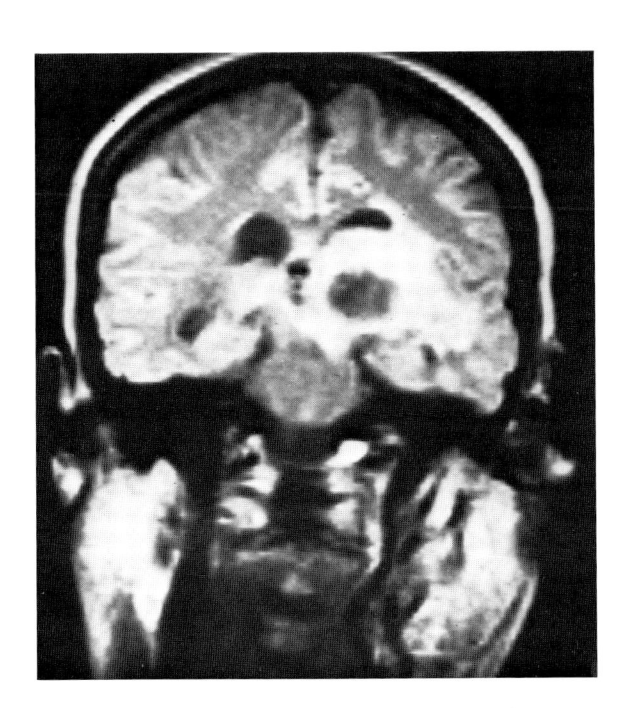

Fig. 6-3. Metastatic gastric carcinoma with abundant vasogenic edema surrounding (SE 1500/28).

4. Nearly all metastases demonstrate a significant component of edema relative to the actual size of tumor deposit (Fig. 6-3). This is particularly true for metastatic adenocarcinoma. Metastatic renal cell carcinoma virtually always has a large area of surrounding edema, and the diagnosis should be in doubt if this is not seen. On the other hand, certain metastatic tumors may demonstrate little if any edema. Melanoma prior to hemorrhage and small cell carcinoma metastasis will often have no detectible edema (Fig. 6-4). Steroid therapy also tends to reduce the component of vasogenic edema surrounding a metastasis.
5. Calcification in metastasis is unusual, but may occur in metastatic sarcomas and mucinous tumors of the colon or breast. Calcification may occur in any metastasis after hemorrhage or therapy. Small calcifications may not be recognized on MRI.
6. Periventricular tumor spread occurs by subependymal tumor invasion. Metastasis may reach the periventricular locations by hematogenous dissemination, direct invasion, or transmission via cerebrospinal fluid. Such modes of dissemination are most commonly seen with lymphoma, melanoma,

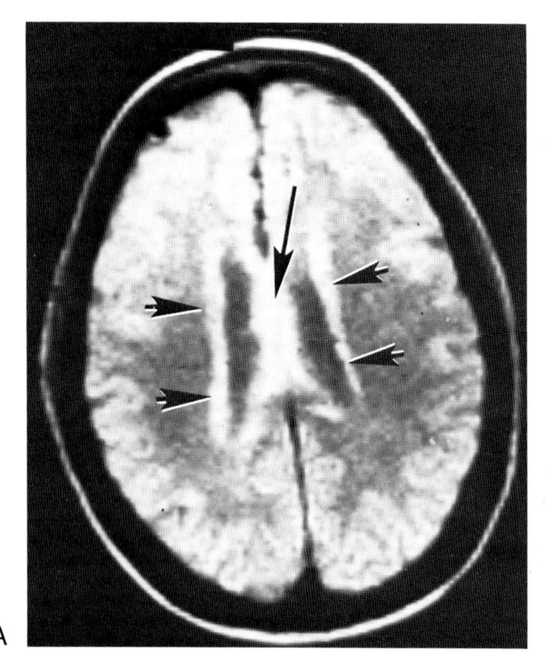

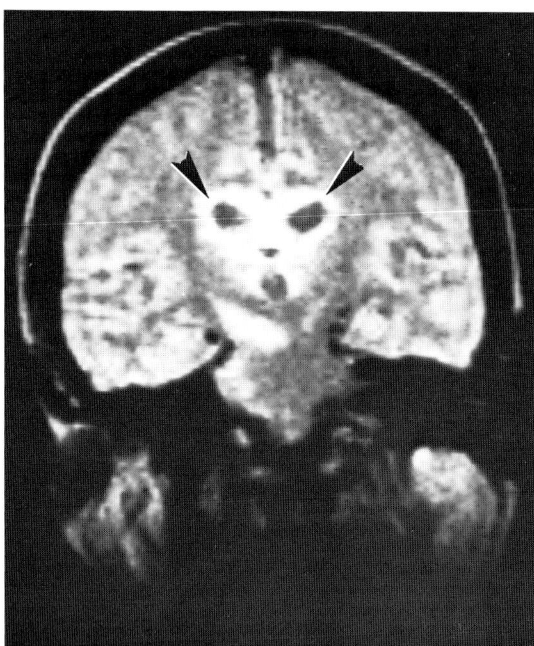

Fig. 6-5. Subependymal spread of breast carcinoma. High periventricular signal is seen. **(A)** SE 2000/28, **(B)** SE 1500/56.

and breast metastases (Figs. 6-5 to 6-7). They are also seen with intracranial spread of primary brain tumors, notably medulloblastoma, ependymoma, glioblastoma, and oligodendroglioma.

7. Calvarial metastases are commonly observed with carcinomas of the prostate, breast, lung, and kidney. The diploic space is involved primarily, and there is usually involvement of both outer and inner tables of the skull (Fig. 6-8). Sometimes only the inner table may be eroded and extension of tumor into the epidural or subdural spaces may be seen. Occasionally direct metastases to the epidural or subdural spaces may occur and be difficult to differentiate from a hematoma or fluid collection. MR signal characteristics may aid in this distinction.
8. The MR appearance of metastatic disease is nonspecific, usually demonstrating multiple long T1/long T2 lesions (Table 6-3). Vasogenic edema may not be separable from tumor. If hemorrhage in the metastasis has occurred, T1 may be short with T2 long. Gd-DTPA improves the detection of very small metastases.[27]

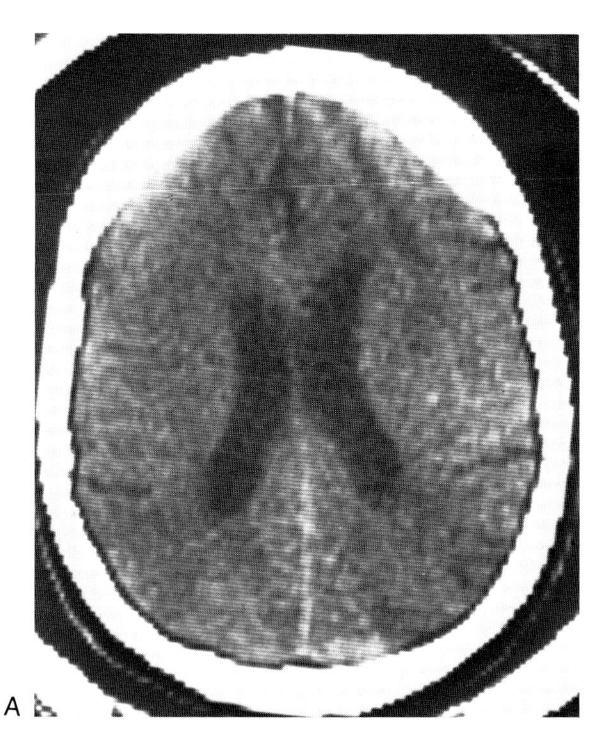

Fig. 6-6. Subependymal and leptomeningeal spread of metastatic adenocarcinoma. **(A)** CT was negative. (*Figure continues*).

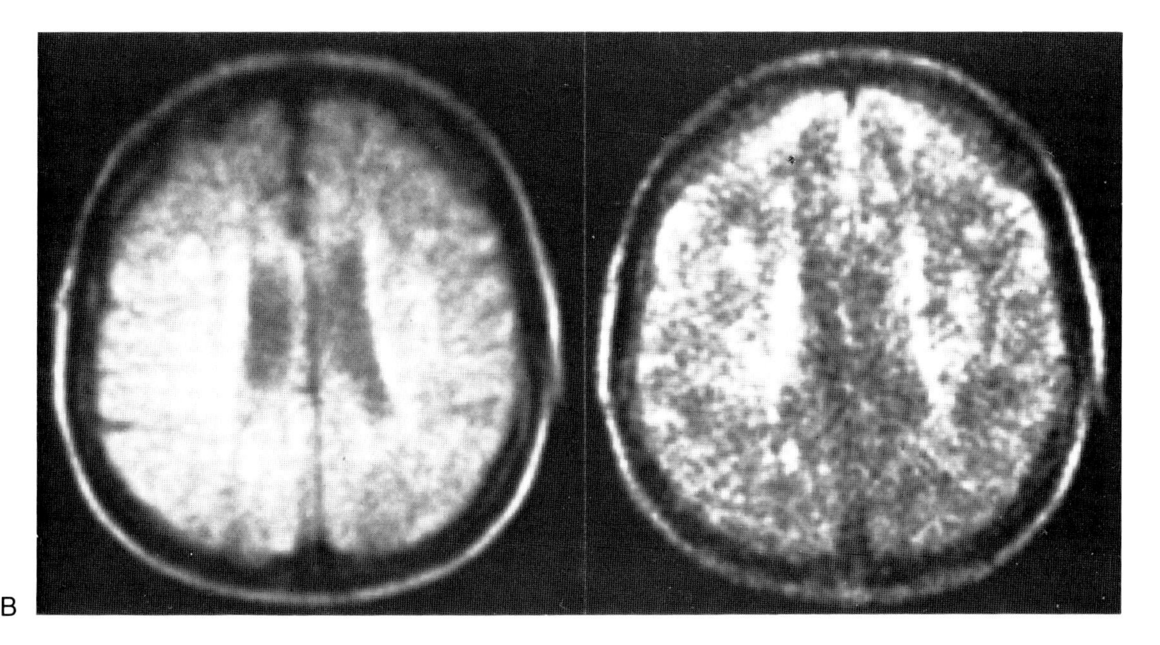

Fig. 6-6 (*continued*). **(B)** MRI shows multifocal high intensity signals from ependymal and subarachnoid spaces.

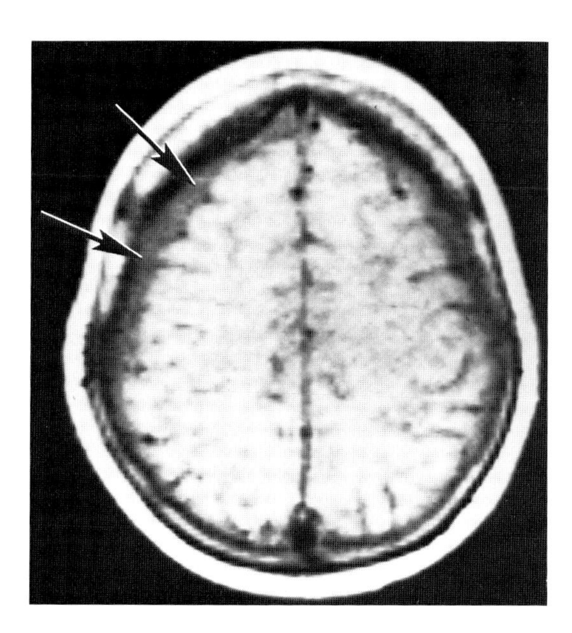

Fig. 6-7. Another patient with meningeal carcinomatosis from breast cancer. CSF protein concentration was greater than 300 mg %. On this SE 500/28 sequence, CSF should be black. Here it has appreciable abnormal signal.

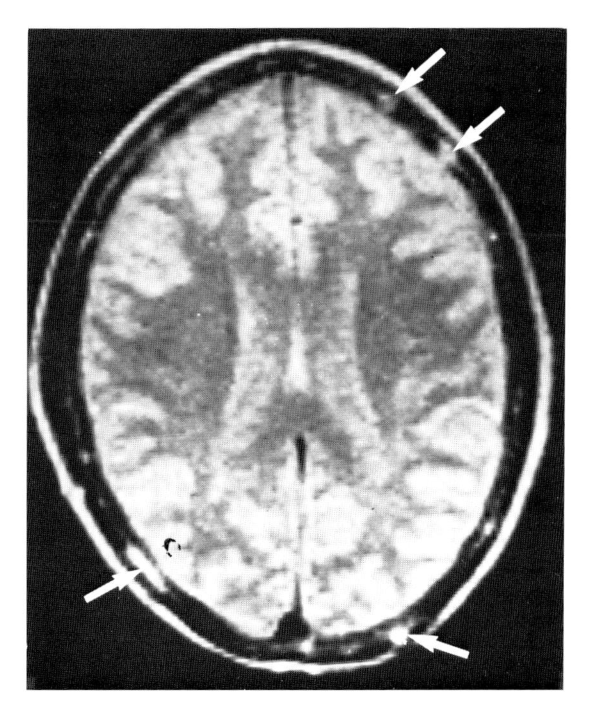

Fig. 6-8. Skull metastases show abnormal high signal form the usually dark calvarium on this SE 2000/56 image.

Table 6-3. Differential Diagnosis of Multifocal High-Intensity Cerebral Lesions on MRI

Congenital
Multiple hamartomas of tuberous sclerosis
Multiple tumors of Neurofibromatosis
Infectious
Abscesses
Septic emboli
Parasitic cysts
Tuberculomas and other granulomas
Multiple areas of encephalitis
Progressive multifocal leukoencephalopathy
Demyelinating Diseases
Vascular
Vasculitis
Binswanger disease
Vascular malformations
Traumatic
Multiple contusions
Shear injuries
Hematomas
Neoplasms
Metastases
Multifocal gliomas
Lymphomas
Other synchronous tumors

INTRA-AXIAL NEOPLASMS

Astrocytomas

Astrocytomas are tumors of glial origin (gliomas) that develop in the white matter of the cerebral hemispheres, cerebellum, or brain stem. In large clinical series, astrocytomas and their varients account for one-third to one-half of all intracranial tumors.[39] The aggressiveness of astrocytomas varies considerably, athough most behave rather poorly because they are unencapsulated and frequently lie deep within the brain. Histologic aggressiveness has been classified by Kernohan as grades I to IV, based on cellularity, mitoses, vascular endothelial proliferation, and necrosis. Low-grade astrocytomas (grades I and II) are relatively slow-growing and often considered "benign." High-grade astrocytomas (grades III and IV) behave malignantly with a correspondingly poor prognosis. Grade IV astrocytomas are also called *glioblastoma multiforme*.

Low-Grade Astrocytomas

Low-grade astrocytomas of the cerebral hemispheres are most commonly seen in young adults. Three histologic varieties (fibrillary, protoplasmic, gemistocytic) are identified, with mixtures being very common.[39] The tumors are usually hypovascular and fairly well circumscribed.[41] Calcification occurs in 10 to 20 percent.[42] Peripheral edema is not usually seen or at most minimal.[43] Tumoral hemorrhage is most uncommon and when present should suggest a higher grade malignancy.[41,42] The tumors may not show contrast enhancement on CT and be difficult to detect or differentiate from infarcts.

Low-grade astrocytomas have been demonstrated in a number of MR reports.[2–6,21,25–28,44–47] While their appearances may vary, most cases are similar to those shown in Figures 6-9 through 6-11. The tumors have elevated T1 and T2 values. They are well circumscribed with little or minimal surrounding edema. Small calcifications seen on CT are not detected by MR. Little if any contrast enhancement with Gd-DTPA is noted. MR may detect low-grade gliomas not recognized by CT.

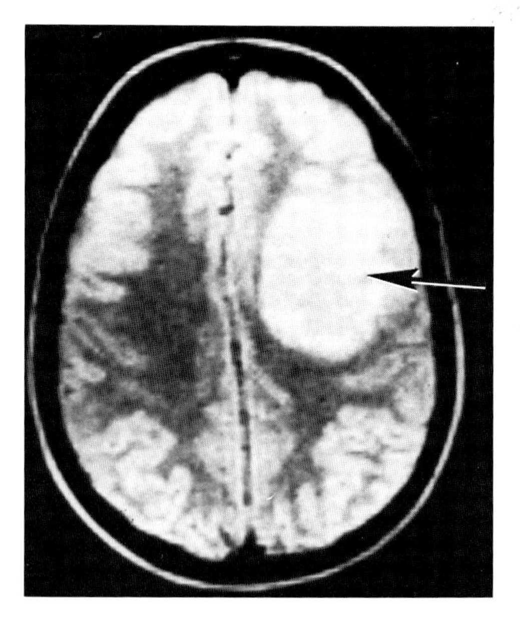

Fig. 6-9. Grade II glioma on this SE 2000/28 axial image. The tumor is well circumscribed and has no vasogenic edema.

Fig. 6-10. Another grade I to II astrocytoma seen on **(A)** SE 2000/56 axial and **(B)** SE 500/28 sagittal images.

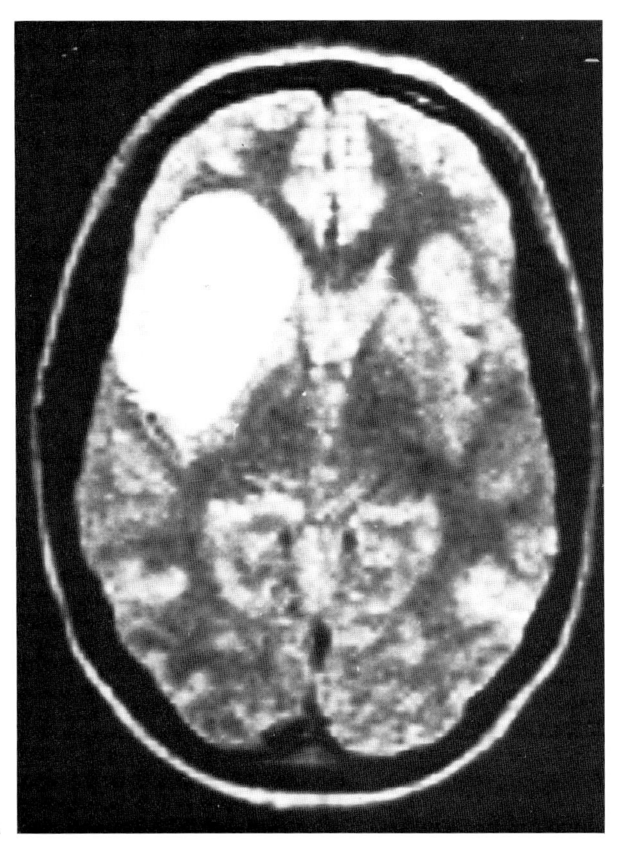

A

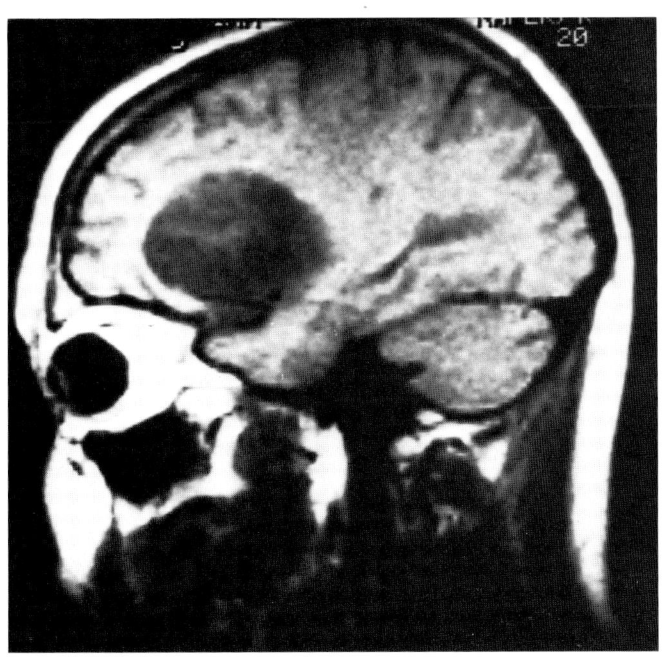

B

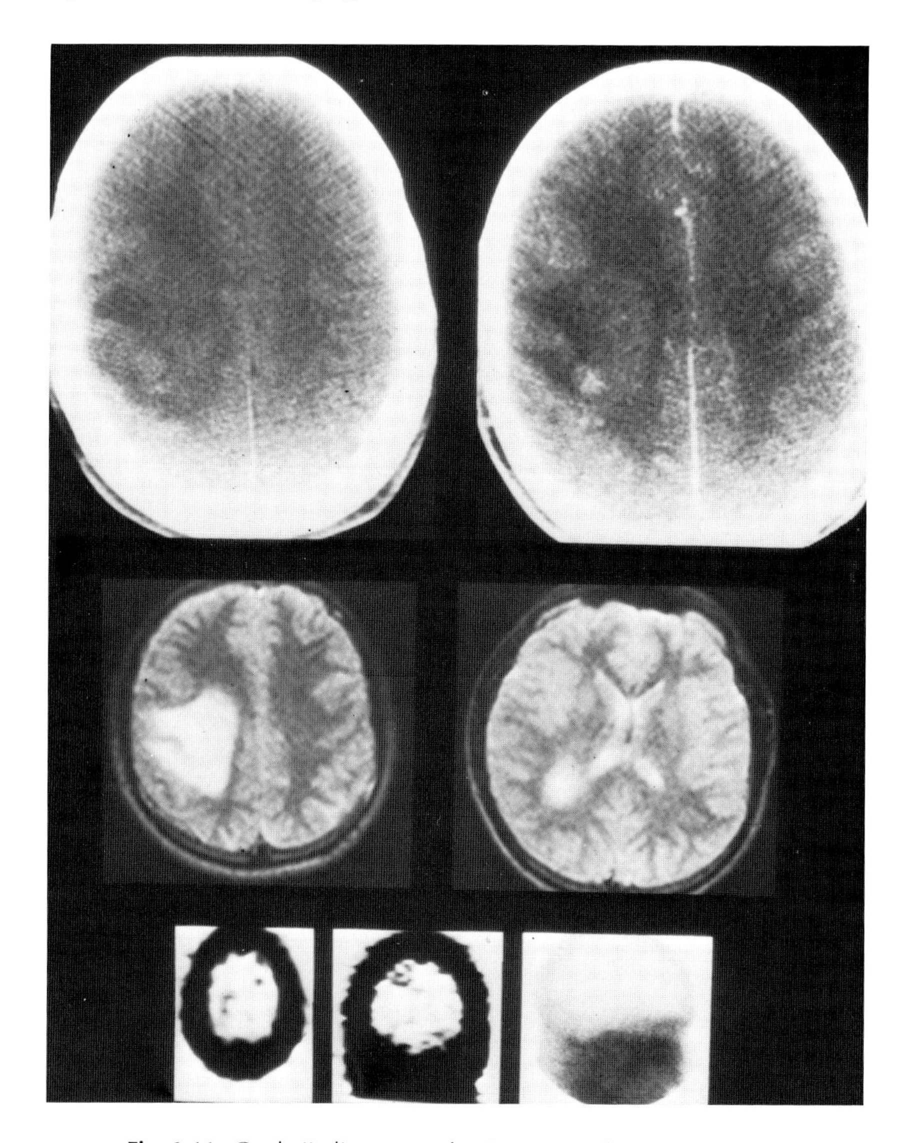

Fig. 6-11. Grade II glioma seen by CT, MR, and SPECT imaging.

Pilocytic Astrocytomas

Pilocytic Astrocytomas (spongioblastomas) are a type of low-grade astrocytoma seen primarily in children.[48] These tumors have a predilection for the brain stem and cerebellar hemispheres. Most are histologically grade I, except for an unusual infiltrative type that occurs in adults. The majority of pilocytic astrocytomas are cystic, some with a mural module. This cyst fluid has high protein content with T1 and T2 values distinct from CSF. Edema is minimal or absent (Fig.6-12 and 6-13).

Brain Stem Astrocytomas

Brain stem astrocytomas are usually pilocytic or fibrillary by histology.[39] Though they are often low-grade, their critical location results in inoperability and eventual fatality. Brain stem astrocytomas are most commonly centered in the pons, with cystic changes in 25 to 50 percent. Exophytic forms are also noted. MRI is superbly qualified to demonstrate the presence and extent of brain stem gliomas—far better than CT in all series to date.[46,49–51] MR is also useful in determining the response of these tu-

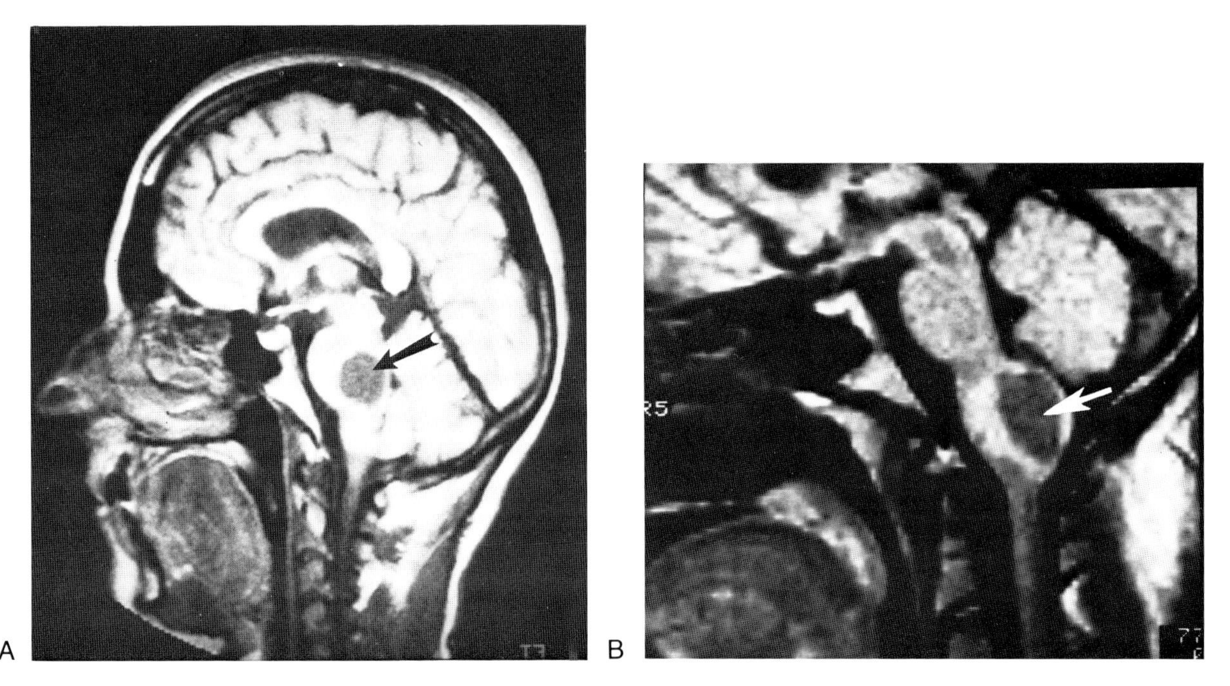

Fig. 6-12. (A) Pilocytic astrocytoma, grade II. These tumors are characteristically cystic and often in the brain stem. (Courtesy of Picker International.) **(B)** The same tumor in another patient.

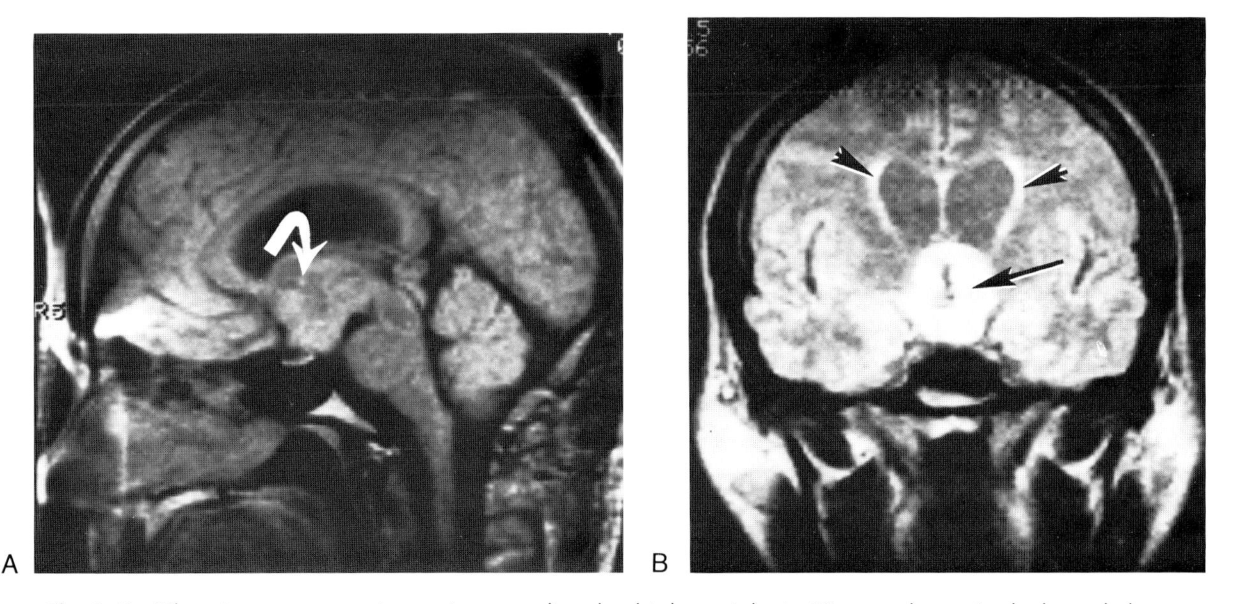

Fig. 6-13. Pilocytic astrocytoma (arrows) surrounding the third ventricle. In **(B)** note obstructive hydrocephalus with transependymal absorption of CSF (arrowheads).

mors to surgery and radiation therapy. Examples of brain stem astrocytomas seen by MRI are shown in Figures 6-12, 6-14 and 6-15.

High-Grade Astrocytomas

High-grade astrocytomas include both anaplastic types (usually grade III), and the very malignant glioblastoma multiforme (grade IV). These tumors are very aggressive, invading both white and gray matter structues. The prognosis is dismal and therapy is only palliative.

High-grade astrocytomas have irregular margins that enhance with Gd-DTPA.[4,5,28] Abundant peritumoral edema is characteristic. Tumor and vasogenic edema may be impossible to separate on noncontrast MR. Necrosis or hemorrhage within the tumor may be seen. MR may better reveal the size and spread of these tumors than CT. Examples of glioblastoma multiforme seen on MR are presented in Figures 6-16 through 6-17.

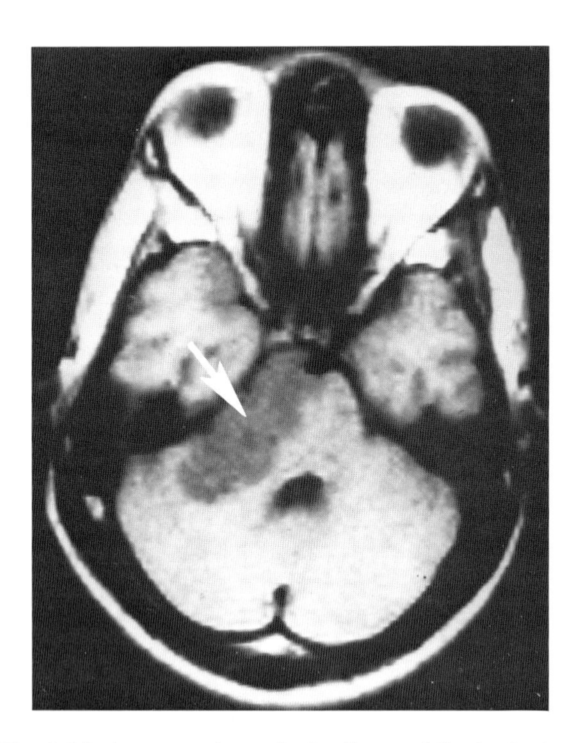

Fig. 6-14. An unusual exophytic glioma of the brain stem (SE 500/28).

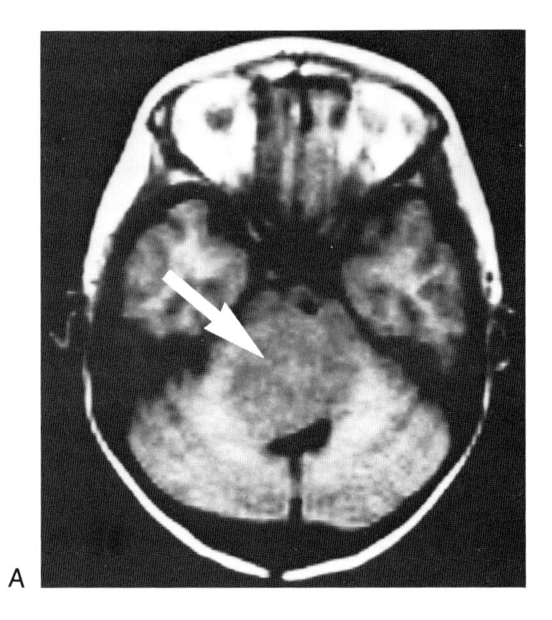

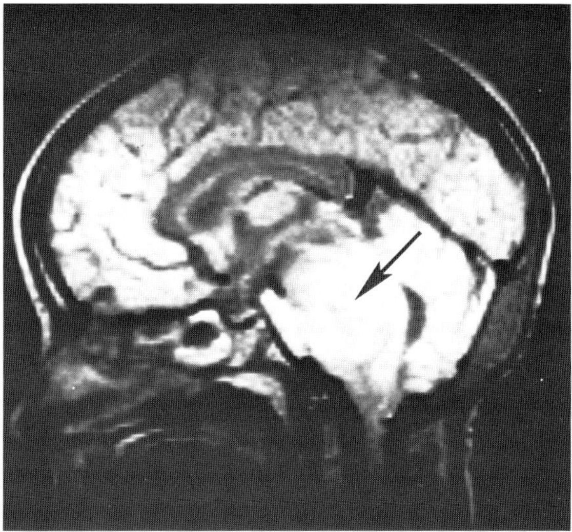

Fig. 6-15. An 8-year-old girl with a large brain stem astrocytoma. **(A)** SE 500/28. **(B)** SE 2000/28.

Multicentric Glioblastoma

Multicentric glioblastoma is seen in about 2.5 percent of patients who die from a high-grade astrocytoma.[52,53] True multicentric glioblastoma implies discrete foci of tumor separate from each other, with no fross or microscopic evidence of connection.[39] The appearance is nonspecific, mimicking multiple ab-

Fig. 6-16. "Butterfly" glioma spreading across the corpus callosum (arrow), well seen on this coronal SE 1500/56 image.

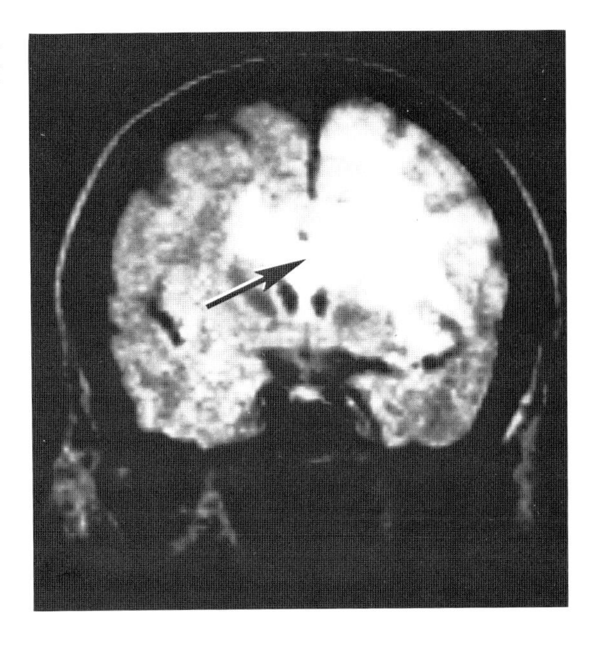

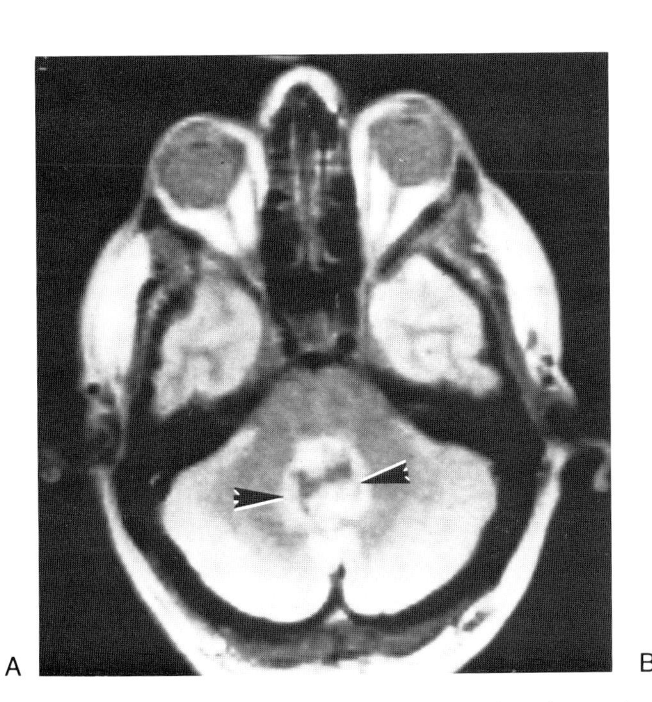

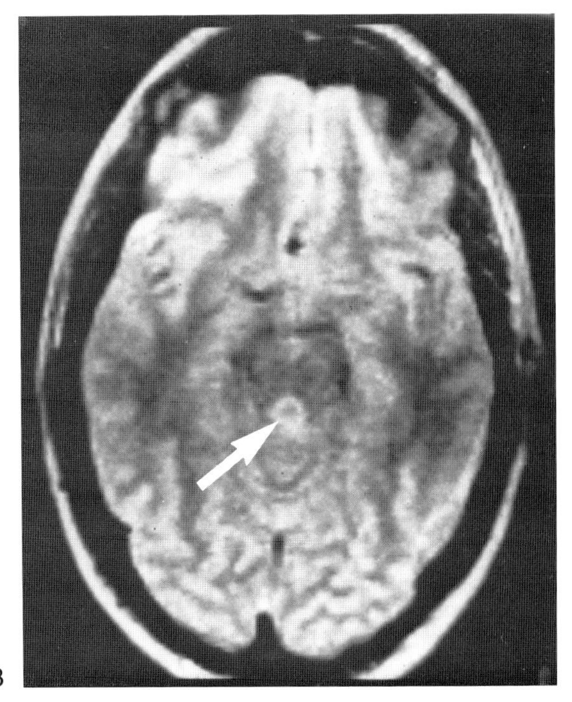

Fig. 6-17. Grade IV glioma, with subependymal spread around the fourth ventricle **(A)** and cerebral aqueduct **(B).**

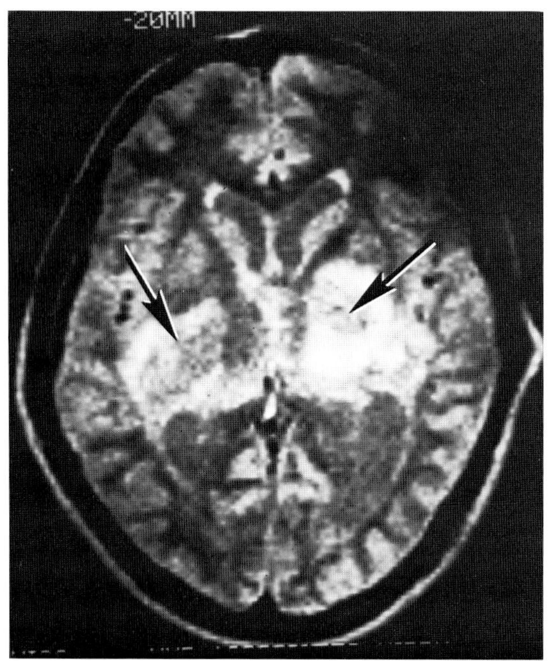

Fig. 6-18. Multifocal gliomas (arrows) in a patient with neurofibromatosis (SE 2000/56).

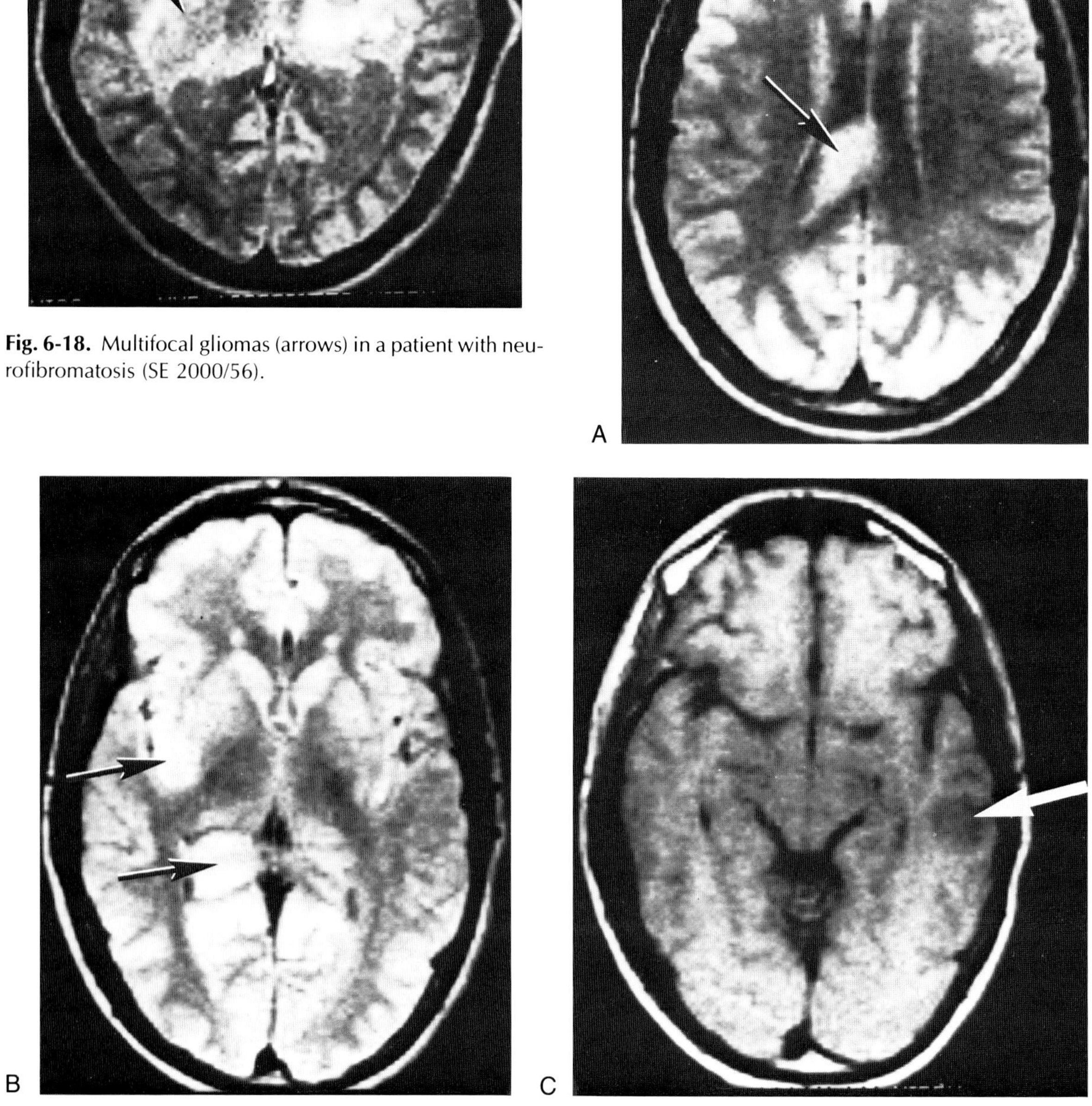

Fig. 6-19. (A–C) Another patient with four separate gliomas (arrows).

scesses or metastases. MR may occasionally show that what appears to be true multifocal glioma is merely a multilobulated glioma with fingerlike extensions resembling discrete tumor foci on axial sectioning.[28] Examples of multicentric glioblastoma on CT and MR are shown in Figures 6-18 and 6-19.

Oligodendrogliomas

Oligodendrogliomas are slow-growing tumors derived from oligodendrocytes of the cerebral white matter. They constitute about 5 to 7 percent of primary intracranial tumors and are seen mostly in adults.[54] The tumors are rare in children and the elderly. Because of their slow growth and location, prognosis is better than for most other gliomas.

Oligodendrogliomas are typically found in the frontal lobes and are very large at the time of diagnosis. Calcification is seen by CT in about 90 percent and constitutes an important diagnostic feature.[54] MR is unable to reliably demonstrate this calcification, and is hence inferior to CT in making the diagnosis.[15] Tumor margins, peripheral edema and contrast enhancement are variable. An example of an oligodendroglioma imaged by CT and MR is shown in Figure 6-20. In another tumor, dense calcification is detected on MR (Fig. 6-21).

Gangliocytomas

Gangliocytomas are relatively rare, slow-growing tumors that occur in children and young adults. They may be located anywhere within the brain, having preference for the temporal and frontal lobes, basal ganglia, and third ventricular region.[55,56] Two major histologic subtypes are identified: ganglioneuroma and ganglioglioma. Malignant degeneration is rare.

Their appearance is nonspecific and may resemble low-grade astrocytomas or gliomas. Calcification occurs in 50 percent, but usually escapes detection by MR. Cystic changes occur in slightly less than half of cases. There is minimal peritumoral edema. An example of a gangliocytoma seen on MR is shown in Figure 6-22.

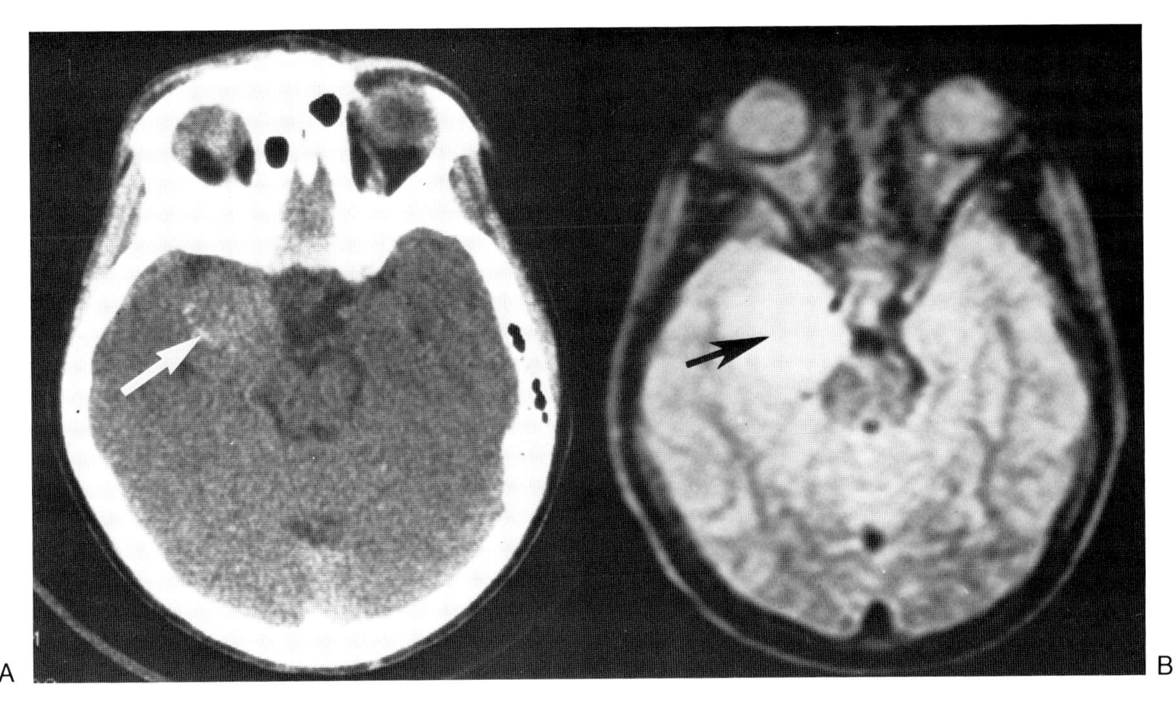

Fig. 6-20. Oligodendroglioma. **(A)** CT shows punctate calcifications, suggesting the diagnosis. **(B)** MRI shows the extent of the tumor clearly, but the calcifications cannot be detected.

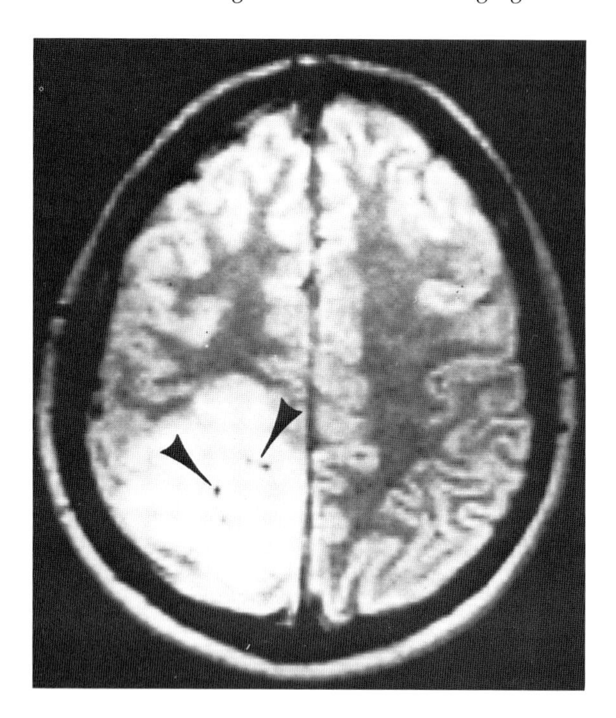

Fig. 6-21. Conglomerate calcifications in this oligodendroglioma were sufficiently large to be identified on MR (SE 2000/28).

Medulloblastomas

Medulloblastomas are the most common primary brain tumors in children presenting usually at ages 5 to 15 years.[57,58] Medulloblastoma also occurs in young adults where a second peak incidence may be seen at 20 to 25 years. Males are affected at least twice as frequently as females. The cellular origin of this tumor is uncertain but may relate to glial or primitive neuroectodermal cell lines.[39] Medulloblastomas occur with increased frequency in the basal cell nevus syndrome (Gorlin syndrome).

The vast majority of medulloblastomas arise from the inferior vermis in the region of the nodulus and are thus midline (axial) masses.[39] The tumor displaces the fourth ventricle superiorly and anteriorly. Moderate to severe obstructive hydrocephalus occurs in 80 to 90 percent as the tumor compresses or invades the fourth ventricle.[58] Very rarely the tumor may enlarge the fourth ventricle. In this situation the differentiation from ependymoma may be very difficult.

Pathologically the usual tumor is round or smoothly ovoid in shape. A rim of surrounding peritumoral

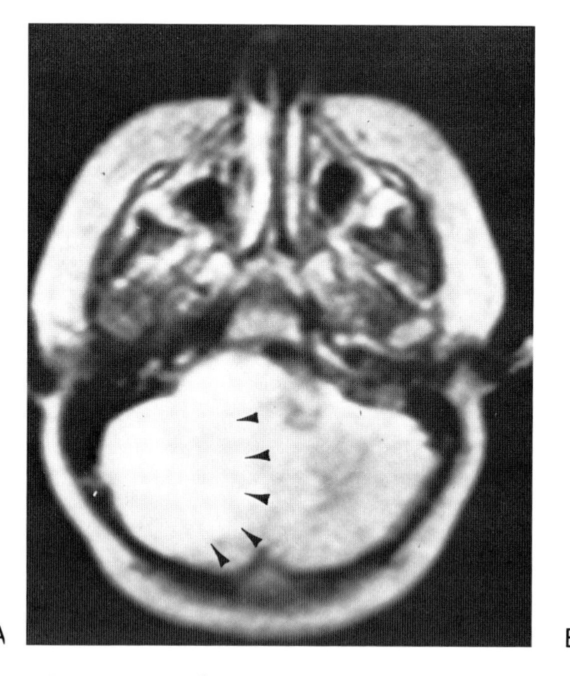

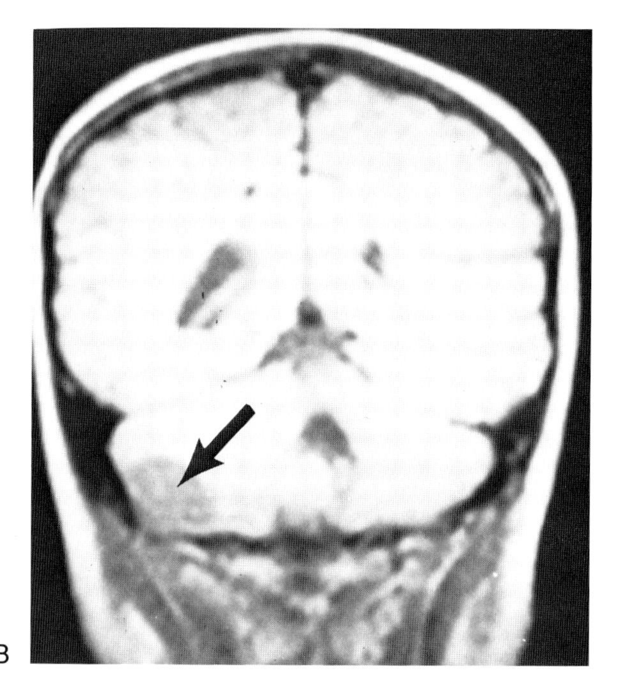

Fig. 6-22. Gangliocytoma. **(A)** Margin of edema is poorly defined on this axial SE 3000/80 image performed on first generation resistive scanner. **(B)** SE 1000/30 coronal image displays cystic component of tumor well.

edema is characteristic. In about 10 percent of cases small cystic areas may be seen, representing true cysts rather than necrosis.[57] About the same number of patients will demonstrate calcifications within the tumor. Usually these calcifications are small, homogeneous, and eccentric and may not be seen by MRI. Small hemorrhages within the tumor are uncommon.

Several isolated examples of medulloblastoma have appeared in the MR literature.[13] In general these tumors display long T1 and T2 values and have a nonspecific appearance. MR may be helpful in defining the epicenter of the tumor to lie in the inferior vermis, helping to suggest the diagnosis (Fig. 6-23). Medulloblastoma metastasizes along the CSF pathways in about 10 percent of cases, and MR studies in these patients should be inspected for ependymal seeding.

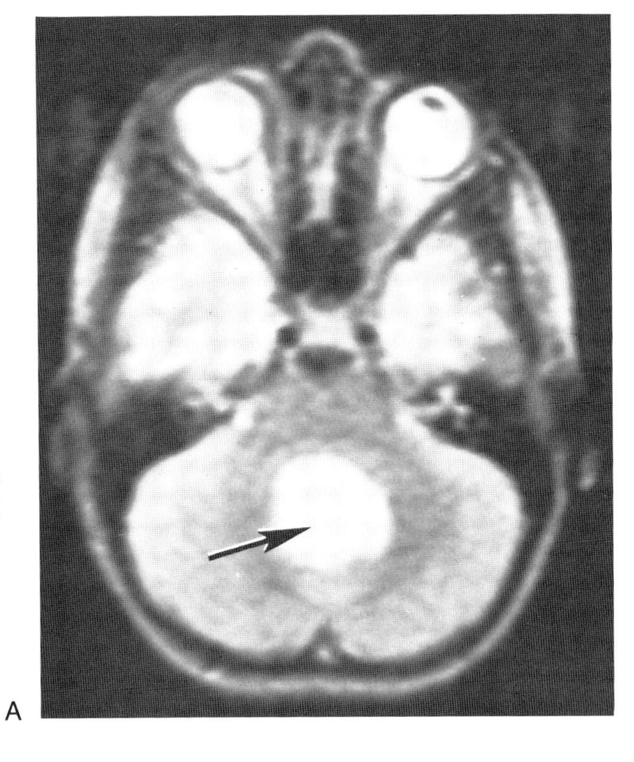

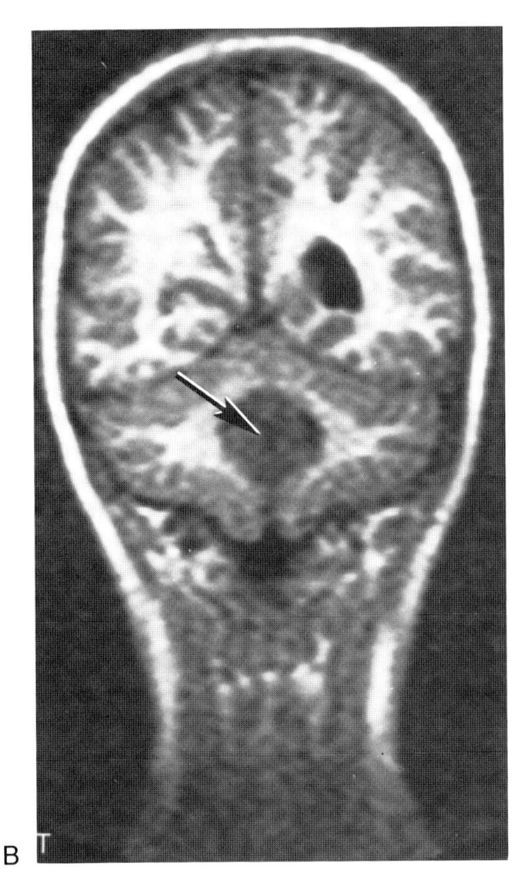

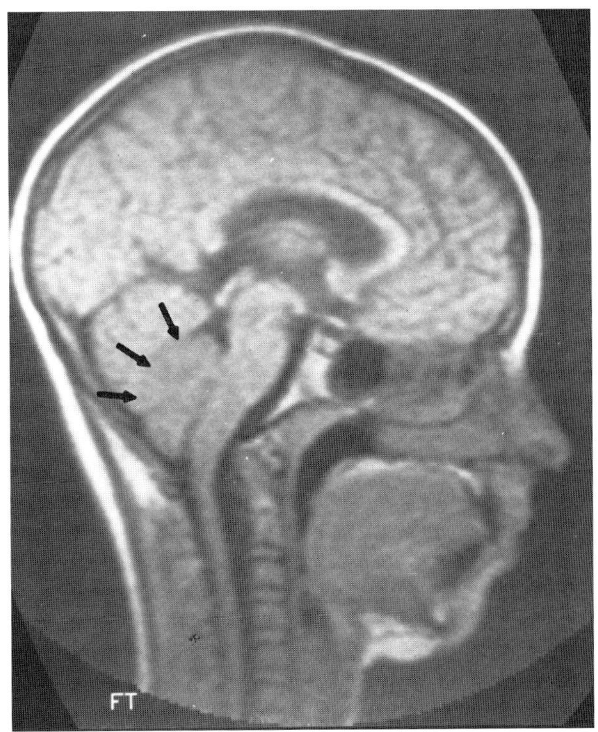

Fig. 6-23. Medulloblastoma. A lesions with long T2 arises from the inferior vermis. Sagittal image is helpful in suggesting site of origin. **(A)** SE 3000/100. **(B)** IR 1400/400/40. **(C)** SE 450/26.

Hemangioblastomas

Hemangioblastomas are histologically benign neoplasms of vascular endothelium that are usually seen in the cerebellum of young and middle-aged adults.[59,60] Macroscopically the tumors are well circumscribed without a capsule. About 60 percent are cystic and contain a highly vascular mural nodule. The tumor may be associated with erythrocytosis and is the primary feature of von Hippel-Lindau syndrome. This latter group of patients frequently have multiple lesions.

MR scanning is nonspecific when the tumor is of the solid variety. When cystic and containing a mural nodule the diagnosis may be strongly suggested, although cystic astrocytomas may have an identical appearance (Fig. 6-24). MRI is useful in von Hippel-Lindau to demonstrate occult spinal and brain stem lesions (Fig. 6-25).

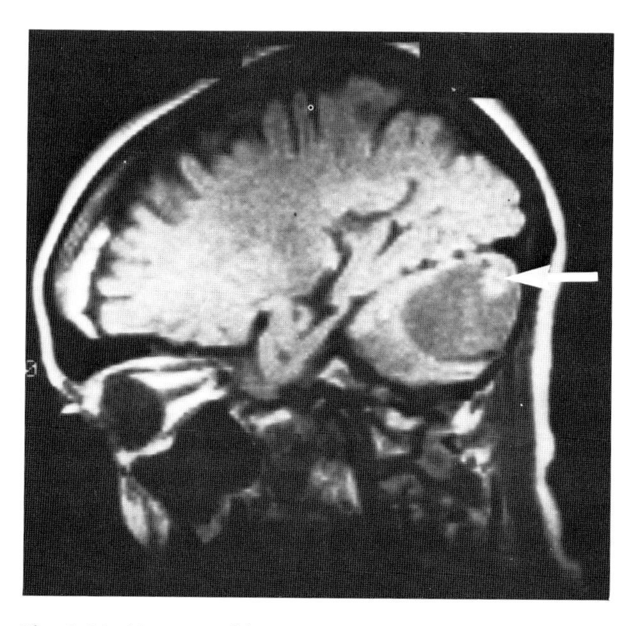

Fig. 6-24. Hemangioblastoma, a cystic tumor with characteristic mural module (arrow). Note, however, that cystic astrocytomas may have a similar appearance.

Lymphoma

Primary lymphoma is an uncommon tumor in the CNS, but its incidence is increasing particularly among allograft recipients and AIDS patients.[61] The cell of origin is uncertain. Diffuse histiocytic (large cell) lymphoma is the most common intracerebral form, while diffuse poorly differentiated lymphocytic lymphoma is the most common meningeal form.[62] Hodgkin disease and other non-Hodgkin lymphoma types are rare intracranially.

Intraparenchymal deposits tend to be located in the periventricular regions, basal ganglia, corpus callosum, thalami, and cerebellar vermis.[62] Cortical and brain stem lesions are also well described. Lymphoma in AIDS patients shows a preference for the frontal lobes.[63] The lesions in CNS lymphoma are relatively large, round or oval masses that are usually sharply demarcated. Perimoral edema is usually moderate but is highly variable. Multiple lesions are seen in 43 percent.[64]

The MR appearance is nonspecific, with focal areas of increased T1 and T2 (Figs. 6-26 to 6-28). Multiplicity of lesions and dimensions of posterior fossa and brain stem lesions may be better appreciated by MR than CT. Diagnosis still requires biopsy, however.

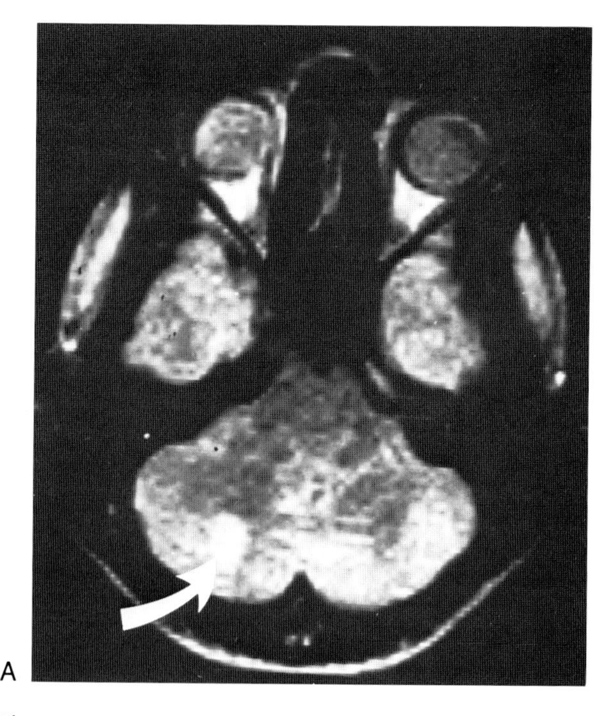

Fig. 6-25. Von Hippel-Lindau disease with multiple hemangioblastomas. The appearance of these lesions are nonspecific. **(A)** First patient, single lesion, SE 2000/56. (*Figure continues*).

Fig. 6-25 (*continued*). **(B)** Second patient, three lesions, SE 3000/100.

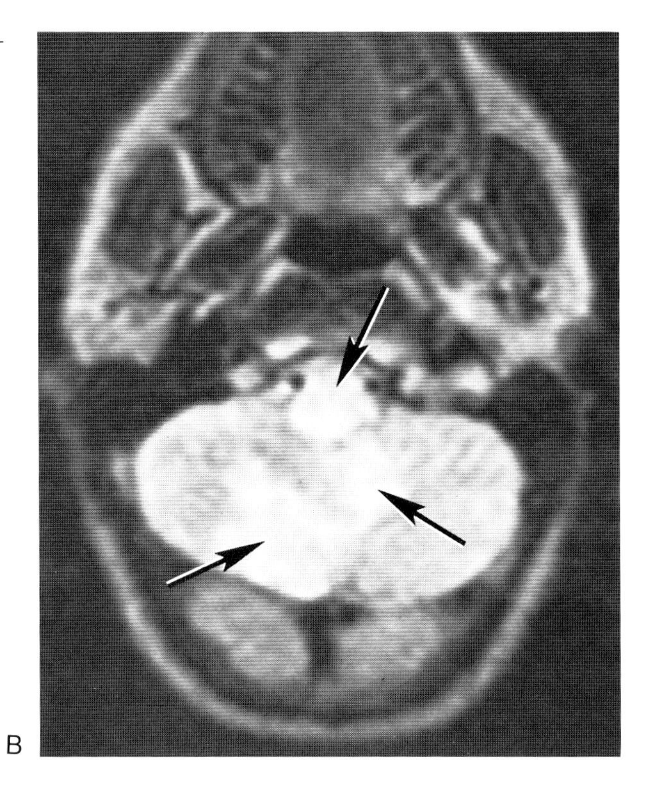

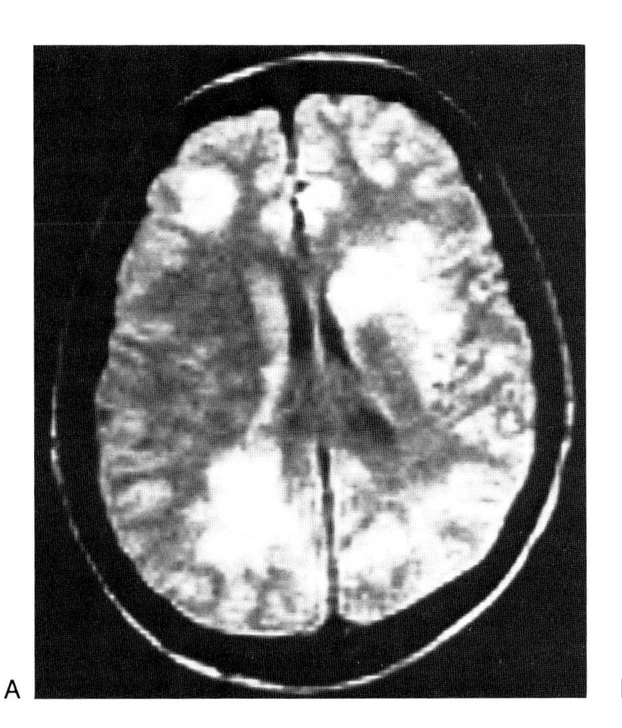

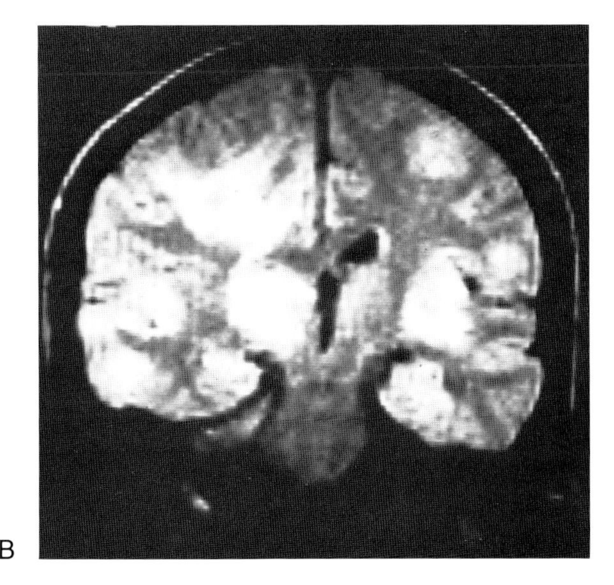

Fig. 6-26. Lymphoma in an AIDS patient. **(A)** SE 2000/28. **(B)** SE 1500/28.

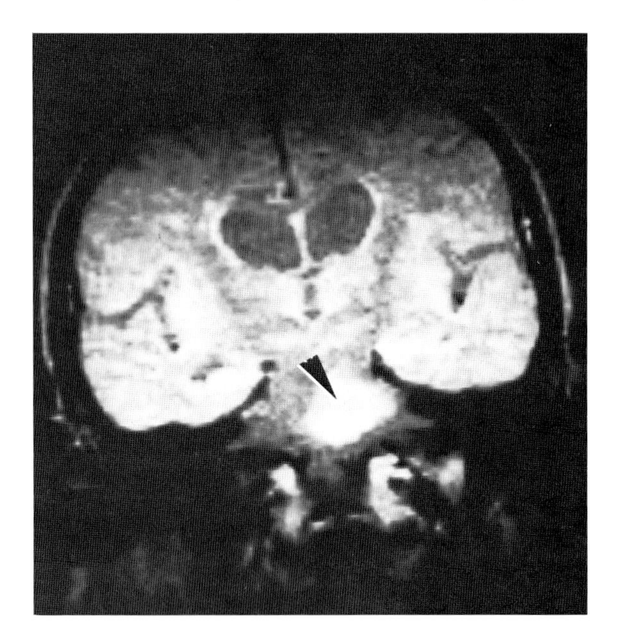

Fig. 6-27. Large cell lymphoma of the pons, SE 1500/56.

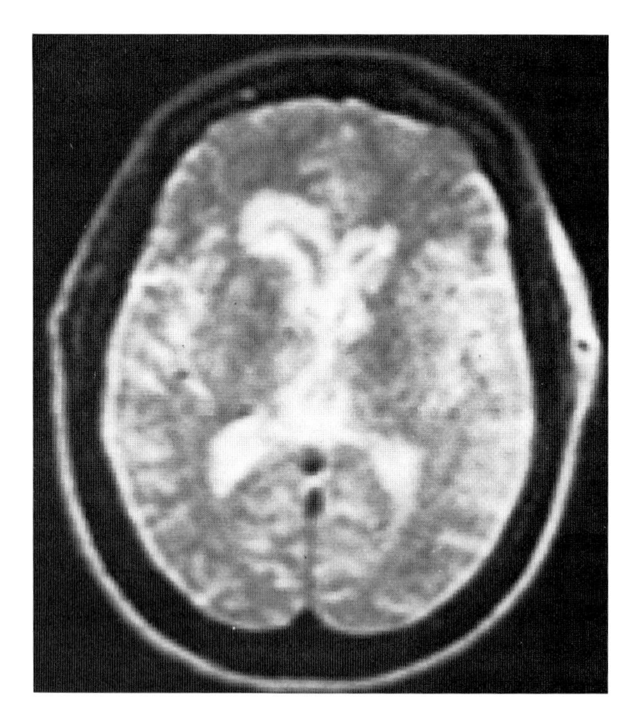

Fig. 6-28. Periventricular lymphoma, SE 3000/100.

TUMORS OF THE PINEAL REGION

The pineal body or epiphysis is a small cone-shaped structure attached to the roof of the third ventricle overlying the tectum of the midbrain. Its function in humans is not fully known, but is believed to be involved with diurmal rhythms, sleep-wake cycles, and gonadal cycling.[65] Microscopically the pineal is composed of glial cells, vascular tissue, large parenchymal cells (pineocytes), and small parenchymal cells (pineoblasts) that resemble lymphocytes and may be immature precursors of pineocytes. Additionally, the pineal body includes midline inclusions of multipotential germinall cells as well as more mature derivatives of germ cell layers (endoderm, ectoderm, and mesoderm). This large number of cell types within the tiny gland account for the varied tumors that may occur in the pineal region.

Tumors of the pineal region are uncommon, comprising less than 1 percent of adult and 5 percent of pediatric brain neoplasms in most large series. The confusing term *pinealoma* has been used in the older literature to denote any pineal region tumor. This nomenclature should be abandoned in favor of the more modern scheme of Russell and Rubinstein based on cell of origin (Table 6-4).

The normal pineal gland is easily visualized on axial, coronal, or sagittal MR scans. The gland is best seen on high resolution, thin section sagittal images. In

Table 6-4. Classification of Pineal Tumors

Tumors of germ cell origin
Germinoma
Teratoma
Malignant teratoid teratomas (embryonal cell carcinoma, choriocarcinoma, endodermal sinus tumor)
Tumors of pineal cell origin
Pineocytoma
Pineoblastoma
Tumors of other cell origin
Cysts (epidermoid, dermoid, nonneoplastic)
Gangliocytoma
Glioma
Lipoma
Melanoma
Meningioma
Metastasis
Primitive neuroectodermal tumor
Vascular tumor (AVM, hemangioma)

Fig. 6-29. (**A** and **B**) The normal pineal on sagittal SE 500/30 image.

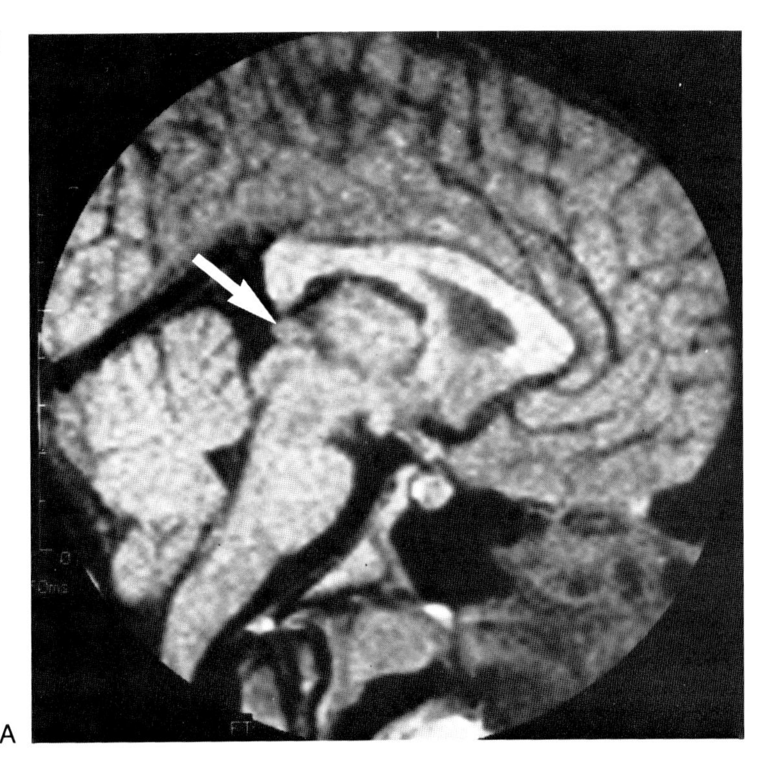

A

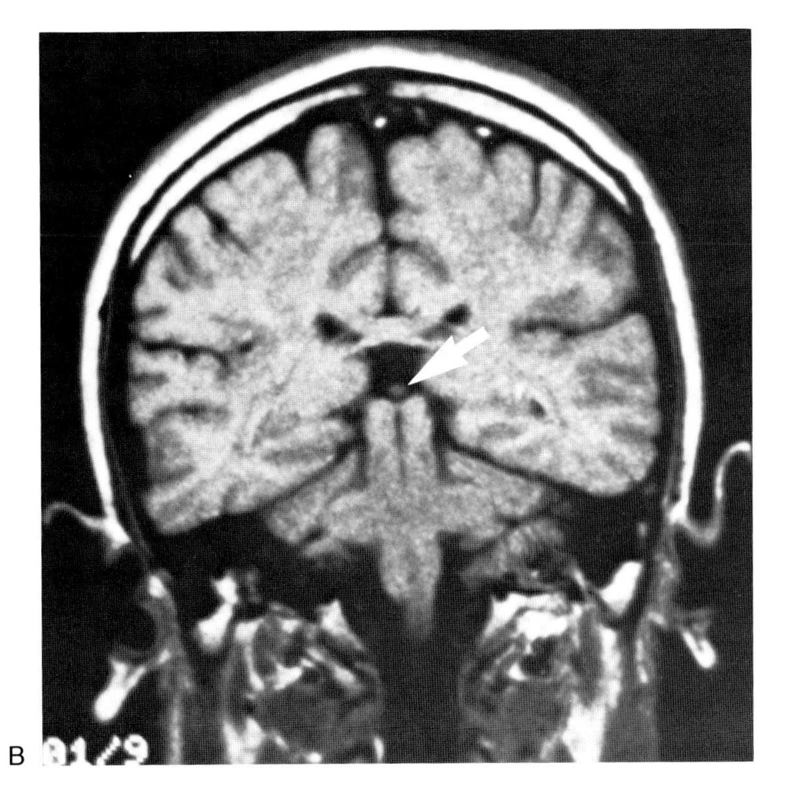

B

about half of cases it is located just anterior to the inferior apex of the spenium and superior to the tectum.[66] In the other half of cases it is more posteriorly located, lying posterior to the superior colliculus.

The normal pineal gland has a rather homogeneous appearance on MRI, with T1 and T2 values similar to gray matter. Even when a gland appears heavily calcified on CT, little or no discernable signal changes can be detected by MR.[66]

In about 10 percent of "normal" glands focal areas of long T1 and T2 may be identified (Fig. 6-29). It has been speculated that these lesions represent benign pineal cysts frequently encountered at autopsy.[66] Such cysts occur in about 10 to 15 percent of patients, contain clear fluid, and are bounded by dense neuroglial fibrils. Because pineal neoplasms also display similar increased T1 and T2 values, clinical history, CT correllation, or follow-up is essential.

Tumors of Germ Cell Origin

Germ cell neoplasms (germinoma, teratoma, and malignant teratoid teratoma) account for about three quarters of pineal region tumors.[68,69] They are most commonly seen in males in the teenage or young adult years. Tumors of identical histology may also occur in the suprasellar-anterior third ventricle area, sometimes coexisting with pineal tumors. These distant tumors are often called *ectopic pinealomas*.

Germinomas

Germinomas (atypical teratomas) are the single most common pineal tumor, accounting for over 50 percent of all enoplasms in the area.[70] They occur almost exclusively in males with a 10:1 male to female predominance. The peak incidence is from 15 to 25 years, but these tumors can occur even in the fifth decade.

Histologically, pineal germinomas are highly cellular neoplasms resembling ovarian dysgerminomas and testicular seminomas.[39] They are well defined and typically present by obstructing the cerebral aqueduct. Hemorrhage, necrosis, and cystic degeneration are uncommon. Peritumoral edema is minimal or absent. Engulfment of a prominently clacified and slightly enlarged pineal is said to be characteristic of germinomas on CT.[70] The tumors are vascular and enhance intensely with intravenous contrast. The tumors are malignant histologically, but show a good response to radiotherapy. Infiltrative spread along the

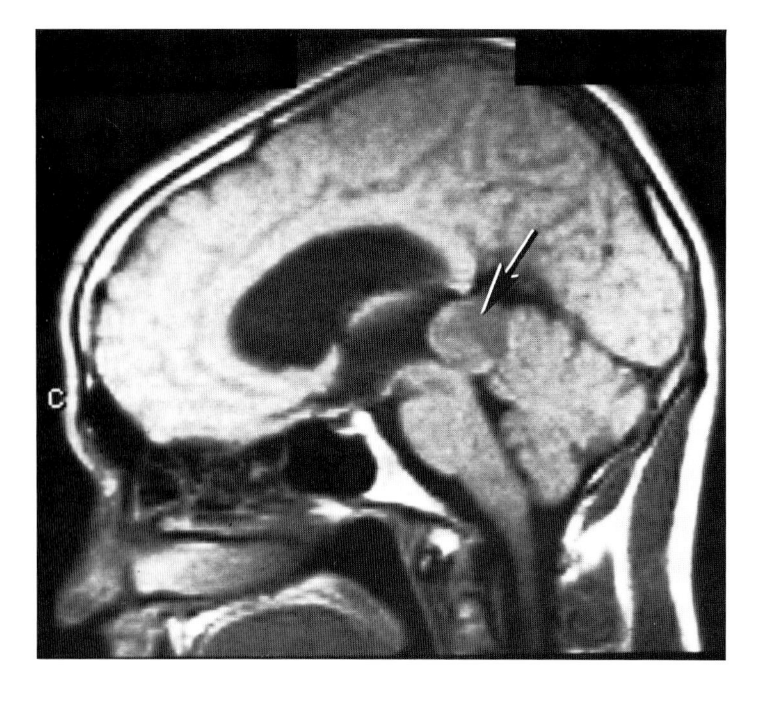

Fig. 6-30. Pineal germinoma. Most have T1 and T2 values similar to gray matter and are distinguished only by shape.

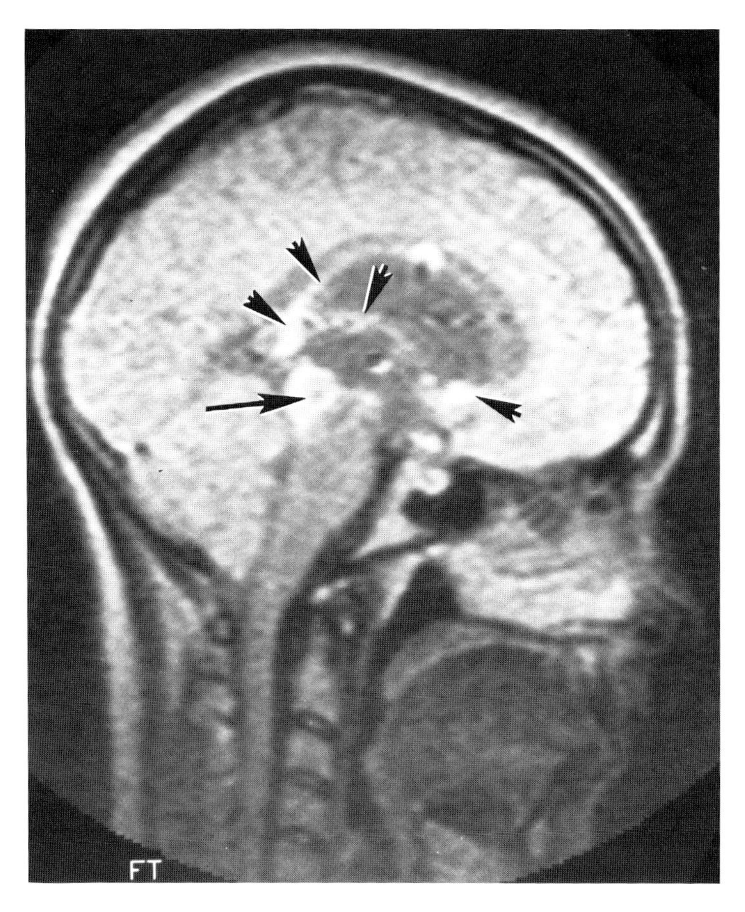

Fig. 6-31. Pineal germinoma, malignant with prolonged T2 values (long arrow). Note subependymal seeding (arrowheads).

floors and walls of the third ventricle is common. Seeding of the subarachnoid spaces may occur.

On MR, the less aggressive pineal germinomas are relatively sharply defined, with T1 and T2 values similar to normal brain (Fig. 6-30). The calcified, engulfed pineal may occasionally be identified as a region of low signal within the tumor mass (Fig. 6-31). Subependymal spread may also be assessed using MRI (Fig. 6-31).

Teratomas

Typical benign teratomas are relatively common pineal region tumors, also most frequent in young males. Their gross morphology may overlap that of germinomas, in which case differentiation by CT or MR may be impossible. Usually, however, several gross features allow these two tumors to be distinguished. Teratomas are more often cystic or multicystic than germinomas, which are usually solid. Calcification may be very prominent in teratomas, and formed elements like teeth are sometimes found. A fatty component is often present and may be dominant. Fat has short T1 and moderate T2 values, with a characteristic high intensity on T1-weighted MR images. When seen on MR fat in a pineal tumor should suggest the diagnosis of teratoma (Fig. 6-32), although lipomas and malignant teratoid tumors may also contain fat, causing confusion.

Malignant teratoid teratomas include *embryonal cell carcinomas*, *choriocarcinomas*, and *endodermal sinus tumors* (yolk sac carcinoma). Histologically, mixtures of the three subtypes may be present within a single tumor. Unlike benign teratomas, the malignant teratoid tumors are less likely to contain fat or be cystic. These tumors are markedly hypervascular. Intratumoral hemorrhage is common. Calcification may occur, especially in embryonal cell carcinoma. No specific MR features have been identified in these tumors.

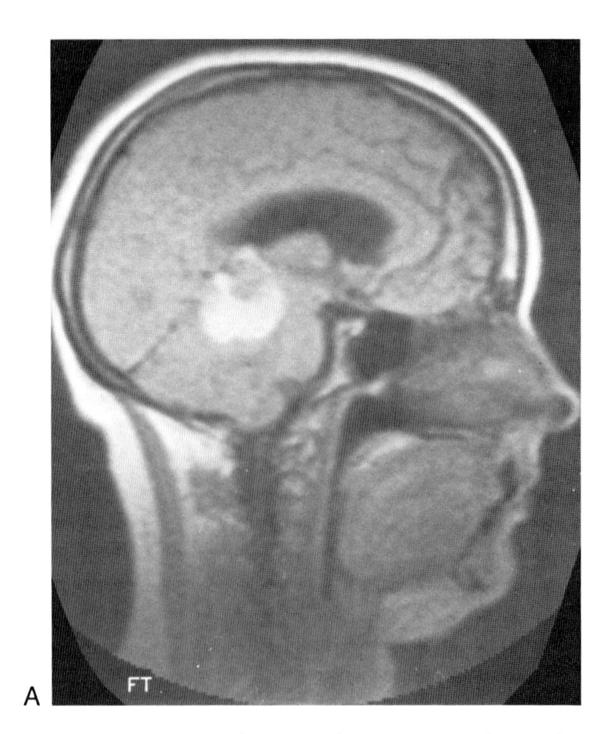

A

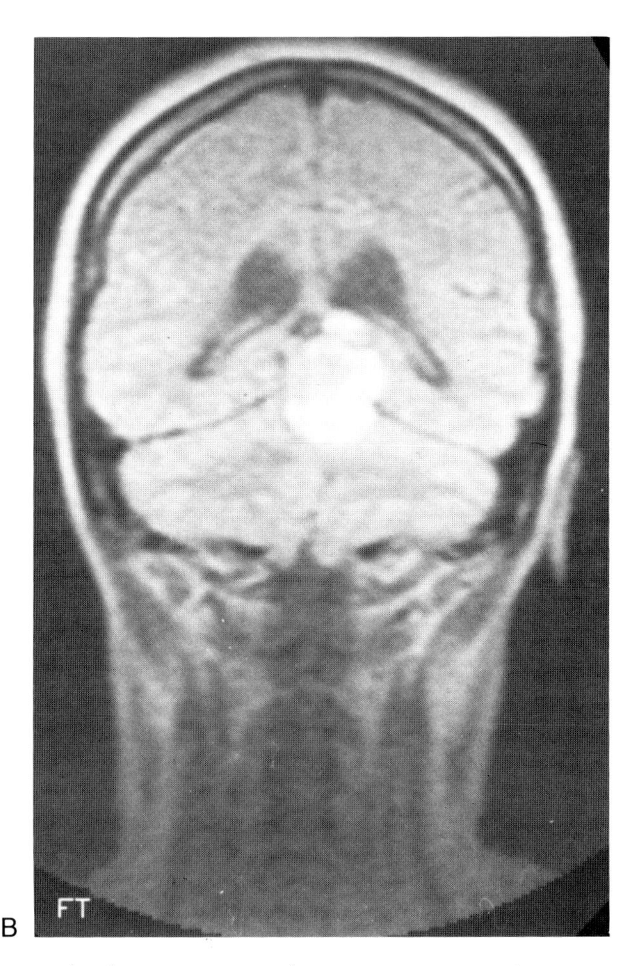

B

Fig. 6-32. (**A** and **B**) Pineal teratoma. High signal on T1-weighted images suggests fatty component, consistent with a teratoma.

Tumors of Pineal Cell Origin

Tumors arising from pineal cells (pineocytoma, pineoblasoma) are only a third as common as tumors of germ cell origin and demonstrate no sex predilection. Histologically, pineal cell tumors resemble neuroblastoma and medulloblastoma, and are considered primitive neuroectodermal tumors. Pineal cell tumors may rarely coexist with bilateral retinoblastomas in a complex called *triangluar retinoblastoma*.

Pineocytomas

Pineocytomas are relatively benign pineal neoplasms that occur at any age beyond childhood. They are usually well circumscribed masses without calcification, necrosis, cystic change, or peritumoral edema. Occasionally they are heavily calcified and have the appearance of an exceptionally large, calcified pineal gland. Because germ cell tumors have such a strong male predilection, a calcified pineal tumor in a female most likely represents pineocytoma.[70]

Pineoblastomas

Pineoblastomas are highly malignant, composed of densely packed primitive cells. Hemorrhage and calcification are rare but central cystic degeneration is occasionally observed.[70] Grossly their appearance and behavior is similar to that of the malignant teratoid teratomas. Pineoblastomas have a tendency to spread along the ependymal surfaces of the third ventricle. Extension to involve the cerebellar vermis can be seen. The tumors are highly vascular, and intense contrast enhancement on CT is the rule. Cases seen on MR have demonstrated prolonged T1 and T2 values (Fig. 6-33).[71]

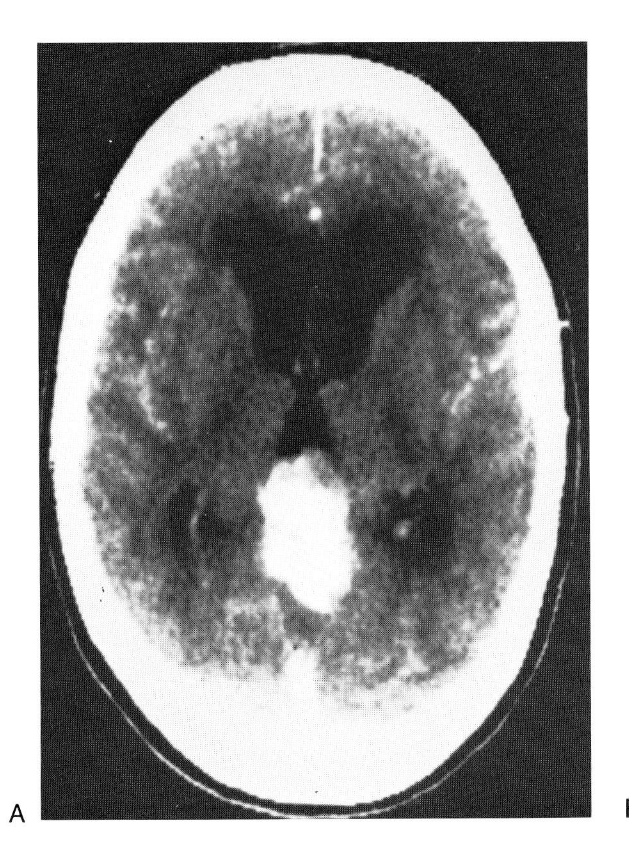

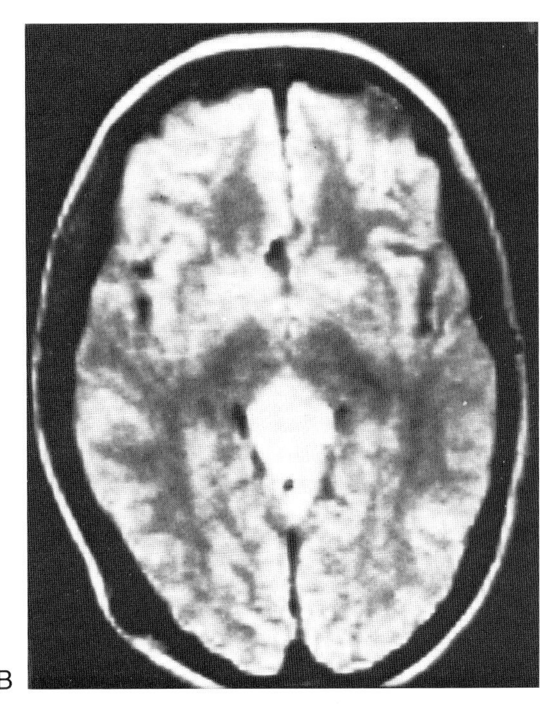

Fig. 6-33. Pineoblastoma. Prolonged T2 values are seen in more malignant and primitive pineal tumors. **(A)** CT scan, **(B)** SE 2000/28, **(C)** SE 1500/56 (*Figure continues*).

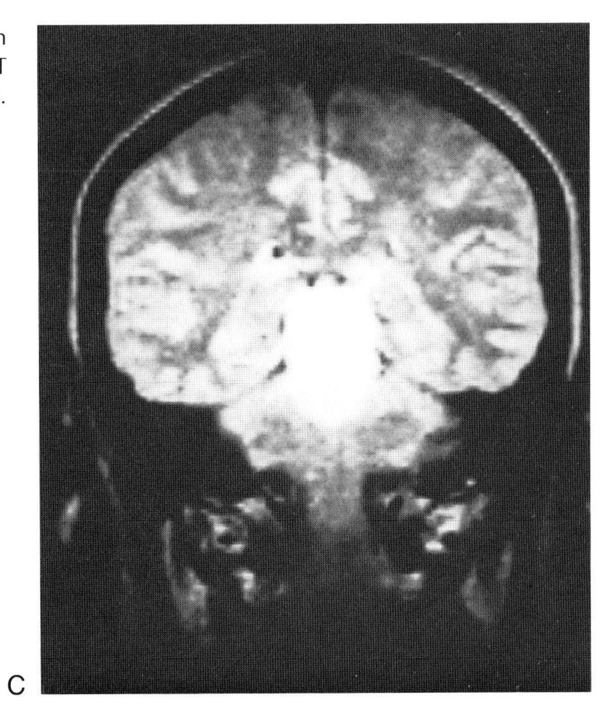

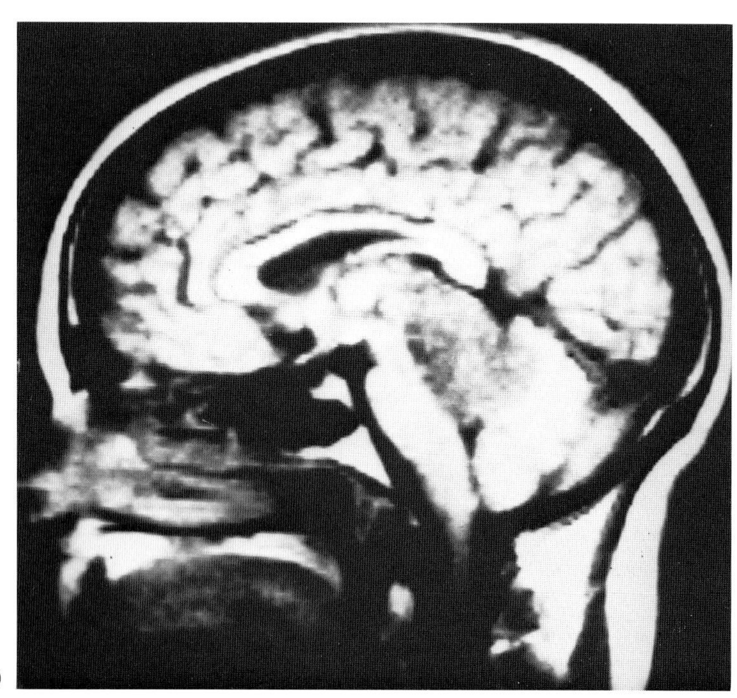

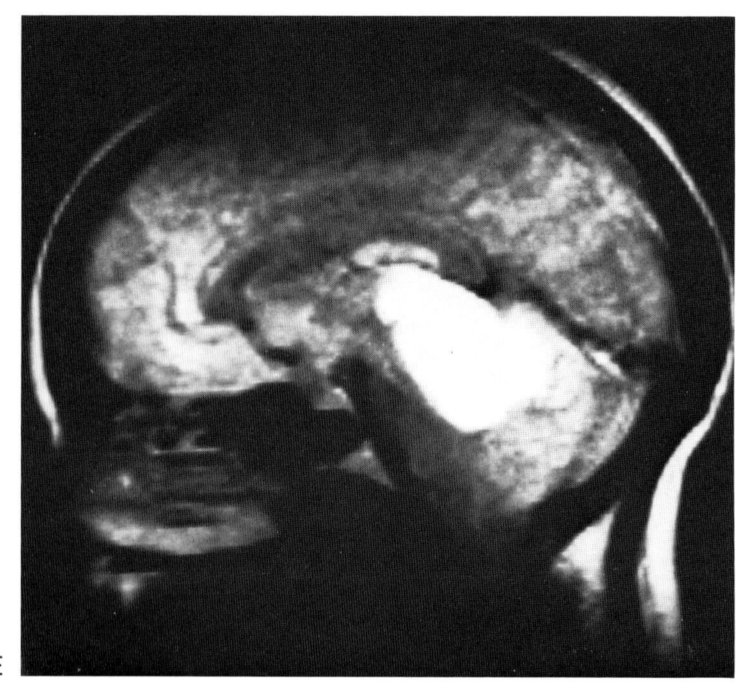

Fig. 6-33 *(continued)*. **(D)** SE 500/28, and **(E)** SE 1500/56.

Role of MR for Pineal Tumors

The inability of MRI to detect pineal calcification consistently limits its primary role in evaluation of pineal region tumors. Particularly in children, abnormal pineal calcification may be the only sign of a pineal tumor.[67] Therefore, MR should not be used as a screening modality for pineal tumors, particularly in the pediatric population.

Magnetic resonance is, however, well suited for delineating certain anatomic relationships of suspected pineal tumors. On axial CT there are times when it is impossible to differentiate tumors arising from the pineal from those of the tectum, splenium, or posterior third ventricle. Sagittal MRI can be most helpful in better defining anatomic relationships and spread of tumor.

The effects of pineal region tumors on adjacent vasculature are better appreciated by MR than by CT. Variations in blood flow alter the MR signal in venous structures near the quadrigeminal cistern. Recognition preoperatively of obstruction of the vein of Galen is important because its occlusion my lead to collateral channels that in turn complicate surgical management.

In summary, therefore, CT should remain the imaging modality of choice for the primary evaluation of pineal tumors. This is true in all age groups, but particularly in children. MR may add significant anatomic and functional information, such as the patency of the surrounding vasculature and of the cerebral aqueduct.

EXTRA-AXIAL NEOPLASMS

Germinal Layer Neoplasms

Inclusion of ectodermal or mesodermal elements within the neural tube during its closure between the 3rd and 5th weeks of gestation may permit development of an intracranial germinal layer neoplasm.[72] If only ectoderm is incorporated, an *epidermoid* tumor results. If both ectoderm and mesoderm are incorporated, a *dermoid* tumor results. Inclusion of only mesodermal tissue may result in an intracranial *lipoma*. If all three germinal layers are incorporated into the neural tube, a *teratoma* may be formed.

Epidermoids

Epidermoids (also called pearly tumors or cholesteatomas) are the most common of the intracranial germinal layer neoplasms, accounting for about one percent of primary brain tumors in most series.[39] Most patients are between 30 and 60 years old at presentation. Males are affected twice as frequently as females.[64]

Epidermoids are slow-growing, lobulated, extra-axial tumors filled with a semisolid fatty or waxy material resembling cottage cheese. This material is composed of epithelial debris, keratin, and cholesterin, the latter being a lipid. Common sites for epidermoid tumors include the cerebellopontine angle, suprasellar region, and middle cranial fossa, in that order. Less common locations include the interhemispheric tissue, cerebral hemisphere, ventricle, and temporal bone. Clinically, epidermoid cysts usually present as a mass causing headaches or seizures. Chemical meningitis may occur if the epidermoid ruptures into the subarachnoid space.[73]

Despite their lipid content, short T1 values are not frequently observed in epidermoids on MRI.[11,74–76] In the usual case the tumors have signals intermediate between brain and CSF (Fig. 6-34). Inhomogeneous texture with variable internal signal is characteristic. Insinuation of the tumor around the brain stem and into the subarachnoid cisterns may be observed (Fig. 6-35). MR with its multiplanar imaging capacity could evaluate mass effect, total extent of tumor and vascular displacement better than CT.

For unclear reasons, some epidermoids have a much fattier composition than the usual type. These may demonstrate very low density on CT (−100 HV) (Fig. 6-36A). These same tumors have a very short T1 and appear bright on T1-weighted images (Fig. 6-36B). When rupture into the subarachnoid space occurs, high intensity fat on T1-weighted images may sometimes be observed in the ventricles or over the cerebral convexities (Fig. 6-37).

Magnetic resonance imaging may aid in distinguishing epidermoids from arachnoid cysts, when both have approximately CSF density on CT. The arachnoid cyst will practically always possess MR signal characteristics identical to CSF, while the T1 values of the epidermoid will nearly always be shorter. However, the differential dilemma of epidermoid and cysticercus cyst may still remain.

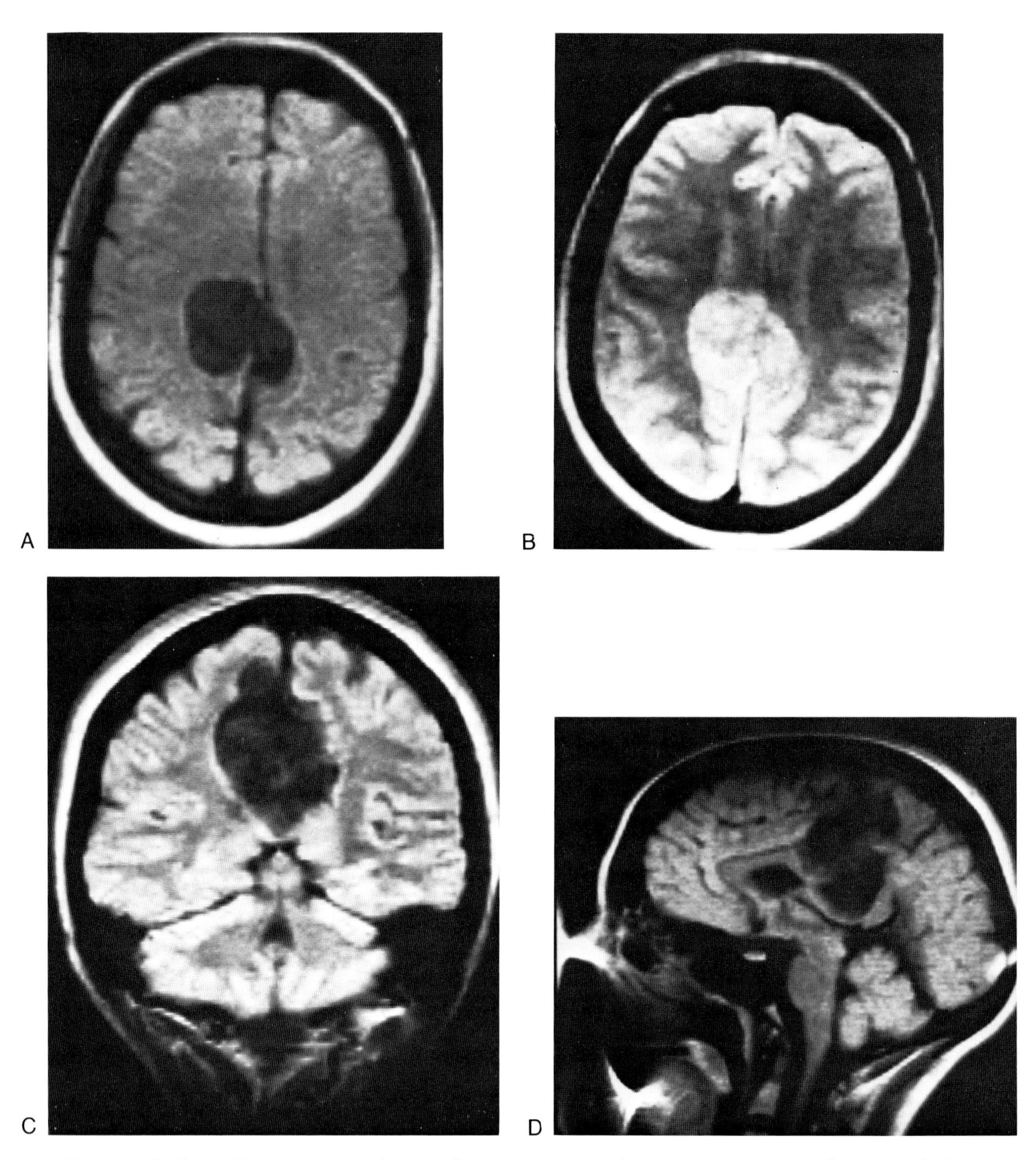

Fig. 6-34. Epidermoid in a 51-year-old man with seizures. Despite fatty appearance on CT, this tumor displayed prolonged T1 and T2 values. **(A)** SE 100/28, **(B)** SE 2000/56, **(C)** SE 1500/28, and **(D)** SE 1000/28.

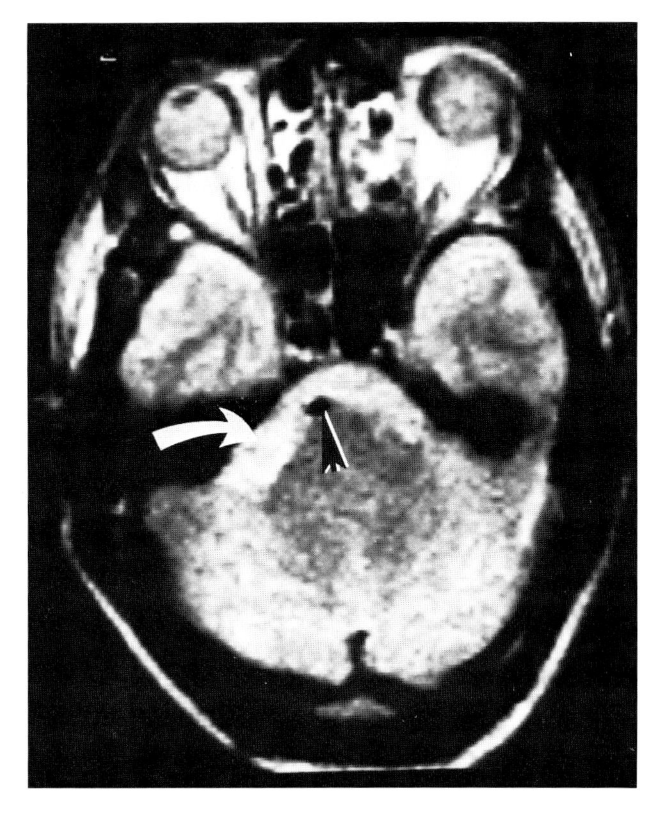

Fig. 6-35. Epidermoid of the right cerebellopontine angle, which drapes around the brain stem, seen on SE 2000/56 image. Note encasement of basilar artery (arrowheads).

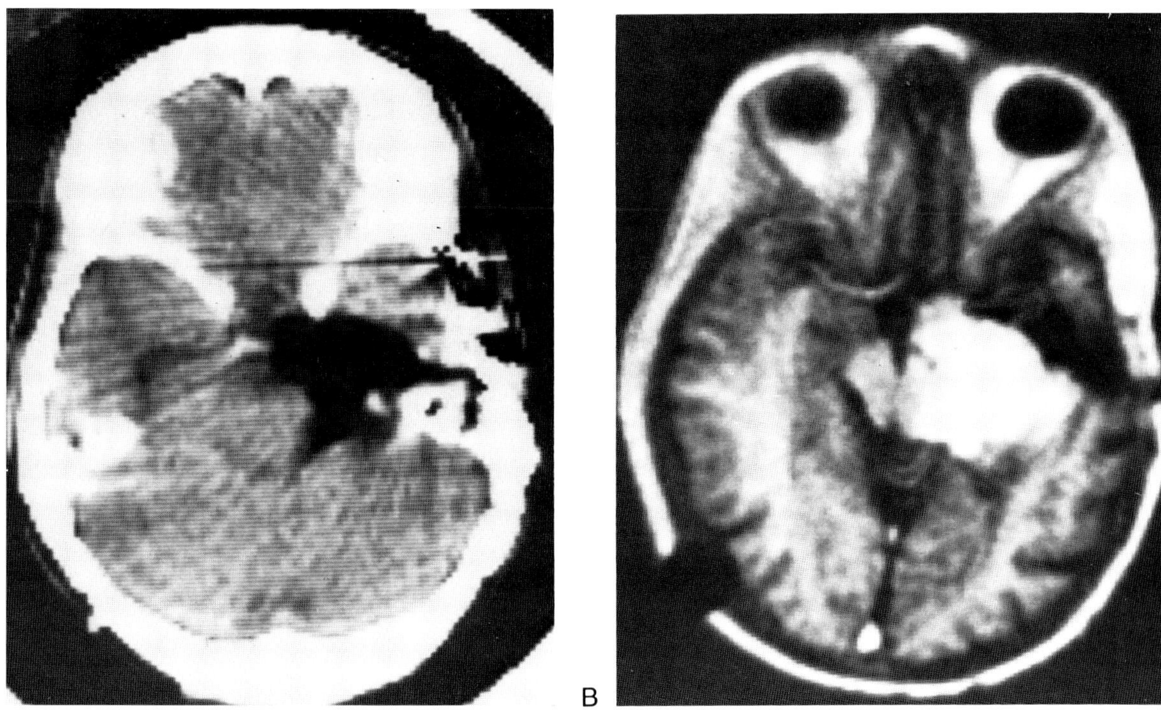

Fig. 6-36. (**A** and **B**) Fatty epidermoid on CT and MRI. This epidermoid has a short T1 consistent with fat.

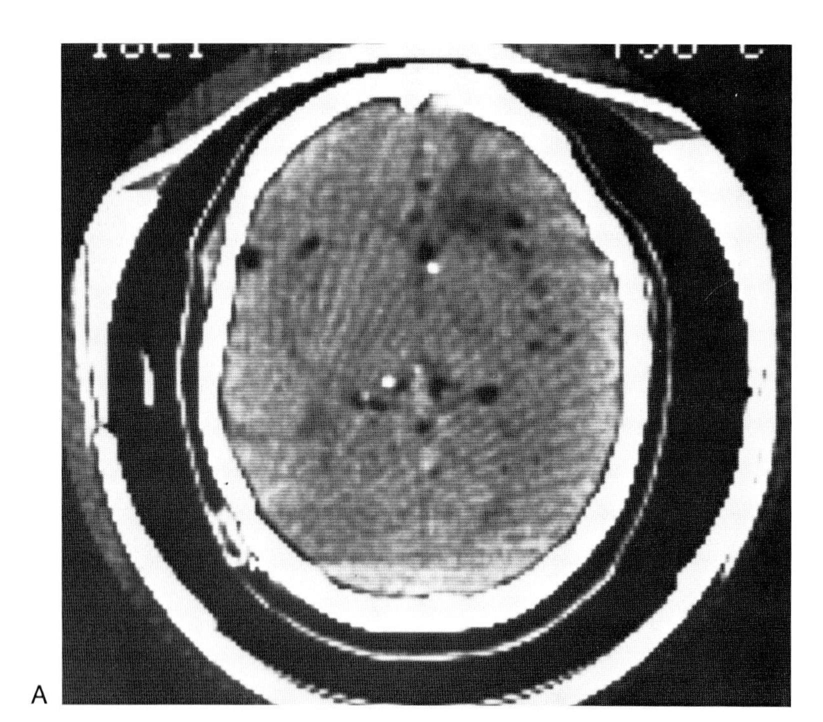

A

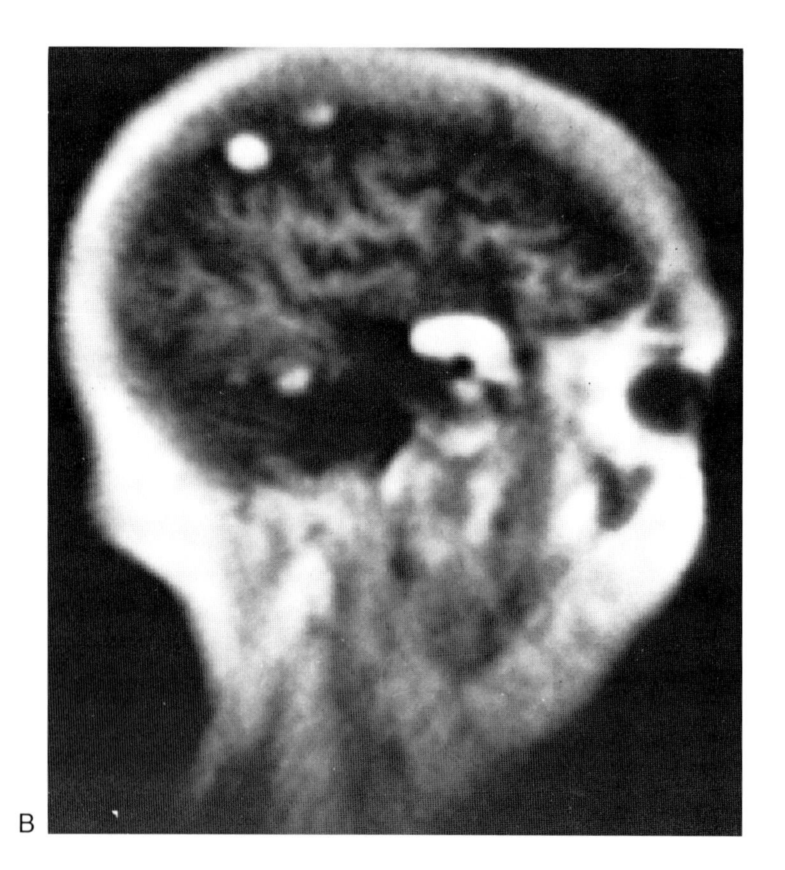

B

Fig. 6-37. (**A** and **B**) Ruptured epidermoid. Fatty material is scattered in the fissures and sulci.

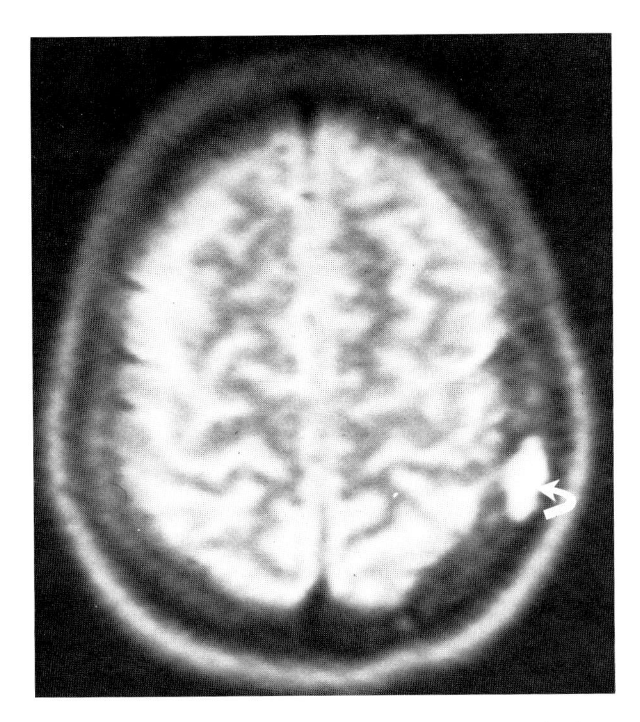

Fig. 6-38. Calvarial epidermoid. Nonspecific MR findings with increased T2 value.

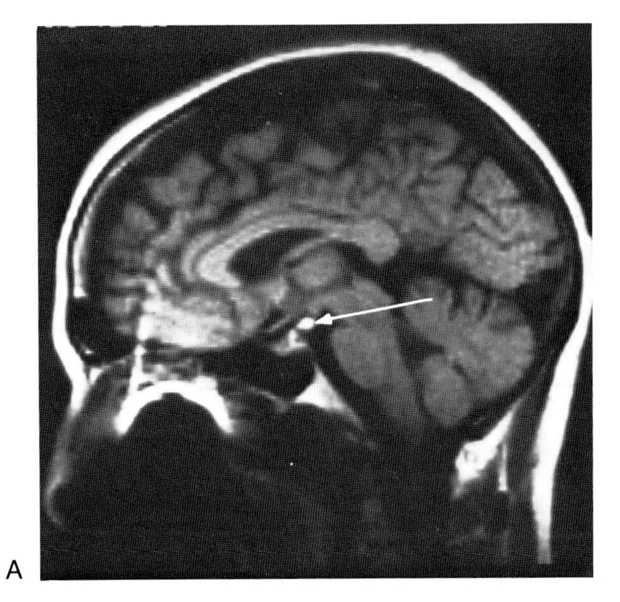

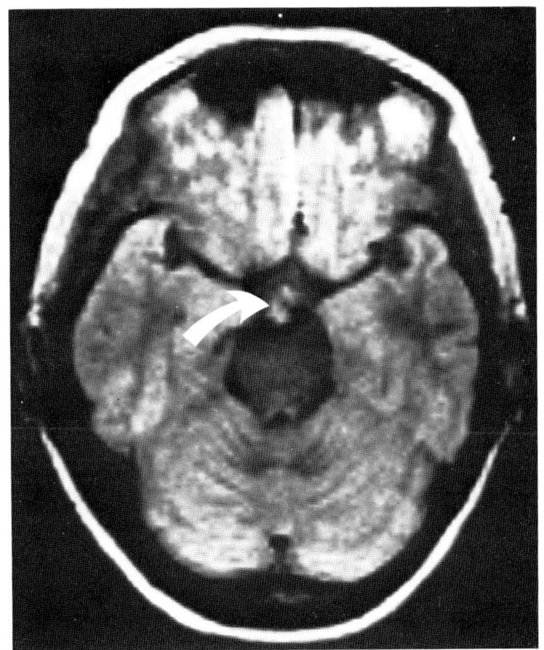

Fig. 6-39. Parasellar dermoid. On T1 weighted images the bright signal of fat is detected. Calcification seen on CT confirmed the diagnosis of dermoid tumor. **(A)** SE 500/28. **(B)** SE 2000/28.

Extradural epidermoids are characteristically well marginated radiolucent defects with sclerotic edges seen on skull x-ray. In the few cases we have seen on MR, these deposits were easily visualized within the diploic space, and possessed a rather high signal on T2-weighted images (Fig. 6-38). The MR appearance, however, was nonspecific requiring confirmation by routine radiography to exclude a significant lesion such as a metastasis.

Dermoids are less common intracranially than epidermoids, and contain both ectodermal and mesodermal elements. Dermoids are seen primarily in the first three decades of life and have a female predominance.[36] The lesions may be asymptomatic or present with mass effect like epidermoids. Frequent locations include the parasellar region, midline subfrontal area, quadrigeminal cistern, and interhemispheric fissure.[72]

In contrast to epidermoids, dermoid tumors frequently contain frank fat and calcium. In nearly all reported cases, dermoids have exhibited short T1 values with a corresponding high signal on T1-weighted images (Fig. 6-39).[11,74] Fatty epidermoids, dermoids, and lipomas may have identical MR appearances with short T1. Therefore, CT is needed to make a reliable diagnosis of dermoid tumor by demonstrating calcification within the lesion.

Lipomas

Lipomas are benign fatty tumors caused by incorporation of mesoderm during neural tube closure. They are exclusively midline lesions and have MR signal characteristics similar to subcutaneous fat.[74] intra-axial lipomas are frequently associated with callosal agenesis and have been presented more completely in Chapter 5. Extra-axial lipomas, presented here, are much less common, and usually are found in the quadrigeminal cistern or interhemispheric fissure (Fig. 6-40). Again, CT is necessary to exclude dermoid tumors, which frequently have MR-invisible calcifications associated with them.

Teratomas

Teratomas are congenital neoplasms that contain all three germ layer elements: ectoderm, mesoderm, and endoderm. They are seen most frequently in males during the first two decades of life. About half of these tumors arise in the pineal gland while about 15 percent are found in the juxtasellar regins. The remainder are widely distributed intracranially. Pineal teratomas have been discussed earlier in this chapter. Teratomas of the juxtasellar region are illustrated in Chapter 11.

Meningiomas

Meningiomas are essentially benign, slowly growing tumors that compress the brain by expansion. After gliomas, meningiomas are the most common primary cerebral neoplasm, accounting for 15 to 20 percent in large series. They are thought to arise from clusters of meningothelial cells imbedded in the dura, often associated with arachnoid granulations.[77]

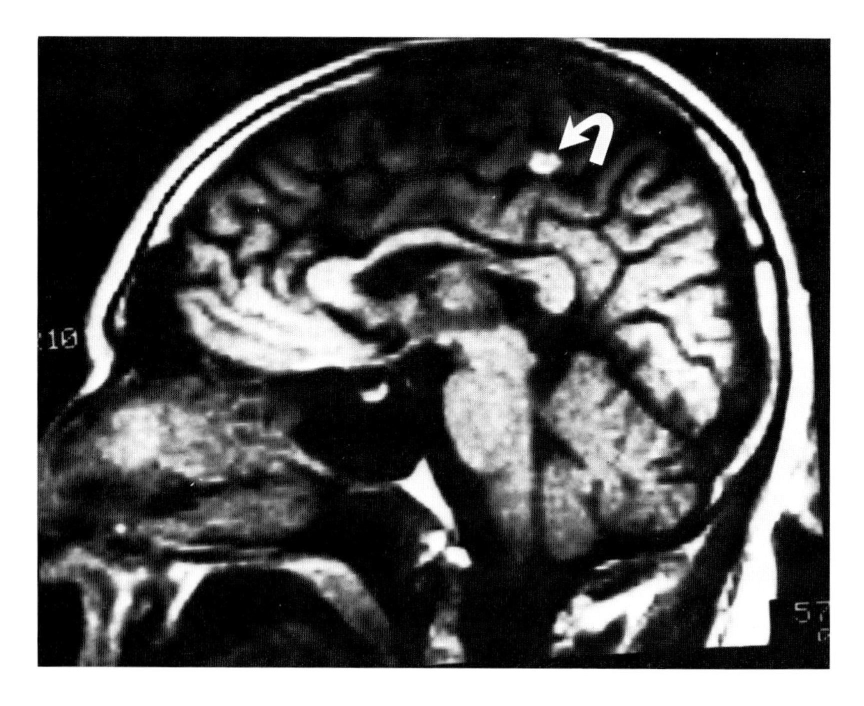

Fig. 6-40. Midline lipoma on this SE 500/28 image.

Meningiomas are primarily tumors of adults. A peak incidence is seen in the fourth through sixth decades, but even children may be affected.[78] Females are affected more often than males. This ratio is especially disproportionate in certain locations, such as the sphenoid ridge, parasellar region, and middle fossa, where meningiomas in women are much more common. Some racial predisposition also exists, because blacks in the U.S. are infrequently affected.

Grossly, meningiomas may present as globular masses or flat, plaquelike lesions. Regardless of shape, they are usually well demarcated and of firm consistency. Meningiomas invade adjacent dura, bones, vessels, and nerves. The brain parenchyma is usually compressed but not invaded. The tumors are usually quite vascular, with a blood supply predominantly or totally of dural origin.[79] Larger lesions may also acquire a pial supply at their periphery while maintaining a dural supply centrally. Predominantly pial supply is rare, but may be seen with extensive invasion of the brain surface. The uncommon intraventricular meningiomas are supplied by choroidal arteries. Arteriovenous shunting is only occasionally seen. Invasion or occlusion of dural sinuses is common.

Most meningiomas are found in the parasagittal regions, over the cerebral convexities, along the falx, or subfrontally (Table 6-5).[80–82] Rarely they may be found in the ventricles or even extradurally. When located in the cerebellopontine angle they may mimic acoustic neuromas. Meningiomas at the foramen magnum are usually located anterior or anterolateral to the neuraxis. Multiple meningiomas are seen in at least 2 percent of patients.[83] Multiple meningiomas as well as the association of meningiomas with neurofibromas or gliomas are common features of von Recklinghausen disease.

Table 6-5. Locations of Meningiomas[a]

Parasagittal and lateral convexity regions	43.4%
Sphenoid ridge	17.7
Subfrontal area	15.5
Posterior fossa	10.9
Falx	6.2
Subtemporal area	2.7
Intraventricular	0.2
Other	1.3

[a] Aggregate of 1241 cases from References 80–82.

Histologically, meningiomas can be classified into several subtypes, in an order corresponding to increasing aggressiveness: fibroblastic, transitional syncytial (meningotheliomatous), angioblastic, and sarcomatous.[39] Tumors containing mixtures of these subtypes may occur. About two-thirds of meningiomas are of the fibroblastic type. These have a heavy connective tissue stroma and frequent psammomatous calcifications. Calcifications may have a number of forms ranging from diffuse and homogeneous to punctate or rimlike. Only moderate edema usually surrounds the less aggressive meningiomas.[84] The more malignant angioblastic and sarcomatous meningiomas tend to be uncalcified, have an inhomogeneous architecture, and produce marked vasogenic edema. Areas of cystic change, necrosis, or hemorrhage are seen in about 20 percent. A wide spectrum of cellular changes may be seen in meningiomas including giant cell infiltration, xanthomatous accumulations as well as bone, cartilage, and melanin formation.

A few meningiomas undergo malignant metaplasia and become invasive.[85,86] They may metastasize within the neuraxis or distantly to the lungs.[87] Pulmonary metastases are most often seen postoperatively after intracranial resection. Meningial sarcomas are highly cellular and malignant throughout their course. A number of sarcomatous subtypes have been distinguished.[88]

The reliable diagnosis of meningiomas by MRI alone has long been fraught with difficulty.[2,12,89] The essential problem is that unlike most other intracranial neoplasms, meningiomas frequently have T1 and T2 values similar to normal brain.[90,91] As a result, intrinsic MR signal contrast between meningiomas and brain may be poor, leading to missed or incorrect diagnoses. Small, plaquelike meningiomas are the most likely to be missed on noncontrast MR, even when their location is known from prior CT. While this error rate depends upon the experience of the reader and numerous technical factors, in good hands, with good equipment, and properly designed MR protocols, fewer than 10 percent of meningiomas should escape detection by MRI.[12] Improved sensitivity and specificity is reported with the use of intravenous gadolinium-DTPA, so even higher detectability rates are expected in the future.[4,29,92]

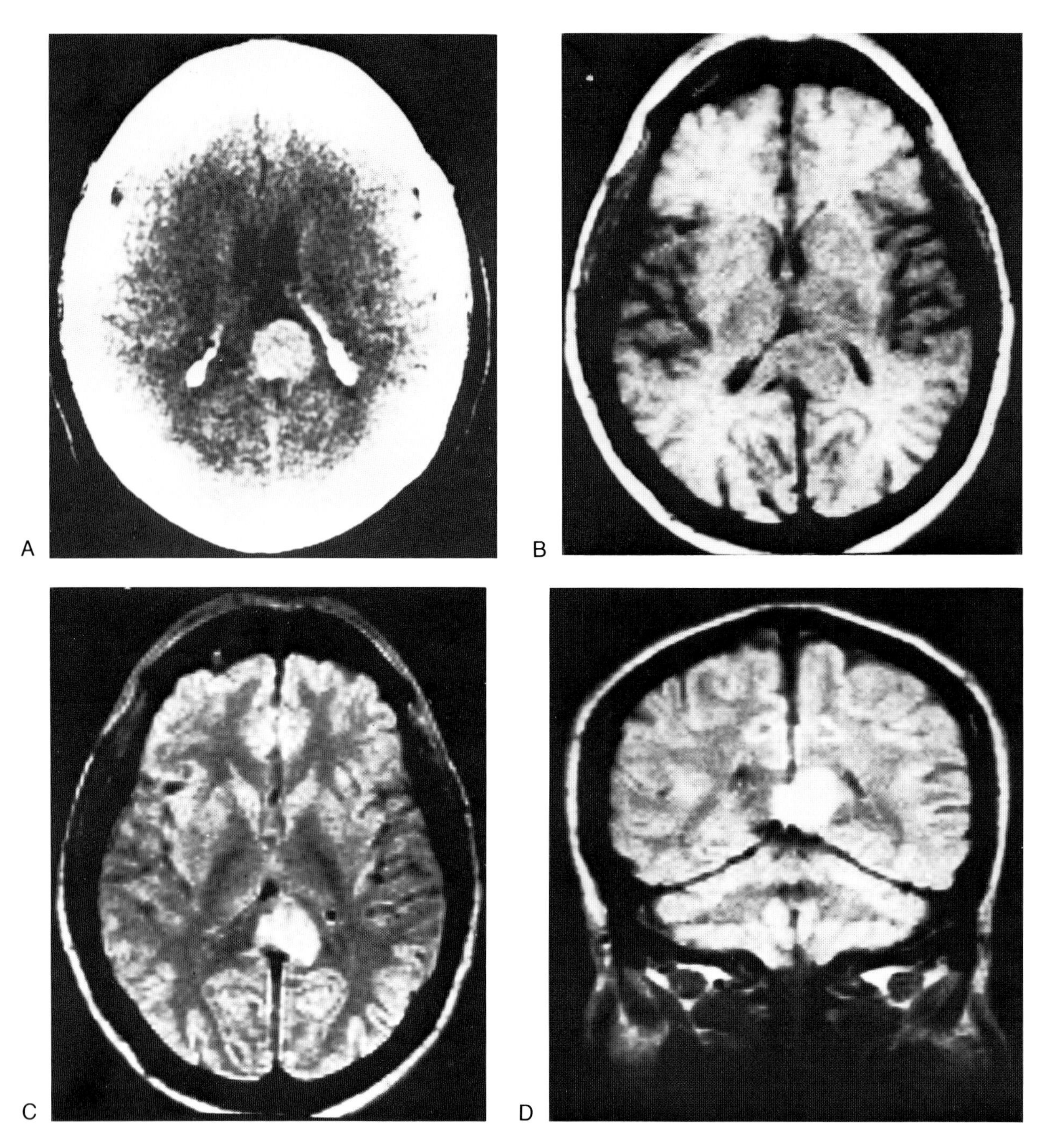

Fig. 6-41. Falx meningioma. **(A)** CT scal, **(B)** SE 500/28, **(C)** SE 2000/28, and **(D)** SE 2000/28.

Computed tomography is clearly superior to MR for evaluating meningiomas in over 50 percent of cases, while in most other situations they are of equal sensitivity. CT is unquestionably better for showing calcification, hyperostosis, and bone erosion, though MR can detect some of these features. However, MR can be superior to CT for evaluating a number of other features of meningioma. First, meningiomas of the posterior fossa, foramen magnum, and skull vertex may be better seen by MR, since bone artifacts and partial volume effects in these regions compromise CT examination. Tumor extent and relationships may be better appreciated by MRI than CT, with the routine use of sagittal and coronal MR images. This is particularly true with high vertex lesions and for tumors along the tentorium.[90] MRI demonstrates vascular encasement, displacement, and occlusion better than CT and as well as digital venous arteriography.[12]

Most meningiomas have T1, T2 and proton density values similar to brain cortex.[6,10,12] On T1-weighted images, about two-thirds will be isointense and one-third hypointense to cerebral gray matter (Fig. 6-41). On heavily T2-weighted images, about half will remain isointense while slightly fewer will become slightly hyperintense to brain (Fig. 6-41). Proton density and mixed T1/T2-weighted images usually show tumors that are isointense with brain.

Another characteristic sign of meningiomas on MR is cortical buckling. This sign, originally described on CT, is indicative of an extra-axial mass lesion.[93] The cortical buckling sign is best noted on inversion recovery images (such as IR 1500/400/30), which are designed to show optimize contrast between gray matter, white matter, and CSF. Cortical buckling may be the only definite evidence for an isointense meningioma on noncontrast MR (Fig. 6-42).

About two-thirds of meningiomas have a hypointense rim at the tumor periphery (Fig. 6-43).[12] This hypointense rim is believed in most cases to represent a CSF–cleft, though in certain situations it may result from buckled dura, a prominent venous capsule, or displaced arterial branches.[91]

Approximately 10 percent of meningiomas depart significantly in appearance from those with the several

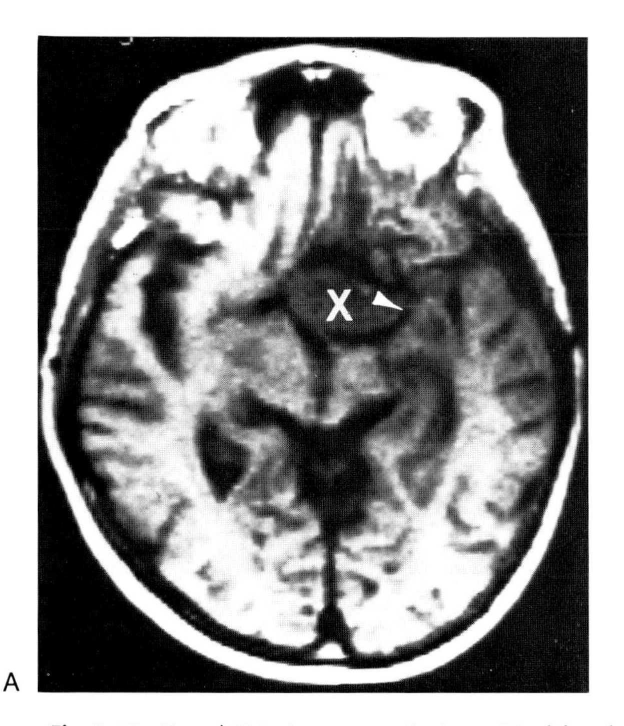

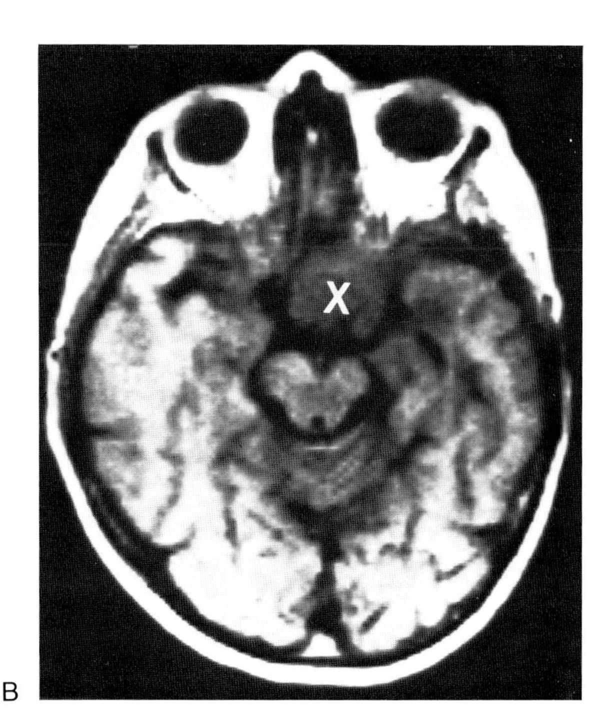

Fig. 6-42. (**A** and **B**) Isointense meningioma (*X*) of the planum sphenoidale, on SE 500/28 images. Note cortical buckling of temporal lobe (arrow).

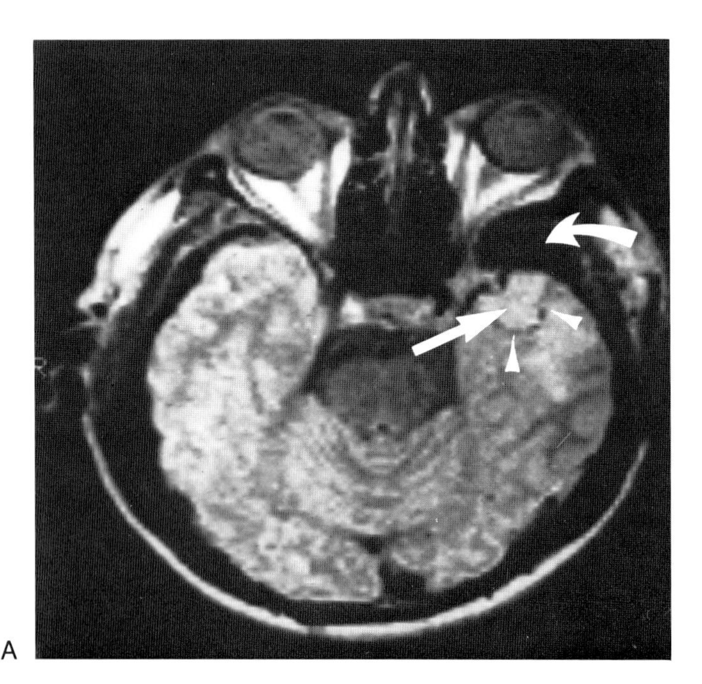

A

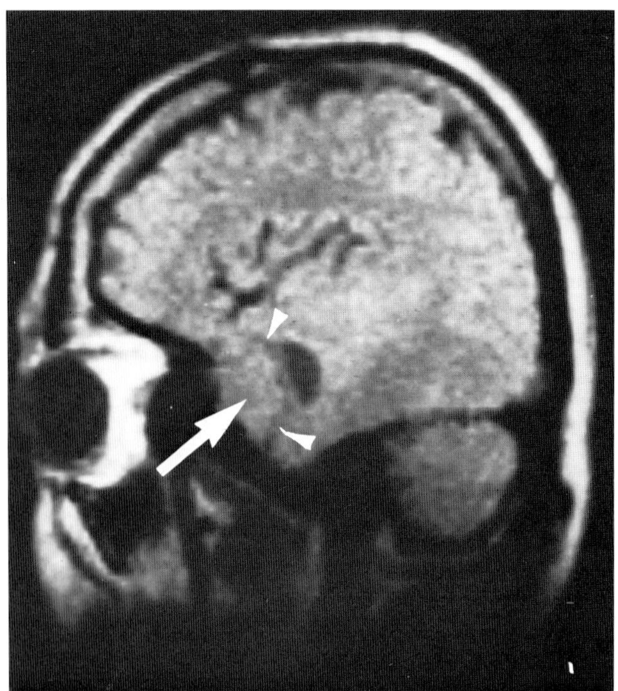

B

Fig. 6-43. Isointense meningioma (straight arrows) of the liddle cranial fossa on **(A)** axial SE 2000/28 and **(B)** sagittal SE 1000/28 images. There is hyperostosis of the adjacent skull (curved arrow). Note also characteristic low intensity band between tumor and brain (arrowheads).

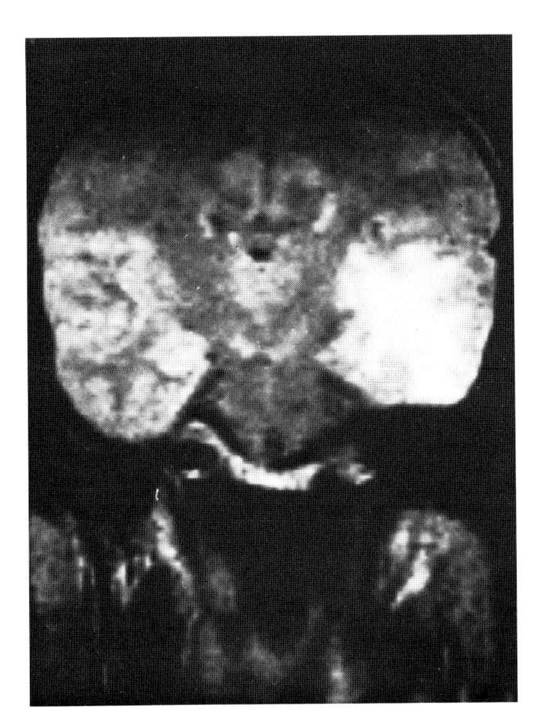

Fig. 6-44. A malignant meningothelial meningioma infiltrating through the left temporal lobe. Was thought to be a glioblastoma on MR (SE 1500/56).

"characteristic" features described above. This second group of meningiomas possess ρ, T1, and T2 values significantly increased relative to brain.[90] It is possible that this second group corresponds to "soft" meningiomas while the first group corresponds to "hard" meningiomas described at surgery and on CT.[94] The distinction between hard and soft meningiomas may be of some surgical importance when the tumors are in close relationship to vital structures. Aggressive meningiomas have ill-defined margins and frequently excite considerable edema in adjacent brain (Fig. 6-44). It may not be possible to separate edema from tumor on noncontrast MR (Fig. 6-45).

Administration of Gd-DTPA significantly improves visualization of meningiomas.[4,29,92] Homogeneous and intense enhancement seen on T1-weighted images is typical. The contrast enhancement of meningiomas on MR may be much greater than their degree of contrast enhancement on CT.[90] In the future, when meningioma is suspected, postcontrast MR following Gd-DTPA administration will likely become routine protocol.

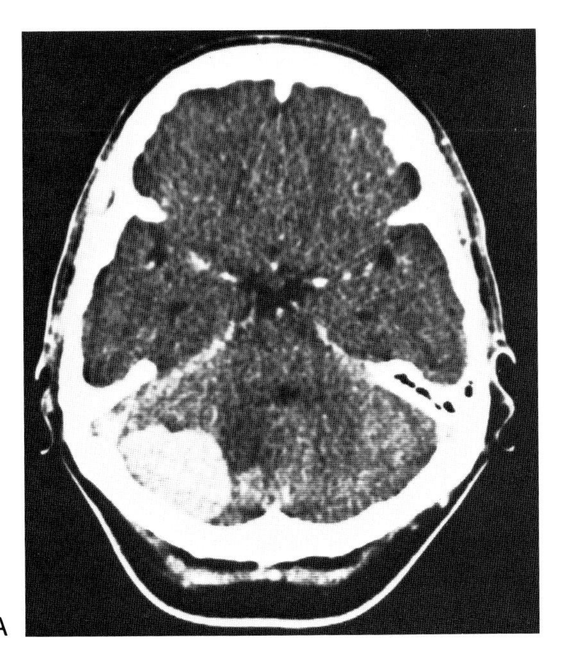

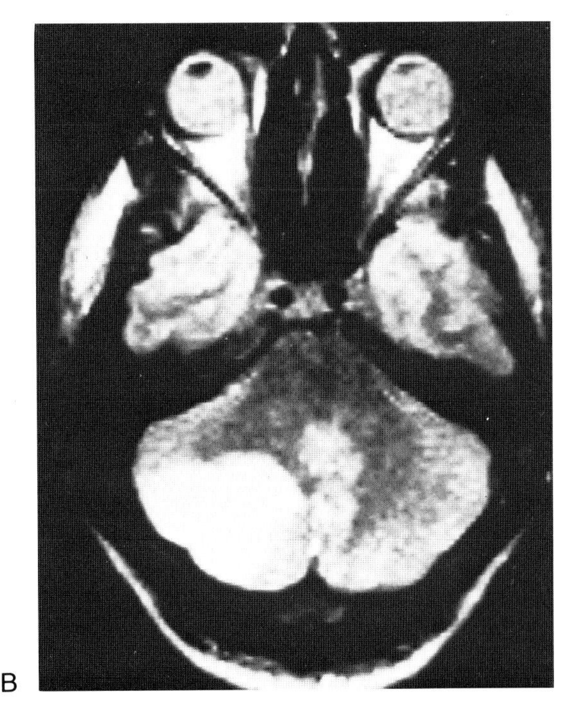

Fig. 6-45. A posterior fossa meningioma with strong T2 signal. This pattern is seen in about 10 percent of meningiomas. **(A)** CT scan. **(B)** SE 2000/56.

REFERENCES

1. Youmans JR: Neurological Surgery. WB Saunders, Philadelphia, 1982
2. Zimmerman RA: Magnetic resonance imaging of cerebral neoplasms. p.127. In Kressel HY (ed): Magnetic Resonance Annual 1985. Raven, New York, 1985
3. Brant-Zawadzki M, Badami JP, Mills CM et al: Primary intracranial tumor imaging: a comparison of magnetic resonance and CT. Radiology 150:435, 1984
4. Felix R, Schorner W, Laniado M et al: Brain tumors: MR imaging with Gadolinium-DTPA. Radiology 156: 681, 1985
5. Graif M, Bydder GM, Steiner RE et al: Contrast enhanced MR imaging of malignant tumors. AJNR 6:855, 1985
6. Mills CM, Crooks LE, Kaufman L et al: Cerebral abnormalities. Use of calculated T1 and T2 magnetic resonance images for diagnosis. Radiology 150:87, 1984
7. Damadian R: Tumor detection by nuclear magnetic resonance. Science 171:1151, 1971
8. Saryan LA, Hollis DP, Economon JS, Eggleston JG: Nuclear magnetic resonance studies of cancer. VI. Correlation of water content with tissue relaxation times. J Natl Cancer Inst 52:599, 1974
9. Cope FW: Nuclear magnetic resonance evidence using D_2O for structured water in muscle and brain. Biophys J 9:303, 1969
10. Araki T, Inouye T, Suzuki H et al: Magnetic resonance imaging of brain tumors: measurement of T1. Radiology 150:95, 1984
11. MacKay IM, Bydder GM, Young IR: MR imaging of central nervous system tumors that do not display increase in T1 or T2. J Comput Assist Tomogr 9:1055, 1985
12. Zimmerman RD, Fleming CA, Saint-Louis LA et al: Magnetic resonance imaging of meningiomas. AJNR 6:149, 1985
13. Randell CP, Collins AG, Young IR et al: Nuclear magnetic resonance imaging of posterior fossa tumors. AJNR 4:1027, 1983
14. Holland BA, Kucharcyzk W, Brant-Zawadzki M et al: MR imaging of calcified intracranial lesions. Radiology 157:353, 1985
15. Oot RF, New PFJ, Pile-Spellman J et al: The detection of intracranial calcifications by MR. AJNR 7:801, 1986
16. Wolf GL, Burnett KR, Goldstein EJ, Joseph PM: Contrast agents for magnetic resonance imaging. p.231. In Kressel HY (ed): Magnetic Resonance Annual 1985. Raven, New York, 1985
17. Slutsky RA, Peterson T, Strich G, Brown JJ: Hemodynamic effects of rapid and slow infusions of manganese chloride and gadolinium-DTPA in dogs. Radiology 154:733, 1985
18. Weinmann HJ, Brasch RC, Press WR, Wesbey GE: Characteristics of gadolinium-DTPA complex: a potential NMR contrast agent. AJR 142:619, 1984
19. Browning E: Toxicity of Industrial Metals, Ed.2. Appleton-Century-Crofts, New York, 1969
20. Strich G, Hagan PL, Gerber KH, Slutsky RA: Tissue distribution and magnetic resonance spin lattice relaxation effects of gadolinium-DTPA. Radiology 154-723, 1985
21. Graif M, Bydder GM, Steiner RE et al: Gd-DTPA in MR imaging of malignant brain tumors. Radiology 157:125, 1985
22. Maravilla KR, Sory C, Mickey B et al: Preliminary clinical experience using intravenous gadolinium-DTPA for MR imaging: diagnostic pulse sequence considerations. Radiology 157:126, 1985
23. Kilgore DP, Breger RK, Daniels DL et al: Cranial tissues: normal MR appearance after intravenous injection of Gd-DTPA. Radiology 160:757, 1986
24. Grossman RI, Wolf GL, Biery D et al: Gadolinium-enhanced nuclear magnetic resonance images of experimental brain abscess. J Comput Assist Tomogr 8:204, 1984
25. Carr DH, Brown J, Bydder GM et al: Clinical use of intravenous gadolinium-DTPA as a contrast agent in NMR imaging of cerebral tumors. Lancet 1:484, 1984
26. Bauer WM, Baierl P, Vogl T et al: Contrast enhancement in intracranial tumors: a comparison of CT and MR. Radiology 157:126, 1985
27. Bradley WG, Brant-Zawadzki, Brasch RC et al: Initial clinical experience with Gd-DTPA in North America. MR contrast enhancement of brain tumors. Radiology 157:125, 1985
28. Brant-Zawadzki M, Berry I, Osaki L et al: Gd-DTPA in clinical MR of the brain: 1. Intraaxial lesions. AJNR 7:781, 1986
29. Berry I, Brant-Zawadzki M, Osaki L et al: Gd-DTPA in clinical MR of the brain. 2. Extraaxial lesions and normal structures. AJNR 7:789, 1986
30. Black P: Brain metastasis: current status and recommended guidelines for management. Neurosurgery 5:617, 1979
31. VanEck JHM, Go KG, Ebels EJ: Metastatic tumors of the brain. Neurol Neurochir 68:443, 1965
32. Baker HI, Houser OW, Campbell JK: National Cancer Institute study: evaluation of computed tomography in diagnosis of intracranial neoplasms. Radiology 136:91, 1980
33. Potts DG, Abbott GF, von Sneider JV: National Cancer Institute study: evaluation of computed tomography in diagnosis of intracranial neoplasms: III metastatic tumors. Radiology 136:657, 1980

34. Posner JB, Chernik NL: Intracranial metastases from systemic cancer. Adv Neurol 19:575, 1978
35. Russcalleda J: Clinical symptomatology and computerized tomography in brain metastases. Comput Tomogr 2:69, 1978
36. Lee SH, Rao KCVG: Cranial Computed Tomography. McGraw-Hill, New York, 1983
37. Chason JL, Walter FB, Landers JW: Metastatic carcinoma in the central nervous system and dorsal root ganglia. A prospective study. Cancer 16:781, 1983
38. Zimmerman RA, Bilaniuk LT: Computed tomography of acute intratumoral hemorrhage. Radiology 135:355, 1980
39. Russell DS, Rubinstein LJ: Pathology of Tumors of the Nervous System. 4th ed. Williams & Wilkins, Baltimore, 1977
40. Kernohan JW, Mabon RF, Svien HJ et al: Symposium on new and simplified concept of gliomas: simplified classification of the gliomas. Proc Staff Meet Mayo Clin 24:71, 1949
41. Butler AR, Horii SC, Kricheff II et al: Computed tomography in astrocytomas. Radiology 129:433, 1978
42. Tans J, De Jongh IE: Computed tomography of supratentorial astrocytoma. Clin Neurol Neurosurg 80:156, 1978
43. Tchang S, Scotti G, Terbrugge K et al: Computerized tomography as a possible aid to histologic grading of supratentorial gliomas. J Neurosurg 46:735, 1977
44. Brant-Zawadzki M, Davis PL, Crooks LE et al: NMR demonstration of cerebral abnormalities: comparison with CT. AJR 140:847, 1983
45. Buonanno FS, Pykett IL, Brady TJ et al: Clinical relevance of two different nuclear magnetic resonance (NMR) approaches to imaging of a low grade astrocytoma. J Comput Assist Tomogr 6:529, 1982
46. Lee BCP, Kneeland JB, Walker RW et al: MR imaging of brain stem tumors. AJNR 6:159, 1985
47. Weinstein MA, Modic MT, Pavlicek W et al: Nuclear magnetic resonance for the examination of brain tumors. Semin Roentgenol 19:139, 1984
48. Zimmerman RA, Bilaniuk LT, Bruno L, Rosenstock J: Computed tomography of cerebellar astrocytoma. AJR 130:929, 1978
49. Peterman SB, Steiner RE, Bydder GM et al: Nuclear magnetic resonance imaging (NMR), (MRI), of brain stem tumors. Neuroradiology 27:202, 1985
50. Bradac GB, Schorner W, Bender A, Felix R: MRI (NMR) in the diagnosis of brain stem tumors. Neuroradiology 27:208, 1985
51. Packer RJ, Zimmerman RA, Luerssen TG et al: Brain stem gliomas of childhood: magnetic resonance imaging. Neurology 35:397, 1985
52. Kieffer SA, Salibi NA, Kim RC et al: Multifocal glioblastoma: diagnostic implications. Radiology 143:709, 1982
53. Batzdorf U, Malamud N: The problem of multicentric gliomas. J Neurosurg 20:122, 1963
54. Vonofakos D, Marcu H, Hacker H: Oligodendrogliomas: CT patterns with emphasis on features indicating malignancy. J Comput Assist Tomogr 3:783, 1979
55. Zimmerman R, Bilaniuk R: Computed tomography of intracerebral gangliogliomas. CT 3:24, 1979
56. Nass R, Whelan M: Gangliogliomas. Neuroradiology 22:67, 1981
57. Zee C, Segall HD, Miller C et al: Less common CT features of medulloblastoma. Radiology 144:97, 1982
58. Zimmerman RA, Bilaniuk LT, Pahlajani H: Spectrum of medulloblastomas demonstrated by computed tomography. Radiology 126:137, 1978
59. Seeger J, Burke D, Knake J et al: Computed tomographic and angiographic evaluation of hemangioblastomas. Radiology 138:65, 1981
60. Ganti S, Silver A, Hilal S et al: Computed tomography of cerebellar hemangioblastomas. J Comput Assist Tomogr 6:912, 1982
61. Snider WD, Simpson DM, Aronyk KE, Nielsen SL: Primary lymphoma of the nervous system associated with acquired immune-deficiency syndrome. N Engl J Med 308:45, 1983
62. Whelan MA, Kricheff II: Intracranial lymphoma. Semin Roentgenol 19:91, 1984
63. Lee Y-Y, Bruner JM, Van Tassel P, Libshitz HI: Primary central nervous system lymphoma: CT and pathologic correlation. AJNR 7:599, 1986
64. Lee SH, Rao KCVG: Cranial Computed Tomography. McGraw-Hill, New York, 1983
65. Erlich SS, Apuzzo MLJ: The pineal gland: anatomy, physiology, and clinical significance. J Neurosurg 63, 321, 1985
66. Riley HK, Maravilla KR, Sory C: MR appearance of the pineal gland. AJNR 7:552, 1986
67. Brazis PW, Masdeu JC, Biller J: Localization in Clinical Neurology. Little, Brown, Boston, 1985
68. Zimmerman RA, Bilaniuk LT, Wood JH et al: Computed tomography of pineal, parapineal, and histologically related tumors. Radiology 157:669, 1980
69. Jooma R, Kendall G: Diagnosis and management of pineal tumors. J Neurosurg 58:654, 1983
70. Ganti SR, Hilal SK, Stein BM et al: CT of pineal region tumors. AJR 146:451, 1986
71. Kilgore DP, Strother CM, Starshak RJ, Haughton VM: Pineal germinoma: MR imaging. Radiology 158:435, 1986
72. Zimmerman RA, Bilaniuk LT, Dolinskas C: Cranial computed tomography of epidermoid and congenital fatty tumors of maldevelopmental origin. CT 3:40, 1979
73. Laster DW, Moody DM, Ball MR: Epidermoid tumors with intraventricular and subarachnoid fat: report of two cases with CT. AJR 128:504, 1977

74. Davidson HD, Ouchi T, Steiner RE: NMR imaging of congenital intracranial germinal layer neoplasms. Neuroradiology 27:301, 1985
75. Kortman KE, Van Dalsem WJ, Bradley WG: MR imaging of epidermoid tumors. Radiology 157:17, 1985
76. Hershey B, Grossman RI, Goldberg HI et al: MR imaging of epidermoid tumors. Radiology 157:17, 1985
77. Bailey OT: Histologic sequences in meningioma with consideration of the nature of hyperostosis cranii. Arch Pathol 30:42, 1940
78. Rosenbaum AE, Rosenbloom SD: Meningiomas revisited. Semin Roentgenol 19:8, 1984
79. Scatliff J, Guinto F Jr, Radcliffe W: Vascular patterns in cerebral neoplasms and their differential diagnoses. Semin Roentgenol 6:59, 1971
80. Olivecrona H: The parasagittal meningiomas. J Neurosurg 4:327, 1947
81. Simpson D: The recurrence of intracranial meningiomas after surgical treatment. J Neurol Neurosurg Psychiatry. 20:22, 1957
82. Cushing H, Eisenhardt L: Meningiomas. Charles C. Thomas, Springfield, 1938
83. Sheehy J, Crockard H: Multiple meningiomas: a long-term review. J Neurosurg 59:1, 1983
84. Vissilouthis J, Ambrose J: Computerized tomography scanning appearance of intracranial meningiomas. J Neurosurg 50:320, 1979
85. Thomas H, Dolman C, Berry K: Malignant meningioma: clinical and pathological features. J Neurosurg 55:929, 1981
86. New P, Hesselink J, O'Carroll C et al: Malignant meningiomas: CT and histologic criteria including a new CT sign. AJNR 3:267, 1982
87. Karasick J, Mullan S: A survey of metastatic meningiomas. J Neurosurg 40:206, 1974
88. Kishikawa T, Namaguchi Y, Fukui M et al: Primary intracranial sarcomas: radiological diagnosis with emphasis on arteriography. Neuroradiology 21:25, 1981
89. Bradley WG, Waluch V, Yadley RA, Wycoff RR: Comparison of CT and MR in 400 patients with suspected disease of the brain and spinal cord. Radiology 152:695, 1984
90. Bydder GM, Kingsley DPE, Brown J et al: MR imaging of meningiomas including studies with and without gadolinium-DTPA. J Comput Assist Tomogr 9:690, 1985
91. Spagnoli M, Goldberg HI, Grossman RI et al: High-field MR imaging of intracranial meningiomas. Radiology 161:369, 1986
92. Bradley WG, Kortman KE, Bucon KA: MR imaging of meningiomas with and without paramagnetic contrast. Radiology 157(P):125, 1985
93. George AJ, Russel EJ, Kricheff II: White matter buckling: CT sign of extraaxial intracranial mass. AJR 135:1031, 1980
94. Kendall BE, Pullicino P: Comparison of consistency of meningioma and CT appearances. Neuroradiology. 18:173, 1979

7

Cerebral Vascular Diseases and Disorders

Disorders of the cerebral vasculature account for a significant portion of neuropathology. In this chapter a wide variety of diseases and vascular abnormalities will be discussed, including aneurysms, vascular malformations, infarction, thrombosis, hemorrhage, and vasculitis. While arteriography and CT continue to serve as mainstays for the diagnosis of vascular diseases, MR may provide equivalent or complementary information. In some instances, MR may obviate arteriography or provide unique data with therapeutic implications.

The clinical presentation of cerebral vascular disorders may range from insidious to cataclysmic. Unruptured aneurysms, vascular malformation, and small vessel occlusion may be asymptomatic or cause nonspecific headache, visual disturbance, or seizures. Occlusion or rupture of larger vessels may cause dramatic symptomatology.

Temporary occlusion or decreased flow in a larger cerebral blood vessel can cause visual loss, paralysis, or altered mental status. If the episode clears completely within 24 hours, it is called a *transient ischemic attack* (*TIA*). If rapid resolution of symptoms does not occur, the patient is said to have suffered a *stroke*. Synonyms for stroke include *apoplexy* and *cerebral vascular accident* (CVA). About three-fourths of strokes are *ischemic*, caused by occlusion or decreased blood flow in a cerebral vessel. The remaining one-fourth of strokes are *hemorrhagic*, resulting from rupture of a cerebral blood vessel. Important causes of stroke are presented in Tables 7-1 through 7-3.

Table 7-1. Causes of Stroke in Adults from the Harvard Cooperative Stroke Registry

Large artery thrombosis	34%
Embolism	31%
Lacunar infarction	19%
Primary intracerebral hemorrhage	10%
Hemorrhage from aneurysm or AVM	6%

(Mohr JP, Caplan LR, Melski JW: The Harvard cooperative stroke registry: a prospective study. Neurology 28:754, 1978.)

Table 7-2. Causes of Ischemic Stroke

- Cerebral embolism
  - of cardiac origin (mural thrombus, endocarditis)
  - of noncardiac origin (air, atherosclerotic plaque from aorta, fat, tumor)
  - Iatrogenic (postarteriography or vascular surgery)
- Arteriolosclerosis (lacunar infarction)
- Systemic hypotension (watershed infarction)
- Vasculitis (infectious, collagen vascular, drugs, idiopathic)
- Venous thrombosis
- Systemic disorders predisposing to thrombosis (malignancies, dehydration)

Table 7-3. Causes of Hemorrhagic Stroke

Hypertensive vascular disease
Aneurysms
Vascular malformations
Hemorrhagic conversion of ischemic infarction
Systemic disorders predisposing to hemorrhage (blood dyscrasias, heparin)
Germinal matrix hemorrhage in premature infants
Hemorrhage into primary and secondary brain tumors
Vascular infections and inflammations

Table 7-4. Cerebral Aneurysm Associations[a]

Coarctation of the aorta
Polycystic kidney disease
Arteriovenous malformation (aneurysm is usually on a feeding vessel)
Renal artery stenosis
Connective tissue disorders (Marfan, Ehlers-Danlos, pseudoxanthoma elasticum)
Moyamoya disease
Congenital cerebral vascular anomalies (hypoplasia on an anterior cerebral artery, persisting fetal anastamoses)
Familial forms

[a] Compiled from References 4, 5, 9, and 10.

INTRACRANIAL ANEURYSMS

An aneurysm is a focal enlargement of an arterial wall, usually composed of only intima and adventitia. Intracranial aneurysms are relatively common in the general population, being detected in 0.25 to 3.0 percent of random autopsies.[1,2] Most are the result of weakness in the arterial wall either on a congenital basis or acquired from atherosclerosis and hypertension.[3] A few develop secondary to trauma or infection (*mycotic aneurysms*).

Cerebral aneurysms may be classified by morphology and size. If the bulging of the arterial wall is lengthy and uniform the aneurysm is termed *fusiform*. If only a portion of the wall is bulging asymmetrically in a saclike fashion the aneurysm is called *saccular*. About 90 percent of cerebral aneurysms are of the saccular type; fusiform dilatation is seen in only 10 percent. Small saccular aneurysms with narrow necks are sometimes referred to as *berry aneurysms*. Aneurysms measuring between 1.0 and 2.5 cm are called *large*, while those greater than 2.5 cm are called *giant aneurysms*.

Most symptomatic cerebral aneurysms are discovered in middle-aged patients, usually in the fifth or sixth decades.[4] Only 5 percent are seen in patients below age 20. Women are affected more frequently than men. Patients with hypertension and atherosclerosis are at higher risk. Certain congenital and hereditary associations are also recognized as risk factors for aneurysm formation (Table 7-4). A second aneurysm is present in 15 to 25 percent of patients with a first aneurysm.[5–8] Multiple aneurysms are particularly common in females.[5]

Most aneurysms are found in the circle of Willis or adjacent proximal main cerebral artery branches. Several large autopsy and clinical series have catalogued the distribution of aneurysms by location and probability of rupture (Table 7-5).[11–18] The discrepancy between clinical and autopsy series largely relates to the fact that many very small (1 to 2 mm) sessile aneurysms exist in the more distal middle cerebral artery territory, which do not present clinically and may not be seen angiographically. For practical purposes one should consider that over 90 percent of clinically significant aneurysms will occur near the circle of Willis, branch points of the internal carotid artery, or proximal middle cerebral artery.

Table 7-5. Location of Cerebral Aneurysms

Location	*Clinical Series[a]*	*Autopsy Series[b]*
Internal carotid artery, including those arising at the junction of anterior choroidal, ophthalmic, and posterior communicating artery branches	37%	26%
Middle cerebral artery	20%	39%
Anterior cerebral artery and anterior communicating artery	32%	24%
Posterior circulation	5%	8%
Other	6%	3%

[a] Compiled from References 12, 13, and 15. Total of 4585 aneurysms.
[b] Compiled from References 11, 14, 15, and 17. Total of 1001 aneurysms.

Clinical Findings

Cerebral aneurysms present clinically in one of two ways: (1) by local mass effect, or (2) by rupture with intracranial hemorrhage. Giant aneurysms and critically positioned smaller aneurysms may press against brain or cranial nerves. Headache, either focal or diffuse, is a common result. Visual disturbance, especially diploia secondary to oculomotor nerve paresis, is the next most frequent symptom due to pressure effects of an aneurysm.[4,19]

The more common presentation of an aneurysm is obtundation, coma, or paralysis, secondary to rupture with intracranial hemorrhage. Five types of hemorrhage may be seen: subarachnoid, intracerebral, intraventricular, subdural, and epidural. The location and type of bleeding depends upon where the aneurysm arose and which direction it was pointing.

Subarachnoid hemorrhage is the most common type of bleeding to accompany aneurysm rupture, occurring in over 50 percent of cases. Intracerebral hematoma is often associated with subarachnoid hemorrhage and is seen in about 25 percent.[6,14] Intracerebral hematoma occurs most frequently following rupture of aneurysms of (1) the anterior communicating artery, (2) the middle cerebral artery, and (3) the internal carotid artery at the origin of the posterior communicating artery, in that order. Intraventricular hemorrhage may occur either by retrograde filling from the subarachnoid space or direct rupture of an intracerebral hemorrhage into the ventricular system. Intraventricular hemorrhage with only minimal cerebral hematoma is seen in about 7 percent.[14] Epidural and subdural hemorrhages following aneurysm rupture are rare.

When primary rupture of an aneurysm occurs, the perforation is usually at the dome or side of the aneurysm, rarely along its neck. The initial rupture of an aneurysm, therefore, usually results in subarachnoid hemorrhage. If the patient survives, hemorrhage from the rupture produces subarachnoid adhesions that bind the sac of the aneurysm more closely to the brain. A second rupture of an aneurysm is more likely to result in intracerebral or intraventricular hematoma. This partially explains the higher mortality seen with rebleeding.[20]

Cerebral infarction can occur after aneurysm rupture, usually a result of vascular spasm about 3 to 10 days following subarachnoid hemorrhage. Direct occlusion of vessels may also occur by extrinsic pressure from a large hematoma. Pre-existing atherosclerosis contributes to the likelihood of infarction. Cerebral infarction is more likely to occur after rupture of proximal aneurysms from the internal carotid artery and circle of Willis.[21] The infarcted area is usually cortical and often in the distribution of the middle cerebral artery. Basal ganglia infarcts more commonly occur after ruptures of posterior communicating artery aneurysms. Bifrontal infarcts may be seen after ruptures of anterior communicating artery aneurysms.

Aneurysms of the internal carotid artery within the cavernous sinus have a different presentation and course than other saccular aneurysms.[22] Expansion of the aneurysm without rupture results in compression of nerves in the cavernous sinus. The usual order of involvement is cranial nerve VI, III, IV, and V. This *cavernous sinus syndrome* has been divided into anterior, middle, and posterior types based upon characteristic patterns of cranial nerve involvement.[23] Rupture of a cavernous aneurysm commonly results in a carotid-cavernous fistula rather than subarachnoid hemorrhage.

Natural History

The natural history of ruptured and unruptured cerebral aneurysms is well documented in several cooperative studies.[4,20] In general the evidence supports surgical intervention in larger berry aneurysms and in those that have bled, provided the patient is a good surgical candidate and the aneurysm is accessible.

The chance rupture of a small asymptomatic aneurysm is often estimated to be about 0.5 to 1.0 percent per year, but this figure must be modified by the size, shape, location, and etiology of the aneurysm. Most ruptured aneurysms found at autopsy are larger than 5 mm in diameter, with an average size of 14 mm.[24] Giant aneurysms greater than 2.4 cm may rupture, but more frequently cause symptoms by pressure effects.

Fusiform aneurysms in any location and saccular aneurysms not arising at bifurcations rarely rupture. Saccular aneurysms that arise at vessel bifurcations,

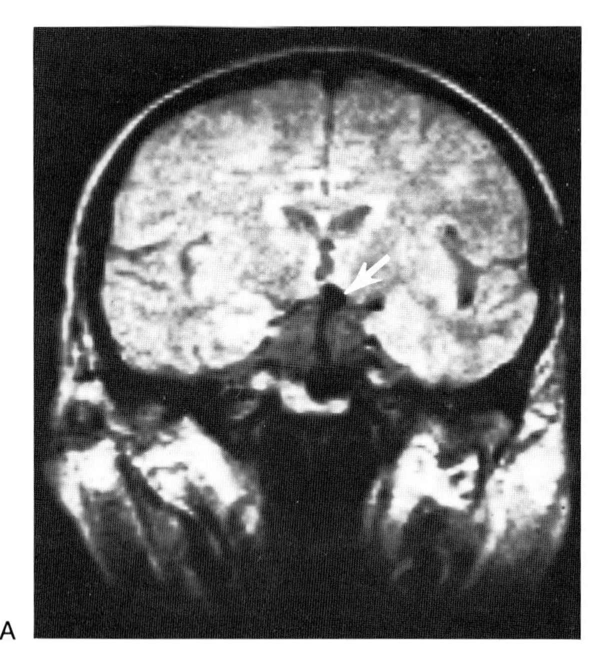

however, are exceedingly prone to rupture. Aneurysms with secondary loculations bleed twice as frequently as smooth-walled lesions. Mycotic and traumatic aneurysms are at high risk for rupture.[25] An aneurysm that has previously bled has a rebleeding rate of 25 to 50 percent in the first year and 4 percent per year thereafter.[4]

A patient with uncontrolled hypertension is at higher risk for aneurysm rupture than a normotensive patient with the same aneurysm. Pregnant patients in gestational weeks 13 through 40 are also at risk, but rupture during actual delivery is rare.[26]

The course and prognosis after aneurysm rupture is well documented, but the figures for mortality and morbidity vary depending on patient population and series selected.[27,28] The following scenario would represent a typical natural history based on the average of several series:

1. Over half of all patient presenting with a ruptured aneurysm will be dead a few months later. Only 15 percent will be alive 10 years later.

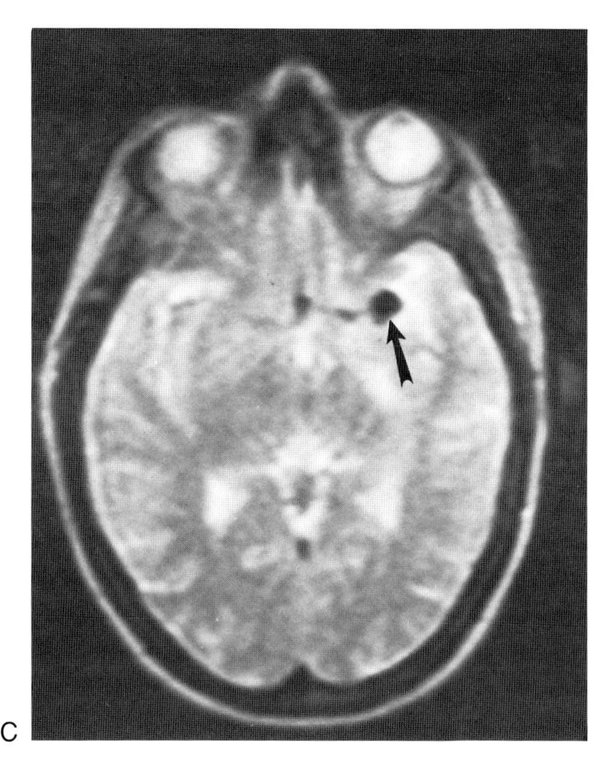

Fig. 7-1. **(A)** Basilar tip aneurysm seen on coronal SE 1500/56 image. **(B)** and **(C)** middle cerebral artery aneurysm on CT and SE 3000/80 image.

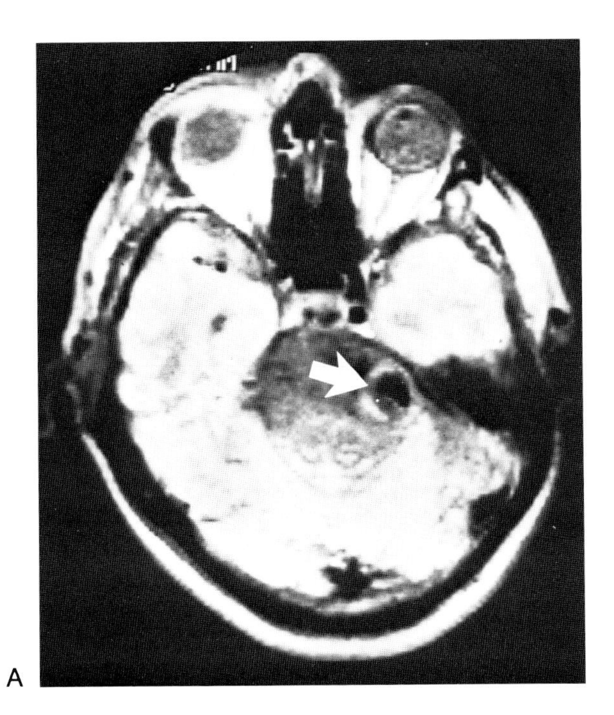

A

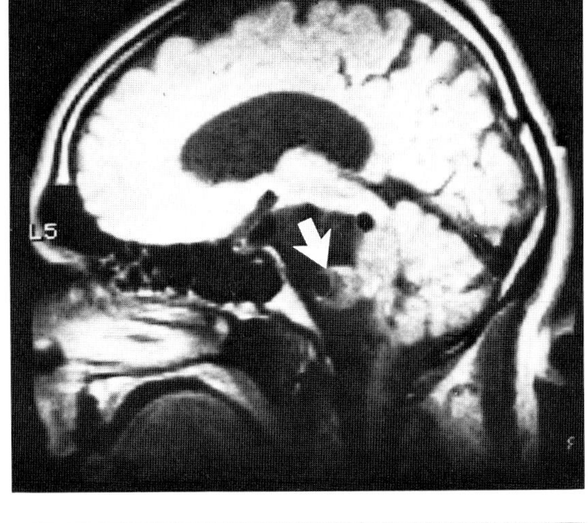

B

C

Fig. 7-2. Large basilar artery aneurysm that distorts brain stem. Note clot within aneurysm and remaining channel of rapidly flowing blood with low signal. **(A)** SE 2000/28. **(B)** SE 1000/28. **(C)** SE 1000/56.

2. Of those who die, about two-thirds will succumb to their original hemorrhage while the other one-third will die from rebleeding.
3. Of those who live and become long-term survivors, about one-third will suffer neurologic deficits, including paralysis, epilepsy, or headache.
4. During the first year there is a 25 to 50 percent chance of rebleeding from the aneurysm. After the first year, the risk of rebleeding is about 4 percent per year. Each rebleed carries with it a 50 percent mortality.
5. Surgical intervention carries a rather high morbidity and mortality, but it is better than the natural history of the disease.

Magnetic Resonance Imaging of Cerebral Aneurysms

Asymptomatic cerebral aneurysms are occasionally encountered in routine MR imaging. In the few small ones we have seen, sufficient flow was present to result in high velocity signal loss so that the aneurysm was rendered dark against higher signal brain or CSF (Figs. 7-1 and 7-2). It is conceivable that with more sluggish flow, paradoxical enhancement of blood in the aneurysm might be noted.

Although arteriography remains the definitive diagnostic modality, some investigators have found MR useful in diagnosis and management of intracranial aneurysms.[29] MRI is particularly helpful in assessing noninvasively the degree of thrombosis in the aneurysm lumen following surgery or balloon occlusion.[29a] MRI could conceivably be used to evaluate those patients who have negative arteriograms after subarachnoid hemorrhage. In this situation, MRI might be able to locate a small thrombosed aneurysm that would not fill at arteriography. In one reported case, MRI established the true site of aneurysmal bleeding in a patient with multiple aneurysms, where CT was inconclusive and arteriography suggested bleeding from the wrong aneurysm.[29b]

A well-known limitation of MRI is its difficulty in detecting very acute intracerebral hematomas and subarachnoid hemorrhages.[30–32] Accordingly CT (not MR) should be performed as the initial step in evaluating patients with suspected aneurysm rupture. Nevertheless, MR is capable of detecting some very acute cerebral hemorrhage especially when high field strength (1.5 T) scanners and heavily T2-weighted images are used.[33] In this setting it is important to review traditional guidelines for predicting the site of aneurysm rupture, given anatomic patterns of hematoma dissection.[19]

1. Anterior communicating artery (ACoA) aneurysms produce subarachnoid hemorrhage in the interhemispheric fissure and suprasellar cisterns. Localized hematoma may be seen in the adjacent frontal lobes. From here the hematoma may dissect upward into the septum pellucidum, and a septal hematoma is said to be highly suggestive of an anterior communicating artery aneurysm (Fig. 7-3). Frontal hematoma may also rupture into the frontal horn of the lateral ventricle. Dissection may occur into the hypothalamus, and ACoA aneurysms are the most likely to injure the hypothalamus.
2. Proximal anterior cerebral artery (ACA) aneurysms behave similarly to ACoA aneurysms. More peripheral aneurysms produce hematomas in the proximity of the lesions, and may dissect into pericallosal sulcus or intercigulate fissure.
3. Aneurysms at the bifurcation of the internal carotid artery (ICA) rupture in a pattern depending upon their direction of pointing. The domes of most of these aneurysms are imbedded in the frontal lobe; rupture then produces a frontal hematoma that may burst into a frontal horn. If the aneurysm extends posteriorly it may rupture instead into the hypothalamus or through the lamina terminalis into the third ventricle.
4. Most aneurysms of the cerebral segment of ICA arise at the origin of the posterior communicating artery (PCoA) and point backward. Subarachnoid hemorrhage is seen in the suprasellar cisterns and may extend into the Sylvian fissure laterally. Forward pointing aneurysms may produce a subfrontal hematoma. In the usual case, hemorrhage is seen in the anterior temporal lobe and may rupture into the temporal horn of the lateral ventricle. Blood may also enter the temporal horn by dissecting along the choroidal fissure above the uncus.
5. Middle cerebral artery (MCA) aneurysms produce subarachnoid hemorrhage in the circular sulcus and Sylvian fissure; suprasellar blood may also be seen. Intracerebral hematoma in the adjacent frontal lobe, temporal lobe, or insula often occurs and may dissect along the external capsule. Forward dissection may result in rupture into the frontal horn. Backward dissection in the external capsule may result in rupture into the ventricular trigone.
6. Basilar artery (BA) aneurysms usually occur at the rostral end and are associated with blood in the interpeduncular and adjacent cisterns. To see blood in the Sylvian or interhemispheric cisterns would be most unusual, except in the case of massive hemorrhage. Basilar tip aneurysms have a propensity to rupture directly into the third ventricle.
7. Posterior inferior cerebellar artery (PICA) aneurysms may produce hemorrhage in the brain stem and basilar cisterns. Blood may dissect retrogradely into the fourth ventricle. When the PICA aneurysm is located more peripherally, a cerebellar hematoma may occur at the site of the lesion.

VERTEBROBASILAR DOLICHOECTASIA

Vertebrobasilar dolichoectasia (VBD) refers to the elongation (Gr. "dolichos") and distension (Gr. "ectasia") of the vertebrobasilar arteries.[34] Other names for this condition include megadolichobasilar artery, fusiform basilar aneurysm, dolichomegavertebralis anomaly, and wandering basilar artery syndrome. The clinical significance of this disorder results from compression effects on the brain stem and cranial nerves by the tortuous and enlarged artery.

Based on CT criteria, the basilar artery will be

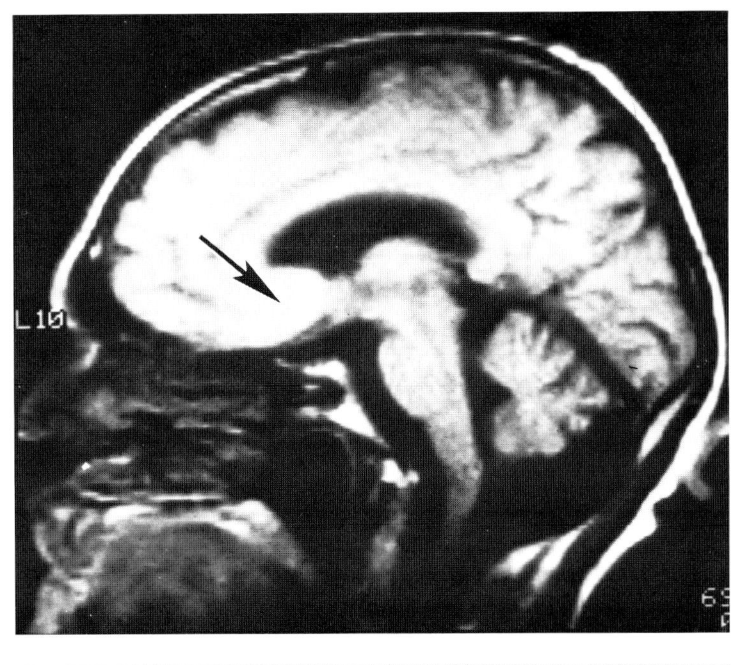

A

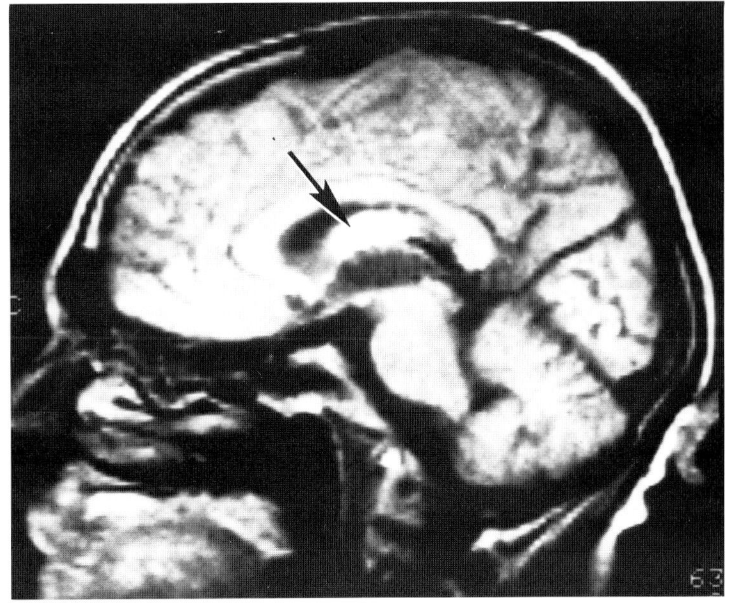

B

Fig. 7-3. Subfrontal **(A)** and septal **(B)** hematoma seen on sagittal SE 500/28 images following hemorrhage from an anterior communicating artery aneurysm.

considered *elongated* if at any point throughout its course it lies lateral to the margin of the clivus or dorsum sellae or if it bifurcates above the plane of the suprasellar cistern.[35] The basilar artery will be considered *ectatic* if its diameter is greater than 4.5 mm. The pathogenesis of this elongation and ectasia is controversial, but likely results from thinning of the arterial media and internal elastic lamina by hypertension and atherosclerosis.[36,37]

Clinical symptomatology of VBD results predomi-

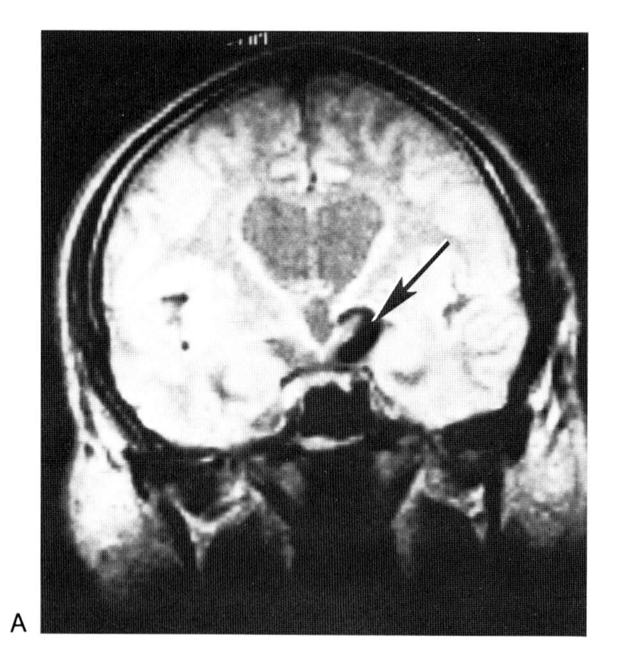

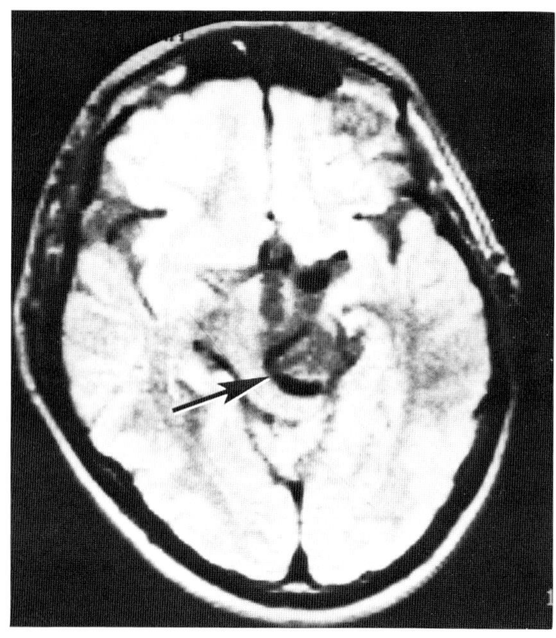

Fig. 7-4. **(A)** Fusiform aneurysm of left internal carotid artery seen on this SE 1500/56 coronal image. **(B)** Vertebrobasilar dolichoectasis noted in the same patient as the brain stem is distorted by the ectactic artery (arrow).

nantly from compressive or ischemic effects on cranial nerves.[38–40] The facial nerve is most frequently affected, and the patient may present with hemifacial spasm.[41] The trigeminal nerve is the next most frequently involved in isolation or in combination with other nerves. Patients with elongated or tortuous, but normal sized, basilar arteries tend to have isolated cranial nerve involvement. Those with dilation (ectasia), however, frequently suffer from multiple neurologic defects caused by brain stem ischemia or compression. VBD may also cause a functional hydrocephalus, usually based on increased CSF pulse pressure or impaired CSF flow by countercurrent pulsations.[42]

Magnetic resonance imaging may be useful in evaluation of patients with symptomatic VBD.[43] Compressive effects on cranial nerves may result in increased signal from them on T2-weighted images. Mass and ischemic effects on the brain stem may be better appreciated by MR than CT (Fig. 7-4).

ARTERIOVENOUS MALFORMATION

Arteriovenous malformations (AVMs) are congenital vascular anomalies representing persistence of direct communications between arteries and veins that develop during embryogenesis. AVMs occur in about 0.14 percent of the general population, being about one-seventh as common as intracranial aneurysms.[44] Men are affected more frequently than women, but not strikingly so.[45] Like cerebral aneurysms, AVMs may be discovered incidentally in asymptomatic patients. The majority of symptomatic AVMs present by age 40, causing hemorrhage, seizures, or headache.[46] Thus, AVMs usually present in a slightly younger age group than do cerebral aneurysms.

Depending upon arterial supply, cerebral AVMs are classified as purely pial (75 percent), purely dural (10 percent), or mixed (15 percent).[47] Nearly all AVMs that occur above the tentorium are purely pial or mixed types. By contrast, most of the purely dural AVMs occur in the posterior fossa or parasellar regions.

Over three-fourths of cerebral AVMs are located supratentorially, often in the temporoparietal region. The middle cerebral artery field is most frequently involved, followed by the anterior cerebral and posterior cerebral artery fields, in that order.[4] Associated saccular aneurysms are reported in 10 percent, often located on a feeding artery.[4,48] Less than one percent of AVMs are multiple.[49]

Pathologically, arteriorvenous malformations have been described as a "bag of worms" composed of grossly dilated, tortuous, and thickened vessels. The tangle of vessels is usually wedge-shaped with its base on the leptomeningeal brain surface and apex directed inward. Associated local cerebral atropy, gliosis, and calcification in the brain parenchyma is often seen and becomes more prominent with each decade of life. Hypoperfusion of normal brain along the margins of the AVM can be demonstrated, probably caused by shunting of arterial blood away from brain capillaries by low resistance malformation channels. This regional hypoperfusion of adjacent brain may account for certain nonhemorrhagic clinical manifestations of AVMs, such as seizures and deteriorating mental status.

Clinical Findings

Arteriovenous malformations usually present in one of three ways: hemorrhage (50 percent), seizures (35 percent), or headaches (15 percent).[46] Hemorrhage is usually subarachnoid or intraventricular, but may be intracerebral. About 12 percent of nontrauma patients presenting with subarachnoid hemorrhage are found to harbor an AVM.[50]

Epilepsy is the second most common manifestation of AVMs, being reported as frequently as hemorrhage in some nonsurgical series. As a general rule, younger patients tend to present with seizures while older patients present with subarachnoid hemorrhage. Seizures tend to be focal, referable to the AVM's locus on the cerebral cortex.

Headaches and progressive neurologic deterioration are relatively common symptoms, although they are generally of little localizing value. Communicating hydrocephalus may occur from repeated small hemorrhages, and this may occasionally be the initial presentation. In infants, large AVMs may produce high output heart failure, but such a presentation in older patients is rare.

The size, density, and location of an AVM correlate well with syptomatology at presentation (Table 7-6).[51] Hemorrhage is more common in small, tenuous temporo-occipital malformations. Seizures are more likely in larger, compact malformations in the frontal or parietal lobes.

Table 7-6. Initial Symptomatogy Based on Size, Density, and Location of an AVM

	Hemorrhage (%)	*Seizure* (%)
Size		
Large (>7 cm^3)	25	75
Small (<7 cm^3)	72	28
Density		
Compact	64	36
Moderate	50	50
Tenuous	14	86
Location		
Frontal	25	75
Parietal	43	57
Temporal	71	29
Occipital	100	0

Dural AVMs have a different clinical course, becoming symptomatic primarily in females over age 40.[52] One mode of presentation is as a carotid-cavernous fistula, which may close spontaneously. Dural AVMs that drain into large leptomeningeal or cortical veins have a propensity to hemorrhage.[53] Those that drain into dural sinuses are more likely to cause brain injury by ischemia or elevated venous pressure.

Natural History

Because AVMs are frequently encountered in MR examinations, it is important to have some idea of the natural history and risk factors associated with them. An unruptured AVM has a risk of future hemorrhage of about 2 percent to three percent per year.[4,46,54,55] In patients who have bled once there is a rebleed rate of about 4 percent per year and may be as high as 6 percent per year for the first four years.[54,56] If an AVM has bled twice, the risk of a third bleed is 25 percent the next year.

Mortality increases with subsequent hemorrhages. The first hemorrhage of an AVM carries a 10 percent mortality, but the second and third hemorrhage carry 13 percent and 20 percent mortalities and high morbidities.[56,57] Long-term follow-up of patients with AVMs treated conservatively are disappointing.[58] Overall, the natural history statistics for AVMs confirm the importance of their diagnosis and treatment in surgical candidates.[46]

Magnetic Resonance Imaging of AVMs

The MRI characteristics of angiographically evident AVMs have been well described.[59,60] The malformation is usually easily recognized as a tangle of serpiginous vessels devoid of MR signal (Fig. 7-5). Because the blood within the channels of most AVMs is rapidly flowing, turbulence and high velocity effects produce signal drop-out on most standard imaging sequences. Slowly flowing portions of the malformation may have high signal on second echo images (even-echo rephasing). Direct coronal or sagittal imaging may make more apparent the characteristic wedge shape of many of these malformations, with their apices directed medially (Fig. 7-6).

Arterial and venous supply may be precisely delineated, with arteries smaller in caliber than the draining veins (Fig. 7-7). MR has been shown superior to CT in imaging the vascular supply of AVMs, particularly in demonstrating cortical veins. The future use of intravenous paramagnetic MR contrast agents may further improve MR sensitivity for these lesions.

Magnetic resonance and CT may play complementary roles in the evaluation of cryptic vascular malformations (i.e., those not demonstrable angiographically). Some cryptic malformations may only be evidenced by tiny (<2 mm) parenchymal calcifications. In this situation CT will detect the calcification while MR is usually normal.[60] The normal MR signal in the region of CT-detected calcification is an exceedingly important finding, however, helping to exclude tumor as the cause of calcification. Thus a "negative" MRI complementarily allows more specificity in CT diagnosis.

On the other hand, MR has been shown superior to CT in detection of cryptic AVMs that have hemorrhaged.[61] Small foci of hemorrhage (with increased T2 and decreased T1) near cryptic AVMs may

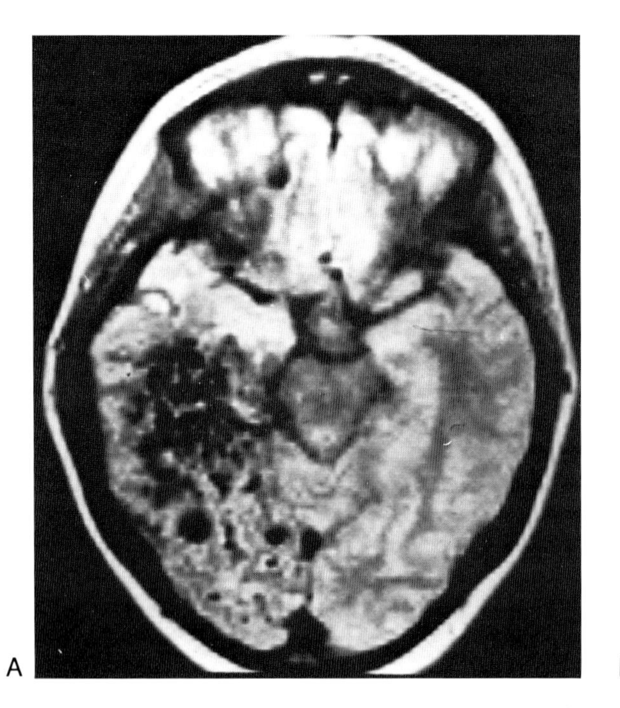

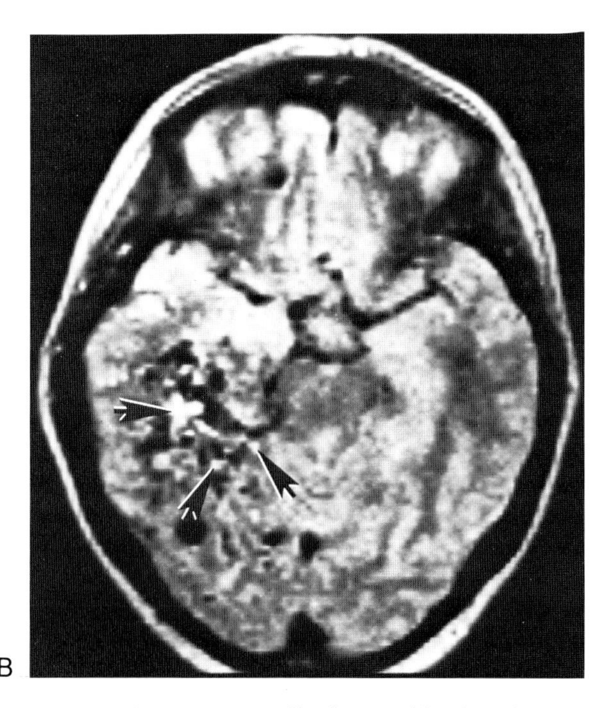

Fig. 7-5. Large AVM seen on **(A)** SE 2000/28 and **(B)** 2000/56 axial images. Rapidly flowing blood in the vascular channels causes signal dropout. In **(B)** blood in some of the more slowly flowing channels undergoes even-echo rephasing (arrowheads).

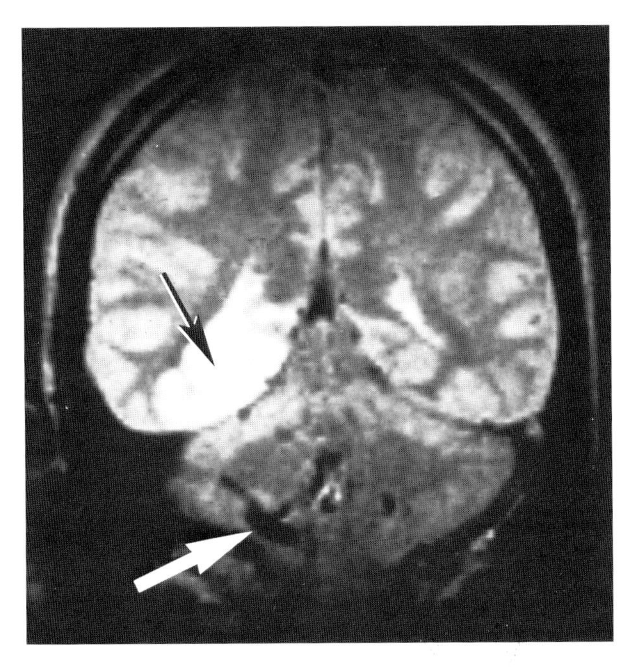

Fig. 7-6. Posterior fossa AVM (white arrow) with infarct right posterior cerebral artery territory (black arrow) secondary to "steal" phenomenon.

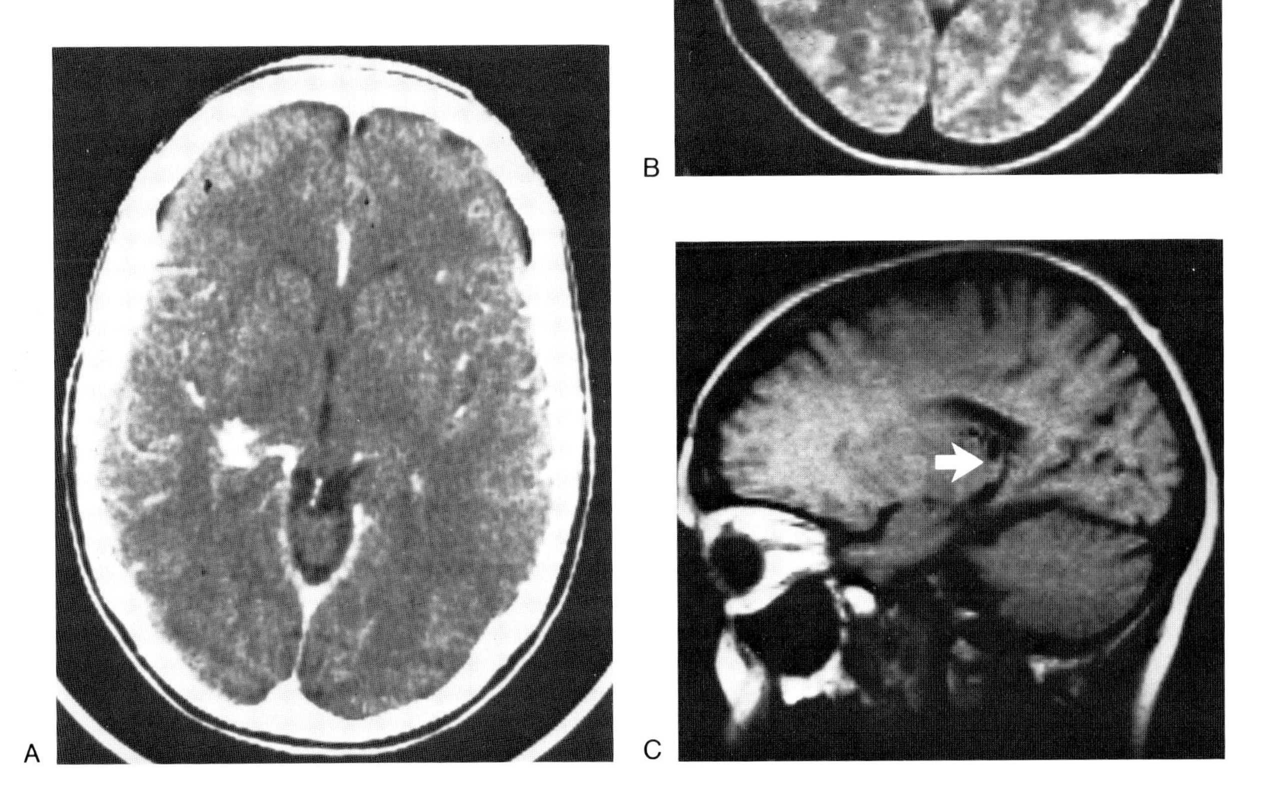

Fig. 7-7. (A-D) AVM with large draining vein; CT, MR, and angiographic correlation. (*Figure continues*).

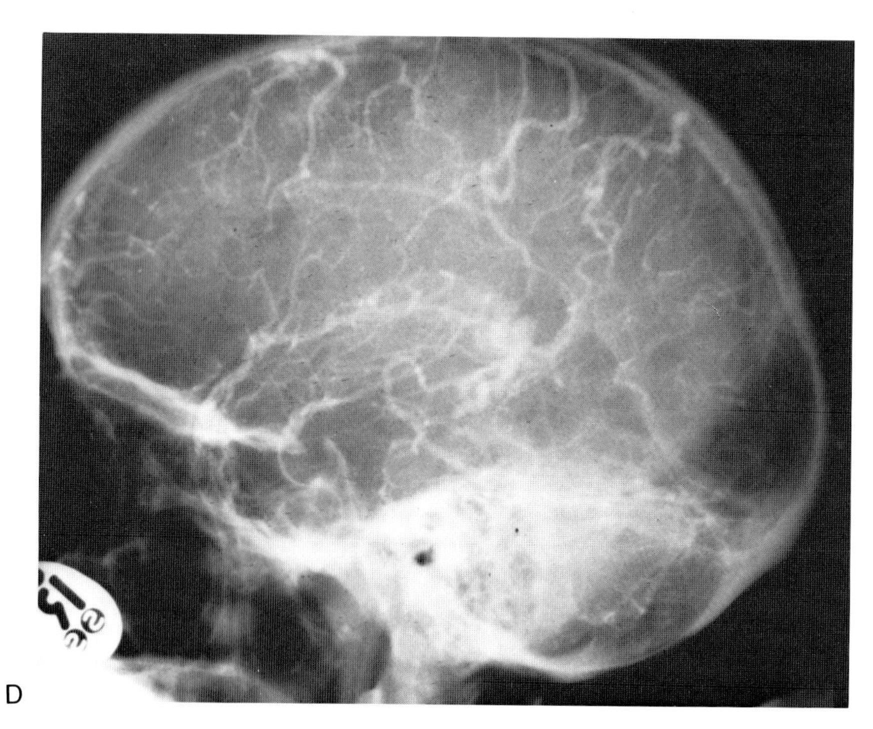

Fig. 7-7 *(continued)*. **D.**

be identified for up to 3 months by MRI. This is a period long after any CT findings would have disappeared. High-field (1.5 T) MR imaging is probably more sensitive and specific than low-field MR or CT in detection of cryptic AVMs.[62] Unusual MR appearances of cryptic AVMs may sometimes result when appreciable gliosis surrounds the lesion (Fig. 7-8).

In summary, therefore, MRI and CT are roughly equivalent at present in their abilities to evaluate AVMs. Nevertheless, certain lesions will escape detectability by both methods, and at other times one modality will be superior to the other. Although angiography is the gold standard for delineation of most AVMs, cryptic (angiographically inapparent) and complicated (thrombosed, infarcted) AVMs may be better evaluated by CT, MR, or both.

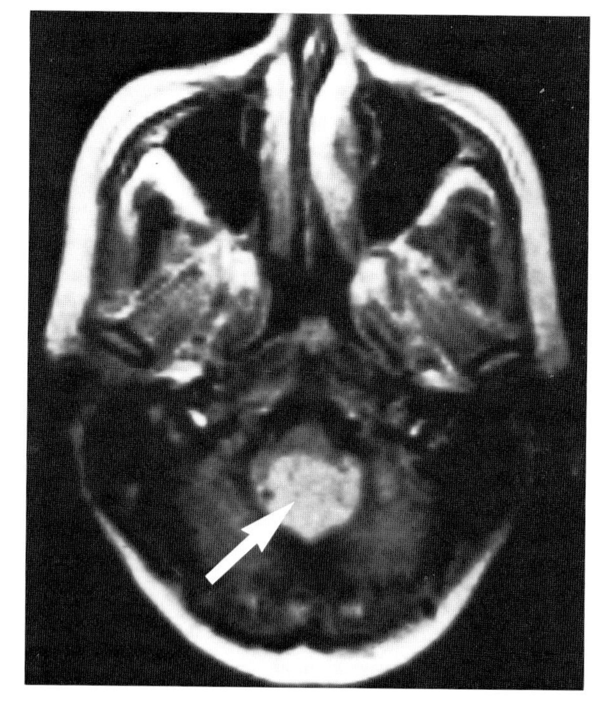

Fig. 7-8. An unusual occult vascular malformation with cystic changes and gliosis. Neither CT nor angiography suggested the diagnosis.

OTHER VASCULAR MALFORMATIONS

Capillary Telangiectasia

Virtually all capillary telangiectasias are radiographically occult and asymptomatic.[65] They are composed of tangles of dilated capillaries separated by normal neural tissue. Most are discovered as incidental findings at autopsy. Rarely one may rupture during life, and is diagnosed by microscopic examination of the wall of the hematoma cavity.

Ataxia-Telangiectasia

Ataxia-telangiectasia (Louis-Bar syndrome) is a rare autosomal recessive disorder characterized by oculocerebrocutaneous telangiectasias and progressive cerebellar ataxia.[66] The telangiectasias are not identifiable radiographically, but cerebellar atrophy can be seen.[67] Associated infections, diminished lymphoid tissue, and hematologic malignancies frequently dominate the clinical picture.

Hereditary Hemorrhagic Telangiectasia

Hereditary hemorrhagic telangiectasia (Rendu-Osler-Weber syndrome) is an autosomal dominant disorder with telangiectasias predominantly involving the skin, mucous membranes, and gastrointestinal tract.[68] Occasionally there may be involvement of the brain and spinal cord.

Cavernous Hemangiomas

Cavernous hemangiomas (cavernomas) consist of tightly clustered, thin-walled vascular sinusoids separated by fibrous tissue.[69] In contrast to other vascular malformations, the vessel walls of cavernomas lack elastic and muscular tissues and no normal brain tissue can be found between the channels. Cavernous hemangiomas are relatively common, being found in every age group, including the neonatal period.[70] Both sexes are equally affected. Cavernous hemangiomas may present clinically with seizures or hemorrhage.

About three-fourths of cavernous hemangiomas are supratentorial, often subcortical in location.[68] They are multiple in 2.5 percent. Calcification is seen in 11 to 30 percent.[69,70] A heavily calcified variant (hemangioma calcificans) is recognized.[46] MR imaging frequently reveals a compact, masslike lesion with calcifications and relatively few vessels compared to AVMs. Adjacent gliosis or evidence of prior hemorrhage may be noted. A very characteristic low-intensity band ("iron rim") is often seen at the lesion's periphery.[70a] This presumably represents the paramagnetic effects of hemosiderin deposition frequently seen adjacent to these lesions (Fig. 7-9).

Intracranial hemangiomas are associated with several syndromes, including *Riley syndrome* (angiomas of skin and CNS, macrocephaly, pseudopapilledema), *Burke syndrome* (angiomas of skin and CNS), *Kasabach-Merritt syndrome* (hemangiomas with thrombocytopenia), *Divry-Van Bogaert disease* (meningocerebral angiomatosis), and *Cobb syndrome* (cutaneous-meningomedullary angiomatosis).[68] A familial form involving the skin, CNS, and retina has also been described, together with its suitability for MR imaging.[71]

Venous Angioma

Venous angiomas are the most common type of intracranial vascular malformation, being found at autopsy in about 2.6 percent of subjects.[72] Pathologically, dilated venous channels containing elastic fibers are seen separated from each other by normal brain parenchyma. The abnormality may be either a single enlarged vein with multiple tributaries or a group of such veins. Venous angiomas are distributed throughout the brain with a slight preference for frontal lobes, deep cerebrum, and cerebellum.[73] The vast majority are asymptomatic.[46] Those that become symptomatic usually do so by age 30, presenting with either seizures or hemorrhage.[74] Cerebellar venous angiomas may be more prone to bleed than venous angiomas in other locations.[75]

Magnetic resonance has been shown useful in the diagnosis of venous angiomas.[76–78] At the current state of technology, however, CT may be slightly superior for those supratentorially.[76] The characteristic appearance is that of a low intensity tubular structure that

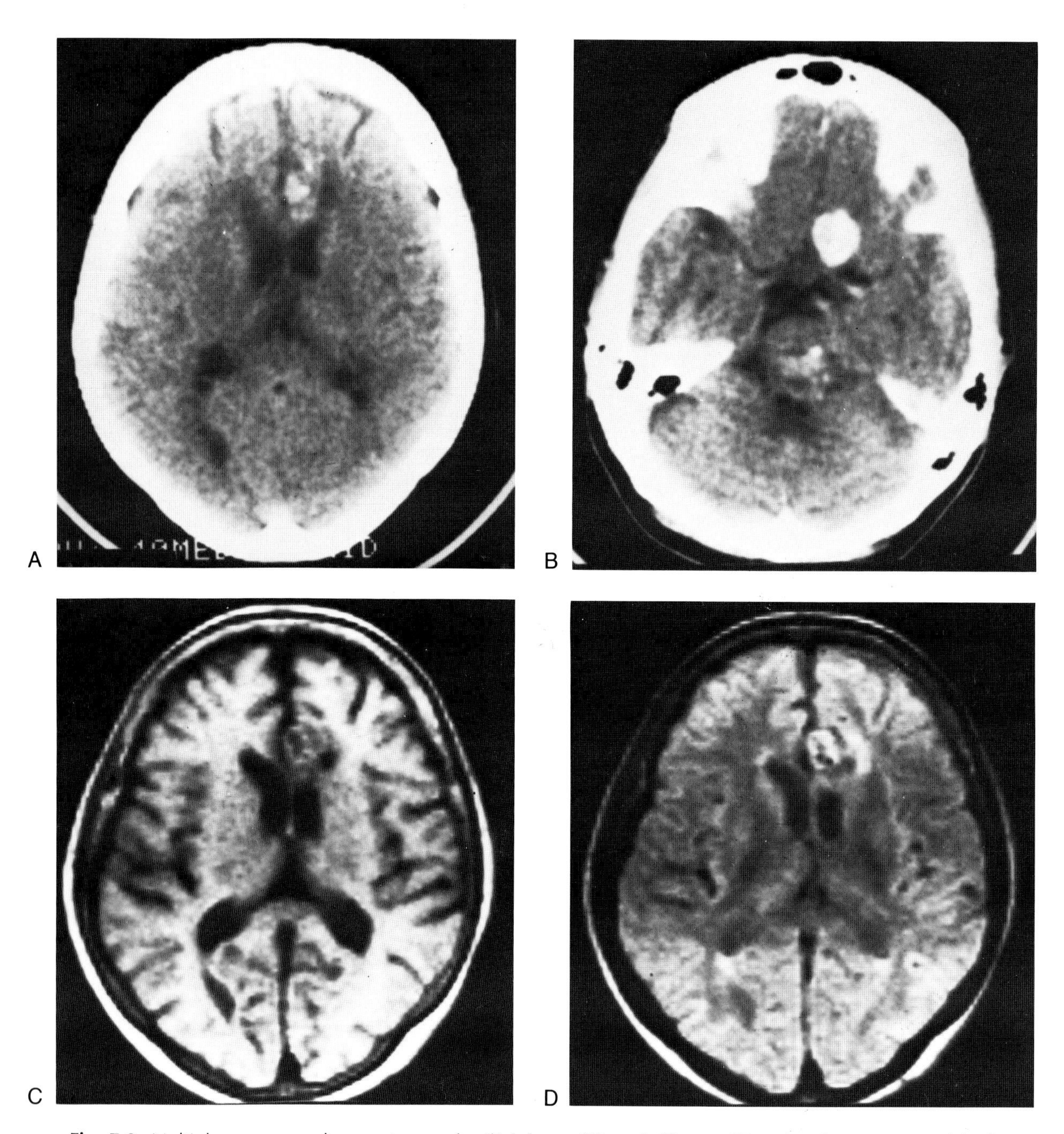

Fig. 7-9. Multiple cavernous hemangiomas, familial form. **(A)** and **(B)** are CT scans showing several high density, contrast-enhancing lesions with small calcifications. **(C)** and **(D)** MR scans show vascular channels and adjacent gliosis. Techniques: **(C)** SE 500/28; **(D)** SE 2000/28. (*Figure continues*).

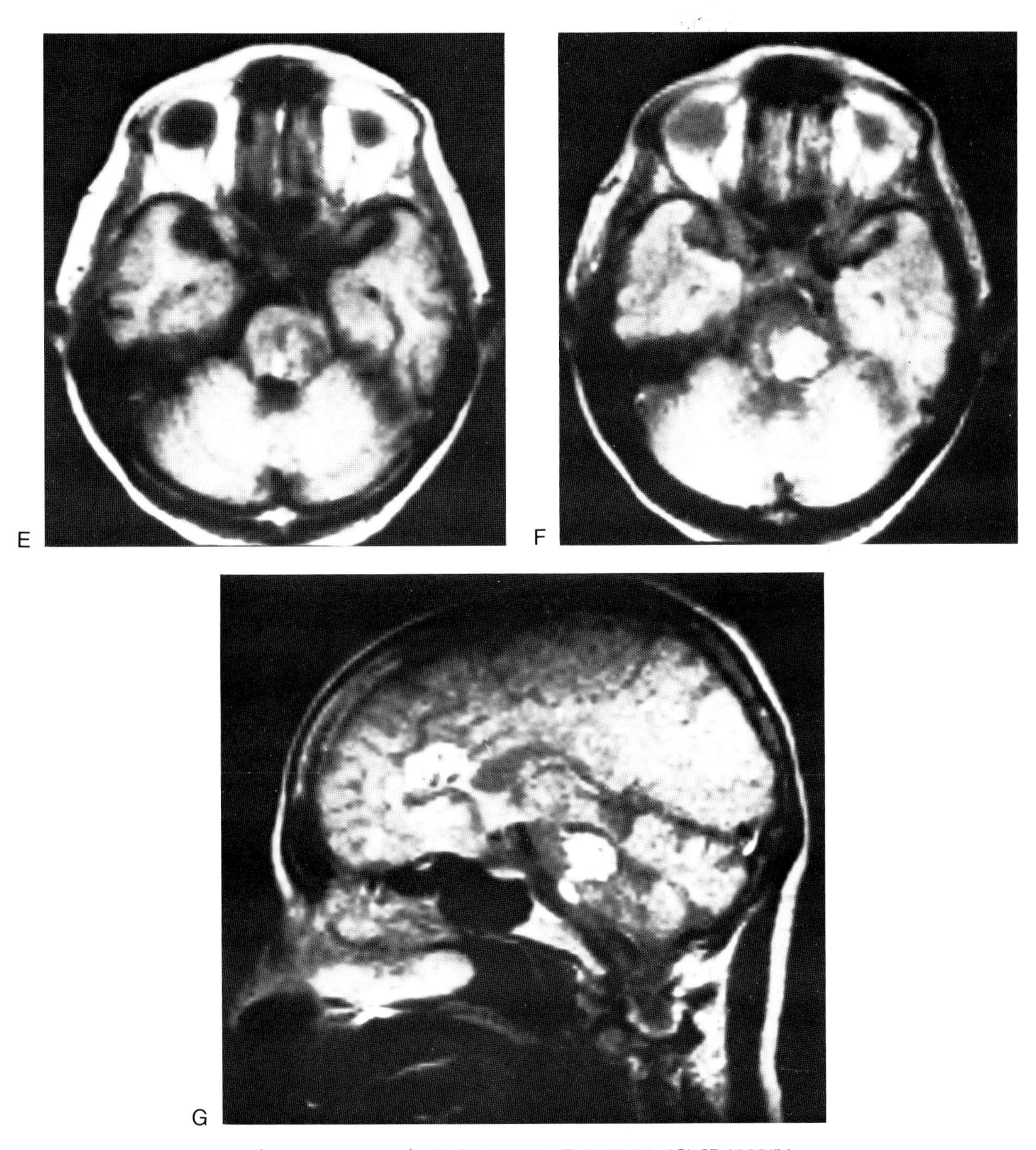

Fig. 7-9 *(continued)*. **(E)** SE 500/28; **(F)** 2000/28; **(G)** SE 1000/56.

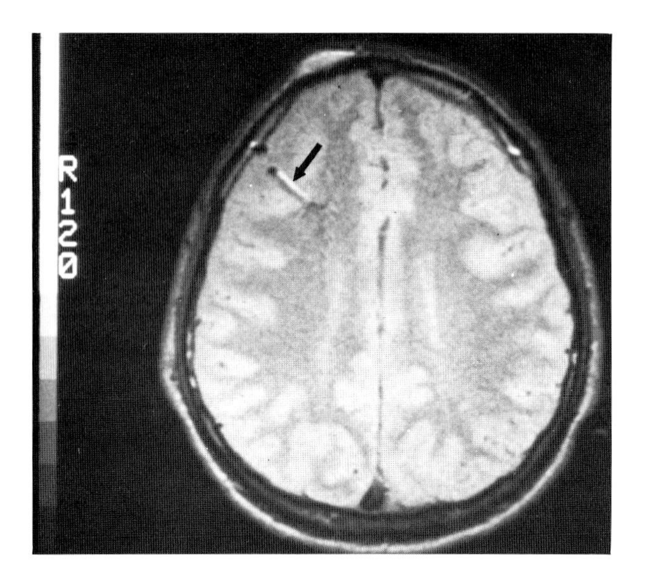

Fig. 7-10. Small venous angioma. Typical "inverted palm tree" appearance and even-echo rephasing.

shows even-echo rephasing due to relatively slow flow (Fig. 7-10). A tuft of deep veins converging to a single large channel ("hydra," "caput medusa," "inverted palm tree") may be noted. The lesion is better seen on T2-weighted than T1-weighted images.[76] Mass effect is absent, but adjacent abnormal signal from the brain parenchyma may be seen particularly in symptomatic patients.[78]

Sturge-Weber Syndrome

The Sturge-Weber syndrome (encephalotrigeminal angiomatosis) is a congenital neurocutaneous vascular anomaly involving the face, eye, and brain. The vascular lesion is primarily a capillary venous angioma, but may also contain components of a true AVM.[19] The following major manifestations comprise the syndrome: (1) port-wine nevus of the face, usually in the territory of the superior branch of the trigeminal nerve, (2) venous angiomas of the leptomeninges, which in the great majority of cases lie in the occipital or occipitoparietal regions ipsilateral to the nevus, and (3) ocular angioma of the choroid with glaucoma.[68] Clinical features of the syndrome include mental retardation, seizures, contralateral hemiplegia, and occasionally cortical blindness.[79]

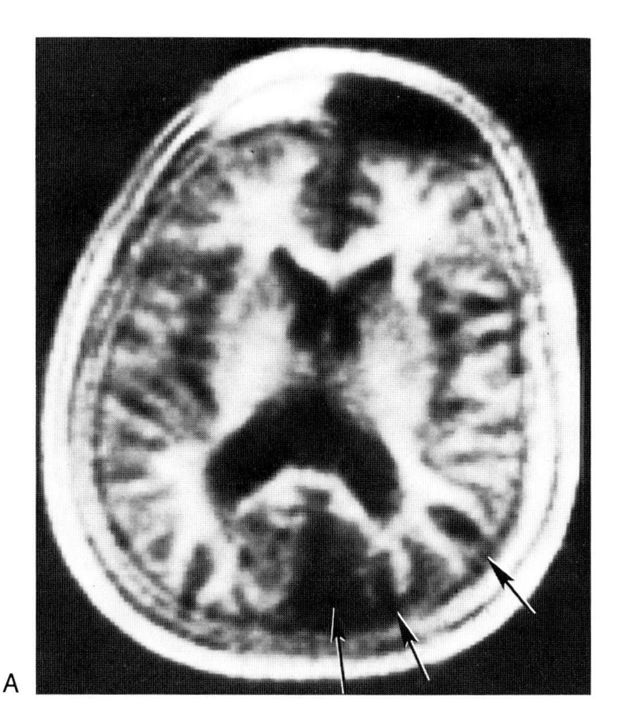

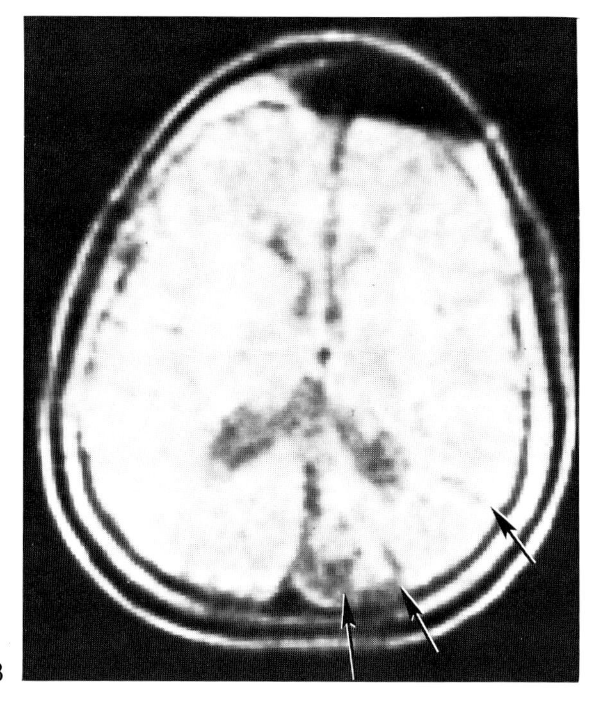

Fig. 7-11. Sturge-Weber syndrome. There is relative atrophy of the left hemisphere especially posteriorly. Characteristic cortical calcifications were seen in the left occipital lobe on CT **(A),** at site of arrows on MR **(B).**

Radiologic manifestations depend upon three features: associated atrophy, gyral calcification, and the vascular channels themselves.[79,80] Hemiatrophy of the involved cerebral hemisphere is usually noted, with smaller hemicranium and secondary thickening of the diploic space. Subarachnoid cisterns and ipsilateral ventricle are often increased in volume. The skull base may be elevated on the affected side and the mastoid air cells may be enlarged. Very rarely, hemihypertrophy instead of hemiatrophy may be present on the side of the nevus.[81] In patients older than 2 years, plain radiographs or CT may reveal parallel, dense curvilinear calcifications in the cerebral cortex. Although once thought to lie within the walls of the vascular malformation, this calcification has clearly been shown to reside in the adjacent superficial cortical gyri. MR is less suited than CT in identifying this calcification at early stages. Contrast enhanced CT or MR may show a blush corresponding to the vascular malformation. Small venous channels may also be seen on noncontrast MR (Fig. 7-11). The usual MR appearance, however, is a nonspecific hemiatrophy, the diagnosis being better made clinically and by CT.

CEREBRAL INFARCTION

Pathophysiologic Principles

Cerebral ischemia means that the blood supply to a part of the brain is insufficient to meet its metabolic needs. Ischemia is not the same as anoxia, which means only lack of oxygen. Ischemia implies that the brain lacks oxygen as well as other nutrients like glucose. Furthermore, the reduced blood flow to ischemic brain allows build up of potentially toxic metabolites such as lactic acid.

If ischemia is temporary, neurologic signs and symptoms may clear with little or no tissue damage. With more severe or prolonged ischemia, neurons and neurologia die, a condition known as cerebral infarction. Important causes of cerebral infarction are listed in Table 7-2 and include atherosclerosis, embolism, and small vessel disease. The physicochemical and pathologic changes induced by ischemia are variable, being governed by several factors:[82–95a]

Process Causing Ischemia

As will be discussed in the ensuing sections, the natural history of cerebral infarction depends to some extent upon the cause of ischemia. About one-third of infarctions are due to thrombosis of an atherosclerotic vessel. These infarctions are typically large and "pale" (nonhemorrhagic). Slightly fewer than one-third of infarctions are due to cerebral embolus from a cardiac or noncardiac source. Emboli typically break up a few days after vascular occulusion subjecting ischemic brain to reperfusion and often resulting in hemorrhage. Small vessel occlusion due to arteriolosclerosis frequently produces small deep infarctions (lacunes). Systemic hypotension results in "watershed infarctions" along border zones of weakly perfused brain between major arterial distributions.

Vascular Supply to the Infarcted Area

The presence or absence of collateral circulation dictates to some extent the evolution of an infarction. Many areas in the cortex may relatively quickly acquire collateral supply at the periphery of an infarct. A patent circle of Willis may allow even occlusion of the internal carotid artery to take place without clinical sequellae. By contrast, many deep perforating arteries such as those to the internal capsule and basal ganglia are essentially end arteries. Occlusion of these often results in complete infarction and necrosis of the involved structures.

Vulnerability of Tissue to Ischemia

Certain cells and regions of the brain are more sensitive than others to ischemia. The neuron is most sensitive, followed in turn by the oligodendrocyte, astrocyte, and other parenchymal cells. In the gray matter of the cerebral cortex, cells in intermediate layers 3, 5, and 6 are most sensitive. In the hippocampus, Sommer's sector is most sensitive. In the thalamus the median nuclei are more sensitive than the anterior nuclei, which in turn are more sensitive than nuclei in the dorsal, posterior, and lateral groups. Selective vulnerability also depends upon the age of the patient (Table 7-7).

Table 7-7. Selective Vulnerability of Brain Regions to Ischemia by Age, in Order of Decreasing Susceptibility

28 Week Fetus	*Infant*	*Adult*
Hippocampus	Hippocampus	Cerebral cortex
Pontine nuclei	Cerebral cortex	Hippocampus
Globus pallidus	Striatum	Cerebellar cortex
Striatum	Thalamus	Globus pallidus
Cerebral cortex	Globus pallidus	Thalamus
Cerebellar cortex	Cerebellar cortex	Striatum
	Pontine nuclei	Pontine nuclei

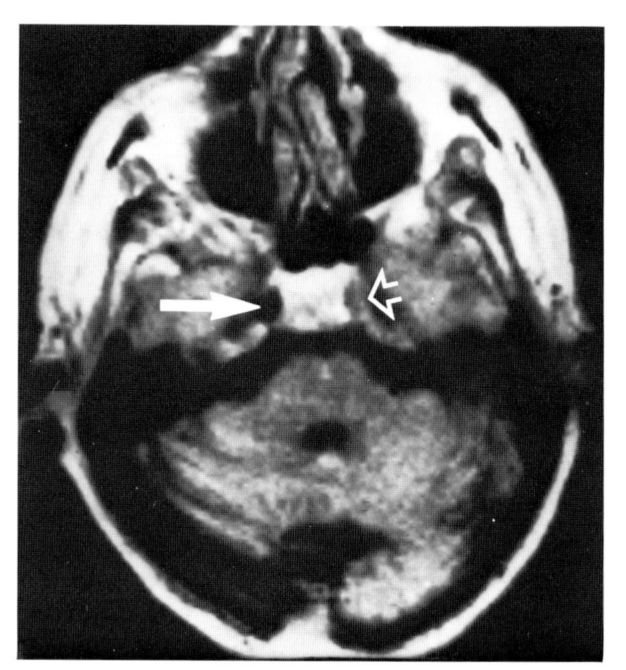

Fig. 7-12. Occlusion of left internal carotid artery shown on axial SE 500/28 image. On the right, rapid flow in a patient right internal carotid artery produces areas of signal void (solid arrow). On the left a relatively high signal is seen in the carotid artery, consistent with occlusion (open arrow).

Underlying Vascular Disease

If vessels in a region are diseased (by vasculitis or atherosclerosis) only small emboli or thrombi may be necessary to occlude them. Likewise the ability of the surrounding diseased vasculature to provide collateral supply may be hindered. Conversely slow or prolonged partial occlusion of a vessel may allow development and maturation of collateral channels, which may relatively protect the ischemic area from further insults.

Thrombic Infarction

Large artery thrombotic occlusion is the most common cause of cerebral infarction.[83] The underlying vessel is nearly always diseased (narrowed) by artherosclerosis. Diminished perfusion pressure and platelet agregation result in thrombus formation, which occludes the vessel. Subsequently distal ischemia and infarction develop in a characteristic pattern corresponding to the distribution of the artery. A knowledge of arterial territories in multiple planes is essential for MR recognition and diagnosis. Lesions that cross arterial boundaries are not likely to represent thrombotic infarctions.[96]

During the first 6 hours following cerebral artery occlusion, only minor changes such as microvacuolization can be observed histiologically. CT is usually normal during this time.[106] However, tissue water has been shown to increase 2 to 3 percent during this period.[86,98] This early cytotoxic edema produces prolongation of brain T1 and T2 values.[99–101] Accordingly, MR may detect cerebral infarction consistently by 4 hours, frequently before 2 hours, and occasionally at 30 minutes.[102–105] In the detection of hyperacute (less than 6 hours old) ischemic infarction, therefore, MR is superior to CT.[106,107] The occluded vessel may sometimes be identified on MR by finding clot within its lumen (Fig. 7-12).

Within the first day following arterial occlusion, progressive edema, cell death, and loss of myelin are noted. Early breakdown of the blood-brain barrier (BBB) occurs. CT will become positive in over 90 percent of patients at this time.[83,97] Contrast enhanced CT will usually not demonstrate breakdown of the BBB unless very high doses are used.[108] MR scanning using Gd-DTPA contrast routinely demonstrates enhancement during the first 24 hours, however.[109] MR can detect tiny amounts of paramagnetic contrast material while relatively larger quanti-

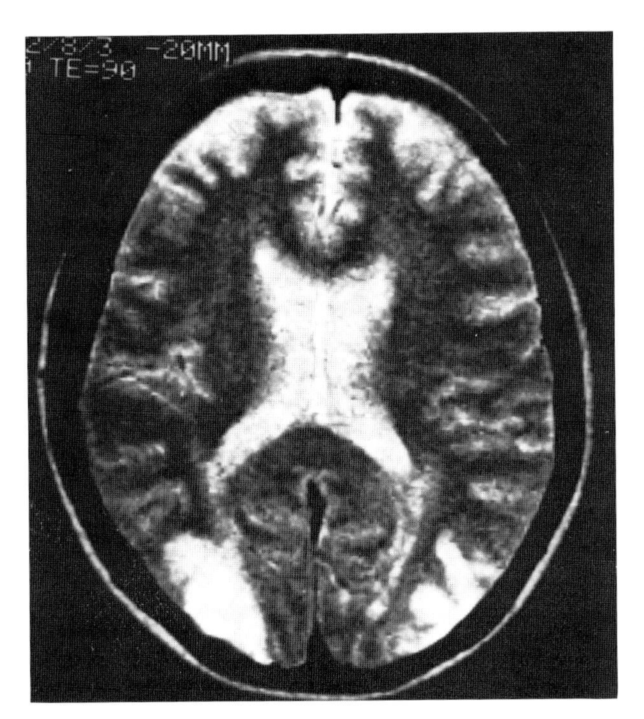

Fig. 7-13. Acute right occipital infarction seen on this SE 2000/90 image. The left occipital stroke is old, but the age of each cannot be determined by MR signal characteristics reliably, unless mass effect or vasogenic edema are present.

ties of iodinated contrast are required for detection by CT.[110]

During the first week of an ischemic stroke, progressive edema and mass effect are noted (Fig. 7-13). Mass effect is maximal on days 3 and 4 corresponding to the time of greatest neurologic deficit.[84,86,97] Brain herniation may occur at this time. Mass effect decreases significantly after the first week and should completely resolve in 12 to 21 days.[83]

During the first week postinfarction, most patients will exhibit CT abnormalities (mass effect, obscured gray-white interface, low density, equivocal enhancement).[83] MR remains abnormal with prolonged T1 and T2 values. However, T1 may actually be slightly shorter than it was during the first 24 hours, although it is still prolonged compared to that of normal brain.[105] This decrese in T1 may result from increased protein content of the brain edema.

In the 2nd and 3rd weeks following occlusion, enzymatic digestion and removal of necrotic tissue is noted pathologically. Neovascularity develops at the periphery of the infarct and maximal permeability of the BBB is seen. Mass effect largely resolves. Maximal contrast enhancement is noted on CT during this period and may persist for several months.[83] MR contrast enhancement is still present, but is not as striking as that seen at 24 to 72 hours.[109]

In the months following an ischemic infarction a spectrum of appearances is possible. At one extreme there can be nearly complete recovery from infarction with only minimal residual changes. At the other extreme there is gliosis, atrophy, and cystic encephalomalacia of the infarcted brain. The leptomeninges become thickened as well. The vessel supplying the region may remain occluded or recanalize.

With revascularization of the ischemic area, T1 and T2 may return to normal or near normal (Fig. 7-14).[120] Alternatively, cystic encephalomalacia shows further prolongation of T1 and T2, resembling CSF (Fig. 7-15). A comparison of pathologic changes with MR and CT appearances in ischemic stroke is presented in Table 7-8.

Embolic Infarction

Arterial embolus is a common cause for cerebral infarction, accounting for 30 to 50 percent of lesions in most series.[83] The embolus may be of cardiac or noncardiac origin (Table 7-2). Cardiac emboli frequently result from intracardiac thrombus, which may form in aneurysms, after myocardial infarction, or in patients with atrial fibrillation. Endocardial and valvular vegetations, from bacterial or marantic endocarditis, are also important sources of emboli. Paradoxical embolism from the right heart may occur with cyanotic congenital malformations. Noncardiac embolism generally results from the showering of thrombus or atheromatous material from the aorta or carotid arteries. Other noncardiac emboli include fat, tumor, and air. Arterial embolus is also a feared complication of cerebral arteriography.

Embolic infarction frequently follows a different clinical and radiographic course than thrombotic infarction. In thrombotic infarction the artery usually remains permanently occluded. With embolic infarction, however, the embolus usually undergoes lysis between the 1st and 5th days.[92] This results in reestablishment of arterial perfusion to the previously

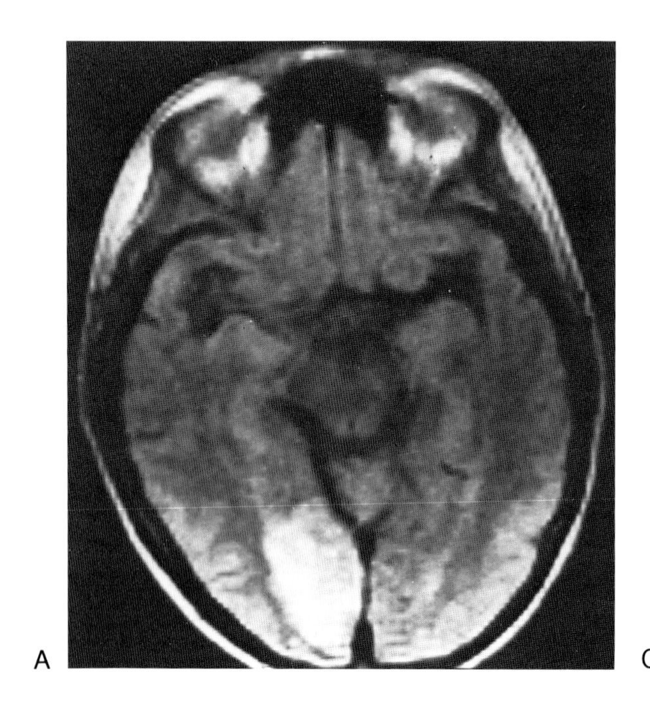
A

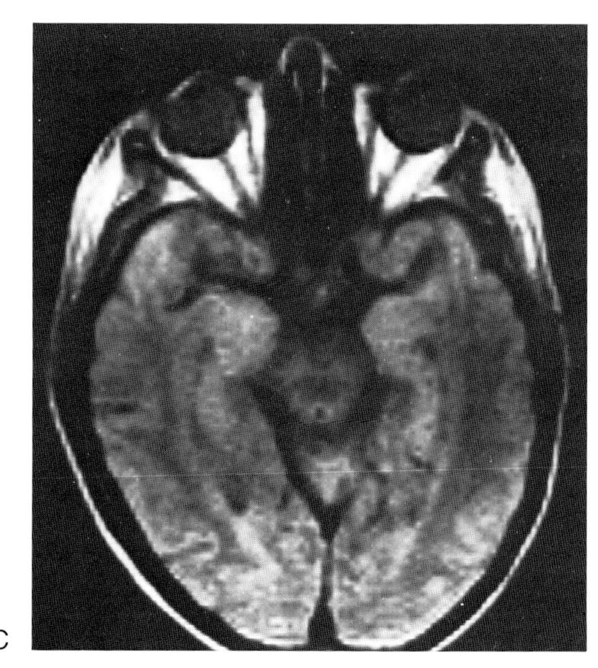
C

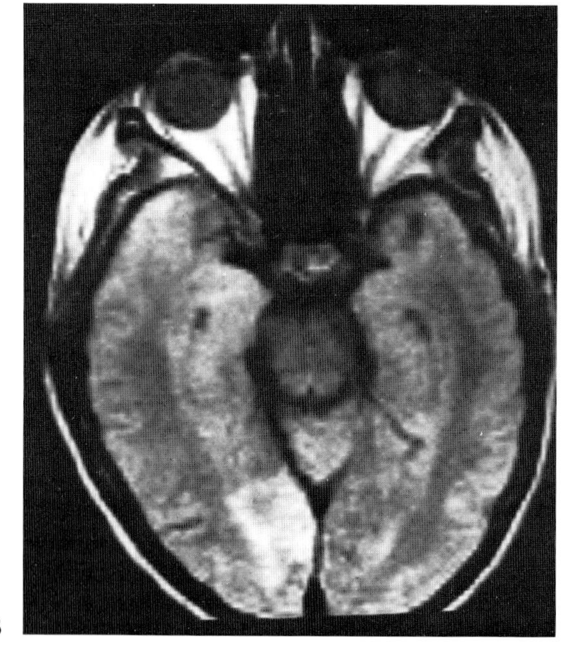
B

Fig. 7-14. Evolution of a cerebral infarction seen on SE 2000/28 MR images. **(A)** Subacute infarction, imaged about 5 days postictus. **(B)** Six weeks later, the area of abnormal signal is smaller and less intense. **(C)** Three months later, the occipital lobe has returned to near normal appearance.

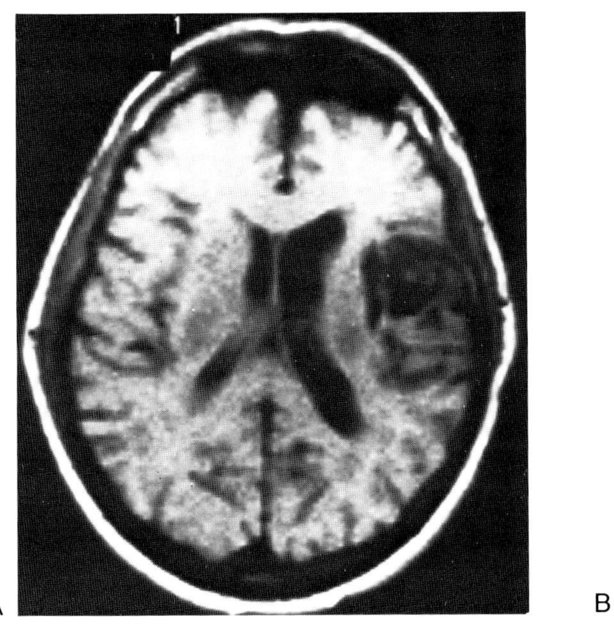

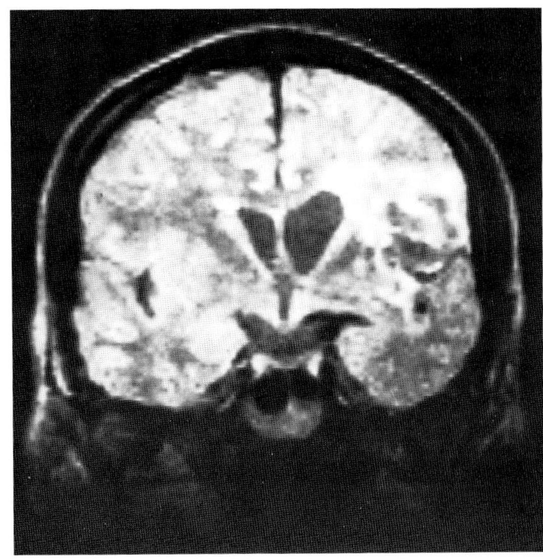

Fig. 7-15. Old left middle cerebral artery infarction that did not heal, but progressed to gliosis and cystic encephalomalacia. **(A)** SE 500/28. **(B)** SE 1500/56.

Table 7-8. Evolution of Thrombic Infarction

Time after Thrombosis	*Pathology*	*CT Findings*	*MR Findings*
0–6 hrs (Hyperacute)	Early cytotoxic edema	Usually none. Occasionally obscured, gray-white interface	↑ T1 and T2 routinely seen at 4 hrs, often sooner
6–24 hrs (Acute)	Progressive edema Loss of myelin Cell death Breakdown of BBB	CT positive in 80% Low density Effaced sulci No contrast enhancement (CE)	Further ↑ T1, T2 Gd-DTPA contrast enhancement observed, maximal 24–72 hrs
1–7 days (Subacute)	Neutrophils/macrophages Mass effect (maximal day 3) Herniation possible	Low density seen 98% Mass effect CE equivocal	T1 may shorten somewhat, but is still abnormally long
7–14 days (Established)	Necrosis centrally Peripheral neovascularity Maximal permeability of BBB	CE seen ↓ Mass effect	T1, T2 stable or longer
Over 14 days (Chronic)	A spectrum ranging from complete recovery to cystic encephalomalacia	CE persists 2–3 months Sharply demarcated low density area	With repair T1 and T2 normalize With gliosis, cystic change T1 and T2 ↑

occluded arterial bed. This arterial bed, damaged by ischemia has lost its normal ability to autoregulate regional blood flow. With clot lysis and loss of autoregulation, the infarcted brain may be exposed to higher perfusion pressures and undergo hemorrhage.[85,87] Infarct hyperemia ("luxury perfusion") is noted especially in the cortex and deep gray matter because these regions have a large capillary bed that dilates passively with reperfusion.[90]

Because of their different pathophysiologies, thrombotic and embolic infarctions may have different natural histories and appearances on MRI (Table 7-9). While thrombotic infarctions tend to remain ischemic and "pale," embolic infarctions frequently undergo hemorrhage. Hemorrhagic transformation may occur with thrombotic infarctions, risk factors being hypertension, large size infarction, and the use of heparin. With hemorrhage T1 values shorten while T2 values remain prolonged (Fig. 7-16). Extracellular (vasogenic) edema seen with embolic infarctions may spread along white matter tracts outside the normal arterial distribution limits seen with thrombotic infarctions, which manifest mostly intracellular (cytotoxic) edema. This may occasionally be helpful in radiographic differentiation of the two types of infarctions.

Lacunar Infarction

Lacunes are small focal infarctions involving the brain stem and deep central structures (basal ganglia, internal capsule, thalamus). They result from arteriolosclerosis involving small penetrating intracranial arteries such as the striates and basal perforators.[112] These vessels are functional end arteries that range in size from 0.2 mm to 0.8 mm at their origins. Because they supply blood to many important nuclei and white matter pathways, occlusion of even one of these small arteries may produce significant neurologic deficit.

Lacunar infarcts typcially occur in older patients with chronic hypertension: large vessel atherosclerosis frequently coexists with the small vessel arteriolosclerosis. Clinical syndromes associated with lacunar infarction include pure motor hemiparesis, pure sensory stroke, dysarthria-clumsy hand syndrome, and ataxia-paresis.[113] Small lacunes in noncritical locations may be asymptomatic. Lacunar infarctions typically

Table 7-9. Comparison of Thrombotic and Embolic Cerebral Infarctions

	Thrombotic Infarction	*Embolic Infarction*
Risk factors	Hypertension Atherosclerosis	Cardiac arrhythmias or infarction
Clinical onset	Often has prodrome	Sudden onset
Pathophysiology	Usually permanent occlusion of cerebral artery	Embolus usually lyses in 1–5 days, subjecting brain to high reperfusion pressures Luxury perfusion seen
Hemorrhagic transformation of infarction	Uncommon	Common
Type of edema	Mostly cytotoxic	Cytotoxic and vasogenic
Maximal leakage of BBB	Late (2–3 weeks) postinfarction, when mass effect gone	Early (1st week) often coexists with mass effect and edema
MR findings	↑ T1 and ↑ T2	↑ T2 but ↓ T1 with hemorrhage

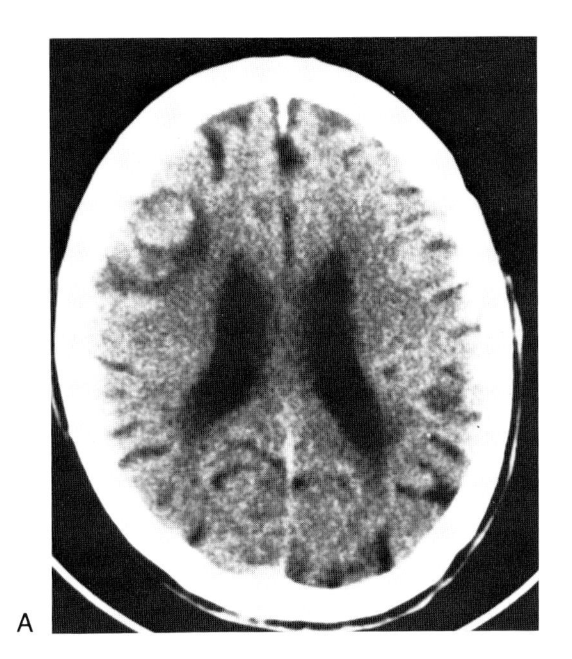

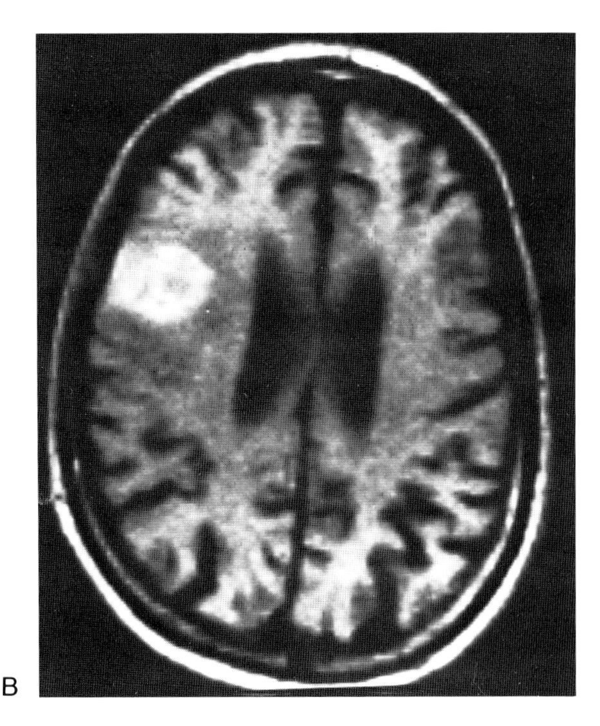

Fig. 7-16. (**A** and **B**) Hemorrhagic infarction secondary to an embolus. Short T1 of methemoglobin results in central high signal on T1 weighted images like this SE 500/28 in **(B).**

evolve into cystic encephalomalacia over 4 to 6 weeks following inception.

On MRI lacunar infarcts are small (2 to 20 mm) oval or comma-shaped lesions demonstrating prolongation of T1 and T2 values (Fig. 7-17).[100] We have observed small lacunes by MRI up to a week prior to their appearance on CT. MRI may detect brain stem lacunes unseen by CT at any stage. The multiplanar format of MRI may aid in precise anatomic localization of a lacune.

Watershed Infarction

Watershed, or hemodynamic, infarctions occur in border zones between major arterial territories (Fig. 7-18). Important watershed regions include: (1) a parasagittal strip over the cerebral convexities between the ACA and MCA distributions, (2) a triangular area in the parieto-occipital region shared by the ACA,

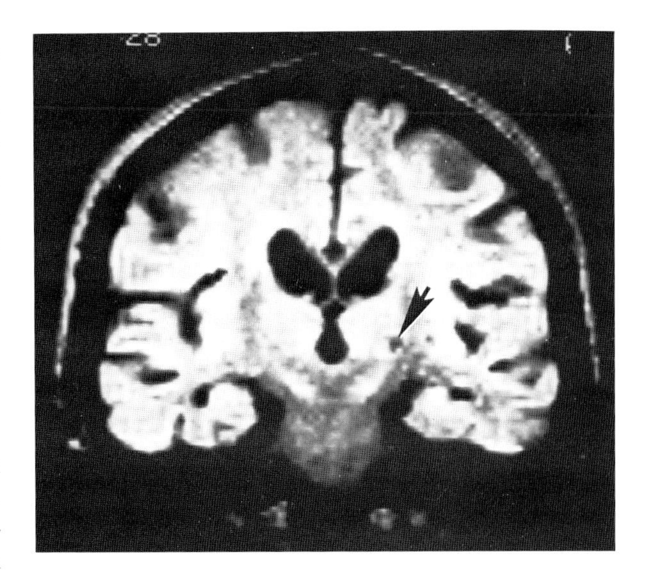

Fig. 7-17. Small lacunar infarct (arrow) in the basal ganglia region is well seen on this SE 1500/28 coronal image.

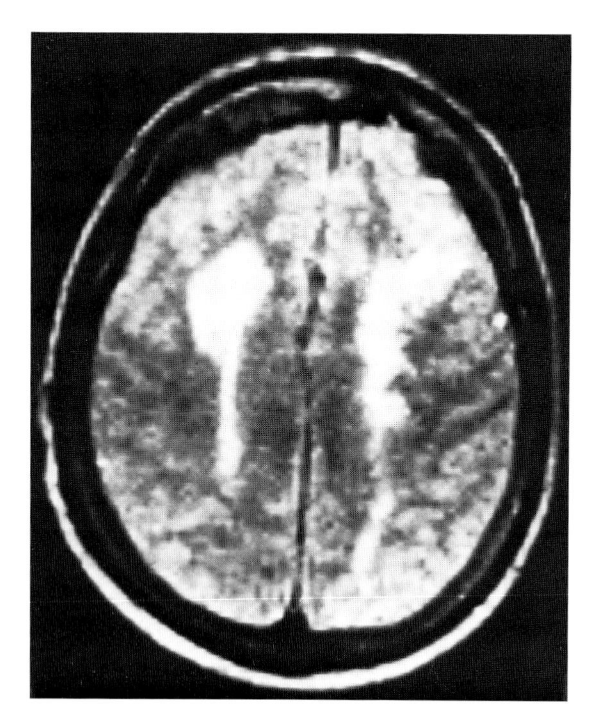

Fig. 7-18. Bilateral watershed infarcts in characteristic locations between anterior and middle cerebral artery distributions.

MAC, and PCA, and (3) a deep curvilinear zone between the superficial and deep branches of the MCA. Other less commonly seen locations for watershed infarcts include the periventricular white matter, the basal ganglia, and the cerebellum.[114]

Because of the baseline marginal perfusion to these watershed areas, small changes in circulatory dynamics can cause ischemia and infarction. Pre-existing occlusive disease in the carotid arteries or their major intracranial branches may contribute to the risk. Watershed infarctions often occur at night, caused by decreased blood pressure and cardiac output during sleep. Watershed infarction may also occur during systemic hypotension from other causes, such as a cardiac arrythmia or sepsis.

Magnetic resonance imaging demonstrates increased T1 and T2 in the infarcted areas, similar to that seen with thrombotic infarction.[100] Watershed infarctions can be diagnosed by their characteristic shapes and locations. Watershed infarctions are often bilateral as shown in Figure 7-18.

INTRACEREBRAL HEMORRHAGE (ICH)

Primary ICH (i.e., that not associated with vascular malformation, tumor, or infarction) accounts for about ten percent of stroke syndromes.[115] Most cases are associated with severe systemic hypertension, although such history may be lacking in up to one-fourth of cases. Hypertensive ICH is thought to be caused by rupture of microaneurysms of deep penetrating arteries. These microaneurysms, originally described by Charcot and Bouchard, are fusiform or saccular dilatations that result from the effects of systemic hypertension and hyaline degeneration in the walls of small penetrating arteries.[116] When these microaneurysms rupture, intracerebral hematoma occurs, often with acute neurologic deficit. Deep brain structures are primarily affected: basal ganglia (52 percent), thalamus (25 percent), cerebral hemisphere (8 percent), cerebellum (6 percent), brain stem (4 percent), and other (5 percent).[113]

Most patients are over 40 at the time of presentation and carry a diagnosis of systemic hypertension or athersclerosis. A clinical ictus (headache or seizure) often occurs immediately prior to hemorrhage. Neurologic deficit is maximal initially or within the first few hours following hemorrhage. Unlike ICH from aneurysm or AVM, continued active bleeding is uncommon after the first day.[117] Mass effect is usually evident in the first 24 hours and may persist for up to 4 weeks.[118] Subfalcine or transtentorial herniation may occur. No change in hematoma size is seen in the first 2 weeks. At the end of the 1st week, peripheral contrast enhancement is noted on CT and MR corresponding to a phase of neovascularity at the hematoma margin.[119]

Magnetic resonance changes in ICH will be described more fully in Chapter 10, Trauma and Surgical Change. Briefly, an ICH presents as a region with relatively normal T1 and T2. In the first 24 hours marked shortening of T2 occurs while T1 remains the same or shortens. With aging T2 lengthens while T1 frequently remains short. The central and peripheral parts of an ICH may vary in their course of MR signal changes. Furthermore, field strength may play a role in determining characteristics as well.

Prognosis after ICH depends upon size and location of the bleed. Medial ganglionic hematomas are worse than lateral ones, because the former frequently affect the internal capsule and cause transtentorial herniation.[113] Rupture of ICH into the lateral or third ventricles is a grave prognostic sign, but cerebellar hematomas that decompress into the fourth ventricle have a better prognosis than those that do not.[120] Cerebellar hemorrhage greater than 3 cm in diameter is a surgical emergency, but smaller posterior fossa hematomas may resolve without surgery, and all less than 1.2 cm do.[121]

CEREBRAL VENOUS THROMBOSIS

Thrombosis of the cerebral veins or dural venous sinuses may occur in a large number of clinical situations (Table 7-11).[122,123] In adults the more important causes include surgery, trauma, infection, and hypercoagulable states (pregnancy, steroids, neoplasia). In infants, children, and debilitated adults, dehydration is an important etiology. The symptoms of veno-occlusive disease depend upon the location, cause, size, and rate of thrombus formation. Accordingly clinical presentation may vary from insidious (headache, blurred vision) to catastrophic (stroke, coma, death).[124]

Angiography and digital subtraction angiography are the most accurate methods for diagnosing venous thrombosis, but are invasive and potentially dangerous. CT is capable of imaging venous thrombosis, but generally requires administration of intravenous contrast and is limited to evaluation of mostly larger vessels and sinuses.[123,125] A number of CT pitfalls are also recognized.[125,126] MRI may become the modality of choice for the diagnosis of venous thrombosis, because it has proved capable of imaging both normal blood flow and its cessation.[127]

The large cerebral veins and sinuses are easily identified on routine MR brain images. In general, flow is relatively swift in the larger veins, which appear dark on most imaging sequences due to high velocity signal loss.[128] In smaller veins and occasionally in the larger venous sinuses blood flow may be sluggish. In this situation high signal within the vessel may be noted as a result of even-echo rephasing or entry phenomenon.[129]

The evolution of venous thrombosis by MR has been studied, and three phases can be identified.[130,130a] Initially (within the first few days after thrombosis) the vessel loses its normal flow void pattern. It becomes isointense with brain parenchyma on T1-weighted images, but hypointense on T2-weighted images. In the intermediate phase (1 to 2 weeks after occlusion) thrombus converts to a high intensity on T1-weighted sequences, followed by high intensity changes on T2-weighted images. This pattern of change occurs first in the periphery and proceeds to fill the center of the vessel. In the late phase (2 weeks post-thrombosis) vascular recanilization may be seen with restoration of flow in the previously occluded vessel. These changes are listed in Table 7-11. Examples of dural sinus thrombosis in various stages of evolution appear in Figures 7-19 through 7-21.

Table 7-10. Factors Associated with Cerebral Venous Thrombosis

Arteriovenous malformation
Arterial occlusion
Collagen vascular disease
Dehydration or rapid diuresis
Diabetes
Drugs (including oral contraceptives, steriods)
Heart disease
Hematologic malignancies (including polycythemia vera)
Hypercoagulable states (including disseminated intravascular coagulation, sickle cell)
Idiopathic
Infection
Pregnancy and puerperium
Surgery
Trauma

Table 7-11. High-Field MR Findings in the Evolution of Venous Thrombosis

0–1 week	Normal flow void not recognized. Vessel becomes isointense with brain on T1-weighted images and hypointense on T2-weighted images.
1–2 weeks	High intensity first noted on T1-weighted images, later on T2. Hyperintensity spreads from peripherally to centrally in larger vessels.
After 2 weeks	Recanalization begins. Flow void noted in recanalized channels.

(Adapted from Macchi PJ, Grossman RI, Gomori JM et al: High-field MR imaging of cerebral venous thrombosis. J Comput Assist Tomogr 10:10, 1986, Raven Press, New York.)

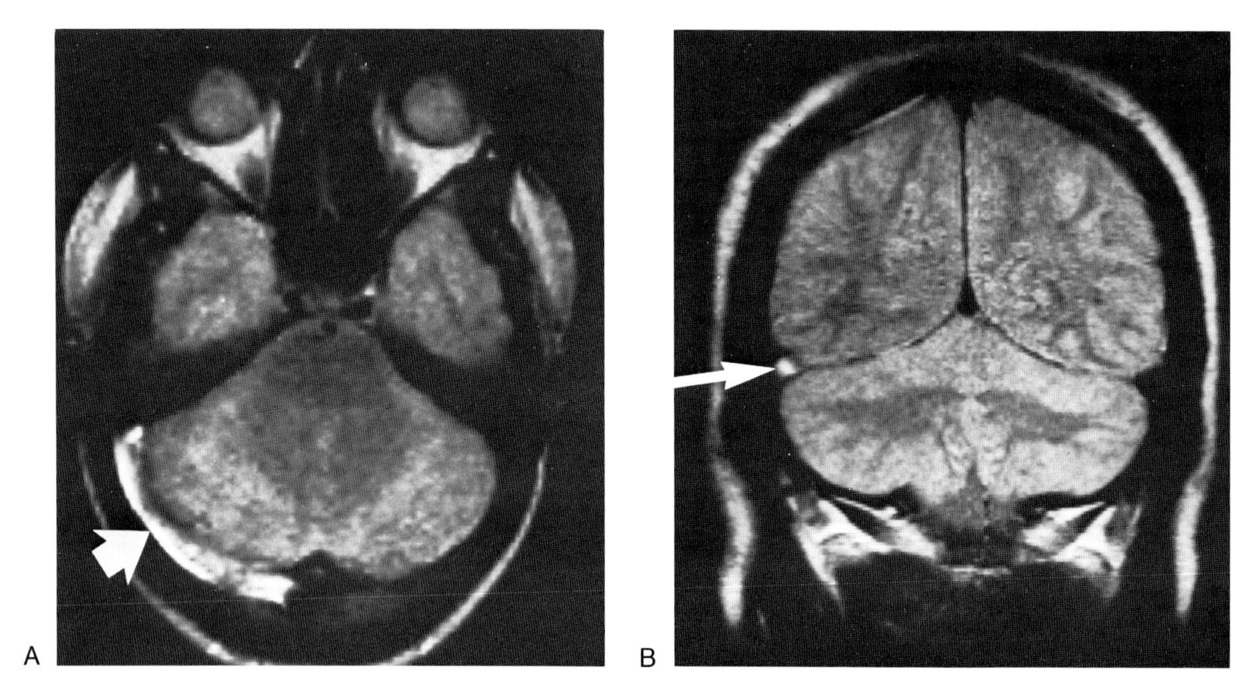

Fig. 7-19. (**A** and **B**) Thrombosis of right transverse sinus, secondary to occlusion by a temporal bone tumor. Abnormal high signal noted on axial and coronal images.

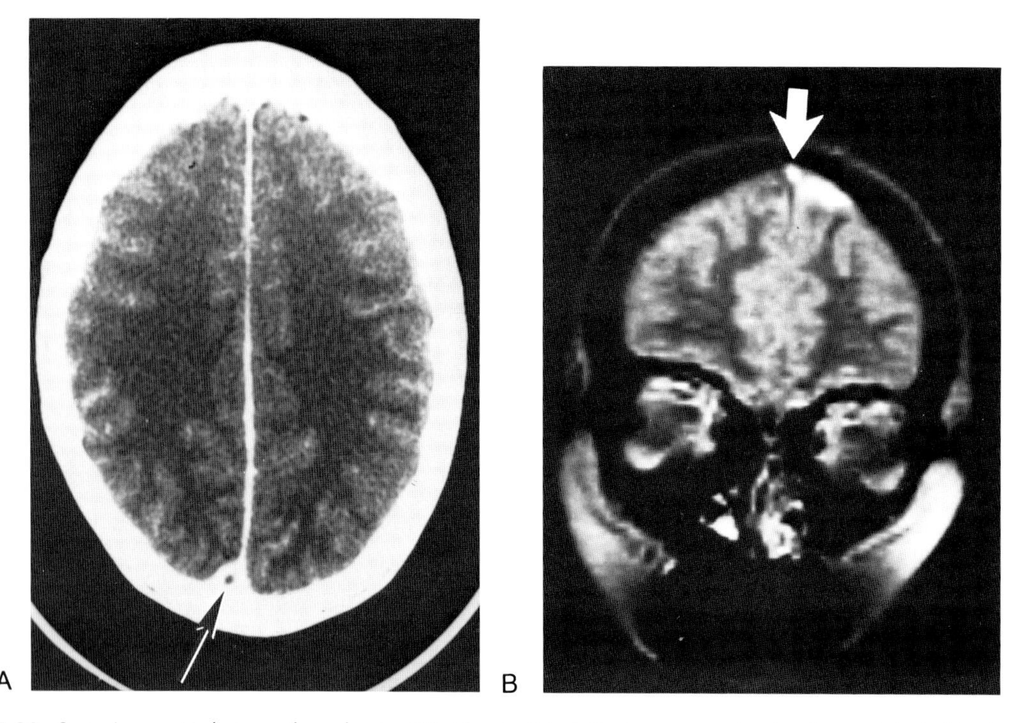

Fig. 7-20. Superior sagittal sinus thrombosis. **(A)** Classic "delta" sign on contrast enhanced CT. **(B)** Coronal SE 2000/60 MR image shows abnormal signal in the superior sagittal sinus consistent with thrombosis (arrow). Adjacent high signal in left frontal lobe was thought to represent a venous infarction.

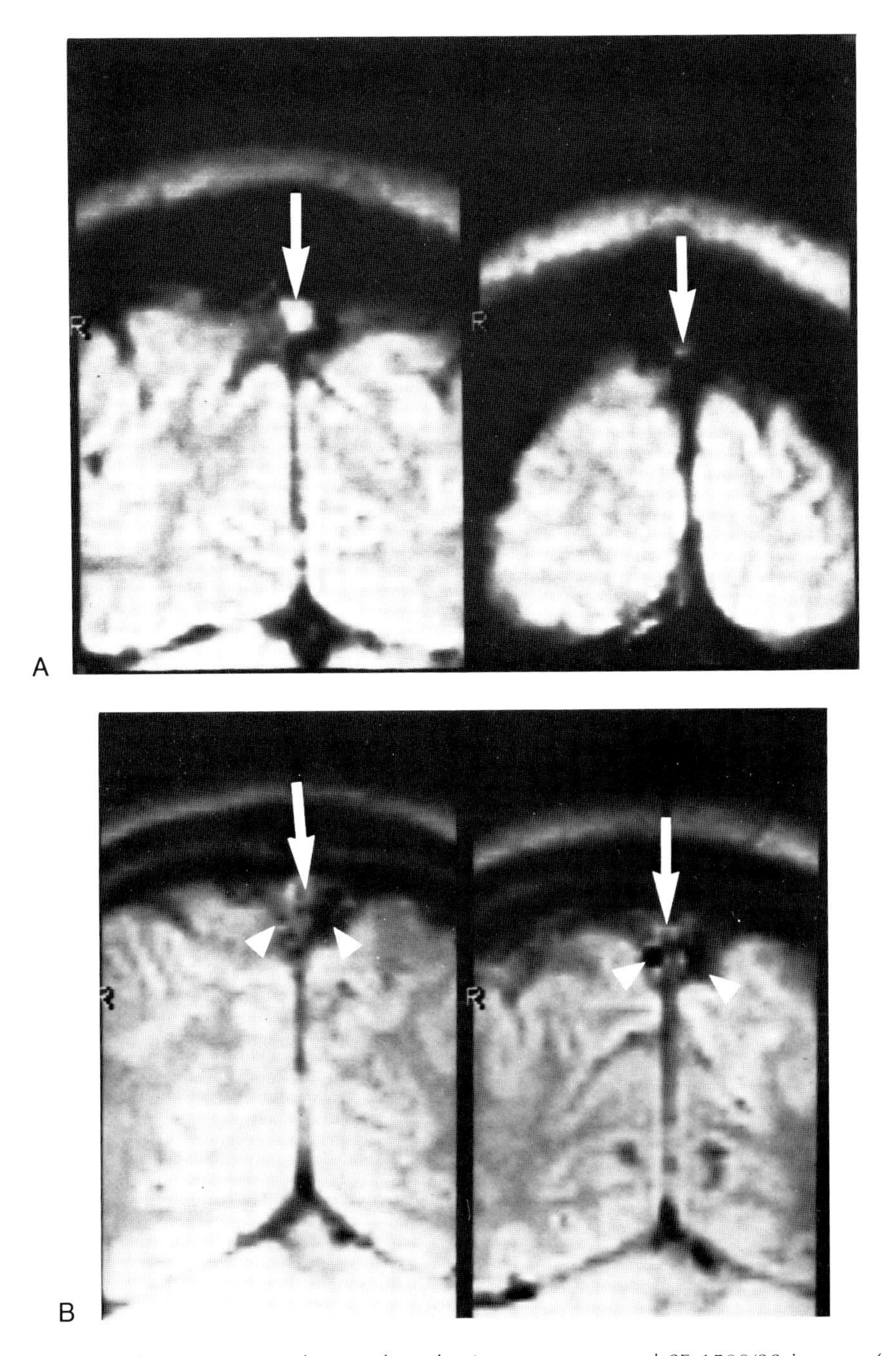

Fig. 7-21. Evolution of superior sagittal sinus thrombosis seen on coronal SE 1500/28 images. **(A)** Within the 1st week after sinus occlusion, high signal (clot) is seen in the sinus and few collaterals. **(B)** A few weeks later the clot signal has decreased in intensity (large arrow). Venous recanalization may be visualized as foci of flow void within the thrombus (arrowheads).

VASCULITIS

Inflammatory changes in the cerebral vasculature may produce neurologic symptoms. Vasculitis may be primary or secondary and involve arteries, veins, or both.[131] Primary vasculitis is frequently one manifestation of a systemic disorder, such as collagen vascular disease, granulomatous arteritis, or tertiary syphilis. Drugs and chemical agents may also cause primary vasculitis. Secondary vasculitis is usually a result of CNS infection such as bacterial or granulomatous meningitis. While primary vasculitis affects arteries predominantly, secondary vasculitis may affect both veins and arteries.

Systemic Lupus Erythematosus

Systemic lupus erythematosus (SLE) is a chronic relapsing inflammatory disease of uncertain etiology. It affects all ages, but more frequently young and middle-aged females. SLE is a multisystem disease that can involve the joints, heart, lungs, kidneys, or CNS. Neurologic abnormalities may be noted in up to 75 percent of patients, usually within the first year of diagnosis, but may occur at any time, especially when the disease is active elsewhere.[132] Cerebral manifestations include dementia, psychosis, seizures, headaches, chorea, and focal neurologic deficits.[132,133] Pathologic findings are based around small vessel vasculopathy with hyalinization, thrombosis, and lymphocytic infiltration. Both microscopic and large infarctions may be seen, as well as hemorrhage. It is uncertain whether cerebral atrophy is a part of the SLE disease process or is a complication of steriod therapy.[134]

MR has proved much more sensitive than CT in detecting cerebral lesions in SLE.[135–137] In each of three reported series, no CNS lesions have been seen by CT and not by MR. Conversely MR has detected many CNS lesions not identified on CT. The presence of focal findings on MR images may be helpful to exclude drug effects or primary psychiatric illness in patients with CNS lupus. Several MR patterns have been noted in SLE including microinfarctions, macroinfarctions, and focal cortical edema.[137] All areas demonstrate prolonged T1 and T2 values. Examples of micro- and macroinfarctions in two patients with SLE are presented in Figure 7-22.

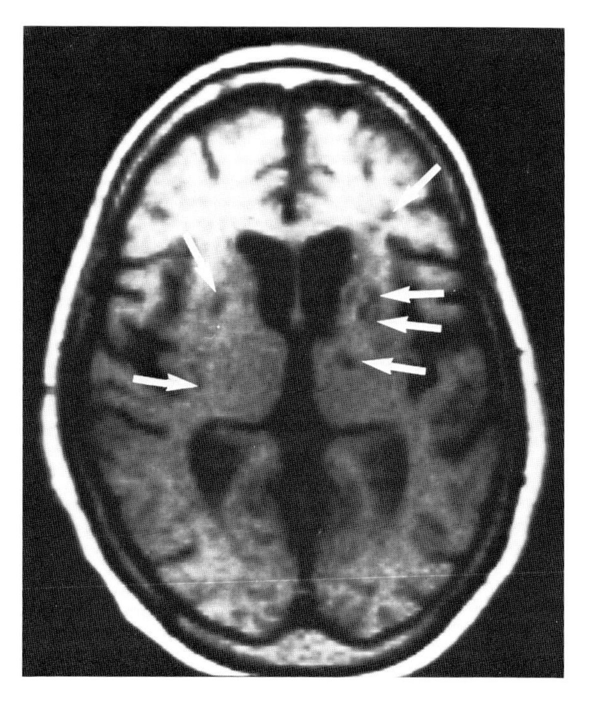

Fig. 7-22. Multiple small infarcts in a young patient with vasculitis secondary to lupus.

Isolated cases of MR findings in cerebral vasculitis secondary to Behçet disease and meningovascular syphilis have been reported.[138,139] Focal areas of increased T1 and T2 corresponding to regions of ischemia were identified in each case. The findings of vasculitis were again nonspecific and MR proved more sensitive than CT in detecting the abnormalities.

REFERENCES

1. Chason JL, Hindman WM: Berry aneurysms of the circle of Willis: results of a planned autopsy study. Neurology 8:41, 1968
2. Housepian EM, Pool JL: A systematic analysis of intracranial aneurysms from the autopsy file of the Presbyterian Hospital, 1914–1956. J Neuropathol Exp Neurol 17:409, 1958
3. Crompton MR: The pathogenesis of cerebral aneurysms. Brain 89:797, 1966

4. Sahs AL, Perret GH, Locksley HB et al (eds): Intracranial Aneurysms and Subarachnoid Hemorrhage: A Cooperative Study. JB Lippincott, Philadelphia, 1969
5. Nehls DG, Flom RA, Carter LP et al: Multiple intracranial aneurysms: determining the site of rupture. J Neurosurg 63:342, 1985
6. Bigelow NH: Multiple intracranial arterial aneurysms. Analysis of their significance. AMA Arch Neurol Psychiatr 73:76, 1955
7. McKissock W, Richardson A, Walsh L et al: Multiple intracranial aneurysms. Lancet 1:623, 1964
8. Poppen JL, Fager CA: Multiple intracranial aneurysms. J Neurosurg 16:581, 1959
9. Bannerman RM, Ingall GB, Graf CJ: The familial occurrence of intracranial aneurysms. Neurology 20:283, 1970
10. Stehbens WE: Cerebral aneurysms and congenital anomalies. Aust Ann Med 11:102, 1962
11. McCormick WF, Nofzinger JD: Saccular intracranial aneurysms. An autopsy study. J Neurosurg 22:155, 1965
12. Locksley HB: Natural history of subarachnoid hemorrhage, intracranial aneurysms, and arteriovenous malformations based on 6,368 cases in the cooperative study. J Neurosurg 25:321, 1966
13. Bull JWD: Contribution of radiology to the study of intracranial aneurysms. Br Med J 2:1701, 1962
14. Stehbens WE: Aneurysms and anatomical variation of cerebral arteries. Arch Pathol 75:45, 1963
15. Parkarinen S: Incidence, etiology, and prognosis of primary subarachnoid hemorrhage. Acta Neurol Scand (Suppl) 29:43, 1967
16. Richardson JCM, Hyland HH: Intracranial aneurysms. Medicine 20:1, 1941
17. Wood EH: Angiographic identification of the ruptured lesion in patients with multiple cerebral aneurysms. J Neurosurg 21:182, 1964
18. McKissock W, Paine KWE, Walsh LS: An analysis of the results of treatment of ruptured intracranial aneurysms, report of 772 consecutive cases. J Neurosurg 17:762, 1960
19. Taveras JM, Wood EH: Diagnostic neuroradiology. 2nd ed. Williams & Wilkins, Baltimore, 1976
20. Jane JA, Kassell NF, Torner JC et al: The natural history of aneurysms and arteriovenous malformations. J Neurosurg 62:321, 1985
21. Crompton MR: Cerebral infarction following the rupture of cerebral berry aneurysms. Brain 87:263, 1964
22. Barr HWK, Blackwood W, Meadows SP: Intracavernous carotid aneurysms. A clinical-pathological report. Brain 94:607, 1971
23. Jefferson G: On saccular aneurysms of the internal carotid artery in cavernous sinus. Br J Surg 26:267, 1938
24. McCormick WF, Acosta-Rua GJ: The size of intracranial saccular aneurysms. An autopsy study. J Neurosurg 33:422, 1970
25. Benoit BG, Wortzman G: Traumatic cerebral aneurysms. J Neurol Neurosurg Psychiatry 36:127, 1973
26. Robinson JL, Hall CS, Sedzimir CB: Arteriovenous malformations, aneurysms, and pregnancy. J Neurosurg 41:63, 1974
27. Alvord EC, Loeser JD, Bailey WL et al: Subarachnoid hemorrhage due to ruptured aneurysms. Arch Neurol 27:273, 1972
28. Sundt TM, Whisant JP: Subarachnoid hemorrhage from intracranial aneurysms. N Engl J Med 299:116, 1978
29. Johnson MH, DeFilipp GJ, Rosenwasser RH et al: The role of MRI in the management of intracranial aneurysms. Magn Reson Imaging 4:162, 1986

29a. Atlas SW, Grossman RI, Goldberg HI et al: Partially thrombosed giant intracranial aneurysms: correlation of MR and pathologic findings. Radiology 162:111, 1987

29b. Hackney DB, Lesnick JE, Zimmerman RA et al: MR identification of bleeding site in subarachnoid hemorrhage with multiple intracranial aneurysms. J. Comput Assist Tomogr 10:878, 1986

30. Bradley WG Jr, Schmidt PG: Effect of methemoglobin formation on the MR appearance of subarachnoid hemorrhage. Radiology 156:99, 1985
31. DeLaPaz RL, New PFJ, Buonanno FS et al: NMR imaging of intracranial hemorrhage. J Comput Assist Tomogr 8:599, 1984
32. Sipponen JT, Sepponen RE, Sivula A: Nuclear magnetic resonance (NMR) imaging of intracerebral hemorrhage in acute and resolving phases. J Comput Assist Tomogr 7:954, 1983
33. Gomori JM, Grossman RI, Goldberg HI et al: Intracranial hematomas: imaging by high field MR. Radiology 157:87, 1985
34. Dorland's Illustrated Medical Dictionary, 26th Ed. WB Saunders, Philadelphia, 1984
35. Smoker WRK, Price MJ, Keyes WD et al: High-resolution computed tomography of the basilar artery: 1. Normal size and position. AJNR 7:55, 1986
36. Sacks JG, Lindenberg R: Dolicho-ectatic intracranial arteries. Symptomatology and pathogenesis of arterial elongation and distension. Johns Hopkins Med J 125:95, 1970
37. Goldstein SJ, Sacks, JG, Lee C et al: Computed tomographic findings in cerebral artery ectasia. AJNR 4:501, 1983
38. Herpers M, Lodder J, Janevski B et al: The symptomatology of megadolichobasilar artery. Clin Neurol Neurosurg 85:203, 1983
39. Resta M, Gentile MA, DiCuonzo F et al: Clinical-

angiographic correlations in 132 patients with megadolichovertebrobasilar anomaly. Neuroradiology 26:213, 1984

40. Smoker WRK, Corbett JJ, Gentry LR et al: High-resolution computed tomography of the basilar artery: 2. Vertebrobasilar dolichoectasia: clinical-pathologic correlation and review. AJNR 7:61, 1986
41. Maroon JC: Hemifacial spasm. Arch Neurol 35:481, 1978
42. Ekbom K, Greitz T, Kugelberg E: Hydrocephalus due to ectasia of the basilar artery. J Neurol Sci 8:465, 1969
43. Hahn FJ, Ong E, McComb R, Leibrock L: Peripheral signal void ring in giant vertebral aneurysm: MR and pathology findings. J Comput Assist Tomogr 10:1036, 1986
44. Michelsen WJ: Natural history and pathothysiology of arteriorvenous malformations. Clin Neurosurg 25:307, 1979
45. Mingrino S: Supratentorial arteriovenous malformations of the brain. Adv Technical Standards Neurosurg 5:93, 1978
46. Wilkins RH: Natural history of intracranial vascular malformations: a review. Neurosurgery 16:421, 1985
47. Newton TH, Cronquist S: Involvement of dural arteries in intracranial arteriovenous malformations. Radiology 93:1071, 1969
48. Okamoto S, Handa H, Hashimoto N: Location of intracranial aneurysms associated with cerebral arteriovenous malformation. Statistical analysis. Surg Neurol 22:335, 1984
49. Ross RT: Multiple and familial intracranial vascular lesions. Can Med Assoc J 81:477, 1959
50. Hayward RD: Intracranial arteriovenous malformations. J Neurol Neurosurg Psychiatry 39:1027, 1976
51. Waltimo O: The relationship of size, density, and localization of intracranial arteriovenous malformations to type of initial symptom. J Neurol Sci 19:13, 1973
52. Leussenhop AJ: Dural arteriovenous malformations. In Wilkins RH, Rengachary SS (eds): Neurosurgery. McGraw-Hill, New York, 1985
53. Malik GM, Pearch JE, Ausman JI et al: Dural arteriovenous malformations and intracranial hemorrhage. Neurosurgery 15:332, 1984
54. Forster DMC, Steiner L, Hankanson S: Arteriovenous malformations of the brain. A long term clinical study. J Neurosurg 37: 562, 1972
55. Parkinson D, Bachers G: Arteriovenous malformations: summary of 100 consecutive cases. J Neurosurg 53:285, 1980
56. Drake CG: Cerebral arteriovenous malformations: considerations for and experience with surgical treatment in 166 cases. Clin Neurosurg 26:145, 1979
57. Perret G, Nishioka H: Arteriovenous malformations. An analysis of 545 cases of craniovertebral arteriovenous malformations and fistulae reported to the cooperative study. J Neurosurg 25:467, 1966
58. Troupp H, Marttila I, Halonen V: Arteriovenous malformations of the brain. Prognosis without operation. Acta Neurochir (Wein) 22:125, 1970
59. Lee BCP, Herzberg L, Zimmerman RD et al: MR imaging of cerebral vascular malformations. AJNR 6:863, 1985
60. Kucharczyk W, Lemme-Pleghos L, Uske A et al: Intracranial vascular malformations: MR and CT imaging. Radiology 156:383, 1985
61. Lemme-Plaghos, Kucharczyk W, Brant-Zawadzki M et al: MR imaging of angiographically occult vascular malformations. AJNR 7:217, 1986
62. Gomori JM, Grossman RI, Goldberg HI et al: Occult cerebral vascular malformations: high field MR imaging. Radiology 158:707, 1986
63. Wyburn-Mason R: Arteriovenous aneurysms of the midbrain and retina, facial naevi, and mental changes. Brain 66:163, 1943
64. Spalline A: Computed tomography in aneurysms of the vein of Galen. J Comput Assist Tomogr 3:779, 1979
65. Rengachary SS, Kalyan-Raman UP: Other cranial intradural angiomas. In Wilkins RH, Rengachary SS (eds): Neurosurgery. McGraw-Hill, New York, 1985
66. Brown LR: Ataxia-telangiectasia (Louis-Bar syndrome). Semin Roentgenol 11:67, 1976
67. Assencio-Ferreira VJ: Computed tomography in ataxia-telangiectasia. J Comput Assist Tomogr 5:660, 1981
68. Taybi H: Radiology of syndromes and metabolic disorders. 2nd Ed. Year Book, Chicago, 1983
69. Bartlett JE, Kishore PRS: Intracranial cavernous angioma. AJR 128:653, 1977
70. Voigt K, Yasargil MG: Cerebral cavernous hemangiomas or cavernomas: Incidence, pathology, localization, diagnosis, clinical features and treatment. Neurochirurgia (Stuttg) 19:59, 1976

70a New PF, Ojemann RG, Davis KR et al: MR and CT of occult vascular malformations of the brain. AJR 147:985, 1986

71. Filling-Katz MR, Katz NNK: Familial systemic cavernous angiomatosis. Neurology 35:309, 1985
72. Sarwar M, McCormick WF: Intracerebral venous angioma: Case report and review. Arch Neurol 35:323, 1978
73. Numaguchi Y, Kitamura K, Fukui M et al: Intracranial venous angiomas. Surg Neurol 18:193, 1982
74. Michels BG, Bentson JR, Winter J: Computed tomog-

raphy of cerebral venous angiomas. J Comput Assist Tomog 1:149, 1977
75. Rothfus WE, Albright AL, Casey KF et al: Cerebellar venous angioma: "benign" entity? AJNR 5:61, 1984
76. Cammarata C, Han JS, Haaga JR et al: Cerebral venous angiomas imaged by MR. Radiology 155:639, 1985
77. Scott JA, Augustyn GT, Gilmor RL et al: Magnetic resonance imaging of a venous angioma. AJNR 6:284, 1985
78. Augustyn GT, Scott JA, Olson E et al: Cerebral venous angiomas: MR imaging. Radiology 156:391, 1985
79. Coulam CM, Brown LR, Reese DF: Sturge-Weber syndrome. Semin Roentgenol 11:56, 1976
80. Welsh, K, Naheedy MH, Abroms IF et al: Computed tomography of Sturge-Weber syndrome in infants. J Comput Assist Tomogr 4:33, 1980
81. Enzmann DR, Hayward RW, Norman D et al: Cranial CT scan appearances of Sturge-Weber disease: unusual presentation. Radiology 122:721, 1977
82. Brust JCM: Stroke: diagnostic, anatomical, and physiological considerations. p. 853. In Kandel ER, Schwartz JH (eds): Principles of Neural Science. Elsevier, New York, 1985
83. Goldberg HI: Stroke. p. 583. In Lee SH, Rao KCVG (eds): Cranial Computed Tomography. McGraw-Hill, New York, 1983
84. Diaz FG, Ausman JI: Experimental cerebral ischemia. Neurosurgery 6:436, 1980
85. Bell BA, Symon L, Branston NM: CBF and time thresholds for the formation of ischemic cerebral edema, and effect of reperfusion in baboons. J Neurosurg 62:31, 1985
86. Hossmann K-A, Schuier FJ: Experimental brain infarcts in cats. I. Pathophysiological observations. Stroke 11:583, 1980
87. Ito U, Ohno K, Nakamura R et al: Brain edema during ischemia and after restoration of blood flow: measurement of water, sodium, potassium content and plasma protein permeability. Stroke 10:542, 1979
88. Alcala H, Gado M, Torack RM: The effect of size, histologic elements, and water content on the visualization of cerebral infarcts. Arch Neurol 35:1, 1978
89. Fishman RA: Brain edema. N Engl J Med 293:706, 1975
90. Hoedt-Rasmussen K, Skinhoj E, Paulson O et al: Regional cerebral blood flow in acute apoplexy: the "luxury perfusion syndrome" of brain tissue. Arch Neurol 17:271, 1967
91. Plum F, Posner JB: Edema and necrosis in experimental cerebral infarction. Arch Neurol 9:563, 1963
92. Klatzo I: Neuropathological aspects of brain edema. J Neuropathol Exp Neurol 26:1, 1967
93. O'Brien MD, Waltz AG, Jordan M: Ischemic cerebral edema. Arch Neurol 30:461, 1974
94. Olsson Y, Crowell RM, Klatzo I: The blood-brain barrier to protein tracers in focal cerebral ischemia and infarction caused by occlusion of the middle cerebral artery. Acta Neuropathol 18:89, 1971
95. Yamaguchi T, Waltz AG, Okazaki H: Hyperemia and ischemia in experimental cerebral infarction: correlation of histopathology and regional blood flow. Neurology 21:565, 1971
95a. Brant-Zawadzki M, Weinstein P, Bartkowski H, Mosely M: MR imaging and spectroscopy in clinical and experimental ischemia: A review. AJNR 8:39, 1987
96. Berman SA, Hayman AL, Hinck VC: Correlation of CT vascular territories with function. 1. Anterior cerebral artery. AJNR 3:259, 1981
97. Inoue Y, Takemota K, Miyamoto T et al: Sequential computed tomography scans in acute cerebral infarction. Radiology 135:655, 1980
98. Schuier FJ, Hossmann K-A: Experimental brain infarcts in cats: II. Ischemic brain edema. Stroke 11:583, 1980
99. Go GK, Hommo T, Edzes MSC: Water in brain edema observations by the pulsed nuclear magnetic resonance technique. Arch Neurol 32:462, 1975
100. DeWitt LD, Buonanno FS, Kistler JP et al: Nuclear magnetic resonance imaging in evaluation of clinical stroke syndromes. Ann Neurol 16:535, 1984
101. Brant-Zawadzki M, Bartkowski HM, Ortendahl DA et al: NMR in experimental brain edema: value of T1 and T2 calculations. AJNR 5:125, 1984
102. Spetzler RF, Zambramski JM, Kaufman B et al: Acute NMR changes during MCA occlusion: a preliminary study in primates. Stroke 14:185, 1983
103. Sipponen JT: Visualization of brain infarction with nuclear magnetic resonance imaging. Neuroradiology 26:590, 1984
104. Bryan RN, Wilcott MR, Schneiders NJ et al: NMR evaluation of stroke in the rat. AJNR 4:242, 1983
105. Brant-Zawadzki M, Pereira B, Weinstein P et al: MR imaging of acute experimental ischemia in cats. AJNR 7:7, 1986
106. Bryan RN, Willcott MR, Schneiders NJ et al: Nuclear magnetic resonance evaluation of stroke. Radiology 149:189, 1983
107. Sipponen JT, Kaste M, Ketonen L et al: Serial nuclear magnetic resonance (NMR) imaging in patients with cerebral infarction. J Comput Assist Tomogr 7:585, 1983
108. Hayman LA, Evans RA, Bastion FO et al: Delayed high dose contrast CT: identifying patients at risk of massive hemorrhagic infarction. AJNR 2:139, 1981

109. McNamara MT, Brant-Zawadzki M, Berry I et al: Acute experimental cerebral ischemia: MR enhancement using Gd-DTPA. Radiology 158:701, 1986
110. Carr DH, Brown J, Bydder GM et al: Gadolinium-DTPA as a contrast agent in MRI: initial clinical experience in 20 patients. AJR 143:215, 1984
111. Kinkel WR, Kinkel PR, Jacobs L et al: Serial magnetic resonance imaging of the evolution of cerebral ischemia and infarction. Neurology 35:136, 1985
112. Fisher CM: Lacunes, small deep cerrebral infarct. Neurology 35:136, 1985
113. Weisberg L, Nice C, Katz M: Cerebral Computed Tomography: a Text-Atlas. p. 105. 2nd Ed. WB Saunders, Philadelphia, 1984
114. Wodarz R: Watershed infarctions and computed tomography. A topographical study in cases with stenosis or occlusion of the internal carotid artery. Neuroradiology 19:245, 1980
115. Mohr JP, Caplan LR, Melski JW: The Harvard cooperative stroke registry: a prospective study. Neurology 28:754, 1978
116. Charcot JD, Bouchard C: Nouvelles recherches sur la pathogénie de l'hémorrhagie cérébrale. Arch Physiol Norm Path (Paris) 1:110, 1868
117. Herbstein DJ, Schaumberg HH: Hypertensive intracerebral hematoma. Arch Neurol 30:412, 1974
118. Messina AV, Chernik NL: The resolving intracerebral hemorrhage. Radiology 118:609, 1975
119. Laster DW, Moody DM, Ball MR: Resolving intracerebral hematoma. Alteration of the "ring sign" with steriods. AJR 130:935, 1978
120. Weisberg LA: Cerebellar hemorrhage in adults. Comput Radiol 6:75, 1982
121. Little JR, Tubman DE, Ethier R: Cerebellar hemorrhage in adults. J Neurosurg 48:575, 1978
122. Yasargil MG, Damur M: Thrombosis of the cerebral veins and dural sinuses. p. 2395. In Newton TH, Potts DG (eds): Radiology of the Skull and Brain: Angiography, Specific Disease Processes. Vol 2. Book 4. CV Mosby, St. Louis, 1974
123. Buonanno FS, Moody DM, Ball MR et al: Computed cranial tomographic findings in cerebral sinovenous occlusion. J Comput Assist Tomogr 2:281, 1978
124. Gabrielsen TO, Seeger JF, Knake JE et al: Radiology of cerebral vein occlusion without dural sinus occlusion. Radiology 140:403, 1981
125. Rao KCVG, Knipp HC, Wagner EJ: Computed tomographic findings in cerebral sinus and venous thrombosis. Radiology 140:391, 1981
126. Segall HD, Ahmadi J, McComb JG et al: Computed tomographic observations pertinent to intracranial venous thrombotic and occlusive disease in childhood. Radiology 143:441, 1982
127. Crooks LE, Mills CM, Davis PI et al: Visualization of cerebral and vascular abnormalities by NMR imaging. The effects of imaging parameters on contrast. Radiology 144:843, 1982
128. Axel L: Blood flow effects in magnetic resonance imaging. AJR 143:1157, 1984
129. Walluch V, Bradley WG: NMR even echo rephasing in slow laminar flow. J Comput Assist Tomogr 8:594, 1984
130. Macchi PJ, Grossman RI, Gomori JM et al: High field MR imaging of cerebral venous thrombosis. J Comput Assist Tomogr 10:10, 1986
130a. McMurdo SK, Brant-Zawadzki M, Bradley WG, et al: Dural sinus thrombosis: study using intermediate field strength MR imaging. Radiology 161:83, 1986
131. Ferris EJ, Levine HI: Cerebral arteritis: classification. Radiology 109:327, 1973
132. Johnson RT, Richardson EP: The neurological manifestations of systemic lupus erythematosus: a clinical pathological study of 24 cases and review of the literature. Medicine 47:337, 1968
133. Ellis SG, Verity MA: Central nervous system involvement in systemic lupus erythematosus: a review of neuropathologic findings in 57 cases, 1955–1977. Semin Arthritis Rheum 8:212, 1977
134. Ostrov SG, Quencer RM, Gaylis NB et al: Cerebral atrophy in systemic lupus erythematosus: steriod- or disease-induced phenomenon? AJNR 3:21, 1982
135. Vermess M, Bernstein RM, Bydder GM et al: Nuclear magnetic resonance (NMR) imaging of the brain in systemic lupus erythematosus. J Comput Assist Tomogr 7:461, 1983
136. Kinkel WR, Green FA, Kinkel PR et al: Magnetic resonance imaging in patients with systemic lupus erythematosus and signs of central nervous system involvement. Neurology 35:216, 1985
137. Aisen AM, Gabrielsen TO, McCune WJ: MR imaging of systemic lupus erythematosus involving the brain. AJR 144:1027, 1985
138. Willeit J, Schmutzhard E, Aichner F et al: CT and MR imaging in neuro-Behcet disease. J Comput Assist Tomogr 10:313, 1986
139. Holland BA, Perrett LV, Mills CM: Meningovascular syphilis: CT and MR findings. Radiology 158:439, 1986

8

Infectious and Inflammatory Diseases

A wide spectrum of infectious and inflammatory diseases may affect the central nervous system. Despite this multitude of potential pathogens, the brain and its coverings can only respond in a limited number of ways. Accordingly, many infections and inflammations of the CNS will have similar gross morphologic and radiographic appearances. Sectional imaging modalities (like MR and CT), therefore, will not generally be able to provide a microbiologic diagnosis. Nevertheless, MR and CT may at times be quite helpful in the diagnosis and management of inflammatory diseases.

While specificity is often lacking, MR can be exquisitely sensitive to the localization of infections and inflammations in the CNS.[1] Although there are as yet no large comparative studies available, in several cases we have found MR superior to CT in detecting focal areas of cerebritis and small subdural effusions. Occasionally, with infections such as herpes or cysticercosis, the MR pattern is so characteristic that a specific microbiologic diagnosis may be entertained.

Perhaps the most important role played by MR (and CT) in the management of intracranial infections is in the detection of complications (Table 8-1). Many of these complications are potentially life-threatening and may require surgical intervention. Examples of these infectious complications appear throughout this chapter.

Table 8-1. Complications of CNS Infection Demonstrable by MRI

Abscess
Arachnoidal adhesions
Atrophy and encephalomalacia
Effusions
Empyema
Granuloma formation
Hydrocephalus
Infarction
Vascular thrombosis
Ventriculitis/ependymitis

PATHOPHYSIOLOGY OF INTRACRANIAL INFECTION

Virtually any microbiologic organism, if implanted in the CNS under appropriate conditions, may induce focal or generalized inflammation. Bacteria commonly cause meningitis, abscesses, and empyema. Viruses cause encephalitis and meningitis. Granulomatous inflammations may result in a wide spectrum of manifestations. Fungal and parasitic infections are important

Table 8-2. External Factors That Increase the Risk of CNS Infection

Acquired immune deficiency syndrome (AIDS)
Agammaglobulinemia and congenital impairments of immunity
Age (very young, very old)
Burns
Cancer (especially leukemia, lymphoma)
Cyanotic congenital heart disease
Debilitating illness
Diabetes
Drugs that impair immune response (cancer chemotherapy, propolged antibiotics, steroids)
Infection elsewhere (especially sinuses, face, scalp)
Malnutrition
Midline bony fusion defects
Radiation
Renal failure, renal transplantation
Sickle cell disease
Splenectomy
Surgery
Trauma

considerations in immunocompromised patients and in those from foreign lands.

Infections can reach the CNS by several routes: (1) hematogenously, from a distant focus of infection, (2) by contiguous spread from an adjacent focus, usually in the sinuses, (3) through direct implantation, or (4) perineurally. Hematogenous spread is most frequently arterial from a distant infected source, but may be from retrograde flow along facial and emissary veins. Contiguous spread of infection is usually from an adjacent suppurative sinusitis, mastoiditis, or otitis. Direct implantation may result from prior surgery or trauma. Perineural entry of infection along nerve sheaths is uncommon, but may occur with certain facial infections and is the predominant mode by which rabies and herpes viruses enter the CNS.

Certain constitutional factors external to the nervous system increase the risk for developing intracranial infection (Table 8-2).[2] Most of these factors relate to impaired immunity or debilitation. Additionally several intrinsic anatomic features may explain why infections become established and grow once the organisms have become implanted intracranially (Table 8-3).[3]

Table 8-3. Intrinsic Factors That May Play a Role in the Pathogensis of Intracranial Infections

Absence of true lymphatics in the CNS
Absence of capillaries in the subarachnoid space
Lack of valves in facial and emissary veins allowing retrograde flow of pathogens from face to brain
Presence of Virchow-Robin arachnoid spaces around veins and large vessels
Propensity of CSF to serve as an excellent culture medium
Phagocytic cells largely excluded by the blood-brain barrier
Low levels of complement and immunoglobulins in normal CSF

BACTERIAL INFECTIONS

Meningitis

Meningitis is an inflammation of the meninges that usually occurs following hematogenous dissemination of microorganisms from a distant focus. Meningeal infections are sometimes subclassified into those affecting the dura (pachymeningitis) and those affecting the pia-arachnoid (leptomeningitis). A superficial inflammation of the brain parenchyma (cerebritis) frequently coexists with meningitis. Virtually any organism may cause meningitis, but most cases are bacterial or viral. Inflammations of the meninges and brain caused by viruses, fungi, and parasites are discussed later in this chapter.

Three bacteria (*Haemophilus influenzae, Neisseria meningitidis, Streptococcus pneumoniae*) cause over 80 percent of cases of purulent meningitis in the United States (Table 8-4).[4] The age and specific clinical situation alter the specific probabilities in a given patient, however.[5,6] In newborns, Group B *Streptococcus* (*Streptococcus agalactiae*) and gram-negative bacilli (*Escherichia coli, Klebsiella*) predominate. In the elderly, *S. pneumoniae* and gram-negatives are more common. *S. pneumoniae* (pneumococci) infections are also frequently associated with otitis media, pneumonia, mastoiditis, basilar skull fractures, postsplenectomy, sickle cell disease, and humoral immune deficiencies. *N. meningitidis* (meningococcus) is the most frequent cause of epidemic meningitis, especially in closed populations such as military camps. *H. influenzae* is most common in children with otitis media or upper respiratory infections; it

Table 8-4. Etiology of Bacterial Meningitis in 13,974 Cases from the National Bacterial Meningitis Surveillance Study, 1978–1981

Organism	*% of Total*
Hemophilus influenzae	48.3
Neisseria meningitidis	19.6
Streptococcus pneumoniae	13.3
Group B Streptococcus	3.4
Listeria monocytogenes	1.9
Other	7.5
Unknown	5.9

(Adapted from Harriman DGE: Greenfield's Neuropathology. p. 238. Adams JH, Corsellis JAN, Duchen LW (eds). Edward Arnold, London, 1976.)

has propensity to cause subdural effusions and frontal cerebritis as well. Extracranial infective foci are likely to seed the CNS with the organism causing that distant infection; frequently this is a gram-negative bacterium from the urinary or biliary tract. Patients who have undergone neurosurgical procedures may become infected by *Staphylococci* (*Staphylococcus epidermidis*, *Staphylococcus aureus*) or *Proprionibacterium acnes*.

The pathologic changes in developing meningitis have been well described.[7–9] Following an episode of septicemia, bacteria lodge in venous sinuses, inducing inflammatory changes. Superficial cerebral and pial vessels become congested. The blood-brain barrier becomes disrupted, and inflammatory cells produce an exudate that fills the subarachnoid space. This purulent exudate is most marked in the basal cisterns and dependent sulci. These inflammatory changes interfere with CSF dynamics leading to stasis or even hydrocephalus. Stagnant CSF can act as an ideal culture medium for bacteria, allowing further invasion of the leptomeninges. In late stages the meninges may become thickened and hemorrhagic. Associated cerebritis may be present.

Computed tomography findings in early or partially treated meningitis are usually normal.[11] We have observed MR signal changes of the CSF early in the course of meningitis when CT was normal (Fig. 8-1). This could represent a potentially useful sign for early diagnosis of meningeal inflammation. In children, distension of the subarachnoid spaces may be a clue of early meningitis. Difuse cerebral edema is also known to occur occasionally very early in meningoencephalitis.[12] This may rarely result in acute ventricular compression or downward transtentorial herniation.

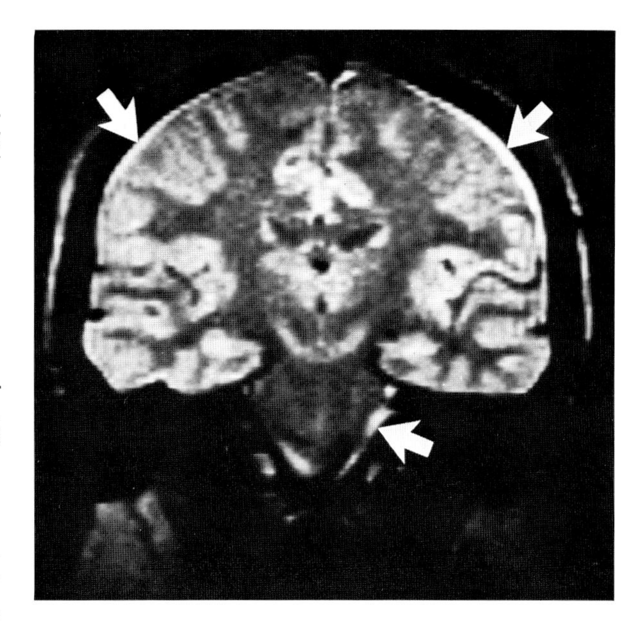

Fig. 8-1. High intensity, presumably proteinaceous, small subdural effusions in a patient with meningitis. SE 1500/56 image. CT was negative.

Within a few days findings on CT and MR are more regularly noted. The meninges and cerebral cortex may show increased intensity on T2-weighted images, corresponding to the phase of meningeal contrast enhancement on CT. Exudate and adhesions may be identified in the basal cisterns (Fig. 8-2).[13] In our limited experience this exudate seems best visualized on MR sequences like SE 1500/30, which are incompletely T2-weighted. Using this technique normal CSF remains dark, but exudate has higher signal intensity and can still be differentiated from nearby brain parenchyma.

Ependymitis (Ventriculitis)

Infection of the ependyma (ependymitis) may be a consequence of leptomeningitis, resulting from retrograde extension of infection into the ventricles from the subarachnoid space. Ependymitis may also occur following rupture of an abscess cavity directly

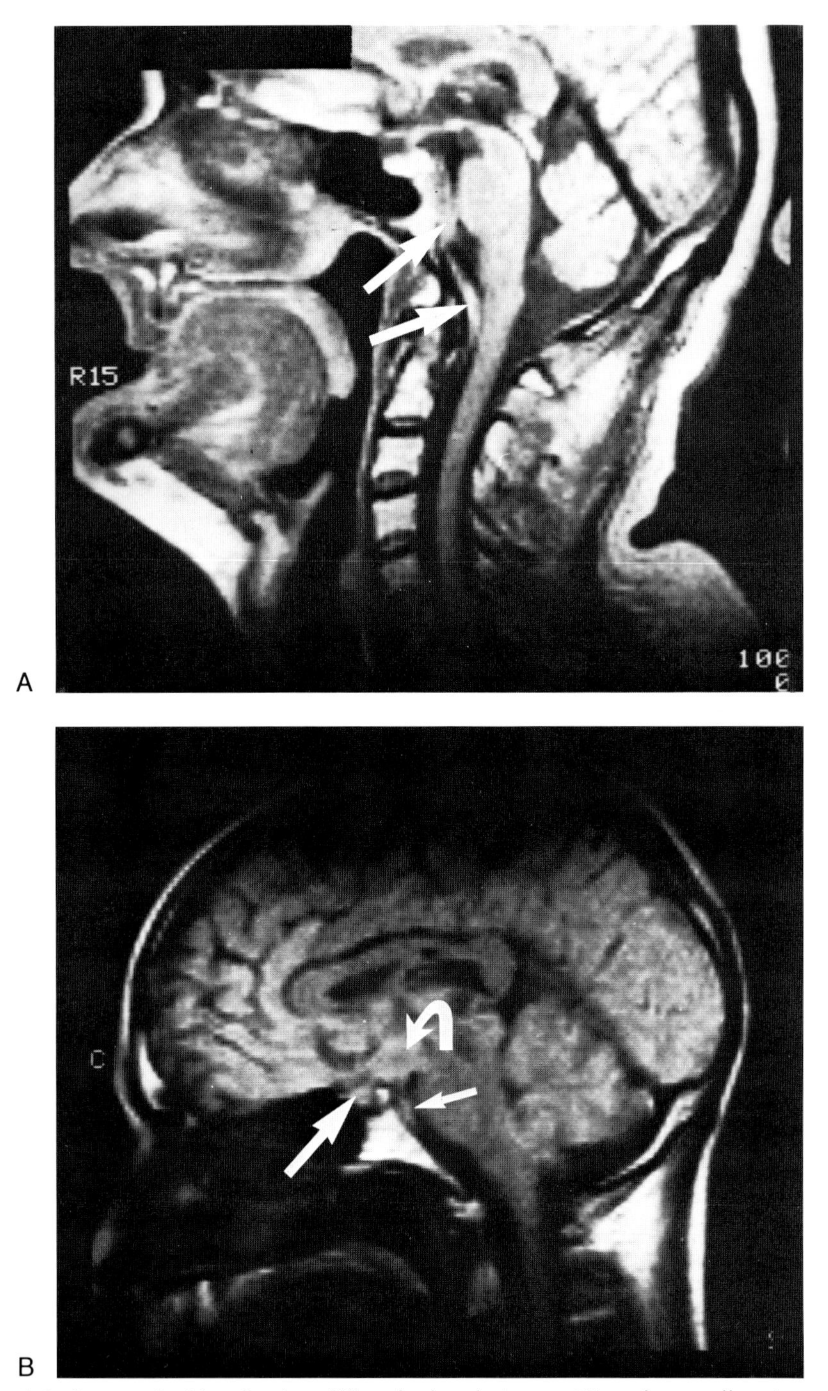

Fig. 8-2. Postmeningitis adhesions filling the basal cisterns **(A)** and parasellar cisterns **(B)**.

into the ventricles. Infected ventriculostomy catheters may induce an ependymitis as well.

Acute bacterial ependymitis is manifested by a periventricular rim of contrast enhancement on CT.[10] MR images reveal a thin rim of prolonged T2 ependymal tissue. This results in a high intensity periventricular signal on T2-weighted images (Fig. 8-3). High signal also involves the septum pellucidum, a feature that may aid in differentiation of ependymitis from periventricular white matter disease.

Intraventricular exudate may organize and form septations, which results in compartmentalization of the ventricular system.[13] This is particularly common in infants. The temporal horn or fourth ventricle may become isolated from the rest of the ventricular system by postinflammatory septations.[14] This "trapped" ventricle may act as an expanding mass, possibly requiring direct shunting (Fig. 8-4). Hydrocephalus may follow meningitis and ependymitis. Depending upon the location of the adhesions the obstruction may be either intraventricular or extraventricular.

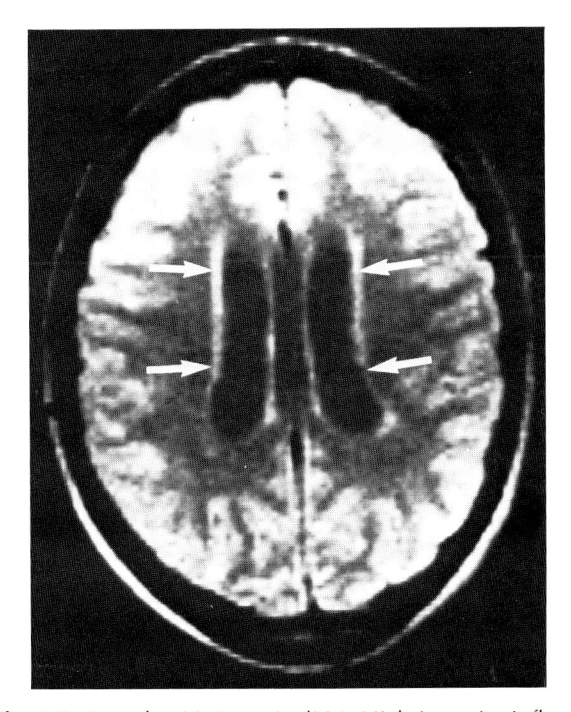

Fig. 8-3. Ependymitis (ventriculitis). High intensity inflammatory tissue coats the inner ventricular surfaces on this SE 2000/28 image.

Empyemas

Subdural and epidural empyemas are collections of pus that usually arise as extension from contiguous infections in the skull or sinuses.[15] Bacteria commonly causing empyema include anaerobic streptococci, *Staphylococcus aureus*, and gram-negative enterics (e.g., *E. coli*).[6] Cerebral empyema, particularly the subdural type, is often considered to be a surgical emergency, because left untreated the mortality is high and the empyema cannot be cured by antibiotics alone.[16]

Subdural empyemas occur with only one-fifth the frequency of parenchymal brain abscesses.[17] In one large series the causes included frontal sinusitis (40 percent), postoperative infection (18 percent), meningitis (14 percent), trauma (14 percent), and presumed hematogenous spread from a distant source (10 percent).[18] Occasionally it may arise from otitis media, mastoiditis, or calvarial osteomyelitis.[19] In the latter case there is usually a coexisting epidural abscess as well.

Subdural empyema usually occurs about 1 to 2 weeks following an episode of sinusitis, having spread intracranially by retrograde thrombophlebitis.[17] The meninges act as a temporary barrier to further spread of infection, which settles in the subdural space. Inflammatory membranes attempt to encapsulate the collection of pus. If this is unsuccessful, the inflammatory process may extend to the brain parenchyma resulting in cortical vein thrombosis, edema, ischemia, or infarction.[20] Empyemas due to otorhinologic infections may occasionally follow a subacute course, and postoperative empyema may be quite indolent, presenting weeks or months after surgery.[21]

Epidural empyema usually results from spread of adjacent infection from frontal sinusitis, mastoiditis, otitis media, or calvanial osteomyelitis.[22,23] It may also be seen after surgery, skull fracture, or penetrating missile injury. Syphilis is one of the few causes of primary epidural abscess (pachymeningitis externa).[24] The tough dura acts as a protective membrane limiting spread of the empyema and minimizing inflammatory effects on adjacent brain.[23] For this reason epidural abscess is often silent clinically, although occasionally patients may present with seizures or focal neurologic defects. This clinical presentation contrasts dramatically with that of subdural empyema, where patients often experience significant neurologic

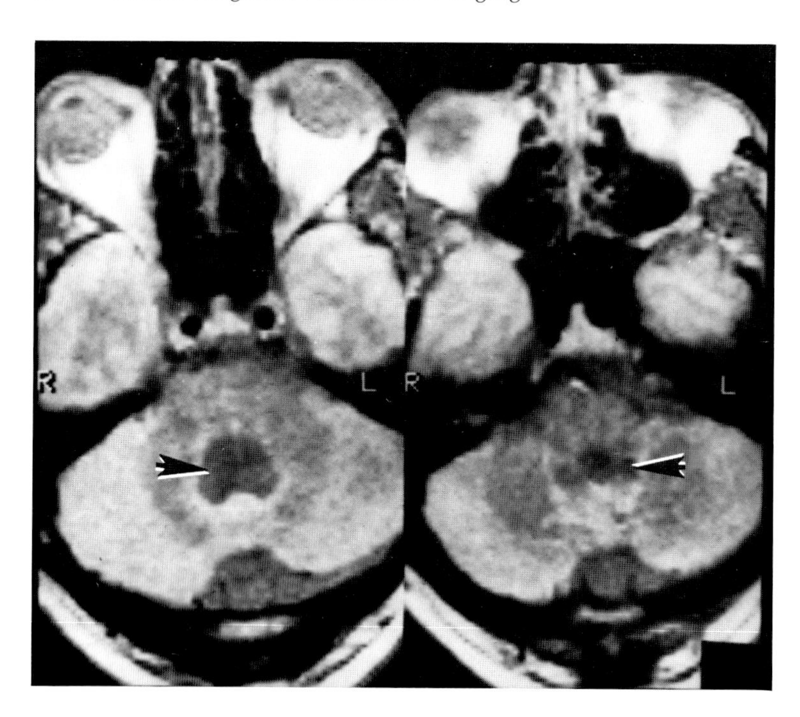

Fig. 8-4. Inflammatory septations intraventricularly may result in a "trapped ventricle."

symptoms due to mass effect and adjacent parenchymal irritation.

Experience worldwide with the MRI of cerebral empyema is limited. Patterns similar to those seen by CT are expected. On CT, subdural empyemas are crescentic or lentiform extra-axial fluid collections with density similar to CSF.[18] Prominent, sharply etched medial rim enhancement, mass effect, and underlying cerebritis are usually noted. Epidural empyemas appear on CT as lentiform, low density fluid collections adjacent to the inner table of the skull, often with calvarial erosion and mass effect on the brain. The medial enhancing membrane is usually thicker in epidural than subdural abscesses (Fig. 8-5).[25] Epidural empyemas may cross the midline, whereas subdural empyemas do not. In our experience and that of others, MRI of empyema shows it to be an extra-axial fluid collection with prolonged T1 and T2 values (Fig. 8-5).[1] Adjacent cerebritis (long T2) may be noted. A thick inflammatory membrane may be visualized in cases of epidural empyema. Both MR and CT are nonspecific in the diagnosis of empyema since the appearance of subdural or epidural hematoma may be identical. The correct diagnosis is frequently suggested by the clinical history. Until MR contrast is available, CT will hold slight advantage in assessing early inflammatory membrane formation. However, MR may be more sensitive than CT in detection of adjacent cerebritis.

Cerebritis

Cerebritis is a regional inflammation of the brain ("focal encephalitis") usually resulting from bacterial or fungal infection. Pathologically cerebritis is characterized by edema, encephalomalacia, hyperemia, petechial hemorrhages, and perivascular inflammatory cell infiltration.[26] If left untreated cerebritis may progress to abscess formation (q.v.) over 10 to 14 days.[27]

Computed tomography scanning in early cerebritis may demonstrate low-attenuation edema and mass effect, but no contrast enhancement.[27] With progressive damage to the blood-brain barrier some gyral and irregular white matter contrast enhancement may be noted. Ringlike enhancement may be present in the absence of a true abscess. MRI demonstrates prolonged T1 and T2 values in regions of brain that are inflammed Fig. 8-6B).[1,28] The use of paramagnetic contrast material (Gd-DTPA) significantly increases the sensitivity of MRI in evaluation of cerebral

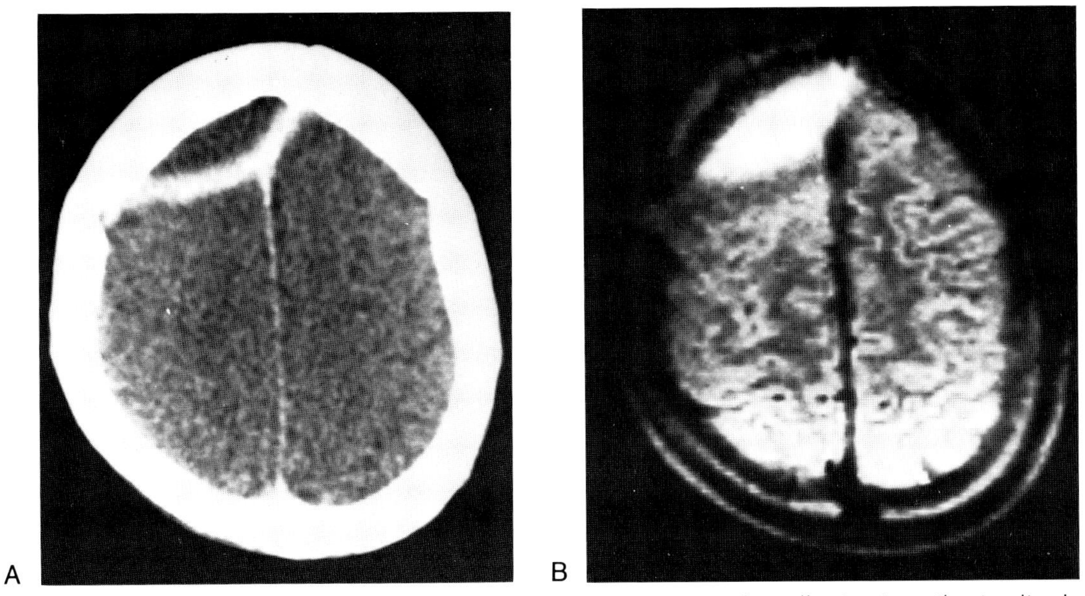

Fig. 8-5. Epidural empyema with thick, enhancing wall seen on CT **(A).** The collection is easily visualized on SE 2000/28 MR image **(B).**

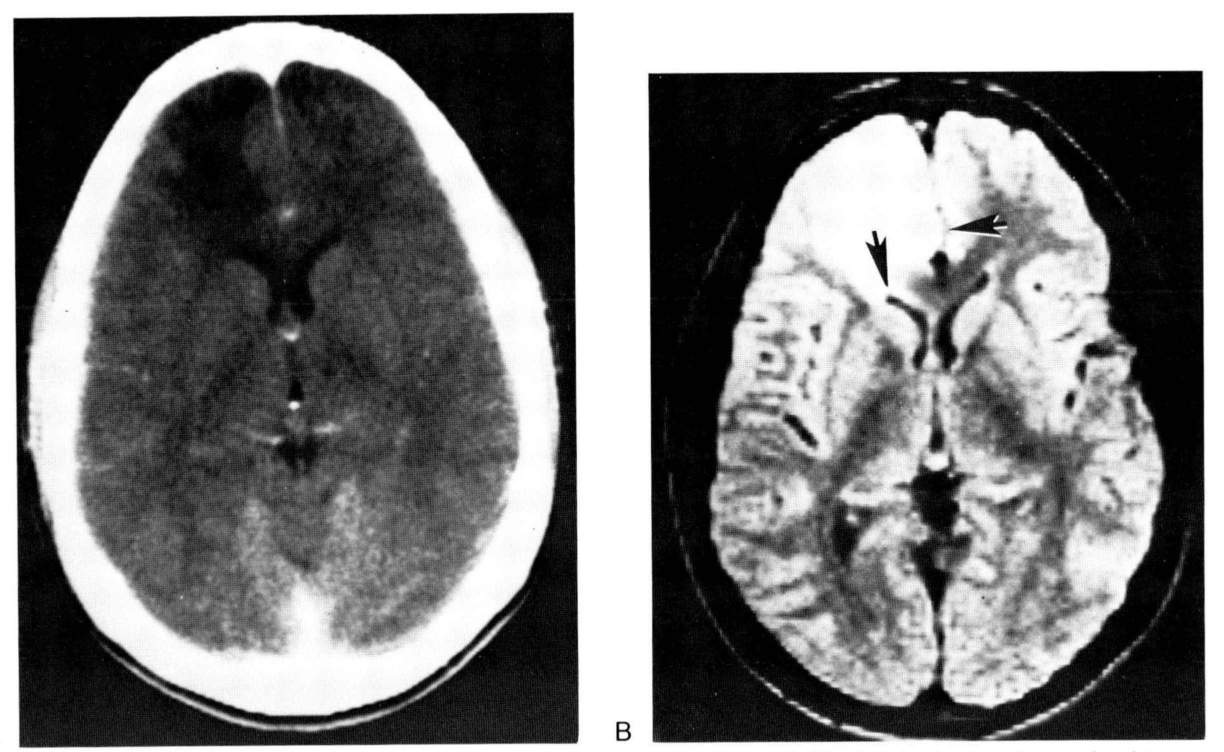

Fig. 8-6. Cerebritis secondary to frontal sinusitis. **(A)** Contrast enhanced CT, day 1, shows edema of right frontal lobe. **(B)** SE 2000/28 image shows high intensity in the same region. Mild mass effect is noted (arrowheads). (*Figure continues*).

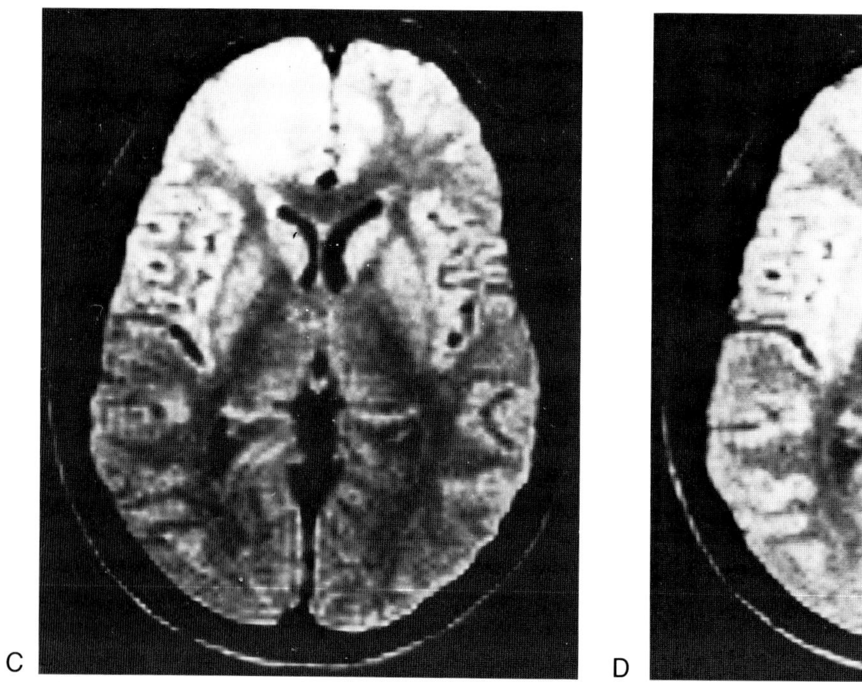

Fig. 8-6 (*continued*). **(C)** After 1 week of antibiotics the mass effect is diminished and the edematous region is smaller. **(D)** Two weeks later there is a nearly normal appearance to the frontal lobe.

inflammations.[29–31] Early cerebritis has been detected on Gd-DTPA enhanced MR images in cases where CT and nonenhanced MR were normal.[29] With resolution of cerebritis, the MR scan returns to normal (Fig. 8-6C,D).

Septic emboli may produce multiple areas of focal cerebritis or microabscess formation (Fig. 8-7).[32] Septic emboli are commonly seen in patients with congenital heart disease, bacterial endocarditis, pulmonary infections, distant abscesses, or a history of intravenous drug abuse. Septic emboli tend to lodge near the corticomedullary junction because of an arteriolar arborization that entraps bacteria.[33] Septic emboli may also result in the formation of larger abscesses, mycotic aneurysms, or regions of septic infarction.

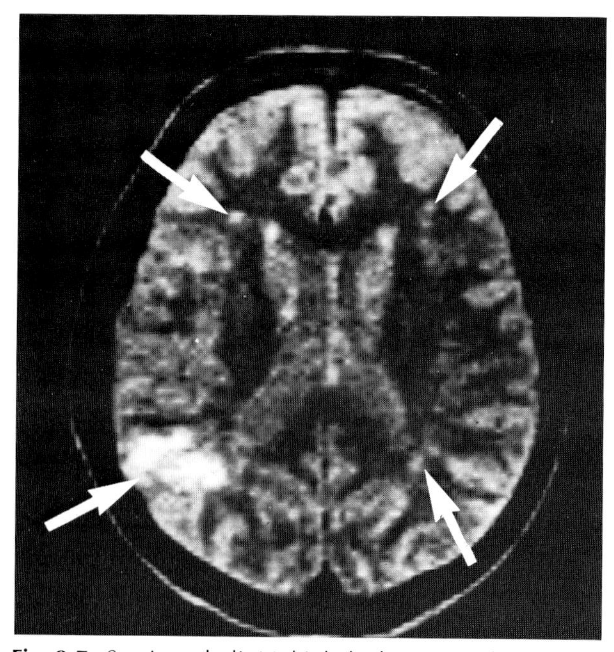

Fig. 8-7. Septic emboli. Multiple high intensity lesions (arrows) at the gray white interfaces on this SE 2000/56 scan in a patient with endocarditis.

Cerebral Abscess

A cerebral abscess is a collection of pus surrounded by a fibrous capsule, which develops in the brain parenchyma as a result of tissue necrosis. Most cerebral abscesses occur following hematogenous dissemination from a distant infectious site.[34] Some are the result of trauma, surgery, or direct extension from sinusitis. Cerebral abscess does not frequently develop after meningitis, but meningitis often develops secondary to an underlying abscess.[35] The causes of cerebral abscess, therefore, largely include most etiologies previously listed for meningitis, cerebritis, and empyema. Common bacterial pathogens include streptococci, staphylococci, and gram-negative species (Table 8-5).[36]

Enzmann has divided abscess formation into four clinicopathologic stages (Table 8-6).[26] The early and late cerebritis phases have been described in the preceeding section. Capsule formation begins about 10 days after implantation of bacteria and matures over the next few weeks. Host resistance, duration of infection, virulence of the organism, and drug therapy (antibiotics, steroids) will affect capsule development and abscess formation.

Computed tomography findings in a developing abscess have been well described.[25–27,32–34)] Several experimental MR studies using Gd-DTPA contrast have also been published.[28–31] In the early stages of cerebritis contrast enhanced MR may detect lesions sooner than CT. Noncontrast MR using T2-weighted images tends to overestimate the extent of cerebritis because abnormally high signal is recorded both from infected brain and areas of reactive (vasogenic) edema. In the early capsulation stage, contrast-enhanced MR may be slightly superior to contrast CT, while noncontrast CT and MR seem largely equivalent. Disruption of the blood-brain barrier is more sensitively detected using contrast-enhanced MR than contrast-enhanced CT. In the late stages of capsular formation, noncontrast MR may differentiate central pus, capsule, surrounding edema, and normal brain. Contrast enhanced MR and CT seem largely equivalent at this stage. Examples of cerebral abscess in the later capsular stages are shown in Figure 8-8.

The differential diagnosis of cerebral abscess includes resolving hematoma, subacute infarction, and primary or metastatic neoplasm. MR may help eliminate resolving hematoma as a consideration because the central region of resolving hematomas usually exhibit decreased T1 values while abscesses demonstrate increased T1 and T2 values.[37] Infarcts typically involve a specific vascular distribution and enhancement is usually more gyral than ringlike. Thickness, irregularity, and nodularity of the capsule wall should raise suspicion of neoplasia or an unusual infection such as a fungus. However, a sufficient number of exceptions make this a somewhat unreliable rule. As always, clinical history will aid in the differential diagnosis.

Table 8-5. Focal Bacterial Infections of the CNS

Location	*Usual Antecedent Events*	*Common Bacteria*
Frontal lobe	Paranasal sinusitis Cavernous sinus thrombosis	Microaerophilic streptococci
Temporal or occipital lobes	Chronic otitis media Mastoiditis	Mixed aerobic-anaerobic flora
Intracerebral	Idiopathic	Microaerophilic streptococci
	Known distant infected focus	Organism from that distant focus
	Postoperative/posttraumatic	Staphylococci Enterobacteriaceae *Pseudomonas* sp. *Bacteriodes* sp.

(Adapted from Butler IJ, Johnson RT: Central nervous system infection. Pediatr Clin North Am 21:649, 1974. Reprinted with permission from WB Saunders Co.)

Table 8-6. Stages of Cerebral Abscess Formation[a]

Stage	*Time*	*Pathologic Features*	*CT Findings*	*MR Findings*
Early cerebritis	1–3 days	Vascular congestion Cytotoxic edema and encephalomalacia Perivascular inflammatory cells seen Petechial hemorrhages	Focal edema (low attenuation) Patchy or gyriform enhancement	Contrast-enhanced MR may show changes before CT or noncontrast MR Increased free water causes prolonged T1 and T2 in area of cerebritis
Late cerebritis	4–9 days	Extracellular edema increases Capillary proliferation Fibroblast migration and early collagen deposition Necrotic center forming	More edema Ring enhancement that may fill in on delayed scan	Noncontrast MR may overestimate extent of cerebritis by imaging entier region of vasogenic edema with cerebritis Contrast MR more sensitive than contrast CT in delineating blood-brain barrier breakdown
Early capsule formation	10–14 days	Central liquefactive necrosis/pus Two-layer capsule: inner wall (granulation tissue and collagen); outer wall (poorly defined glial reaction)	Low density center (pus) Usually thin, uniform rim that shows contrast enhancement Peripheral vasogenic edema	Noncontrast MR: T2-weighted images show high intensity surrounding edema and lower intensity abscess Contrast MR T1-weighted images show high intensity central abscess
Late capsule formation	After 14 days	Capsule thicker and well defined Three-layer capsule: inner wall (granulation tissue); middle layer (collagen); outer layer (astroglial) Capsule is thicker laterally than medially	Ring enhancement, thicker possibly irregular Multilocularity may be seen (daughter abscesses)	Noncontrast MR may distinguish central cavity, wall, and peripheral edema Contrast MR similar to CR

[a] Compiled from References 25–30.

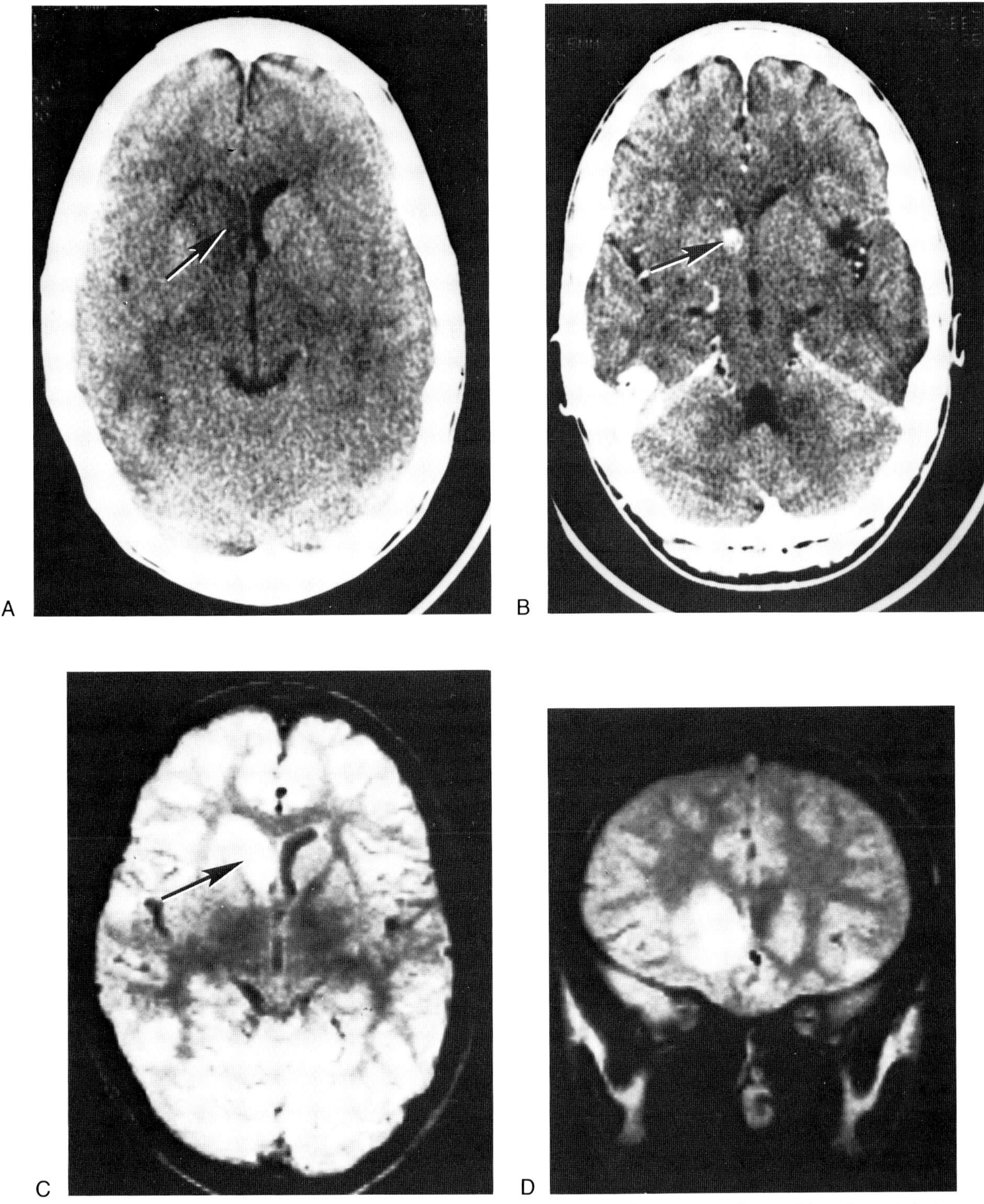

Fig. 8-8. Abscess in the caudate nucleus. Aspiration at surgery revealed 3 to 4 ccs of pus, although no definite capsule was seen on CT or MR. **(A)** Precontrast CT; **(B)** postcontrast CT; **(C)** axial SE 1000/28 MR image; **(D)** coronal SE 1500/28 image.

Complications of cerebral abscess include pressure effects, intraventricular rupture, and spontaneous hemorrhage. Mass effect in rapidly growing infections may be life-threatening, often related more to surrounding edema than to the actual abscess cavity itself. Cerebellar abscesses may potentially cause significant morbidity by mass effects, but if recognized and treated early their prognosis is better than abscesses elsewhere.[38] Intraventricular rupture with ventriculitis is a feared complication, with mortality of 40 to 60 percent.[39] The medial wall of abscesses is typically thin and weak due to the relatively poor vascular supply of deep white matter. This may predispose to intraventricular rupture or intraparenchymal rupture with formation of abscess daughters. Finally, bleeding is occasionally seen as an unusual complication of cerebral abscess. When it occurs, it is usually caused by hemorrhagic infarction secondary to venous thrombosis.[40]

GRANULOMATOUS DISEASES

Tuberculosis

Although relatively uncommon in the United States, tuberculosis is the most frequent granulomatous infection of the CNS world-wide.[41,42] Young children and the elderly are primarily affected, though the disease may occur at any age.[43,44] Two forms of CNS tuberculosis are identified: tuberculous meningitis and tuberculoma. The diagnosis may be difficult because CNS tuberculosis occurs without evidence of associated extracranial disease in a significant minority of cases.[41] If undiagnosed or untreated, the morbidity and mortality are high.[43]

Tuberculous Meningitis

Tuberculous meningitis is seen predominantly in children following hematogenous dissemination from a focus in the chest, abdomen, or urinary tract. The critical event in the development of meningitis is rupture of a juxtaependymal tubercle (Rich focus) into the subarachnoid space.[45] This indirect pathway of CNS contamination explains why meningitis is not usually seen until several weeks into the course of miliary tuberculosis.

Once implanted in the subarachnoid space, the tubercle bacilli incite a proliferative meningitis. A purulent, fibrinous exudate accumulates especially in the basal cisterns. With time this exudate may consolidate into a thick, gelatinous mass, possibly compromising the function of cranial or spinal nerves. Vasculitis with spasm or thrombosis may develop in vessels that traverse this basilar exudate. Hydrocephalus develops in a majority of untreated cases.[46]

No MR studies of tuberculous meningitis have been published, but the CT appearance is well known.[46–51] Basilar exudate is commonly seen having preference for the chiasmatic and prepontine cisterns. Cerebral infarctions are noted in about 25 percent of untreated cases, involving primarily cortex in the middle cerebral artery distribution and the basal ganglia.[46] Hydrocephalus occurs in a high percentage, and is seen in nearly all patients who survive 4 to 6 weeks without therapy. Coexisting tuberculomas are present in 10 percent. Meningeal calcification may be observed in nearly 50 percent of patients within 3 years after onset of the disease.[52]

Tuberculoma

Tuberculoma represents a focal infection of the brain parenchyma or meninges by *Mycobacterium tuberculosis*. A spectrum of closely related processes can be identified, including focal tuberculous cerebritis, tuberculoma, and tuberculous brain abscess.[41] These may represent a continuity of disease ranging from clustered or conglomerate masses of tubercles to pus-containing cavities. In general it is not possible to distinguish reliably these focal processes from one another by CT or MRI.

A tubercle is a small (1 to 3 mm) core of epithelioid cells surrounded by lymphocytes. Focal tuberculous cerebritis represents an area of inflamed brain containing several small tubercles. A true tuberculoma is formed by the conglomeration of multiple tubercles. The center of the tuberculoma becomes necrotic, filled with caseous debris. Peripherally, a capsule of fibrous tissue and reactive gliosis is formed. Edema is usually seen in the normal brain surrounding the lesion. Progression to abscess formation, with a thick wall and frank pus, is rare.[53,54]

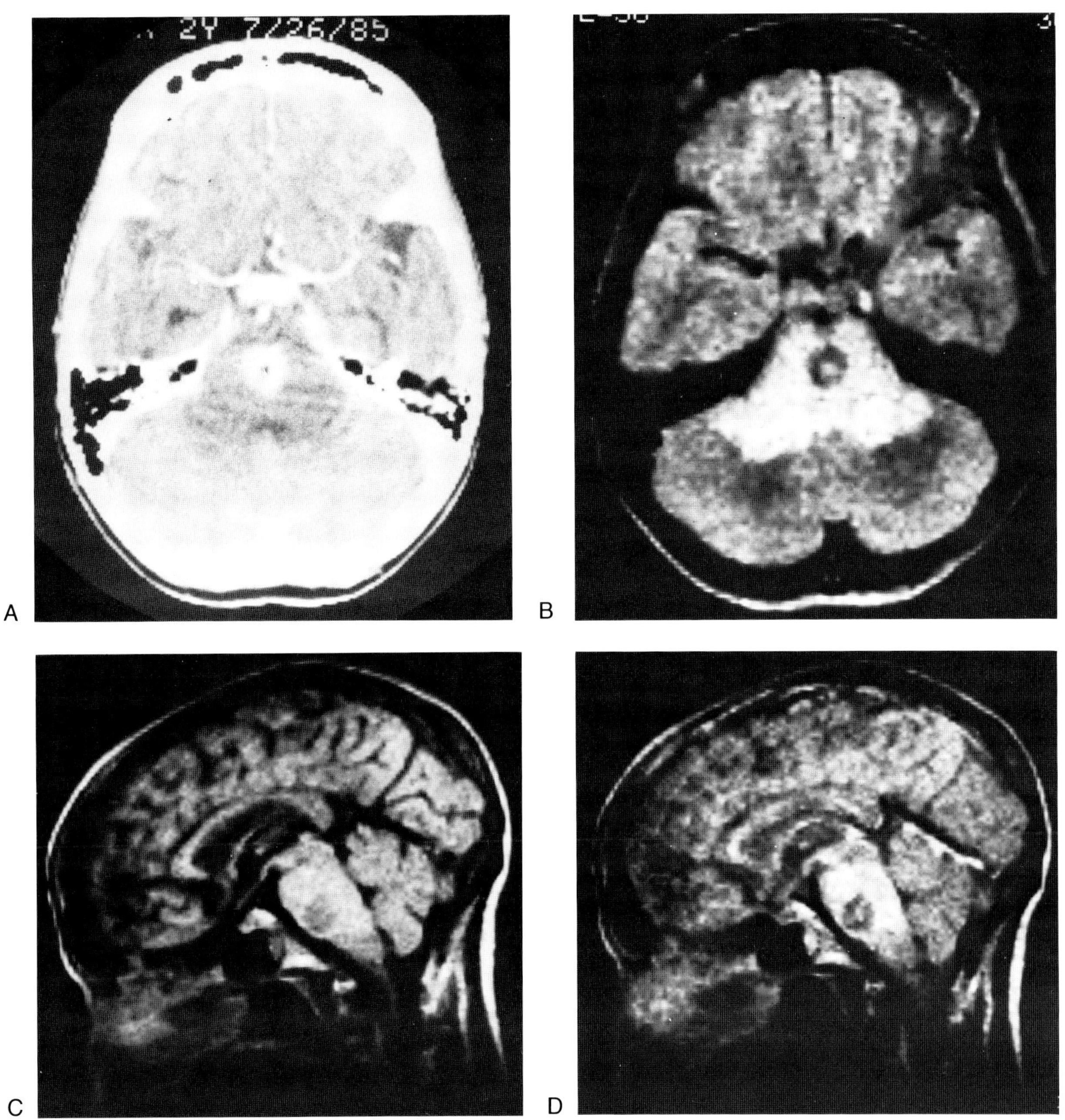

Fig. 8-9. Tuberculoma. **(A)** Contrast CT shows an enhancing lesion in the pons with central low attenuation region and surrounding vasogenic edema. **(B)** These same features are shown to good advantage on this axial SE 2000/56 MR image. **(C)** The lesion is less well defined on this sagittal SE 1000/28 image, although brain stem swelling can be well appreciated. **(D)** Sagittal SE 1000/56 image increases contrast between edema and the lesion.

The diagnosis of tuberculoma may be difficult because its appearance is often indistinguishable from other space-occupying lesions. In children intracranial tuberculomas are often infratentorial and coexist with progressive primary disease.[54] Most tuberculomas in the United States occur in young adults, supratentorially, during reactivation TB.[56,57] About 42 percent of these patients will have no evidence of active extracranial tuberculosis.[57]

Old inactive tuberculomas are small parenchymal nodules, calcified in 1 to 13 percent.[52] Active tuberculomas vary in size from a few millimeters to several centimeters in diameter, and are multiple in 60 percent.[49,58] On CT, a "micro-ring" appearance of tuberculomas has been described, consisting of an enhancing peripheral rim of granulation tissue surrounding a lower density central region of caseous necrosis (Fig. 8-9A).[56] We have observed a corresponding pattern in a tuberculoma imaged by MR (Fig. 8-9B–D). In this case, the central necrotic region contained prolonged T1 and T2 material, while the rim showed decreased relaxation times compared to brain. Edema with prolonged T1 and T2 was present surrounding the lesion, seen also on CT.

Other patterns of tuberculomas on CT include (1) homogeneously enhancing nodules, (2) nodules with peripheral enhancement and central calcification, and (3) densely calcified nodules.[59,60] Dural tuberculomas may cause calvarial hyperostosis and resemble meningiomas.[61] Further research and experience with paramagnetic contrast agents should refine our knowledge of the appearances of tuberculomas on MR.

Sarcoidosis

Sarcoidosis is a systemic disease of unknown etiology that can present intracranially as a diffuse granulomatous meningitis or mass lesion. The disease is most common in females during the third and fourth decades, but any age may be affected. At autopsy up to 27 percent of patients with sarcoidosis have microscopic evidence of CNS involvement.[62] However, only about 5 percent of patients with systemic sarcoidosis have neurologic symptoms. Only rarely is CNS involvement the sole manifestation of the disease.[63]

A

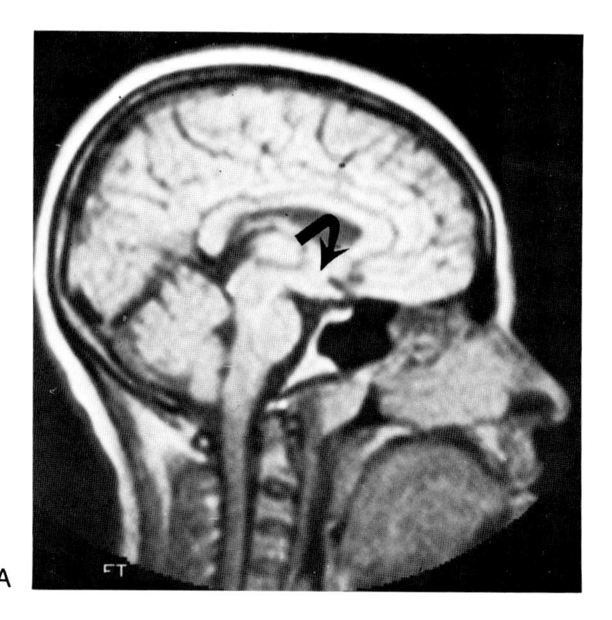

B

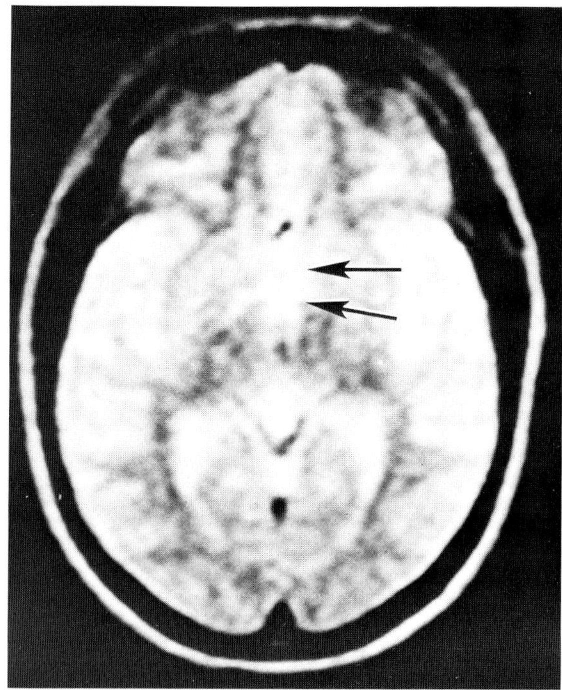

Fig. 8-10. CNS sarcoidosis. **(A)** There is granulomatous infiltration in the hypothalamus and around the optic chiasm, which is thickened on this sagittal SE 500/26 image. **(B)** On axial SE 3000/100 image there is subtle abnormal increased signal in the same region.

Chronic granulomatous leptomeningitis is the most common pattern of CNS involvement.[64–69] This meningitis may occur diffusely or as a localized process, often in the basilar cisterns. Involvement of the pituitary, hypothalamus, optic chiasm, and other cranial nerves (particularly VII) may be seen. Communicating hydrocephalus often results. The granulomatous meningitis frequently spreads along the Virchow-Robin spaces; causing small vessel angiitis and thrombosis.[65]

Granulomatous masses in the brain similar to tuberculomas may be seen in sarcoidosis.[66–70] The masses may be solitary or multiple and associated arachnoiditis is usually present. The sarcoid nodules create problems by mass effect, and obstructive hydrocephalus has been reported by a periaqueductal lesion.[71] Unlike tuberculomas, sarcoid nodules do not develop caseous centers and are not associated with edema. These features may assist in the differential diagnosis by CT or MRI. Both sarcoid nodules and granulomatous meningitis enhance significantly on CT.[66–69]

Only a single case of neurosarcoidosis imaged by MR has been reported.[72] Diffuse areas of increased T2 and proton density, that resolved after 8 weeks of steroid therapy were identified in the hypothalamic and basal ganglia regions. The MR abnormalities were more extensive than those noted on CT. In the cases we have seen, similar findings were observed. (Fig. 8-10).

FUNGAL INFECTIONS

The number of CNS fungal infections has increased dramatically in recent years owing to the increased use of steroids, antibiotics, and immunosuppressive chemotherapy.[73] A useful clinical classification (Table 8-7) divides fungal diseases into those that affect normal patients and those that affect primarily the immunocompromised.[74] Alternatively, a classification scheme based on histopathologic patterns of CNS involvement (Table 8-8) may be more useful for radiographic differential diagnosis.[75]

Most fungi gain access to the CNS by hematogenous dissemination from a distant infected focus, usually in the lungs, bones, or lymph nodes. Certain fungi (e.g., *Mucor*) may directly invade the brain from adjacent sinus infection. The gross appearance of intracranial mycosis depends on the type of fungus as well as the dominant infecting form, i.e., yeast or hyphae.[75,76] Yeast typically elicits a diffuse leptomeningitis with intraparenchymal masses (granulomas, abscesses). Frequently the fungus can be recovered from the CSF. When the hyphal form predominates (seen with aspergillosis and mucormycosis) hyphae may block blood vessels, commonly resulting in hemorrhagic infarction. Further fungal erosion through the vascular wall into ischemic brain parenchyma may produce focal cerebritis or mycotic aneurysms.

Table 8-7. CNS Fungal Diseases Classified by the Patient Population They Primarily Affect[a]

Infections in Normal Hosts
Cryptococcosis (torulosis)
Coccidiomycosis
Histoplasmosis
Blastomycosis
Actinomycosis[b]
Infections in Immunocompromised Hosts[c]
Candidiasis
Aspergillosis
Cryptococcosis
Mucormycosis
Nocardiosis[b]

[a] Listed in order of decreasing frequency for each group.
[b] Diseases caused by bacteria that behave like fungi.
[c] Note that all fungal infections (especially cryptococcosis) that affect normal patients may also affect the immunocompromised.

Table 8-8. CNS Fungal Diseases Classified by Their Predominant Pathophysiologic Pattern

Granulomatous Pattern (meningitis, granulomas)
Cryptococcosis
Coccidiomycosis
Histoplasmosis
Blastomycosis
Discrete Abscess Pattern
Candidiasis
(Actinomycosis)
(Nocardiosis)
Hemorrhagic Infarction Pattern
Aspergillosis
Mucomycosis

Candidiasis

Candidiasis accounts for nearly 50 percent of autopsy proven cases of cerebral mycosis, usually caused by the yeast form of *Candida albicans*.[76] Most patients are immunocompromised or debilitated, often with systemic (disseminated) candidiasis. CNS infection with *Candida* has also been reported in premature infants, intravenous drug abusers, patients with prosthetic heart valves, and patients with longstanding central venous catheters. The typical pathology seen consists of multiple microabscesses scattered especially in the middle cerebral artery distribution.[77] Other presentations include meningitis, noncaseating granulomas, septic infarction, and larger cavitations, resembling pyogenic abscess. The course is frequently fulminant and fatal.

Cryptococcosis

Cryptococcosis (torulosis) is caused by the budding yeast, *Cryptococcus neoformans*, whose polysaccharide capsule distinctively stains with India ink. *C. neoformans* is a ubiquitous fungus that grows especially well on window ledges or in soil contaminated with pigeon droppings.[78] Presentation of the disease is indeed "cryptic" in that nearly half of infected patients lack fever or meningeal signs.[79] Normal and immunosuppressed patients are affected with equal frequency, although the course of the disease is more indolent in the former.

Cryptococcus usually reaches the CNS by dissemination from an inapparent pulmonary focus. Granulomatous meningitis with a thick basal exudate resembling tuberculosis is the common mode of presentation.[80] Abscesses or large parenchymal granulomas (torulomas) may be present and simulate tuberculomas.[81] Frequently these are multiple. A latex agglutination test on CSF is the most rapid and reliable laboratory method for diagnosis.[79]

Histoplasmosis

Histoplasmosis is caused by the dimorphic fungus *Histoplasma capsulatum*, which is endemic throughout the central and eastern United States, especially in soil contaminated by bat or bird droppings.[74] Although subclinical pulmonary infection is common, CNS histoplasmosis is unusual even in disseminated disease.[82] Most reported cases have presented as a granulomatous meningitis, although mass lesions (histoplasmomas) are also known.[73] We have observed a single thick-walled histoplasma abscess resembling a pyogenic one on MRI (Fig. 8-11).[82a]

Coccidiomycosis

Coccidiomycosis is caused by *Coccidioides immitis*, a fungus endemic to the arid regions of the southwestern United States (San Joaquin Valley) and northern Mexico. Only a small percentage of patients with cutaneous or pulmonary coccidiomycosis develop CNS involvement. An exudate granulomatous meningitis with preference for the basilar cisterns is the most common CNS manifestation.[83] Ependymitis and communicating hydrocephalus are often seen.[84] Vasculitis with occlusion is rare, but reported.

North American Blastomycosis

North American blastomycosis is caused by the airborne yeast, *Blastomyces dermatiditis*, found primarily in the Ohio and Mississippi River valleys of the eastern United States. CNS infection is uncommon and most always a late manifestation of disseminated blastomycosis.[74] Granulomatous meningitis seems to be the most common manifestation, but focal cerebritis, granulomas, and abscesses are reported.[85] We have observed a single case on MRI, manifest by focal parenchymal inflammation and basilar meningitis (Fig. 8-12).

Aspergillosis

Aspergillosis is caused by hyphal forms of the genus *Aspergillus*, usually *A. fumigatus*. After *Candida*, *Aspergillus* is the most common CNS fungal infection in the immunocompromised host.[75] *Aspergillus* usually enters the CNS by hematogenous spread from

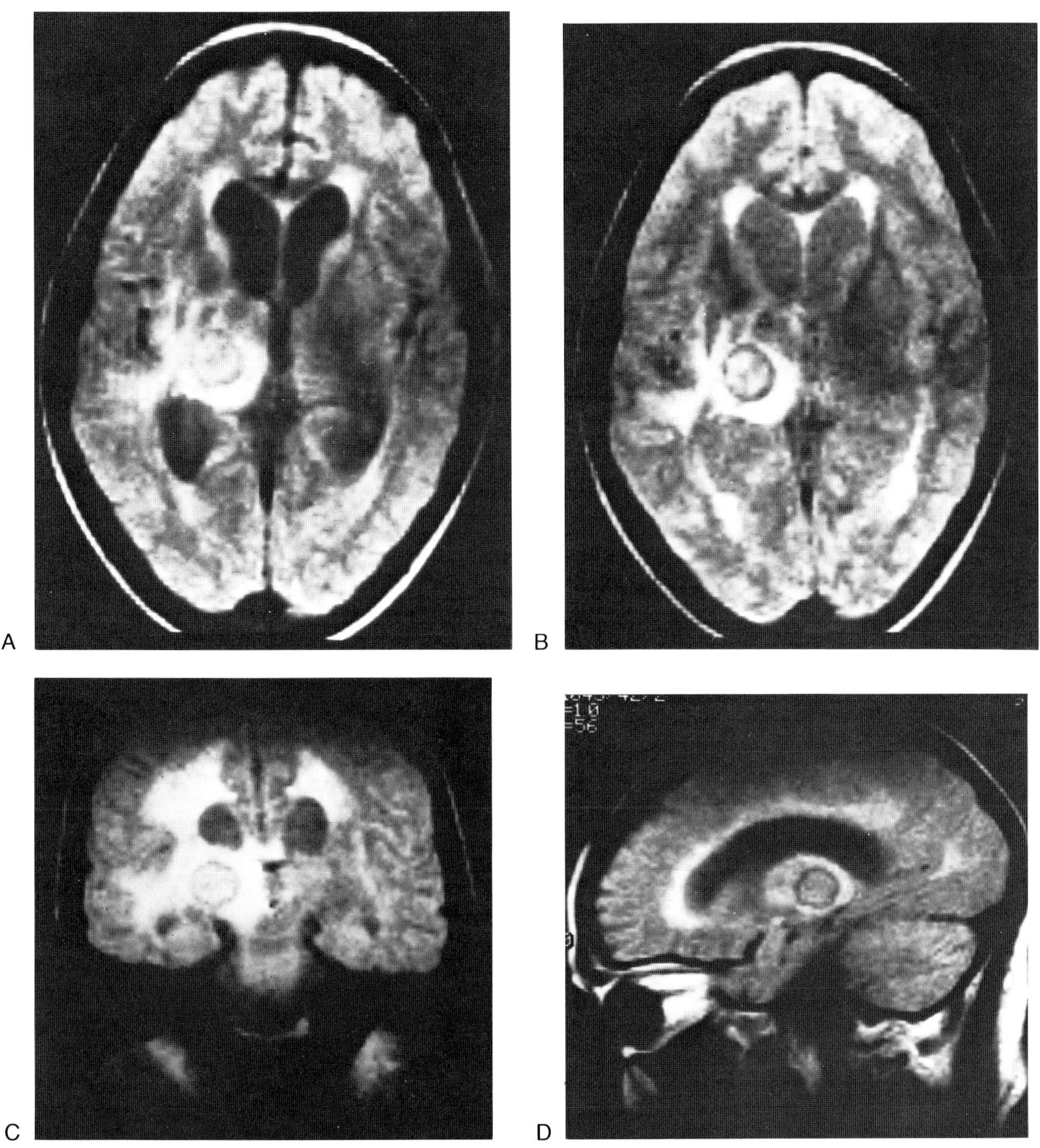

Fig. 8-11. Histoplasmosis abscess in the right thalamus. MR scans demonstrate capsule, abscess, and surrounding edema to good advantage. **(A)** Axial SE 2000/28; **(B)** axial SE 2000/56; **(C)** coronal SE 1500/56; **(D)** sagittal SE 1000/56.

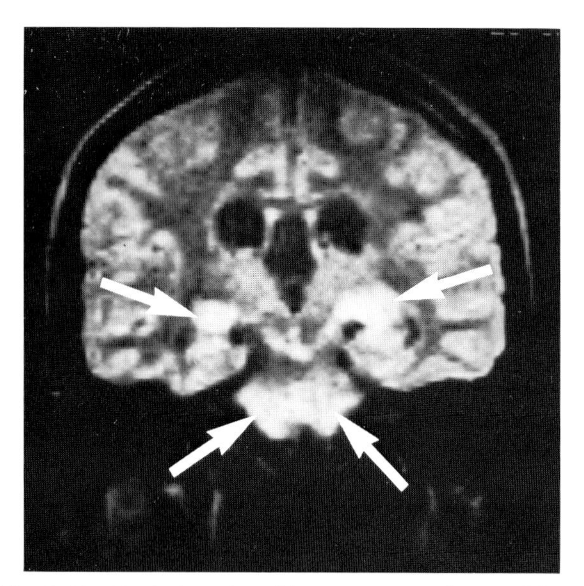

Fig. 8-12. North American blastomycosis. Abnormal high signal around the pons and medial temporal lobes seen on this coronal SE 1500/56 image.

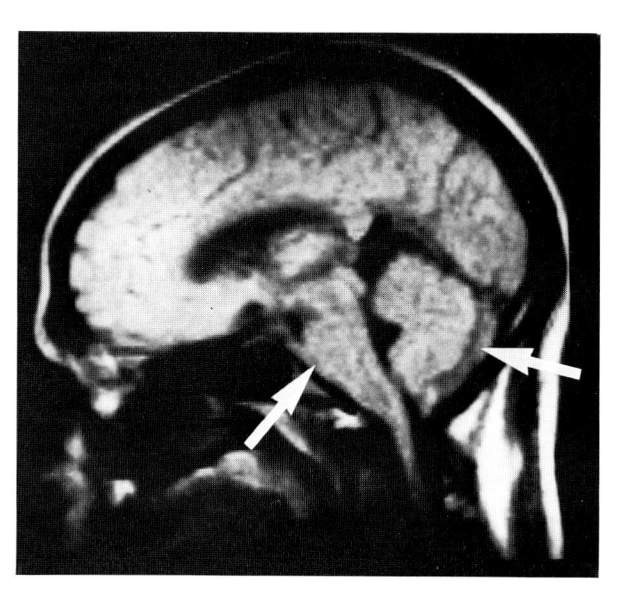

Fig. 8-13. Mucormycosis of the meninges in an intravenous drug abuser. Abnormal inflammatory tissue is seen around the pons and cerebellum on this sagittal SE 1000/28 image.

the lung, but direct extension from the sinuses or orbit has also been reported.[74,86] Patients usually have minimal fever and present with symptoms of a mass lesion.[87] An initial focal area of cerebral inflammation is often poorly defined.[88] The hyphae may block or invade blood vessels, leading to infarction and hemorrhage.[76] *Aspergillus* is the most common cause of fungal mycotic aneurysm.[89]

Mucormycosis

Mucormycosis is produced by various moldlike fungi of the class Zygomycetes. Like *Aspergillus*, these fungi grow in hyphal forms, causing vascular thrombosis, tissue destruction, and abscess formation.[90] The disease is usually seen in diabetics with ketoacidosis and in debilitated patients suffering from burus, uremia, or malnutrition. A slightly less aggressive form has been described in intravenous drug abusers.[92] The fungus is usually inhaled and rapidly destroys the nasal mucosa forming black crusts. It then extends into the paranasal sinuses, orbit, and CNS. Vascular thrombosis, necrosis, and abscess formation are prominent features. Unlike most other fungal infections, which often run a chronic course, untreated mucormycosis usually leads to death in a few days. This aggressive form of mucormycosis has not yet been recorded by MRI, but we have seen a case of chronic mucormycosis of the meninges in an intravenous drug abuser (Fig. 8-13).

Actinomycosis

Actinomycosis is an uncommon intracranial infection caused by the pleomorphic bacterium, *Actinomyces israelii*. In tissues *A. israelii* exists in a mycelial form resembling a fungus, resulting in its pathologic similarities and classification with fungal diseases. However, research has firmly established *A. israelii* as a bacterium, which exists in a bacillary form in the mouth, particularly in patients with dental caries. Extracerebral abscess is the most likely manifestation of CNS actinomycosis, which is most uncommon.[93] The organism gains access to the CNS by (1) hematogenous dissemination from lung, abdomen (especially appendix), or mandible ("lumpy jaw"), or (2) direct extension from infection of the head and neck.

Identification of "sulphur granules" in sinus tracts or sputum aids in identification. The organism responds to penicillin therapy.

Nocardiosis

Nocardiosis resembles actinomycosis in that it is also caused by a bacterium (*Nocardia asteroides*) that resembles a fungus pathologically and morphologically. While actinomycosis frequently occurs in immunocompetent patients, *Nocardia* is usually seen in those with compromised immunity, often on steroid therapy.[94] Nocardiosis coexists with a number of chronic conditions including pulmonary alveolar proteinosis, sarcoidosis, ulcerative colitis, intestinal lipodystrophy, and diseases of the reticuloendothelial system.[95] Humans acquire the organism only from exposure to contaminated soil. It reaches the CNS by hematogenous dissemination, usually from the lung. The CNS is the most common system affected in disseminated disease.[96] This usually results in parenchymal brain abscess although meningitis is rarely seen. CT scans have shown an enhancing capsule with multiple loculations.[97] No MR reports are available. Because *Nocardia* responds to sulfonamides the diagnosis should be considered in the appropriate clinical setting.

Table 8-9. Viral Infections of the Central Nervous System

Adenovirus	Multiple serotypes
Arbovirus (Togavirus, Bunyavirus, Orbivirus)	Equine encephalomyeoitis [a]St. Louis encephalitis Japanese encephalitis Tick-borne encephalitis [a,b]Rubella California encephalitis Colorado tick fever
Arenavirus	Lymphocytic choriomeningitis
Herpesvirus	[a]Herpes simplex Zoster-varicella Cytomegalovirus Epstein-Barr (infectious mononucleosis)
Orthomyxovirus	Influenza
Papovavirus	Progressive multifocal leukoencephalopathy
Paramyxovirus	[a]Measles (subacute sclerosing panencephalitis) [a,b]Mumps
Picornavirus	Poliovirus [b]Coxsackievirus [b]Echovirus Enteroviruses 70,71
Poxvirus	Vaccinia
Rhabdovirus	Rabies
"Slow viruses"	Kuru, Creutzfeldt-Jakob

[a] Common cause of encephalitis
[b] Common cause of meningitis

VIRAL INFECTIONS

A large number of viruses may affect the CNS, usually producing meningitis or encephalitis (Table 8-9). Only a few viral diseases (including those caused by herpes and papovaviruses) have characteristic MR features. The remainder present with nonspecific findings related to meningeal inflammation or diffuse brain edema.

While MR is very nonspecific in diagnosing viral diseases it can be helpful in detecting the sequellae of such infections. MR may detect altered patterns of myelination following inflammatory disease in infants.[98] It may also show persisting abnormalities in brain regions after CT has returned to normal and the patient has clinically recovered. An example of residual inflammation and cystic encephalomalacia developing in an infant after viral encephalitis is shown in Figure 8-14.

Herpes Simplex

Herpes simplex, an encapsulated DNA virus, is the most common cause of fatal endemic encephalitis in the United States.[99] Two viral types may be distinguished by differences in their capsular antigens. The type 1 strain causes oral lesions and encephalitis in adults. The type 2 strain causes genital lesions and neonatal meningoencephalitis.

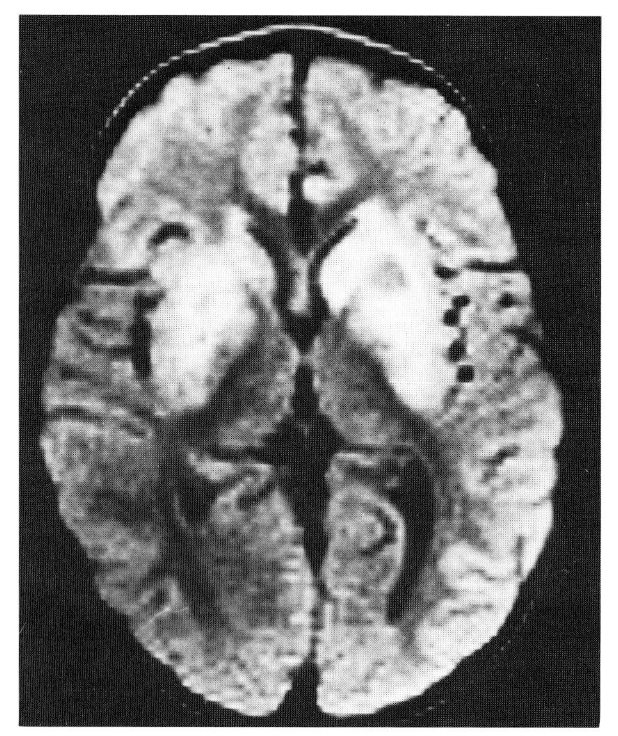

Fig. 8-14. Abnormal signals from the basal ganglia of a child, which remain several weeks following a viral encephalitis.

Herpes simplex encephalitis in adults is nearly always caused by the type 1 strain, although type 2 has been isolated in several patients with altered cellular immunity.[100] Herpes simplex produces a focal necrotizing encephalitis preferring the medial temporal lobes, insula, and subfrontal regions. This characteristic distribution is thought to be the result from the entry of virus to the CNS along the trigeminal nerve.[101]

The clinical picture is usually a subacute onset of fever, headache, and altered mentation. Hallucinations, memory loss, and findings of limbic dysfunction result from temporal lobe involvement. Untreated, mortality is about 70 percent, with severe neurologic sequellae occurring in the survivors. Mortality and morbidity are significantly reduced with early antiviral therapy (acyclovir).[99] Brain biopsy is frequently required for diagnosis.[102]

Computed tomography findings are well known, and include temporal lobe edema, mass effect and nonhomogeneous contrast enhancement.[103–107] Isolated frontal or occipital lobe involvement is uncommon.[107] Abrupt transition to normal brain at the external capsule is considered characteristic.[106] Hemorrhagic infarction may occur, and cystic encephalomalacia may be seen as a late sequella.[104,106] CT may be normal early in the infection.[106]

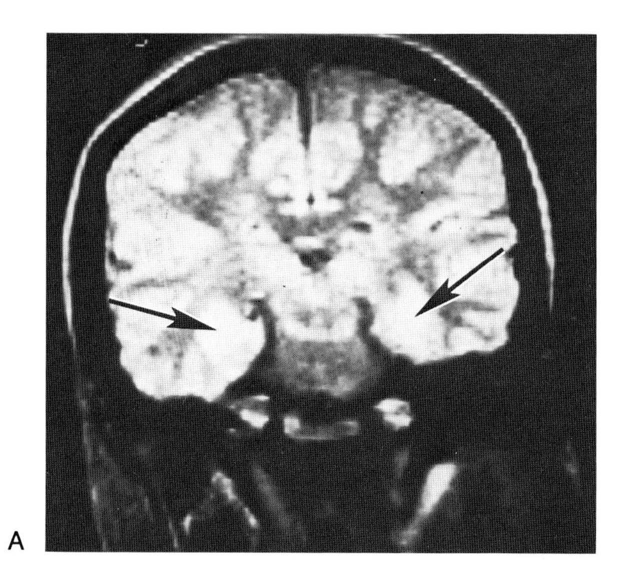

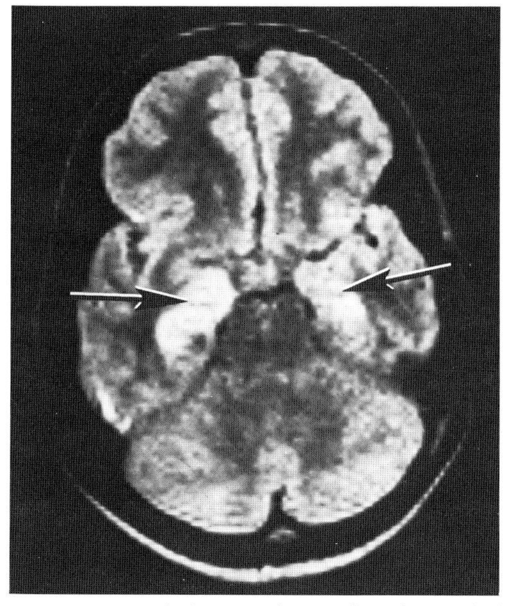

Fig. 8-15. Herpes encephalitis. High signal in the medial temporal lobes seen on **(A)** coronal SE 1500/56 and **(B)** axial SE 2000/56 images. CT was normal.

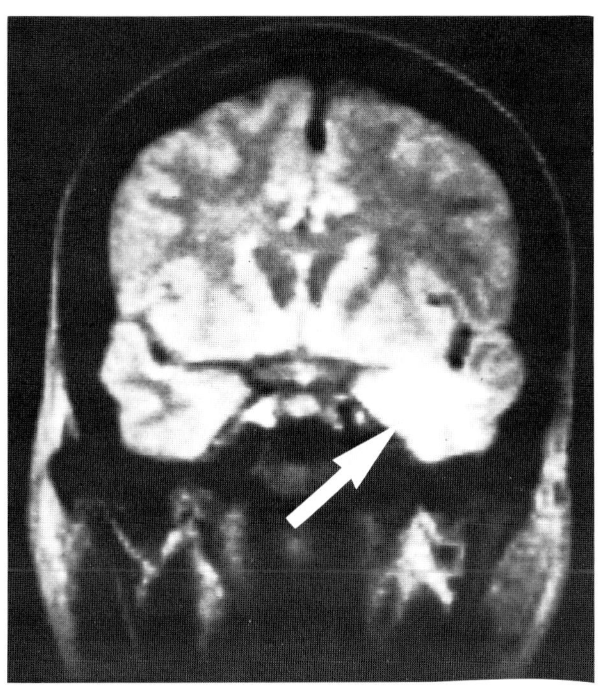

Fig. 8-16. Asymmetric involvement of the temporal lobes in another patient with herpes encephalitis.

Several isolated examples of herpes encephalitis seen by MR have appeared in the literature, but no comprehensive or comparative studies have yet been published. In these cases and in our own experience MR demonstrates areas of abnormal parenchymal signal corresponding to, but more extensive than those seen on CT. Direct coronal MR imaging is useful to show medial temporal lobe involvement. Two examples of documented herpes simplex encephalitis appear in Figures 8-15 and 8-16.

Neonatal infection with herpes simplex nearly always involves the type 2 strain acquired from a mother with genital lesions.[99] Early intrauterine infection acquired transplacentally results in micrencephaly, intracranial calcification, mental retardation, microphthalmia, and retinal dysplasia.[108] Postnatal infection is acquired by the infant's passage through an infected birth canal. These patients present with skin lesions and disseminated viremia, often with diffuse meningoencephalitis and high mortality.[109]

Other Viral Infections

Cytomegalovirus

Cytomegalovirus (*CMV*), like herpes simplex, may infect the CNS in a congenital or acquired form.[110] Congenitally acquired CMV infections result in atrophy, microencephaly, and periventricular calcifications. Adult acquired CMV infections are important considerations in the immunocompromised host.[111] Here they may present as diffuse for focal encephalitis. An example of localized CMV cerebritis in an AIDS patient not detected by CT is shown in Figure 8-17.

Varicella-Zoster Virus

Varicella-Zoster virus is a herpesvirus that causes both primary varicella (chickenpox) and the recurrent disease herpes zoster (shingles). Meningitis or meningoencephalitis complicates fewer than 1 in 1000 cases of chickenpox and has no distinguishing radiographic features, although its clinical presentation with cerebellar ataxia is quite characteristic.[112] Encephalitis complicating herpes zoster is usually seen in debilitated or immunosuppressed adults with disseminated disease.[99] A distinctive pattern of presentation in about one-third of cases is an initial zoster ophthalmicus with development of contralateral hemiplegia over the next several weeks.[113] The pathologhysiology of zoster encephalitis may differ from varicella encephalitis in that the former demonstrates a significant granulomatous angiitis.[114]

Subacute Sclerosing Panencephalitis

Subacute sclerosing panencephalitis (SSPE) is a rare, usually fatal disease of children and adolescents that occurs 3 to 10 years following an innocent measles (rubeola) infection.[115] Progressive dementia, ataxia, myoclonus, and focal neurologic deficits develop as the latent measles virus slowly destroys brain parenchyma. Findings include diffuse cortical atrophy, periventricular demyelinization, and degenerative changes in the caudate nuclei.[116] A similar disease to SSPE occurs years following rubella, called *progressive rubella panencephalitis PRPE*). Both SSPE

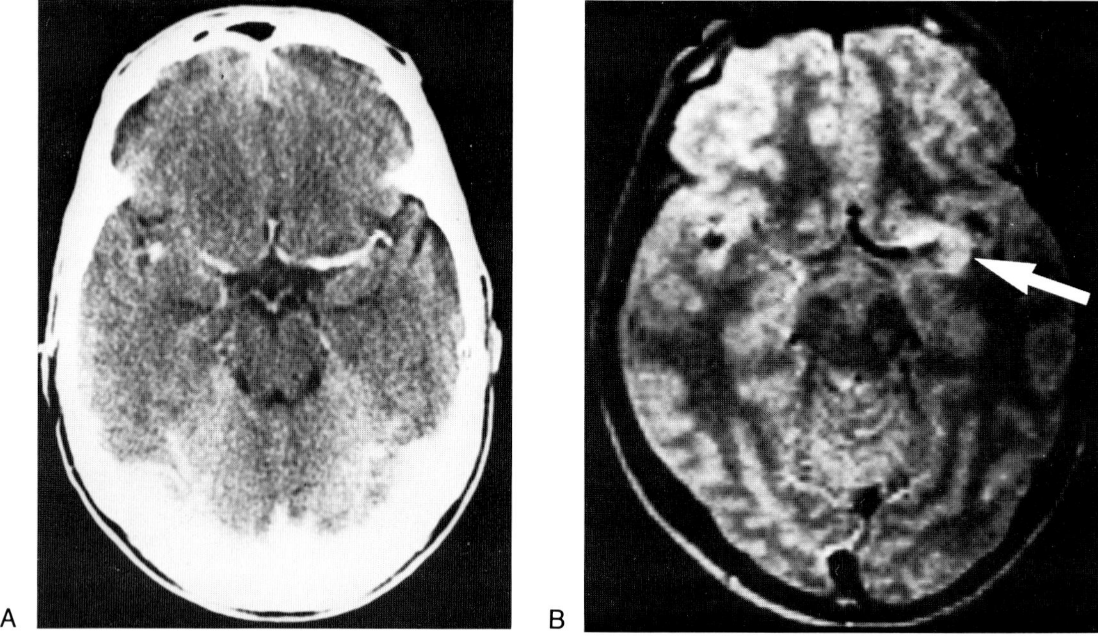

Fig. 8-17. Cytomegalovirus infection in a patient with AIDS. **(A)** Contrast CT, initially read as normal. **(B)** Abnormal signal along the proximal left middle cerebral artery seen on this SE 2000/56 image.

and PRPE should be considered in the differential diagnosis of diffuse demylinating disease in children.

Progressive Multifocal Leukoencephalopathy

Progressive multifocal leukoencephalopathy (PML) is a subacute demylinating disease of immunocompromised patients caused by a papovavirus. It will be discussed more fully in Chapter 9. *Creutzfeldt-Jacob disease*, a fatal brain degeneration caused by a slow virus, will also be discussed at that time.

PARASITIC INFECTIONS

Cysticercosis

Cysticercosis is an infestation by larvae of the pork tapeworm, *Taenia solium*. Although relatively infrequent in the United States, cysticercosis is the most common parasitic infestation of the CNS worldwide. With increasing immigration from endemic areas, especially Mexico and South America, its incidence in the U.S. is increasing.[117] The diagnosis of cysticercosis should be seriously considered in patients from endemic areas who present with seizures or have cystic intracranial lesions demonstrated by MR or CT.

Man is the definitive host for the adult tapeworm, which resides in the small bowel with its head (scolex) held on by suckers and hooklets.[118] The mature tapeworm sheds its eggs (ova) in human stool. When humans ingest ova from contaminated soil or water, the ova hatch into oncospheres that invade the gut wall. These oncospheres spread hematogenously throughout the body, especially to skeletal muscle and brain. In about two months the oncospheres form into encysted larvae, called cysticerci.

Swine may become intermediate hosts for the parasites when they ingest *Taenia* ova. Hematogenous dissemination then occurs in the swine, with cysticerci forming in its skeletal muscle. Humans eating undercooked infected pork may develop an intestinal tapeworm infestation, but not widespread cysticercosis. Only human ingestion of the ova passed by these tapeworms poses the threat of whole body infestation, including neurocysticercosis.

Cysticerci are distributed throughout the CNS with

a preference for the subarachnoid space, cortical sulci, and cortical gray matter.[119] Classified by location, cysticerci are meningeal (39 percent), parenchymal (20 percent), intraventricular (17 percent), mixed (23 percent), or intraspinal (1 percent).[120] Only 4 percent of lesions are solitary.[121]

Active parenchymal cysticerci are cystic lesions, usually about 1 cm in diameter, with a range in size of 0.5 to 4.0 cm.[122] A 2 to 4 mm mural nodule containing the scolex is commonly seen internally at one end of the cyst. The cyst fluid is transparent in live cysticerci, but turns turbid or jellylike with the death of the organism.[123] A thin capsule and layer of compressed gliotic brain surround the cyst. A striking feature of the disease at this stage is a lack of inflammatory reaction in the surrounding brain.[124] Lesions may appear "punched out" with surprisingly little distortion of neighboring structures.[117]

After a period of years, parenchymal brain cysts slowly expand and the larvae die. At this time a marked inflammatory response may be noted in the surrounding brain, associated with edema and contrast enhancement on CT.[125] Calcifications occur in dead larvae and may require 10 or more years to develop.[126] Calcification is the most common CT manifestation of cysticercosis.[121] The calcifications are rounded or oval, 7 to 11 mm in size, and are usually located in the gray matter or in the gray-white matter junction.[120] In cases of reinfection, calcified (dead) larvae may coexist with active lesions.

Some cysticerci that grow in the subarachnoid space degenerate into large, multibolular cysts called *racemose cysts*.[118] These cysts may reach several centimeters in diameter and are frequently located in the basilar cisterns near the cerebellopontine angle or sella. Although sterile and nonviable in that they lack a scolex, racemose cysts may grow by proliferation of their walls, and may induce adjacent inflammatory reactions.[122]

Symptoms of neurocysticercosis depend upon the age, number, and location of the lesions. Parenchymal and juxtaparenchymal larvae may cause seizures, headaches, paralysis, focal neurologic defects, or dementia.[127] In endemic areas, cysticercosis is the most common identifiable cause of seizures in young adults. Symptoms may appear or exacerbate during the phase of parasite death, when adjacent brain parenchymal inflammation is at a maximum. Meningeal involvement may result in arachnoditis, ependymitis, vasculitis, or vascular thrombosis. Intraventricular cysts may cause ventricular obstruction chronically or acutely, the latter of which may produce sudden death.[128]

On MR cysticerci image as spherical cystic masses containing fluid with signal properties closely paralleling CSF (Fig. 8-18). A higher intensity 2 to 4 mm mural nodule containing the scolex is frequently identified (Fig. 8-18). Racemose cysts are identified by their larger size, characteristic location in the basal cisterns, and lack of a mural nodule (Fig. 8-19). A rim of parenchymal inflammation with prolonged T1 and T2 values may surround dying or degenerating cysts.[129]

Magnetic resonance and computed tomography seem to be complementary in the evaluation of neurocysticercosis. MR is superior in demonstrating some cysts over the convexities, particularly those near the skull vertex. MR may reveal intraventricular cysts not identified by CT; MR may even show intraventricular lesions when complete obstruction prevents entry of metrizamide.[130] MR is probably more sensitive than CT in identifying surrounding edema, and may show internal signal changes within the cyst indicative of parasite death.[129] By comparison, CT is much better able to detect calcified cysticerci. Until MR contrast agents become available, CT will continue to hold an advantage in assessment of blood-brain barrier integrity. MR may obviate the need for CT ventriculography in certain situations, and is potentially useful for monitoring therapy of cysts and edema.

Toxoplasmosis

Cerebral toxoplasmosis is an infection caused by the obligate intracellular protozoan *Toxoplasma gondii*. A congenital form is transmitted transplacentally to the fetus from an infected mother. An acquired form, usually seen in adults, occurs with high frequency in those with impaired immunity (underlying malignancy, AIDS, collagen vascular disease, organ transplantation, or chemotherapy).

The cat is an important intermediate host for the parasite, and infection can be acquired by ingestion of oocysts from cat feces or contaminated soil.[131] Eating unwashed vegetables or undercooked meat is another mode of transmission. Once ingested the organ-

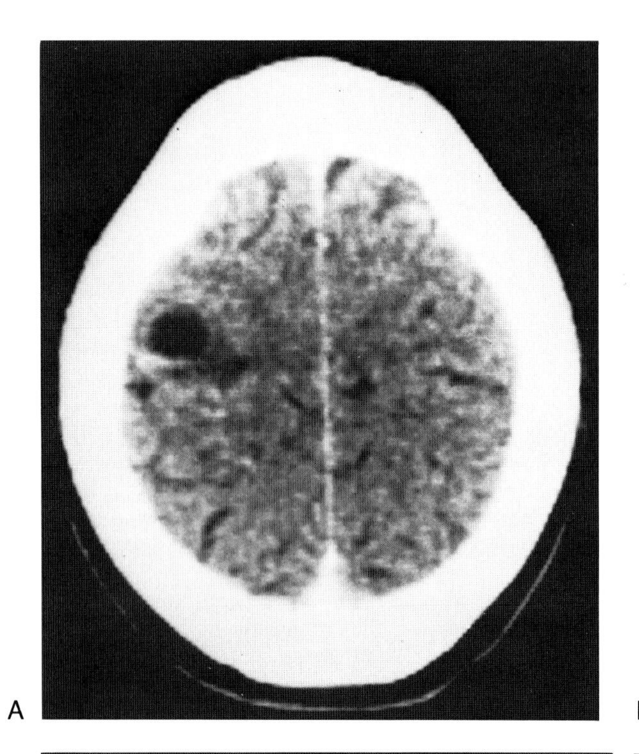
A

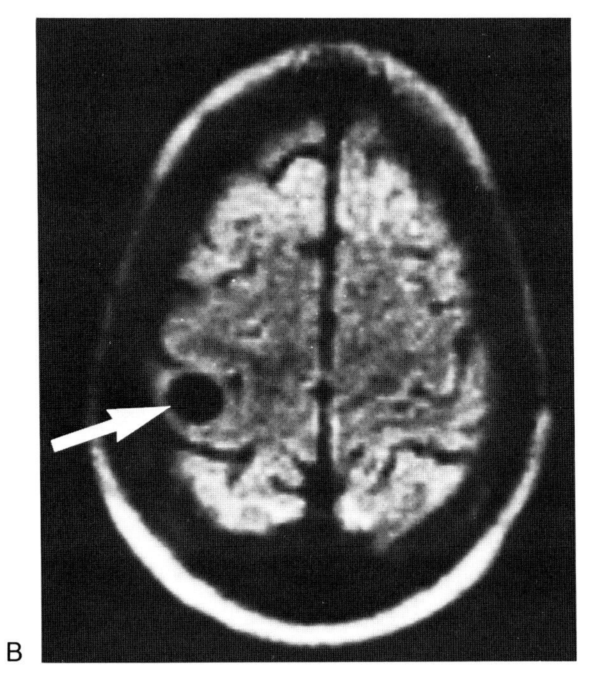
B

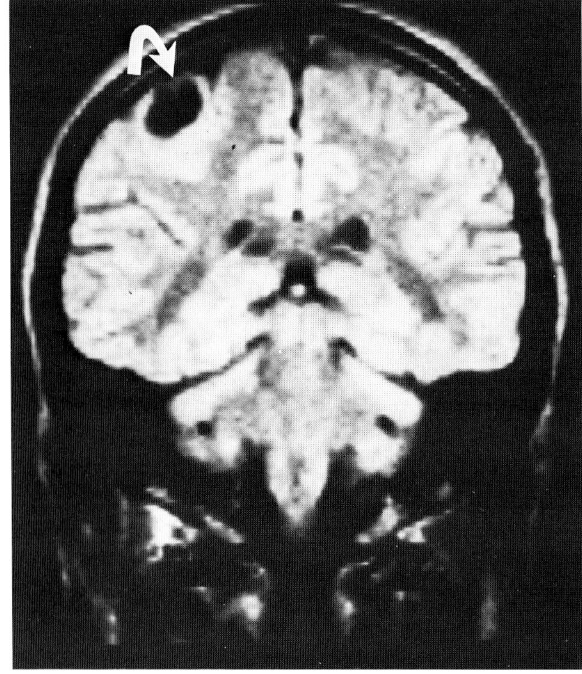
C

Fig. 8-18. Cysticercosis. Cortical cyst seen on **(A)** CT and **(B)** axial SE 1000/28 image. **(C)** Coronal SE 1500/28 image provides important diagnostic clue, the presence of a scolex (arrow) within the cyst.

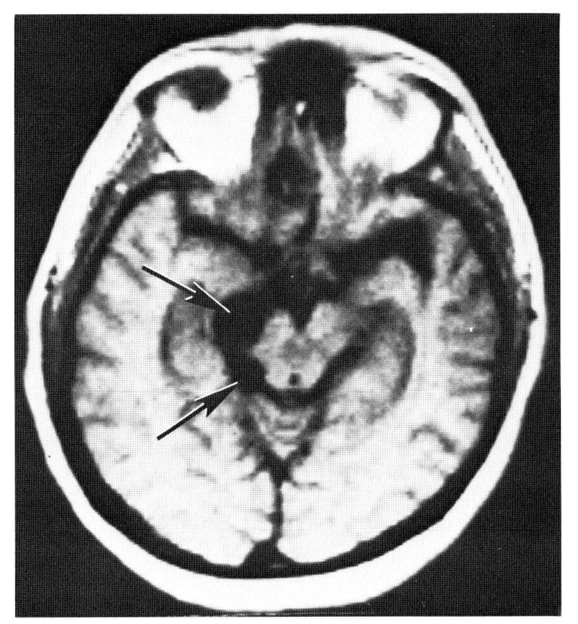

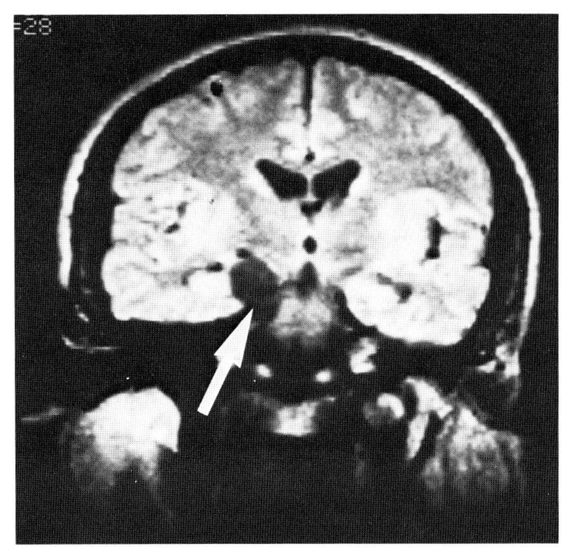

A B

Fig. 8-19. Cysticercosis. Racemose cysts in the perimesencephalic cistern. **(A)** On SE 500/28 the signal from these cysts cannot be distinguished from CSF. **(B)** Noticeably different signal from cysts is seen on this coronal SE 1500/28 MR image. Virtually all arachnoid cysts are isointense with CSF on any pulse sequence.

ism invades the bloodstream and is disseminated throughout the body including the CNS. Most adult acquired infections are asymptomatic, but initial dissemination may produce a syndrome with lymphadenopathy resembling acute mononucleosis.[132] If the patient is pregnant transmission to the fetus may occur at this time. Otherwise, the acute phase resolves and tissue cysts (containing bradyzoites) form in multiple organs (including the brain), representing the latency stage of the disease. With subsequent immunosuppression the cysts may rupture, releasing invasive trophozoites. These trophozoites destroy adjacent host cells, producing necrotic foci of infection.[131]

In the congenital form of the disease, symptoms and signs are usually recognized within the first few days of life.[133] Common manifestations include microcephaly, seizures, mental retardation, spasticity, chorioretinitis, microphthalmus, and hepatosplenomegaly. Periventricular calcifications can be seen on CT, less frequently on plain skull radiographs or MRI (Fig. 8-20). Prognosis is poor with over 50 percent of affected infants dying within the first few weeks. A subacute early childhood form of the disease has also been described.[134]

Healthy adults normally quickly overcome acute exposure to *Toxoplasma*, save those few who experience a viral-like syndrome. Immunocompromised patients are at significant risk to develop reactivation disease of great severity, however.[135] Patients at high risk include those with malignancies, especially leukemia and lymphoma, and AIDS.[136,137] The risk of an AIDS patient for developing cerebral toxoplasmosis is estimated to be as high as 6 to 12 percent.[138] Between one-fourth and one-third of all CNS infections in patients with AIDS are caused by *Toxoplasma gondii*.[139] Despite appropriate antibiotic therapy, mortality is high.[131]

The CT findings are nonspecific but the diagnosis may be suggested in the appropriate clinical setting with characteristic findings.[139–141] CT usually shows multiple round, low-density lesions in both hemispheres, often subcortically. The basal ganglia or brain stem are involved in over half of cases. Intravenous contrast administration results in ringlike or nodular enhancement. CT is known to underestimate the extent of disease.[140,141]

Several examples of cerebral toxoplasmosis imaged by MR have been published.[111,142] In most cases MR

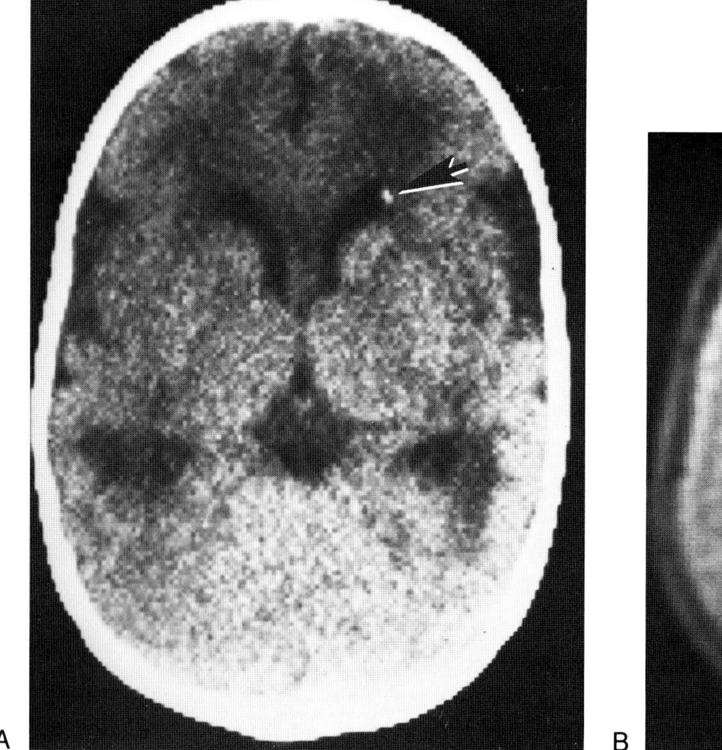

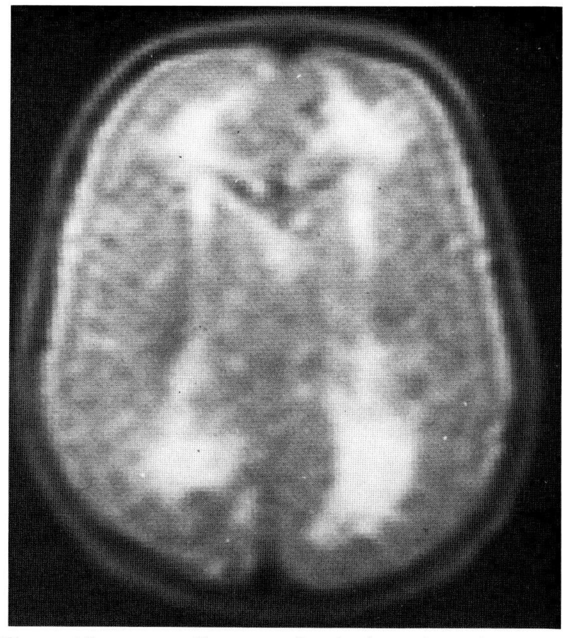

Fig. 8-20. Congenital toxoplasmosis. **(A)** CT scan revealed diffuse white matter disease and a single periventricular calcification adjacent to the left frontal horn (arrow). **(B)** Axial SE 3000/80 MR scan reveals extensive white matter disease, but the calcification is not detected.

showed closer resemblance than CT to the lesions seen on the pathologic specimen. All lesions demonstrated had long T1 and T2 values, and were round and multiple. Cerebellar and brain stem lesions were better appreciated by MR. These features are illustrated in Figure 8-21.

Other Parasitic Infestations of the CNS

Several other parasites that may invade the CNS are mentioned for completeness. However, no documented cases of these infestations as imaged by MR have been published, and neither the author nor his colleagues have had personal experience with these diseases.

Primary Amebic Meningoencephalitis

Primary amebic meningoencephalitis is usually caused by the protozoan *Naegleria fowleri.*[143] The disease occurs in healthy children and adults who were recently swimming in freshwater lakes. The ameba penetrate the nasal mucosa and enter the CNS through the cribiform plate. The amebae incite a rapidly progressive and purulent meningoencephalitis with marked edema particularly in the subfrontal region. Most patients die within 72 hours.

Amebic Brain Abscess

Amebic brain abscess is rare but occurs most frequently after hemotogenous dissemination of

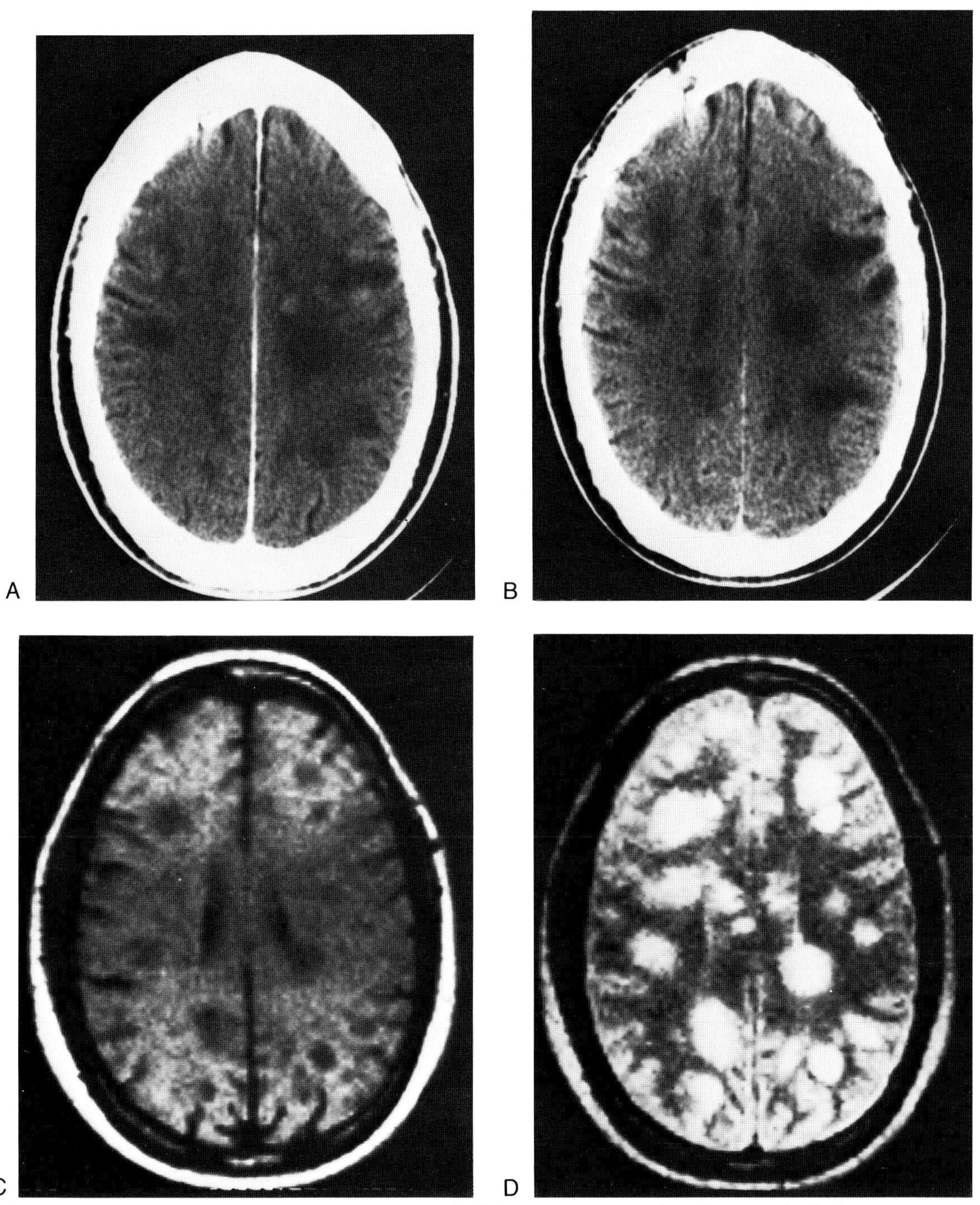

Fig. 8-21. Toxoplasmosis in an immunosuppressed adult. CT scan (**A** and **B**) shows multiple low attenuation lesions, some with vague contrast enhancement. These lesions can be seen on SE 500/28 images fairly **(C)**. Excellent delineation of the lesions is seen on this T2-weighted SE 2000/56 image **(D)**.

Entameba histolytica.[144] The brain abscess is nearly always accompanied by a coexisting liver abscess. Frequently, multiple small brain abscesses are seen.

Hydatid Disease

Hydatid disease of the CNS may occur after ingestion of eggs from a dog tapeworm, *Echinococcus granulosus.*[145] The disease is rare in the United States, being seen mainly in immigrants from rural endemic areas like the Near East where humans and dogs live closely with sheep and cattle (which are intermediate hosts for the parasite). In a life cycle analogous to that of cysticercosis, oncospheres released from ingested ova hematogenously disseminate to multiple organs including the CNS. Only about two percent of ecchinococcal infestations invade the CNS. Children account for about 75 percent of cases.[146]

Hydatid cysts of the brain are usually solitary and large.[145] Most lie a few millimeters beneath the cortex, but may be extradural. Cyst contents are similar to CSF in density. Occasionally daughter cysts within the larger cyst may be noted. The cyst can produce mass effect and obstructive hydrocephalus, but there is no surrounding edema.

Worms

Certain worms or their larvae occasionally invade the CNS. These include: (1) pork tapeworm (*Trichinella spiralis*), which produces trichinosis; (2) the human nematode *Strongyloides stercoralis;* (3) the rat lungworm (*Angiostrongylus cantonensis*), which causes eosinophilic meningitis; (4) the dog and cat roundworms (*Toxocara canis* and *cati* called larva migrans); and (5) trematodes of the species *Schistosoma*, which cause bilharziasis. These worms may cause meningoencephalitis or present with widespread or localized areas of granulomatous reaction.[131]

Paragonimiasis

Paragonimiasis is an infestation caused by the oriental lung fluke, *Paragonimus westermani.*[147] Humans are infected by eating raw freshwater crabs or crayfish. Both pulmonary and cerebral hematogenous dissemination are common. Mostly cases occur in children below the age of 10. Three forms of reaction of the CNS to invading flukes have been described: chronic arachnoiditis, granuloma, and encapsulated abscess. Multiple parenchymal and intraventricular cysts may be seen. In the later stages shell-like and conglomerate calcifications are characteristic.[148]

REFERENCES

1. Davidson HD, Steiner RE: Magnetic resonance imaging in infections of the central nervous system. AJNR 6:499, 1985.
2. Moore GA, Thomas LM: Infections including abscesses of the brain, spinal cord, intraspinal, and intracranial lesions. Surg Ann 6-413, 1974
3. Harriman DGE: Greenfield's Neuropathology. p. 238–247. Year Book, Chicago, 1976
4. Schlech WF III, Ward JI, Band JD et al: Bacterial meningitis in the United States, 1978–1981. JAMA 253:1749, 1985
5. Edberg SC: Conventional and molecular techniques for the laboratory diagnosis of infections of the central nervous system. Neurol Clin 4:13, 1986
6. Ovetuf GD: Pyogenic bacterial infections of the CNS. Neurol Clin 4:69, 1986
7. Moxon ER, Smith Al, Averill DR et al: *Haemophilus influenzae* meningitis in infant rats after intranasal inoculation. J Infect Dis 129:154, 1974
8. Scheld WM, Long WJ, Brodeur JM: Effects of experimental *E. coli* meningitis on the blood-brain barrier: in vitro and in vivo studies. Clin Res 31:375A, 1983
9. Täuber MG, Brooks-Fournier RA, Sande MA: Experimental models of CNS infections. Neurol Clinics 4:249, 1986
10. Zimmerman RA, Patel S, Bilaniuk L: Demonstration of purulent bacterial intracranial infections by computed tomography. AJR 127:155, 1976
11. Auh YH, Lee SH, Toglia JU: Excessively small ventricles on cranial CT: Clinical correlation in 75 patients. J Comput Tomogr 4:325, 1980
12. Cockrill HH Jr, Dreisbach J, Lowe B et al: Computed tomography in leptomeningeal infections. AJR 130:511, 1978
13. Schultz P, Leeds NE: Intraventricular septations complicating neonatal meningitis. J Neurosurg 38:620, 1973
14. Zimmerman RA, Bilaniuk LT, Gallo E: CT of the trapped fourth ventricle. AJR 130:503, 1978

15. Smith H, Hendrick E: Subdural empyema and epidural abscess in children. J Neurosurg 58:392, 1983
16. Bannister GB, Williams B, Smith S: Treatment of subdural empyema. J Neurosurg 55:82, 1981
17. Bhandari YS, Sarkari NB: Subdural empyemas. A review of 37 cases. J Neurosurg 32:35, 1970
18. Zimmerman RD, Leeds NE, Danziger A: Subdural empyema: CT findings. Radiology 150:417, 1984
19. Sadhu VK, Handel SF, Pinto RS et al: Neuroradiologic diagnosis of subdural empyema and CT limitations. AJNR 1:39, 1980
20. Courville CB: Subdural empyemas secondary to purulent frontal sinusitis; clinico-pathologic study of 42 cases verified at autopsy. Arch Otolaryng Otolaryngol 39:211, 1944
21. Post EM, Modesti LM: "Subacute" postoperative subdural empyema. J Neurosurg 55:761, 1981
22. Handel SF, Klein WC, Kim YW: Intracranial epidural abscess. Radiology 111:117, 1974
23. Sharif HS, Ibrahim A: Intracranial epidural abscess. Br J Radiol 55:81, 1982
24. Galbraith JG, Barr VW: Epidural abscess and subdural empyema. Adv Neurol 6:257, 1974
25. Lott T, El Gammal T, Dasilva R et al: Evaluation of brain and epidural abscesses by computed tomography. Radiology 122:371, 1977
26. Enzman DR, Britt RH, Yeager AS: Experimental brain abscess evolution: computed tomography and neuropathologic correlation. Radiology 133:113, 1979
27. Enzman DR, Britt RH, Placone R: Staging of human brain abscess by computed tomography. Radiology 146:703, 1983
28. Brant:Zawadzki M, Enzmann DR, Placone RC Jr et al: NMR imaging of experimental brain abscess: comparison with CT. AJNR 4:250, 1983
29. Runge VM, Clanton JA, Price AC et al: Evaluation of contrast-enhanced MR imaging in a brain-abscess model. AJNR 6:139, 1985
30. Grossman RI, Wolf G, Biery D et al: Gadolinium enhanced nuclear magnetic resonance images of experimental brain abscess. J Comput Assist Tomogr 8:204, 1984
31. Grossman RI, Joseph PM, Wolf G et al: Experimental intracranial septic infarction: magnetic resonance enhancement. Radiology 155:649, 1985
32. Lee SH: Infectious diseases. p. 505. In Lee SH, Rao KCVG (eds): Cranial Computed Tomography. McGraw-Hill, New York, 1983
33. Stevens EA, Norman D, Kramer RA et al: Computed tomographic brain scanning in intraparenchymal pyogenic abscesses. AJR 130:111, 1978
34. Moussa AH, Dawson BH: Computed tomography and the mortality rate in brain abscess. Surg Neurol 10:301, 1978
35. Butler IJ, Johnson RT: Central nervous system infection. Pediatr Clin North Am 21:649, 1974
36. DeLouvois J: The bacteriology and chemotherapy of brain abscess. J Antimicrob Chemother 7:395, 1978
37. Gomori JM, Grossman RI, Goldberg HI et al: Intracranial hematomas: imaging by high-field MR. Radiology 157:87, 1985
38. Morgan H, Wood MW: Cerebellar abscesses. A review of 7 cases. Surg Neurol 3:93, 1975
39. Reed JE, Williams JP, Cooper MD: Intraventricular abscess rupture. Neuroradiology 7:261, 1974
40. Whalen MA, Hilal SK: Computed tomography as a guide in the diagnosis and follow-up of brain abscesses. Radiology 135:663, 1980
41. Sheller JR, Des Prez RM: CNS tuberculosis. Neurologic Clinics 4:143, 1986
42. Traub M, Colchester ACF, Kingsley DPE et al: Tuberculosis of the central nervous system. Q J Med 209:81, 1984
43. Kennedy DH, Fallon RJ: Tuberculosis meningitis. JAMA 241:264, 1979
44. Stead WW, Lofgren JP, Warren E et al: Tuberculosis as an endemic and nosocomial infection among the elderly in nursing homes. N Engl J Med 312:1483, 1985
45. Rich AR, McCordock HA: Pathogenesis of tuberculous meningitis. Bull Johns Hopkins Hosp 52:5, 1933
46. Bhargava S, Gupta AK, Tandon PN: Tuberculous meningitis: A CT study. Br J Radiol 55:189, 1982
47. Rovira M, Romero F, Torrent U et al: Study of tuberculous meningitis by CT. Neuroradiology 19:137, 1980
48. Chu N-S: Tuberculous meningitis. Arch Neurol 37:458, 1980
49. Price HJ, Danziger A: Computed tomography in cranial tuberculosis. AJR 130:769, 1978
50. Armitsu T, Jabbari B, Buckler RE et al: CT in verified cases of tuberculous meningitis. Neurology 29:384, 1979
51. Casselman ES, Hasso AN, Ashwal S et al: CT of tuberculous meningitis in infants and children. J Comput Assist Tomogr 4:211, 1980
52. Lorber J: Intracranial calcification following tuberculous meningitis in children. Acta Radiol 50:204, 1958
53. Whiterier DR: Tuberculous brain abscess. Report of a case and review of the literature. Arch Neurol 35:148, 1978
54. Reichenthal E, Cohen ML, Schujman E et al: Tuberculous brain abscess and its appearance on computed tomography. J Neurosurg 56:597, 1982
55. Sibley WA, O'Brien JL: Intracranial tuberculomas: a review of clinical features and treatment. Neurology 6:157, 1956
56. Whalen MA, Stern J: Intracranial tuberculoma. Radiology 138:75, 1981

57. Mayers MM, Kaufman DM, Miller MH: Recent cases of intracranial tuberculomas. Neurology 28:256, 1978
58. Capon A, Noterman J, Huber JP: Multiple tuberculoma of the brain. Acta Neurochir 32:303,1975
59. Welchman JM: CT of intracranial tuberculomata. Clin Radiol 30:567, 1979
60. Bhargava S, Tandon PN: Intracranial tuberculomas: A CT study. Br J Radiol 53:935, 1980
61. Elisevich K, Arpin EJ: Tuberculoma masquerading as a meningioma. J Neurosurg 56:435, 1982
62. Manz HJ: Pathobiology of neurosarcoidosis and clinicopathologic correlation. Can J Neurol Sci 10:50, 1983
63. Delaney P: Neurologic manifestations in sarcoidosis: review of the literature with a report of 23 cases. Ann Intern Med 87:336, 1977
64. Post MJD, Quencer RM, Tabe SZ: Intracranial sarcoidosis: CT demonstration of the optic nerve, frontal lobes, and falx cerebri. AJNR 3:523, 1982
65. Mirfakhraee M, Crofford MJ, Guinto FC Jr et al: Virchow-Robin space: a path of spread in neurosarcoidosis. Radiology 158:715, 1986
66. Bahr AL, Krumholz A, Kristt D et al: Neuroradiological manifestations of intracranial sarcoidosis. Radiology 127:713, 1978
67. Kendall BE, Tateler GLV: Radiologic findings in neurosarcoidosis. Br J Radiol 51:81, 1978
68. Morehouse H, Dazinger A: CT findings in intracranial neurosarcoid. J Comput Tomogr 4:267, 1980
69. Brooks BS, Gammal TE, Hungerford GD et al: Radiologic evaluation of neurosarcoidosis: role of computed tomography. AJNR 3:513, 1982
70. Griggs RC, Manesberry WR, Condemi JJ: Cerebral mass due to sarcoidosis: regression during corticosteroid therapy. Neurology 23:981, 1973
71. Kumpe DA, Rao CVGK, Garcia JH et al: Intracranial neurosarcoidosis. J Comput Assist Tomogr 3:324, 1979
72. Reed LD, Abbas S, Markivee CR et al: Neurosarcoidosis responding to steriods. AJR 146:819, 1986
73. Salaki JS, Louria DB, Chmel H: Fungal and yeast infections of the central nervous system: a clinical review. Medicine 63:108, 1984
74. Lyons RW, Andriole VT: Fungal infections of the CNS. Neurologic Clinics 4:159, 1986
75. Whelan MA, Stern J, deNapoli RA: The computed tomographic spectrum of intracranial mycosis: correlation with histopathology. Radiology 141:703, 1981
76. Parker JC, McCloskey JJ, Lee RS: The emergence of candidosis. The dominant postmortem cerebral mycosis. Am J Clin Pathol 70:31, 1978
77. Lipton SA, Hickey WF, Morris JH et al: Candidal infection in the central nervous system. Am J Med 76:101, 1984
78. Mishra SK, Damodaran VN: Observations on the natural habitats of *Cryptococcus neoformans* and *Nocardia asteriodes*. Ind J Chest Dis 15:263, 1973
79. Sabetta JR, Andriole VT: Cryptococcal infection in the nervous system. Med Clinics North Am 69:333, 1985
80. Everett BA, Kusske JA, Rush JL et al: Cryptococcal infection of the central nervous system. Surg Neurol 9:157, 1978
81. Cornell S, Jacoby C: The varied computed tomographic appearance of intracranial cryptococcosis. Radiology 143:703, 1982
82. Goodwin RA Jr, Shapiro JL, Thurman GH et al: Disseminated histoplasmosis: clinical and pathologic correlations. Medicine 59:1, 1981
82a. Dion FM, Venger BH, Landon G, Handel SF: Thalamic histoplasmoma: CT and MR imaging. J Comput Assist Tomogr 11:193, 1987
83. Dublin A, Phillips H: Computed tomography of disseminated cerebral coccidiomycosis. Radiology 135: 361, 1980
84. McGahan J, Graves D, Palmer P et al: Classic and contemporary imaging of coccidiomycosis. AJR 136:393, 1981
85. Gonyea EF: The spectrum of primary blastomycotic meningitis: a review of central nervous system blastomycosis. Ann Neurol 3:26, 1978
86. McCormick WF, Schochet SS, Weaver PR et al: Disseminated aspergillosis. Arch Pathol 100:353, 1975
87. Beal M, O'Carroll C, Kleinman G et al: Aspergillosis of the nervous system. Neurology 32:473, 1982
88. Grossman RI, Davis KR, Taveras JM et al: Computed tomography of intracranial aspergillosis. J Comput Assist Tomogr 5:646, 1981
89. Ahuja GK, Jain N, Vijayaraghavan M et al: Cerebral mycotic aneurysm of fungal origin. J Neurosurg 49:107, 1978
90. Lehrer RI, Howard DH, Sypherd PS et al: Mucormycosis. Ann Intern Med 93:93, 1980
91. Centeno RS, Bentson JR, Mancuso AA: CT scanning in rhinocerebral mucormycosis and aspergillosis. Radiology 140:383, 1981
92. Pierce CF Jr, Solomon SL, Kaufman L et al: Zygomycetes brain abscesses in narcotic addicts with serological diagnosis. JAMA 248:2681, 1982
93. Fetter B, Klintworth G, Hendry W: Mycoses of the Central Nervous System. Williams & Wilkins, Baltimore, 1967
94. Smith PW, Steinkraus GE, Henricks BW et al: CNS nocardiosis. Arch Neurol 37:729, 1980
95. Case 20–1980: Case records of the Massachusetts General Hospital. N Engl J Med 302:1194, 1980
96. Curry WA: Human nocardiosis. A clinical review with selected case reports. Arch Intern Med 140:818, 1980

97. Tyson GW, Welch JE, Butler AB et al: Primary cerebellar nocardiosis. J Neurosurg 51:408, 1979
98. Johnson MA, Pennock JM, Bydder GM et al: Clinical NMR imaging of the brain in children: normal and neurologic disease. AJR 141:1005, 1983
99. Barnes DW, Whitley RJ: CNS diseases associated with varicella zoster virus and herpes simplex virus infection. Neurologic Clinics 4:265, 1986
100. Nahmias AJ, Whitley RJ, Visintine AN et al: Herpes simplex virus encephalitis: laboratory evaluations and their diagnostic significance. J Infect Dis 145:829, 1982
101. Davis LE, Johnson RT: An explanation for the localization of herpes simplex encephalitis. Ann Neurol 5:2, 1979
102. Barza M, Pauker SG: The decision to biopsy, treat, or wait in suspected herpes encephalitis. Ann Intern Med 92:641, 1980
103. Davis JM, Davis KR, Kleinman GM et al: Computed tomography of herpes simplex encephalitis, with clinical pathological correlation. Radiology 129:409, 1978
104. Enzmann DR, Ranson B, Norman D et al: Computed tomography of herpes simplex encephalitis. Radiology 129:419, 1978
105. Dublin AB, Merten DF: Computed tomography in the evaluation of herpes simplex encephalitis. Radiology 125:133, 1977
106. Zimmerman RD, Russell EJ, Leeds NE et al: CT in the early diagnosis of herpes simplex encephalitis. AJR 134:61, 1980
107. Ketonen L, Koskiniemi ML: CT appearance of herpes simplex encephalitis. Clin Radiol 31:161, 1980
108. South MA, Tompkins WA, Morris CR et al: Congenital malformation of the central nervous system associated with genital type (type 2) herpes virus. J Pediatr 75:13, 1969
109. Whitley RJ: Herpes simplex virus infections of the central nervous system in children. Semin Neurol 2:87, 1982
110. Overall J Jr, Glasgow L: Virus infections of the fetus and newborn infant. J Pediatr 77:315, 1970
111. Post MJD, Sheldon JJ, Hensley GT et al: Central nervous system disease in acquired immunodeficiency syndrome: prospective correlation using CT, MR imaging and pathologic studies. Radiology 158:141, 1986
112. Johnson R, Milbourn PE: Central nervous system manifestations of chickenpox. Can Med Assoc J 102:831, 1970
113. Hedges TR III, Albert DM: The progression of the ocular abnormalities of herpes zoster. Ophthalmology 39:165, 1982
114. Rosenblum WI, Hadfield MG: Granulomatous angiitis of the nervous system in cases of herpes zoster and lymphosarcoma. Neurology 22:348, 1972
115. Modlin JF, Halsey NA, Eddius DL et al: Epidemiology of subacute sclerosing panencephalitis. J Pediatr 94:231, 1979
116. Duda EE, Huttenlocher PR, Patronas NJ: CT of subacute sclerosing panencephalitis. AJNR 1:35, 1980
117. Schultz TS, Ascherl GF Jr: Cerebral cysticercosis: occurrence in the immigrant population. Neurosurg 3:164, 1977
118. Marquex-Monter H: Cysticercosis. p. 592. In Marcial-Rojas R (ed): Pathology of Protozoal and Helminthic Diseases with Clinical Correlation. Williams & Wilkins, Baltimore, 1971
119. Byrd SE, Locke GE, Biggers S et al: The computed tomographic appearance of cerebral cysticercosis in adults and children. Radiology 144:819, 1982
120. Rodriquez-Carbajal J, Palacios E, Azar-Kia B et al: Radiology of cysticercosis of the central nervous system including computed tomography. Radiology 125:127, 1977
121. Minguetti G, Ferreira MVC: Computed tomography in neurocysticercosis. J Neurol Neurosurg Psychiatry 46:936, 1983
122. Rabiela-Cervantes MT, Rivas-Hernandez A, Rodriguez-Ibarra J et al: Anatomo-pathological aspects of human brain cysticercosis. p. 179. In Flisser A, Wilms K, Laclette JP et al (eds): Cysticercosis: Present State of Knowledge and Perspectives. Academic Press, New York, 1982
123. Escobar A, Nieto D: Parasitic diseases. p. 2503. In Minckler J (ed): Pathology of the Nervous System. Vol 3. McGraw-Hill, New York, 1972
124. Enzmann DR: Imaging of Infections and Inflammations of the Central Nervous System: Computed Tomography, Ultrasound, and Nuclear Magnetic Resonance. p. 103. Raven, New York, 1984
125. Handler LC, Mervis B: Cerebral cysticercosis with reference to the natural history of parenchymal lesions. AJNR 4:709, 1983
126. Dixon HBF, Lipscomb FM: Cysticercosis: An analysis and follow-up of 450 cases. Privy Council, Medical Research Council Report, No. 229. p. 1. London, Her Majesty's Stationery Office, 1961
127. McCormick GF, Zee C-S, Heiden J: Cysticercosis cerebri: review of 127 cases. Arch Neurol 39:534, 1982
128. Zee C-S, Segall HD, Apuzzo MLJ et al: Intraventricular cysticercal cysts: further neuroradiologic observations and neurosurgical implications. AJNR 5:727, 1984
129. Suss RA, Maravilla KR, Thompson J: MR imaging of intracranial cysticercosis: comparison with CT and anatomopathologic features. AJNR 7:235, 1986
130. Waluch V, Solti-Bohman LG, Wade CT et al: MR imaging of intraventricular cysticercosis. Radiology 157(P):212, 1985

131. Bia FJ, Barry M: Parasitic infections of the central nervous system. Neurologic Clinics 4:171, 1986
132. McCabe RE, Remington JS: Toxoplasmosis. p. 281. In Warren KS, Mahmoud AAF (eds): Tropical and Geographical Medicine. McGraw-Hill, New York, 1984
133. Wilson CB, Remington JS, Stagno S et al: Development of adverse sequellae in children born with subclinical toxoplasma infection. Pediatrics 66:767, 1980
134. Sabin AB: Toxoplasmic encephalitis in children. JAMA 116:801, 1941
135. Krick JA, Remington JS: Current concepts in parasitology. Toxoplasmosis in the adult—an overview. N Engl J Med 298:550, 1978
136. Hakes TB, Armstrong D: Toxoplasmosis: Problems in diagnosis and treatment. Cancer 52:1535, 1983
137. Vietzke WM, Gelderman AH, Grimley PM et al: Toxoplasmosis complicating malignancy. Experience at the National Cancer Institute. Cancer 21:816, 1968
138. Luft BJ, Brooks RG, Conley FK et al: Toxoplasmic encephalitis in patients with acquired immune deficiency syndrome. JAMA 252:913, 1984
139. Levy RM, Bredesen DE, Rosenblum ML: Neurological manifestations of the acquired immunodeficiency syndrome (AIDS): Experience at UCSF and review of the literature. J Neurosurg 62:475, 1985
140. Post MJD, Chan JC, Hensley GT et al: Toxoplasma encephalitis in Haitian adults with acquired immunodeficiency syndrome: a clinical-pathologic CT correlation. AJR 140:861, 1983
141. Menges HW, Fischer E, Valavanis A et al: Cerebral toxoplasmosis in the adult. J Comput Assist Tomogr 3:413, 1979
142. Zee C-S, Segall HD, Rogers C et al: MR imaging of cerebral toxoplasmosis: correlation of computed tomography and pathology. J Comput Assist Tomogr 9:797, 1985
143. John DT: Primary amebic meningoencephalitis and the biology of *Naegleria fowleri*. Ann Rev Microbiol 36:101, 1982
144. Lombardo L, Alonso P, Arroyo LS et al: Cerebral amebiasis. J Neurosurg 21:704, 1964
145. Ozgen T, Erbengi A, Bertan V et al: The use of computed tomography in the diagnosis of cerebral hydatid cysts. J Neurosurg 50:339, 1979
146. Hamza R, Touibi S, Jamoussi M et al: Intracranial and orbital hydatid cysts. Neuroradiology 22:211, 1982
147. Kim SK, Walker AE: Cerebral paragonimiasis. Acta Psychiatr Neurol Scand 36:153, 1961
148. Sim BS: CT findings of parasitic infestations of the brain in Korea. J Korean Neurosurg Soc 9:7, 1980

9

Degenerative Brain Diseases

Degenerative brain diseases result in the loss or alteration of one or more components of the brain. A wide spectrum of metabolic, toxic, inherited, and acquired processes may produce cerebral degeneration. Some of these disorders will produce relatively specific MR findings, while many will be indistinguishable by imaging criteria alone. For the purposes of this chapter, degenerative brain diseases will be divided into those manifest primarily by atrophy, those affecting the basal ganglia, and those characterized by destruction or defective formation of white matter.

DISEASES CHARACTERIZED PRIMARILY BY ATROPHY

The term *atrophy* implies loss of brain parenchyma, which is usually accompanied by a secondary enlargement of surrounding CSF-containing spaces. Almost all degenerative brain diseases will eventually result in atrophy.[1] The diseases discussed in this section, however, will be those whose primary imaging manifestation is atrophy. For the purposes of differential diagnosis, atrophic brain disorders will be divided into three groups, based upon patterns of atrophy (Table 9-1). Generalized cerebral atrophy may be a normal phenomenon of aging, and is also seen in Alzheimer disease, malnutrition, slow virus infections, and toxic insults. Cerebellar atrophy, which may also involve the brain stem, may be seen with certain hereditary conditions, drugs, and as a paraneoplastic syndrome. Focal brain atrophy is usually a sequela to local brain injury by trauma or vascular insufficiency, but rare metabolic causes are also known.

Table 9-1. Patterns of Cerebral Atrophy on MRI

Generalized Atrophy
- Normal aging
- Alzheimer disease
- Creutzfeldt-Jacob disease
- Substance abuse (alcohol, barbiturates, amphetamines)
- Cancer
- Steroids
- Dehydration, starvation
- The final stage of any longstanding degenerative brain disease

Cerebellar Atrophy
- Normal aging
- Primary cerebellar degenerations (Table 9-2)
- Paraneoplastic syndrome
- Phenytoin therapy

Focal Atrophy
- Trauma
- Infarction
- Inflammatory
- Developmental
- Associated with vascular anomalies
- Pick disease

The Aging Human Brain

Numerous postmortem and CT studies have demonstrated atrophic involution of the brain, which is seen with normal aging.[2–9] Gray matter is lost in the earlier decades (20 to 50 years), while white matter is lost in later life (70 to 90 years).[3] Selective reduction in the numbers of certain neurons occur in different brain regions.[4] Neurofibrillary tangles, senile plaques, and granulovacuolar degeneration are seen histologically with increasing frequency from the fifth decade onward.[4] Age-related depletion of dopaminergic pathways has been documented, and similar neurochemical depletions may be present in cholinergic systems.[4] While there are conflicting reports in the literature, as yet there seems to be no convincing relationship between dementia and measurements of ventricular or sulcus size in patients older than 65 years.[5]

Generalized cerebral atrophy from aging or other causes has an appearance on axial images that has been likened to a "cracked walnut" (Fig. 9-1).[10]

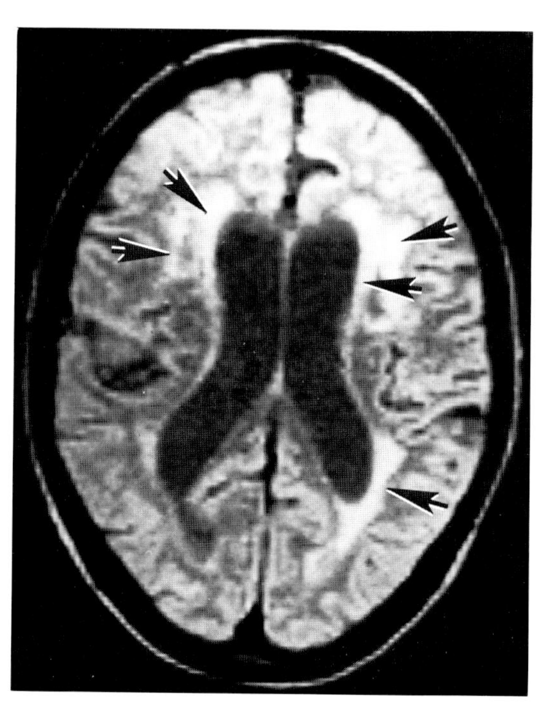

Fig. 9-2. Patchy periventricular and deep white matter disease, presumably on a vascular basis are frequently noted in the elderly with little cognitive impairment.

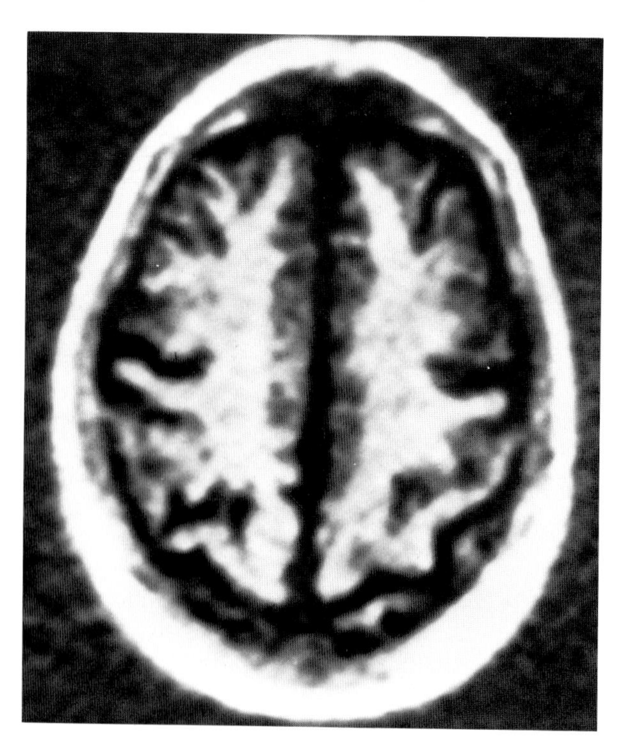

Fig. 9-1. The "cracked walnut" appearance of generalized cerebral atrophy.

Patchy, predominantly periventricular white matter lesions of prolonged T1 and T2 have been noted in elderly patients with no appreciable cognitive impairment (Fig. 9-2).[11–14] At least some of these patients have subcortical vascular encephalopathy (Binswanger disease, q.v.) to account for their lesions, but most have no definite history of hypertension or atherosclerosis elsewhere. There is no consistent correlation of MR detected white matter lesions with dementia or aging, although preliminary data suggest more lesions in demented and older patients.[11,13] Some authorities have recommended that the noncommittal term "leuko-araiosis" be applied to such white matter lesions whose pathogenesis is unclear.[14a]

Alzheimer Disease

Alois Alzheimer (1907) first described the pathologic changes seen in the brains of patients with presenile dementia.[15] For many years, the term Alzheimer disease was applied only to specific dementias occurring in patients below the age of 65. It is now recognized, however, that Alzheimer disease and senile dementia may share common biochemical and histologic features; the term *dementia of the Alzheimer type* is now applied to both.[16]

Dementia of the Alzheimer type is a progressive disorder of insidious onset, characterized by memory loss, confusion, and a variety of cognitive impairments. The disease may occur as early as age 40, but onset after age 60 is more typical. Disturbance of speech is an early symptom, and myoclonus is an early sign. The disease is relentlessly progressive, with the terminal stage characterized by weakness, vegetative mentation, and loss of bowel and bladder control.[17]

The etiology of the disease is unknown. It has been considered by some to be merely an acceleration of normal aging, because its pathologic features (neurofibrillary tangles, senile plaques, and granulovacuolar degeneration) are found in nondemented elderly persons also. Genetic predisposition is also a factor, because there is a 60 percent concordance in identical twins, and over one-third of patients with Alzheimer disease have one or more first-degree relatives who are also affected.[17] Degeneration of the brains of patients with Down syndrome (which occurs commonly in the fourth and fifth decades) has an identical course and histology to Alzheimer disease.[17]

Computed tomography and magnetic resonance findings in Alzheimer disease are nonspecific, most commonly demonstrating generalized cerebral and cerebellar atrophy. Atrophy may sometimes be more severe in the frontal and temporal lobes. Periventricular white matter lesions were noted in seven of eight patients in one study with clinically suspected Alzheimer disease, while CT could detect abnormalities in only three of these patients.[13] The Alzheimer group also seemed to have more severe white matter changes than a control group, but pathologic confirmation was not obtained in this small series.

Creutzfeldt-Jacob Disease

Creutzfeldt-Jacob disease is an extremely rare, rapidly progressive degeneration of the cortex, basal ganglia, and spinal cord that occurs in middle-aged and elderly adults.[18,19] The disease is caused by a spongiform virus, which may be transmitted via corneal transplants or inadequately sterilized surgical instruments.[20] The incidence of the disease worldwide is about 1/1,000,000, with a higher risk noted among certain families and in Libyan Jews.[21]

Clinical features include mental deterioration (100 percent), movement disorder (90 percent), extrapyramidal signs (60 percent), and cerebellar signs (56 percent).[19] Pathologic findings are similar to other spongiform encephalopathies with neuronal loss, astrocytosis, and cytoplasmic vacuoles in neurons and astrocytes.[18] The gray matter of the cerebral cortex and basal ganglia are primarily affected, although white matter lesions may occur late. The disease runs a rapid and inexorable course, with death occurring nearly always within a year of onset.

Magnetic resonance and computed tomography imaging have revealed rapidly progressive and symmetric cortical atrophy as the only abnormalities seen in a few proven cases.[22,23] In our single documented case (Fig. 9-3), no abnormal white matter signals could be detected either. The pattern seen cannot be distinguished from cortical atrophy of other etiology. Since white matter lesions are known to occur by pathology late in the disease, it is possible that

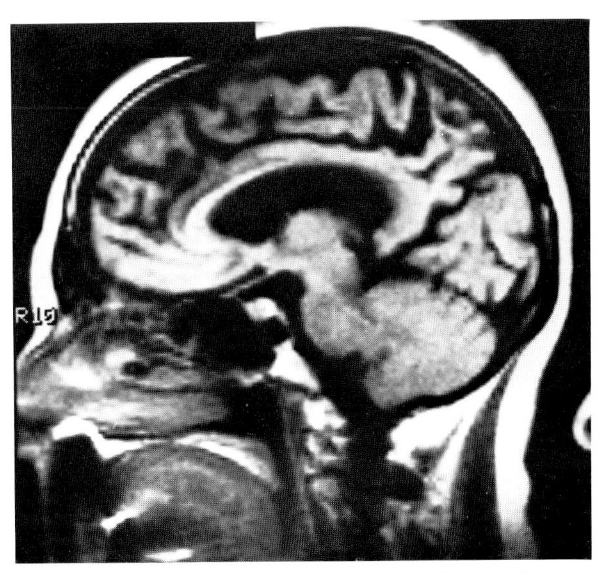

Fig. 9-3. A patient with Creutzfeldt-Jacob disease showing rapidly progressive atrophy over a 6 month period, but no abnormal MR signal changes.

prolonged T2 lesions might be seen in some cases using spin-echo sequences.[23]

Alcohol-Induced Parenchymal Degeneration

Generalized cerebral or cerebellar atrophy may occur in patients who abuse alcohol.[24] Alcoholics have a larger mean ventricular size than age-matched controls. The age of the patient and length of drinking history correlate well with the degree of cerebral atrophy.[25] However, there is poor correlation between the degree of cognitive impairment and the extent of cerebral atrophy noted on CT or MR. It is uncertain whether the parenchymal degeneration results from the toxic effects of alcohol exclusively, or whether coexisting nutritional deficiencies and occult head trauma also play important roles.[26]

Discrete cerebellar atrophy may occur in alcoholics, being much more common in men than women.[27] These patients usually relate a long history of alcohol abuse and often display an associated peripheral neuropathy. Because the anterior vermis is first affected, the initial symptoms are of a gait disorder, with staggering, stumbling, and truncal instability. Pathologically, all cell layers of the cerebellum are degenerated, especially the Purkinje cells. The olivary nuclei are also frequently involved.

Alcoholic patients with *Wernicke disease* may have degeneration of their mamillary bodies, structures that may be seen on sagittal MRI. No significant change in size of the mamillary bodies with age has been observed. However, their size on MRI has been shown to be markedly reduced in patients with chronic Wernicke disease.[28]

Marchiafava-Bignami disease is a primary degeneration of the corpus callosum resulting from the toxic effects of crude wine or moonshine liquor. In a single case, MR demonstrated atrophy and a high intensity linear signal in the middle lamina of the corpus callosum, together with scattered lesions in the centra semiovale.[28a]

Paraneoplastic Degeneration

Cerebral and especially cerebellar atrophy may be associated with a remote neoplasm, being part of a *paraneoplastic syndrome*. These patients can present clinically with dementia, dysarthria, gait disturbance, opsoclonus, or myoclonus. In a large group of cancer patients the sizes of the ventricles and sulci were significantly larger than in age-matched controls.[29] Paraneoplastic cerebral and cerebellar degeneration has been reported with Hodgkin disease, bronchogenic carcinoma, ovarian carcinoma, breast carcinoma, uterine carcinoma, and neuroblastoma.[30,31] The remote effects of tumors on the brain occur by unknown mechanisms, possibly autoimmune. Pathologically, there is widespread loss of Purkinje cells, demyelination of the superior cerebellar peduncles, and brain stem nuclear degeneration.

Atrophy Induced by Drugs

Cerebral atrophy may be induced by a number of medicinal and recreational drugs. Barbiturate or amphetamine abuse may produce generalized cerebral atrophy, although it is not usually as severe as that seen with alcohol abuse.[26] Very little atrophy seems to be induced by heroin or cannabis abuse,

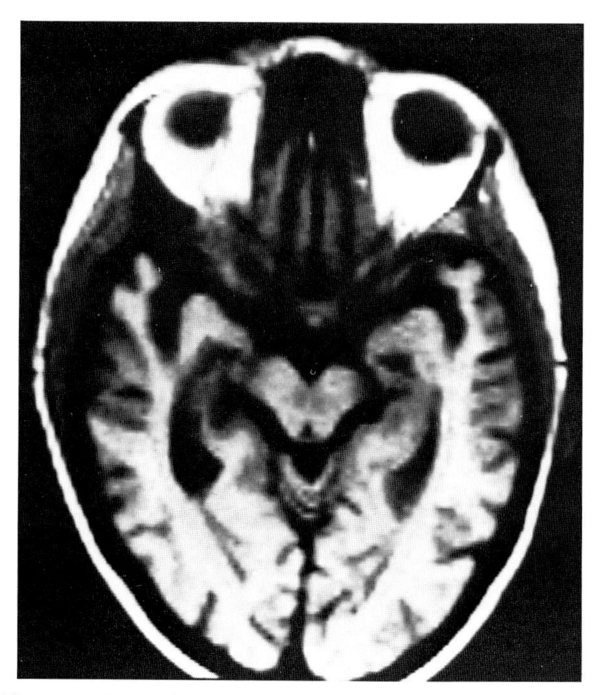

Fig. 9-4. Generalized cerebral and cerebellar atrophy induced by steroids.

however.[9] Corticosteroid or ACTH therapy is accompanied by a high incidence of cerebral atrophy, which is at least partially reversible following cessation of therapy (Fig. 9-4). Mechanisms that may explain this apparent atrophy include steroid-induced protein catabolism and loss of brain volume secondary to water loss.[9] A correlation between steroid dosage and degree of atrophy can be noted.[32] Finally, phenytoin therapy, commonly used to control seizures in a variety of neurologic diseases, will itself produce isolated cerebellar atrophy.[33]

Primary Cerebellar Degenerations

A number of mostly hereditary degenerations of the cerebellum, brain stem, and spinal cord have been described (Table 9-2).[34] While classification of many of these is still in dispute and clinical manifestations are variable, most present as chronic, slowly progressive ataxias that usually begin in the legs. In addition, there may be signs of lesions elsewhere in the brain such as in the basal ganglia or pyramidal tracts. Histologically, these disorders demonstrate selective neuronal system degeneration with reactive gliosis and demyelinization. Mixed types of these disorders are observed clinically and pathologically.

Friedreich Ataxia

Friedreich ataxia is perhaps the most common and well-known of the hereditary ataxias, but primarily involves the dorsal half of the spinal cord. Degeneration rarely extends above the medulla and the cerebellum is usually normal. Cranial MR has been normal in the few cases we have examined.

Table 9-2. Primary Cerebellar Degenerations (Hereditary Ataxias)

Hereditary spinocerebellar ataxia (Friedreich)
Hereditary ataxia with muscular atrophy (Levy-Roussy)
Hereditary cerebellar ataxias
Resembling olivopontocerebellar atrophy
Cerebello-olivary degeneration (Holmes)
Olivopontocerebellar atrophy (Dejerine-Thomas)
Joseph disease
Others

Hereditary Cerebellar Ataxias

Hereditary cerebellar ataxias comprise a rather diverse group of cerebellar and brain stem degenerations. Greenfield has classified them into two major groups: type A, resembling olivopontocerebellar atrophy; and type B, cerebello-olivary degeneration of Holmes.[35] We have observed a single case of this latter type (Fig. 9-5). The cerebellar atrophy was somewhat characteristic in that it involved the cortex more than the medulla.[36,37] The vermis was also markedly atrophic. Little change was noted in the brain stem other than degeneration of the olives.

Olivopontocerebellar Atrophy

Olivopontocerebellar atrophy, first described in 1900 by Dejerine and Thomas, is a chronic progressive ataxia that begins in middle age.[38] As many as five clinical subtypes have been defined.[39] Although most cases are sporadic, a few seem to be recessively inherited and possibly associated with a deficiency of the enzyme glutamate dehydrogenase.[40] Symptoms include ataxia of the limbs and trunk, dysphasia, impaired equilibrium, and extrapyramidal rigidity or tremor. Pathologically, there is severe atrophy of the cerebellum, middle cerebellar peduncle, ventral pons, and olivary nuclei.[39,41] The cerebellar nuclei, pontine tegmentum, and inferior cerebellar peduncle are usually spared.

Magnetic resonance is especially well suited for diagnosing and imaging olivopontocerebellar atrophy (Figs. 9-6, 9-7). Cerebellar atrophy is easily appreciated in any plane, while brain stem degeneration is best noted on sagittal images. Depending on the severity and duration of illness, brain stem atrophy may vary from mild (Fig. 9-6) to severe (Fig. 9-7). We have not observed increased T1 or T2 values in the areas of demyelination or degeneration, but conceivably this could occur.

Pick Disease

Pick disease is a rare cerebral degeneration characterized by symmetrical focal atrophy predominantly involving the frontal lobes.[42] The disease occurs

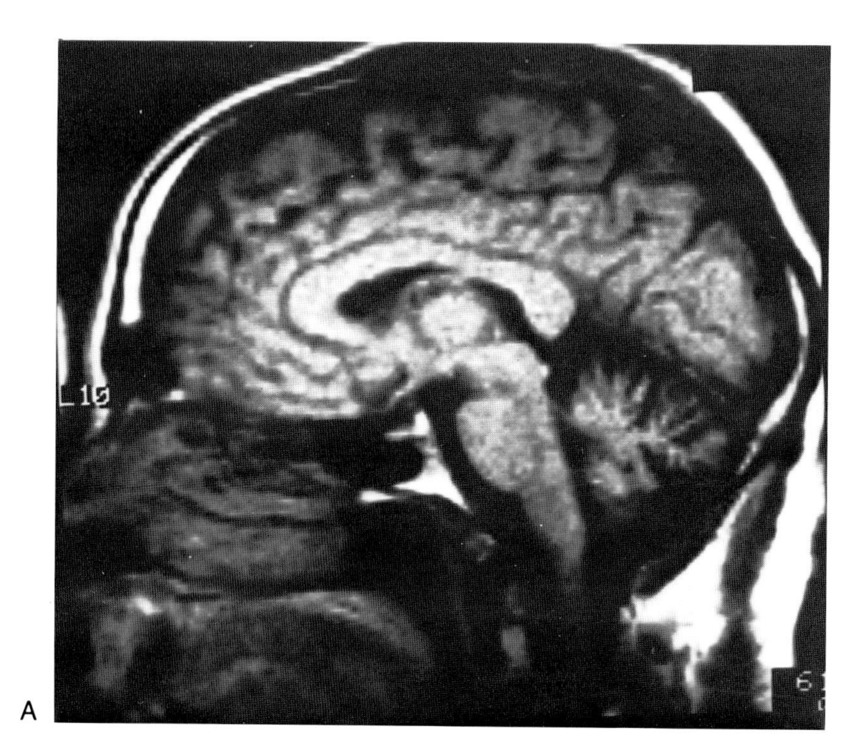

A

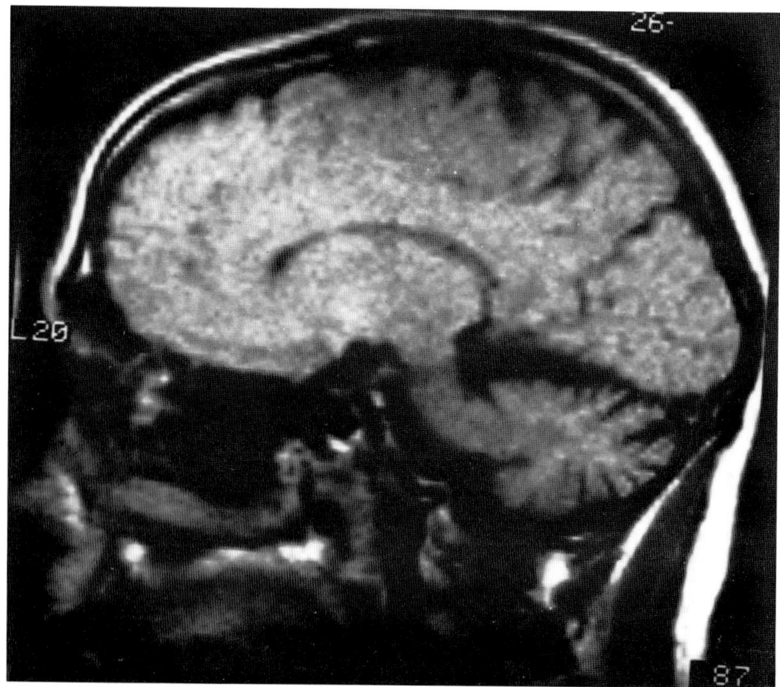

B

Fig. 9-5. (**A** and **B**) A familial cerebellar atrophy resembling type B cerebello-olivary degeneration (of Holmes). The cerebellar cortex is more affected than the medulla.

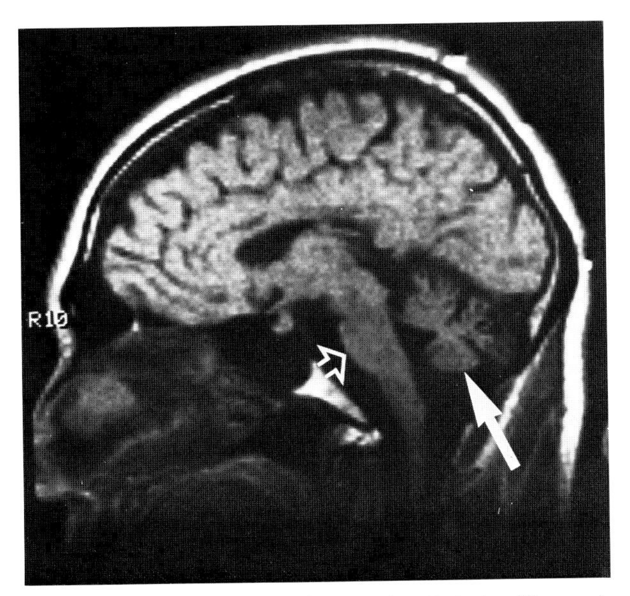

Fig. 9-6. Olivopontocerebellar atrophy (Dejerine-Thomas). Note cerebellar atrophy and flattening of the pons (arrows).

almost exclusively in the presenium, with clinical presentation similar if not identical to Alzheimer disease.[43] Because Alzheimer disease may occasionally present with focal brain atrophy, definitive diagnosis requires pathologic demonstration of *Pick bodies* (argentophilic intraneuronal inclusions) at autopsy.[44]

In a single proven case we have observed, classic gross pathologic findings were noted on MRI (Fig. 9-8). Severe symmetrical atrophy of both frontal lobes was observed. Characteristic sparing of the motor and sensory cortex and the proximal superior temporal and angular gyri were noted. No white matter lesions of significance could be identified.

DEGENERATIONS OF THE BASAL GANGLIA SYSTEM

The basal ganglia consist of three deep masses of gray matter, the *caudate nucleus*, *globus pallidus*, and *putamen*. The *claustrum* and *amygdaloid nuclei* are sometimes also included in the basal ganglia system. Two other subcortical nuclei, the *subthalamic*

A

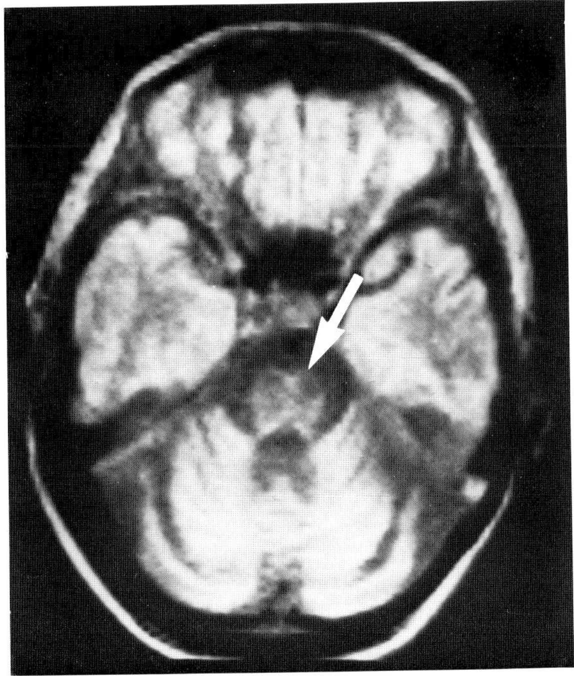

B

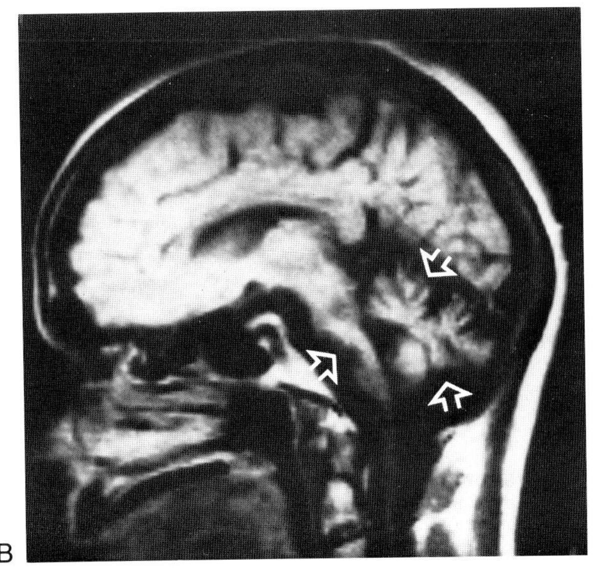

Fig. 9-7. (**A** and **B**) Severe olivopontocerebellar atrophy.

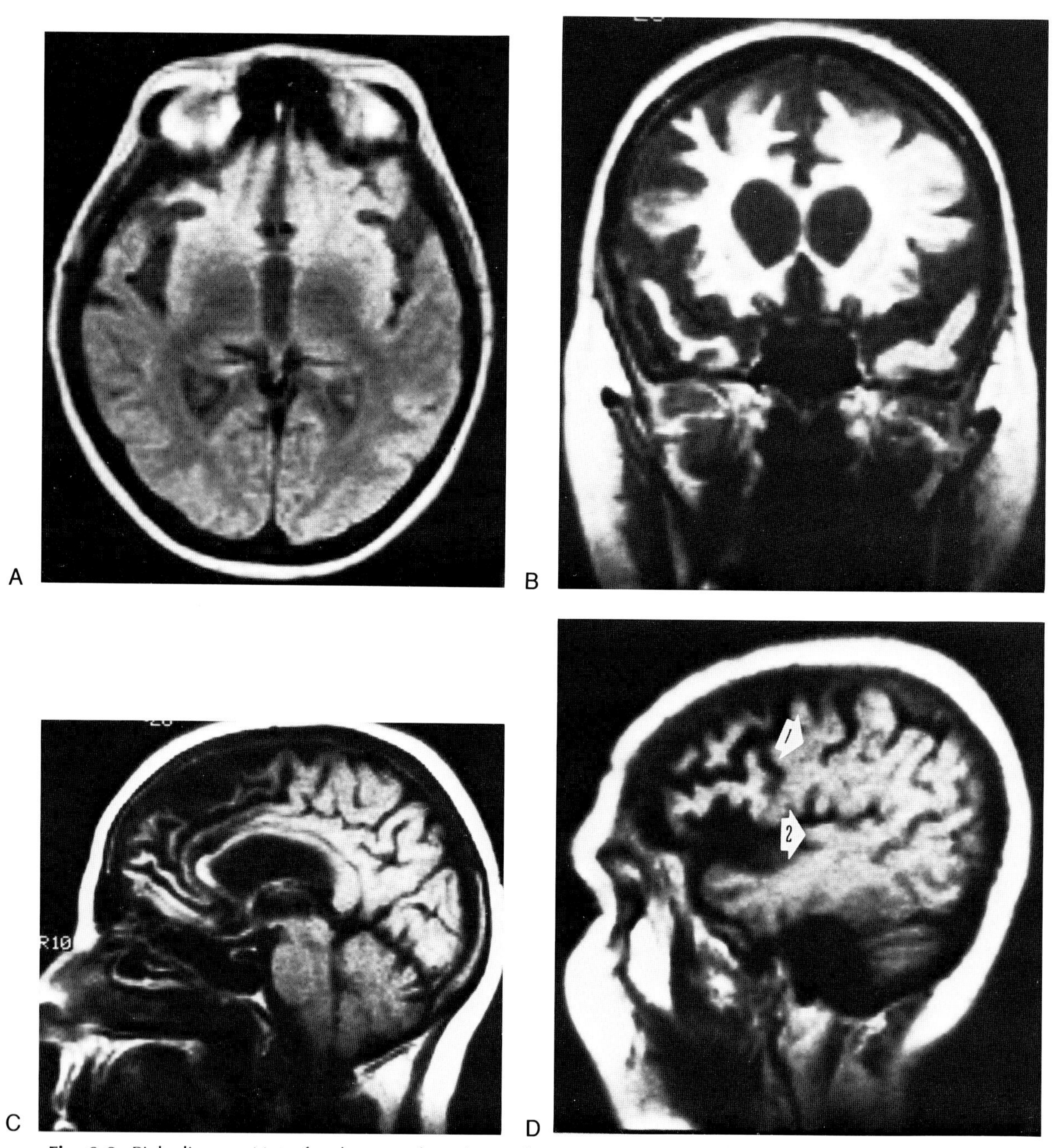

Fig. 9-8. Pick disease. Note focal severe frontal atrophy, as well as atrophy of the anterior temporal lobes **(A–C)**. Parasagittal image **(D)** demonstrates sparing of the precentral gyrus (1) and posterior portion of the superficial temporal gyrus (2), said to be quite characteristic of Pick disease.

nucleus and *substantia nigra*, are not strictly considered to be basal ganglia; nevertheless, they are functionally related and form important components in the basal ganglia system. The globus pallidus and putamen together are called the *lentiform nucleus*, while the putamen and caudate together are called the *striatum*.

The basal ganglia form the basis for the extrapyramidal motor system. While the precise functions of this system are not fully known, it has been suggested that the basal ganglia are responsible for smoothly carrying out complex previously learned motor acts. Lesions of the basal ganglia may be subclinical or may result in profoundly disabling movement disorders such as choreoathetosis, dyskinesia, or rigidity.

Several degenerative disorders of the basal ganglia system have been investigated by MRI. In general MR has been found to be at least as sensitive and often superior to CT in delineating abnormalities in the basal ganglia. Two exceptions to this general principle exist. First, patients with uncontrolled chorea or movement disorders cannot hold still for MR image acquisition and are better evaluated by CT. Secondly, diseases primarily manifest by basal ganglionic calcification are usually better seen on CT, because small or moderate degrees of calcification will not be detected by MR.

A pattern approach to the MR differential diagnosis of basal ganglionic lesions may be helpful (Table 9-3). Two of the more common diseases, Parkinson's and Huntington's, usually show only atrophy or degeneration of normal structures without obvious MR signal changes. In Huntington disease the caudate nuclei become atrophic, while in Parkinson disease the substantia nigra may become hard to visualize.

Several basal ganglia disorders are characterized by hypodense lesions on CT and long T1/long T2 lesions on MRI. On T2-weighted images, these lesions appear bright. Examples of such disorders include Wilson disease, Leigh disease, and carbon monoxide poisoning. Here cell death and reactive gliosis account for the MR signal characteristics.

Some basal ganglia degenerations result in decreased MR signal. Diseases such as lead poisoning, Fahr disease, and hypoparathyroidism may result in such dense basal ganglionic calcifications that the number of available hydrogen protons to produce an MR signal is diminished. A second mechanism by which basal ganglia signal may be decreased is by the pathologic deposition of paramagnetic substances (especially iron), which reduces tissue T1 and T2. Disorders in which paramagnetic metals are deposited in the basal ganglia include Hallervorden-Spatz disease and the Parkinson-plus syndromes. It should be noted that while copper deposition in the basal ganglia occurs in Wilson disease, cell death and reactive gliosis dominate the pathologic picture. The T1 and T2 values of the basal ganglia in Wilson disease are therefore lengthened, and paramagnetic effects of copper do not account for the imaging characteristics observed.

Table 9-3. MR Signal Characteristics of Diseased Basal Ganglia on T2-Weighted Images

Atrophy or Degeneration without Obvious Signal Change
- Parkinson disease (most cases)
- Huntington disease (most cases)

High Signal (Long T2)
- Wilson disease
- Anoxia
- Carbon monoxide poisoning
- Leigh disease
- Cytoplasmically inherited striatal degeneration

Low Signal
- Short T2 caused by paramagnetic substance deposition
  - Hallervorden-Spatz disease
  - Parkinson-plus syndromes (Shy-Drager, Multiple system atrophy)
  - Huntington disease (bradykinetic/rigid form only)
- Decreased [hydrogen] caused by dense calcification
  - Fahr disease
  - Lead poisoning
  - Hypoparathyroidism

Parkinson Disease

James Parkinson first described in 1817 a syndrome characterized by tremor, muscular rigidity, and loss of postural reflexes.[45] Since that time Parkinson disease has become recognized as the most frequent basal ganglia degeneration and as a leading cause of neurologic disability in patients over 60. The fundamental pathology involves loss of dopamine-containing neurons in the substantia nigra and striatum.[46] *Primary parkinsonism* (shaking palsy, paralysis agitans) is of unknown etiology. *Secondary parkinsonism* occurs when basal ganglia degeneration follows other diseases, such as encephalitis, intoxications (carbon monoxide, manganese), drugs (Aldomet, phenothiazines), or vascular disorders.

In the early 1960s, Hornykiewicz and colleagues demonstrated that the loss of pigmented neurons, particularly in the pars compacta of the substantia nigra, was a common feature in all forms of parkinsonism.[47] Subsequent research revealed that dopamine, a neurotransmitter substance, was found in the same structures affected by Parkinson disease (neostriatum, substantia nigra, locus ceruleus, and dorsal vagal nucleus). Degeneration of these structures correlated with the degree of clinical disability.[46] Pathologically, intracellular inclusion bodies (Lewy bodies) were noted in the involved areas.

Parkinson disease is more common in patients over 60. All races and both sexes are affected. The classic clinical triad includes tremor, rigidity, and akinesia.[48] The tremor, often referred to as "pill-rolling," is usually worse at rest. Rigidity of muscles (resistance to passive motion) is seen in nearly all cases. A festinating gait with stooped body posture is characteristic. A peculiar akinesia ("paucity of movement") is often present and may be the most disabling feature of the disease. A number of autonomic and cognitive dysfunctions are also commonly encountered.

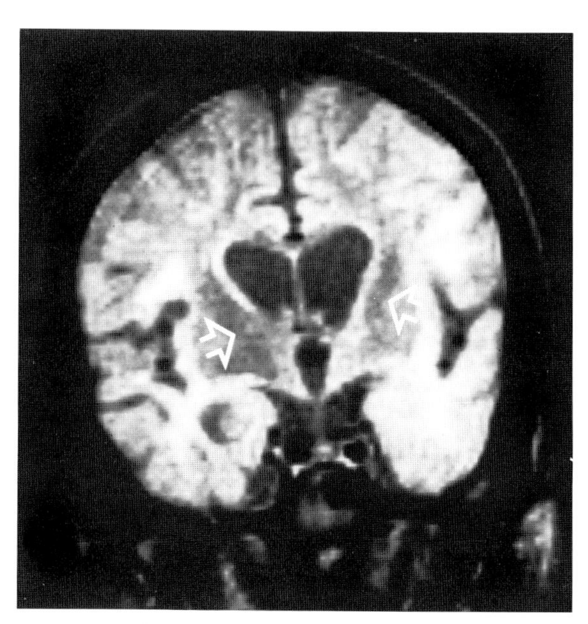

Fig. 9-10. Parkinson-plus syndrome. Abnormal decreased signal in the basal ganglia on this SE 1500/56 image. This pattern is consistent with abnormal iron accumulation in the striatum.

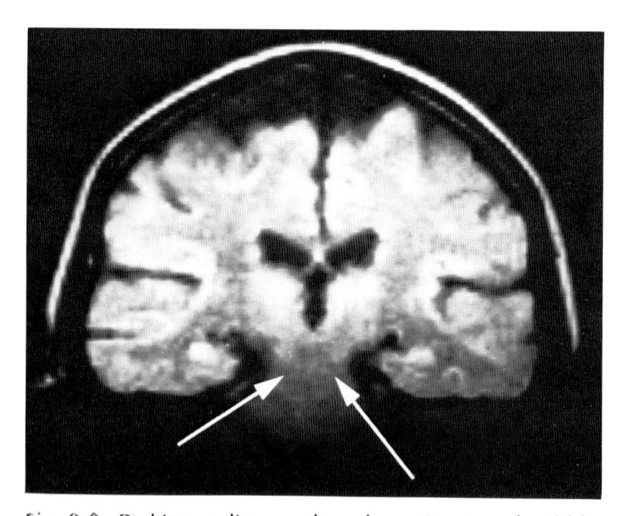

Fig. 9-9. Parkinson disease, the substantia nigra should be seen on this coronal SE 1500/28 image at the site of the arrows. Here it is not visualized, a finding sometimes associated with longstanding parkinsonism.

No specific CT findings are characteristic of Parkinson disease. Cerebral atrophy is commonly noted, out of proportion to that in age-matched controls.[49] The atrophy is better correlated with cognitive and emotional symptoms than with tremor or akinesia. Basal ganglia abnormalities are seldom seen.

Comprehensive studies of MR imaging of patients with Parkinson disease have not yet been published. A single early report suggested that primary and secondary forms could be differentiated by MR signal characteristics, but this has not yet been confirmed.[50] At low field strengths (<0.5 T) we have usually observed no significant abnormalities, save atrophy, in most patients with primary parkinsonism. Occasionally in cases of longstanding disease, the substantia nigra may fail to visualize (Fig. 9-9).

At high field strengths certain Parkinsonlike diseases may have interesting imaging characteristics due to deposition of paramagnetic substances in the basal ganglia. These disorders include the "Parkinson-plus syndromes": Shy-Drager, multiple system atrophy, and progressive supranuclear palsy.[51,52] Mostly iron, but also copper, manganese, and neuromelanin are known to accumulate in the basal ganglia of patients with these disorders.[53] Paramagnetic substances induce more rapid MR tissue relaxation. As a result, the T1 and T2 values are shortened. On T2-weighted images, therefore, tissues that accumulate iron and other paramagnetic substances will appear less intense than normal. In the Parkinson-plus syndromes the putamina, and to a lesser extent the other basal ganglia, show decreased T2-signal.[54,55] An example is shown in Figure 9-10.

Huntington Disease

Huntington disease is an inherited neurodegenerative disorder characterized by choreiform movements, progressive dementia, and emotional disturbances. It is transmitted as an autosomal dominant trait with complete penetrance, whose locus lies on chromosome 4.[56] The disease occurs worldwide, but pockets of high prevalence exist in western Scotland and Venezuela where large families with disease reside.[57]

Although a juvenile variant of the disease may present in childhood, the typical case becomes apparent in the 40 to 60 year age range. Symptoms include chorea, emotional lability and progressive dementia. Two clinical forms of the disease have been distinguished: choreic-hypotonic and bradykinetic-rigid. Because of the relatively late onset of the disease most patients have already produced offspring by the time of diagnosis, making reliable genetic counseling difficult.

Pathologically there is atrophy that initially involves the basal ganglia. The caudate and putamen are particularly affected demonstrating neuronal loss and gliosis. The clinical appearance of chorea correlates with caudate atrophy.[58] Patients with the bradykinetic-rigid form have more gliosis in the putamen than those with the choreic-hypotonic form.[59] Cortical atrophy sets in after basal ganglia degeneration. It usually begins in the frontal lobes and progresses posteriorly. The degree of cortical atrophy is strongly correlated with functional and cognitive impairments.[58]

Primary diagnosis by CT or MR requires documentation of basal ganglia atrophy. Ratios of frontal horn to bicaudate diameters[60] and bicaudate to outer table diameters[61] have been traditionally utilized by CT

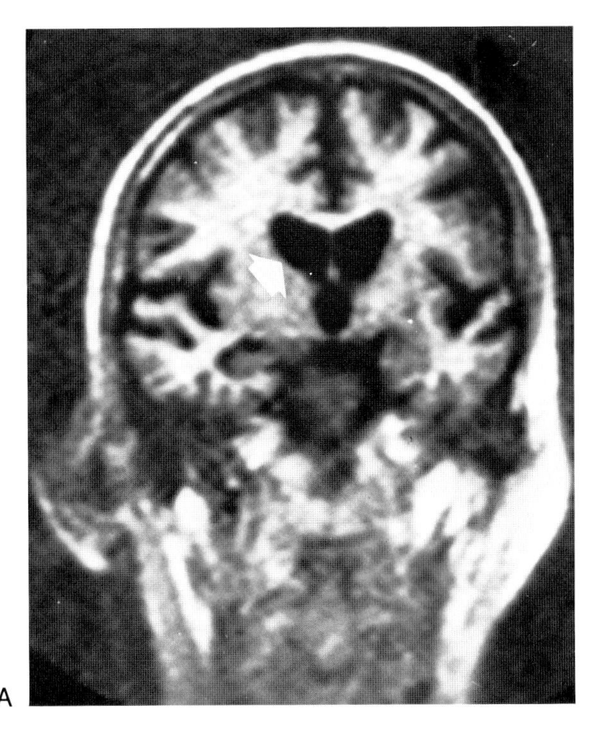

A

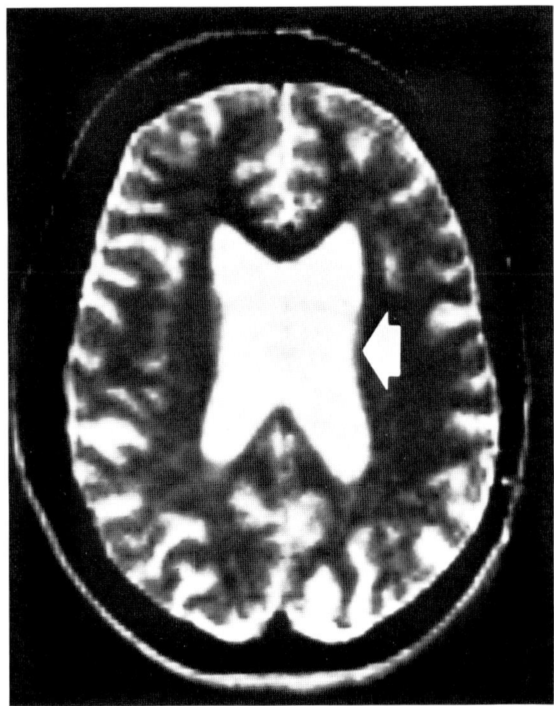

B

Fig. 9-11. Huntington disease. Marked atrophy of the caudate nucleus head and body seen on **(A)** coronal IR 1400/40 and **(B)** axial SE 3000/80 MR images.

to assess this atrophy. MR has proven at least as sensitive as CT in assessing these atrophic changes.[62] MR demonstrates the putamen and caudate with better definition than CT. Furthermore, coronal views aid in the three-dimensional assessment of their sizes (Fig. 9-11). On SE 2000/32 decreased signal has been noted in the anterior portion of the globus pallidus and posterior portion of the internal capsule.[63] This may result from iron deposition known to occur in these areas in patients with Huntington disease. In patients with the bradykinetic-rigid form of the disease, prolonged T2 may be noted in the putamina, corresponding to the known gliosis.[59] Positron emission tomography may actually prove more sensitive than either MR or CT in assessing patients with Huntington disease since it is capable of demonstrating hypometabolic areas in the caudate and putamen before there is bulk tissue loss.[64]

Wilson Disease

Wilson disease (hepatolenticular degeneration) is an inborn error of copper metabolism associated with degeneration of the basal ganglia and cirrhosis of the liver.[65] The precise biochemical defect is unknown, but likely involves inadequate synthesis of ceruloplasmin, a serum α_2-globulin for copper transport.[66] Defective copper transportation and utilization results in abnormal deposition of the metal in the liver, brain, and other organs. Here the copper is sequestered into lysosomes where it may disrupt membrane lipids and cause cellular degeneration.

In the brain copper deposition takes place mainly in the basal ganglia, which appear brick-red at autopsy. The putamen, globus pallidus, and caudate are most involved, in that order. Neuronal loss occurs, but the pathologic picture is dominated by a marked gliosis, mainly composed of astrocytes.[67] It is this gliosis rather than paramagnetic effects of copper that account for the imaging changes noted on MRI.

Copper deposition and lesser degenerative changes also occur in the brain stem, cerebellum, substantia nigra, and convolutional white matter. In the cornea metal accumulates peripherally resulting in the well-known Kayser-Fleischer rings observed clinically. Liver involvement is characteristically a pattern of postnectrotic cirrhosis. Toxic changes in renal tubular cells may also be observed.

Wilson disease is slightly more common in males than females. The age of onset is usually 10 to 40 years. It is transmitted as an autosomal recessive disorder, and a history of consanguinity is often present. While the disease occurs in all races there is a higher incidence in eastern European Jews, southern Italians, and in isolated populations where inbreeding is common.[66]

Neurologic manifestations are varied and insidious.[68] Dysarthria is the most common symptom. Choreoathetosis and a parkinsonlike rigidity are often seen. A peculiar "wing-beating" arm tremor has been described. Personality and cognitive changes also occur. Liver involvement is frequently asymptomatic. Progression is slow but definite and death usually occurs 10 to 15 years following onset.

Diagnosis is suspected clinically when tremor, chorea, incoordination, and rigidity occur in children and young adults. A Kayser-Fleischer ring may be seen

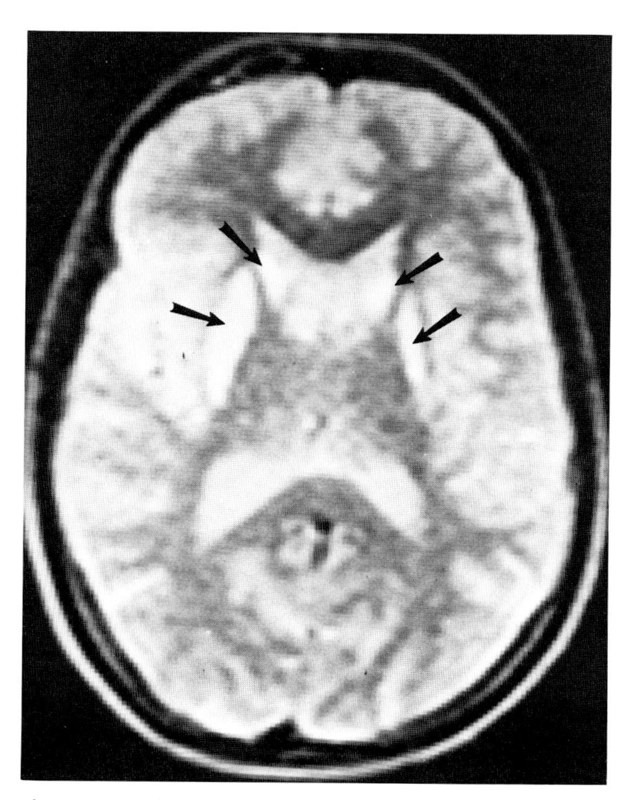

Fig. 9-12. Wilson disease. Symmetrical areas of increased signal from the caudate and lenticular nuclei on this SE 3000/80 axial MR scan.

in 90 percent of older patients, although may be absent in children. Serum ceruloplasmin and copper levels are nearly always low. Treatment is with D-penicillamine, a copper chelating agent.[69]

Computed tomography abnormalities are seen in about two-thirds of patients with Wilson disease.[70–72] These include atrophy (focal or diffuse) and areas of hypodensity (in the basal ganglia, brain stem, and white matter). CT has not proven particularly useful in correlating clinical symptoms, in assessing prognosis, or evaluating response to therapy.[70–73]

Several reports of MR imaging in Wilson disease have been published.[73–75] About 80 percent of patients will have abnormal scans, making MR more sensitive than CT.[74,75] Symmetric areas of increased signal on T2-weighted images were seen in the lenticular, thalamic, caudate, and dentate nuclei. A smaller number of patients showed brain stem and asymmetric white matter lesions. Clinical correlation was generally good. Dystonia correlated with putamen, caudate, and nigra lesions; bradykinesia with putamen lesions; and tremor with dentate and red nucleus lesions.[75] The lesions detected on MR likely represent gliosis and edema rather than paramagnetic effects of copper, which would serve to decrease signal intensity. An example of a patient with Wilson disease shown on MRI is presented in Figure 9-12.

Leigh Disease

Leigh disease, also called *subacute necrotizing encephalomyelopathy*, is a rare, rapidly progressive neurodegenerative disorder of infancy and early childhood.[76] Although the clinical picture may vary, a typical presentation would be in an infant who initially fails to thrive and later becomes floppy.[77] Progressive impairment of vision and hearing occurs, with ataxia, nystagmus, hypotonia, and psychomotor retardation. Affected patients usually die a few months to several years after onset. More chronic forms of the disease exist, with survival to the third decade.[78] An adult variety with predilection for the brain stem has also been reported.[79]

Histopathologically, the lesions of Leigh disease are identical to those of Wernicke disease, but encompass a different distribution.[80] Leigh disease commonly involves the basal ganglia, brain stem, optic nerves, and optic tracts. The white matter of the centrum semiovale is occasionally affeged. Leigh disease typically spares the hypothalamus and mamillary bodies, which are involved in over 95 percent of patients with Wernicke disease. Wernicke disease only occasionally involves the basal ganglia, while this is a relative common occurrence with Leigh disease.

The similarity of Leigh and Wernicke diseases led investigators to propose that a congenital metabolic defect of thiamine or pyruvate metabolism, possibly on a hereditary basis, might be involved. Deficiencies of pyruvate carboxylase, cytochrome-c oxidase, and thiamine triphosphate have been reported in patients with Leigh disease.[80–82] Additionally, an inhibitor of the phosphoryl-transferase enzyme necessary to convert thiamine pyrophosphate to thiamine triphosphate has been found in their urine. It is possible that several different metabolic defects involving regulation of the pyruvate dehydrogenase complex may have clinical presentations that are all classified as Leigh disease.

Several CT reports of Leigh disease have noted characteristic symmetrical areas of low attenuation in the basal ganglia, especially the putamina.[83–85] In some cases, only the putamina are involved, while in others, more extensive lesions including the caudate and cerebral white matter have been reported. In general the juvenile form of the disease seems confined to the basal ganglia, while the infantile form may be much more extensive.[84]

Figures 9-13 show CT and MR findings in a case of Leigh disease in a 2-year-old patient. The MR images demonstrate high-intensity signals from the basal ganglia on T2-weighted sequences that correspond exactly to regions of low attenuation seen on CT. MRI seems more sensitive than CT in evaluating patients with possible Leigh disease, although in our case the two were of equal value. In two of five reported cases, however, MR was judged superior to CT for characterizing the full extent of the disease, finding necroses in the mesencephalon and olives while CT was normal in these regions.[86]

The differential diagnosis of Leigh disease by MR or CT must include bilateral infarction of the basal ganglia, carbon monoxide poisoning, striatonigral atrophy, and Wilson disease. Other rarer conditions causing symmetric pallidal necrosis must also be considered, including barbiturate intoxication, trauma, cyanide poisoning, hypoglycemia, and hydrogen sul-

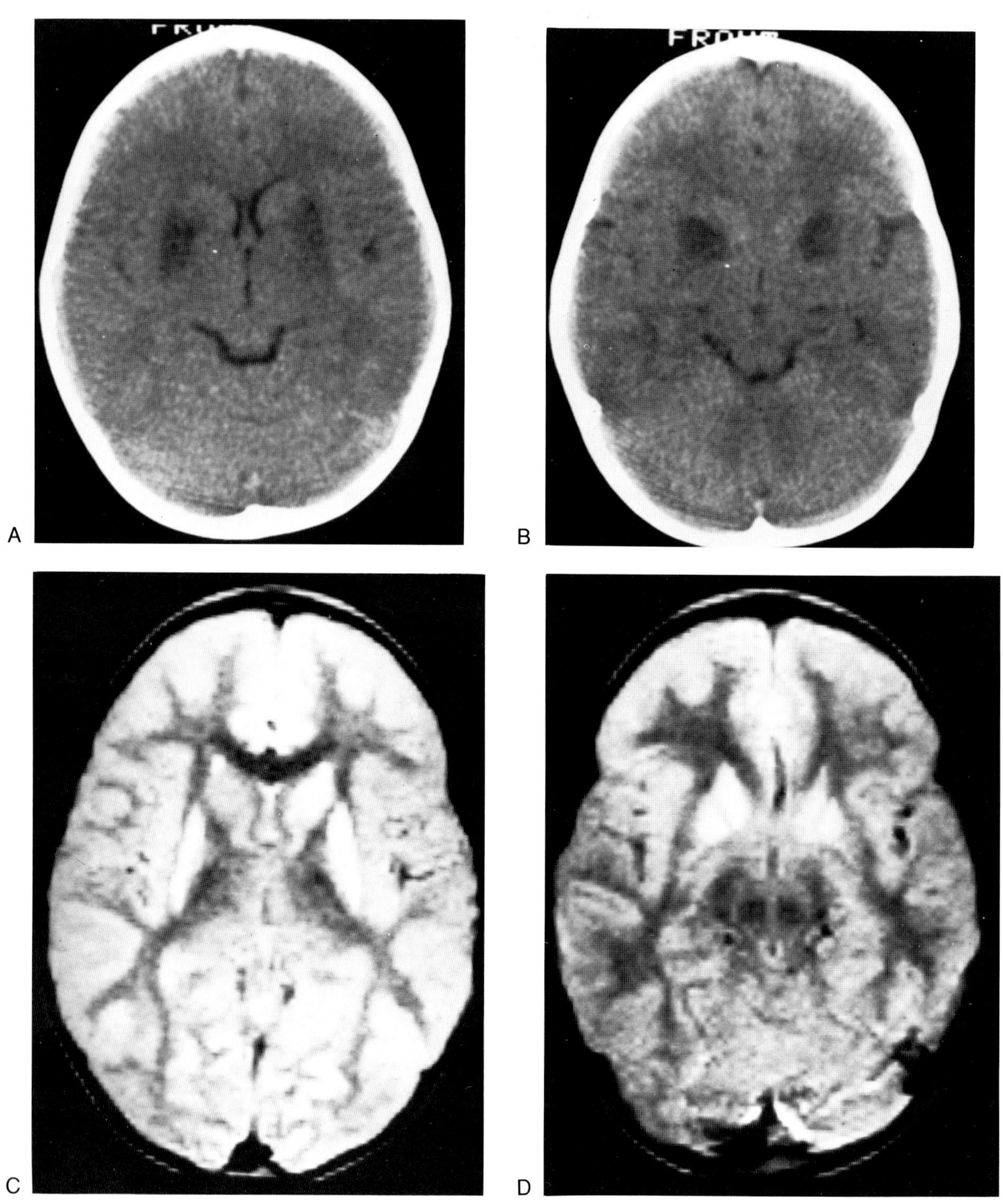

Fig. 9-13. Leigh disease. **(A,B)** CT scans show low density areas of both lentiform nuclei. **(C,D)** T2-weighted MR images show high signal from the same areas.

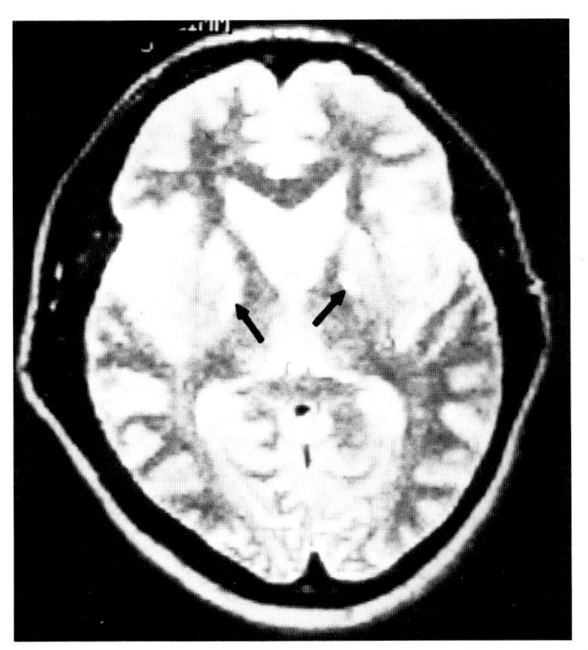

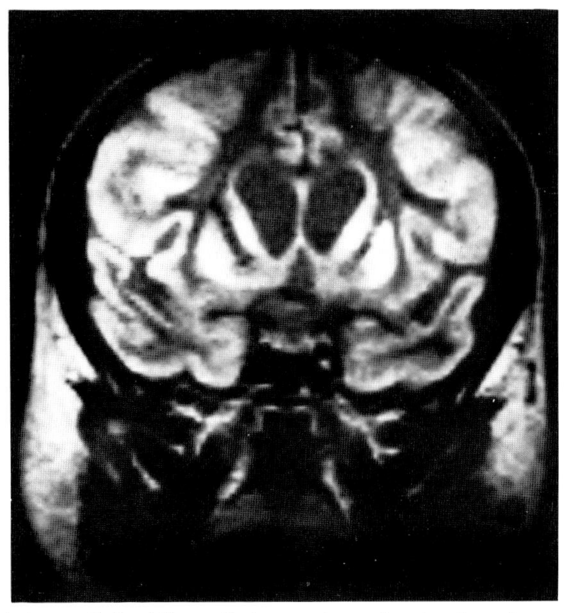

Fig. 9-14. Basal ganglia infarcts. **(A)** Small infarctions of the globus pallidus bilaterally in a patient who developed arm atoxia after a hypotensive episode. **(B)** Severe global hypoxia that developed during surgery 2 weeks prior to this scan. Bilateral high intensity lesions seen in the basal ganglia on this coronal SE 2674 scan.

fide poisoning. Usually clinical history will allow differentiation of these diseases to be made (Fig. 9-14).

Hallervorden-Spatz Disease

Hallervorden-Spatz disease (HSD) is an insidiously progressive, autosomal recessive movement disorder of childhood and adolescence.[87,88] It is characterized by abnormal accumulation of iron and other metals in the globus pallidus, reticular zone of the substantia nigra, and red nucleus.[89] Diffuse disordered myelination also occurs and is most marked in the tracts between the striatum and pallidum. The disease may be related to infantile neuroaxonal dystrophy.[90]

The disease usually presents with stiffness of gait, with pes equinovarus and toe-walking.[87,88] There is progressive spasticity and rigidity to involved arm and face musculature. Some children become dystonic and assume bizarre postures. Pigmentary degeneration of the retina is sometimes observed. Intellectual function usually remains intact. Death from intercurrent illness usually occurs about 10 to 15 years after onset. Definitive diagnosis requires autopsy, although the clinical findings are sufficiently characteristic to suggest the diagnosis premortem.

Computer tomography findings are minimal and nonspecific with atrophy of the caudate nuclei, brain stem, and cerebellum sometimes observed.[91] (Fig. 9-15) In three reported cases as well as our own, MR revealed decreased T2 values in the lentiform nuclei and surrounding white matter.[92,93] This may be due to the accumulation of endogenous paramagnetic substances (e.g., iron) in these regions. However, disordered myelination may produce similar shortening of T1 and T2, so the appearance is nonspecific.[92] Nevertheless, the pattern of decreased T1 and T2 in the basal ganglia is sufficiently unusual that it may support the diagnosis of HSD in the appropriate clinical setting. In the reported cases periventricular white matter showed increased signal intensity on T2-weighted sequences. The pathologic correlate of this finding may relate to disordered myelination, since altered CSF dynamics to explain this signal have been reported with HSD.

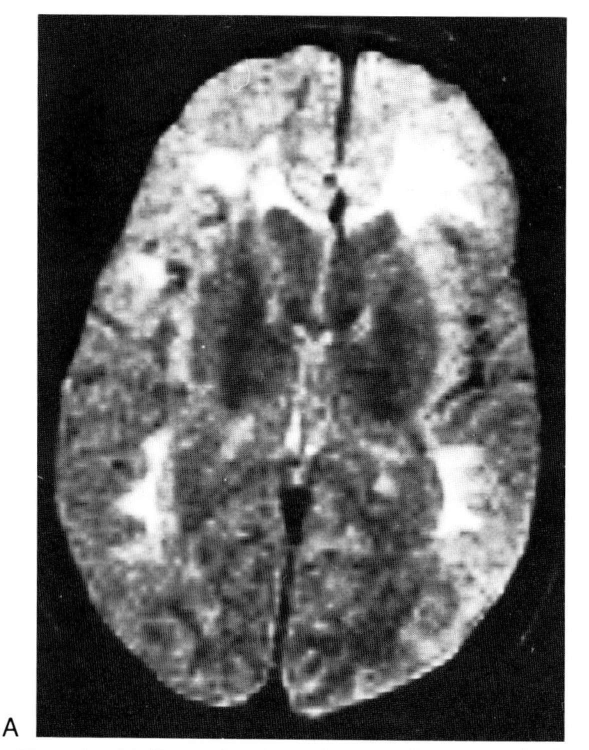

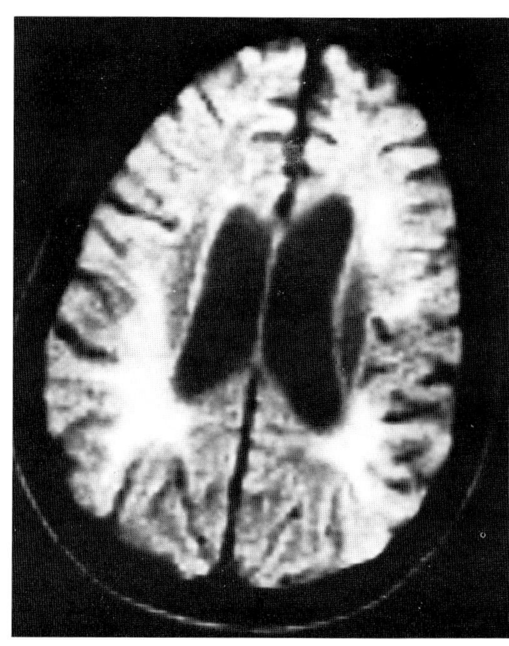

Fig. 9-15. Hallervorden-Spatz disease. There is markedly decreased signal in the basal ganglia, probably secondary to iron deposition **(A)**. Diffuse white matter high signal, reflecting dysmyelination, is also noted **(B)**.

WHITE MATTER DISEASES

A large number of disease processes may involve the white matter, including infections, tumors, vascular disorders, and primary degenerations. This section concerns itself almost totally with primary disorders of white matter *(leukoencephalopathies)*.

White matter diseases may be subdivided into three types: *demyelinating (myelinoclastic), dysmyelinating*, and *mixed*.[94] Demyelinating diseases cause destruction of myelin, which has been normally formed. Examples include multiple sclerosis, toxic insult, malnutrition, and progressive multifocal leukoencephalopathy. In dysmyelinating diseases there is abnormal formation or maintenance of myelin as a result of an enzymatic disturbance. Examples include Alexander disease and metachromatic leukodystrophy. Adrenoleukodystrophy has features of both demyelination and dysmyelination, and may be considered a mixed disease. A classification of the various types of leukoencephalopathy is presented in Table 9-4.

Table 9-4. White Matter Diseases

Demyelinating
Multiple sclerosis
Progressive multifocal leukoencephalopathy
Disseminated necrotizing leukoencephalopathy
Acute disseminated encephalomyelitis
Secondary types (anoxic, infectious, traumatic, ischemic, etc.)
Adrenoleukodystrophy
Dysmyelinating
Metachromatic leukodystrophy
Adrenoleukodystrophy[a]
Krabbe disease
Canavan disease
Alexander disease
Pelizaeus-Merzbacher disease
Inborn errors of metabolism

[a] Has features of both demyelinating and dysmyelinating disease.

Patterns of White Matter Degeneration

Although there is considerable overlap between the appearances of various white matter diseases on MRI, some patterns are sufficiently characteristic to suggest one diagnosis over another. A summary of various patterns of white matter degeneration is presented in Table 9-5.

The *focal symmetric* pattern is exemplified by adrenoleukodystrophy. In the typical case, adrenoleukodystrophy spreads from posteriorly to anteriorly in a systematic fashion, with contiguity over the corpus callosum. An enhancing anterior margin may be noted with the administration of intravenous contrast. Disseminated necrotizing leukoencephalopathy may also exhibit a focal symmetric pattern of myelin breakdown. This disorder, which results from intrathecal methotrexate therapy, shows symmetric involvement of the white matter around the frontal horns. DNL typically progresses posteriorly to involve both centra semiovale.

The *focal asymmetric* pattern commonly results from a localized insult, such as trauma, infection, or ischemia. There may be involvement of gray matter as well. Typical locations include the temporal poles, frontal poles, and occipital lobes. Radiation therapy, if applied with narrow collimation, may result in focal areas of white matter involvement. Progressive multifocal leukoencephalopathy, an infection caused by a papovavirus, commonly presents with focal asymmetric white matter lesions. Toxoplasmosis may have a similar appearance. Metastases and multifocal glioma should also be considered in the differential diagnosis of this pattern.

The *periventricular* pattern of white matter involvement is typified by multiple sclerosis (MS). MS plaques frequently form along the margins of the lateral ventricles. They may spread as thin periependymal lesions or coalesce into a lumpy-bumpy pattern. Involvement at the angles of the frontal and occipital horns may occur, and is more irregular than ependymitis granularis, which is an anatomic variant found in the same regions. Subcortical arteriosclerotic encephalopathy (Binswanger disease) also occurs in the periventricular regions. However, the lesions of Binswanger disease are often not immediately adjacent to the ventricular surface as are the lesions of MS. Furthermore, involvement of the corpus callosum does not appear to occur in Binswanger disease while it is quite common in MS; callosal lesions tend to exclude the diagnosis of Binswanger disease. Additionally, MS may demonstrate brain stem or spinal cord lesions, which are unusual in Binswanger disease.

Transependymal absorption of CSF may cause high intensity lesions in the periventricular areas. Hydrocephalus is nearly always present, and the lesions are typically smooth and symmetric.

A number of white matter diseases characteristically involve the *centrum semiovale* regions primarily. These include Binswanger disease, multiple sclerosis, metachromatic leukodystrophy, progressive multifocal leukoencephalopathy, Krabbe disease, and Canavan disease. Often clinical history will aid in the differential diagnosis.

Lesions in the immediate subcortical white matter frequently are caused by blood-borne distant pro-

Table 9-5. Patterns of White Matter Degeneration

Focal Symmetric
Adrenoleukodystrophy
Disseminated necrotizing leukoencephalopathy
Focal Asymmetric
Localized insult (trauma, infection, ischemia)
Radiation therapy
Progressive multifocal leukoencephalopathy
Toxoplasmosis
Metastasis
Multifocal glioma
Periventricular
Multiple sclerosis
Binswanger disease
Transependymal absorption of CSF
Centrum Semiovale
Binswanger disease
Metachromatic leukodystrophy
Multiple sclerosis
Progressive multifocal leukoencephalopathy
Krabbe disease
Canavan disease
Subcortical White Matter
Emboli
Metastasis
Diffuse
Radiation therapy
Alexander disease
Acute disseminated encephalomyelitis
Pelizaeus-Merzbacher disease
End stage of many other white matter diseases

cesses. Metastases and septic emboli should be considered when this pattern is observed.

Finally, *diffuse* white matter disease may be seen in a number of settings. First, it may be an end stage of many congenital and acquired leukodystrophies. It is a common pattern following whole brain irradiation. Alexander disease, acute disseminated encephalomyelitis, and the very rare Pelizaeus-Merzbacher disease may present in this pattern initially.

Multiple Sclerosis

Multiple sclerosis is a chronic relapsing disease of unknown etiology producing demyelinization and sclerosis in the central nervous system. The disease typically affects young adults, with a broad spectrum of clinical signs and symptoms. Characteristically, MS produces multiple focal neurologic deficits "separated in space and time."[95] Episodes of exacerbation and remission are common, but overall the trend is one of slow progression.

Even early in its development, MR demonstrated extraordinary sensitivity for visualizing the plaques of multiple sclerosis.[96] MR is now considered to be the imaging modality of choice for MS and other demyelinating diseases. Because of the historic and continued importance of MRI in the evaluation of MS, it is important to have a detailed understanding of the etiology, pathology, clinical course, and diagnostic modalities for MS. Several excellent reviews are available.[97–99]

A striking geographic distribution of multiple sclerosis has been thoroughly documented.[100] Inhabitants of the higher latitudes in each hemisphere are affected by MS up to eight times more frequently than those living closer to the equator. Migrants who move from a low-risk to a high-risk area before the age of 15 are affected as frequently as those born in the high-risk area.[101] Conversely, people migrating in childhood from a high-risk to a low-risk area acquire the lower risk. These studies of geographic distribution and migration effects indicate that MS is acquired from an environmental factor. Some genetic predisposition does exist, because there is an increased incidence of the disease associated with certain antigens of the major histocompatibility complex on chromosome 6.[102] There is also a slightly increased risk of MS in relatives of affected patients, but it is uncertain whether this is based on genetic or environmental factors.[97]

It is quite possible that the environmental factor responsible for MS is an infectious agent, such as a slow virus. Support for this theory may be found in a "multiple sclerosis epidemic" that occurred in the Faroe Islands between 1943 and 1960.[103] Prior to the occupation of the islands by British troops in World War II, no cases of MS had occurred in the native population. The emergence of this disease in natives in the postwar years suggests that the troops brought with them an infectious agent causing MS. Nevertheless, despite many attempts to isolate an infectious agent or characterize it, no causative organism has been found.

An alternate pathogenesis for MS involves the possibility that it represents an autoimmune disease. Several types of experimental allergic encephalomyelitis resembling, but not identical to MS, may be induced by injection of brain extract into animals.[98] The autoimmune theory is further strengthened by the association of MS with certain histocompatibility types (HLA-A_3, -B_7, -DW_2, -DR_2). Perhaps both the infectious and autoimmune theories have elements of truth. For example, it has been suggested that MS represents the effects of a slow virus on certain genetically susceptible individuals, which creates an immunoregulatory state allowing an autoimmune attack upon central myelin.

Pathology

Pathologically, the lesions of MS begin as local areas of demyelination with surrounding edema scattered throughout the white matter. The lesions, known as plaques, vary in size from less than a millimeter to several centimeters in diameter. Particularly common sites of involvement include the optic nerves, spinal cord, brain stem, periventricular white matter, corpus callosum, and centrum semiovale. White matter anywhere may be affected; gray matter involvement is usually by extension from a white matter focus.

Microscopically, the lesions of MS show primary breakdown of the myelin sheath, with relative sparing of axons. A perivenous distribution of macrophages, lymphocytes, and plasma cells around the plaques are seen. Breakdown products of myelin, largely lipids, are found free and in macrophages. With age and subsidence of activity, astrocytic hyperplasia (gliosis) occurs. Oligodendroglia, the myelin-producing cells, are reduced in number and frequently absent.[97]

Clinical Manifestations

Clinical manifestations of multiple sclerosis depend upon the number, location, and size of the demyelinating lesions.[104] The typical onset of the disease is between 20 and 40 years of age. Occasionally children and adults over 50 may be affected. No documented case of onset over age 64 has been reported. The disease is characterized by episodes of exacerbation and remission.

About 40 percent of patients will present with optic neuritis as their initial manifestation of disease. A rapid onset of pain and vision loss in one eye occurs over several days, which usually improves over the following several weeks. The remaining 60 percent of patients with MS present with a lesions in the spinal cord or brain stem. The symptoms from a lesion in one of these locations include numbness, weakness, diplopia (internuclear ophthalmoplegia), nystagmus, cerebellar ataxia, and bladder dysfunction.

Natural History

The natural history of MS is variable, depending on the number of lesions and the severity of each attack.[104] Early remissions are often complete, and the remission period after an attack may last several years. With each acute recurrence, however, the chances of complete recovery are diminished. An acute, fulminant form of multiple sclerosis may lead to death within weeks or months. Typically, however, the average survival of patients with MS after diagnosis is about 27 years. End-stage disease is characterized by dementia, incontinence, blindness, quadriplegia, and ataxia.

Clinical Diagnosis

Clinical diagnosis of MS requires documentation of neurologic lesions that have occurred on more than one occasion and at more than one site, not explainable by other etiologies. More recently, electrophysiologic studies and cerebrospinal fluid analysis have served as extensions of the neurologic examination. Based upon clinical findings and ancillary tests, a number of criteria for the definitive, probable, or possible diagnosis of MS have been proposed.[95,105–109] Of these many criteria, those by McAlpine et al.[95] and Poser et al.[105] seem to have gained the widest popularity. The McAlpine criteria are based solely on clinical assessment, while the Poser criteria include clinical, paraclinical, and laboratory data. A summary of the Poser criteria is presented in Table 9-6.

Several electrophysiologic tests are in use to detect white matter lesions in the cord or brain stem that may not be clinically or radiographically apparent.[110] These tests are called evoked-response tests, because they measure brain wave activity in response to visual, auditory, or somatosensory stimuli. The visual evoked response test can reveal involvement of the optic nerve or tract. The brain stem auditory evoked response test can reveal lesions in the acoustic nerves or brain stem. The somatosensory evoked response test detects white matter abnormalities from the spinal cord level to the sensory cortex. These tests can

Table 9-6. Diagnostic Criteria for Multiple Sclerosis[a]

Definite Multiple Sclerosis
- Two attacks at least 1 month apart with:
  - Clinical evidence[b] of one lesion and at least paraclinical evidence of a second lesion, OR
  - Clinical or paraclinical evidence[c] of one lesion and positive CSF

OR

- One attack with positive CSF[d], clinical evidence for one lesion, and at least paraclinical evidence for a second lesion

Probable Multiple Sclerosis
- Two attacks at least 1 month apart with:
  - Clinical evidence of only one lesion, OR
  - Positive CSF only

OR

- One attack with clinical evidence for one lesion, and at least paraclinical evidence for a second lesion

[a] Adapted from Poser CM, Paty DW, Scheinberg L et al: New diagnostic criteria for multiple sclerosis: Guidelines for research protocols. Ann Neurol 13:227, 1983.)

[b] A focal neurologic deficit explainable by a single white matter lesion.

[c] Abnormal urodynamic or evoked-response testing.

[d] Elevated cerebrospinal fluid IgG, myelin basic protein, or the presence of oligoclonal bands.

be complementary to other types of evaluation, because they may reveal the presence of lesions too small to be seen by MRI and not large enough to cause detectable clinical symptoms.[111] At least one of the evoked response tests will be abnormal in 88 percent of patients with clinically definite MS.

Urodynamic testing involves cystometric filling and emptying of the bladder. The entire micturation reflex arc is thus evaluated including the sacral plexus, spinal cord, pons, and cerebrum. The examination is frequently abnormal, even in patients without symptoms of bladder dysfunction. Up to 96 percent of patients with clinically definite MS have abnormal urodynamics.[112]

Abnormalities in the spinal fluid can be seen in about 80 percent of patients with clinically definite MS.[110] Immunoglobulins, particularly IgG, are increased in at least two-thirds of patients with MS. In about 90 percent with elevated IgG, immunoelectrophoresis will detect oligoclonal bands representing high concentration of a few species of IgG.[113] Though a sensitive test, oligoclonal bands are also found in many other diseases, such as acute or chronic infections, tumors, cerebrovascular diseases, and peripheral neuropathies.[114]

Myelin degradation products, such as myelin basic protein, are often found in the cerebrospinal fluid in patients with MS.[115] Myelin basic protein can be detected in about 90 percent of patients for a week following an acute clinical episode. In the chronic situation, however, most patients tend to have normal levels of myelin basic protein in their spinal fluid. As with oligoclonal bands, myelin basic protein may be found in a number of other conditions besides MS, including infarction, encephalitis, leukodystrophies, metabolic encephalopathies, and methotrexate myelopathy.[97,114]

Radiographic Imaging

Radiographically, CT has been used reliably over the last decade for the evaluation of patients with suspected MS. Perhaps more important than the identification of MS plaques has been CT's ability to exclude other significant pathology such as tumor, which might clinically mimic MS. The optimal CT technique should probably include high dose contrast and delayed scanning.[116]

The CT appearances of MS have been fully described.[117–120] Old, inactive plaques typically appear as sharply defined areas of decreased attenuation in the periventricular white matter.[117] Acute lesions may exhibit contrast enhancement secondary to breakdown of the blood-brain barrier.[121] Mass effect may be noted in acute lesions, presumably due to edema. [122] Depending upon degree of clinical suspicion for MS, 60 percent to 80 percent of patients will have abnormal CT scans.[118]

A number of early studies illustrated the sensitivity and usefulness of MRI in the detection of MS lesions.[96,123,124] The lesions detected by MRI match closely with MS plaques found at autopsy in the same patients.[125] The T1 and T2 values of MS plaques are lengthened significantly compared to normal brain.[126–128] This is a likely result of edema and demyelination. When extensive parenchymal cellular infiltration (encephalitis) accompanies inflammation and demyelination, the T1 and T2 values of the affected white matter may be normalized.[129] This could potentially result in false negative MR scans in some patients with active MS.

Both T1- and T2-weighted sequences are effective in detecting the lesions of MS.[128,130,131] T1-weighted sequences show MS plaques to be low intensity and rather well-defined. T2-weighted sequences show MS plaques to be high intensity (Fig. 9-16). Many more lesions are identified on T2-weighted sequences than on those that are T1-weighted.[124,132] Sometimes multiple small lesions on T1-weighting will fuse into a large confluent region on T2-weighting.[130] The MS lesions are usually distributed in a typical fashion corresponding to common locations seen at autopsy: the periventricular white matter, centrum semiovale, brain stem, cerebellar hemispheres, and spinal cord. Because of limited spatial resolution at present, optic nerve lesions are not routinely visualized.

In several comparative studies, MR has been found superior to CT in detecting the plaques of MS.[126,130,131] No lesions detected by CT are missed by MR. Many more lesions are seen by MR than CT, especially in the posterior fossa, brain stem, and spinal cord. At least 75 percent of MS lesions detected by MR are clinically "silent"; only about 5 percent can be definitely related to clinical symptoms and signs.[130] These clinically silent lesions are usually located in the centra semiovale, temporo-occipital, or periventricular white matter.

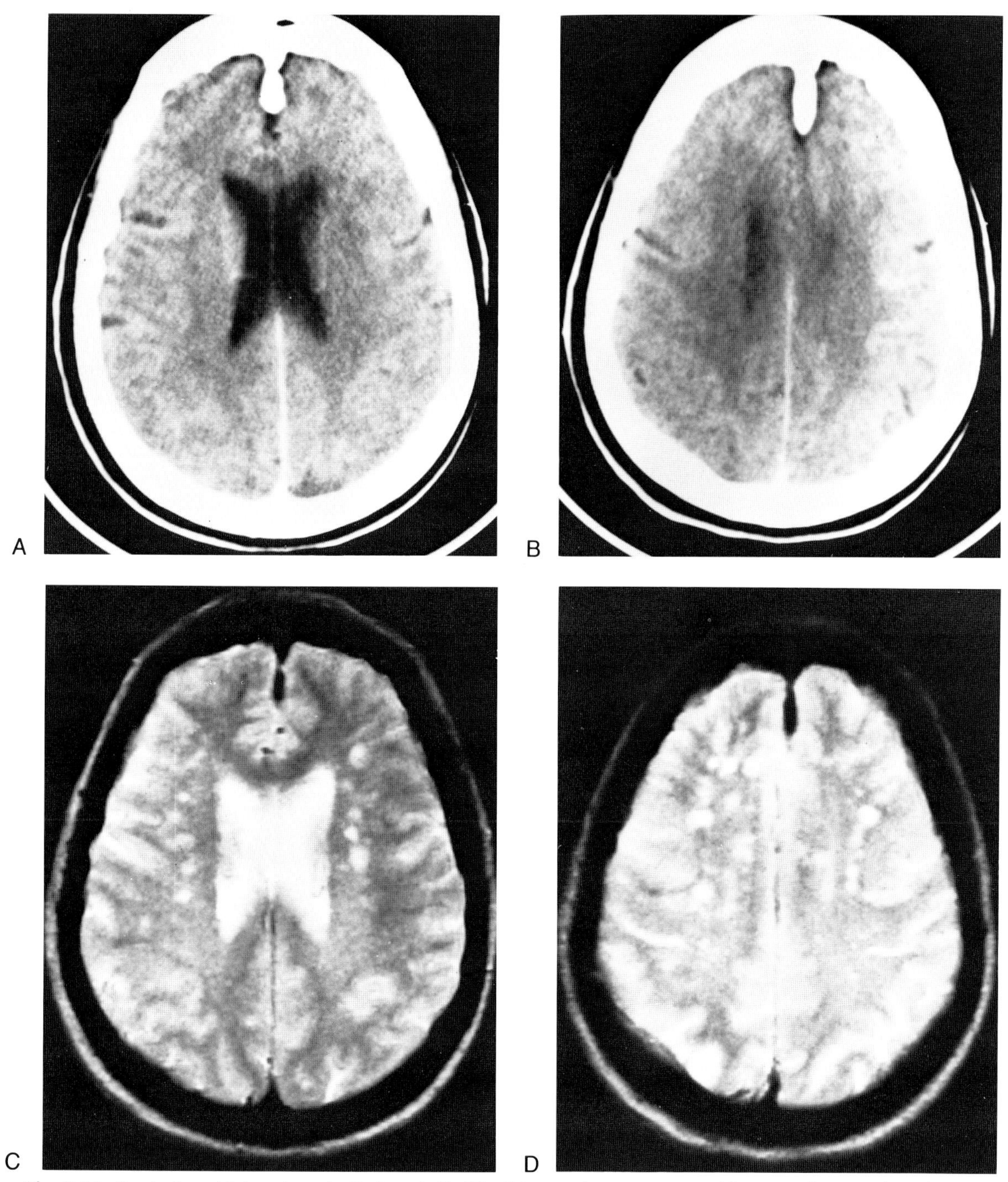

Fig. 9-16. Typical multiple sclerosis. Patient 1 **(A–D):** CT scan shows questionable, nonenhancing low density areas in the white matter, a difficult call to make. MR images (SE 2500/100) show multiple high intensity MS lesions in the centra semiovale bilaterally. (*Figure continues*).

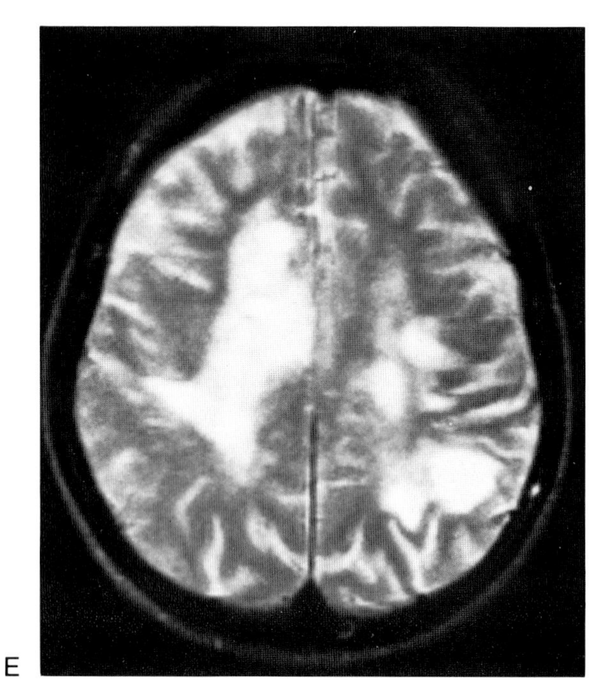

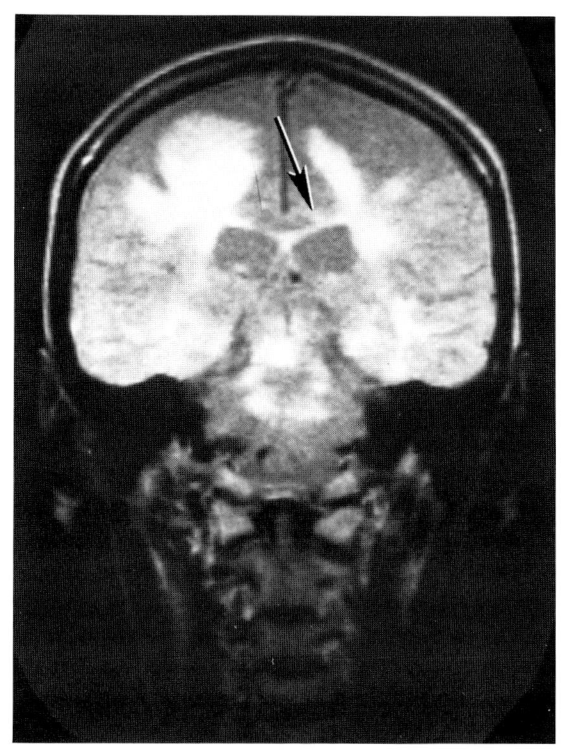

Fig. 9-16 (*continued*). Patient 2 **(E,F):** More severe involvement seen on axial and coronal MR images. Note involvement of the corpus callosum in **(F),** a finding not seen in Binswanger disease, but common in MS.

In general, MRI is complementary to evaluation by clinical examination, evoked potentials, urodynamics, and spinal fluid analysis.[131] Generally, severely symptomatic patients tend to have many lesions detected by MRI, while mildly symptomatic patients have fewer lesions.[132] However, because many MS plaques are clinically silent, the extent of disease demonstrated by MRI does not necessarily correlate with clinical severity or disability.[133,134] Progression of disease based on the appearance of new lesions may be more accurately assessed by MRI than by clinical examination. Evoked potential tests are more sensitive than MRI in detecting small lesions, but MR may detect lesions not included in the pathways tested by the evoked potentials.[111] Cerebrospinal fluid testing is less sensitive overall than MRI.[131]

MRI is most likely to show lesions in the following situations: (1) if the patient can be classified clinically as having definite MS, (2) the disease is active at the time of scanning, (3) the duration of disease is greater than 2 years, and (4) if multiple neurologic deficits are present indicating multiple lesions.[131] By contrast, MR scan is often normal in short duration disease, in patients with optic neuritis only, and in those with only clinically probable MS.

The relaxation times of acute and chronic MS lesions overlap, so it is not possible on a single scan to assess disease activity.[126] The demonstration of a plaque's contrast enhancement on CT has been said to be an indicator of disease activity, because it implies disruption of the blood-brain barrier.[118,121] However, other investigators have shown somewhat poor correlation between CT enhancement and disease activity.[130,135] Thus it is unclear whether CT holds any advantage to MR in measuring physiologic activity of plaques. In any case, equivalent MR information about the integrity of the blood-brain barrier will be available with future use of paramagnetic contrast agents.[136]

The course of MS by serial MR studies has been documented.[137] In general, prolonged T1 and T2 values may be seen in plaques for many years after

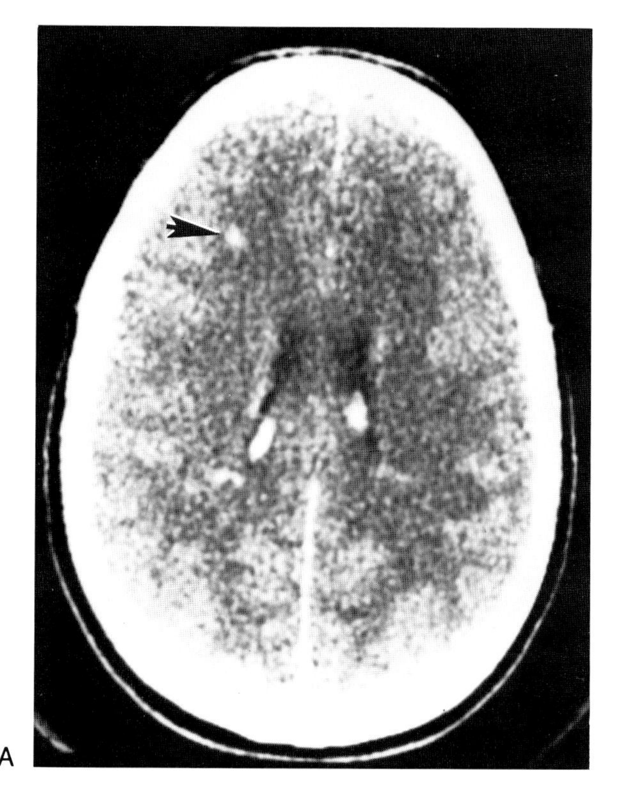

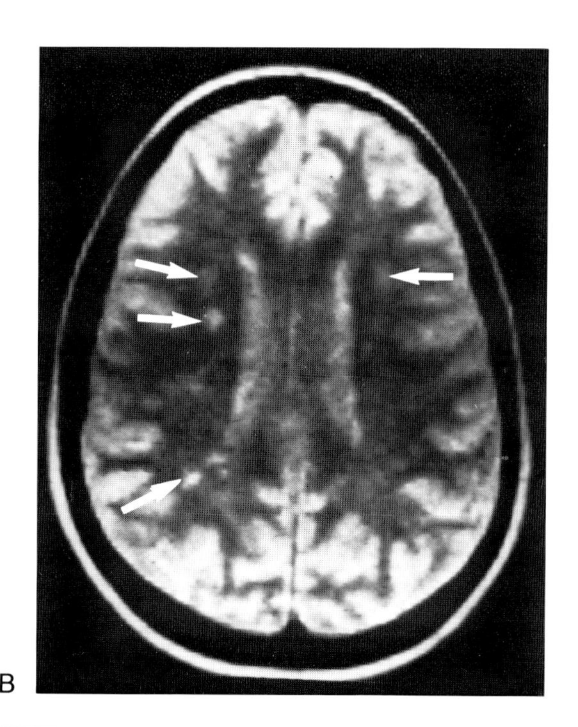

Fig. 9-17. Evolution of MS lesions. **(A,B)** CT and MR scans (SE 2000/56) in a patient with acute MS attack. Single enhancing lesions seen on CT while multiple lesions are noted on MR. **(C)** Follow-up MR scan 6 months later. The lesions are much less apparent. However, many lesions may not change their MR appearance with time, so persisting signal does not necessarily imply active disease.

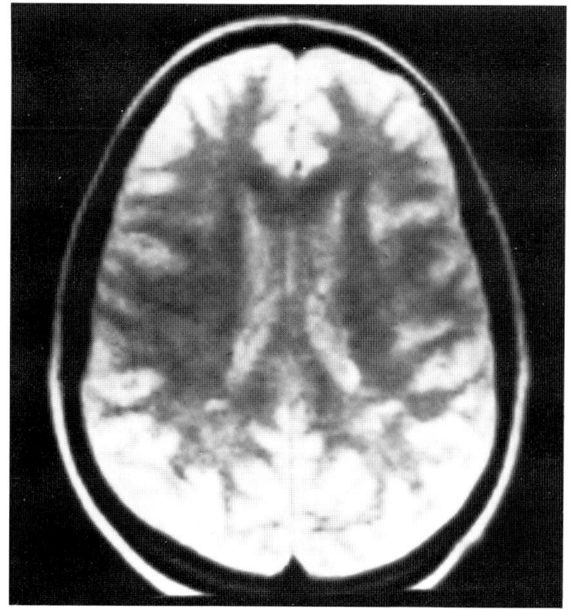

they become inactive. On a single scan, active lesions cannot be distinguished from inactive lesions. Some lesions, however, do tend to fade with time and healing (Fig. 9-17). Lesions in some patients treated with immunosuppressive agents have been observed to decrease in size and become more sharply marginated.[134] Active disease can be recognized on serial MR scans if the detected lesions increase in size or number.

Because routine cranial MR imaging commonly includes the upper cervical cord, a considerable number of MS patients will be shown to have lesions here.[138] Spinal cord plaques are characterized by an elongated configuration, and may extend longitudinally along the cord for several centimeters. Preferential occurrence in the lateral and dorsal segments has been described.[95] An example of incidental detection of MS plaques in the cervical cord is shown in Figure 9-18.

Occasionally, unusual MR presentations will be seen. In the Schilder's (diffuse sclerosis) variant of MR, large, confluent areas may be involved (Fig. 9-19) Occasionally, in very acute lesions, mass effect may be seen. Sometimes only a few large lesions are identified (Fig. 9-20).

Longstanding MS is associated with atrophic

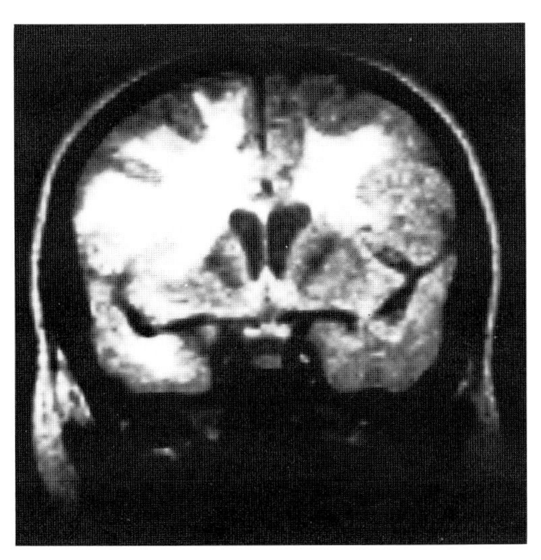

Fig. 9-19. Severe MS, confluent, also involving the gray matter.

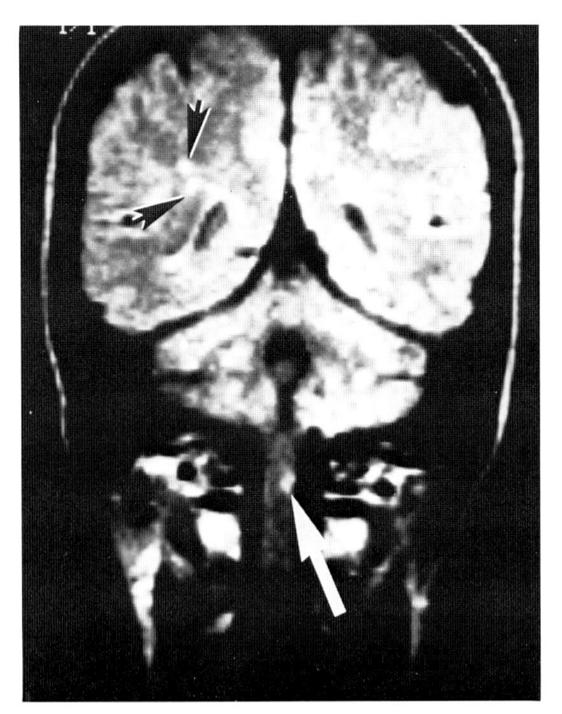

Fig. 9-18. MS of the cord. An elongated cord plaque is noted on this coronal SE 1500/28 image (white arrow). Note other lesions in the right centrum semiovale (black arrows).

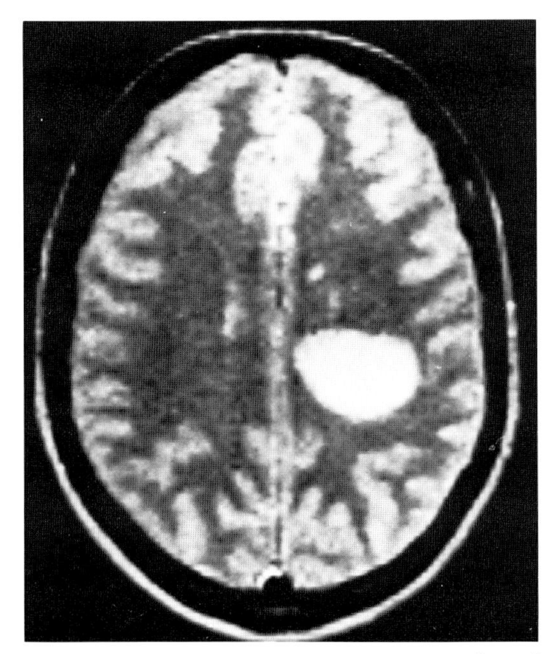

Fig. 9-20. Single large MS plaque, though initially to be a low-grade glioma. Patient subsequently developed other lesions.

changes, observed in 80 percent of patients followed in one large series up to 19 years.[139] This has been attributed to loss of white matter from periventricular MS lesions.[140] Chronic steroid administration may also play a role. The atrophy may be diffuse or involve the cerebellum selectively, particularly the vermis.[117]

Subcortical Arteriosclerotic Encephalopathy

Subcortical arteriosclerotic encephalopathy (SAE, Binswanger disease) is a degenerative disorder of deep white matter resulting from arteriolar insufficiency. First described by Binswanger in 1894, this entity was originally thought to be rare.[141] When strict clinical and pathologic criteria are applied, however, SAE may be identified at autopsy in about 4 percent of patients over the age of 60.[142]

The pathologic changes in SAE are thought to be secondary to vasculopathy of the long, penetrating arterioles that extend from the brain surface to the deep white matter.[143] Hyalinization of these medullary arterioles is a constant finding, with thickening of the media, elastosis, and dilatation of the perivascular space (état criblé).[143] Focal or diffuse areas of demyelination and necrosis are observed. Multiple small cystic white matter infarcts are commonly noted. The white matter disease occurs mainly in the centra semiovale and periventricular areas and is bilateral but not necessarily symmetrical. Subcortical U-fibers and the corpus callosum are not involved.[143] Cortical atrophy and lacunar infarcts in the basal ganglia are common associated findings.

Clinically, most patients are over 60 with a history of hypertension and other signs of arteriosclerotic vascular disease.[144] Most show patchy mental lapses characterized by memory disturbance, confusion, and labile emotions. Urinary incontinence is very common.[143] Small focal neurologic abnormalities consistent with lacunar infarction may be present. SAE or a similar process may be responsible in many cases of the clinical syndrome known as multi-infarct dementia.[143]

White matter disease on CT or MRI is seen in 16 to 30 percent of elderly patients.[11,14] This is a higher percentage than the autopsy incidence of SAE. The term *leukoariaosis* has been applied by pathologists to account for some of these "soft areas" in the deep

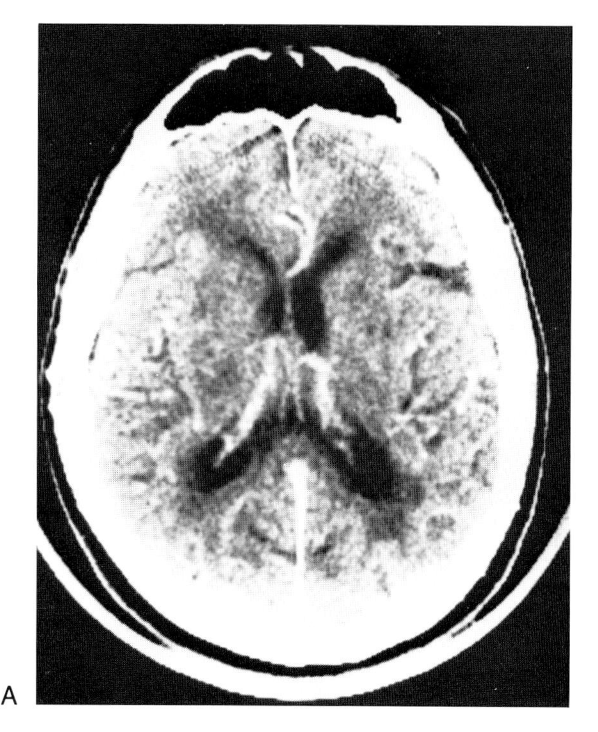

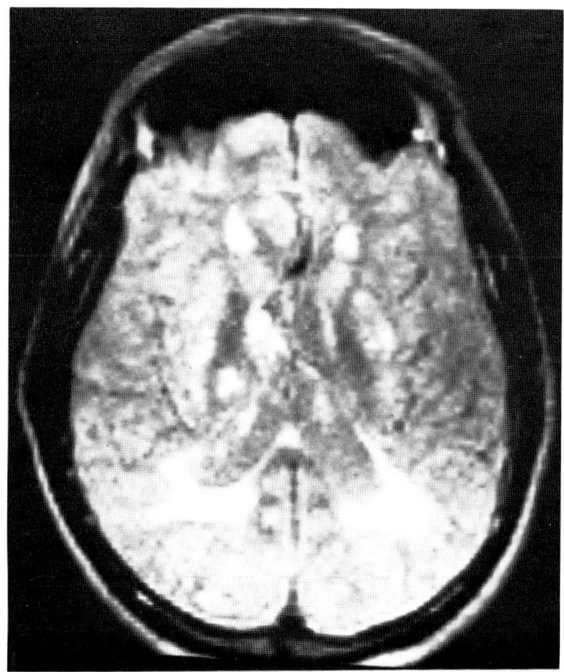

Fig. 9-21. Binswanger disease. **(A)** Multiple low density areas in white matter seen on CT. **(B)** The abnormalities are much more apparent on this axial MR image (SE 2000/56).

white matter.[14a] Ependymitis granularis and local collections of periependymal fluid may account for some white matter lesions adjacent to the ventricular margins.[145] Other benign etiologies include small vascular malformations and ventricular diverticula.[144a]

Magnetic resonance imaging is more sensitive than CT in identifying the white matter lesions of SAE.[145a] The lesions are characterized by prolonged T1 and T2 values making them appear dark on T1-weighted images and bright on T2-weighted images. The lesions of SAE are frequently discrete but may be confluent.[146] They are best seen in the centra semiovale and periventricular regions.[145a] In contrast to the plaques of MS, the periventricular lesions of SAE are not usually immediately adjacent to the ependymal surface.[145] The corpus callosum is characteristically spared in SAE while it is involved in up to 30 percent of patients with MS.[146a] Examples of white matter lesions in patients with pathologically proven SAE are shown in Figure 9-21.

Progressive Multifocal Leukoencephalopathy

Progressive Multifocal Leukoencephalopathy (PML) is a rare viral infection of the immunocompromised host, which produces progressive white matter demyelination. Initially described in patients with leukemia and lymphoma, PML has also been reported in patients with other malignancies, tuberculosis, lupus, renal transplants, AIDS, sprue, sarcoidosis, macroglobulinemia, and in those on immunosuppressive therapy.[147,148] The cause of PML is a papovavirus of the JC or SV40 strain.[149]

Pathologically, multiple, partially confluent areas of demyelination are seen scattered throughout the white matter, with relative sparing of neurons.[150] It is thought that the demyelination is due to destruction of oligodendroglia by the virus. Areas of demyelination are most prominent in the subcortical white matter with less involvement of the cerebellum, brain stem, and spinal cord. Rarely PML may present as a discrete or expansive mass, resembling a glioma or hematoma.[151] Pathology in these cases has revealed marked tissue necrosis, inflammatory infiltration, and gliosis. In the later stages of PML, progression to cystic atrophy may be seen.

Clinical manifestations are diverse, related to the number and location of the lesions.[150] Characteristically the onset is sudden, with mental deterioration, hemiplegia, ataxia or cranial nerve palsies. The clinical course is usually progression to death within 3 to 6 months, although there are some long-term survivors reported. Therapy is largely unsatisfactory. The final diagnosis depends upon biopsy, because serology is unreliable.

Corresponding features of the disease process are demonstrated by CT and MR.[152] The lesions usually begin in the parietooccipital white matter manifest by edema (low x-ray attenuation; prolonged T1 and T2 relaxation times). Usually these areas show no mass effect and progress by coalescence.[148] The lesions often demonstrate a scalloped, sharply marginated outer border. MR may be more sensitive in monitoring the development and progression of PML. This may be helpful because CT is known to underestimate the amount of white matter disease clinically.[148] Examples of PML in immunosuppressed patients imaged by MR are shown in Figures 9-22 and 9-23.

Central Pontine Myelinolysis

Central pontine myelinolysis (CPM) is an uncommon disorder characterized by destruction of myelin sheaths in the basis pontis. The disease usually occurs in middle aged alcoholic or malnourished patients with electrolyte disturbances, especially hyponatremia.[153] CPM is frequently associated with diseases of other organ systems, including the liver (cirrhosis, Wilson disease), brain (Wernicke disease, tumors), and kidney (nephropathy, transplant).[154–157] Less common associations include diabetes, amyloidosis, leukemia, infections, and Marchiafava-Bignami disease.

Clinical presentation is usually as a rapidly progressive corticospinal and corticobulbar syndrome.[153–155] Findings include facial weakness, quadriparesis, hyperreflexia, and extensor plantar response. A "locked-in syndrome" may be present, where the patient is mute and paralyzed but not comatose. Brain stem auditory evoked responses are abnormal. Most patients die within days or weeks following onset. Some patients may survive with careful correction of their electrolyte disturbances.

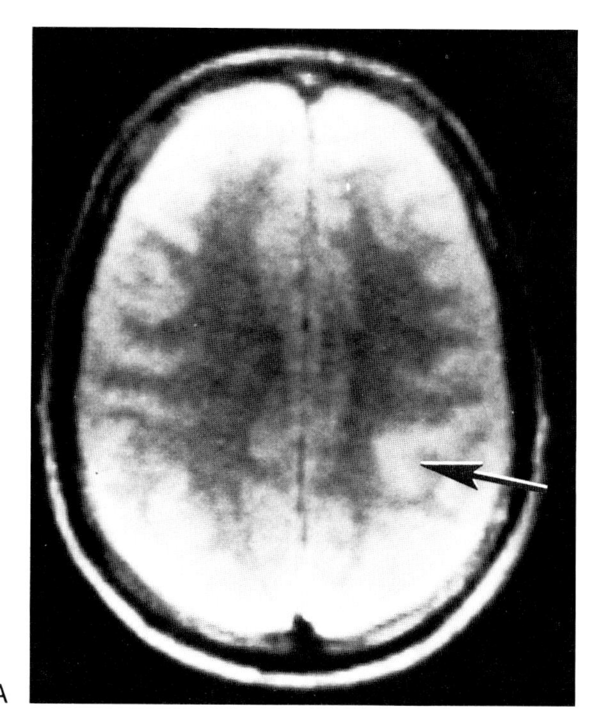

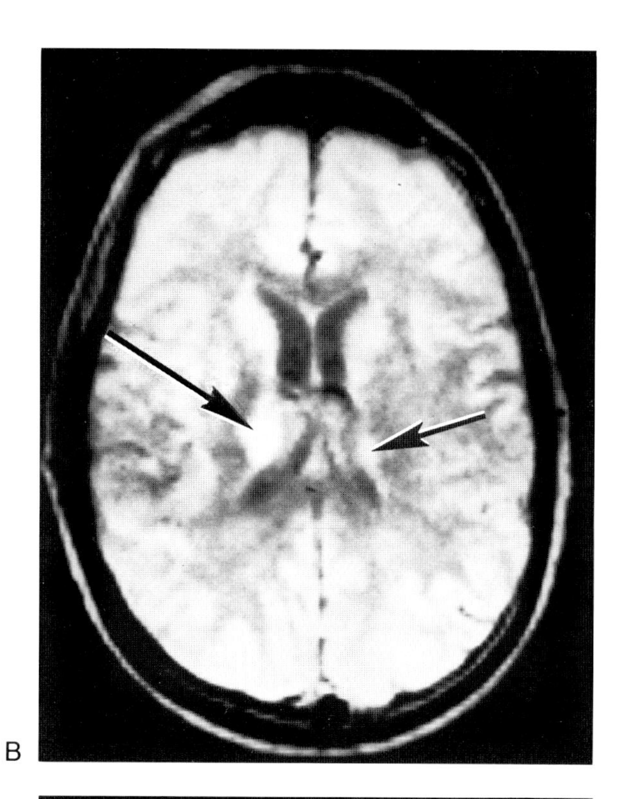

Fig. 9-22. Progressive multifocal leukoencephalopathy. Several rounded, nonspecific foci of high signal scattered throughout the brain seen on T2-weighted MR images **(A–C).**

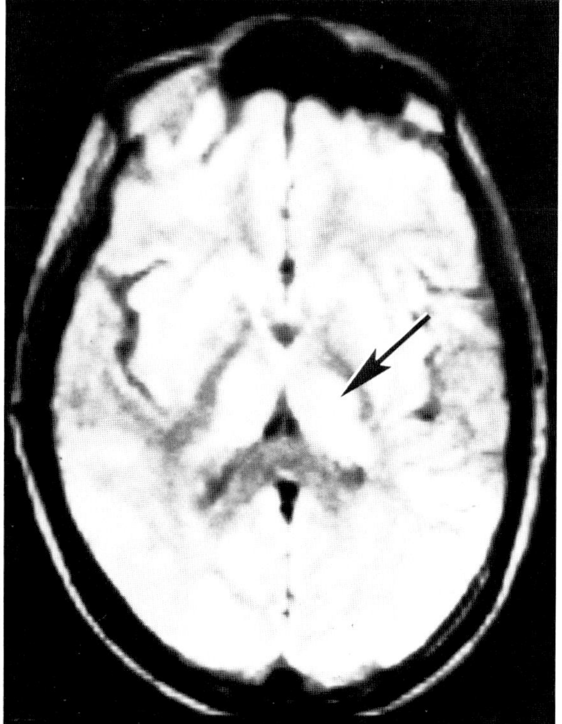

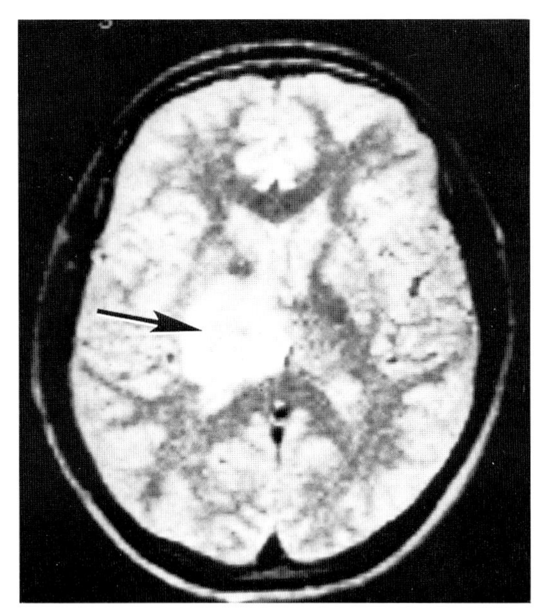

Fig. 9-23. Progressive multifocal leukoencephalopathy. Single large lesion, nonspecific MR findings.

Pathologically, CPM demonstrates symmetrical demyelinization in the basis pontis, which usually begins in the median raphe and advances laterally.[158,159] Although myelin sheaths are destroyed, the nerve cells and axis cylinders remain fairly well preserved. There is no inflammation. The lesion may spread to the pontine tegmentum or upward to the mesencephalon. Extrapontine lesions are relatively common, and may occur in the subcortical white matter, thalamus, striatum, corpus callosum, centrum semiovale, cerebellum, or cervical cord.[154]

Increasing clinical and experimental evidence suggests that CPM is caused by rapid changes in serum sodium, which subjects the brain to focal severe osmotic pressures.[160–162] Pontine and extrapontine myelinolysis has been produced in animals by rapid elevation of serum sodium following a period of chronic hyponatremia. Some cases in humans have arisen following correction of hyponatremia with hypertonic saline in an injudicious manner.[163,164] It is possible, therefore, that CPM may largely be an iatrogenic disorder caused by too rapid correction of electrolyte disturbances.

The diagnosis of CPM has occasionally been made by CT scan in the appropriate clinical setting when low attenuation lesions are noted in the pons.[165–167] Several examples of CPM have appeared in the MR literature, showing symmetrical areas of prolonged T1 and T2 in the midpons.[168–170] MR should be much more sensitive than CT in detecting these lesions. Extrapontine involvement may also be noted on MR. An example of CPM with only pontine lesions is shown in Figure 9-24.

Adrenoleukodystrophy

Adrenoleukodystrophy is an x-linked recessive disease manifest by adrenal insufficiency and cerebral demyelination. Most cases occur in males between the ages of 5 and 14 years. A typical clinical picture is frequently seen:[171] a boy, previously in perfect health begins to experience behavior problems and poor grades in school. Over the next several years there is progressive mental deterioration. The patient may experience visual loss, hearing difficulties, cerebellar, ataxia, and quadriparesis. Adrenal insufficiency may be a prominent feature, but can be subclinical.[172] Progression without remission is the rule. Several variant forms of the disease in adults (including adrenomyeloneuropathy and adrenoleukomyeloneuropathy) have been described.[173,174]

The underlying metabolic defect in adrenoleukodystrophy is not known. An abnormal accumulation of very long chain fatty acids associated with impaired oxygenation has been reported.[175,176] Adrenal biopsy is diagnostic, revealing atrophy of the zona reticularis and fasciculata, ballooned and striated cells in the adrenal cortex, and periodic acid-Schiff (PAS) positive cytoplasmic inclusions.[177] Brain biopsy may be misleading, frequently demonstrating demyelination and inflammatory reaction.[178]

A rather characteristic pattern of cerebral demyelination has been described, which when seen virtually assures the diagnosis.[171,172,179–182] (It should be noted, however, that atypical patterns particularly early in the disease are not infrequent, and may make the diagnosis difficult by imaging modalities alone.) In the usual case, demyelination typically begins in the subcortical white matter of the posterior cerebrum. At the time of diagnosis it usually involves portions of the occipital, parietal, and posterior temporal lobes. Contiguity of spread across the splenium of the corpus callosum is characteristic, but the remain-

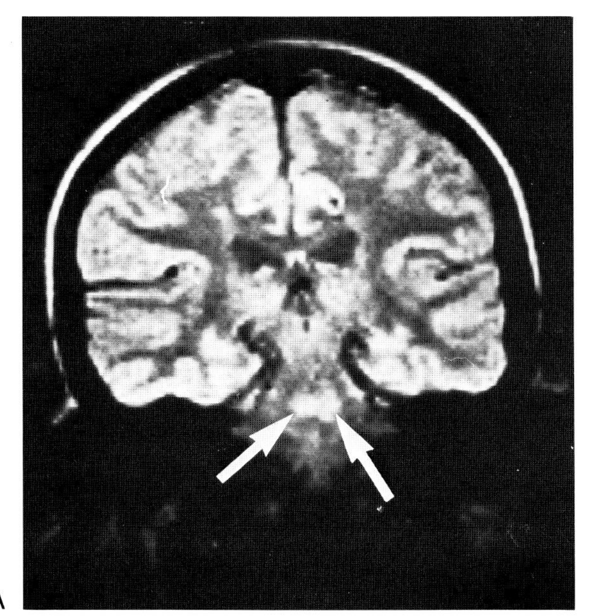

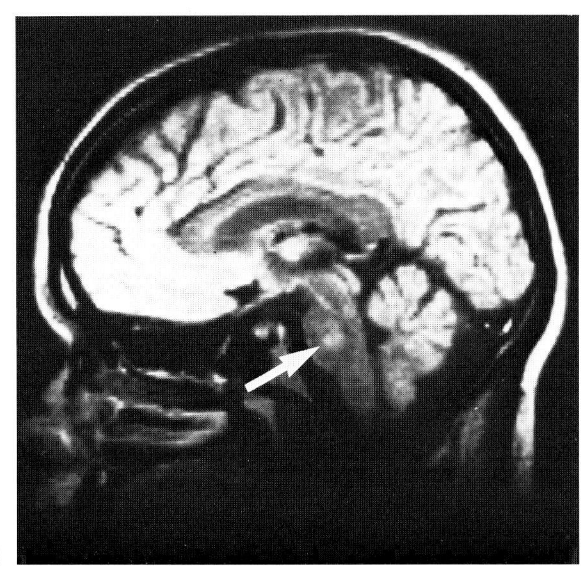

Fig. 9-24. Central pontine myelinolysis. An alcoholic in hyposmolar coma (**A** and **B**). Symmetric high intensity lesions in the central pons (arrows) seen on coronal SE 2500/56 and sagittal SE 1500/28 images.

der of the corpus callosum is not involved. Contiguous spread from posteriorly to anteriorly through the deep white matter then occurs. The leading edge of the demyelination is often serpiginous, demonstrating inflammatory changes and breakdown of the blood-brain barrier. More posteriorly the disease appears inactive with glial fibrosis and dystrophic calcifications observed pathologically. The final stages of the disease are characterized by generalized cerebral and cerebellar involvement, progressing to diffuse atrophy.[180,181]

The pathologic progression described above has been well documented by CT (Fig. 9-25).[171,179–182] The leading edge frequently shows contrast enhancement. A few cases of adrenoleukodystrophy and its variants seen by MR have been published.[183,184] On inversion recovery sequences there is loss of gray matter-white matter contrast posteriorly. Areas of demyelination demonstrate prolonged T1 and T2 values; these areas correspond to low attenuation lesions seen on CT. MR is able to detect focal areas of involvement not seen on CT.[185] MR is thus more sensitive and should be considered the imaging modality of choice in following these patients.

Metachromatic Leukodystrophy

Metachromatic leukodystrophy (MLD) is an autosomal recessive disease of white matter caused by a deficiency in the enzyme cerebroside sulfatase A.[186] As a result of this enzyme defect, sulfated lipids (sulfatides) increase and the membranes of myelin sheaths deteriorate in both the central and peripheral nervous systems. The strong negative charges of these sulfatides alter the normal histological staining characteristics of dyes such as cresyl violet. This phenomenon is called metachromasia and gives the disease its name.[187]

Pathologically, sulfatides accumulate in the plasma membranes of many cells, including Schwann cells, oligodendroglia, renal tubular cells, and gall bladder epithelium. In the CNS profound demyelination, which is usually diffuse and symmetric occurs, but spares the arcuate fibers.[186] A sulfatide lipidosis may also involve the dentate and brain stem nuclei. Peripheral nerves are also involved, and slowing of nerve conduction velocity is often an early and important sign.[188]

Infantile, juvenile, and adult forms of MLD are

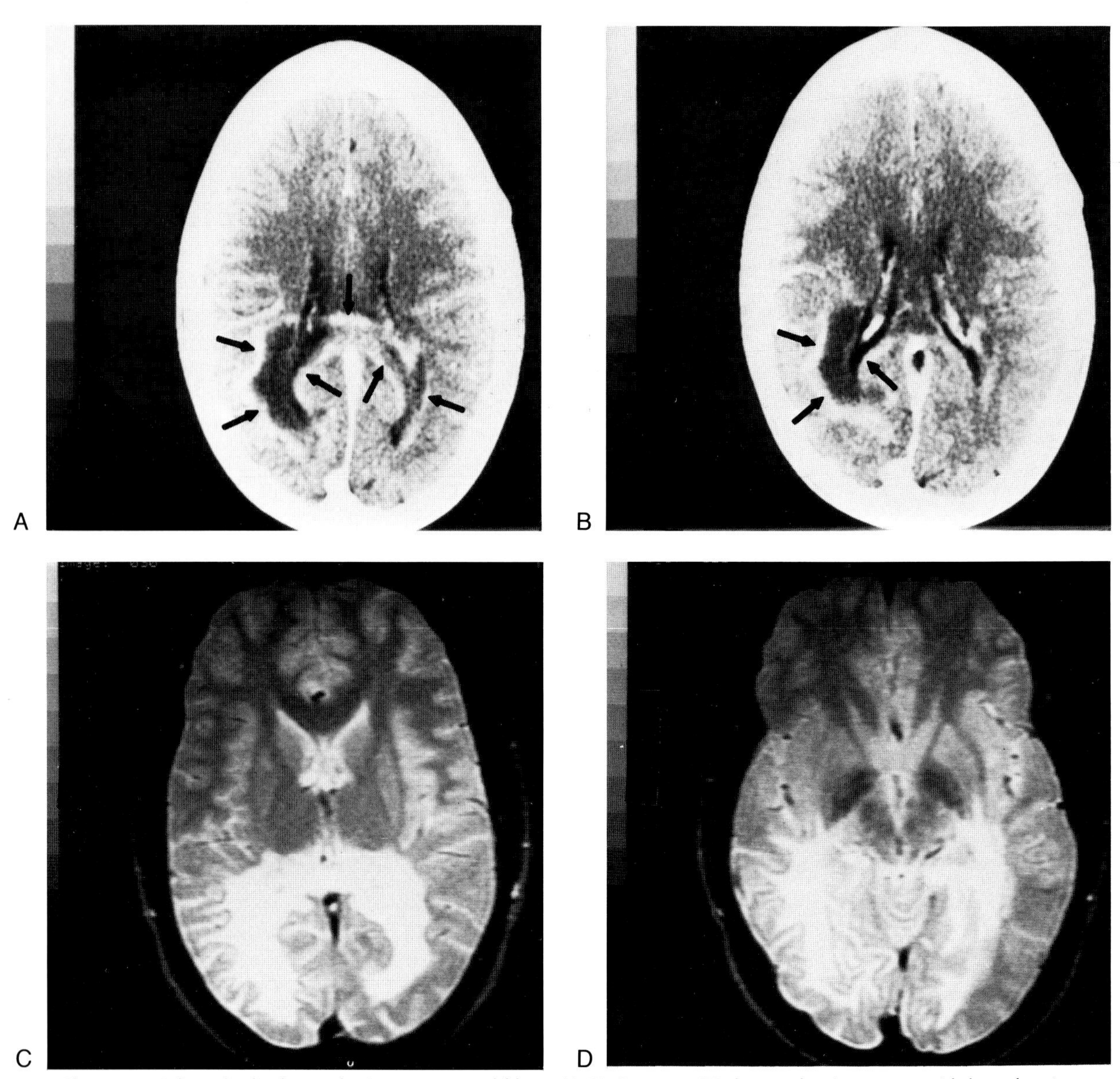

Fig. 9-25. Adrenoleukodystrophy in a 12-year-old boy. **(A,B)** Contrast CT shows classic pattern with low density in posterior white matter and enhancing serpiginous edge (arrows). **(C,D)** Axial MR images (SE 2000/100) show the white matter disease superbly. (*Figure continues*).

Fig. 9-25 (*continued*). **(E,F)** Sagittal SE 900/30 images show spread across the spelium of the corpus callosum (low signal).

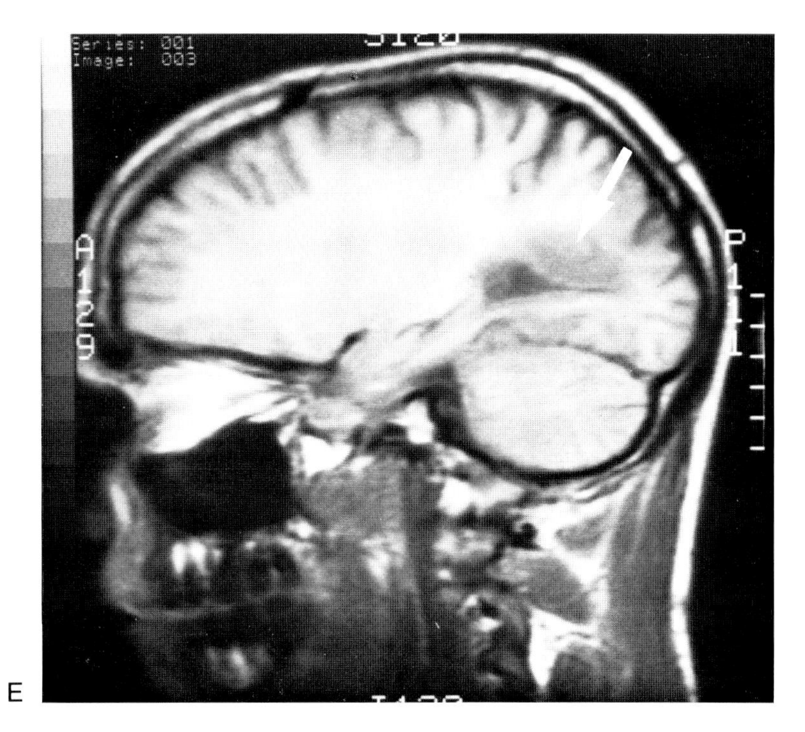

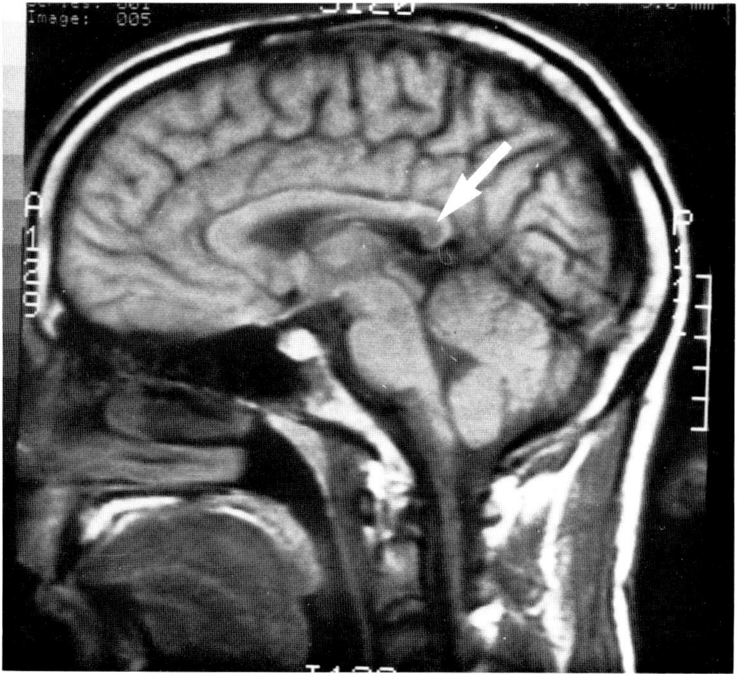

recognized.[189] A multiple sulfatase deficiency variant has also been described. About two-thirds of cases of MLD are of the infantile variety and present before the age of 3 years. The child is normal at birth, but begins to experience gait difficulties between 1 and 2 years of age. There is weakness and hypotonia of the legs with decreased or absent reflexes and positive Babinski signs bilaterally. Later there is involvement of the arms and face. Swallowing and speech difficulties ensue. A dementia emerges, and death occurs within 7 to 8 years following onset.

The juvenile form usually occurs between ages 3 and 10 years. Heralding symptoms include emotional lability and poor school performance. Pyramidal and cerebellar signs then occur and progress as in the infantile variety.

The adult form of MLD presents insidiously, with symptoms of emotional lability, poor concentration, memory loss or frank psychosis. Seizures may occur, and some patients have unusual spasmotic movements. The prognosis in the adult is better than in the juvenile or infantile forms, although death usually occurs about 14 years after onset.

Computed tomographic findings are nonspecific with patchy but generally symmetric diffuse low attenuation seen throughout the white matter.[188–190] The immediate subcortical white matter may be spread. In the cases we have seen by MR, MR was far more sensitive than CT in detecting areas of leukodystrophy (Fig. 9-26).

Globoid Cell Leukodystrophy

Globoid cell leukodystrophy (Krabbe disease, galactosylceramide lipidosis) is a rare, autosomal recessive disorder of neuronal lipid metabolism caused by a deficiency in the enzyme galactocerebrodise-β-galactoside.[191] This enzyme has two substrates, galactocerebroside and galactosylsphingosine (psychosine), both of which accumulate abnormally when the enzyme is defective. Galactosylsphingosine is highly cytotoxic and is found in concentrations at least 100 times higher than in normal brains.[192] As a consequence of the enzyme defect and cytotoxic substrates that accumulate there is widespread deficiency or total lack of myelination.

Although subacute and late-onset forms of the disease are known, globoid leukodystrophy nearly always presents during the first year of life, usually between 3 and 6 months of age.[193] Affected infants are initially hyperirritable and later suffer developmental delay. Progressive retardation, spasticity, seizures, and blindness then occur. Two unusual clinical features are characteristic: rapid spontaneous nystagmus and

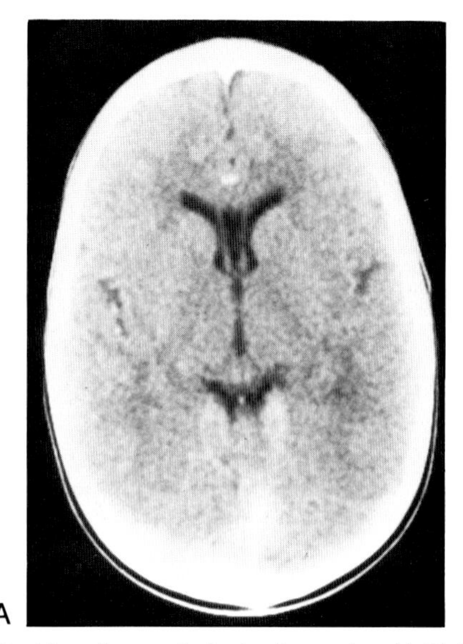

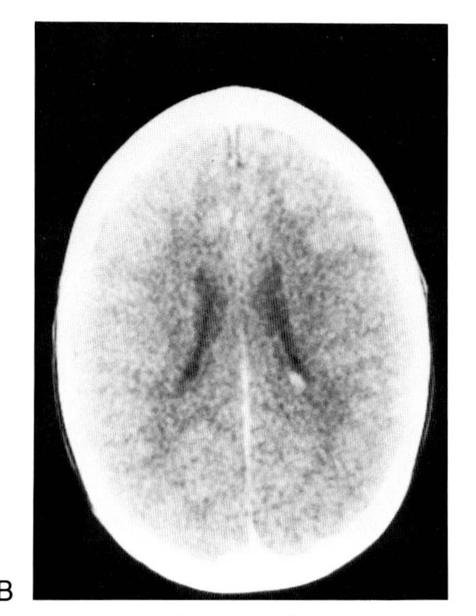

Fig. 9-26. Metachromatic leukodystrophy. **(A,B)** CT scan interpreted as normal. (*Figure continues*).

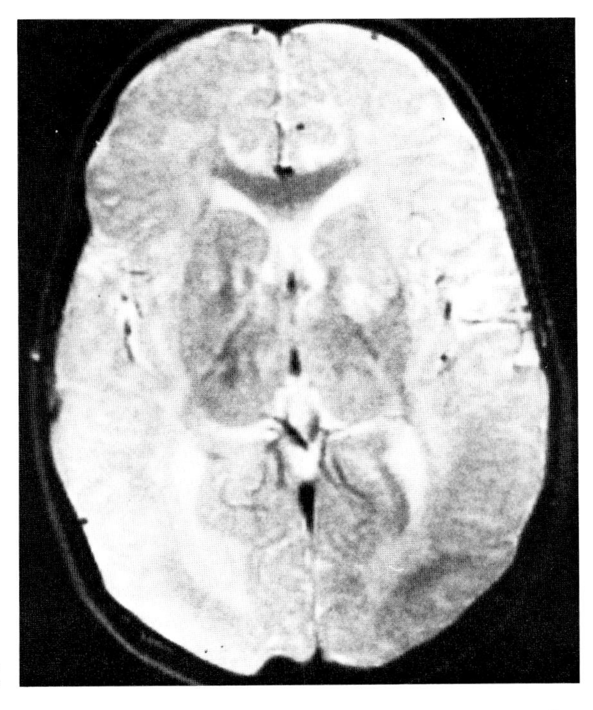

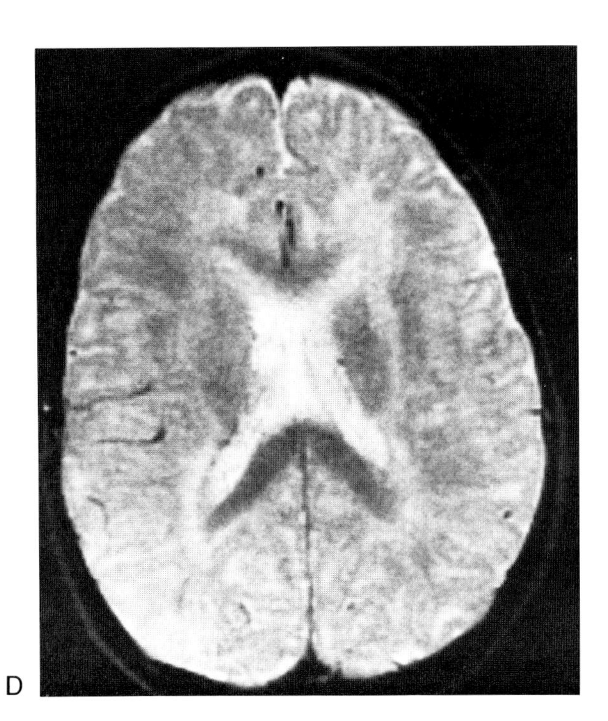

Fig. 9-26 (*continued*). **(C,D,E)** MR scan (SE 2000/100) shows symmetric, ill-defined, multiple high intensity regions in the white matter consistent with a dysmyelinating process.

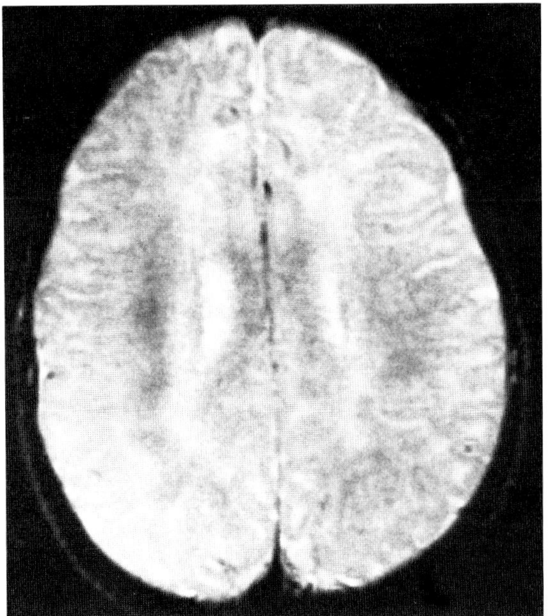

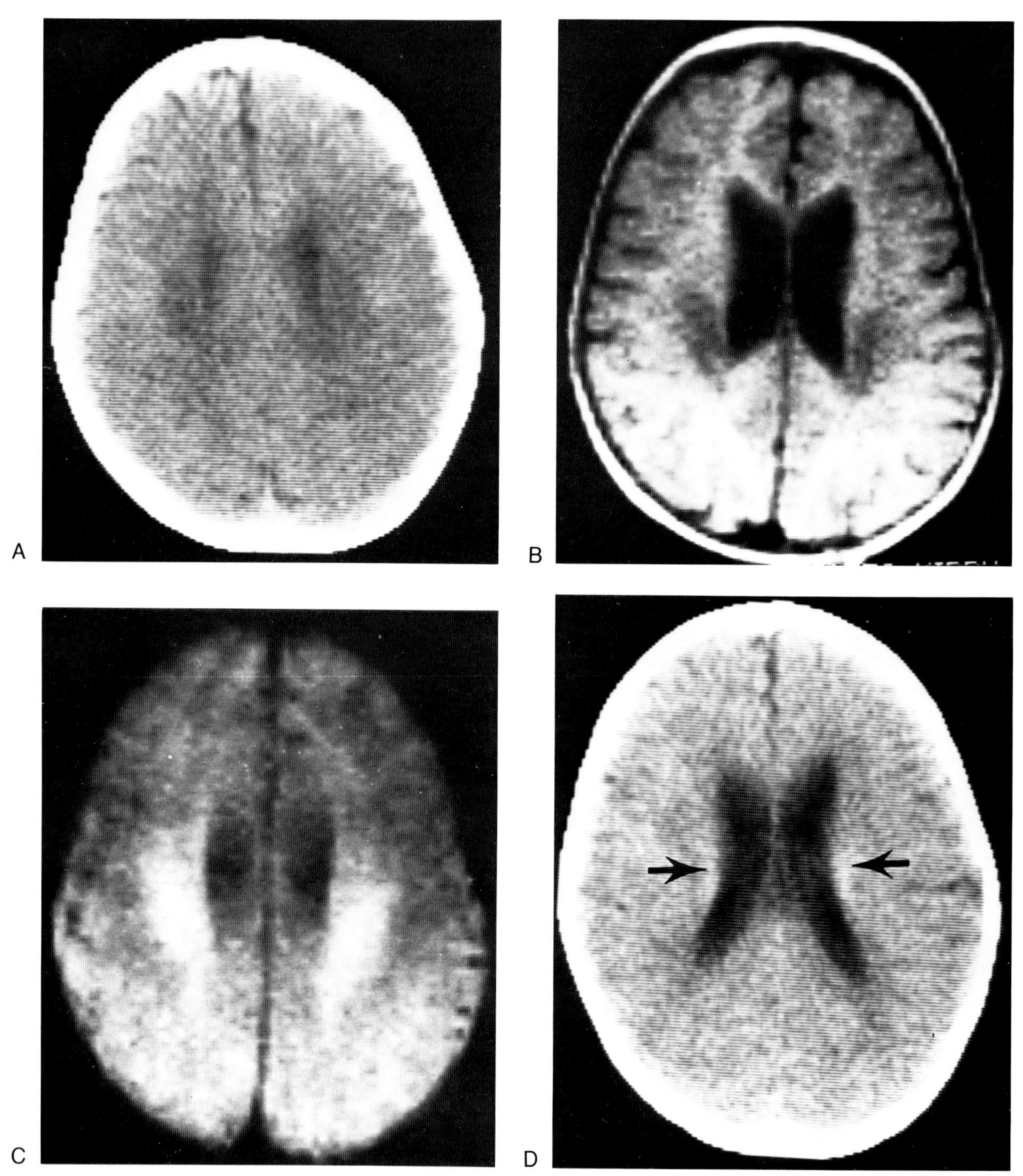

Fig. 9-27. Krabbe disease. CT **(A,D)** showed low attenuation lesions in the centra semiovale and dense areas (arrowheads) along the lateral ventricles. T1-weighted MR images **(B,E)** were not especially helpful but T2-weighted images **(C,F)** showed extensive white matter disease. (*Figure continues*).

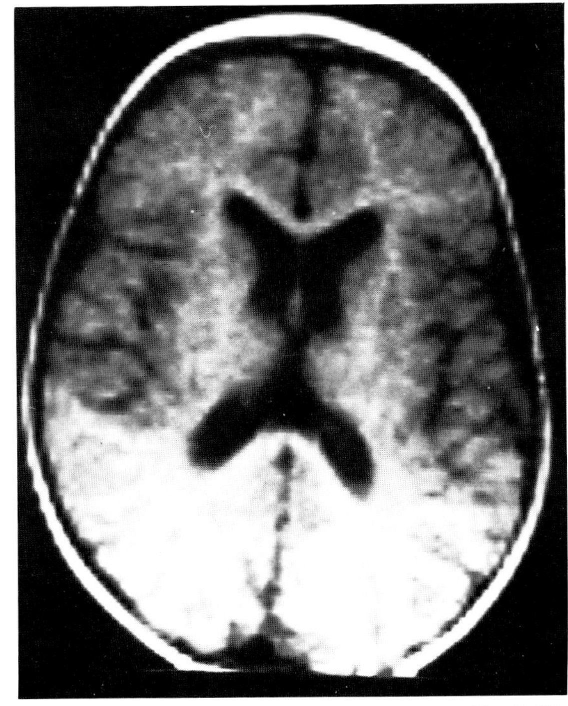

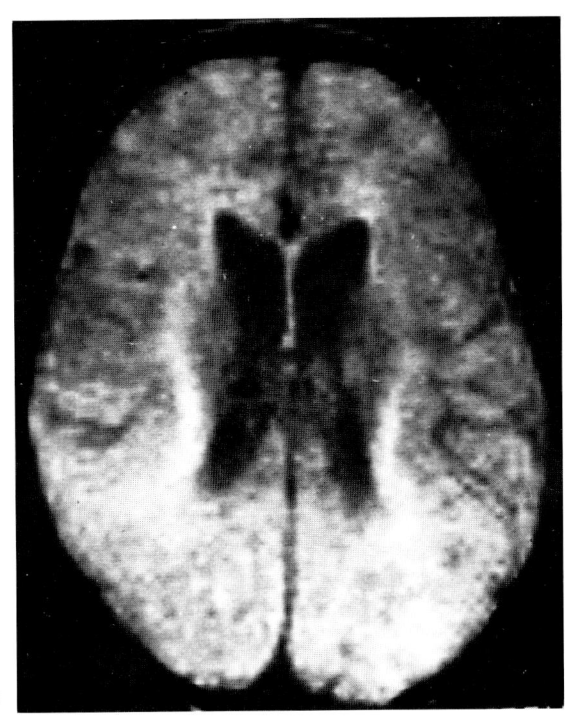

Fig. 9-27. *(continued)*.

poikilothermia.[194] The variations in body temperature may mimic sepsis. Progressive deterioration and death within a year is the rule.

Pathologically, clusters of large histocytes (globoid cells) filled with galactocerebroside are scattered throughout the white matter.[195] Initially, prominent involvement of the cerebellum, periventricular white matter, and centrum semiovale is noted. Gray matter becomes progressively involved with the pons, dentate nucleus, and thalamus affected more than cortical areas. In the late stages, profound loss of myelin, axonal degeneration, gliosis, fibrosis, and atrophy are seen.

Computed tomography findings may initially be normal.[196] Later there are white matter lucencies that are fairly symmetric and involve large areas of the centra semiovale.[197,198] Discrete and symmetric high attenuation areas may be seen in the deep gray matter and periventricular white matter, a finding with unknown pathologic correlate.[197–199] In the final stages, diffuse cerebral atrophy is noted.[200]

In a single proven case, MRI revealed white matter abnormalities far better than CT (Fig. 9-27).[199] The demyelinated areas possess prolonged T1 and T2 values. The dense or calcified deep gray matter lesions exhibited decreased signal on T1- and T2-weighted pulse sequences. This may relate to accumulation of endogenous paramagnetic substances such as iron in these regions.

REFERENCES

1. Barnes DM, Enzmann DR: The evaluation of white matter disease as seem on computed tomography. Radiology 138:379, 1981
2. Dekaban AS, Sadowsky D: Changes in brain weights during the span of human heights and body weights. Ann Neurol 4:345, 1978
3. Miller AKH, Alston RL, Corsellis JAN: Variation with age in the volumes of gray and white matter in the cerebral hemispheres of man: measurements with an image analyses. Neuropathol Appl Neurobiol 6:119, 1980
4. Creasey H, Rapoport SI: The aging human brain. Ann Neurol 17:2, 1985

5. Hughes CP, Gado M: Computed tomography and aging of the brain. Radiology 139:391, 1981
6. Meese W, Kluge W, Grumme T et al: CT evaluation of the CSF spaces of healthy persons. Neuroradiology 19:131, 1980
7. Gonzalez CF, Lantieri RL, Nathan RJ: CT scan appearance of the brain in the normal elderly population: A correlative study. Neuroradiology 16:120, 1978
8. Jacoby RJ, Levy R, Dawson JM: Computed tomography in the elderly: normal population. Br J Psychiatry 136:249, 1980
9. Huckman MS: Computed tomography in the diagnosis of degenerative brain disease. Radiol Clin N Am 20:169, 1982
10. Mann AH: Cortical atrophy and air encephalography: a clinical and radiological study. Pschol Med 3:374, 1973
11. Brant-Zawadzki M, Fein G, Dyke CV et al: MR imaging of the aging brain: patchy white matter lesions and dementia. AJNR 6:675, 1985
12. Bradley WG, Waluch V, Brant-Zawadzki M et al: Patchy, periventricular white matter lesions in the elderly: common observation during NMR imaging. Noninvasive Med Imaging 1:35, 1984
13. George AE, deLeon MJ, Kalnin A et al: Leukoencephalopathy in normal and pathologic aging: 2. MRI of brain lucencies. AJNR 7:567, 1986
14. Zimmerman RD, Fleming CA, Lee BCP et al: Periventricular hyperintensity as seen by magnetic resonance: prevalence and significance. AJNR 7:13, 1986
14a. Hachinski VC, Potter P, Merskey H: Leuko-araiosis. Arch Neurol 44:21, 1987
15. Alzheiner A: Über eine eigenartige Erkrankung der Hirnrinde. Zbl Nervenheilk 30:177, 1907
16. Terry RD, Davies P: Dementia of the Alzheimer type. Ann Rev Neurosci 3:77, 1980
17. Khachaturian ZS: Diagnosis of Alzheimer's disease. Arch Neurol 42:1097, 1985
18. Siedler H, Malmud N: Creutzfeldt-Jacob's disease. Clinicopathologic report of 15 cases and review of the literature. J Neuropathol Exp Neurol 22:381, 1963
19. Brown P, Cathala F, Sadowsky D et al: Creutzfeldt-Jacob disease in France. II Clinical characteristics of 124 consecutive verified cases during the decade 1968–1977. Ann Neurol 6:430, 1979
20. Baringer JR, Gajdusek DC, Gibbs CJ Jr et al: Transmissible dementias: current problems in tissue handling. Neurology 30:302, 1980
21. Masters CL, Harris JO, Gajdusek DC et al: Creutzfeldt-Jacob disease: patterns of worldwide occurrence and the significance of familial and sporadic clustering. Ann Neurol 5:177, 1979
22. Rao CVGK, Brennan TG, Garcia JH: Computed tomography in the diagnosis of Creutzfeldt-Jacob disease. J Comput Assist Tomogr 1:211, 1977
23. Kovanen J, Erkinjuntti T, Iivanainen M et al: Cerebral MR and CT imaging in Creutzfeldt-Jacob disease. J Comput Assist Tomogr 9:125, 1985
24. Fox JH, Ramsey RG, Huckman MS et al: Cerebral ventricular enlargement: chronic alcoholics examined by computerized tomography. JAMA 236:365, 1976
25. Lusins J, Zimberg S, Smokler H et al: Alcoholism and cerebral atrophy: a study of 50 patients with CT scan and psychologic testing. Alcoholism 4:406, 2980
26. Rumbaugh CL, Fang HCH, Wilson GH et al: Cerebral CT findings in drug abuse: clinical and experimental observations. J Comput Assist Tumogr 4:330, 1980
27. Victor M, Adams RD, Mancall EL: A restricted form of cerebellar cortical degeneration occurring in alcoholic patients. Arch Neurol 1:579, 1959
28. Charness ME, DeLaPaz RL, Diamond I et al: MR imaging of atrophic mamillary bodies in chronic Wernicke disease. Radiology 157(P):344, 1985
28a. Holland BA: Diseases of white matter. p. 259. In Brant-Zawadzki M, Norman D (eds): Magnetic Resonance Imaging of the Central Nervous System. Raven, New York, 1986
29. Huckman MS, Ramsey RG, Shenk GI: CT scanning in patients with suspected cerebral metastases. J Comput Assist Tomogr 2:511, 1978
30. Brain WR, Wilkinson M: Subacute cerebellar degeneration associated with neoplasms. Brain 88:465, 1965
31. Brazis PW, Biller J, Fine M et al: Cerebellar degeneration with Hodgkins disease: computed tomographic correlation and literature review. Arch Neurol 38:253, 1981
32. Bentson JR, Reza M, Winter J et al: Steroids and apparent cerebral atrophy on computed tomographic scans. J Comput Assist Tomogr. 2:16, 1978
33. Selhorst JB, Kaufman B, Horwitz SJ: Diphenylhydantion-induced cerebellar degeneration. Arch Neurol 27:453, 1972
34. Sorbi S, Blass JP: Hereditary ataxias. Curr Neurol 4:37, 1982
35. Greenfield J: The Spinocerebellar Degenerations. Charles C. Thomas, Springfield, 1954
36. Holmes G: A form of familial degeneration of the cerebellum. Brain 30:466, 1907
37. Schut JW: Hereditary ataxia. Arch Neurol Psychiatry 63:535, 1950
38. Dejerine J, Thomas A: L'atrophie olivo-ponto-cérébelleuse. Nouv Iconog Salpêtriere. 13:330, 1900
39. Konigsmark BW, Weiner LP: The olivopontocerebellar atrophies: a review. Medicine 49:277, 1970
40. Plaitakis A, Nicklas WJ, Desnick RJ: Glutamate dehydrogenase deficiency in three patients with spinocerebellar syndrome. Ann Neurol 7:297, 1980

41. Critchley M, Greenfield JG: Olivo-ponto-cerebellar atrophy. Brain 71:343, 1948
42. Pick A: Über die Beziehungen der senilen Hirnatrophie zur Aphasia. Prog Med Wochenschr 17:165, 1892
43. Wisniewski HM, Coblentz JM, Terry RD: Pick's disease: a clinical and ultrastructural study. Arch Neurol 26:97, 1972
44. Weschler AF, Verity A, Rosenchein S et al: Pick's disease. A clinical, computed tomographic, and histologic study with Golgi impregnation observations. Arch Neurol 39:287, 1982
45. Parkinson J: An Essay on the Shaking Palsy. Neely and Jones, London, 1817
46. Alvord EC, Forno LS, Kusske JA et al: The pathology of parkinsonism. A comparison of degenerations in cerebral cortex and brain stem. Adv Neurol 5:175, 2974
47. Hornykiewicz O: Neurohumoral interactions and basal ganglia function and dysfunction. Res Publ Assoc Res Nerv Ment Dis 55:269, 1976
48. Yahr MD: Parkinsonism. p.526. In Rowland LP (ed): Merritt's Textbook of Neurology. Lea & Febiger, Philadelphia, 1984
49. Adam P, Fabre N, Guell A et al: Cortical atrophy in Parkinson's disease: correlation between clinical and CT findings with special emphasis on prefrontal atrophy. AJNR 4:442, 1983
50. Rutledge JN, Hilal SK, Defendini R, Silver AJ: MR imaging of movement disorders. Radiology 157:290, 1985
51. Shy GM, Drager GA: A neurological syndrome associated with orthostatic hypotension. Arch Neurol 2:511, 1960
52. Polinsky RJ: Multiple system atrophy: clinical aspects pathophysiology and treatment. Neurol Clin 2:487, 1984
53. Borit A, Rubinstein LJ, Urich H: The striato-nigral degenerations: putaminal pigments and nosology. Brain 98:101, 1975
54. Pastakia B, Polinsky R, Di Chiro G et al: Multiple system atrophy (Shy-Drager syndrome): MR imaging. Radiology 159:499, 1986
55. Drayer BP, Olanow W, Burger P et al: Parkinson plus syndrome: diagnosis using high field MR imaging of brain iron. Radiology 159:493, 1986
56. Grisella JF, Wexler NS, Conneally PM et al: A polymorphic DNA marker genetically linked to Huntington's disease. Nature 306:234, 1983
57. Martin JB, Gusella JF: Huntington's Disease. N Engl J Med 315:1267, 1986
58. Sax DS, O'Donnell B, Butters N et al: CT, neurologic and neuropsychological correlates of Huntington disease. Int J Neurosci 18:21, 1983
59. Sax DS, Buonanno FS: Putaminal changes in spin-echo magnetic resonance imaging signal in bradykinetic/rigid forms of Huntington's disease. Neurology 36:311, 1986
60. Terrence CF, Delaney JF, Alberts MC: CT for Huntington's disease. Neuroradiology 13:173, 1977
61. Barr AN, Heinze WJ, Dobben GD et al: Bicaudate index in CT of Huntington's disease and cerebral atrophy. Neurology 28:1196, 1978
62. Simmons JT, Pastakia B, Chase TN et al: Magnetic resonance imaging in Huntington disease. AJNR 7:25, 1986
63. Kozachuk W, Salange V, Conomy J, Smith A: MRI (magnetic resonance imaging) in Huntington's disease. Neurology 36:310, 1986
64. Kuhl DE, Phelps ME, Markham CH et al: Cerebral metabolism and atrophy in Huntington's disease determined by 18-FDG and CT scan. Ann Neurol 12:425, 1982
65. Wilson SAK: Progressive lenticular degeneration: a familial nervous disease associated with cirrhosis of the liver. Brain 34:295, 1912
66. Menkes JH: Disorders of metal metabolism. p.426. In Rowland LP (ed): Merritt's Textbook of Neurology, 7th Ed. Lea & Febiger, Philadelphia, 1984
67. Schulman S: Wilson's disease. p. 1133. In Minckler J (ed): Pathology of the Nervous System. McGraw-Hill, New York, 1968
68. Scheinberg IH, Sternlieb I: Wilson's Disease. WB Saunders, Philadelphia, 1984
69. Sternlieb I, Scheinberg IH: Penicillamine therapy for hepatolenticular degeneration. JAMA 189:748, 1964
70. Kvicala V, Vymazal J, Nevsimalova S: Computed tomography of Wilson's disease. AJNR 4:429, 1983
71. Kendall BE, Pollock SS, Bass NM, Valentine AR: Wilson's disease. Clinical correlation with cranial computed tomography. Neuroradiology 22:1, 1981
72. Williams JB, Walsche JM: Wilson's disease: an analysis of the cranial computerized tomographic appearances found in 60 patients and the changes in response to treatment with chelating agents. Brain 104:735, 1981
73. Aisen AM, Martel W, Gabrielsen TO et al: Wilson disease of the brain: MR imaging. Radiology 157:137, 1985
74. Lawler GA, Pennock JM, Steiner RE et al: Nuclear magnetic resonance (NMR) imaging in Wilson disease. J Comput Assist Tomogr 7:1, 1983
75. Starosta-Rubenstein S, Young AB, Kluin K et al: Quantitative clinical assessment of 25 Wilson's patients: correlation with structural changes on MRI. Neurology 35:175, 1985
76. Leigh D: Subacute necrotizing encephalomyelopathy in an infant. J Neurol Neurosurg Psychiatry 14:216, 1951
77. Pincus JH: Subacute necrotizing encephalomyelopa-

thy (Leigh disease): a consideration of clinical features and etiology. Dev Med Child Neurol 14:87, 1972

78. Plaitakis A, Whetsell WO Jr, Cooper JR, Yahr MD: Chronic Leigh Disease: a genetic and biochemical study. Ann Neurol 7:304, 1980
79. Sipe JC: Leigh's syndrome: the adult form of subacute necrotizing encephalomyelopathy with predilection for the brain stem. Neurology 23:1030, 1973
80. DeVivo DC: Necrotizing encephalomyelopathy and lactic adidosis. p.444. In Rowland LP (ed): Merritt's Textbook of Neurology. Lea & Febiger, Philadelphia, 1984
81. DeVivo DC, Haymond MW, Obert K et al: Defective activation of the pyruvate dehydrogenase complex in subacute necrotizing encephalomyelopathy (Leigh disease). Ann Neurol 6:483, 1979
82. Willems JL, Monnens LAH, Trijbels JMF et al: Leigh's encephalomyelopathy in a patient with cytochrome *c* oxidase deficiency in muscle tissue. Pediatrics 60:850, 1977
83. Hall K, Gardner-Medwin D: CT scan appearances in Leigh's disease (subacute necrotizing encephalomyelopathy). Neuroradiology 16:48, 1978
84. Chi JG, Yoo HW, Chang KH et al: Leigh's subacute necrotizing encephalomyelopathy: possible diagnosis by CT scan, Neuroradiology 22:141, 1981
85. Patiel H, O'Gorman AM, Meagher-Villemure K et al: CT study of subacute necrotizing encephalomyelopathy (Leigh disease). Radiology 162:115, 1987
86. Geyer CA, Sartor K, Prensky AJ et al: Leigh's disease (subacute necrotizing encephalomyelopathy): CT and MR findings in 5 cases. AJNR 7:558, 1986
87. Elejalde BR, De Elejalde MMJ, Lopez F: Hallervorden-Spatz disease. Clin Genet 16:1, 1979
88. Vakili S, Drew AL, Von Schuching S et al: Hallervorden-Spatz syndrome. Arch Neurol 34:729, 1977
89. Goldberg W. Allen N: Nonspecific accumulation of metals in the globus pallidus in Hallervorden-Spatz disease. Trans Am Neurol Assoc 104:106, 1979
90. Indravasu S, Dexter RA: Infantile neuroaxonal dystrophy and its relationship to Hallervorden-Spatz disease. Neurology 18:693, 1968
91. Dooling EC, Richardson EP Jr, Davis KR: Computed tomography in Hallervorden-Spatz disease. Neurology 30:1128, 1980
92. Johnson MA, Pennock JM, Bydder GM et al: Clinical NMR imaging of the brain in children: normal and neurologic disease. AJR 141:1005, 1983
93. Littrup PJ, Gebarski SS: MR imaging of Hallervorden-Spatz disease. J Comput Assist Tomogr 9:491, 1985
94. Heinz ER, Drayer BP, Haenggeli CA et al: Computed tomography in white matter disease. Radiology 130:371, 1979
95. McAlpine D, Lumsden CE, Acheson ED: Multiple Sclerosis: A Reappraisal. Churchill-Livingstone, London, 1972
96. Young IR, Hall AS, Pallis CA et al: Nuclear magnetic resonance imaging of the brain in multiple sclerosis. Lancet 2:1063, 1981
97. McFarlin DE, McFarland HF: Multiple sclerosis. N Engl J Med 307:1183, 1982
98. McFarlin DE, McFarland HF: Multiple sclerosis. N Engl J Med 307, 1246, 1982
99. Ellison GW (moderator): Multiple sclerosis. Ann Intern Med 101:514, 1984
100. Kurtzke JF: Epidemiologic contribution to multiple sclerosis: an overview. Neurology 30:61, 1980
101. Alter M, Kahana E, Loewenson R: Migration and risk of multiple sclerosis. Neurology 28:1089, 1978
102. Batchelor JR, Compston A, McDonald WI: The significance of the association between HLA and multiple sclerosis. Br Med Bull 34:279, 1978
103. Kurtzke JF, Hyllested K: Multiple sclerosis in the Faroe Islands: I. Clinical and epidemiological features. Ann Neurol 5:6, 1979
104. Poskanzer DC, Adams RD: Multiple sclerosis and other demyelinating diseases. p.1900. In Thorn GW, Adams RD, Braunwald E et al (eds): Harrison's Principles of Internal Medicine. 8th Ed. McGraw-Hill, New York, 1977
105. Poser CM, Paty DW, Scheinberg L et al: New diagnostic criteria for multiple sclerosis: Guidelines for research protocols. Ann Neurol 13:277, 1983
106. Schumacher GA, Beebe GW, Kibler RF et al: Problems of experimental trials of therapy in multiple sclerosis. Ann NY Acad Sci 122:552, 1965
107. Rose AS, Ellison GW, Nyers LW et al: Criteria for the clinical diagnosis of multiple sclerosis. Neurology 26:20, 1976
108. Bauer H: Concerning the diagnostic criteria for multiple sclerosis. p.555. In Bauer H, Poser R, Ritter G (eds): Progress in Multiple Sclerosis Research. Springer, New York, 1980
109. McDonald WI, Halliday AM: Diagnosis and classification of multiple sclerosis. Br Med Bull 33:4, 1977
110. Bartel DR, Markand ON, Kolar OJ: The diagnosis and classification of multiple sclerosis: evoked responses and spinal fluid electrophoresis. Neurology 33:611, 1983
111. Tramo MJ, Schneck MJ, Lee BCP et al: Evoked potentials and MRI in the diagnosis of multiple sclerosis. Neurolory 35 (Suppl 1):105, 1985
112. Andersen JT, Bradley WE: Abnormalities of detrusor and sphincter function in multiple sclerosis. Br J Urol 48:193, 1976
113. Johnson KP, Nelson BJ: Multiple sclerosis: diagnostic usefulness of cerebrospinal fluid. Ann Neurol 2:425, 1977

114. Arnason BGW: Immunology of multiple sclerosis. Clin Immunol Update 4:235, 1983
115. Cohen SR, Herndon RM, McKhann GM: Radioimmunoassay of myelin basic protein in spinal fluid: an idnex of active demyelination. N Engl J Med 295:1455, 1976
116. Speigel SM, Vinuela F, Fox AJ et al: CT of multiple sclerosis: reassessment of delayed scanning with high doses of contrast material. AJNR 6:533, 1985
117. Cala LA, Mastaglia FL, Black JL: Computerized tomography of brain and optic nerve in multiple sclerosis. J Neurol 36:411, 1978
118. Barrett L, Drayer B, Shin C: High-resolution computed tomography in multiple sclerosis. Ann Neurol 17:33, 1985
119. Vinuela FV, Fox AJ, Debrum GM et al: New perspectives in computed tomography of multiple sclerosis. AJNR 3:277, 1982
120. Hershey LA, Gado MH, Trotter JL: Computerized tomography in the diagnostic evaluation of multiple sclerosis. Ann Neurol 5:32, 1979
121. Sears ES, Tindall RSA, Zarnow H: Active multiple sclerosis. Enhanced computerized tomographic imaging of lesions and the effect of corticosteroids. Arch Neurol 35:426, 1978
122. Van der Velden, Bots AM, Endtz LJ: Cranial CT in multiple sclerosis showing a mass effect. Surg Neurol 12:307, 1979
123. Buonanno FS, Kistler JP, Lehrich JR et al: ^{1}H Nuclear magnetic resonance imaging in multiple sclerosis. Neurol Clin 1:757, 1983
124. Lukes SA, Crooks LE, Aminoff MJ et al: Nuclear magnetic resonance imaging in multiple sclerosis. Ann Neurol 13:592, 1983
125. Stewart WA, Hall LD, Berry K et al: Correlation between NMR scan and brain slice data in multiple sclerosis. Lancet 2:412, 1984
126. Jackson JA, Leake DR, Schneiders NJ et al: Magnetic resonance imaging in multiple sclerosis: results in 32 cases. AJNR 6:171, 1985
127. Bailes DR, Young IR, Thomas DJ et al: NMR imaging of the brain using spin-echo sequences. Clin Radiol 33:395, 1982
128. Runge VM, Price AC, Kirshner HS et al: Magnetic resonance imaging of multiple sclerosis: a study of pulse-technique efficacy. AJNR 5:691, 1984
129. Noseworthy JH, Strejan G, Gilbert JJ et al: Comparison of in vitro nuclear magnetic resonance (NMR) properties and histopathology in experimental allergic encephalomyelitis (EAE). Neurology 35(Suppl 1):259, 1985
130. Jacobs L, Kinkel WR, Polachini I et al: Correlations of nuclear magnetic resonance imaging, computerized tomography, and clinical profiles in multiple sclerosis. Neurology 36:27, 1986
131. Sheldon JJ, Siddharthan R, Tobias et al: MR imaging of multiple sclerosis: comparison with clinical and CT examinations in 74 patients. AJNR 6:683, 1985
132. Edwards MK, Farlow MR, Stevens JC: Multiple sclerosis: MRI and clinical correlation AJNR 7:595, 1986
133. Paty DW, Bergstrom M, Palmer M et al: A quantitative magnetic resonance image of the multiple sclerosis brain. Neurology 35(Suppl 1)137:1985
134. Crisp DT, Kleiner JE, DeFillip GJ et al: Clinical correlations with magnetic resonance imaging in multiple sclerosis. Neurology 35 (Suppl 1)137:1985
135. Ebers GC, Vinuela FV, Feasby T et al: Multifocal CT enhancement in MS. Neurology 34:341, 1984
136. Gongalez-Scarano F, Grossman RI, Galetta SL et al: Enhanced magnetic resonance images in multiple sclerosis. Neurology 36:285, 1986
137. Johnson MA, Li DKB, Bryant DJ et al: Magnetic resonance imaging: serial observations in multiple sclerosis. AJNR 5:495, 1984
138. Maravilla KR, Weinreb JC, Suss R: Magnetic resonance demonstration of multiple sclerosis plaques in the cervical cord. AJR 144:381, 1985
139. Glydensted C: Computed tomography of the brain in multiple sclerosis: a radiologic study of 110 patients with special reference to demonstration of cerebral plaques. Acta Neurol Scand 53:386, 1976
140. Brownell B, Hughes JT: The distribution of plaques in the cerebrum in multiple sclerosis. J Neurol Neurosurg Psychiatry 25:315, 1962
141. Binswanger O: Die Abgrenzung der allgemeinen progressiven Paralyse. Berl Klin Wochenschr 31:1103, 1894
142. Tomonaga BM, Yamamouchi H, Tohgi H, Kameyama M: Clinical pathologic study of progressive subcortical vascular encephalopathy (Binswanger type) of the elderly. J Am Geriatr Soc 30:524, 1982
143. Goto K, Ishii N, Fukasawa H: Diffuse white-matter disease in the geriatric population. Radiology 141:687, 1981
144. Caplan LR, Schoene WC: Clinical features of subcortical arteriosclerotic encephalopathy (Binswanger disease). Neurology 28:1206, 1978

144a. Kirkpatrick JB, Hayman LA: White matter lesions in MR imaging of clinically healthy brains of elderly subjects: possible pathologic basis. Radiology 162:509, 1987

145. Sze G, De Armond SJ, Brant-Zawadzki M et al: Foci of MRI signal (pseudo lesions) anterior to the frontal horns: histologic correlations of a normal finding. AJNR 7:381, 1986

145a. Kinkel WR, Jacobs L. Polachini I et al: Subcortical arteriosclerotic encephalopathy (Binswanger's disease). Computed tomographic, nuclear magnetic resonance, and clinical correlations. Arch Neurol 42:951, 1985

146. Kortman KE, Tsuruda JS, Price J et al: MR imaging patterns in multiple sclerosis and white matter infarction. Radiology 157:290, 1985

146a. Simon JH, Holtas SL, Schiffer RB et al: Corpus callosum and subcallosal periventricular lesions in multiple sclerosis: detection with MR. Radiology 160:363, 1986

147. Astrom KE, Mancall EL, Richardson EP Jr: Progressive multifocal leukoencephalopathy: a hitherto unrecognized complication of chronic lymphatic leukemia and Hodgkin's disease. Brain 81:93, 1958

148. Krupp LB, Lipton RB, Swerdlow ML et al: Progressive multifocal leukoencephalopathy: clinical and radiolographic features. Ann Neurol 17:344, 1985

149. Norkin LC: Papovaviral persistent infections. Microbiol Rev 46:384, 1982

150. Richardson EP, Webster H: Progressive multifocal leukoencephalopathy: its pathological features. Prog Clin Biol Res 105:183, 1983

151. Preskorn SH, Watanabe I: Progressive multifocal leukoencephalopathy: cerebral mass lesions. Surg Neurol 12:231, 1979

152. Carroll BA, Lane B, Norman D et al: Diagnosis of progressive multifocal leukoencephalopathy by computed tomography. Radiology 122:137, 1977

153. Adams RD, Victor M, Mancall EL: Central pontine myelinolysis: a hitherto undescribed disease occurring in alcoholic and malnourished patients. Arch Neurol Psychiatry 81:154, 1959

154. Wright DG, Laureno R, Victor M: Pontine and extrapontine myelinolysis. Brain 102:361, 1979

155. Messert B, Orrison WW, Hawkins MJ, Quaglieri CE: Central pontine myelinolysis. Neurology 29:147, 1979

156. Shurtliff CF, Ajax ET: Central pontine myelinolysis and cirrhosis of the liver. Am J Clin Pathol 46:239, 1966

157. McCormick WF, Dannell CM: Central pontine myelinolysis. Arch Int Med 119:444, 1967

158. Goebel HH, Herman-Ben Zur P: Central pontine myelinolysis: a clinical and pathological study of 10 cases. Brain 95:495, 1972

159. Endo Y: Central pontine myelinolysis: a study of 37 cases in 1,000 consecutive autopsies. Acta Neuropathol 53:145, 1981

160. Kleinschmidt-Demasters BK, Norenberg MD: Rapid correction of hyponatremia causes demyelination: relation to central pontine myelinolysis. Science 211:1068, 1971

161. Laureno R: Experimental pontine and extrapontine myelinolysis. Trans Am Neurol Assoc 105:354, 1980

162. Holliday MA: Factors that limit brain volume changes in response to acute and sustained hyper- and hyponatremia. J Clin Invest 47:1916, 1968

163. Laureno R. Central pontine myelinolysis following rapid correction of hyponatremia. Ann Neurol 13:232, 1983

164. Norenberg MD, Leslie KO, Robertson AS: Association between rise in serum sodium and central pontine myelinolysis. Ann Neurol 11:128, 1982

165. Thompson DS, Hutton JT, Stears JC, et al: Computerized tomography in the diagnosis of central and extrapontine myelinolysis. Arch Neurol 38:243, 1981

166. Hazratji SMA, Kim RC, Marasigan AV: Evolution of pontine and extrapontine myelinolysis. J Comput Assist Tomogr 7:356, 1983

167. Gerber O, Geller M, Stiller T, et al: Central pontine myelinolysis: resolution shown by computed tomography. Arch Neurol 40:116, 1983

168. DeWitt LD, Buonanno FS, Kistler JP, et al: Central pontine myelinolysis: demonstration by nuclear magnetic resonance. Neurology 34:570, 1984

169. Takeda K, Sakuta M, Saeki F: Central pontine myelinolysis diagnosed by magnetic resonance imaging. Ann Neurol 17:310, 1985

170. Bydder GM: Magnetic resonance imaging of the posterior fossa, p. 5. In Kressel HY (ed): Magnetic Resonance Annual 1985. New York, Raven Press, 1985

171. Aubourg P, Diebler C: Adrenoleukodystrophy—its diverse CT appearances and an evolute or phenotypic variant. Neuroradiology 24:33, 1982

172. Schaumburg HH, Powers JM, Raine CS et al: Adrenoleukodystrophy. A clinical and pathological study of 17 cases. Arch Neurol 32:577, 1975

173. Griffin JW, Goren E. Schaumburg H et al: Adrenomyeloneuropathy: a probable variant of adrenoleukodystrophy. I. Clinical and endocrinologic aspects. Neurology 27:1107, 1977

174. O'Neill BP, Marmion LC, Feringa ER: The adrenoleukomyeloneuropathy complex: expression in four generations. Neurology 31:151, 1981

175. Singh I, Moser AB, Moser HW et al: Adrenoleukodystrophy: impaired oxidation of very long chain fatty acids in white blood cells, cultured skin fibroblasts, and amniocytes. Pediatr Res 18:286, 1984

176. Molzer B, Bernheimer H, Heller R et al: Detection of adrenoleukodystrophy by increased C26:0 fatty acid levels in leukocytes. Clin Chem Acta 125:299, 1982

177. Powers JM, Schaumburg HH: The adrenal cortex in adrenoleukodystrophy. Arch Pathol 96:305, 1973

178. Powell H, Tindall R, Schultz P et al: Adrenoleukodystrophy: electron microscopic findings. Arch Neurol 32:250, 1975

179. Fernandez RE, Kishore PRS: White matter disease of the brain. P.659. In Lee SH, Rao CKVG (eds): Cranial Computed Tomography. McGraw-Hill, New York, 1983

180. Quisling RG, Andriola MR: Computed tomographic

evaluation of the early phase of adrenoleukodystrophy. Neuroradiology 17:285, 1979

181. Greenberg HS, Halverson D, Lane B: CT scanning and diagnosis of adrenoleukodystrophy. Neurology 27:884, 1977
182. Lane B, Carroll BA, Pedley TA: Computerized cranial tomography in cerebral diseases of white matter. Neurology 28:534, 1978
183. Young IR, Randell CP, Kaplan PW et al: Nuclear magnetic resonance (NMR) imaging in white matter disease of the brain using spin-echo sequences. J Comput Assist Tomogr 7:290, 1983
184. Bewermeyer H, Bamborschke S, Ebhardt G et al: MR imaging in adrenoleukomyeloneuropathy. J Comput Assist Tomogr 9:793, 1985
185. O'Neill BP, Forbes GS, Gomez MR et al: A comparison of magnetic resonance imaging (MRI) and computed tomography (CT) in adrenoleukodystrophy. Neurology 35:83, 1985
186. Austin J: Metachromatic leukodystrophy. p.418. In Rowland JP (ed): Merritt's Textbook of Neurology. Lea & Febiger, Philadelphia, 1984
187. Dulaney J, Moser H: Sulfatide lipidosis: metachromatic leukodystrophy. p.770. In Stanbury J, Wyngaarden L, Fredrickson D (eds): Metabolic Basis of Inherited Disease. 4th Ed. McGraw-Hill, New York, 1978
188. Carlin L, Roach ES, Riela A et al: Juvenile metachromatic leukodystrophy: evoked potentials and computed tomography. Ann Neurol 13:105, 1983
189. MacFaul R, Cavanagh N, Lake B et al: Metachromatic leuko-dystrophy: review of 38 cases. Arch Dis Child 57:168, 1982
190. Buonanno F, Ball M, Laster DW et al: Computed tomography in late infantile metachromatic leukodystrophy. Ann Neurol 4:43, 1978
191. Suzuki K, Suzuki Y: Globoid leukodystrophy (Krabbe disease) deficiency of galactocerebroside-beta-galactosidase. Proc Natl Acad Sci (USA) 66:302, 1970
192. Svennerholm L, Vanier MT, Mansson JG: Krabbe disease: a galactosylsphingosine (psychosine) lipidosis. J Lipid Res 21:53, 1980
193. Hagberg B, Kollberg H, Sourander P et al: Infantile globoid cell leukodystrophy (Krabbe's disease): a clinical, morphological and genetical study of 32 Swedish cases. Neuropadiatre 1:74, 1970
194. Malone MJ, Szoke MC, Looney GL: Globoid leukodystrophy: clinical and enzymatic studies. Arch Neurol 32:606, 1975
195. Cromel, Stern J: Inborn lysosomal enzyme deficiencies. p.541. In Blackwood W. Corsellis JAN (eds): Greenfield's Neuropathology. Year Book Medical Publishers, Chicago, 1976
196. Barnes DM, Enzmann DR: The evolution of white matter disease as seen on computed tomography. Radiology 138:379, 1981
197. Ieshima A, Eda S, Matsui A et al: Computed tomography in Krabbe's disease: comparison with neuropathology. Neuroradiology 25:323, 1983
198. Kwan E, Drace J, Enzmann DR: Specific CT findings in Krabbe's disease. AJR 143:665, 1984
199. Baram TZ, Goldman AM, Percy AK: Krabbe disease: specific MRI and CT findings. Neurology 36:111, 1986
200. Lane B, Carroll BA, Pedley TA: Computerized cranial tomography in cerebral diseases of white matter. Neurology 28:534, 1978

10

Hemorrhage, Trauma, and Therapeutic Change

The significant contribution made by CT to the diagnosis and management of patients with intracranial hemorrhage, trauma, or therapy needs no elaboration. Early reports suggested that MRI could be quite useful in similar clinical settings.[1,2] However, it was soon recognized that acute hemorrhage, especially the subarachnoid type, was difficult or impossible to detect on routine MR images.[3,4] This observation, together with MR's well known failure to image bone detail raised serious questions about the role of MRI for evaluating acutely ill or traumatized patients.

With the passage of time and with accrued MR experience, a more optimistic role is seen for MR in patients who have experienced trauma, intracranial hemorrhage, surgery, or radiotherapy. MR can provide significant information about trauma patients, particularly those in the subacute and chronic stages.[5–8] The varied MR appearances of cerebral hemorrhage and their physical bases are just now beginning to be appreciated.[3,4,9–12] It is conceivable that such data may have important prognostic and therapeutic implications that are complimentary to information obtained by CT.

This chapter will initially focus upon intracranial hemorrhage, including parenchymal, subarachnoid, and extra-axial types. Considerable attention will be directed to the physical and chemical bases responsible for the observed MR signals within and around a hematoma. Next the discussion will turn to trauma with some tentative judgments made concerning the roles of CT and MR imaging in this setting. The chapter will conclude by showing examples of various postoperative and post-therapeutic changes of the brain and skull that may be encountered in MRI. Effects of radiation therapy and chemotherapy for cancer will also be considered at that time.

INTRACRANIAL HEMORRHAGE

Physical Properties of Hemoglobin and Its Degradation Products

In order to understand the variable appearance of hemorrhage on MR images, the structure of hemoglobin and its breakdown products must be considered in some detail. Circulating hemoglobin alternates between oxy and deoxy forms depending on its state of oxygen saturation. To bind oxygen, the iron in the hemoglobin must be maintained in the reduced ferrous (Fe^{+2}) state.[13] Both oxyhemoglobin and deoxyhemoglobin contain iron in this state and may bind oxygen reversibly. A number of metabolic pathways in the RBC exist to maintain heme iron in the ferrous state. When these pathways fail (e.g., during the evolution of a hemorrhage or following red cell lysis) hemoglobin is converted to one of several inactive forms. These inactive forms include methemoglobin, hemichromes, hemosiderin, and ferritin. Iron in these inactive forms exists in the ferric (Fe^{+3}) state.

Heme iron normally resides suspended in a crevice in the center of the hemoglobin molecule. A free coordination site allows for reversible attachment of an oxygen molecule. In the deoxygenated state heme iron in the ferrous state has four unpaired electrons and is strongly paramagnetic.[13] When oxygen attaches, however, electron sharing occurs resulting in conversion of the complex to a low spin form. Accordingly while deoxyhemoglobin is strongly paramagnetic, oxyhemoglobin possesses no particular paramagnetic properties.[14–16]

The presence of deoxyhemoglobin in an intracerebral hematoma results in a marked decrease in T2-relaxation time.[9] This effect is known as *proton relaxation enhancement*, and relates to the paramagnetic properties of the deoxyhemoglobin molecules. This T2-relaxation enhancement effect is proportional to the square of the magnetic field, to the square of the deoxyhemoglobin concentration, and inversely to a function of the hematocrit.[17] This T2-relaxation enhancement phenomenon probably explains the low intensity from the center of an intracerebral hematoma seen during the first few days on T2-weighted images.[9]

Although deoxyhemoglobin is strongly paramagnetic in intact RBCs or in the center of a hematoma, such properties do not occur in aqueous solution. Oxy- and deoxyhemoglobin in aqueous solution (such as in the CSF) exhibit no differences in their proton paramagnetic relaxation effects.[18–20] This observation helps explain why acute subarachnoid hemorrhage is so difficult to detect on MRI—oxy- and deoxyhemoglobin, the predominant forms in hyperacute subarachnoid hemorrhage, cannot sufficiently alter relaxation times of CSF to insure reliable detection.[21]

With RBC metabolic failure and lysis (as occurs in an aging hematoma), the hemoglobin molecule begins to undergo oxidative denaturation. Ferrous heme iron is oxidized to the ferric state and methemoglobin is formed. The coordination site previously occupied by oxygen in oxyhemoglobin is now filled by water or hydroxylate, depending on pH. Methemoglobin both in intact RBCs and in aqueous solution possesses significant paramagnetic properties. The presence of small quantities of methemoglobin result in marked shortening of T1 with little or no effect on T2.[20,21] These relaxation effects of methemoglobin help explain certain MR signal changes in aging hematomas and in subarachnoid hemorrhage.

With continued oxidative denaturation, heme iron maintains its ferric state but progressive alteration of the structure of the globin molecule occurs. Methemoglobin is thus converted into a number of derivatives known as hemichromes.[15] During phagocytosis iron is stripped from the hemichrome complexes and

Table 10-1. Forms of Hemoglobin and Its Degredation Products

Oxyhemoglobin	Iron in ferrous (Fe^{+2}) state Carries oxygen No significant paramagnetic effects
Deoxyhemoglobin	Iron in ferrous state Significant paramagnetic relaxation of T2 in intact RBC
Methemoglobin	Iron in ferric (Fe^{+3}) state Significant paramagnetic relaxation of T1 especially in aqueous solution
Hemichromes	Iron in ferric (Fe^{+3}) state Changes in tertiary structure of globin molecule
Hemosiderin	Ferric iron stripped from globin complex
Ferritin	Stored promarily in macrophages Significant parallel paramagnetic relaxation of T1 and T2

Table 10-2. Effect of Altered Blood Components on T1 and T2 in Vitro[a]

Blood Component Alteration	*Effect on T1*	*Effect on T2*
Increased hematocrit	Linear decrease	Linear decrease
Lysis of oxygenated RBCs	No change	No change
Decreased serum pH	Increased	Increased
Decreased serum osmolarity (to produce swollen but intact RBC's)	Increased	Increased
Deoxygenation of intact RBC	Slight increase	Marked decrease
Deoxygenation of lysed RBC	Slight increase	Slight decrease or no change
Methemoglobin formation	Marked decrease	No change

[a] Data compiled from references 14–26.

is stored in macrophages as hemosiderin or ferritin. Both ferritin and hemosiderin exist primarily with iron in the Fe^{+3} states.[13] They produce proton relaxation enhancement by proton-electron dipole-dipole interactions, resulting in parallel shortening of T1 and T2.[9] Because iron-laden macrophages accumulate at the margin of a hematoma, these paramagnetic effects of hemosiderin and ferritin may explain the low intensity "iron ring" seen in the MRI of some resolving intracerebral hemorrhages.[9]

A summary of these metabolic pathways of hemoglobin and effects of these hemoglobin derivatives on T1 and T2 are presented in Tables 10-1 and 10-2.

Critical Analysis of the Existing Literature

Considerable confusion and contradictory findings exist in the literature concerning the MR imaging characteristics of hemorrhage. The conflicts and paradoxes that exist in this volume of data arise from several sources.

Clinical Reporting

In many examples reported in the literature there is difficulty in knowing the precise age of the hemorrhage. Furthermore, various authors use different definitions of acute and chronic. Pathologic confirmation is seldom available in reported cases. Failure to differentiate true hematoma from hemorrhagic tissue may also cause confusion.

Physical Characteristics of the Clot

A number of physical and chemical changes take place in an evolving hematoma that may occur at different rates in various sites throughout the brain. For example, fresh hematoma may clot rapidly, slowly, or not at all. Clot retraction, red cell lysis, serum production, and liquefaction occur at different rates. Adjacent tissue pH, oxygenation, and edema effect the evolution of hematomas as well. Mixing of clot with CSF may have significant effects on proton relaxation times.

Different Scanning Methods

Magnetic resonance studies of hematoma evolution have been performed on a variety of different scanners at different field strengths. Pulse sequences frequently vary between institutions. For example, what are commonly accepted as T1-weighted pulse sequences often have considerable T2-weighting as well. Actual measurement of tissue T1 and T2 are also quite methodology-dependent.

Bias in Experimental Models

A number of experimental models for studying the evolution of hemorrhage on MRI have been created. Each may be biased in various ways, however. Such biases, which may introduce errors include the age of the blood, whether the clot is isolated from air, ambient temperature, whether the clot was agitated prior to scanning, and whether methemoglobin formation occurred naturally or was induced by exogenous chemicals.

In summary, hematomas are complex mixtures of whole clot, lysing clot, plasma, and serum influenced in varying ways by adjacent normal or diseased tissue. The physical and chemical properties of this mixture change constantly. Although we may make general conclusions regarding MR imaging of hematomas, we should not be surprised when an individual case shows significant deviance from what we may call a classic pattern of evolution.

Intracerebral Hematomas (ICH)

While much controversy still remains, a clearer understanding of the MR signal changes in ICH has now emerged. While some of the imaging findings reported here may not be applicable for very low field strength scanners (i.e., those with fields <0.2 T), the majority of findings to be described are applicable for all magnets. The conclusions drawn herein are from our own data and from several of the more credible reports available in the literature.[4,9,12] In summary these conclusions are presented in Figure 10-1.

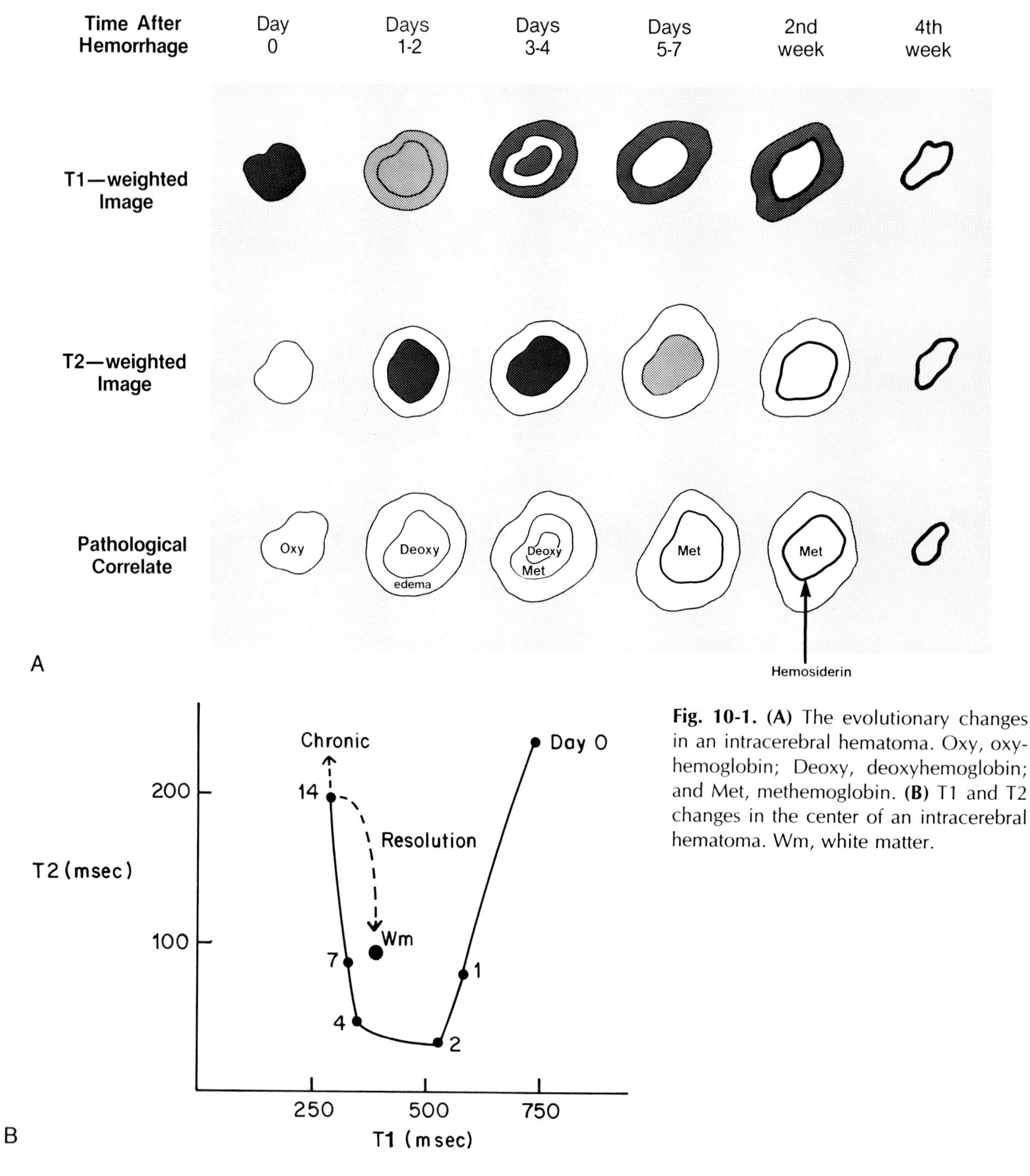

Fig. 10-1. (A) The evolutionary changes in an intracerebral hematoma. Oxy, oxyhemoglobin; Deoxy, deoxyhemoglobin; and Met, methemoglobin. **(B)** T1 and T2 changes in the center of an intracerebral hematoma. Wm, white matter.

Acute blood clot has long T1 and T2 values.[11,12] Therefore, immediately after an ICH forms one would expect to see a high intensity signal on T2-weighted images (T2WI) and a low signal on T1-weighted images (T1WI). However, very few patients are examined by MRI in the first 3 or 4 hours after ICH.[12a] Nearly all cases of clinically "acute" ICH published in the literature were scanned 12 to 72 hours posthemorrhage. However, even within the first 24 hours, significant changes in T1 and T2 are occurring. It is not surprising, therefore, that a variety of appearances of acute hemorrhage on MRI have been reported in the literature.

Within the first 24 to 48 hours following ICH a

dramatic shortening of T2 occurs in the center of the hematoma.[9,12] It has been postulated that this phenomenon is due to high deoxyhemoglobin concentration in intact erythrocytes, which shortens T2 by a paramagnetic relaxation enhancement.[9] This effect is more marked at high fields, but may be seen to some degree in lower strength scanners (Fig. 10-2). T1 during this period remains prolonged relative to that of white matter, but is shortened compared to its high value in acute clot. At some stage of hematoma evolution in the first day, T1 and T2 effects may balance so that the hematoma appears isodense to brain on some pulse sequences. However, it is unlikely that isodensity will remain on all images when several different pulse sequences and timing parameters are used. Vasogenic edema commonly occurs in the white matter surrounding the hematoma for the first 2 weeks.[27] This may be observed as a region of increased intensity on T2WI (Fig. 10-3).

Between about days 2 and 5 posthemorrhage, several changes occur in and around the hematoma. These changes occur from the periphery of the hematoma inward as deoxyhemoglobin is denatured into methemoglobin. The formation of methemoglobin is accompanied by a decrease in T1, while T2 remains low or increases slightly. As a result, the ICH will appear to fill in with high signal on T1WI from the periphery inward. The central region will remain low intensity on all imaging sequences (Fig. 10-3).

Simultaneous with this central filling in of the hematoma signal, changes are occurring in the brain immediately adjacent to the hematoma. With the peripheral breakdown of hemoglobin into ferric derivatives such as hemosiderin, an "iron rim" is deposited in the surrounding brain parenchyma.[28] This hemosiderin deposition results in marked shortening of T2 with little change in T1. As a result, an iron rim of low intensity on T2WIs may often be observed in

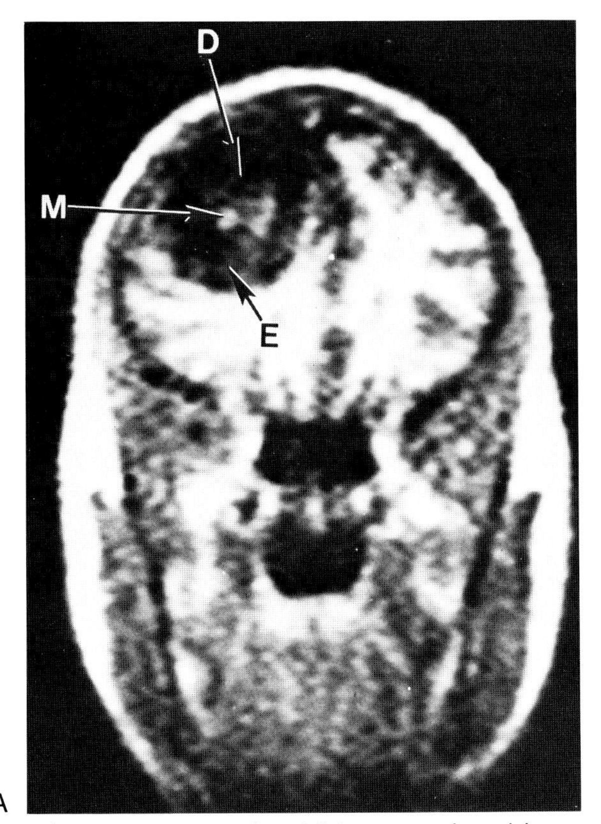

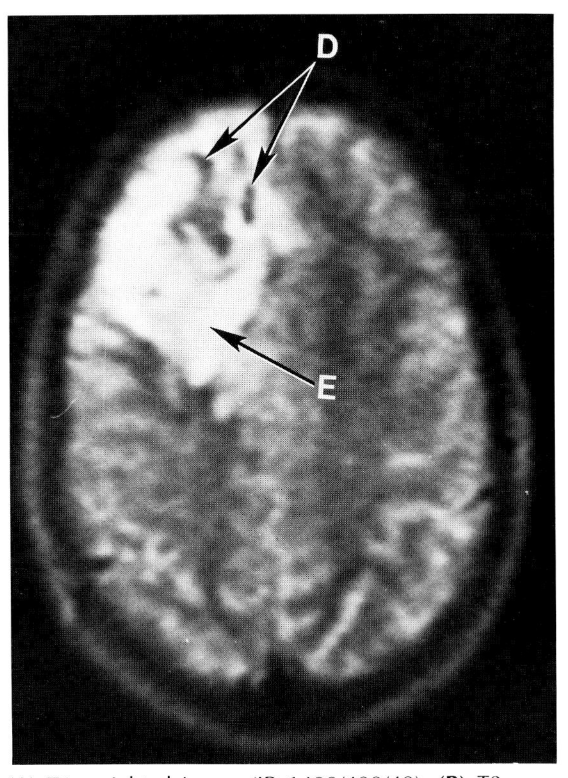

Fig. 10-2. Acute (2-day-old) intraparenchymal hematoma. **(A)** T1-weighted image (IR 1400/400/40). **(B)** T2-weighted image (SE 3000/100). E, edema; D, deoxyhemoglobin; M, methemoglobin.

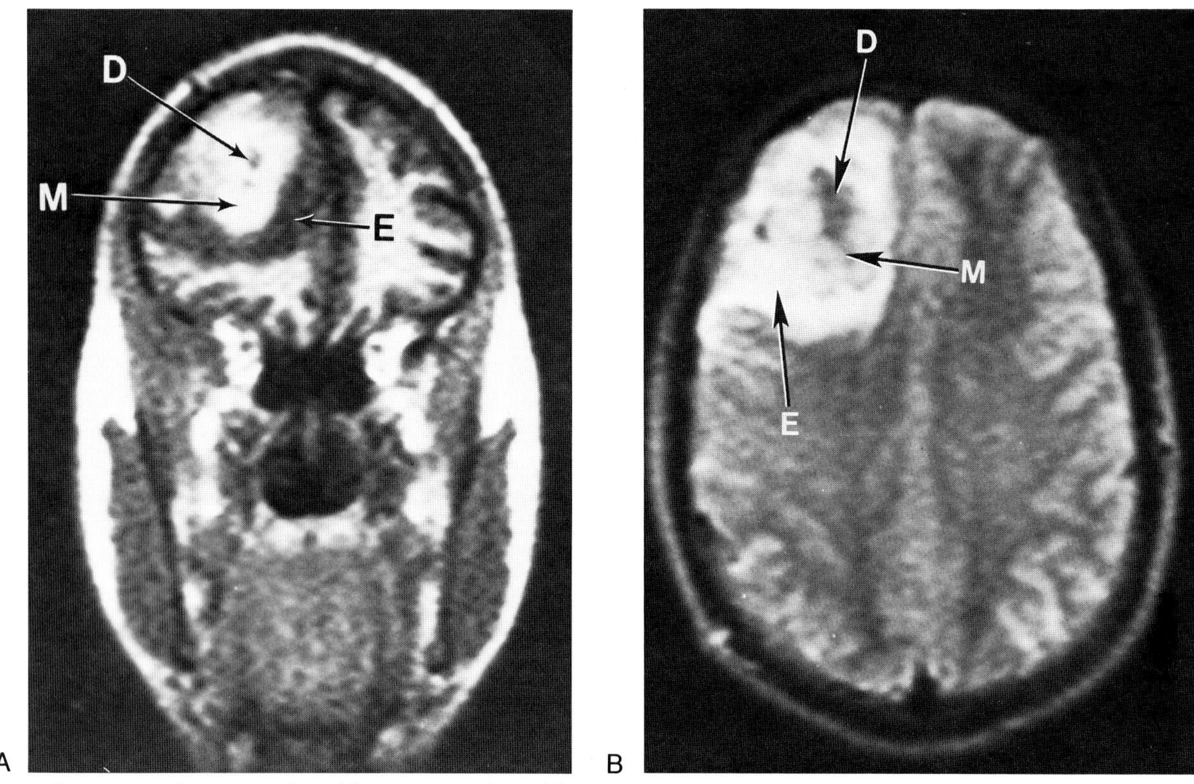

Fig. 10-3. Subacute (5-day-old) intraparenchymal hematoma **(A)** T1-weighted image (IR 1400/400/40). **(B)** T2-weighted image (SE 3000/100). Only a small region of deoxyhemoglobin (*D*) present centrally. The bulk of the hematoma is now composed of methemoglobin (*M*). Peripheral edema (*E*) is still noted.

the brain parenchyma surrounding a resolving hematoma in the subacute and chronic stages. This phenomenon is more marked with high filed strength scanners but can still be observed at all fields (except possibly those <0.15 T). This rim of hemosiderin laden macrophages may persist indefinitely, even after the ICH has totally resolved, serving as an MR marker of distant hemorrhages (Fig. 10-4).

By the end of the first week posthemorrhage the entire hematoma has generally filled in completely and possesses high signal on T1WI's (Fig. 10-5). The T2 relaxation time in the hematoma has gradually increased due to increasing liquefaction and complete disappearance of the deoxyhemoglobin moiety. Between about 6 and 12 days the T2 of the hematoma may be similar to that of white matter, making the ICH isointense on T2WIs. Peripheral vasogenic edema of the white matter may persist, but is usually gone by the end of the second week.[27] Leakage at the blood-brain barrier may occur in the first several days and last 2 to 3 months.[29] It is expected that a similar process should occur with Gd-DTPA, when this becomes available. Steroid administration may alter this contrast enhancement in the early but not later stages.[29]

The MR appearance of the chronic phase of ICHs depends upon what happens to the hematoma. In the usual case liquefaction occurs with gradual resorption of fluid over several weeks to months. A residual orange-walled slit lined with hemosiderin-ladened macrophages is often the only late persisting evidence for old ICH. MRI here may reveal nothing or an area of decreased T1 due to the hemosiderin deposits. Sometimes, however, the area of hematoma may remain as a fluid filled intraparenchymal cyst. This is particularly true if hemorrhage has occurred into a brain laceration or cystic tumor. When an ICH persists as a chronic fluid filled cavity, T2 remains pro-

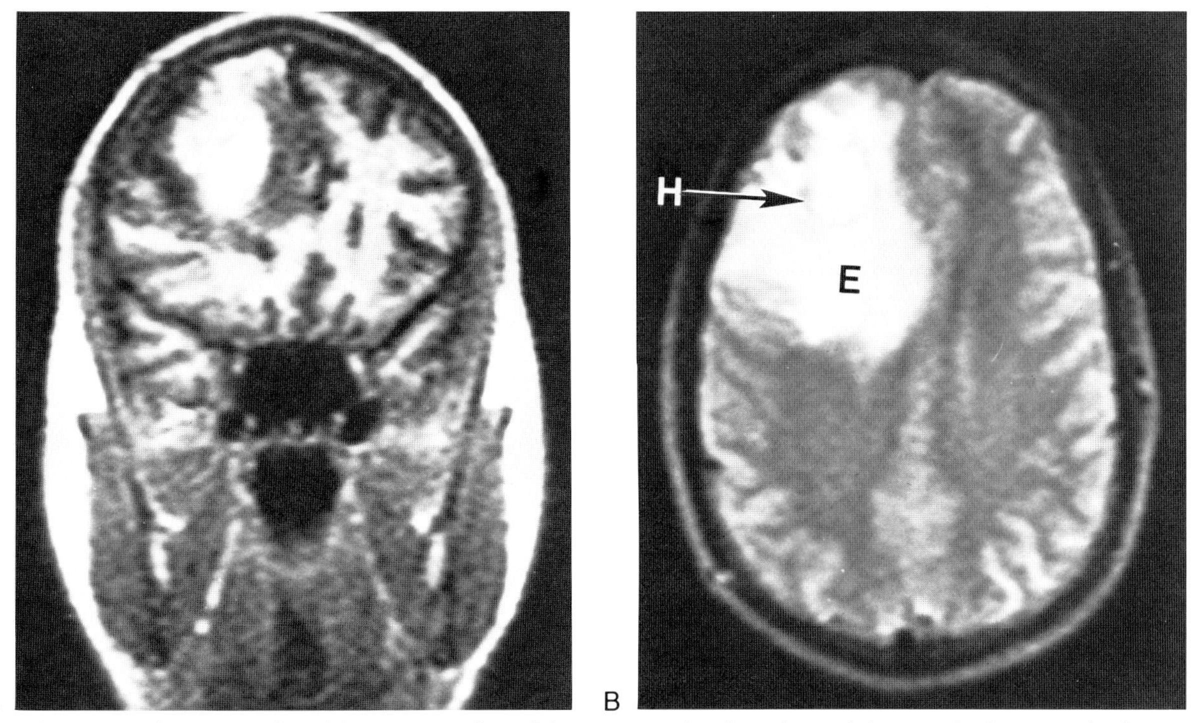

Fig. 10-4. Chronic (10-day-old) intraparenchymal hematoma. The deoxyhemoglobin moeity has completely disappeared, being replaced by methemoglobin and other hemoglobin degradation products. An early hemosiderin (*H*) ring ("iron rim") is noted on the T2-weighted image. **(A)** IR 1400/400/40. **(B)** SE 3000/100.

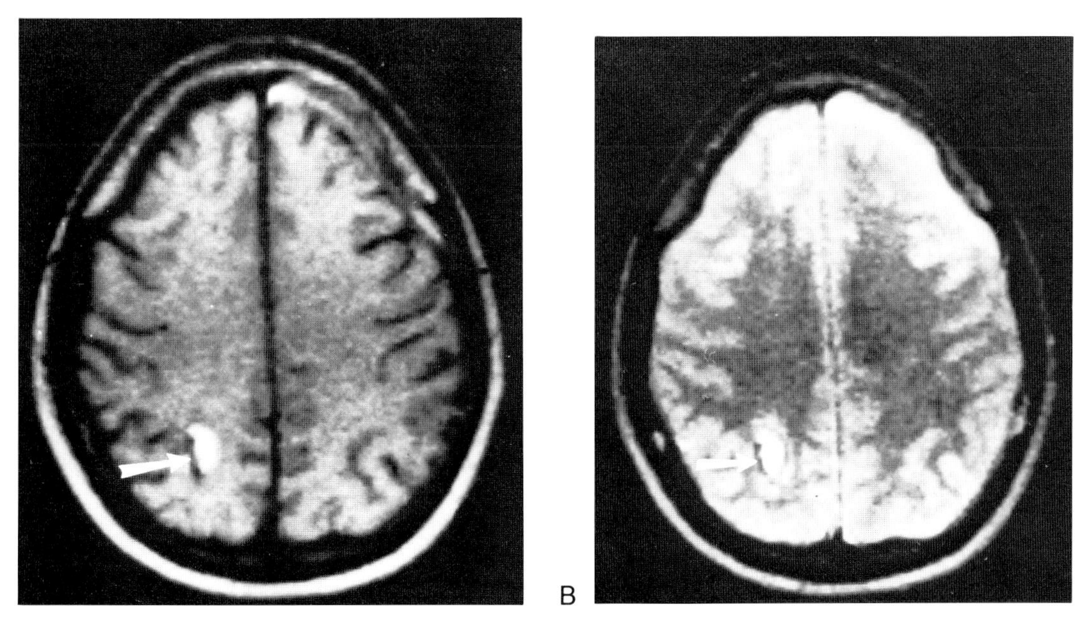

Fig. 10-5. Two-week-old intraparenchymal hematoma demonstrating hemosiderin rim (arrow) on T1 and T2 weighted sequences. **(A)** SE 500/28. **(B)** SE 2000/28.

longed. T1 may remain slightly shortened for some time, but will often gradually lengthen so that the collection becomes more like water in signal characteristics with the passing of time. Peripheral calcifications in the wall of the old hematoma cavity may be noted as areas of signal drop out only if they are very large.

Epidural Hematoma

An epidural hematoma (EDH) is an extra-axial collection of blood that most frequently develops following laceration of the middle meningeal artery or one of its branches.[30] Occasionally the bleeding may originate from a meningeal vein or venous sinus. An associated skull fracture is present in about 85 percent of cases.[31] In the usual case blood under high pressure from meningeal artery laceration dissects the dura from the inner table of the skull. Because this outer layer of dura is fused with the periosteum of the inner table, an EDH may be thought of as a subperiosteal hemorrhage.

The classic clinical presentation of an EDH involves (1) initial loss of consciousness due to concussion, (2) a 4 to 6 hour lucid internal where consciousness is regained but EDH is developing, and (3) rapid neurologic deterioration and coma, when the EDH reaches a critical size for compressive damage to the brain. However, many patients present with severe symptoms immediately. Chronic and delayed EDH's may also develop remote from trauma, though they are rare.

The classic appearance of an EDH on CT is well recognized.[30] A collection of high density (40 to 90 HU) blood assuming a biconvex (lentiform) shape is seen adjacent to the inner table, usually in the frontal, parietal, or temporal regions. Compression of underlying brain parenchyma is commonly noted. Because the dura is more strongly attached to the calvarium at the sutures, EDHs usually do not spontaneously dissect across a suture line unless a fracture also extends across that line.

These morphologic criteria used on CT may be applied to the diagnosis of EDH by MRI. Acute EDH has long T1 and short T2 similar to freshly clotted blood (Fig. 10-6).[7] Subacute and chronic EDHs may have a short T1 due to the presence of methemoglobin and other ferric degradation products (Fig. 10-7). The

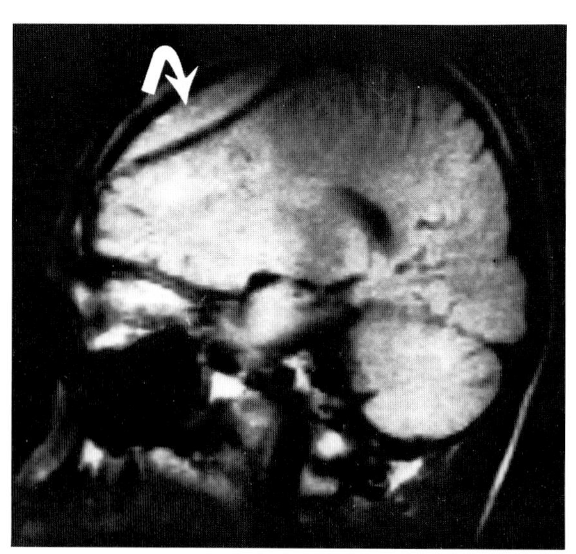

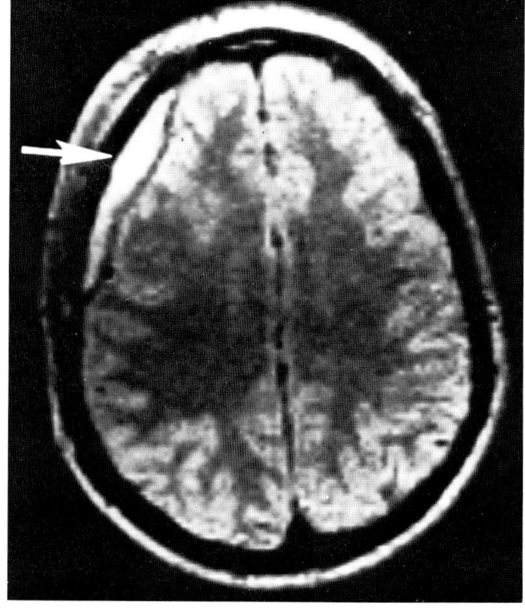

Fig. 10-6. Subacute (3-day-old) epidural hematoma that developed postoperatively. **(A)** T1-weighted image. **(B)** T2-weighted image.

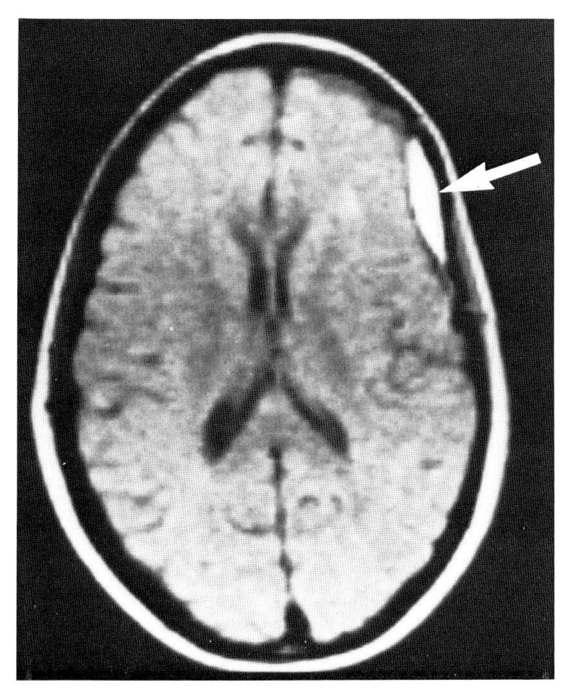

Fig. 10-7. Six-day-old epidural hematoma (arrow) shows shortened T1, presumably due to methemoglobin formation.

evolution of EDH signal by MRI has not been studied prospectively because most cases are evacuated surgically at the time of diagnosis. CT and MRI seem largely equivalent in their abilities to diagnose EDH. However, MR may be superior to CT in the diagnosis of subtemporal and high convexity hematomas.[7]

Subdural Hematoma

A subdural hematoma (SDH) most commonly results from tearing of bridging veins that traverse the space between the cerebral cortex and venous sinuses.[30] Other etiologies include direct rupture of arteries or veins at the site of brain contusion or tears in small cortical arterial branches. While often a sequela of severe trauma, SDHs may also develop after minor trauma or spontaneously. This latter situation is especially encountered in patients with bleeding diatheses, on steroids, on dialysis, in infants with hydrocephalus, and in the elderly with atrophy (Fig. 10-8). While considerable attention is frequently drawn to the differentiation of SDH and EDH, it should be noted that blood is found in both spaces at surgery or autopsy in up to 20 percent of cases.[31]

Several morphologic features serve to distinguish SDH and EDH.[30] Because blood more easily dissects in the subdural space, an SDH is usually more diffuse than an EDH. While an EDH is usually biconvex, an SDH is typically crescenic. It should be noted, however, that chronic SDHs may have a biconvex shape and resemble acute EDHs.

The age of an SDH may be estimated by its CT attenuation value. In one study 100 percent of acute (<7-day-old) SDHs were hyperdense compared to brain; 70 percent of subacute (7 to 21 days old) hematomas were isodense; and 76 percent of chronic (>21-day-old) SDHs were isodense.[34] It should be noted, however, that anemic patients may have hypodense or isodense lesions acutely.[35] Rebleeding in chronic hematomas may alter their appearance as well. While considerable attention in the early days of CT was drawn to the problems in identifying isodense SDHs, with high resolution scanners today, there should seldom be any such problem.[30]

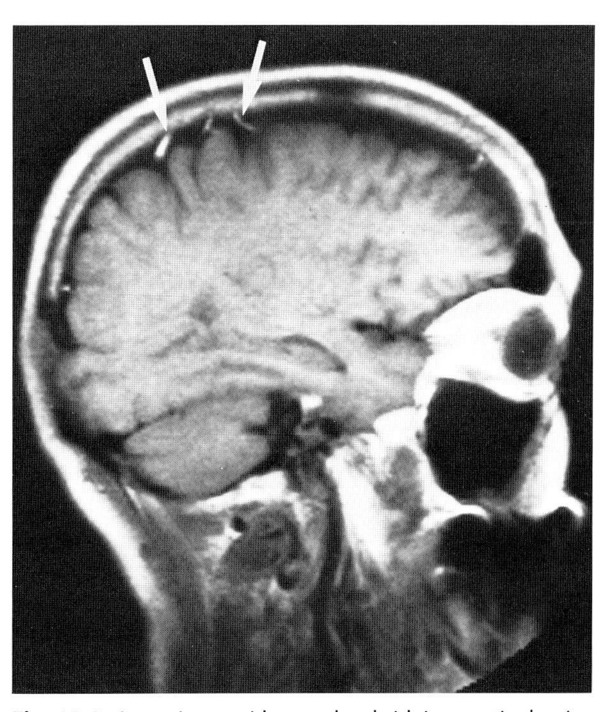

Fig. 10-8. In patients with atrophy, bridging cortical veins (arrows) are at risk for shearing injury that may result in a subdural hematoma.

A number of investigators have observed SDHs my MRI at various stages of evolution.[3,7–9,36–38] In general, the evolution of MR signal in an SDH is similar to that seen in the center of an ICH.[9,36] However, mixed density within an SDH is not uncommon, making for varied appearances that depart from the classic scheme.[39] Mixed composition of an SDH may result from unclotted blood, extruded serum, or leakage of CSF into the subdural space from an arachnoid tear. Admixture of CSF in the hematoma diminishes the T2-shortening effect of deoxyhemoglobin.[21] This could conceivably alter the appearance on MRI. Absorption of water across the membrane into chronic SDHs may cause both T1 and T2 values to be elevated. Examples of SDH at varying stages on MRI are seen in Figures 10-9 and 10-10.

Subarachnoid Hemorrhage

It has long been recognized that acute subarachnoid hemorrhage (SAH) is difficult or impossible to detect on routine MR images.[3,4,7] As such, MR has taken a somewhat secondary role in the initial evaluation of acutely ill or traumatized patients. Recent research has begun to clarify at which stages SAH may be identified by MR, and has better defined the role MR may play in patients with suspected SAH.[21,26]

When blood of increasing concentration is mixed with CSF in vitro, shortening of T1 and T2 that is directly related to the amount of blood added occurs.[26] A simultaneous linear increase in CT attenuation has long been recognized in the same experimental setting.[40] Formation of high density "microclots" in clinical SAH may explain the rather easy detectability of such hemorrhages by CT. With standard MR imaging techniques, however, the relatively subtle alterations in T1 and T2 relaxation times induced by acute SAH usually do not produce visually recognizable signal changes. Although at least one 17-hour-old SAH has been demonstrated by MRI, this is clearly the exception, not the rule.[21]

Whereas acute changes in MR signal at the center of an intraparenchymal hematoma may be explained by conversion of oxyhemoglobin to deoxyhemoglobin, such an effect does not occur in aqueous solution.[20] Methemoglobin formation, however, results in significant T1 shortening in aqueous solution.[19] This phenomenon is thought to be due to a combination of "inner sphere" and "outer sphere" effects. (Inner sphere effects refer to the relaxation that occurs be-

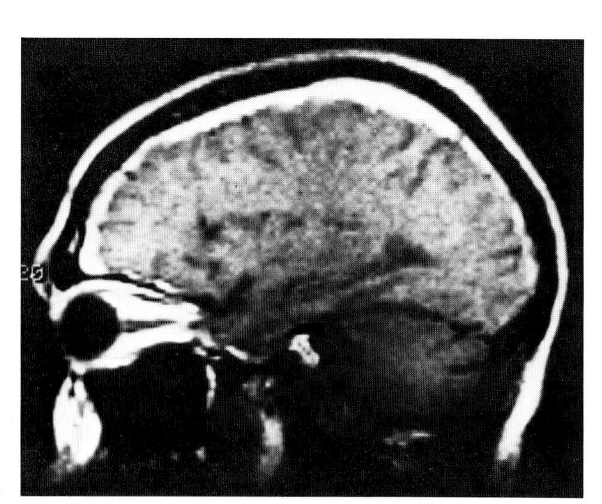
A

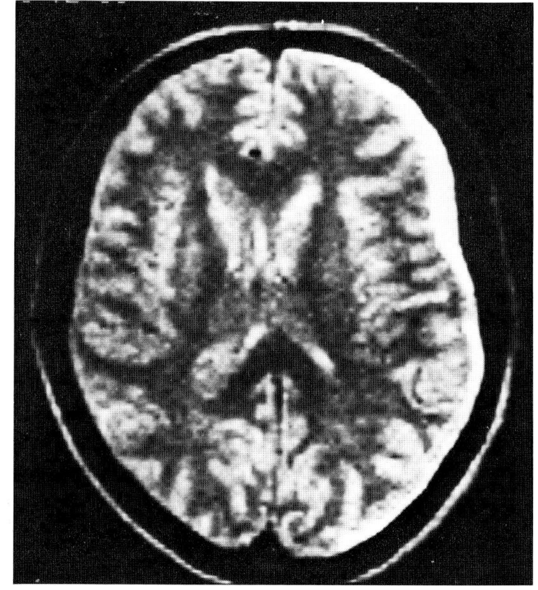
B

Fig. 10-9. Twelve-day-old subdural hematoma illustrates short T1 and long T2 values. **(A)** SE 500/28. **(B)** SE 2000/56.

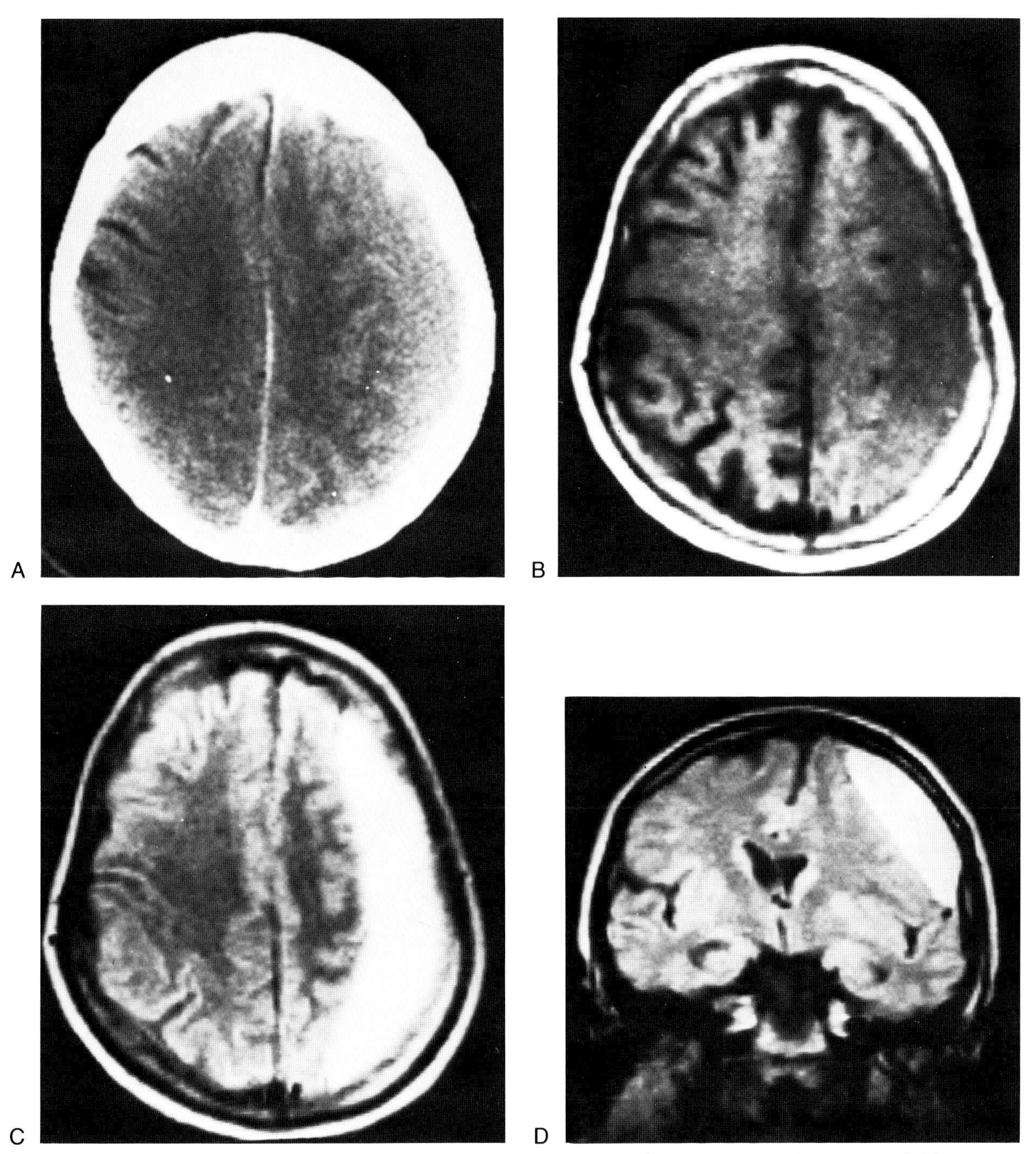

Fig. 10-10. Three-week-old subdural hematoma. **(A)** The SDH is isodense on CT. **(B)** SE 500/28 axial image shows an extra-axial fluid collection whose T1 is shorter than that of CSF. **(C)** T2-weighted image (SE 2000/28) shows the SDH to good advantage, by virtue of its long T2 value. **(D)** Coronal SE 1500/28 image shows the mass effect to good advantage.

tween ligand water at the methemoglobin binding site. Outer sphere effects refer to diffusional processes where solvent water protons gain access to the ferric ion through the nonpolar crevice of the methemoglobin molecule.)[19]

The very short T1 value of methemoglobin in solution has implications for the MR imaging of SAH. Although acute SAH may be MR-invisible, subacute (>1-week-old) SAH can often be demonstrated as high subarachnoid signal on T1-weighted images.[21] In this time frame, CT evidence for SAH is usually absent. MRI may be of significant value, therefore, in demonstrating evidence for prior SAH in patients examined 1 week or more postictus.[21]

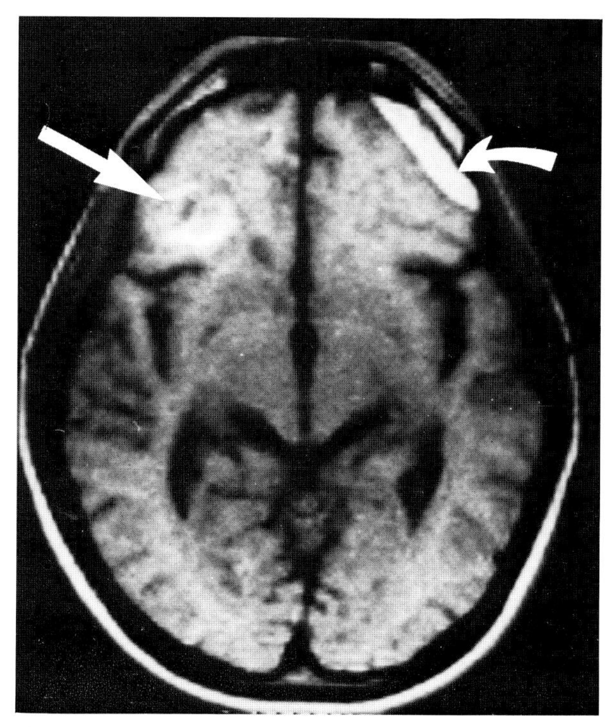

Fig. 10-11. Hemorrhagic parenchyma (straight arrow) with mottled signal intensity on this T1-weighted (SE 500/28) image. No true hematoma was seen by CT in this same area. Note also the extra-axial hematoma on the left side (curved arrow).

Hemorrhagic Parenchyma

In discussing signal changes that occur in intracranial hemorrhage, it is important to distinguish hematoma from hemorrhagic tissue. The term hematoma implies a focal collection of blood without intervening brain parenchyma. Hemorrhagic tissue implies blood mixed with brain parenchyma and edema, not clearly separated into a loculated space. Hemorrhagic parenchyma is seen with brain contusion, shear injuries, hemorrhagic cortical infarction, and hemorrhage into tumors.

Cerebral contusion is defined as bruising or crushing of brain tissue without cortical disruption. It is characterized by punctate hemorrhages, edema, and tissue necrosis near the site of impact.[30] Depending on the relative amounts of blood, edema, necrotic tissue, and normal brain, the appearance on MR may vary. The appearance is often dominated by edema with prolonged T1 and T2 values in the acute phase.[4] Subacute and chronic lesions may have short T1 dues to hemosiderin deposition, although T2 may remain prolonged due to edema and gliosis (Fig. 10-11).

Hemorrhagic transformation is a relatively common sequel of embolic, watershed, and venous infarctions. This hemorrhagic conversion, known pathologically as "red softening," occurs in about 20 percent of infarcts and is usually confined to the cortex.[41] Blood in the pia, perivascular space, and overlying subarachnoid space may also be seen.

The MR imaging characteristics of hemorrhagic cortical infarction (HCI) differ from those seen with other forms of parenchymal hemorrhage or hematoma.[42] It has been postulated that the hemoglobin in acute HCI is subjected to relatively higher local PO_2 than with ICH because of early vascular recanalization and luxury perfusion.[42,43] As a result of these factors, less deoxyhemoglobin is formed in the acute stage of HCI. Therefore, the low signal on T2-weighted images seen in the first several days of ICH is absent or not nearly so marked in HCI. The subacute stage (1 to 3 weeks) of HCI may show shortening of T1 values due to methemoglobin accumulation. Edema is surrounding brain parenchyma with long T1 and T2 times is present in the acute and subacute stages. Shortening of T1 and T2 may occur in the chronic stage of HCI with hemosiderin deposition in the tissues. An example of HCI in the acute and subacute stages seen by CT and MRI is presented in Figure 10-12.

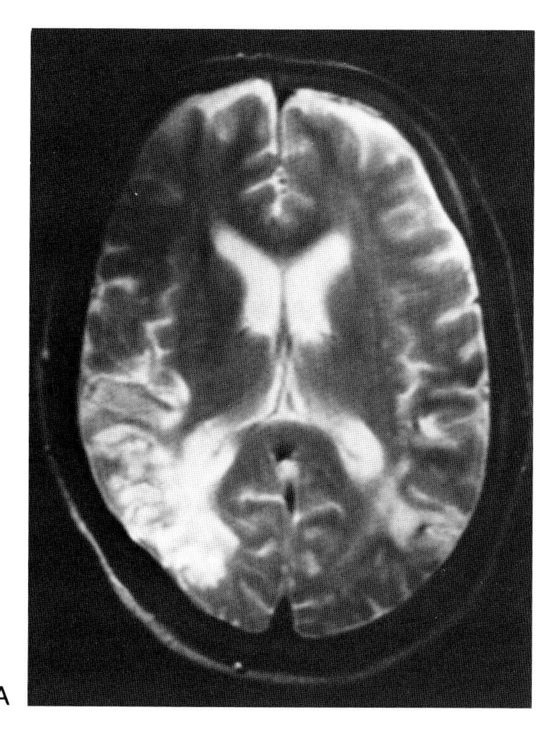

Fig. 10-12. Hemorrhagic cortical infarction (HCI).

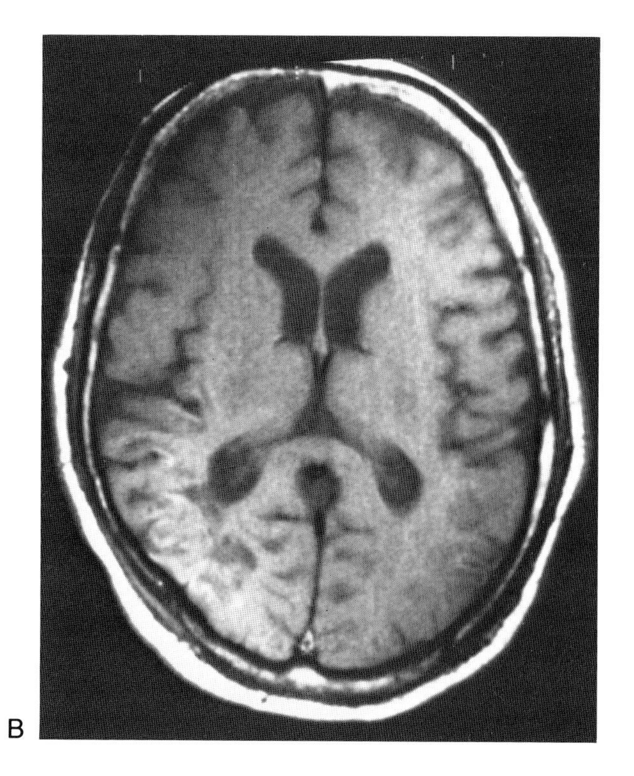

MR IMAGING OF TRAUMA AND ITS SEQUELAE

The enthusiasm of early investigators for the use of MRI in head trauma was initially damped by the realization that acute hemorrhage, especially subarachnoid, was difficult or impossible to detect on routine MR images.[3,4] Additionally bone detail needed to detect skull fractures could not be obtained from MR signal data. Because head trauma patients are frequently uncooperative and agitated, motion artifacts frequently degrade their MR scans. Clinical instability also makes MR scanning more difficult than CT, because few MR scanning suites are currently supplied with nonferromagnetic equipment for monitoring, ventilating, or resuscitating these patients. These factors have made CT the unquestionable modality of choice in imaging acutely traumatized patients.

However, it has also become clear that CT has its own limitations, even acutely. CT findings are only about 70 percent accurate in predicting the prognosis after head injury.[44] CT may underestimate the extent of pathology, especially with regard to shearing and brain stem injuries. Subfrontal and subtemporal hematomas may be obscured by partial volume effects with the skull base. An occasional isodense subdural hematoma may escape detection by CT.

The relative merits of CT and MR in head trauma patients have been compared in several investigations, but none of the studies to date are either prospective or randomized. Below are some tentative conclusions based upon available data.[5–8a]

Extra-Axial Lesions

Magnetic resonance demonstrates all SDHs and EDHs seen by CT as well as additional ones not seen by CT. Those detected by MR and not CT are usually very small and do not affect surgical management, however. High parietal, subfrontal, and subtemporal extra-axial collections are easily visualized on coronal MRI while they may be overlooked or difficult to quantitate on axial CT. Isodense SDHs on CT are all easily seen on MRI. MR, with its multiplanar capacity, better characterizes the size and extent of extra-axial collections than CT. Subdural hygromas can be differentiated from chronic SDHs easily by MR, but not by CT.

Intracerebral Lesions

Magnetic resonance demonstrates nonhemorrhagic contusions much better than CT, although surgical management does not necessarily change from this added information. Shearing injuries in white matter and brain stem injuries may be better appreciated by MR than by CT. This knowledge may potentially affect prognosis and management. The extent of apparent encephalomalacia in the chronic stages of injury is better evaluated by MR than CT, which often underestimates the extent of injury (Fig. 10-13).

In conclusion, CT is clearly preferred to MRI in at least the first 72 hours following head trauma. This is based upon CT's documented ability to detect SAH, skull fractures, and all surgically significant extra-axial collections. MRI may add additional information during this period, but it is uncertain whether this knowledge can be applied to determine prognosis or aid in management. In the subacute and late stages following head injury MR is at least equal to CT in most regards and superior to it in several. MRI may better delineate and characterize post-traumatic gliosis and extra-axial fluid collections.

POST-THERAPEUTIC CHANGES

Radiation Injury

Both transient and permanent adverse effects may be recognized following therapeutic radiation of the central nervous system.[45–47] Transient effects include headache, nausea, and acute cerebral edema. Permanent effects include radiation-induced demyelination and radionecrosis. Radiation injury has been well studied by CT, but relatively few MR reports are available.[48–50]

The primary pathology responsible for CNS radiation injury appears to be vascular damage, first manifest by arteriolar thickening and hyalinization.[51,52] Later, demyelination occurs in these same areas supplied by the damaged arterioles. The deep white matter of the periventricular regions and centra semiovale are most effected. Usually a period of several months to several years elapses before symptoms occur.[45,47] Most patients have focal neurologic signs, although in others nonspecific symptoms may be seen includ-

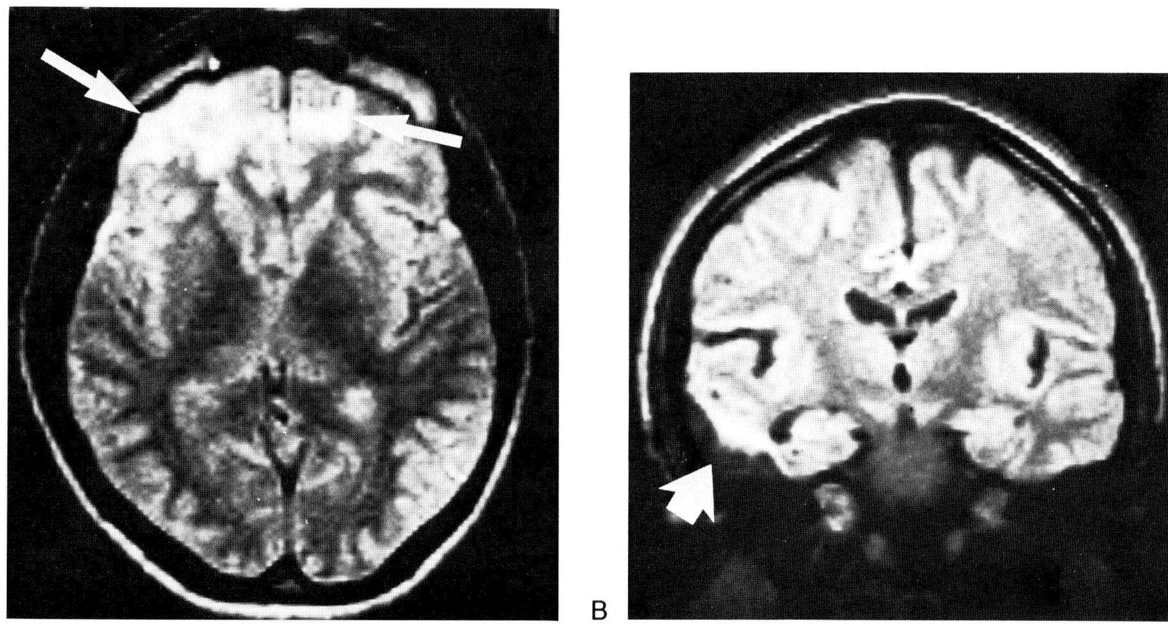

Fig. 10-13. Utility of MRI in evaluating this distant effects of trauma. A man who fell 16 feet 1 year ago, with persisting neuro-behavioral changes **(A)** Extensive gliosis in both frontal lobes, not appreciated by CT. **(B)** Temporal lobe atrophy is also well appreciated on this direct coronal image.

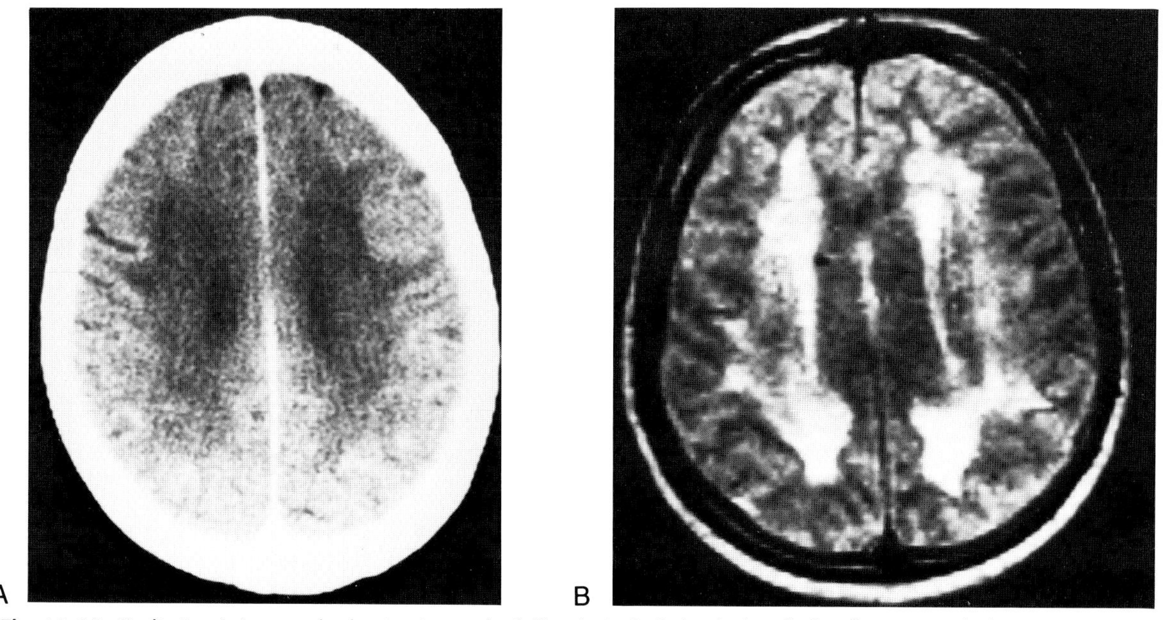

Fig. 10-14. Radiation injury to the brain, 6 months following whole brain irradiation for metastatic lung cancer. **(A)** CT image. **(B)** SE 2000/56 MR image shows extensive, symmetrical white matter disease.

ing confusion, lethargy, or dementia. Late alterations that are irreversible but relatively asymptomatic include atrophy and mineralizing microangiopathy. Radiation-induced neoplasms, vasculopathy, and necrosis are late changes that are both irreversible and life-threatening.

The brain will usually tolerate doses of up to 5000 Rad (50 Gy) administered in fractionated doses over 5 weeks.[53] Radiation changes are not usually expected with doses below 2000 Rad (20 Gy) but have been reported rarely in children receiving only 100 Rad (1 Gy). Doses in excess of 6000 Rad (60 Gy) significantly increase the risk of injury.[46,53] Concomitant chemotherapy, especially with methotrexate, vincristine, or actinomycin-D, has a synergistic effect posing increased risk for brain injury.

Several reports of MR imaging in CNS radiation injury have now appeared.[55–58] In general MR seems to be more sensitive than CT in revealing the presence and extent of radiation lesions.[56] Focal radionecrosis cannot be reliably differentiated from recurrent tumor with MR any more easily than with CT.[55,56] MR without contrast may occasionally be inferior to contrast enhanced CT in the evaluation of postirradiation changes. This was demonstrated in a published case where subependymal tumor spread was seen on contrast CT but obscured on MR by high signal from white matter radiation lesions.[55]

Cerebral radiation lesions demonstrate prolonged T1, T2, and proton density values compared to normal white matter.[55] They are best seen on coronal or axial T2- and proton density-weighted images such as SE 1000/80 or SE 2000/56. With whole brain irradiation a characteristic appearance may be noted (Fig. 10-14). Fairly symmetric periventricular and deep white matter hyperintensity, which in severe cases extends to involve the subcortical U-fibers, is seen. The margins are irregular and flame-shaped. The corpus callosum is frequently not involved with whole brain irradiation, perhaps due to its rich vascular supply by short penetrating arteries. Local areas of brain (including the corpus callosum) may be selectively involved, however, when the radiation portal is more limited or regional doses are administered.

Radiation necrosis is a focal region of severe injury that presents as a mass lesion months to years following radiotherapy. The differentiation of radiation necrosis from recurrent tumor is difficult or impossible without biopsy. Because radiation necrosis frequently

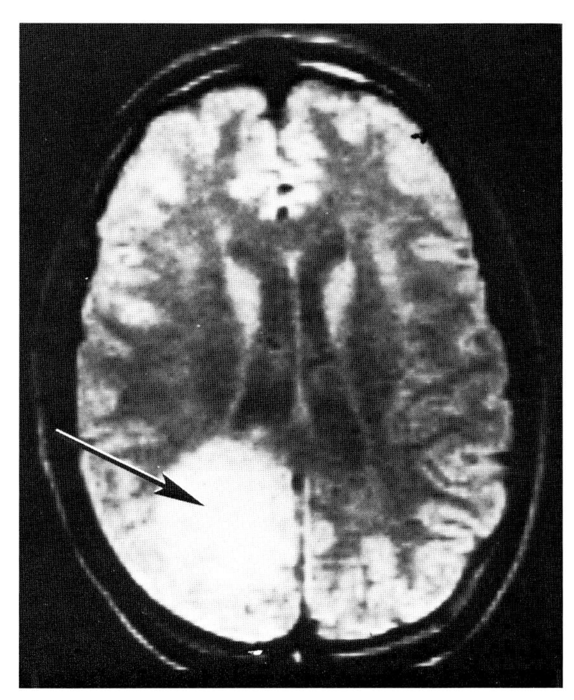

Fig. 10-15. Radiation necrosis presenting as a mass lesion. Biopsy was necessary to exclude recurrent tumor.

enhances, CT has been shown to be of little value in the differential diagnosis. Unfortunately, MR appears to offer no more specificity in regard to this diagnosis.[58] An example of radionecrosis resembling recurrent tumor is shown by MR in Figure 10-15.

Chemotherapy-Induced Leukoencephalopathy

Chemotherapeutic agents for cancer, especially methotrexate, are known to produce toxic effects on cerebral vessels and white matter, similar pathologically to radiation injury.[54,59–61] Although uncommon, drug-induced encephalopathy most frequently develops in leukenic patients who receive prophylaxis with intrathecal methotrexate. Occurrence following intravenous administration of the drug has also been reported.[62] Associated cranial irradiation may produce additive effects, and the individual contributions of radiotherapy and methotrexate injury may be difficult to separate.

The underlying pathology for methotraxate en-

Fig. 10-16. Drug-induced leukoencephalopathy. **(A)** This leukemic 8-year-old received intrathecal methotrexate plus radiation. Note extensive white matter disease that spares the corpus callosum. **(B)** This adult leukemic received chemotherapy and radiation 18 months prior to this scan which was performed to evaluate progressive weakness and dementia. Note extensive white matter disease. (SE 1000/56).

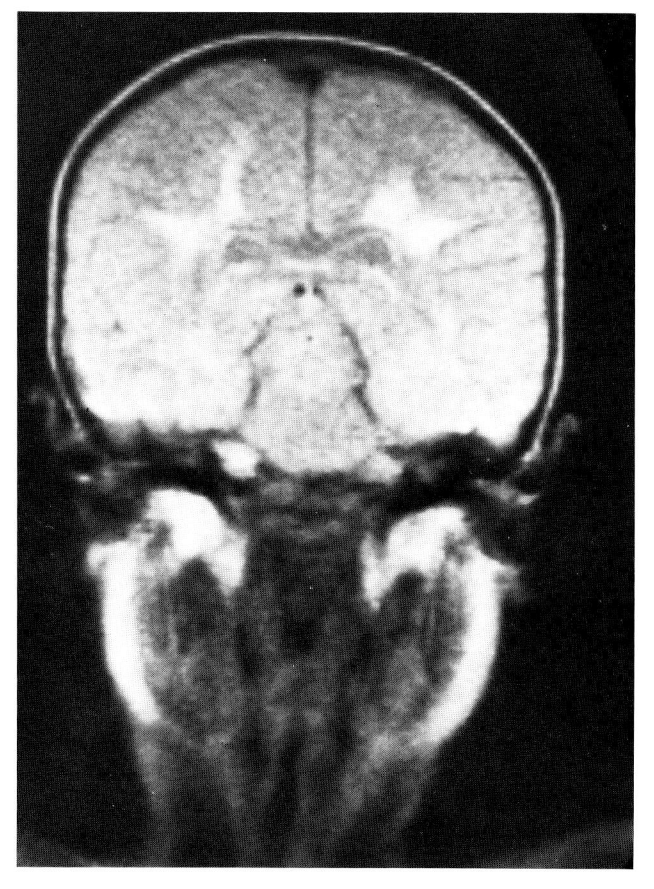
A

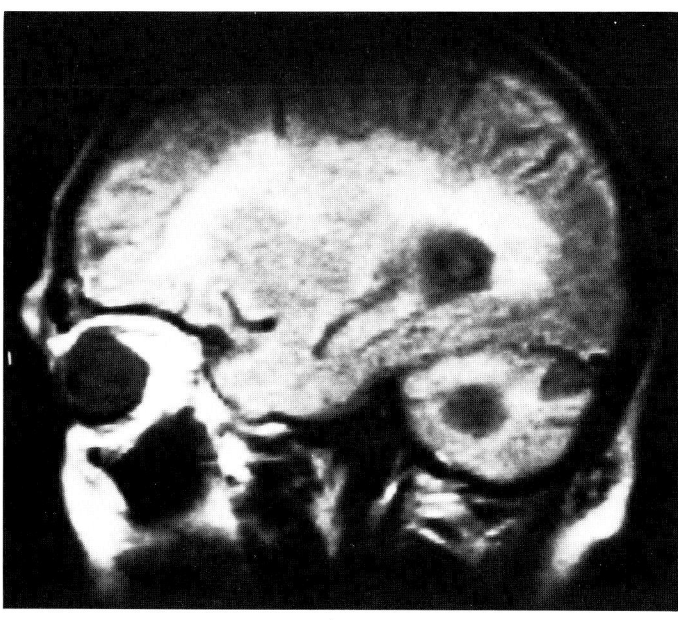
B

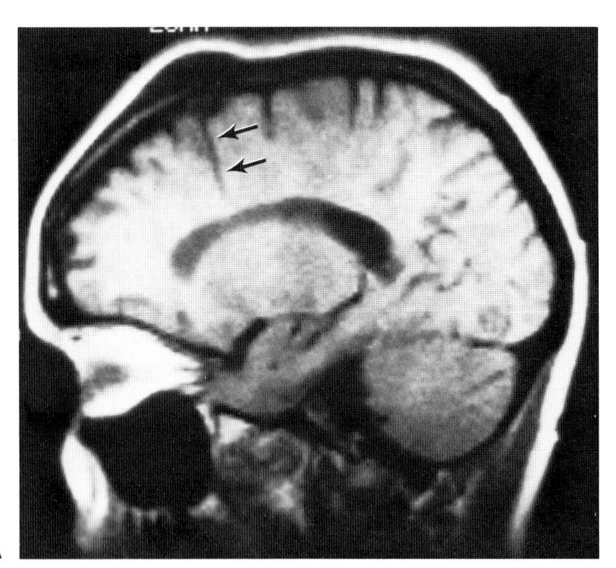

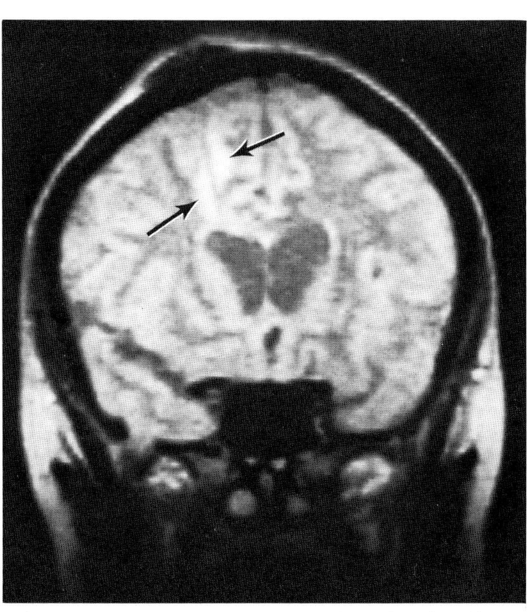

Fig. 10-17. Leakage of chemotherapeutic agent along the tract of an Omaya reservoir may result in localized leukoencephalopathy. **(A)** T1-weighted image shows catheter only (arrows). **(B)** T2-weighted image shows localized white matter disease along the catheter tract (arrows). A traumatic surgical insertion of any ventriculostomy catheter may also produce such an effect.

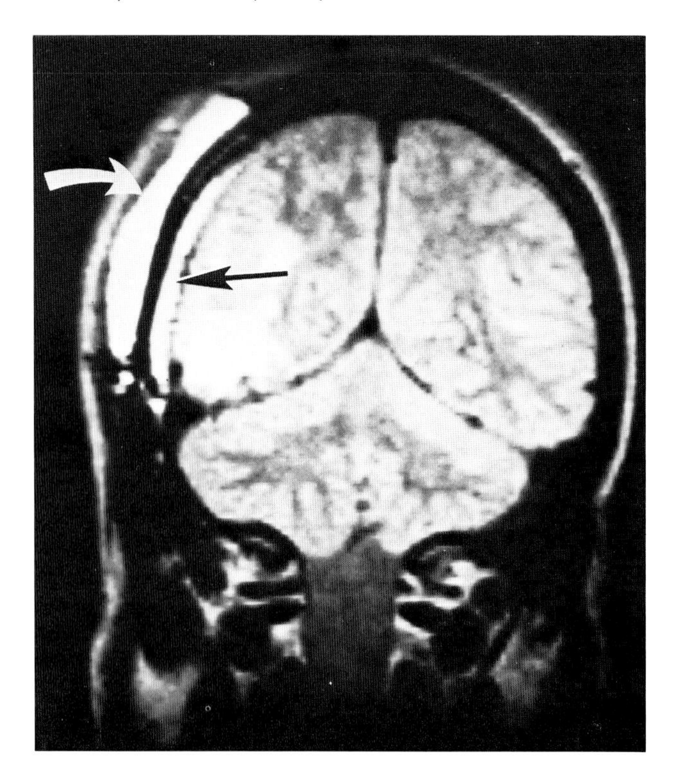

Fig. 10-18. Subgaleal (curved arrow) and extraaxial (straight arrow) postoperative hematomas.

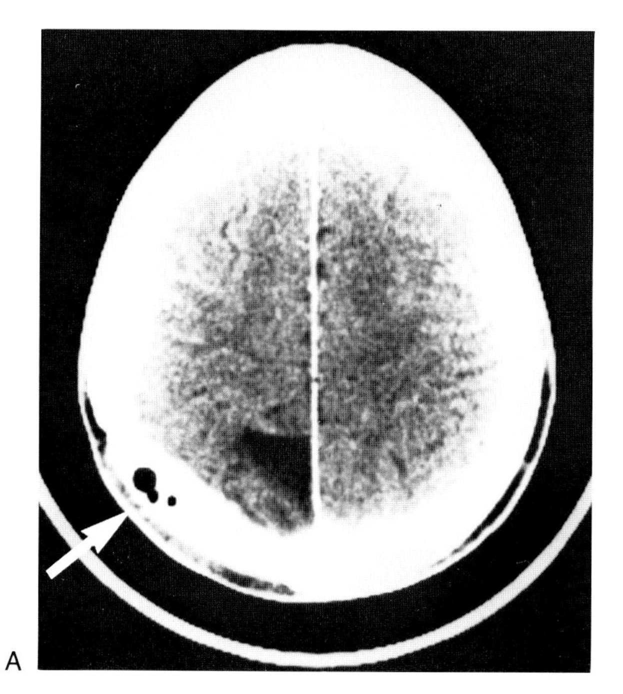

A

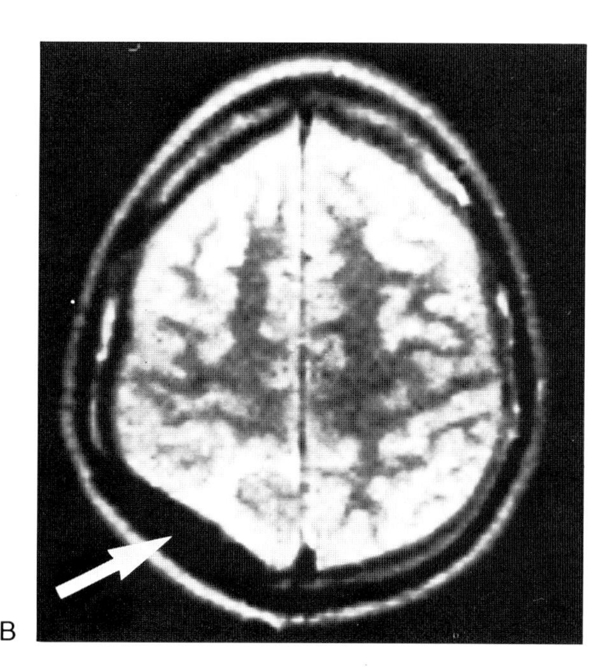

B

Fig. 10-19. Methachrylate skull flap. **(A)** Bubbly appearance on CT, not to be mistaken for abscess. **(B,C)** No signal from flap on MRI.

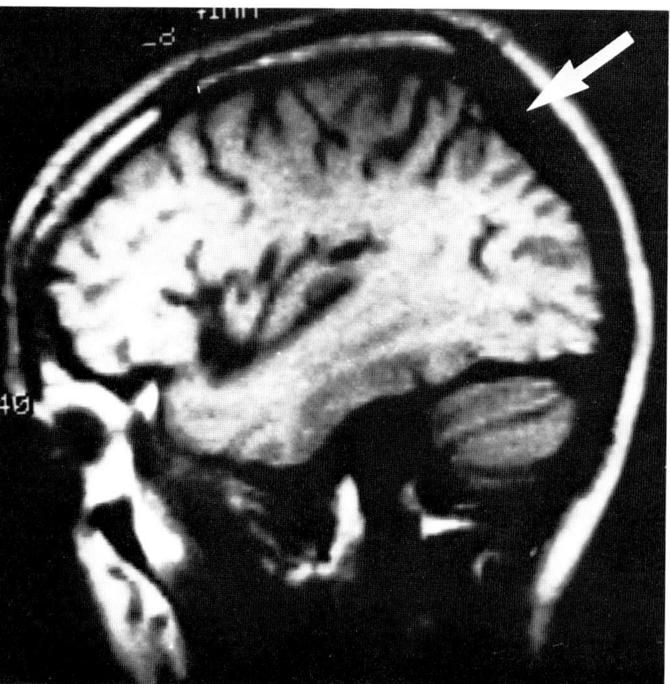

C

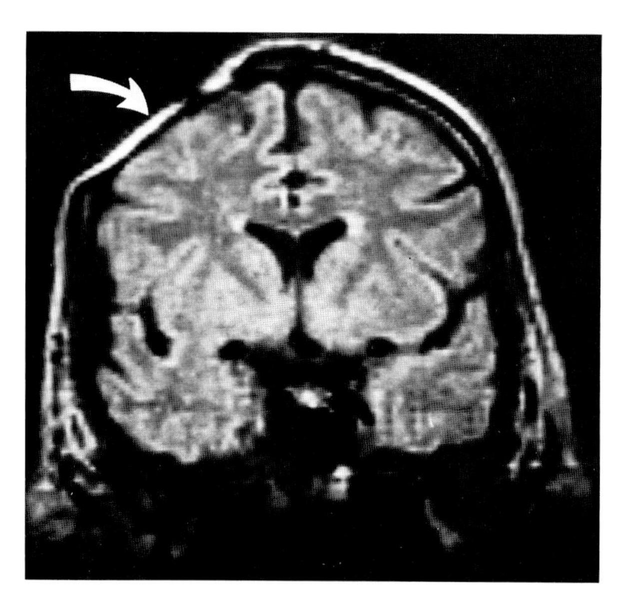

Fig. 10-20. Large craniotomy defect.

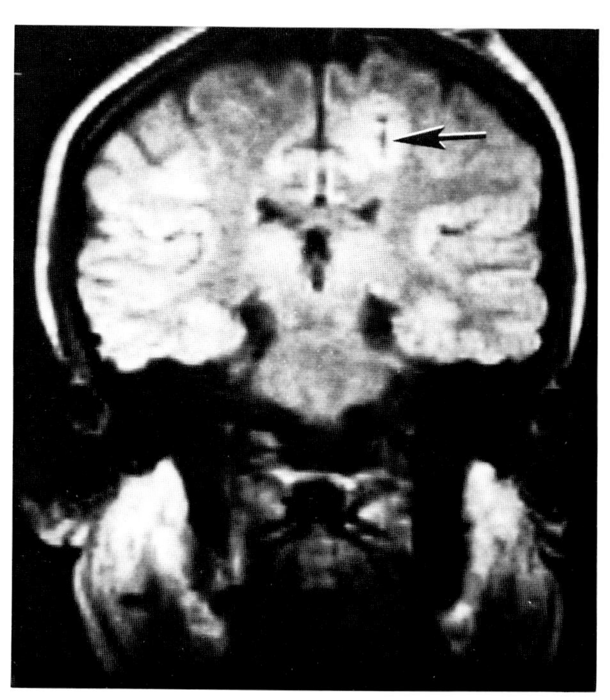

Fig. 10-21. Needle biopsy tract (arrow).

cephalopathy is not clear, but may relate to toxic accumulations of the drug in regions of brain previously damaged by vascular insufficiency.[61] Multifocal areas of coagulative necrosis, axonal swelling and fragmentation are seen microscopically. Symptoms usually begin a few months following therapy and include confusion, ataxia, seizures, dementia, and, rarely, coma and death. The appearance of these lesions on MR is nonspecific and cannot be separated from those of simple postirradiation change (Fig. 10-16).

Transient cortical and subcortical lesions corresponding to clinical deficits have been observed by MR in several leukemic patients following intrathecal or systemic chemotherapy.[63] The etiology of these lesions is unknown but none was observed with CT and all resolved on MRI. Thus MRI seems more sensitive than CT in evaluating the response of the brain to chemotherapeutic agents.

Intrathecal administration of chemotherapeutic agents is often performed through an indwelling ventriculostomy device such as an Omaya Reservoir (Fig. 10-17). Leakage of toxic chemotherapeutic agent may occur along the tract of the catheter, causing local necrosis and gliosis. (Y.Y. Lee, personal communication.) This finding has been observed by MR in several of our patients (Fig. 10-17).

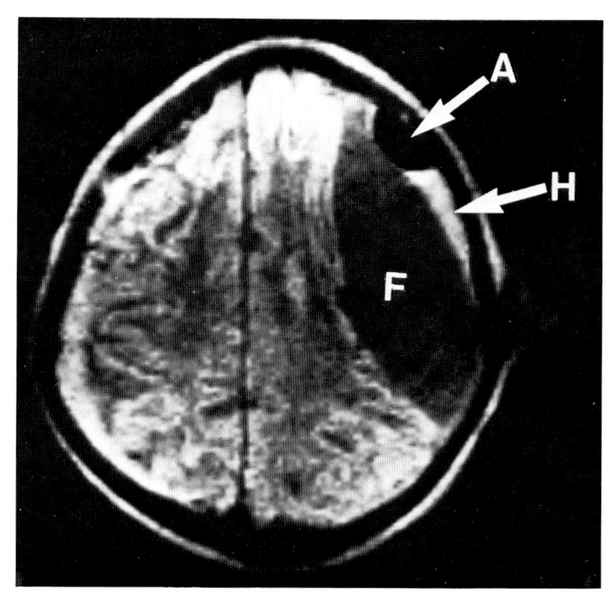

Fig. 10-22. Postoperative extra-axial fluid collections (septated). A, air; H, hematoma; F, fluid.

Postoperative Changes

A number of postoperative changes of the skull and brain may be encountered on MRI. While they generally produce few diagnostic challenges with appropriate history, occasionally they may cause confusing appearances on MR images. Because their appearances are somewhat different than on CT examples of miscellaneous postoperative changes are presented here for completeness. These changes include craniectomy defects, skull flaps, fluid collections, pneumocephalus, needle tracts, and postoperative parenchymal defects (Figs. 10-18 to 10-22).

REFERENCES

1. Bailes DR, Young IR, Thomas DJ et al: NMR imaging of the brain using spin-echo sequences. Clin Radiol 33:395, 1982
2. Bydder GM, Steiner RE, Young IR et al: Clinical NMR imaging of the brain: 140 cases. AJR 135:215, 1982
3. Sipponen JT, Sepponen RE, Sivula A: Nuclear magnetic resonance (NMR) imaging of intracerebral hemorrhage in acute and resolving phases. J Comput Assist Tomogr 7:954, 1983
4. DeLaPaz RL, New PFJ, Buonanno FS et al: NMR imaging of intracranial hemorrhage. J Comput Assist Tomogr 8:599, 1984
5. Gandy SE, Snow RB, Zimmerman RD, Deck MDF: Cranial nuclear magnetic resonance imaging in head trauma. Ann Neurol 16:254, 1984
6. Langfitt TW, Obrist WD, Alavi A et al: Computerized tomography, magnetic resonance imaging, and positron emission tomography in the study of brain trauma. J Neurosurg 64:760, 1986
7. Han JS, Kaufman B, Alfidi RJ et al: Head trauma evaluated by magnetic resonance and computed tomography: a comparison. Radiology 150:71, 1984
8. Snow RB, Zimmerman RD, Gandy SE, Deck MDF: Comparison of magnetic resonance imaging and computed tomography in the evaluation of head injury. Neurosurgery 18:45, 1986

8a. Jenkins A, Teasdale G, Hadley MDM et al: Brain lesions detected by magnetic resonance imaging in mild and severe head injuries. Lancet 9:445, 1986

9. Gomori JM, Grossman RI, Goldberg HI et al: Intracranial hematomas: imaging by high-field MR. Radiology 157:87, 1985
10. Zimmerman RA, Bilaniuk LT, Grossman RI et al: Resistive NMR of intracranial hematomas. Neuroradiology 27:16, 1985
11. Swensen SJ, Keller PL, Berquist TH et al: Magnetic resonance imaging of hemorrhage. AJR 145:921, 1985
12. DiChiro G, Brooks RA, Girton ME et al: Sequential MR studies of intracerebral hematomas in monkeys. AJNR 7:193, 1986

12a. Nose T, Enomoto T, Hyodo A et al: Intracerebral hematoma developing during MR examination. J Comput Assist Tomogr 11:184, 1987

13. Wintrobe MM, Lee GR, Buggs DR et al: Clinical Hematology. Lea & Febiger, Philadelphia, 1981
14. Pauling L, Coryell C: The magnetic properties and structure of hemoglobin oxyhemoglobin and carbonmonoxyhemoglobin. Proc Nat Acad Sci (USA) 22:210, 1936
15. Pauling L, Coryell C: The magnetic properties and structure of the hemochromogens and related substances. Proc Nat Acad Sci (USA) 22:159, 1936
16. Brooks RA, Battocletti JH, Sances A Jr et al: Nuclear magnetic relaxation in blood. IEEE Trans Biomed Eng 22:12, 1975
17. Thulborn KR, Waterton JC, Matthews PM, Radda GK: Oxygenation dependence of the transverse relaxation time of water protons in whole blood at high field. Biochem Biophys Acta 714:265, 1982
18. Bloembergen N, Purcell EM, Pound RV: Relaxation effects in nuclear magnetic resonance absorption. Phys Rev 73:679, 1948
19. Koenig SH, Brown RD, Lindstrom TR: Interactions of solvent with the heme region of methemoglobin and fluoromethemoglobin. Biophys J 34:397, 1981
20. Singer JR, Crooks LE: Some magnetic studies of normal and leukemic blood. J Clin Eng 3:237, 1978
21. Bradley WG Jr, Schmidt PG: Effect of methemoglobin formation on the MR appearance of subarachnoid hemorrhage. Radiology 156:99, 1985
22. Zipp A, James TL, Kuntz ID, Shohet SB: Water proton magnetic resonance studies of normal and sickle erythrocytes: temperature and volume dependence. Biochem Biophys Acta 428:291, 1976
23. Sandhu HS, Friedmann GB: Proton spin-spin relaxation time and hemoglobin content. J Clin Eng 3:237, 1978
24. Chuang AH, Waterman MR, Yamaoka K, Cottam GL: Effect of pH, carbamylation and other hemoglobins on deoxyhemoglobin S aggregation inside intact erythrocytes as detected by proton relaxation rate measurements. Arch Biochem Biophys 167:145, 1975
25. Cohen MD, McGuire W, Cory DA, Smith JA: MR appearance of blood and blood products: an in vitro study. AJR 146:1293, 1986
26. Chakeres DW, Bryan RN: Acute subarachnoid hemorrhage: in vitro comparison of magnetic resonance and computed tomography. AJNR 7:223, 1986

27. Dolinskas CA, Bilaniuk LT, Zimmerman RA, Kuhl DE: Computed tomography in intracerebral hematomas. II Radionuclide and transmission CT studies of the perihematoma region. AJR 129:689, 1977
28. Som PM, Patel S, Nakagawa H, Anderson PJ: The iron rim sign. J Comput Assist Tomogr 3:109, 1979
29. Laster DW, Moody DM, Ball MR: Resolving intracerebral hematoma. Alteration of the "ring sign" with steroids. AJR 130:935, 1978
30. Williams AL, Haughton VM: Cranial Computed Tomography: A Comprehensive Text. CV Mosby, St. Louis, 1985
31. Jamieson KG, Yelland JDN: Extradural hematoma: report of 167 cases. J Neurosurg 29:13, 1968
32. Kishore PRS, Lipper MH, Becker DP et al: The significance of CT in the management of patients with severe head injury: correlation with ICP. AJNR 2:307, 1981
33. Lipper MH, Kishore PRS, Girevendulis AK et al: Delayed intracranial hematoma in patients with severe head injury. Radiology 133:645, 1979
34. Scotti G, Terbrugge K, Melancon D et al: Evaluation of the age of subdural hematomas by computerized tomography. J Neurosurg 47:311, 1977
35. Smith WP Jr, Batnitzky S, Rengachary SS: Acute isodense subdural hematomas: a problem in anemic patients. AJNR 2:37, 1981
36. Dooms GC, Uske A, Brant-Zawadzki M et al: Spin-echo MR imaging of intracranial hemorrhage. Neuroradiology 28:132, 1986
37. Sipponen JT, Sepponen RE, Sivula A: Chronic subdural hematoma: demonstrated by magnetic resonance. Radiology 150:79, 1984
38. Moon RL, Brant-Zawadzki M, Pitts PH, Mills CM: Nuclear magnetic resonance imaging of CT-isodense subdural hematomas. AJNR 5:319, 1984
39. Reed D, Robertson WD, Graeb DA et al: Acute subdural hematoma: atypical CT findings. AJNR 7:417, 1986
40. Norman D, Price D, Boyd D et al: Quantitative aspects of computed tomography of the blood and cerebrospinal fluid. Radiology 123:335, 1977
41. Fazio C: Red softening of the brain. J Neuropathol Exp Neurol 8:43, 1949
42. Hecht-Leavitt C, Gomori JM, Grossman RI et al: High-field MRI of hemorrhagic cortical infarction. AJNR 7:581, 1986
43. Hayman LA, Evans RA, Bastion FO, Hinck VC: Delayed high dose contrast CT: identifying patients at risk of massive hemorrhagic infarction. AJR 136:1151, 1981
44. Lipper MH, Kishore PRS, Enas GG et al: Computed tomography in the prediction of outcome in head injury. AJNR 6:7, 1985
45. Kramer S, Lee KF: Complications of radiation therapy: the central nervous system. Semin Roentgenol 9:75, 1974
46. Allen JC: The effects of cancer therapy on the nervous system. J Pediatr 93:903, 1978
47. Sheline GE, Wara WM, Smith V: Therapeutic irradiation and brain injury. Int Radiat Oncol Biol Phys 6:1215, 1980
48. Mikhael MA: Radiation necrosis of the brain: correlation between computed tomography, pathology, and dose distribution. J Comput Assist Tomogr 2:71, 1978
49. Mikhael MA: Radiation necrosis of the brain: correlation between patterns on computed tomography and dose of radiation. J Comput Assist Tomogr 2:241, 1978
50. Kingsley DPE, Kendall BE: CT of the adverse effects of therapeutic radiation of the central nervous system. AJNR 2:453, 1981
51. Courville CB, Myers RO: The process of demyelination in the central nervous system. II. Mechanism of demyelination and necrosis of the cerebral centrum incident to x-irradiation. J Neuropathol Exp Neurol 17:158, 1958
52. DeReuck J, Van der Eecken H: The anatomy of late radiation encephalopathy. Eur Neurol 13:481, 1975
53. Marks JE, Baglan RJ, Prassad SC et al: Cerebral radionecrosis: incidence and risk in relation to dose, time, fractionation, and volume. Int J Radiat Oncol Biol Phys 7:243, 1981
54. Phillips TL, Fu KK: Quantification of combined radiation therapy and chemotherapy effects on critical normal tissues. Cancer 37:1186, 1976
55. Dooms GC, Hecht S, Brant-Zawadzki M et al: Brain radiation lesions: MR imaging. Radiology 158:149, 1986
56. Curnes JT, Laster DW, Ball MR et al: Magnetic resonance imaging of radiation injury to the brain. AJNR 7:389, 1986
57. DiChiro G, Patronas NJ, Oldfield EH: PET, CT, and NMR of cerebral necrosis following radiation or intra-arterial chemotherapy for cerebral tumors. AJNR (in press)
58. Harris MI, Bleck TP: Magnetic resonance imaging of CNS radionecrosis. Neurology 35:172, 1985
59. Liu HM, Maurer HS, Vongsvivut S et al: Methotrexate encephalopathy. A neuropathologic study. Hum Pathol 9:635, 1978

60. Bjorgen JE, Gold LHA: Computed tomographic appearance of methotrexate-induced leukoencephalopathy. Radiology 122:377, 1977
61. Shalen PR, Ostrow PT, Glass PJ: Enhancement of the white matter following prophylactic therapy of the central nervous system for leukemia. Radiology 140:409, 1981
62. Allen JC, Thaler HT, Deck MDF et al: Leukoencephalopathy following high-dose intravenous methotrexate chemotherapy: quantitative assessment of white matter attenuation using computed tomography. Neuroradiology 16:44, 1978
63. Patronas NJ, Reinig JW, Seibel NL et al: MR imaging detection of transient cerebral lesions following prophylactic chemotherapy in leukemic patients. Radiol 157:247, 1985

11

The Skull Base

THE SELLA AND PITUITARY GLAND

The pituitary gland is classically divided into two parts, separable by gross anatomic features, physiology, and embryologic origin. The larger anterior lobe or *adenohypophysis* develops as an upward outpouching of the foregut endothelium (Rathke's pouch). The smaller posterior lobe or *neurohypophysis* develops as a downward displacement of brain tissue from the primitive hypothalamus. The two lobes are separated by a thin band of epithelial cells, the *pars intermedia*, and a narrow cleftlike potential space where certain cysts and cystic tumors may develop.

The anterior pituitary is responsible for secretion of a number of important hormones including prolactin (PRL), growth hormone (GH), thyroid stimulating hormone (TSH), adrenocorticotropic hormone (ACTH), follicle stimulating hormone (FSH), luteinizing hormone (LH), and melanocyte stimulating hormone (MSH). Secretion of these hormones is regulated largely by hypothalamic releasing factors that flow from the hypothalamus along a portal venous system to the adenohypophysis.

The posterior pituitary is responsible for the secretion of two important hormones, antidiuretic hormone (ADH) and oxytocin (OXY). The cells of origin of these hormones lie in the supraoptic and paraventricular nuclei of the hypothalamus. Long axons from these neurons of origin extend along the pituitary stalk or *infundibulum* into the posterior pituitary. From here their hormones are released directly into the systemic circulation.

The pituitary gland lies nestled in a bony depression in the body of the sphenoid, the *sella turcica*. The sella is covered by a dural membrane, the *diaphragma sella*, which is fenestrated centrally to transmit the pituitary stalk. Otherwise the diaphragma sella forms a boundary between the basal subarachnoid cisterns and the pituitary fossa. Superior to the pituitary lie the optic chiasm and tracts. Inferiorly may be found the sphenoid sinus. Laterally to the pituitary lie the cavernous sinuses, which contain the internal carotid arteries as well as cranial nerves III through VI.

Normal MR Anatomy (Figs. 11-1 to 11-3)

The normal pituitary gland is best visualized on coronal or sagittal thin section MR images that are T1-weighted. Depending upon aeration of the sphenoidal sinus, the bony sellar floor is variably identified. The dorsum sella is commonly filled with fatty marrow and its cortex bordering the pituitary fossa may often be seen as a low-intensity line. Posteriorly within the pituitary fossa a high-intensity signal on T1-weighted images can be identified in about 95 percent of patients. Initial research suggested that this high signal was the result of an intrasellar fat pad behind the true gland.[1] However, others have

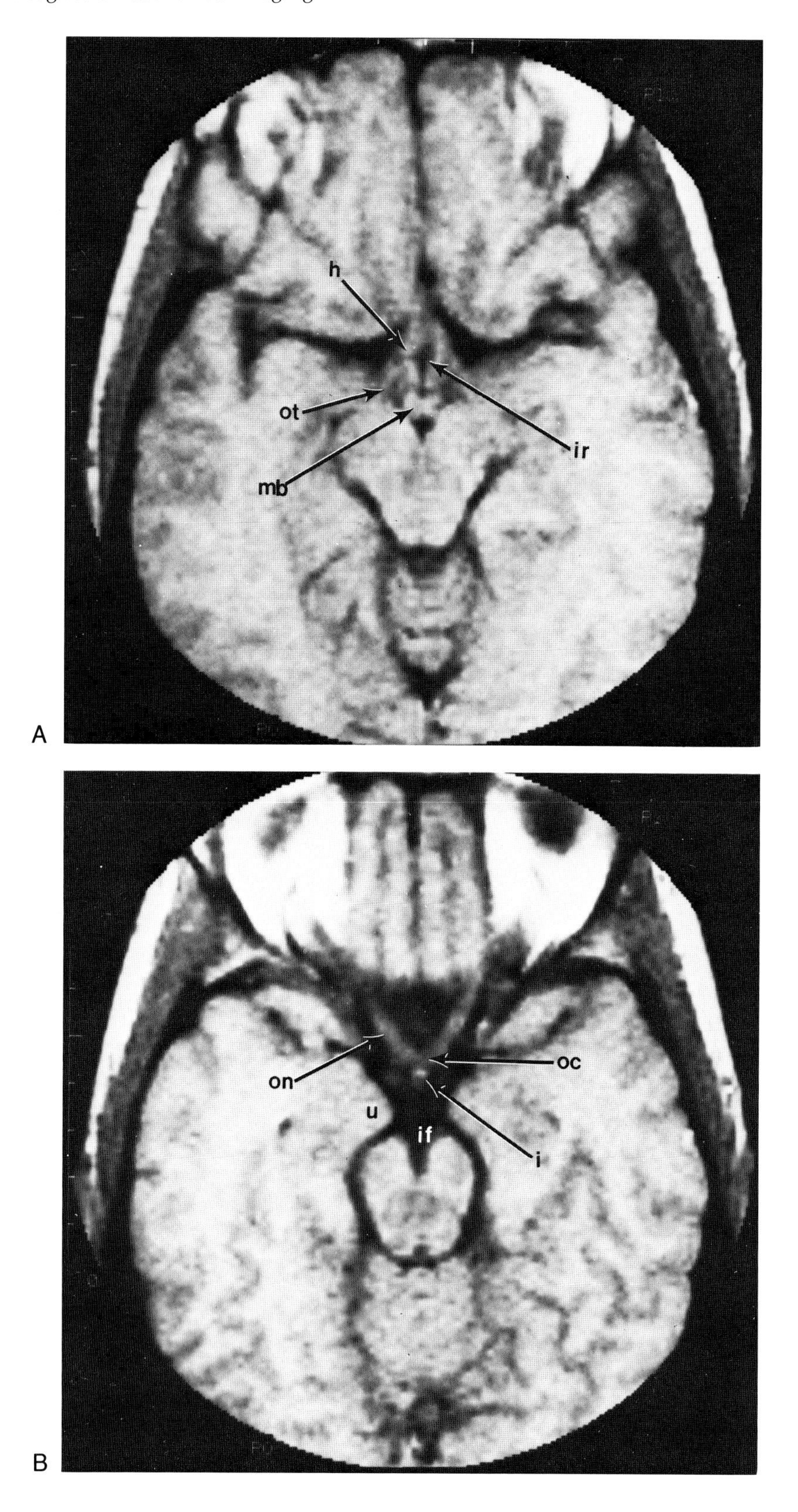
h
ot
ir
mb
A
on
oc
u
if
i
B

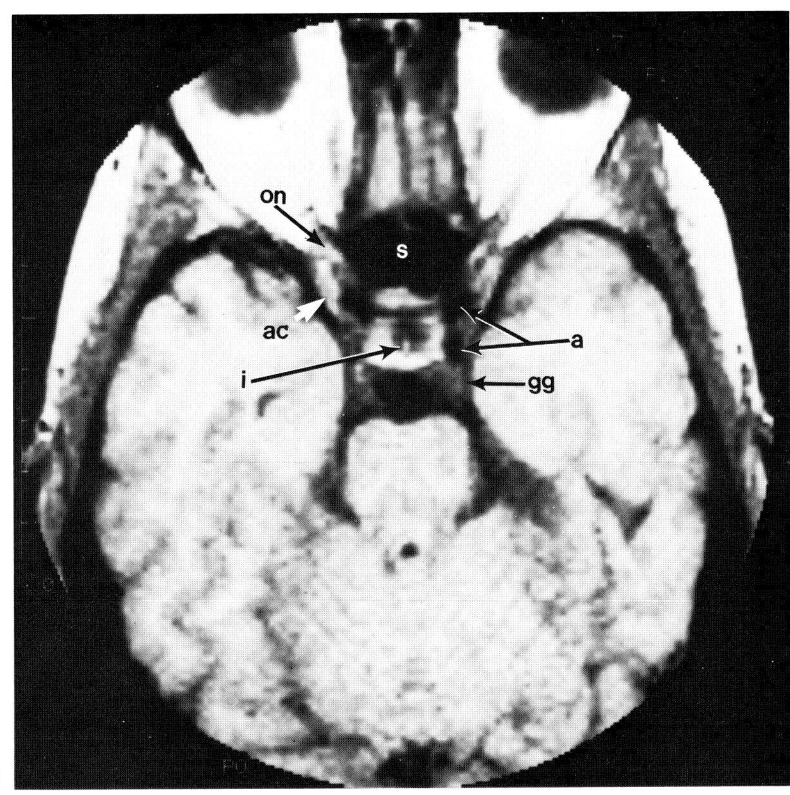

Fig. 11-1. (A–E) Axial SE 450/26 images of the sella and parasellar region from above downward. (See legend for explanation of symbols.) (*Figure continues*).

a, carotid artery;
ac, anterior clinoid process;
am, amygdala;
cp, cerebral peduncle;
cs, cavernous sinus, venous portion;
fp, fat in tip of petrous pyramid;
gg, Gasserian (semilunar) ganglion;
h, hypothalamus;
i, infundibulum;
icv, internal cerebral veins;
if, interpeduncular fossa;
io, inferior olive;
ir, infundibular recess of third ventricle;
j, jugular bulb;
mb, mamillary bodies;
oc, optic chiasm;
on, optic nerve;
ot, optic tract;
p, pituitary;
pca, posterior communicating artery;
po, pons;
ppf, pterygopalatine fossa;
s, sphenoidal sinus;
sof, superior orbital fissure;
u, uncus;
V, trigeminal nerve;
V_1, ophthalmic division of V;
V_2, maxillary division of V;
V_3, mandibular division of V;
III, oculomotor nerve;
3, third ventricle.

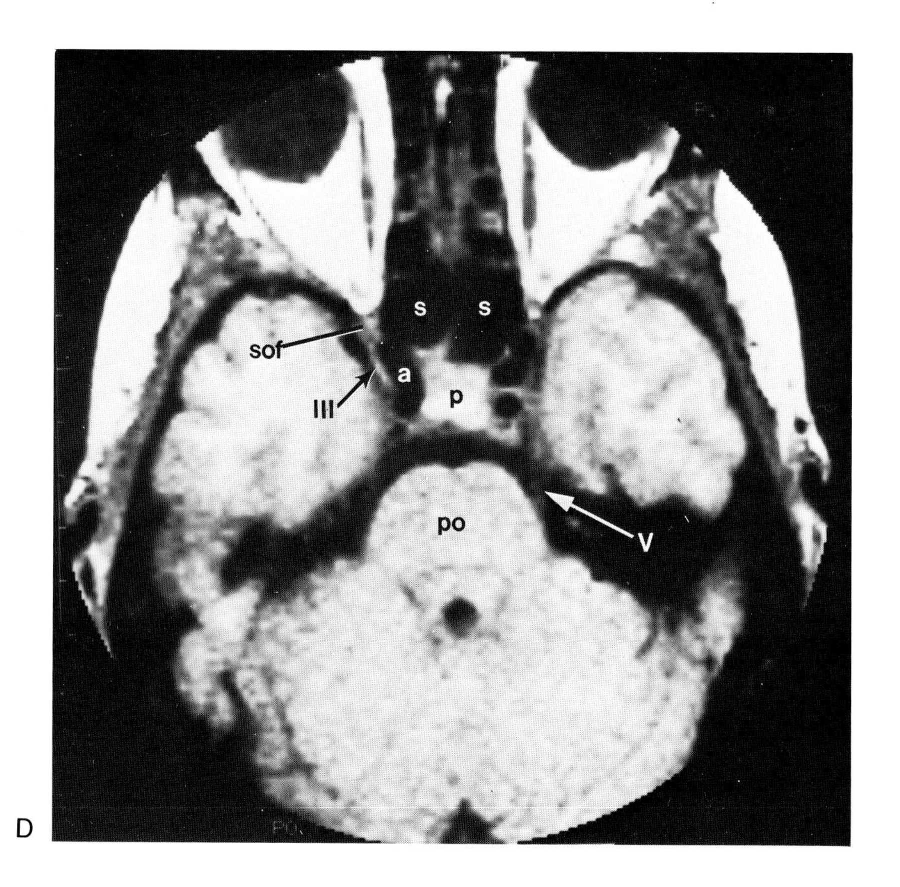
s
s
sof
a
III
p
po
V
D

E

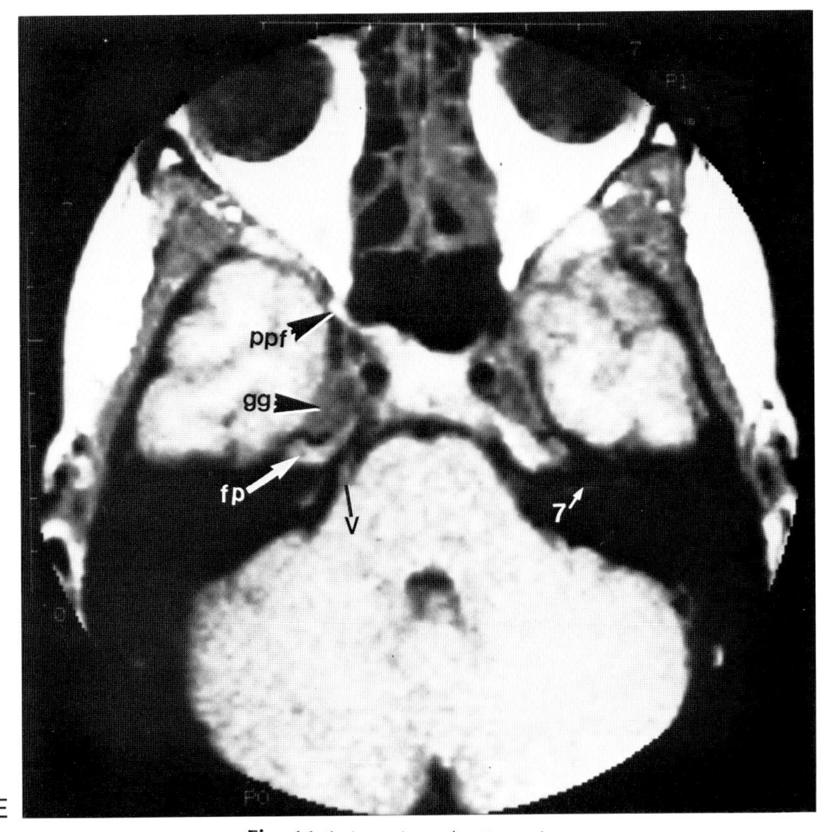

Fig. 11-1 (*continued*). (**D** and **E**)

a, carotid artery;
ac, anterior clinoid process;
am, amygdala;
cp, cerebral peduncle;
cs, cavernous sinus, venous portion;
fp, fat in tip of petrous pyramid;
gg, Gasserian (semilunar) ganglion;
h, hypothalamus;
i, infundibulum;
icv, internal cerebral veins;
if, interpeduncular fossa;
io, inferior olive;
ir, infundibular recess of third ventricle;
j, jugular bulb;
mb, mamillary bodies;
oc, optic chiasm;
on, optic nerve;
ot, optic tract;
p, pituitary;
pca, posterior communicating artery;
po, pons;
ppf, pterygopalatine fossa;
s, sphenoidal sinus;
sof, superior orbital fissure;
u, uncus;
V, trigeminal nerve;
V_1, ophthalmic division of V;
V_2, maxillary division of V;
V_3, mandibular division of V;
III, oculomotor nerve;
3, third ventricle.

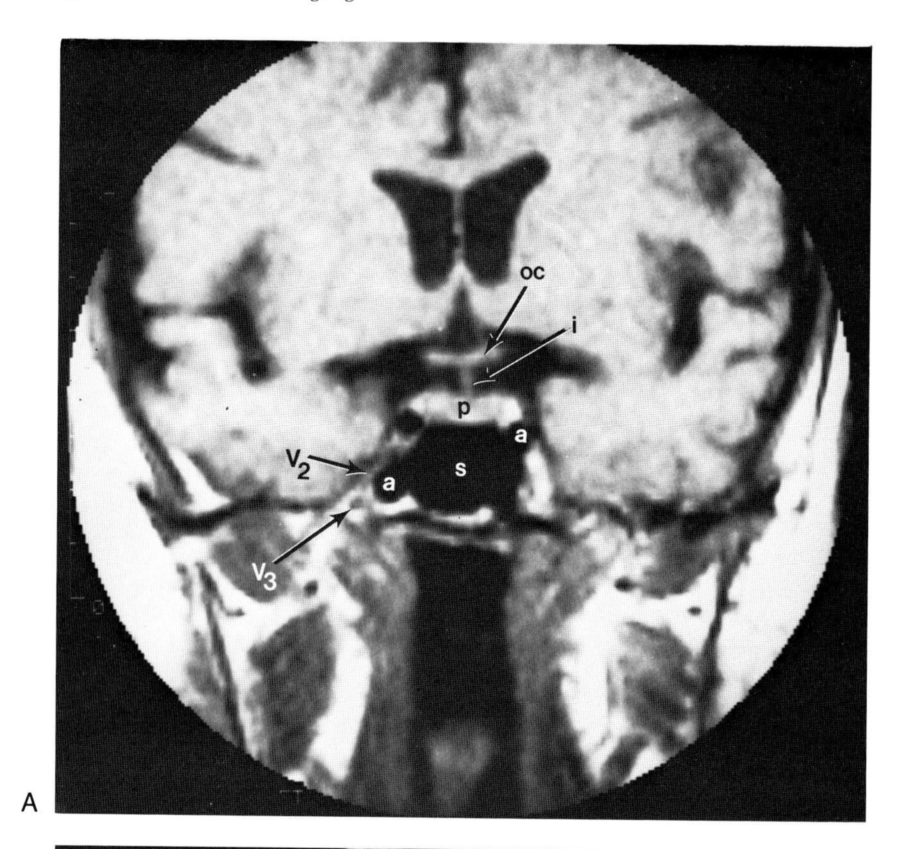
oc
i
p
a
V_2
s
a
V_3
A

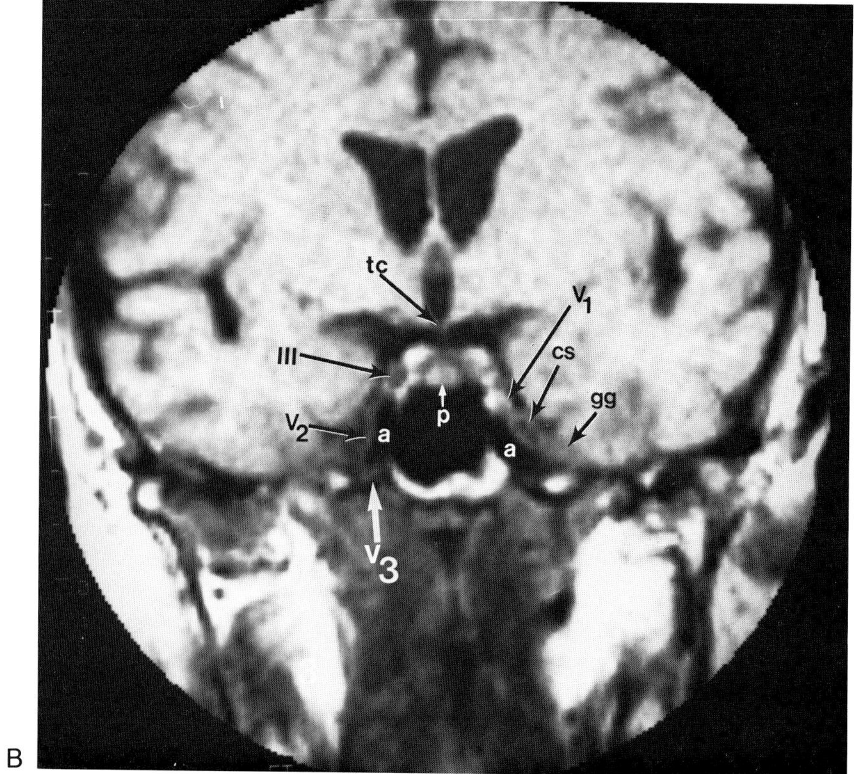
tc
V_1
III
cs
gg
p
V_2
a
a
V_3
B

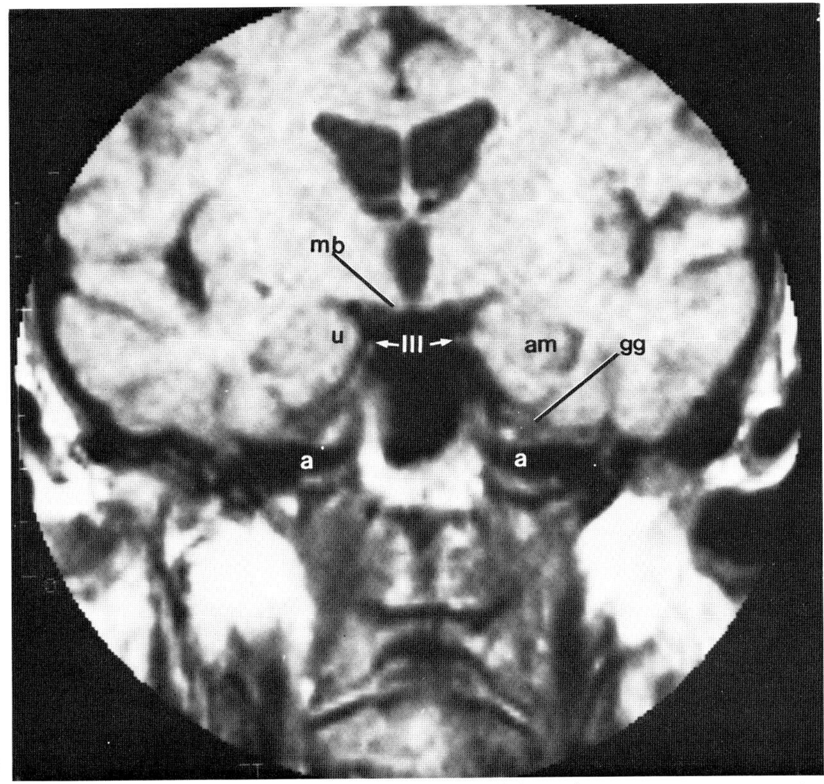

Fig. 11-2. (A-F) Coronal SE 450/26 images of the sella and skull base from anterior to posterior. (See legend for explanation of symbols.) (*Figure continues*).

a, carotid artery;
ac, anterior clinoid process;
am, amygdala;
cp, cerebral peduncle;
cs, cavernous sinus, venous portion;
fp, fat in tip of petrous pyramid;
gg, Gasserian (semilunar) ganglion;
h, hypothalamus;
i, infundibulum;
icv, internal cerebral veins;
if, interpeduncular fossa;
io, inferior olive;
ir, infundibular recess of third ventricle;
j, jugular bulb;
mb, mamillary bodies;
oc, optic chiasm;
on, optic nerve;
ot, optic tract;
p, pituitary;
pca, posterior communicating artery;
po, pons;
ppf, pterygopalatine fossa;
s, sphenoidal sinus;
sof, superior orbital fissure;
u, uncus;
V, trigeminal nerve;
V_1, ophthalmic division of V;
V_2, maxillary division of V;
V_3, mandibular division of V;
III, oculomotor nerve;
3, third ventricle.

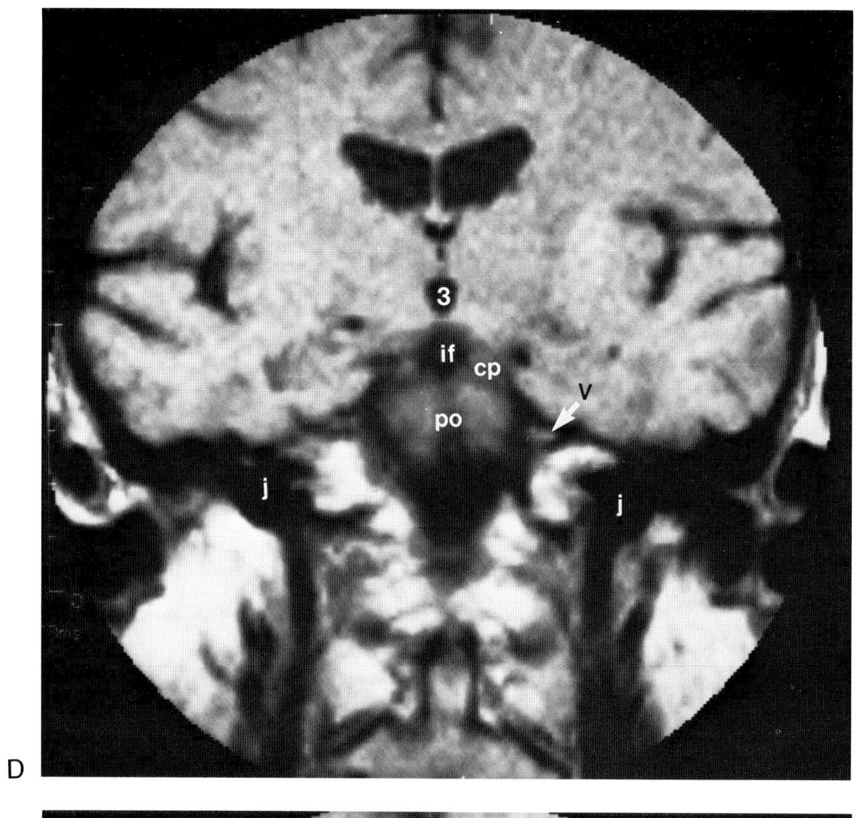

D

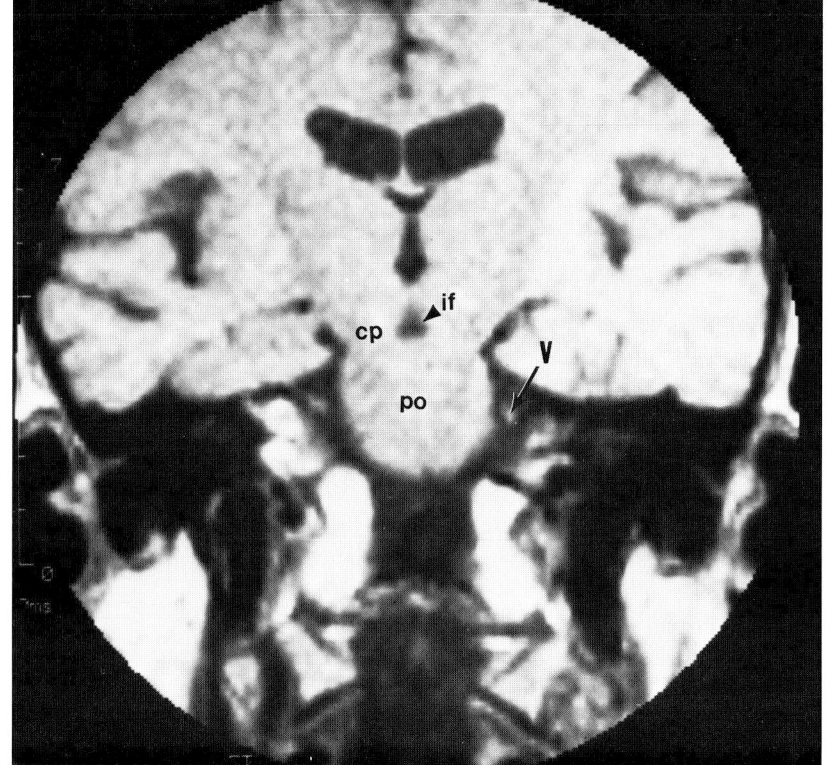

E

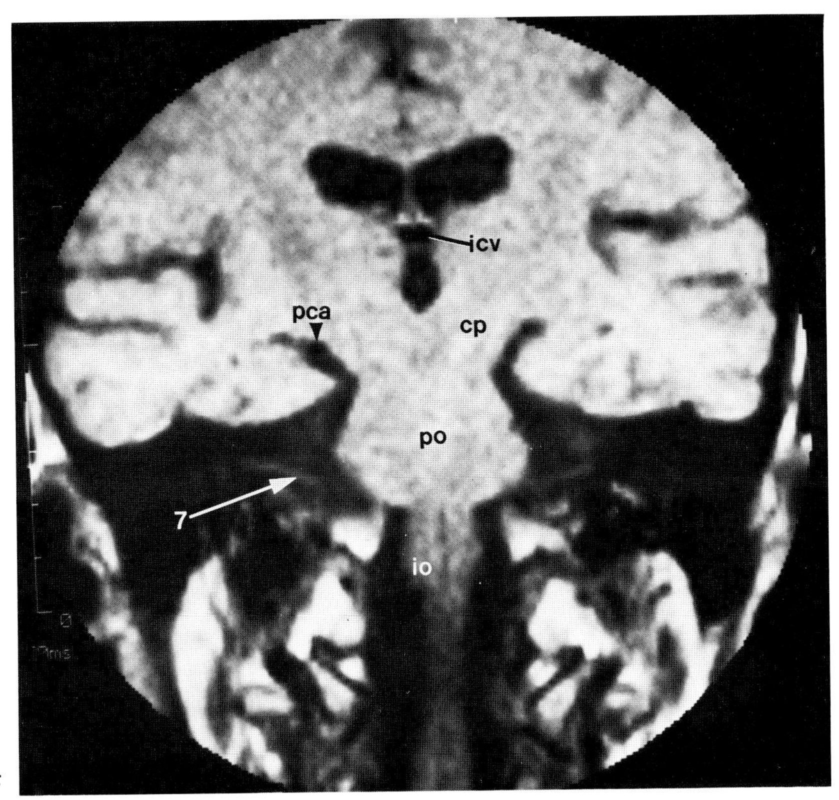

Fig. 11-2 (*continued*). **D, E, F.**

a, carotid artery;
ac, anterior clinoid process;
am, amygdala;
cp, cerebral peduncle;
cs, cavernous sinus, venous portion;
fp, fat in tip of petrous pyramid;
gg, Gasserian (semilunar) ganglion;
h, hypothalamus;
i, infundibulum;
icv, internal cerebral veins;
if, interpeduncular fossa;
io, inferior olive;
ir, infundibular recess of third ventricle;
j, jugular bulb;
mb, mamillary bodies;
oc, optic chiasm;
on, optic nerve;
ot, optic tract;
p, pituitary;
pca, posterior communicating artery;
po, pons;
ppf, pterygopalatine fossa;
s, sphenoidal sinus;
sof, superior orbital fissure;
u, uncus;
V, trigeminal nerve;
V_1, ophthalmic division of V;
V_2, maxillary division of V;
V_3, mandibular division of V;
III, oculomotor nerve;
3, third ventricle.

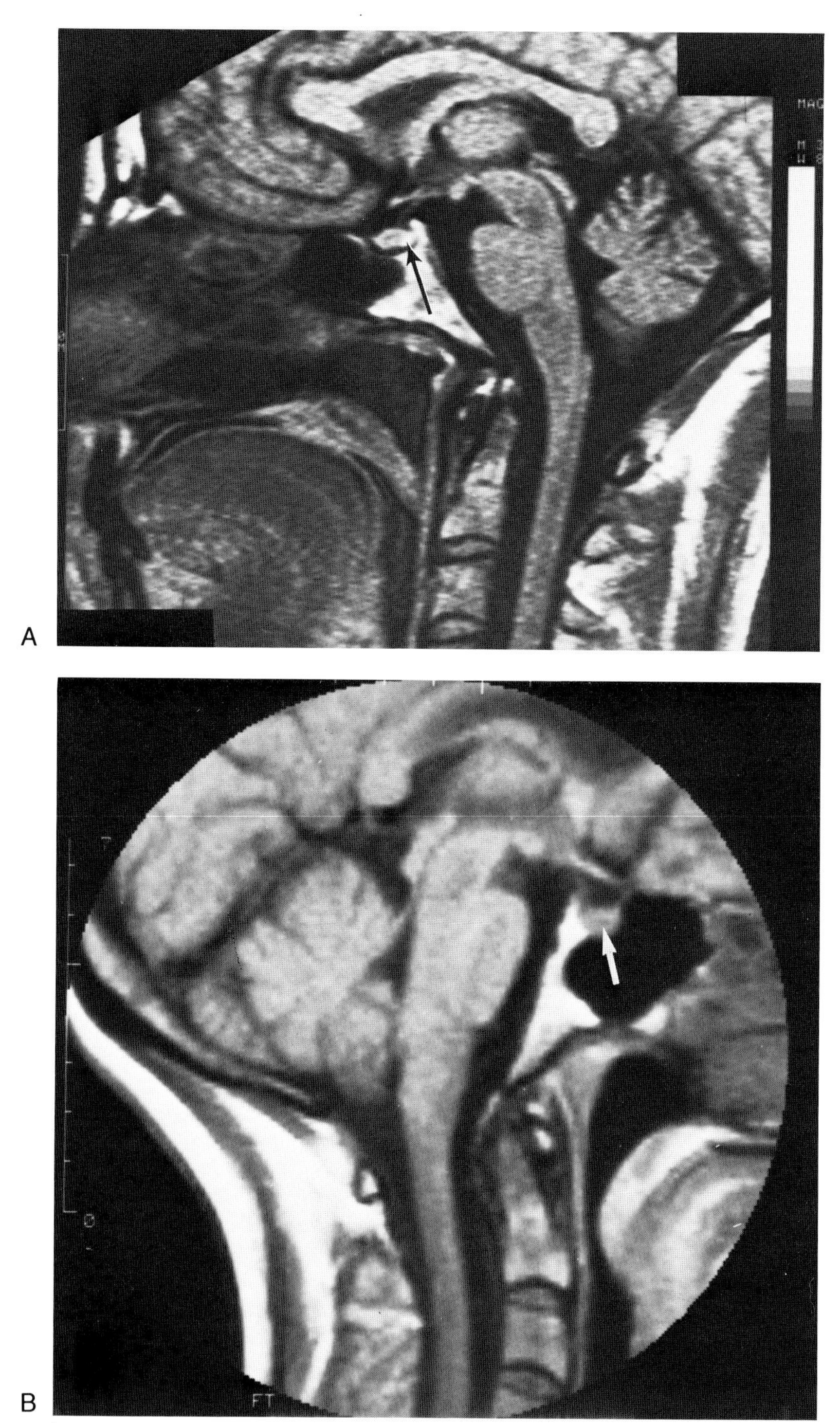

Fig. 11-3. Sagittal SE 500/30 images of the pituitary. **(A)** Neurohypophysis (arrow) has high intensity on this T1-weighted pulse sequence. **(B)** Low intensity signal (arrow) commonly observed within the gland, possible a partial volume effect with parasellar vascular channels.

now demonstrated conclusively that the high signal arises from the posterior pituitary lobe itself.[2]

The normal pituitary gland has MR signal intensity similar to gray matter elsewhere in the brain. The infundibulum is visualized as a vertically oriented tubular structure on sagittal or coronal images (Figs. 11-2, 11-3A). On axial images the infundibulum appears as a dot (Fig. 11-1C).

Pituitary gland height on sagittal SE 500/40 images have been shown to correlate well with coronal CT measurements both in normal and abnormal glands.[3] Because some controversy over normal gland height still exists in the CT literature, it is unlikely that the question will be soon settled in MRI. Mark et al. (1984) found the average height of normal pituitaries by MRI to be 5.7 mm, with a range of 3.3 to 8.3 mm. Weiner et al. (1985) reported an average height of 5.4 mm with a range of 3.0 to 9.0 mm. As a general guideline one may choose to consider that the normal pituitary gland should not exceed 7 mm in men and postmenopausal women, keeping in mind that some women in their childbearing years may have normal glands up to 9.7 mm in height.[4]

General CT criteria may also be followed concerning the configuration of the upper margin of the pituitary. Usually the pituitary gland is concave or flat on its upper surface. Gentle upward convexity may occasionally be noted as a normal variant. A more significant finding is focal convexity of the diaphragma sella, which by CT has a 91 percent correlation with the site of a microadenoma.[5]

The pituitary stalk or infundibulum is normally 4.0 mm or less in diameter. A useful reference for size of the pituitary stalk is the basilar artery. The basilar artery is larger than the pituitary stalk in over 90 percent of normal patients.[6] The pituitary stalk is usually near the midline of the sella although minor deviations occur normally. Displacement of the pituitary stalk may be a useful secondary indicator of pituitary pathology when seen on MRI.[7] Enlargement of the pituitary stalk should arouse suspicion of a possible pathologic condition such as histiocytosis X, pituitary adenoma, infection, hypothyroidism, sarcoidosis, metastases, Rathke cleft cysts, or hypothalamic lesion (Fig. 11-4).

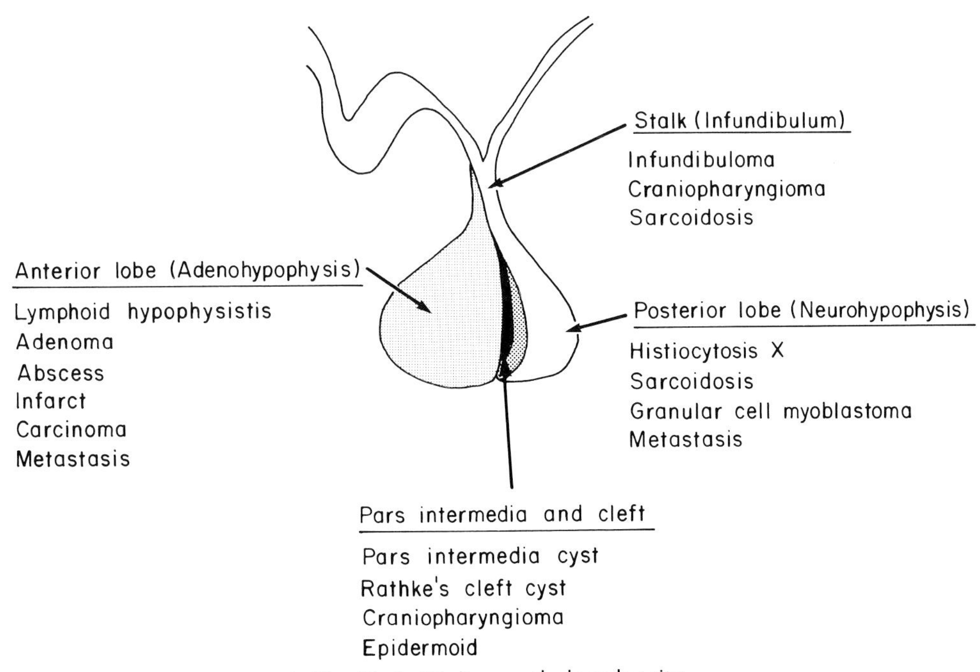

Fig. 11-4. Pituitary pathology by site.

"Incidental" Pituitary Pathology

Any discussion concerning the imaging of pituitary pathology must take into account the surprisingly frequent occurrence of small asymptomatic mass lesions noted within otherwise normal glands. A number of autopsy series have focused attention on the pathologic findings in pituitary glands in random patients without evidence of pituitary disease. Microadenomas have been reported in 14 percent to 27 percent, with a figure of 20 percent being typical.[8–11] In the largest of these series most of the incidental adenomas were less than 2 mm in diameter and all were less than 4 mm.[9] However, a second group of investigators found a significant percentage of incidental microadenomas to measure at least 3 mm.[8]

Incidental pars intermedia cysts have been encountered in 13 percent to 22 percent of random autopsies.[12,13] Most of these cysts are less than 3 mm in diameter, but almost a third may be larger.[8] Small areas of pituitary infarction a few millimeters in diameter may also be seen in 1 percent to 3 percent of patients without clinical evidence of hypopituitarism.[14] Incidental foci of metastasis or abscess may occasionally be found in patients without pituitary dysfunction who have carcinoma or sepsis.[15,16]

In conclusion, the results of several autopsy series suggest the frequent occurrence of small mass lesions a few millimeters in size in otherwise normal pituitary glands. Incidental microadenomas and pars intermedia cysts may each occur in about 20 percent of patients, and a few percent may harbor small infarcts, metastases, or foci of hyperplasia in their glands. A relatively high incidence (20 percent) of low density lesions greater than 3 mm has been reported in "normal" pituitaries studied by CT.[8] Although no comparable MR series has been reported using high resolution, it is anticipated that a similar distribution of small asymptomatic lesions will be detected. The significance of small pituitary lesions detected by MR or CT must always be correlated with clinical and laboratory data, because they will be more frequently incidental than pathologic.

Table 11-1. Classification of Pituitary Adenomas: Correlation of Traditional Light Microscopy and Functional Schemes.

Staining Pattern	*Functional Cell Types*
Eosinophilic (Acidophilic)	Most are GH-secreting Some are PRL-secreting or mixed PRL- and GH-secreting
Basophilic	Nearly all are ACTH-secreting
Chromophobe	60% are PRL-secreting 20% are Nonfunctional 20% are ACTH-, TSH-, FSH-, or FSH/LH-secreting

(Turski PA, Damm M: The role of computed tomography in the evaluation of pituitary disease. Semin Ultras CT MR 6:276, 1985.)

Pituitary Adenoma

Pituitary adenoma is one of the more common primary intracranial tumors and is the single most common tumor arising within the sella. Nearly all adenomas arise within the anterior lobe (adenohypophysis), possibly from pre-existing foci of hyperplasia.[17] His-

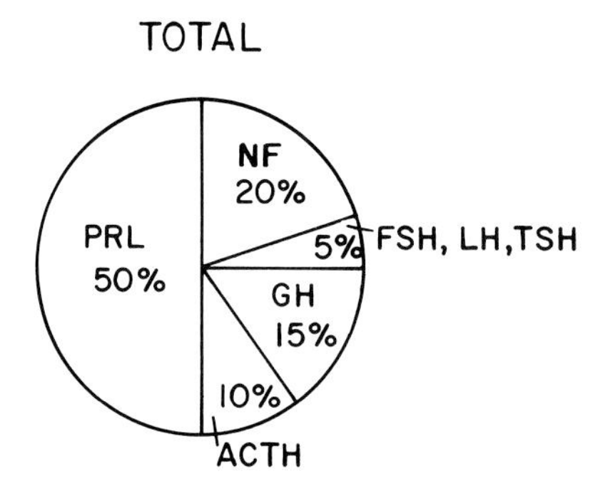

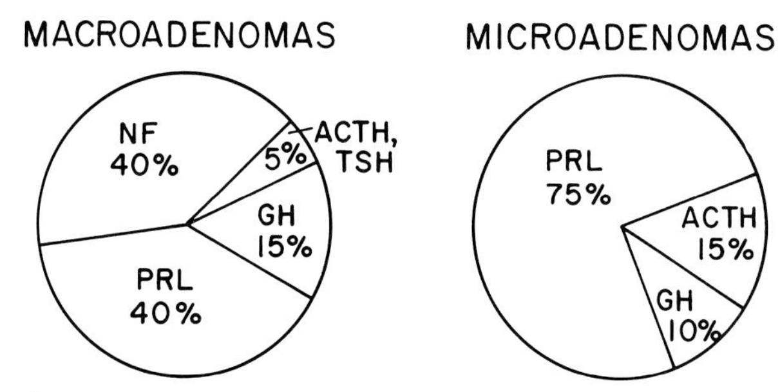

Fig. 11-5. Distribution of pituitary adenoma types by size. NF, nonfunctioning.

torically, pituitary adenomas have been classified into three histologic types based on staining seen by light microscopy: eosinophilic, basophilic, and chromophobic (Table 11-1). A more modern classification scheme is based upon type of hormone secretion and cytoarchitecture by electron microscopy. Pituitary adenomas are also classified grossly by size: tumors less than 1 cm in diameter are called *microadenomas* while those larger are called *macroadenomas* (Fig. 11-5).

Hormone-secreting adenomas characteristically present clinically in well-known syndromes relating to excess hormone production (Table 11-2). Because they present relatively early with hormonal symptomatology, hormone-secreting adenomas tend to be smaller than nonfunctioning tumors. Nonfunctioning adenomas and hormone-secreting macroadenomas may often present by virtue of their mass effects. Pressure effects may induce dysfunction of the optic nerves or chiasm, nerves within the cavernous sinuses, or glandular tissue in the remaining normal portion of the pituitary. Of these visual disturbances are the most common. Bitemporal hemianopsia is classically described, but a wide spectrum of visual field defects may exist.

Prolactin-Secreting Tumors

Prolactin-secreting tumors (prolactinomas) constitute over half of all pituitary adenomas removed surgically and an overwhelming majority of incidentally detected microadenomas.[9,18] They occur most commonly in females of reproductive age and present with a syndrome of amenorrhea and galactorrhea. Headache, infertility, and visual disturbances may occasionally be seen. Most prolactinomas in females are only a few millimeters in size at the time of presentation. In contrast, prolactinomas in males have a more insidious onset clinically and are often macroadenomas with suprasellar extension at the time of detection. The clinical syndrome in males includes decreased libido, gynecomastia, and infertility, as well as symptoms from intracranial mass effect.

Measurement of serum prolactin level will often allow the diagnosis of a prolactinoma to be made before it is large enough to be detected radiographically. Normal serum prolactin levels in most laboratories are less than about 25 ng/ml. Prolactinomas may elevate serum levels from 100 to several thousand ng/ml. With very high serum prolactin levels the diagnosis of prolactinoma is assured. However, some PRL-secreting adenomas may elevate serum prolactin levels only minimally. In this situation the chemical diagnosis may be unclear and confused with drug effects and other clinical conditions that may moderately elevate serum prolactin.[19] Here, precise evaluation by MR or CT may be critical for proper diagnosis.

ACTH-Secreting Tumors

ACTH-secreting tumors are typically microadenomas (<5 mm diameter) that present with Cushing syndrome (obesity, hirsuitism, hypertension). The

Table 11-2. CT-Surgical-Clinical Correlation of Pituitary Adenomas in 113 Patients

Functional Type	*Microadenomas/ Macroadenomas*	*Total Number*	*Average Gland Size by CT*	*Typical Symptoms*
PRL-secreting	45/18	63 (56%)	10.9 mm	Amenorrhea, galactorrhea, infertility, headache, visual disturbances
GH-secreting	8/11	19 (17%)	13.6	Acromegaly, headache
ACTH-secreting	9/2	11 (10%)	10.3	Weight gain, hirsuitism, diabetes mellitus, hypertension, (Cushing syndrome, Nelson syndrome)
TSH-secreting	0/2	2 (2%)	14.5	Hyperthyroidism
Nonfunctioning	0/17	17 (15%)	26.0	Visual distrubance, headache, hypopituitarism

(Davis PC, Hoffman JC Jr, Tindall GT et al: CT-surgical correlation in pituitary adenomas: evaluation in 113 patients. AJNR 6:711, 1985.)

term Cushing disease is used when the Cushing syndrome is caused by an ACTH-secreting pituitary tumor. The diagnosis of Cushing disease may be made by serum assay for ACTH and the decadron suppression test. Nelson syndrome, characterized by hyperpigmentation, may develop in patients with unrecognized ACTH-secreting pituitary adenomas who have undergone adrenalectomy. Nelson syndrome may also occur secondary to hypersecretion of melanocyte stimulating hormone (MSH) by the rare *MSH-secreting pituitary adenomas.*

GH-Secreting Adenomas

GH-secreting adenomas present with gigantism or acromegaly, depending upon whether the tumor occurs before or after puberty. Because acromegaly may take many years to develop and be recognized clinically, these adenomas are often quite large at the time of diagnosis.

Nonfunctioning Adenomas

Nonfunctioning adenomas are usually not discovered until they are quite large and produce symptoms by pressure on the optic chiasm or cavernous sinus. Although nonfunctioning adenomas account for only about 15 percent of all symptomatic adenomas, they represent over one-third of all macroadenomas. Nonfunctioning adenomas may also present as a syndrome of secondary hypopituitarism by their compression of adjacent normal gland. A nonfunctioning macroadenoma may falsely elevate serum prolactin level by compressing the pituitary stalk, and even present as Forbes-Albright syndrome (ammenorrhea-galactorrhea). Such tumors have been called *pseudoprolactinomas.*[20]

Pituitary Apoplexy

Pituitary apoplexy refers to sudden infarction, necrosis, or hemorrhage within a pre-existing pituitary adenoma.[21,22] This may occur spontaneously or following radiation therapy. Hemorrhage and infarction induce rapid swelling of the gland with resultant pressure effects on nearby cranial nerves. Common symptoms of pituitary apoplexy include severe headache, vomiting, diplopia, and visual loss. Subarachnoid hemorrhage is often seen.

Role of MRI in the Diagnosis of Pituitary Adenomas

Any imaging modality used to investigate pituitary pathology seeks to answer several questions: (1) Is a lesion present? (2) If a lesion is present, what is the probable histologic type? (3) How large is it? (4) Does it invade or affect adjacent structures (e.g., cavernous sinuses, carotid arteries, visual system)? The relative merits of CT and MR may be assessed by how well each modality can answer these questions.

Magnetic resonance and computed tomography are largely equivalent in their ability to detect macroadenomas.[23,24] However, MR is clearly inferior to CT in the detection of microadenomas.[25] Despite recent technologic improvements, the spatial resolution of MR still lags slightly behind CT. Thus partial volume effects still limit detection of very small microadenomas. Secondly, MR is unable to detect subtle erosions of the sellar floor, which may be important in CT diagnosis.[23] Finally, at least some microadenomas seem to have identical T1 and T2 values as the normal gland, making MR contrast differentiation impossible.[25]

When visualized, microadenomas usually have prolonged T1 and T2 values.[25] Focal upward convexity of the gland may be noted (Figs. 11-6, 11-7). The coronal plane seems most sensitive for detection, because partial volume and chemical shift artifacts are minimized in this projection.[25] Three-dimensional limited volume techniques may provide superior signal to noise ratio and resolution to 2DFT methods, and may be useful for detection of microadenomas.[26]

Unfortunately, the best high resolution techniques on MR for diagnosing pituitary microadenomas are still inferior to the best CT. In one comparison study only 6 of 11 prolactin-secreting microadenomas detected by CT were also seen on MR.[25] Additionally, it has been recognized that while most microadenomas have elevated T1 and T2 values some have shortened or normal relaxation times. The future use of paramagnetic intravenous contrast agents may improve the accuracy of MR diagnosis.

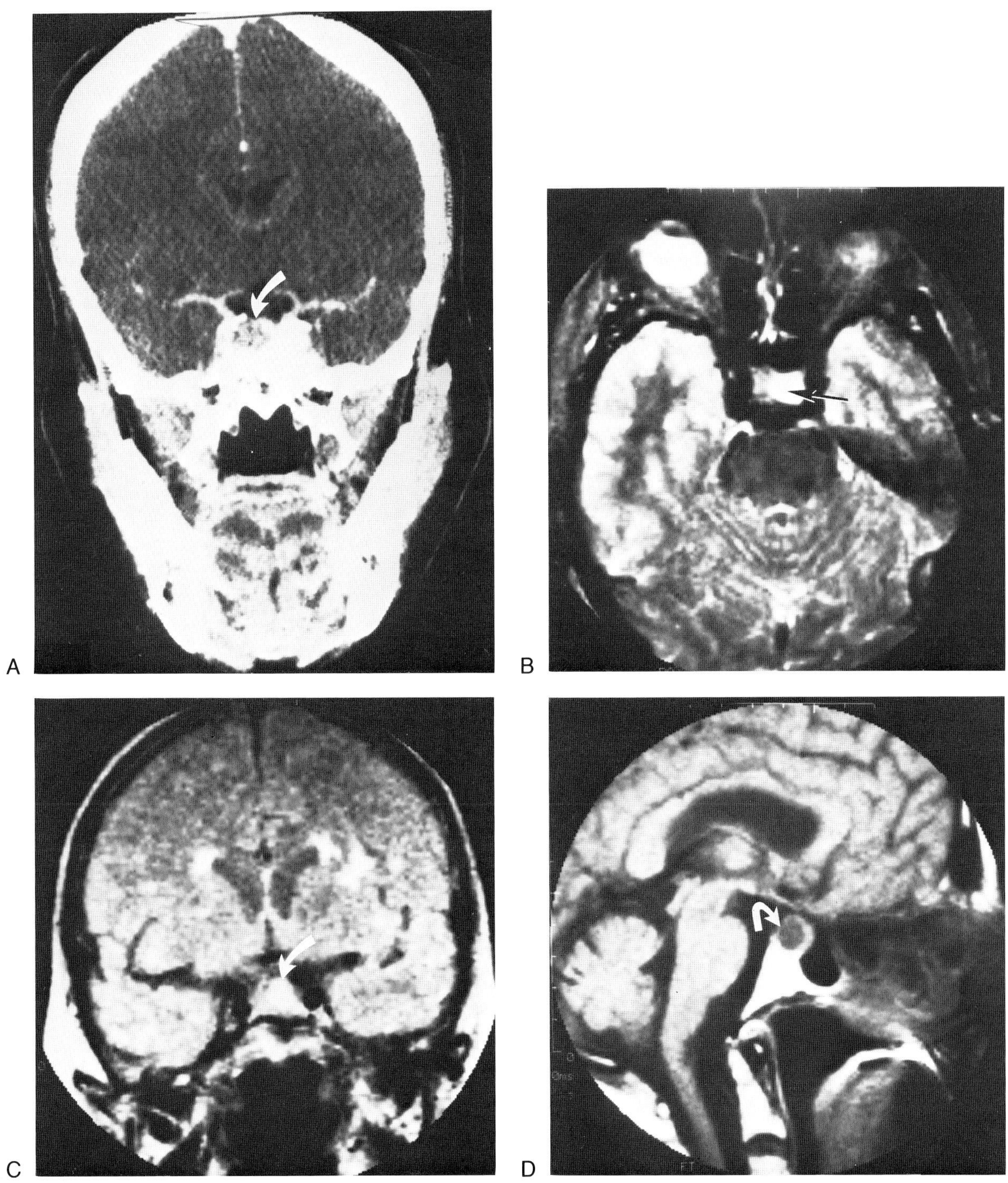

Fig. 11-6. Typical pituitary microademona. **(A)** CT reveals upward convexity of gland. **(B–D)** MRI shows well a discrete lesion with long T2 and slightly prolonged T1 consistent with an ademona.

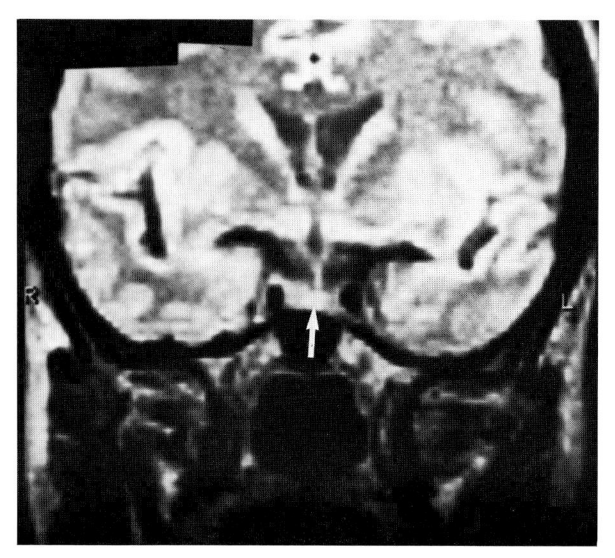

Fig. 11-7. A tiny prolactinoma seen on this coronal SE 2000/28 scan. Serum prolactin was over 200 ng/dl.

Preliminary reports have suggested that MR signal changes may be observed in up to one-third of prolactinomas clinically responsive to bromocriptine.[26a] The apparent signal changes in such treated tumors has been increased intensity on T1-weighted images (i.e., T1 has shortened). T2 values usually also decrease. The explanation for these changes is unclear, and future studies are under way to elucidate the prognostic value of these MR signal changes.

While MR is generally inferior to CT for the evaluation of microadenomas, it is probably equal or superior to CT for evaluating macroadenomas.[24,27] In particular, extrasellar extension is well demonstrated on MRI. The relationship of the tumor to the optic nerves, chiasm, and tract is appreciated on sagittal or coronal MR images (Fig. 11-8). Sagittal images may show displacement of the anterior cerebral arteries. Invasion of the cavernous sinuses is particularly well demonstrated on coronal MR images (Fig. 11-9).

Invasive Adenomas and Pituitary Carcinoma

Pituitary tumors may be classified on the basis of their biologic behavior into expanding adenomas, invasive adenomas, and carcinoma. Expanding adenomas are slow-growing and tend to maintain a relatively globular configuration. They usually grow upward into the suprasellar cistern but may also erode the sella and compress the cavernous sinuses laterally. Occasionally expanding adenomas may have an asymmetric configuration. A waistlike indentation may be produced between the intrasellar and suprasellar components of the adenoma by the diaphragma sella.[29] Focal nodular outgrowths likened to the buds of a potato may occasionally burst through the mesenchymal envelope of the tumor and invade adjacent structures.

Invasive pituitary adenomas directly invade adjacent bony and soft tissue structures while often maintaining benign histologic characteristics. Considerable diversity of opinion exists in the pathology literature concerning the definition of pituitary carcinoma. Some feel that anaplastic histologic features are sufficient for the diagnosis of pituitary carcinoma.[30] Others feel that conclusive diagnosis of pituitary carcinoma cannot be made microscopically, but only by the documentation of distant metastases.[31] It would be presumptuous to assume that a clear distinction between invasive adenomas and carcinoma without metastasis can be made by MRI, when it has not yet been clearly made pathologically.

Expansion or invasion of a pituitary adenoma may occur in a number of directions, and the multiplanar capacity of MRI may exquisitely document the limits of invasion (Fig. 11-10). Lateral invasion may occur into the cavernous sinus, with engulfment of the cavernous carotid artery and displacement or encirclement of cranial nerves III through VI (Fig. 11-9). Coronal MRI may be helpful in distinguishing this from two other types of lateral extension that do not invade the cavernous sinus: extradural subcavernous extension and intradural supracavernous extension.[32] The recognition of these types of lateral extension has considerable surgical importance (Fig. 11-11).

Superior extension of pituitary tumor frequently involves the optic nerves or chiasm. Very rarely, the tumor may indent the third ventricle anteriorly and even obstruct the foramen of Monro.[33] Superior and anterior extension may occur beneath the frontal lobes while superior and posterior extension may involve the interpeduncular cistern and impinge on the midbrain.

Posterior extension through the dorsum sellae may

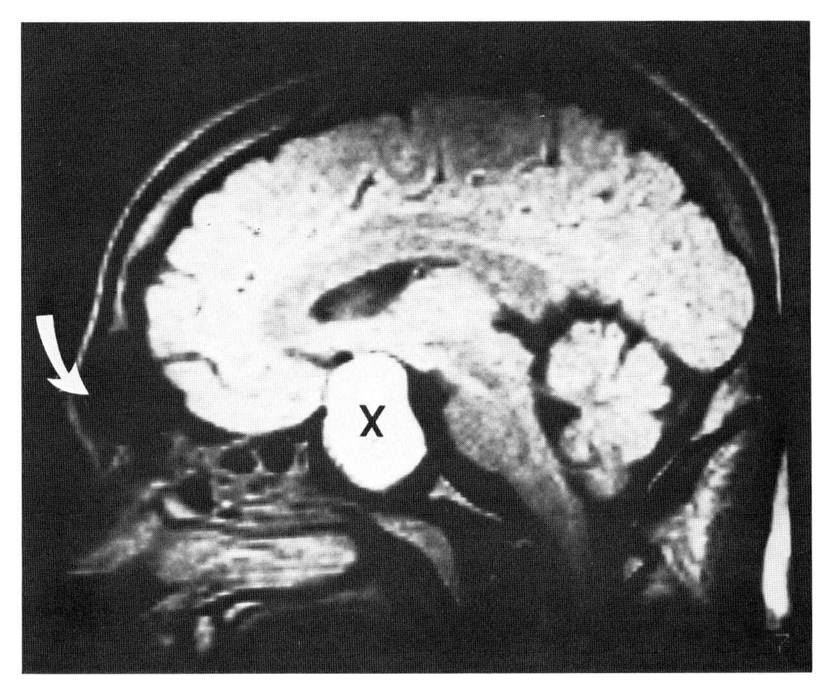

Fig. 11-8. GH-secreting macroademona (*X*) in a patient with acromegaly. Also note enlarged frontal sinus (arrow) characteristic of acromegaly.

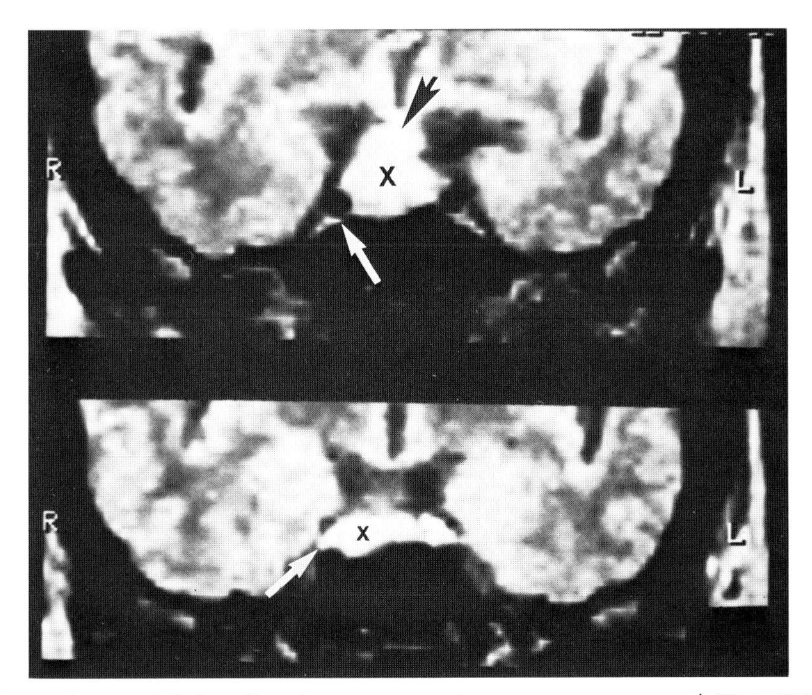

Fig. 11-9. Pituitary adenoma (*X*) invading the cavernous sinuses seen on coronal SE 1500/56 image. The tumor is wrapping around and over the right cavernous carotid artery (white arrows). Also note suprasellar extension (black arrowhead).

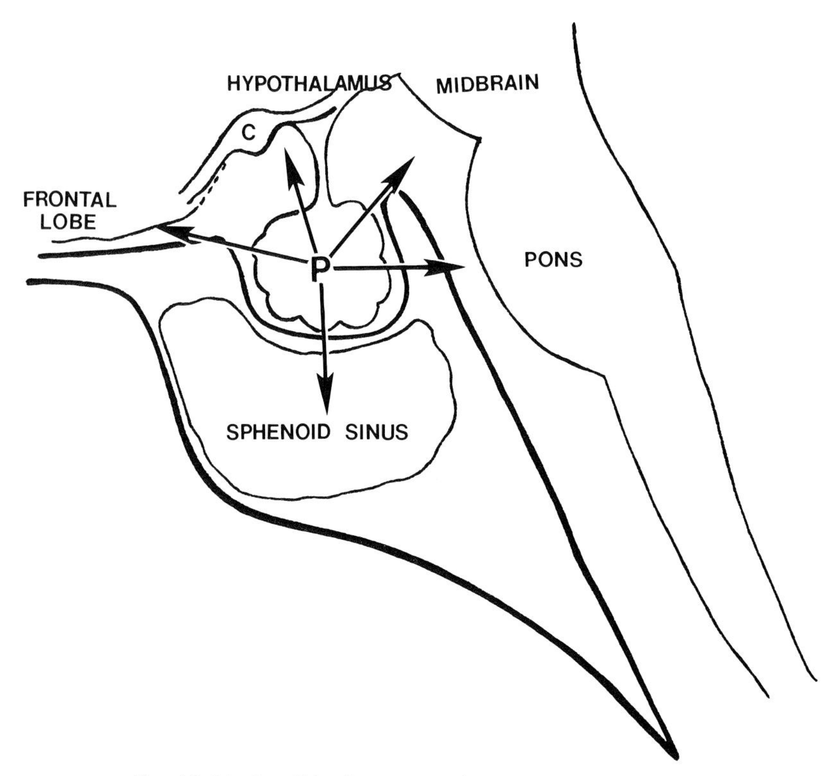

Fig. 11-10. Possible directions of pituitary tumor spread.

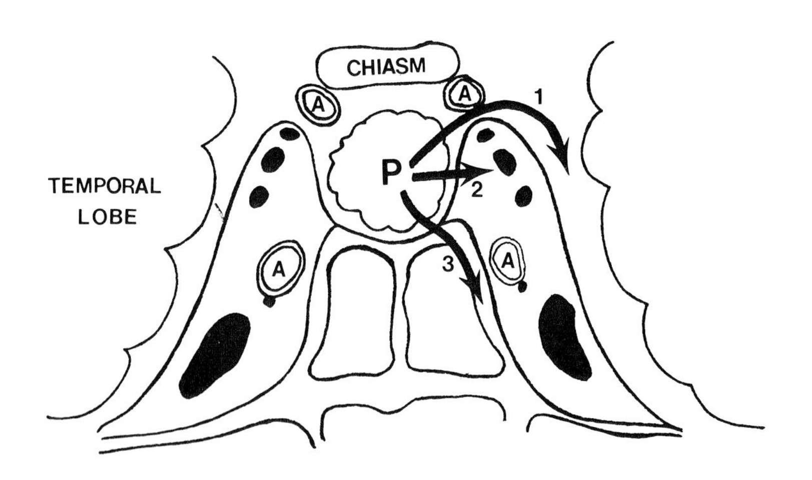

Fig. 11-11. Modes of lateral extension. (1) Intradural supracavernous extension. (2) Direct invasion. (3) Extradural subcavernous extension.

occur, and the tumor may enter the pontine or cerebellopontine angle cisterns. Even more rarely, anterolateral extension of tumor through the superior orbital fissure into the orbit may be seen.

Inferior extension of tumor is common through the bony floor of the sella and into the sphenoid sinus. With very large tumors, invasion of the nasopharynx or nose may also occur (Fig. 11-12).

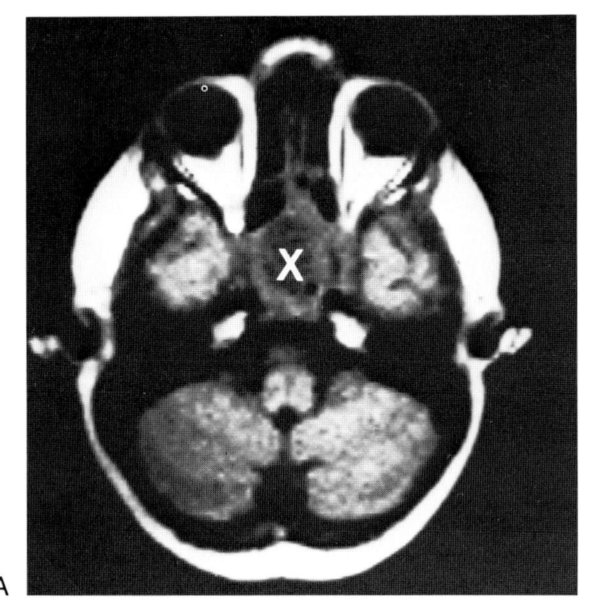

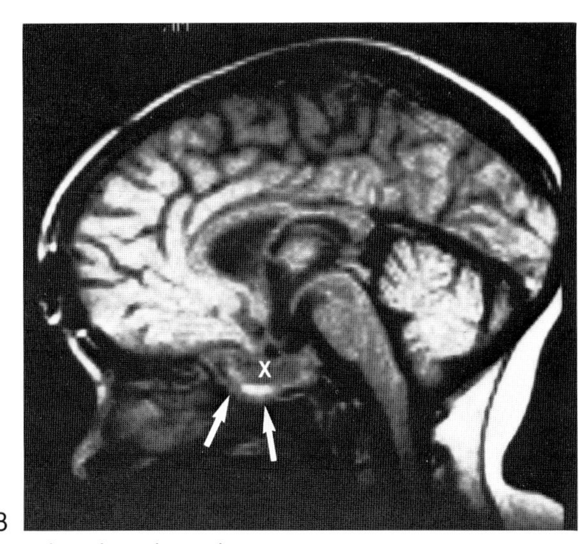

Fig. 11-12. (**A** and **B**) Pituitary tumor (*X*) invading the sphenoid sinus (arrows).

Rare malignant carcinomas may occasionally arise within the sella that are not of adenomatous origin. These include diffuse histiocytic lymphoma, juvenile pilocytic astrocytoma, meningioma, melanoma, and fibrosarcoma. Of these unusual malignancies, fibrosarcomas are the most significant because they typically occur as complications of radiotherapy for pituitary adenoma. Pituitary fibrosarcoma occurs with a latent period of 3 to 20 years (median 10 years) after radiation therapy where the mean dose was 7220 R.[34] No MR reports of these unusual pituitary neoplasms have yet been published.

Metastasis to the Pituitary

Metastases to the pituitary occur frequently in patients with disseminated carcinoma. The most common primary tumor sites are the lung, breast, and prostate, although conceivable any hematogenously borne metastasis could reach the pituitary. Metastases may be found in 25 percent of women who undergo hypophysectomy for advanced breast cancer.[35] Fortunately, metastases to the pituitary rarely produce clinical symptoms and are more likely to represent incidental findings on CT, MR, or at autopsy.

Infarction

As mentioned previously, small areas of infarction a few millimeters in diameter are found in pituitary glands of 1 percent to 3 percent of asymptomatic patients.[8,14] More massive necrosis of the pituitary may occur postpartum in the setting of shock, sepsis, or hemorrhage (Sheehan syndrome).[36] In this situation the anterior gland is affected far more extensively than the posterior lobe. Eventually, the anterior lobe becomes a thin collagenous scar. Acute or subacute pituitary necrosis has also been reported in a number of other conditions including sinus thrombosis, temporal arteritis, diabetes mellitus, and raised intracranial pressure.[37] These infarctions of a normal pituitary gland should not be confused with pituitary apoplexy, which represents infarction within a pituitary adenoma.

Empty Sella

The term empty sella refers to an abnormal extension of the subarachnoid space into the sella turcica. Normally the pituitary fossa is separated from the subarachnoid space of the suprasellar cistern by the fibrous diaphragma sellae. If the diaphragma sellae is deficient, herniation of subarachnoid space into the sella may occur. Such incompetence of the diaphragma sellae may be congenital or acquired.

Empty sella in children is extremely rare, although congenital deficiencies in the diaphragma sellea occur in over 70 percent.[38] Empty sella is relatively common in adults, found in at least six percent. Females are affected nine times more often than males.[39] It has been hypothesized that the periodic enlargement and atrophy of the female pituitary gland with menses and pregnancy causes ischemic injury to the diaphragma sellae. Potentially, therefore, areas of congenital weakness in the diaphragma may develop into frank perforations with ischemia, and herniation of subarachnoid space downward into the sella may occur.

A secondary type of empty sella that is clearly acquired can develop from a pathologic process within the pituitary gland. Most commonly this process is a large pituitary adenoma that presses upward against the diaphragma sellae and compromises its blood supply. Ischemic necrosis of a portion of the diaphragma sellae may again result. As long as the sella remains filled by tumor, destruction of the diaphragma may pass unnoticed. If involution of the adenoma occurs, subarachnoid fluid may now herniate through the perforated diaphragma and fill the intrasellar space formerly occupied by tumor. Secondary or acquired empty sella may also develop after rupture of an intrasellar cyst, ischemic necrosis of the pituitary gland, or postoperatively.

Empty sella is often an incidental finding on CT or MRI studies of patients examined for other reasons (Fig. 11-13). Because cerebrospinal fluid (CSF) pulsations are transmitted into the pituitary fossa, remodeling and enlargement of the sella turcica may occur. Empty sella is considered to be a benign cause of sellar enlargement seen on plain radiographs of the skull.

A few patients, particularly those with acquired empty sella, may become symptomatic. Mild or subclinical pituitary hypofunction may be present.[40] Herniation of the optic nerves into the sella or compression of a displaced optic chiasm against the fundibulum may produce visual disturbances (Fig. 11-14). Thinning of the walls of the sella from cerebrospinal fluid pulsations may result in CSF-rhinorrhea.[42]

Abscess

Pituitary abscess may develop in a normal gland or in the necrotic center of a pituitary adenoma.[43] The infection may occur by hematogenous dissemination from a distant focus or direct extension of contiguous source such as sphenoid sinusitis, meningitis, or cavernous sinus thrombophlebitis. Pituitary dysfunction may occur. The pituitary is often elevated superiorly by the abscess cavity.[44] The MR findings are nonspecific, but the diagnosis can be suggested in the presence of clinical meningitis, sinusitis, or abscesses elsewhere in the brain.

Granular Cell Myoblastoma (Pituicytoma, Choristoma)

Myoblastomas are exceedingly rare benign neoplasms thought to arise from pituicytes of the neurohypophysis. These tumors occur in adults with a 2:1 preference for females.[45] Many are incidental small lesions found at autopsy. Larger lesions are found in the posterior sella or suprasellar regions and present with mass effect (visual disturbances, hypopituitarism). No myoblastomas have been described by MRI. CT and angiographic appearance may be suggestive because in contrast to most pituitary adenomas myoblastomas are hypervascular and enhance intensely with intravenous contrast.[46]

Pars Intermedia Cysts

As mentioned previously, pars intermedia cysts are encountered in 13 percent to 22 percent of random autopsies. These benign cysts are fibrous lined and filled with a colloid material. They occur at the junction of the anterior and posterior pituitary gland. Most are less than 3 mm in size, although occasionally they may become quite large (Fig. 11-15). Pars intermedia cysts are nearly always asymptomatic. Their significance is in their simulation of microadenomas from which differentiation can be difficult. Pars intermedia cysts are strictly midline. On CT they do not show any contrast enhancement. MR experience is limited,

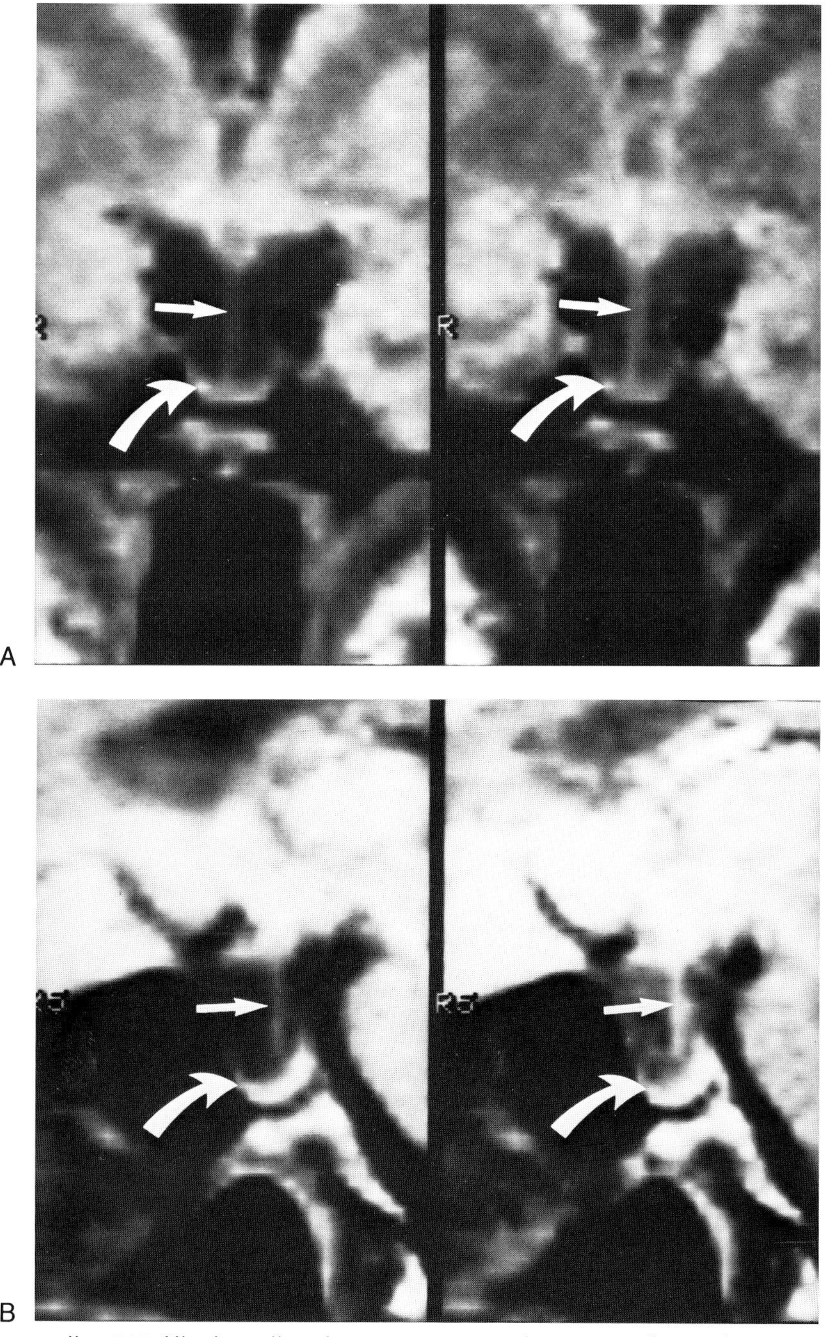

Fig. 11-13. Empty sella. CSF fills the sella. The pituitary (curved arrow) is flattened against the floor. The infundibulum is stretched and elongated (straight arrow). **(A)** Coronal SE 1500/28 and 1500/56 images. **(B)** Sagittal SE 1000/28 and 1000/56 images.

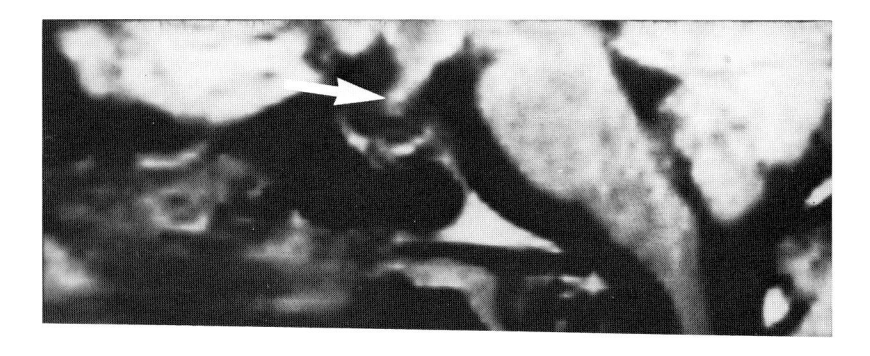

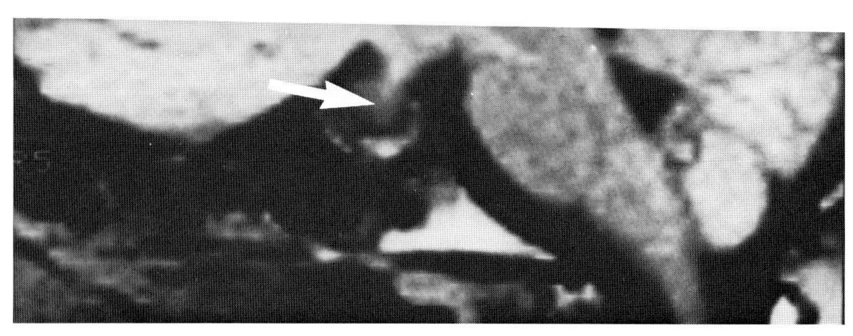

Fig. 11-14. Prolapse of the optic chiasm (arrow) into an empty sella posthypophysectomy.

but it is thought that they may have signal characteristics more fluidlike (i.e., longer T1 and T2) than the typical adenoma.[47]

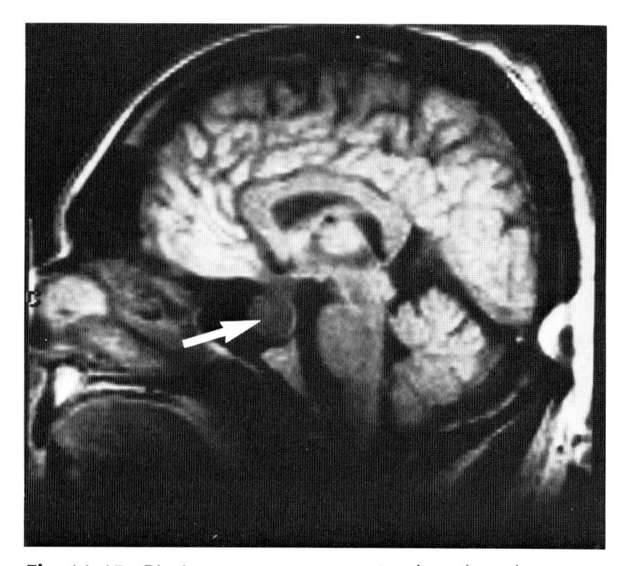

Fig. 11-15. Pituitary cyst seen on MRI, thought to be empty sella on CT. The cyst has slightly different T1 and T2 values from CSF.

Rathke Cleft Cyst

Rathke cleft cysts are generally believed to be derived from remnants of the Rathke pouch.[48] They are found in the sella in the same location as pars intermedia cysts, but are lined with cuboidal or columnar epithelium. Some controversy about their true origin exists.[49] The cysts are filled with fluid or thick mucoid material. An aseptic inflammatory process may occur in the cyst and adjacent pituitary, raising its CT density and causing contrast enhancement.[50] Occasionally, the cysts may be suprasellar.

While small cysts may be asymptomatic, most symptomatic cysts tend to be large (>1 cm) and present with visual impairment or hypophyseal dysfunction. Rarely the cyst may be so large that it may

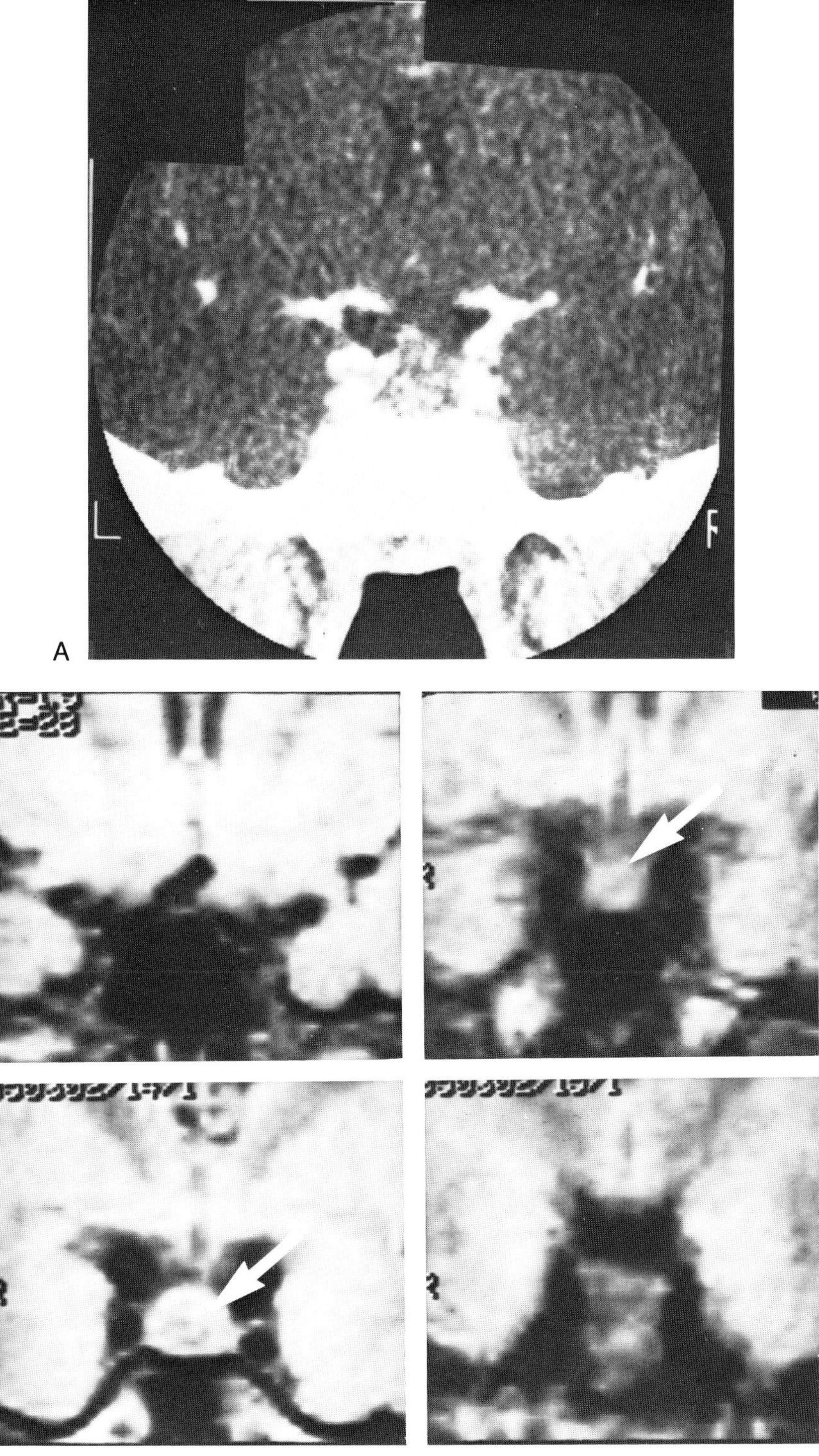

Fig. 11-16. Rathke Cleft cyst on **(A)** CT and **(B)** SE 1500/28 coronal MR. Diagnosis made by pathologist, because radiographically this resembled an adenoma.

indent the third ventricle, compress the foramen of Monro, and produce hydrocephalus.

In the ideal case, a Rathke cyst should have low density contents on CT and enhance either not at all or only peripherally. MR should reveal a fluid-filled (long T1 and T2) cystic lesions with little surrounding mass. However, considerable pathologic variation exists with regard to these cysts. Inflammatory reaction, proteinaceous cyst contents, and peripheral calcification may occur, causing them to resemble pituitary adenomas or craniopharyngiomas. In our single MR case (no others are in the literature), the Rathke cleft cyst was indistinguishable from a pituitary adenoma both by MR and CT (Fig. 11-16).

Pituitary Hyperplasia

The pituitary gland may undergo hyperplasia in a variety of conditions, including overstimulation by hypothalmic releasing factors, pregnancy, target endocrine organ failure, and regeneraion after partial destruction of the gland. Histologically, areas of hyperplasia may be indistinguishable from those of adenoma. Grossly, hyperplasia is usually diffuse whereas adenoma represents a more circumscribed process. Pituitary hyperplasia should be suspected in the appropriate clinical setting with diffuse homogeneous enlargement of the gland on MRI.[51]

Lymphocytic Adenohypophyitis of Pregnancy

Late in pregnancy a poorly understood autoimmune inflammation may affect the anterior pituitary. Histologically, the adenohypophysis becomes infiltrated with lymphocytes and antipituitary antibodies may be demonstrated. This infiltration initially results in enlargement of the gland, which may simulate a pituitary adenoma.[52] Later in pregnancy the gland may atrophy with resultant hypopituitarism.[53] The diagnosis should be suspected when a pregnant patient presents with a pituitary mass and hypopituitarism in the last trimester of pregnancy. Simple decompression of the chiasm and subtotal removal of the mass appears to be adequate therapy, and symptoms normally decrease when pregnancy ends.

THE SUPRASELLAR REGION

MRI is particularly valuable in assessing lesions that arise in the suprasellar region. In one comparison study of suprasellar mass lesions MR was judged equal to CT in 75 percent and better than CT in 20 percent of cases.[54] The advantage of MR largely relates to its ability to image the regional pathology in coronal and sagittal planes. This provides important diagnostic information concerning the relationship of masses here to the optic apparatus and hypothalamus.[24] Nevertheless, it must be cautioned that certain tumors, such as tuberculum meningiomas may be missed on noncontrast MR scans.[54]

Craniopharyngioma

Craniopharnygiomas are locally aggressive sellar or suprasellar tumors that arise from squamous epithelial remnants of the craniopharyngeal duct. The craniopharyngeal duct is a constriction of Rathke's pouch, which develops embryologically as an upward extension of foregut to form the anterior pituitary gland. These epithelial remnants are usually found in the suprasellar region near the infundibulum. They may also be found within the sella or, rarely, within the sphenoid bone or third ventricle. This corresponds to the spatial distribution of craniopharyngiomas observed clinically. Overall, about 90 percent are suprasellar and 10 percent are intrasellar.[55]

Craniopharyngiomas may occur at any age, but they are more common in older children near puberty and in middle-aged adults. Males are more frequently affected than females, with a ratio of about 2 to 1. Craniopharyngioma is the most common juxtasellar tumor seen in children. It is also reasonably frequent in the adult population.

Craniopharyngiomas may be cystic, solid, or mixed.[56] The cystic tumors are often called *adamantinomatous* because they are composed of columnar epithelium arranged in stellate forms resembling the enamel organ. Most cystic craniopharyngiomas occur in children and teenagers. They have thick walls and are filled with a material resembling motor oil in which floats cholesterol crystals. Curvilinear, plaque-like calcifications are seen along the cyst wall in a

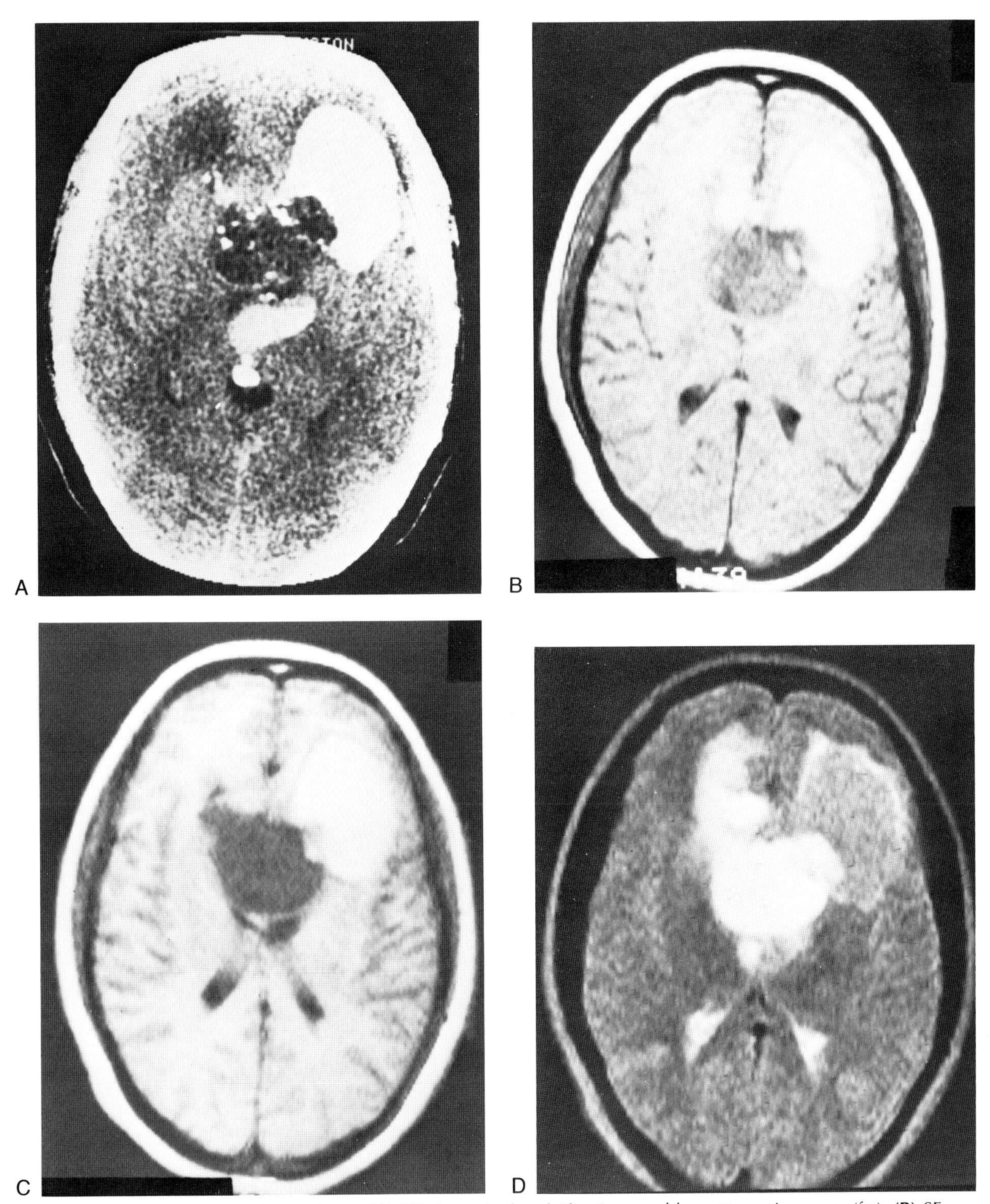

Fig. 11-17. Huge craniopharyngioma. **(A)** CT shows both calcifications and low attenuation areas (fat). **(B)** SE 700/30. **(C)** IR 1700/450/30. **(D)** SE 2000/120. (*Figure continues*).

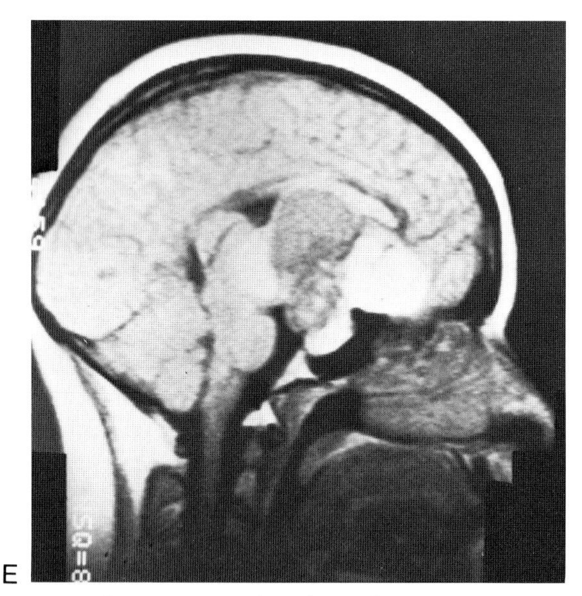

Fig. 11-17 *(continued)*. **(E)** SE 700/30.

high percentage of cases. Cystic craniopharyngiomas may become quite large and extend upward to invade the third ventricle.

The solid and mixed varieties contain simple squamous epithelium. They occur in all age groups and carry a better prognosis than the purely cystic types. The solid types tend to be smaller and less frequently calcified than the cystic tumors. Calcification, when it occurs, may either be peripheral or central (within degenerated solid portions). Calcification in all forms of craniopharyngioma is more frequent in children than in adults; it occurs in the tumors of 20 percent of patients under 2 years old, 80 percent of children over 2 years, and in about 40 percent of adults.[57]

Sellar enlargement is seen with craniopharyngioma, usually in the form of an elongated, shallow ("open") sella caused by shortening of the dorsum sellae. This appearance is characteristic for the larger cystic and mixed tumors, but the smaller solid tumors may produce a symmetrically rounded ("ballooned") sella. The sella walls may be thickened with craniopharyngioma in contrast to pituitary adenoma where thinning is often seen.[58] This may be helpful in differential diagnosis on standard radiographs. Rarely, a corticated channel in the sellar floor representing persistence of the craniopharyngeal duct may be seen. Its presence, together with a juxtasellar mass should suggest the diagnosis of craniopharyngioma.[58] When large, this persistent duct can be visualized on sagittal MR images.[59]

In arriving at an initial tissue diagnosis, CT is probably superior to MRI. This is because CT can detect the characteristic calcifications that tend to distinguish craniopharyngiomas from other suprasellar tumors. (MRI is very poor at detecting these calcifications.) Nevertheless, the cystic portions of most craniopharyngiomas contain material with short T1 and long T2 values, a relatively unique MR signature.[24,60] This cyst material, possibly containing cholesterol or hemorrhage, appears hyperintense to white matter on both T1- and T2-weighted sequences (Figs. 11-17, 11-18).

In one comparison study, tumor extent was as well or better evaluated by MR than CT in 11 of 12 cases.[60] This was largely due to the ability to image directly with MR in sagittal or coronal planes. In another study, MR was judged to be superior to CT in demonstrating the relationship of tumor mass to the major vascular structures and optic apparatus.[61] Furthermore, MR was also excellent for small tumors, visualizing cyst septation, or loculation, and in evaluating the postoperative patient for recurrent or residual tumor.[61]

Hypothalamic Hamartomas

A hamartoma is not a true neoplasm but rather a congenital rest or normal tissue lying in an abnormal location. Hypothalamic hamartomas are extremely rare collections of neural and glial tissue that arise in the region of the tuber cinereum. Most lesions are pedunculated "collar button" masses measuring a few millimeters up to 2 cm in diameter.[62] The mass is usually connected by a stalk to the tuber cinereum and extends downward into the prepontine and interpeduncular cisterns. Occasionally the mass may lie entirely within the hypothalamus and contain calcification.[63]

Hypothalamic hamartomas usually present in the first decade of life with precocious puberty, seizures, diabetes insipidus, or mental changes. Males have a slightly higher incidence than females. The etiology of precocious puberty is unclear but may relate to

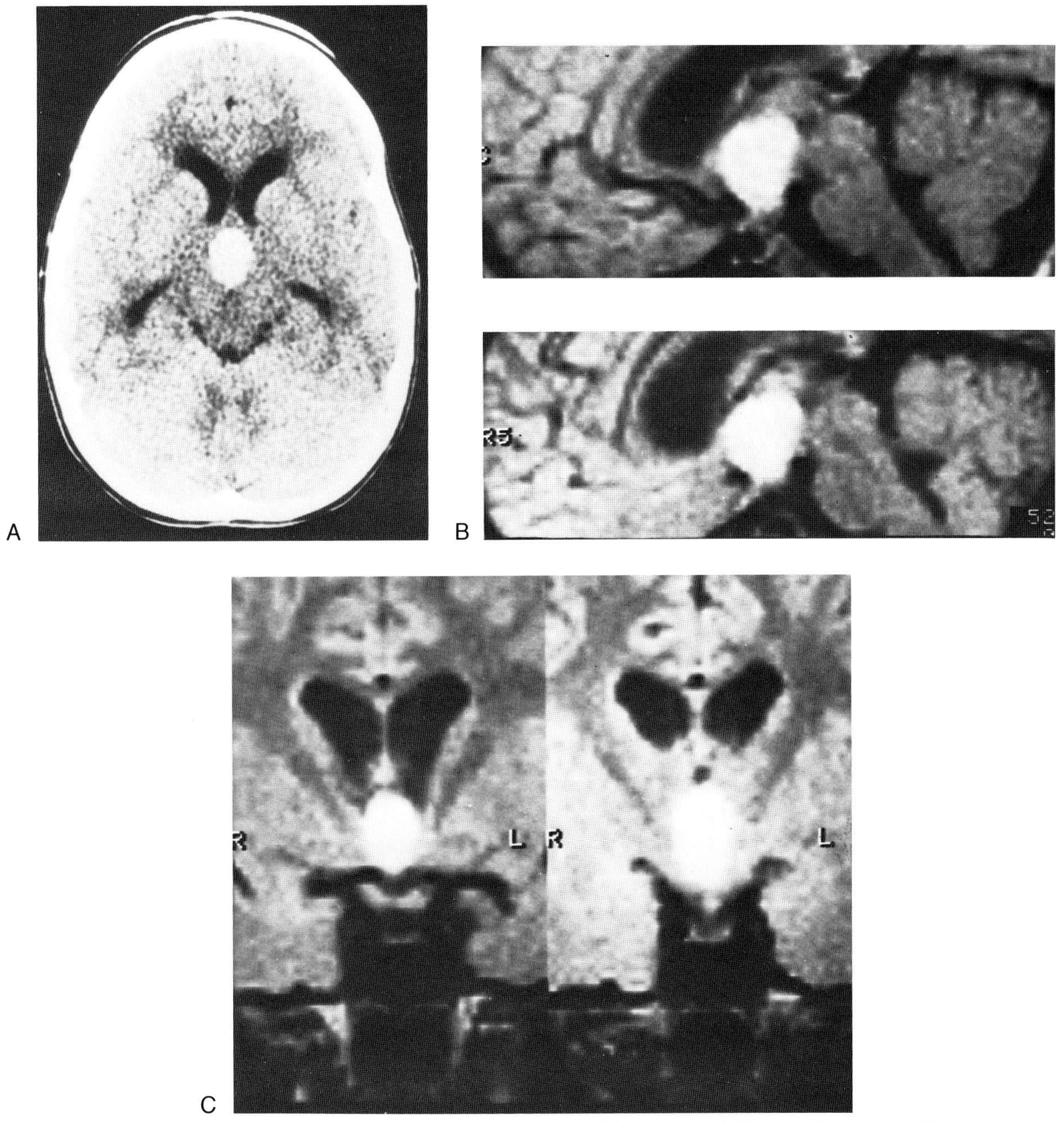

Fig. 11-18. (A–C) Smaller craniopharyngioma in a 7-year-old boy. CT and SE 1500/28 images. Other than location, there are no characteristic CT or MR features in this tumor to suggest the diagnosis.

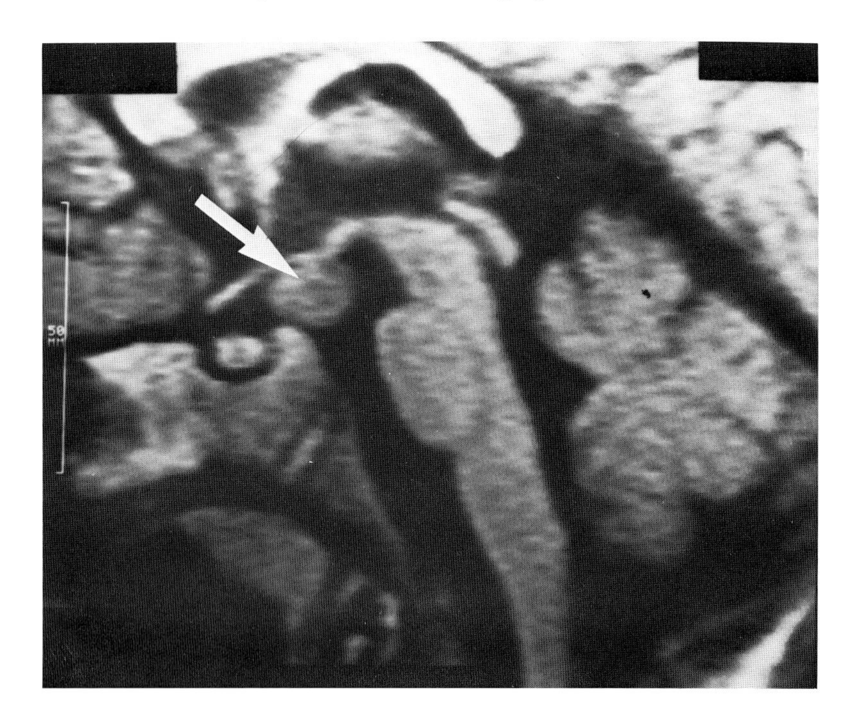

Fig. 11-19. Hamartoma of tuber cinereum. Characteristic mushroom-shaped mass in a boy with precocious puberty. (Courtesy Fonar Corporation, 110 Marcus Drive, Melville, NY.)

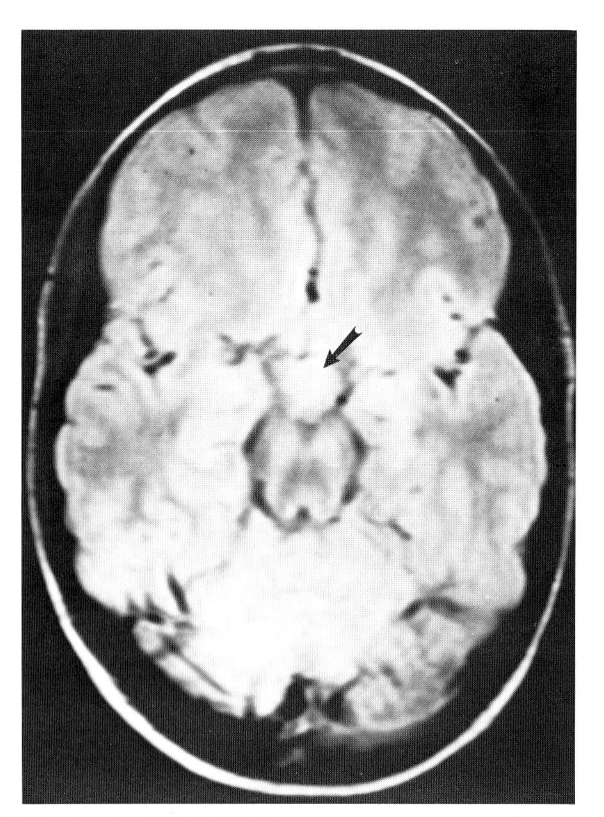

Fig. 11-20. Hypothalamic hamartoma seen in another patient. Signal characteristics are similar to normal brain.

direct neural stimulation of the hypothalamus or autonomous production of gonadotropin releasing factors.[64]

No specific MR findings have been reported. In one case the mushroom-like tumor had T1 and T2 values similar to brain and was well seen on sagittal MR images (Fig. 11-19). In another case the tumor was entirely within the hypothalamus, manifest by distortion of the third ventricle but otherwise invisible because it was isointense to brain (Fig. 11-20). In a single case reported in the literature, the tumor showed high intensity on T2-weighted images.[65]

GENERALIZED DISEASES AFFECTING THE SKULL BASE

Metastatic Carcinoma

Metastases may affect any area of the skull base but are particularly common in the clivus and temporal bones. On plain radiographs, metastases may be osteolytic, permeative, or osteoblastic. However, no correlation between plain radiographic patterns of involvement and MR signal changes has yet been pos-

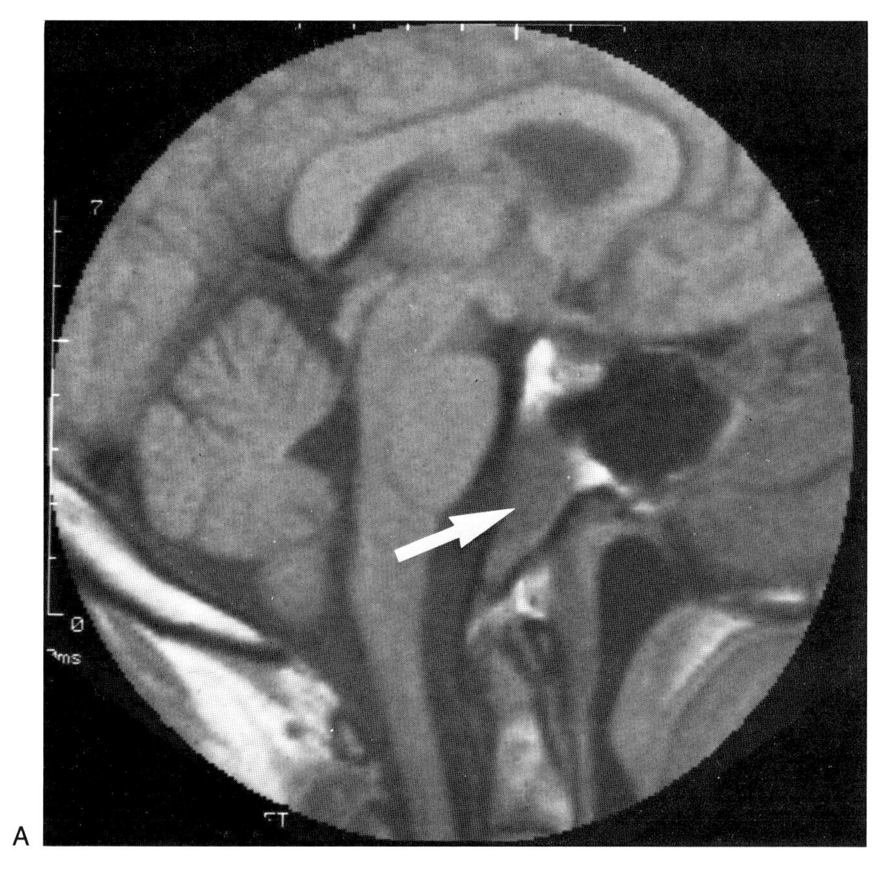

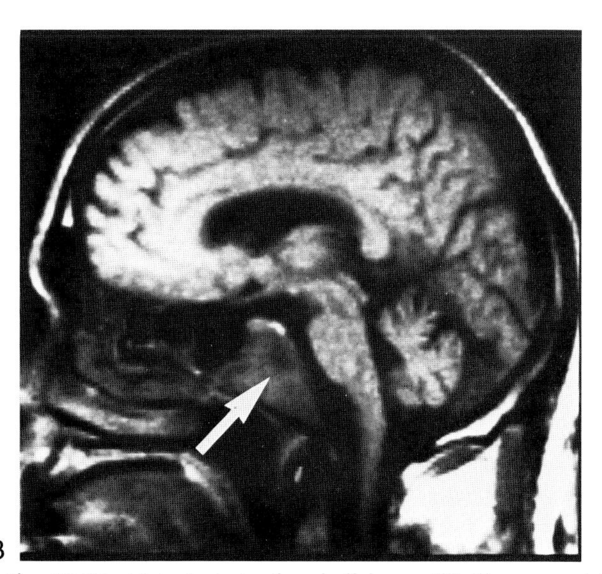

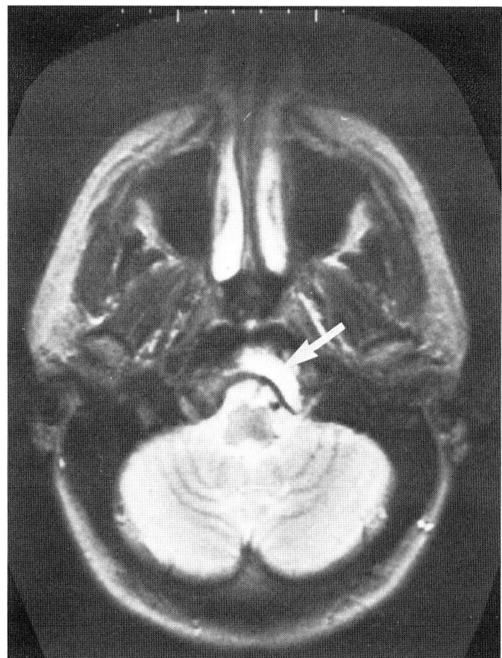

Fig. 11-21. Metastasis to the skull base. **(A)** The normal fat in the clivus is replaced by tumor (arrow). **(B)** Metastatic prostate cancer in another patient. **(C)** High signal from tumor on T2-weighted image.

ited. As demonstrated in Chapter 6, most bone metastases have long T1 and T2 values. On T1-weighted images, metastases may be identified as low intensity areas where they have replaced high intensity marrow (Fig. 11-21A,B). On T2-weighted images metastases usually have high signal (Fig. 11-21C). In the sinuses or temporal bone such foci of high signal cannot be separated from benign air-cell opacification by fluid. In general, CT or radionuclide bone scan will be superior to MR for identifying skull base metastases. MR may be helpful in equivocal cases being more sensitive than CT in some parts of the skull base and having superior spatial resolution to nuclear bone scan in this same region.

Invasive Carcinoma

The skull base is frequently invaded by tumors that arise in the sinuses or nasopharynx. With direct sagittal and coronal imaging MR may elegantly display the anatomic patterns of invasion, thus aiding surgical approach or other therapy. Examples of resectable and nonsectable sinus tumors evaluated by MRI are shown in Figures 11-22 and 11-23.

Fibrous Dysplasia

Fibrous dysplasia is a bone disorder of unknown etiology where medullary cavities are variably replaced by fibrous tissue and poorly calcified primitive new bone.[66] The lesions are not present at birth, but appear several years before puberty and progress through the life of the patient. Both monostotic and polyostotic forms have been described. The bone disease may be associated with irregularly outlined ("coast of Maine") cafe-au-lait spots and precoccious puberty, a complex known as McCune-Albright syndrome.[67]

In the skull fibrous dysplasia has a propensity for the skull base, but may also involve the calvarium. Cystic expansions tend to spare the inner table.[67] There may be encroachment of the neural foramina and basilar invagination. At the skull base involvement of adjacent facial bones in common.[67]

The appearance is usually quite characteristic on plain radiographs. On MRI we have observed diffuse expansion of the marrow cavities with a homogeneous material of intermediate intensity on both T1- and T2-weighted images (Fig. 11-24). The MR appearance, however, is nonspecific, with calvarial lesions resembling other disorders such as Paget's disease.

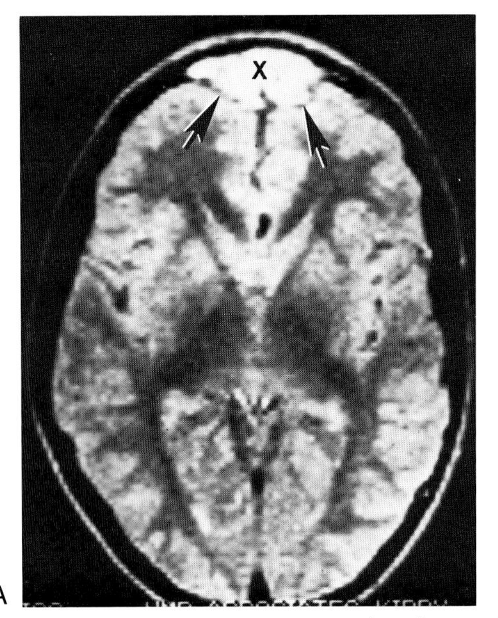

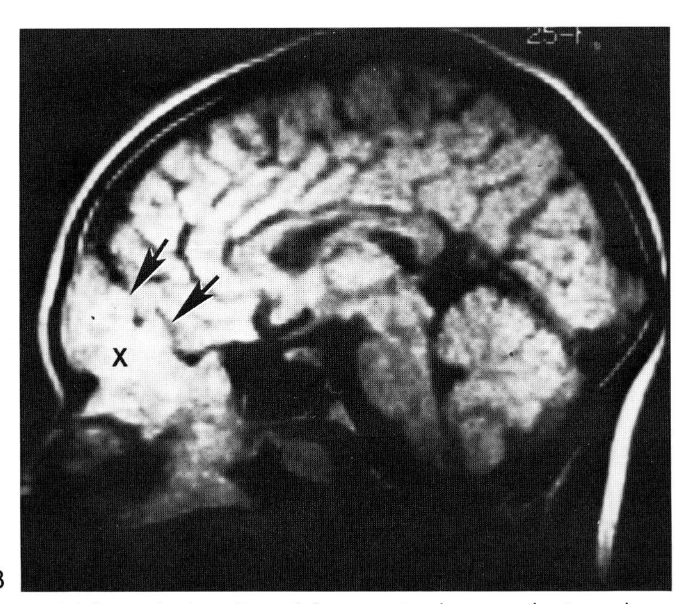

Fig. 11-22. Sinus tumor (*X*) invading the anterior cranial fossa. A clear line of demarcation between brain and tumor is identified (arrows); resection is therefore possible.

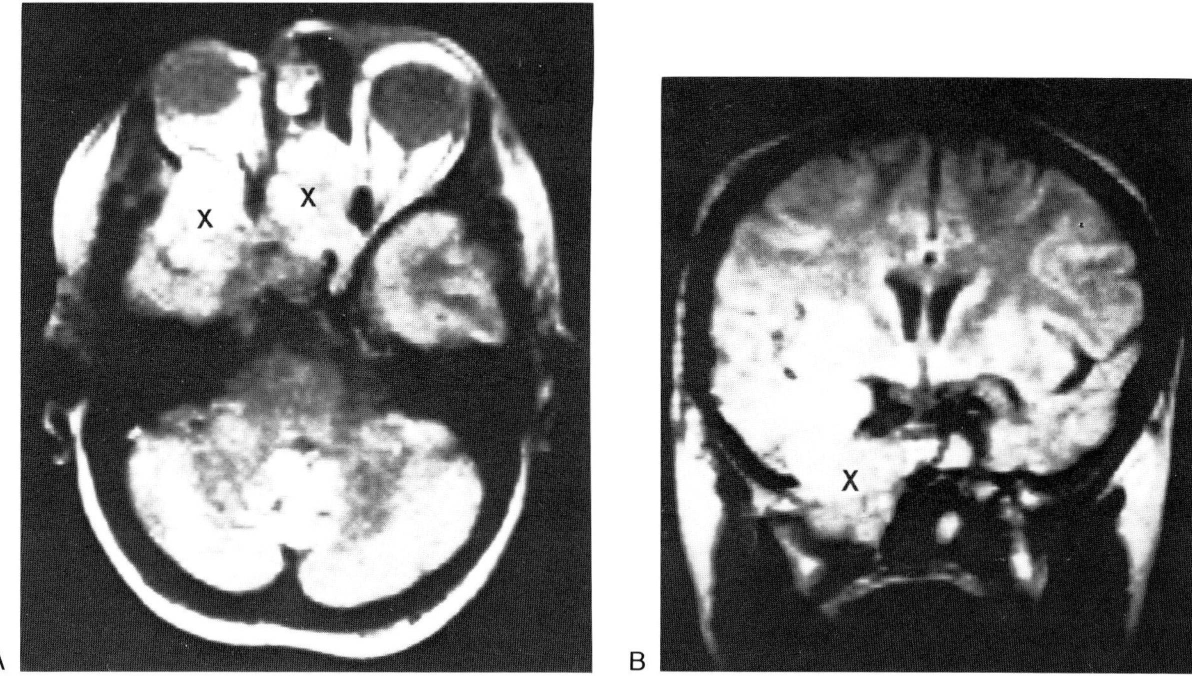

Fig. 11-23. (**A** and **B**) This adenocarcinoma (*X*) has invaded the temporal lobe and skull base and is unresectable.

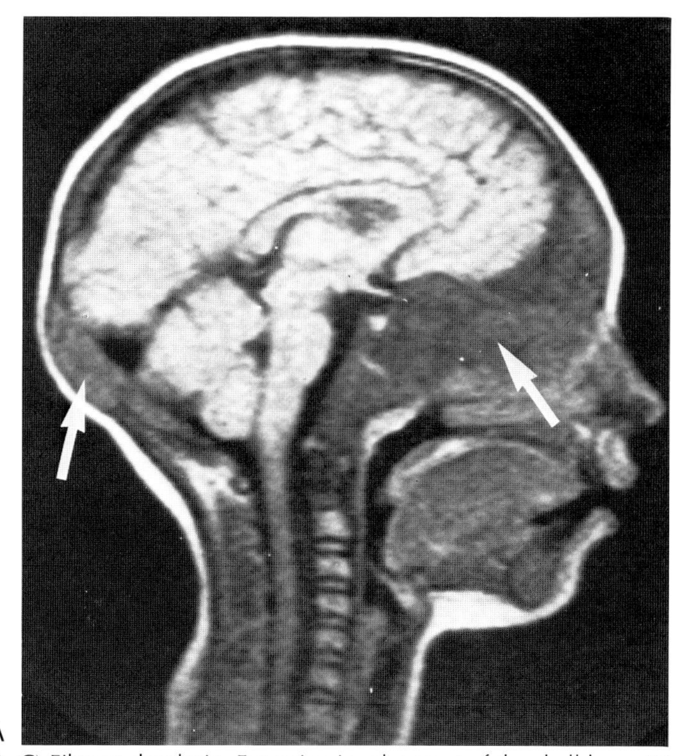

Fig. 11-24. (**A–C**) Fibrous dysplasia. Extensive involvement of the skull base. (*Figure continues*).

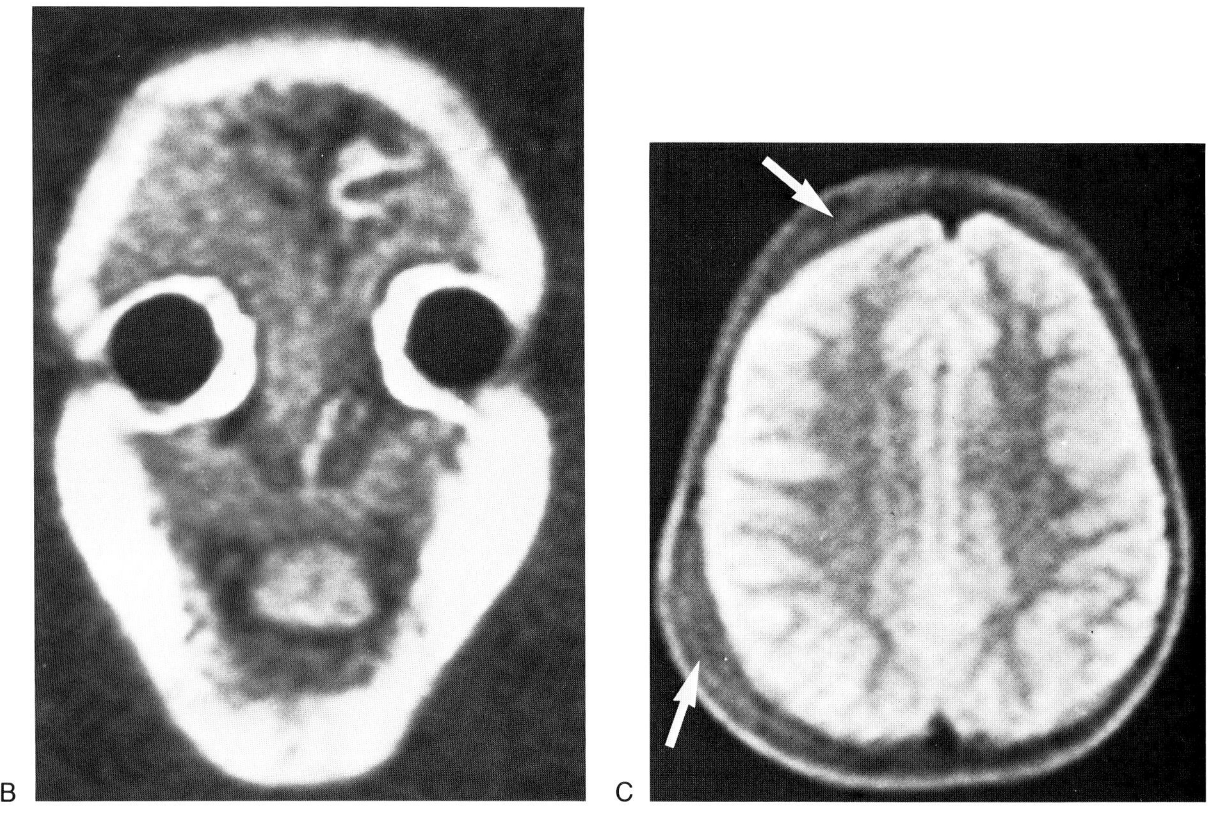

Fig. 11-24 *(continued)*.

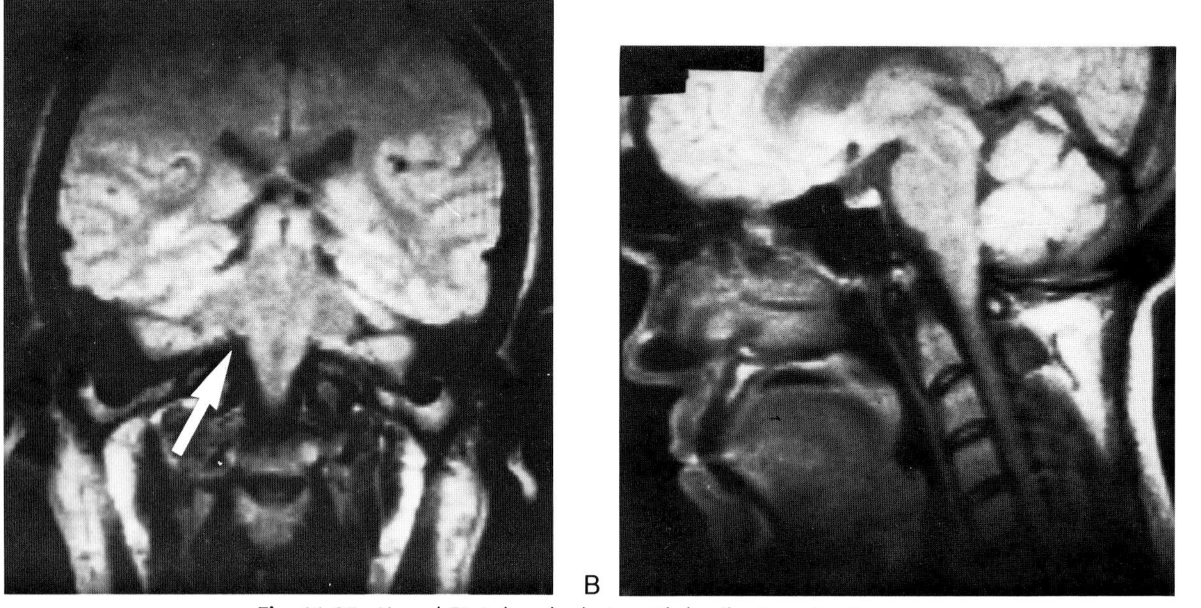

Fig. 11-25. (**A** and **B**) Achondoplasia with basilar invagination.

Achondroplasia

Achondroplasia is a hereditary disorder of endochondral bone formation, arising as a spontaneous mutation or inherited as an autosomal dominant trait. Extracranial manifestations include short-limbed dwarfism, trident hands, champagne glass pelvis, tracheal cartilage deficiency, and spinal stenosis. In the head, the skull base is primarily affected because it arises form endochondral bone formation. The calvarium is spared, because it is formed in membrane. With growth the skull base remains small while the calvarium enlarges. The foramen magnum is narrow and elongated in an anteroposterior direction. Basilar invagination is characteristic. Hydrocephalus may develop from impaired venous or CSF outflow. MRI may be useful for demonstrating brain stem distortion secondary to basilar invagination (Fig. 11-25).[67]

Neurofibromatosis

The skull base is frequently deformed in neurofibromatosis. A characteristic finding is orbital dysplasia with a markedly widened superior orbital fissure. On one side there is unilateral absence of a large part of the greater wing of the sphenoid as well as hypoplasia and elevation of the lesser wing.[67] The temporal lobe may protrude through this large defect in the posterolateral wall of the orbit, resulting in pulsating exophthalamus. It should be noted that no neurofibroma is associated with this skull defect; it is purely a bone dysplasia. An example of sphenoid dysplasia associated with neurofibromatosis appears in Figure 11-26. More complete discussion of the intracranial manifestations of this disorder are presented in Chapter 4.

PRIMARY TUMORS OF THE SKULL BASE

Chordomas

Chordomas are rare tumors of the skull base and spine that arise from remnants of the embryonic notochord. The notochord is a mesodermal derivative that forms the primitive axial skeleton of all vertebrates. In the fully developed human notochord remnants persist as cell rests in the basiocciput and coccyx

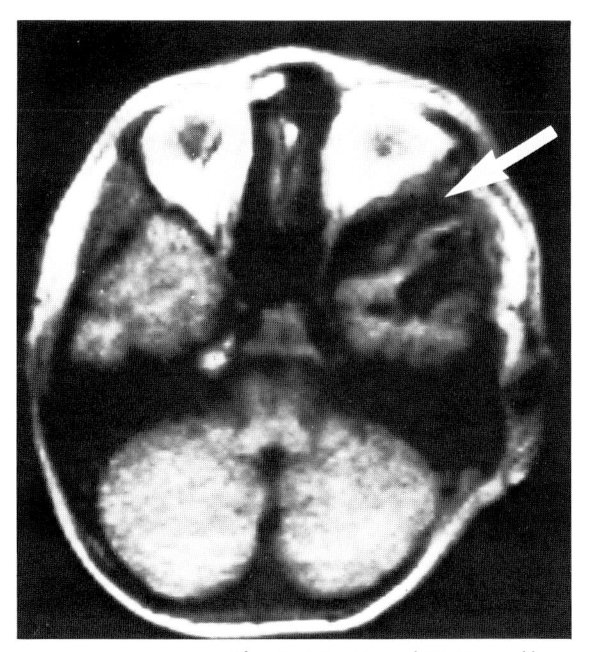

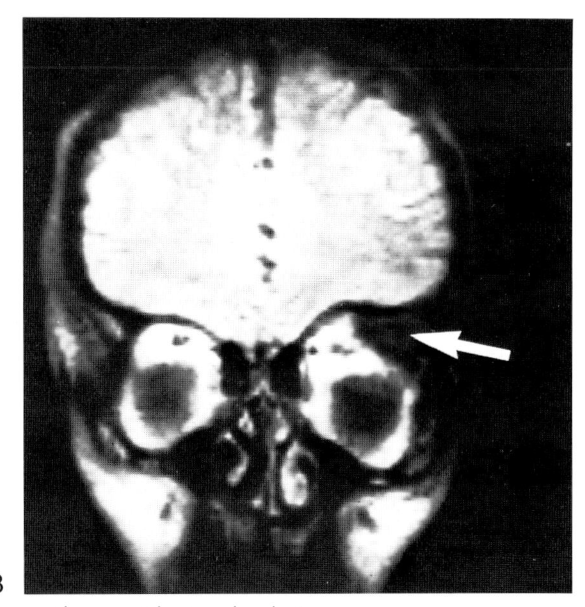

Fig. 11-26. (**A** and **B**) Neurofibromatosis with spenoid wing dysplasia.

and become the nuclei pulposi of intervertebral disks.

Chordomas may occur at any age, but are unusual in children. Cranial chordomas are most common in the third and fourth decades of life, while sacral chordomas are more frequent in the fifth and sixth decades. Males are affected more often than females by a ratio of 2 or 3 to 1.[68] Over half of chordomas are found in the sacrum and about one-sixth occur in the spine. The remaining one-third of chordomas occur at the base of the skull, in and around the clivus. The most typical intracranial location is near the spheno-occipital synchondrosis.[69] Chordomas may also arise in the nasopharyngeal tissues anterior to the clivus and in the epidural space posterior to the clivus. Occasionally they are found at the tip of the clivus near the foramen magnum. Rarely they may arise within the sella and mimic a pituitary tumor.

Grossly, chordomas are firm, bulky tumors consisting of lobules of gelatinous material surrounded by fibrous septa and a well-defined capsule. Although they are poorly vascularized, hemorrhage and tumor necrosis are common. They are relatively slow growing but behave malignantly because of their location and persistent expansion.

Nearly every cranial chordoma will involve the clivus. Destruction is typically lytic and sharply defined, but reactive sclerosis is seen in 10 percent.[70] Sequestration of bone fragments from the clivus may occur and when seen is a useful diagnostic sign to suggest chordoma. Additionally, areas of calcification may be identified by CT in 50 percent.[70] The larger of these may be visualized by MRI.

From the clivus, chordomas may extend in several directions: anteroinferiorly to the nasopharynx or spenoidal sinus, laterally to the petrous apex, superiorly to the sella, or posteriorly into the pontine and medullary cisterns. Anterior extension is the most common, and a nasopharyngeal mass is seen in about 40 percent of cases.[70] Posterior extension results in displacement of the brain stem, but the brain stem itself is seldom invaded. A thin CSF-containing space may be identified on sagittal MR images between the tumor and brain stem. Chordomas may engulf and invade cranial nerves. Those arising near the sella may compress the optic chiasm and engulf cranial nerves 3 to 6 in the cavernous sinuses. Those arising in the basiocciput may affect cranial nerves 7 to 12.[71] Chordomas may also encircle and occlude the internal carotid arteries and this complication can be suggested on MRI.

Such large, aggressive chordomas should be distinguished from tiny, asymptomatic, lesions called "benign chordomas" or *chordal ectopias*. These small, soft, mushroomlike tumors are attached to the sphenoid bone and may protrude through the dura on stalks. They are always benign and totally asymptomatic. They are found in 1.5 percent to 2 percent of random necropsy specimens.[72]

Several chordomas have been illustrated in the recent MR literature.[73,74] The marrow cavity of the clivus is normally seen as a region of high signal intensity on T1-weighted images because of its fat content. Replacement of this marrow cavity by lower intensity tumor is characteristic for chordoma. A sharp margination between tumor and normal clivus may be seen.[5] Large areas of calcification or bone sequestration may be identified by signal drop-out independent of pulse sequence chosen. The sagittal and coronal planes are most useful for evaluating local extension of tumor. Examples of chordomas seen on MRI are presented in Figures 11-27 through 11-29.

Cranial Cartilaginous Tumors

Cranial cartilaginous tumors are uncommon neoplasms thought to arise from embryonic cartilage rests, usually near the base of the skull. Most patients are between 20 and 60 years old at the time of diagnosis. Benign chondromas are most common in the younger age groups, while malignant transformation (to chondrosarcoma) occurs in the elderly. Intracranial cartilage tumors may be a feature of multiple enchondromatosis or Maffucci syndrome.[75]

Most cranial cartilage tumors are extradural and found at the skull base. Over half arise in or adjacent to the body of the spenoid bone. The cerebellopontine angle is the next most common site. The calvarium may be the site of origin in a few percent of cases. Intradural origin is seen in less than one-fourth cartilage tumors. These are usually along the falx, over the frontoparietal convexities or intracerebral.[76]

Cartilaginous tumors may contain widely different proportions of bone and cartilage. Accordingly, the benign forms are called either chondromas or osteochondromas depending on their histology. Malignant

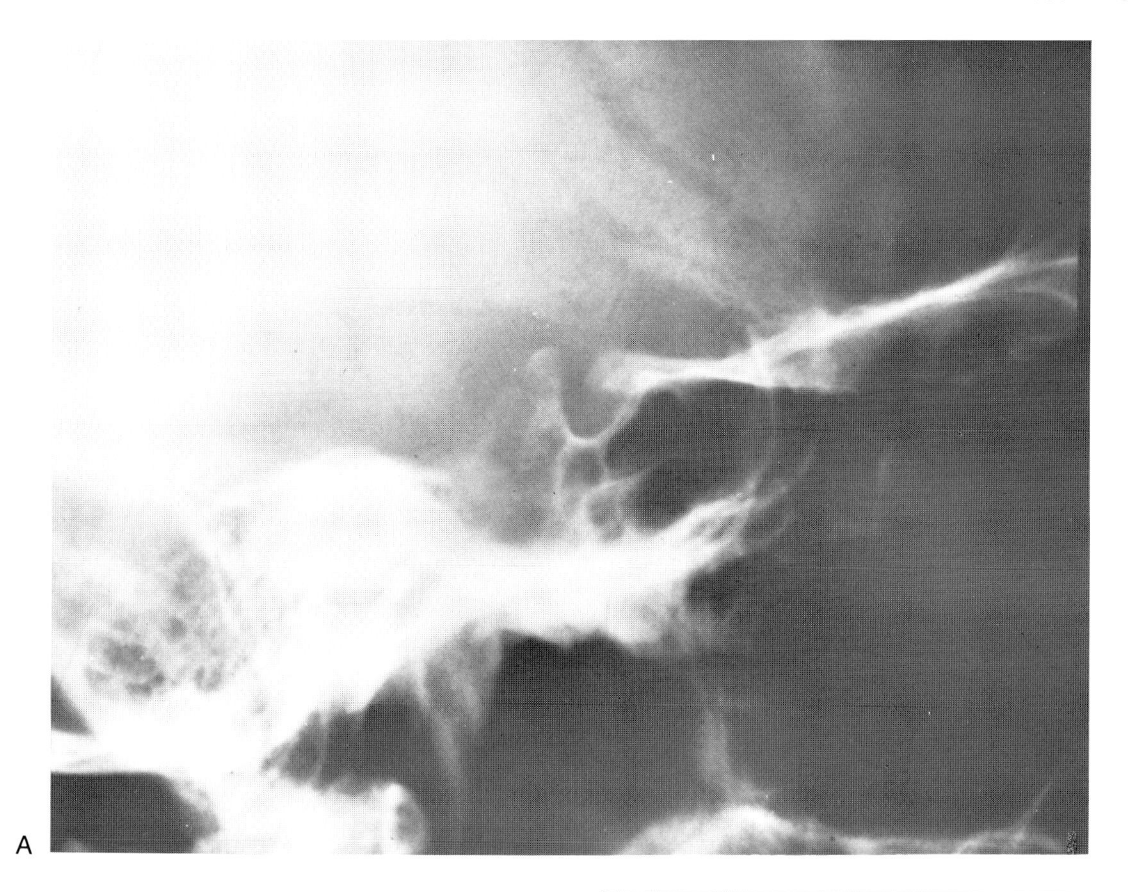

A

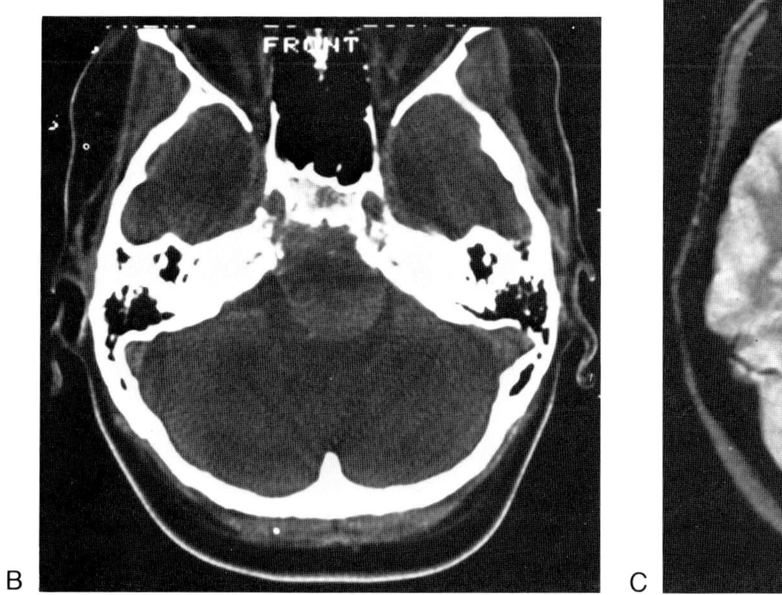

B

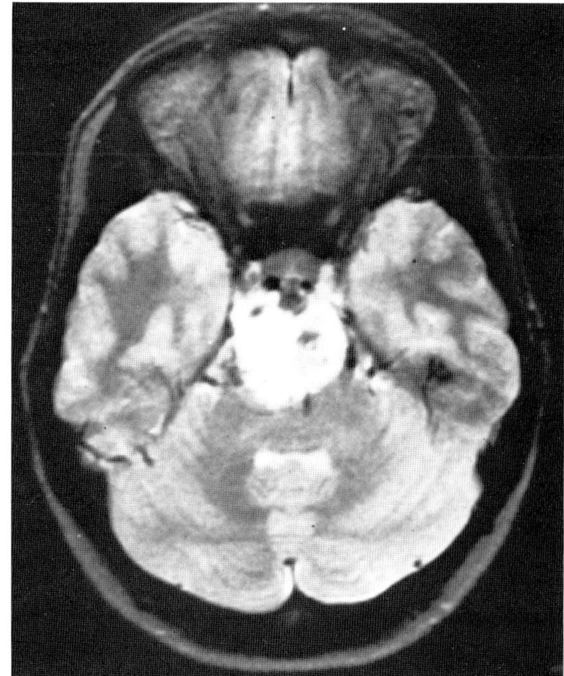

C

Fig. 11-27. Chordoma. **(A)** Plain x-ray shows clival destruction. **(B)** Mass seen on CT. (*Figure continues*).

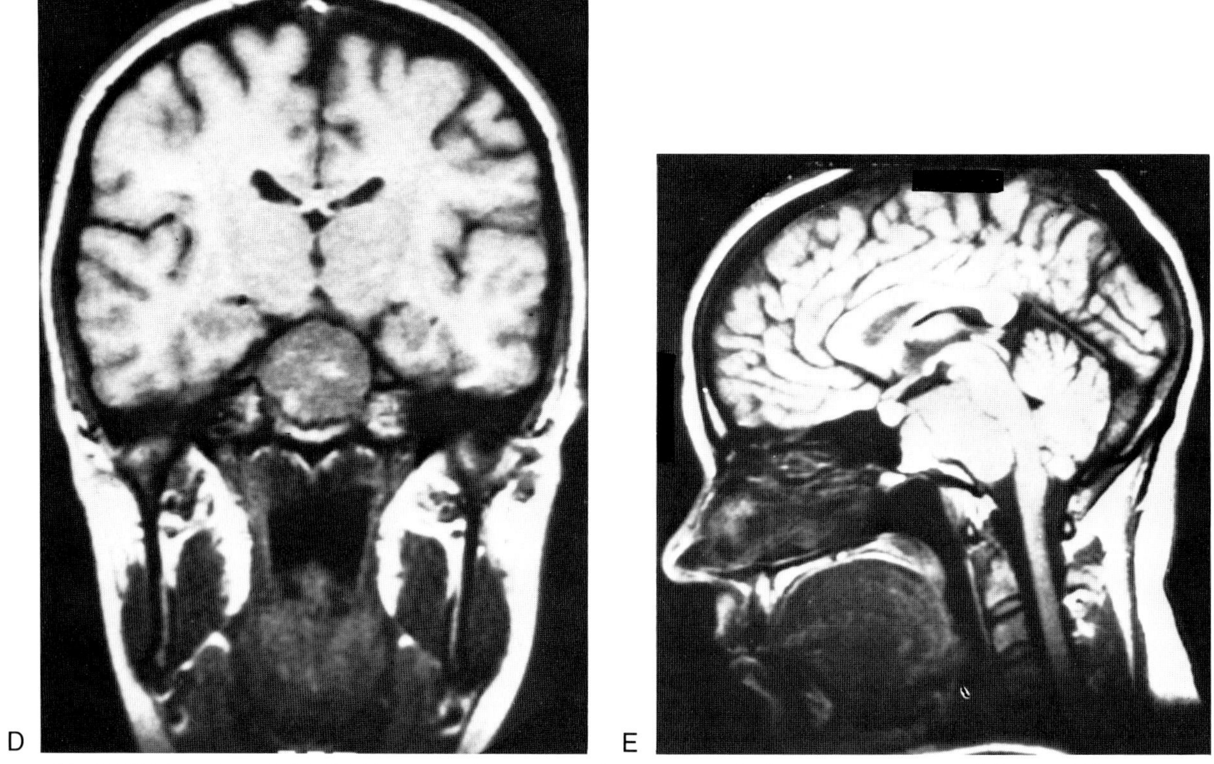

Fig. 11-27 (*continued*). **(C–E)** MR images show extent of mass to good advantage. (Courtesy of Thomas Wiggans, M.D.)

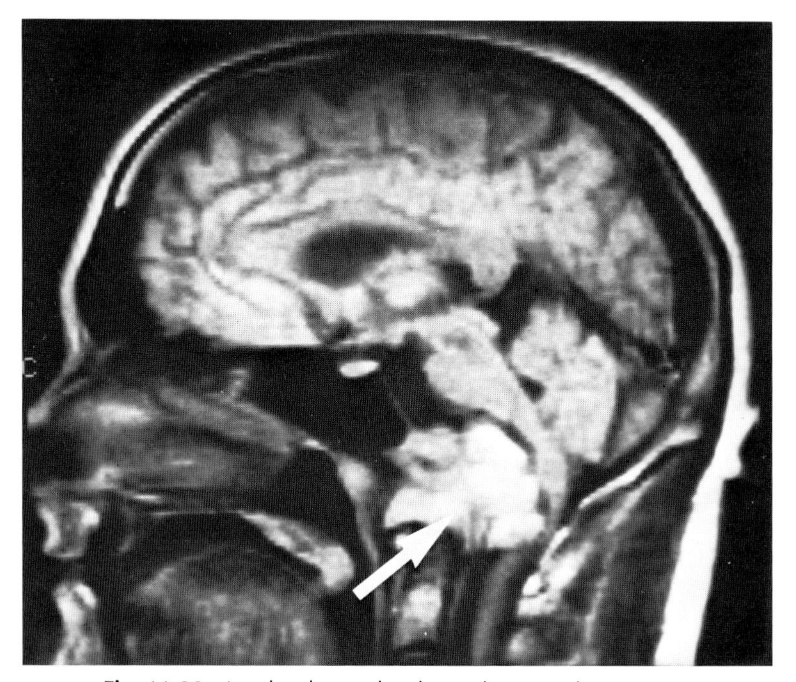

Fig. 11-28. Another large chordoma distorting brain stem.

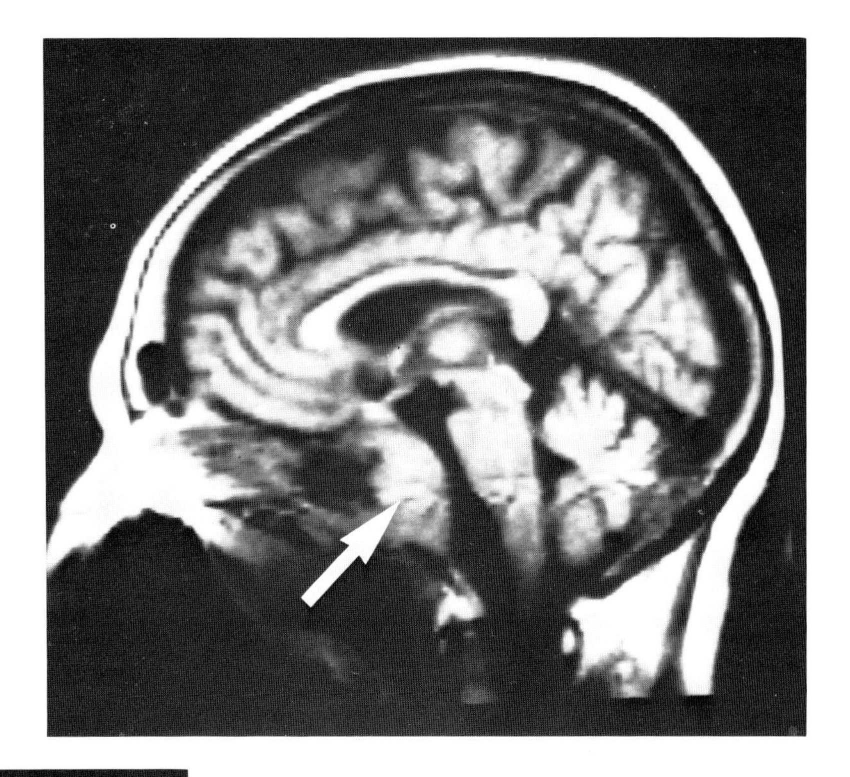

Fig. 11-29. Smaller chordoma (arrow).

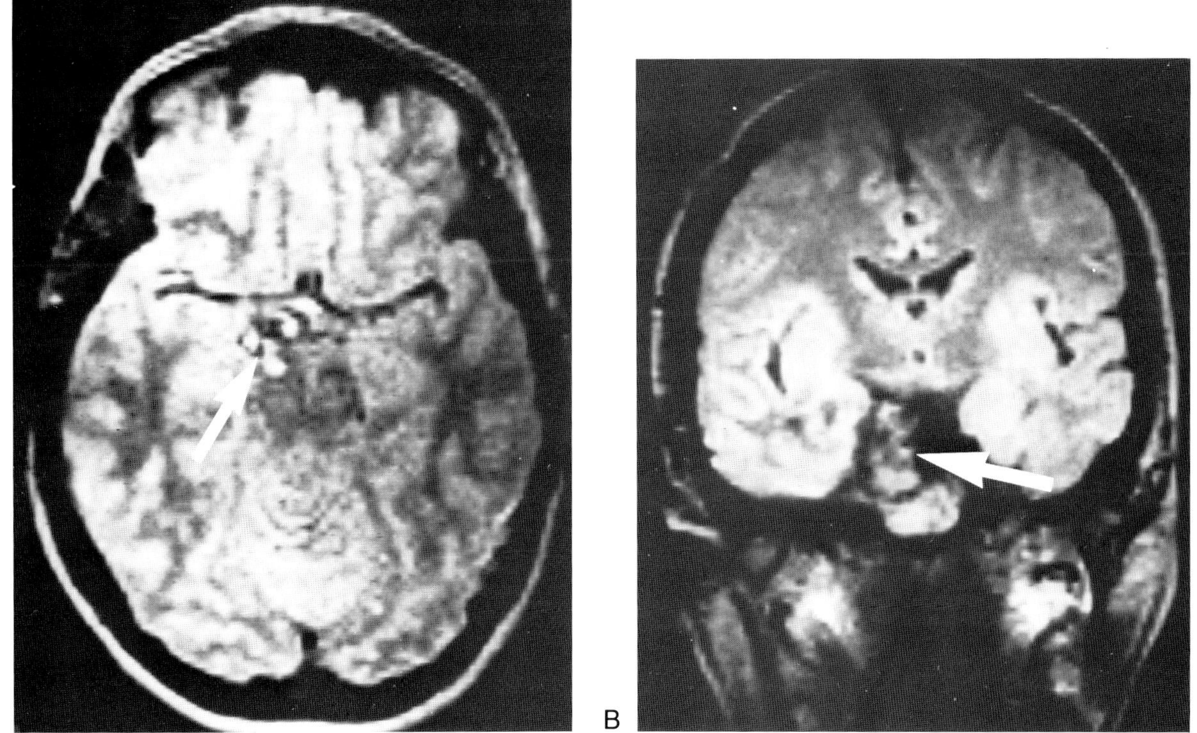

Fig. 11-30. (**A** and **B**) Chondrosarcoma of the skull base. Darker areas of tumor were shown on CT to be calcifications.

transformation occurs in less than 1 to 3 percent.[76] Perhaps more commonly chondrosarcoma may originate as a malignant tumor.

Chondromas grow slowly, causing pressure effects on adjacent cranial nerves, particularly those in the cavernous sinus. Bone erosion may be seen in about 50 percent of cases. Calcification (mottled, stippled, or ringlike) occurs in 60 percent, but is curiously rare in those that originate in the cerebellopontine angle.[77]

Chondrosarcomas may also grow slowly if low grade. Rapid growth may be present in more malignant lesions. Radiographically the distinction between chondromas, chondrosarcomas, and chordomas may be impossible. In our case, as well as several scattered in the literature, dense curvilinear calcifications could be identified within the tumor as areas of signal dropout (Fig. 11-30). Otherwise, the appearance was nonspecific on MR resembling chordomas and osteochondromas (Fig. 11-31).

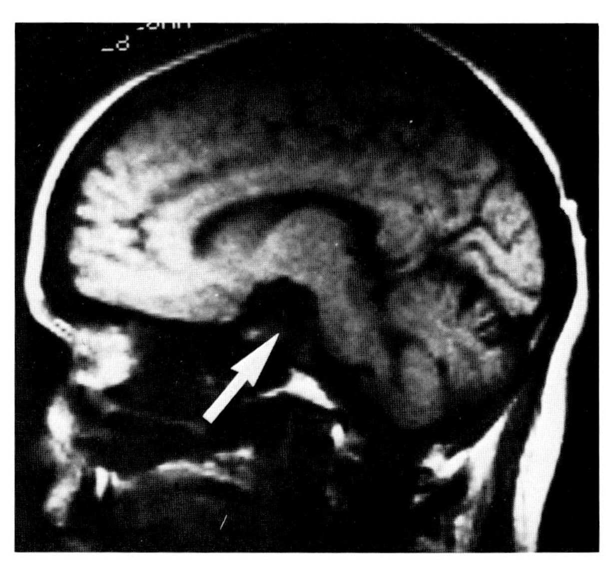

Fig. 11-31. Large osteochondroma (arrow) casts virtually no MR signal due to its heavy calcification (low hydrogen density).

THE CEREBELLOPONTINE ANGLE AND TEMPORAL BONE

Normal Anatomy

The cerebellopontine angle (CPA) is that region of the posterior fossa bounded medially by the pons, posteriorly by the cerebellum, and anterolaterally by the petrous temporal bone. The CPA cistern contains cranial nerves VII and VIII, the petrosal veins, and distal portions of the anterior inferior cerebellar arteries. The flocculus of the cerebellum projects into the posterior portion of the CPA cistern. Superiorly the trigeminal nerve emerges from the ventral pons to pass forward into Meckel's cave. Inferiorly, the rootlets of cranial nerves IX, X, and XI pass downward to exit the skull through the jugular foramen. MRI provides excellent visualization of the nervous structures in and around the CPA.

By contrast the temporal bone is a remarkably complex osseous structure well suited to CT examination. The ossicles, bony canals, and walls of the middle ear cavity are poorly defined by MR. However, the cochlea, nerves, and vestibular apparatus are exceedingly well visualized by MR and free of bone artifact.

Figures 11-32 through 11-34 demonstrate CPA and temporal bone anatomy using surface coil MR imaging.[78,79] Note particularly the excellent visualization of the separate components of the vestibulocochlear nerve, the clarity of anatomic relationships, and the course of the more distal facial and greater superficial petrosal nerves. Unfortunately, high signal from a few fluid-filled mastoid air cells may hide this beautiful anatomy and presents a major limitation to MRI of the temporal bone (Fig. 11-35).

The medial portions of the seventh and eighth cranial nerves are best seen on T1-weighted pulse sequences.[80] If a too T2-weighted pulse sequence is used, the bright signal from CSF can easily obscure the contour and signal from these nerves proximally. The lateral portions of these nerves are best seen with slight T2-weighting, perhaps using a technique such as SE 1500/60. Disease in the middle ear and mastoids must be examined using both T1- and T2-weighted sequences. Mastoid inflammatory disease will appear bright on T2-weighted images and less bright on T1 images. By contrast lesions such as cholesteatoma may appear bright on T1-weighted images due to the short T1 of their fatty components.

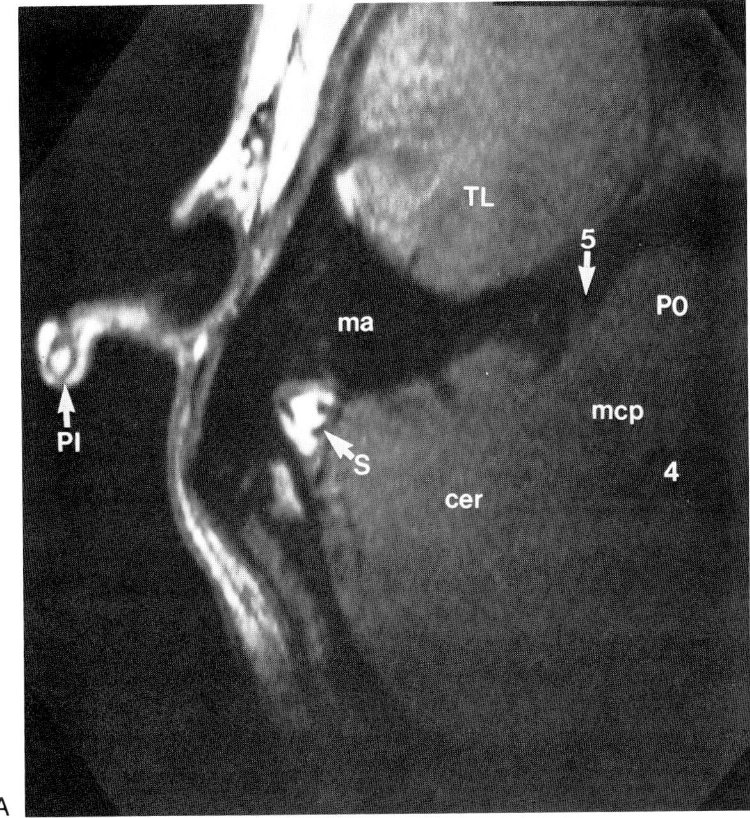

Fig. 11-32. (A–D) Axial SE 500/30 surface coil images of the right temporal bone and CPA cistern. (See legend for explanation of symbols.) (Courtesy of Fonar Corporation, 110 Marcus Drive, Melville, NY.) (*Figure continues*).

4, fourth ventricle;
5, trigeminal nerve;
7, facial nerve;
7h, facial nerve, horizontal segment;
7d, facial nerve, descending limb;
8, vestibulocochlear nerve;
8c, cochlear division;
8i, inferior vestibular division;
8s, superior vestibular division;
9, glossopharyngeal nerve;
10, vagus nerve;
11, spinal accessory nerve;
12, hypoglossal nerve;
EX, external auditory canal;
LP, lateral pterygoid muscle;
M, medulla;
PE, petrous apex;
PI, pinna;
PO, pons;
S, sigmoid sinus;
TL, temporal lobe;
cer, cerebellum;
co, cochlea;
fl, flocculus;
gg, geniculate ganglion;
gsp, greater superficial petrosal nerve;
ica, internal carotid artery;
ima, internal maxillary artery branches;
jb, jugular bulb;
jv, jugular vein;
lsc, lateral semicircular canal;
ma, mastoid air cells;
mcp, middle cerebellar peduncle;
smf, stylomastoid foramen;
ssc, superior semicircular canal;
ves, vestibule.

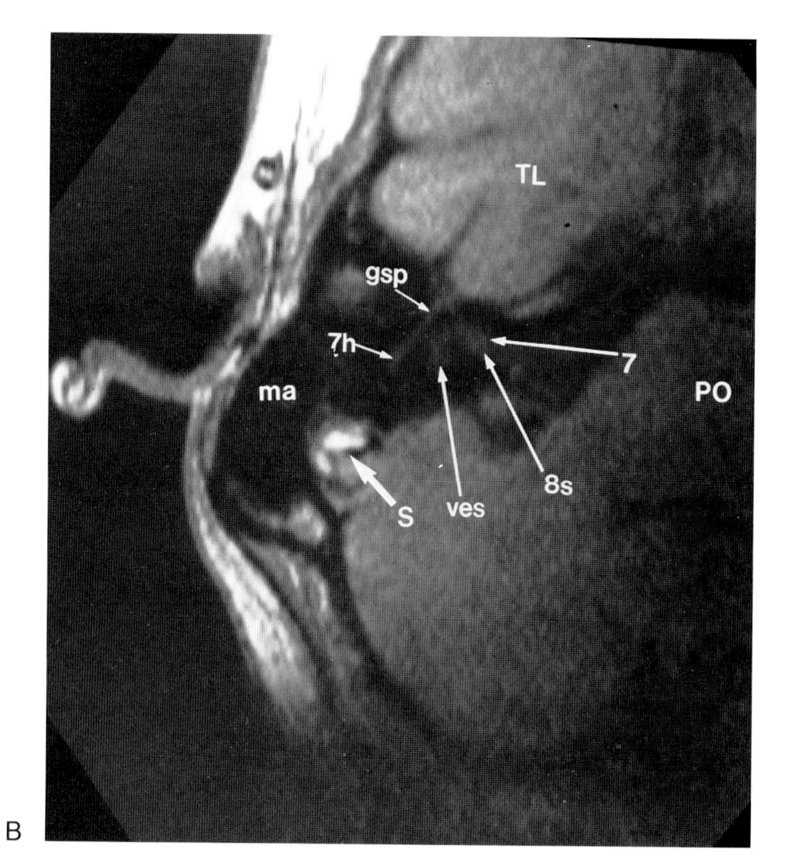

B

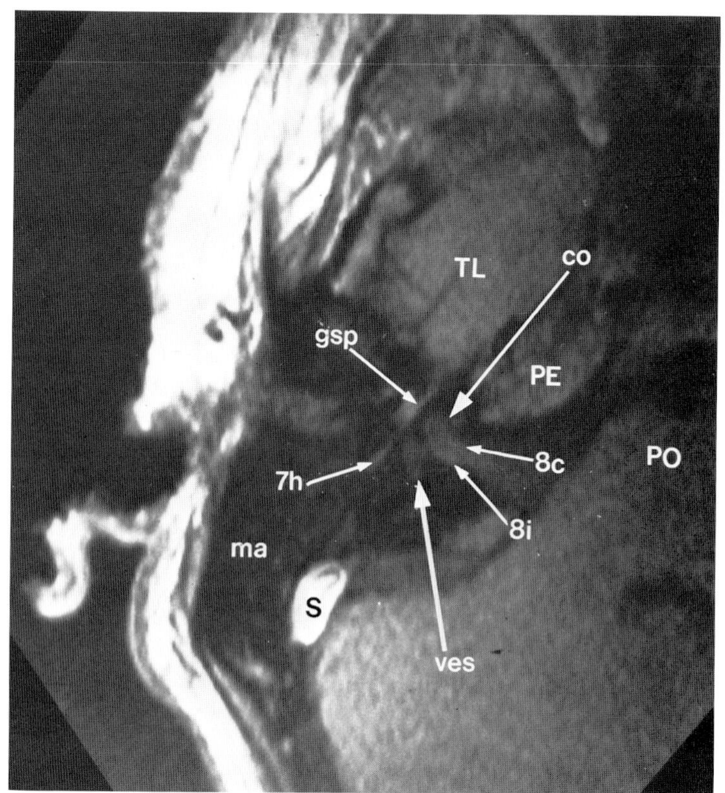

C

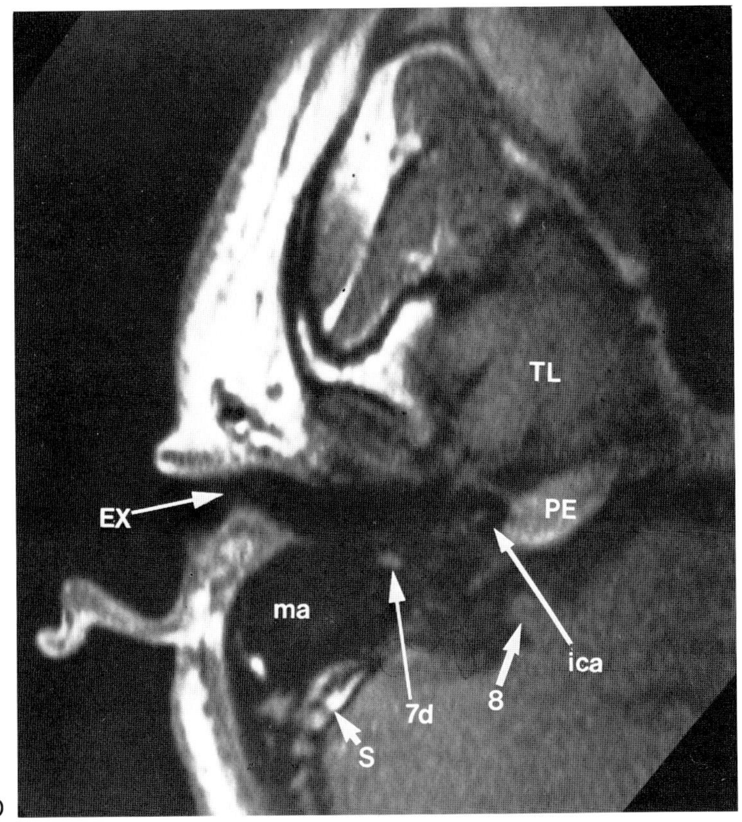

Fig. 11-32 *(continued)*. **B, C, D.**

4, fourth ventricle;
5, trigeminal nerve;
7, facial nerve;
7h, facial nerve, horizontal segment;
7d, facial nerve, descending limb;
8, vestibulocochlear nerve;
8c, cochlear division;
8i, inferior vestibular division;
8s, superior vestibular division;
9, glossopharyngeal nerve;
10, vagus nerve;
11, spinal accessory nerve;
12, hypoglossal nerve;
EX, external auditory canal;
LP, lateral pterygoid muscle;
M, medulla;
PE, petrous apex;
PI, pinna;
PO, pons;
S, sigmoid sinus;
TL, temporal lobe;
cer, cerebellum;
co, cochlea;
fl, flocculus;
gg, geniculate ganglion;
gsp, greater superficial petrosal nerve;
ica, internal carotid artery;
ima, internal maxillary artery branches;
jb, jugular bulb;
jv, jugular vein;
lsc, lateral semicircular canal;
ma, mastoid air cells;
mcp, middle cerebellar peduncle;
smf, stylomastoid foramen;
ssc, superior semicircular canal;
ves, vestibule.

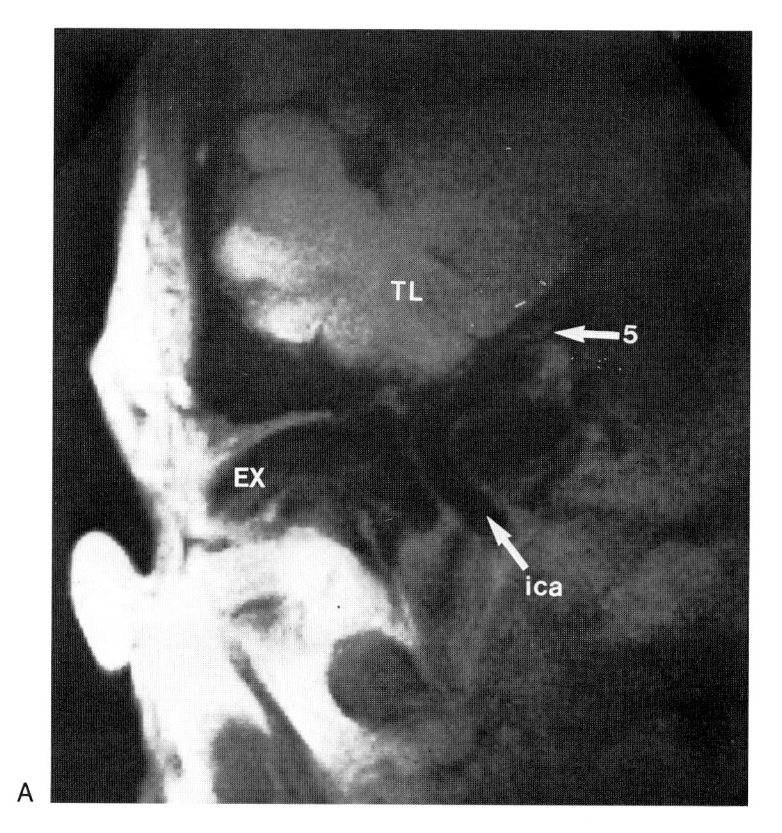
TL
5
EX
ica
A

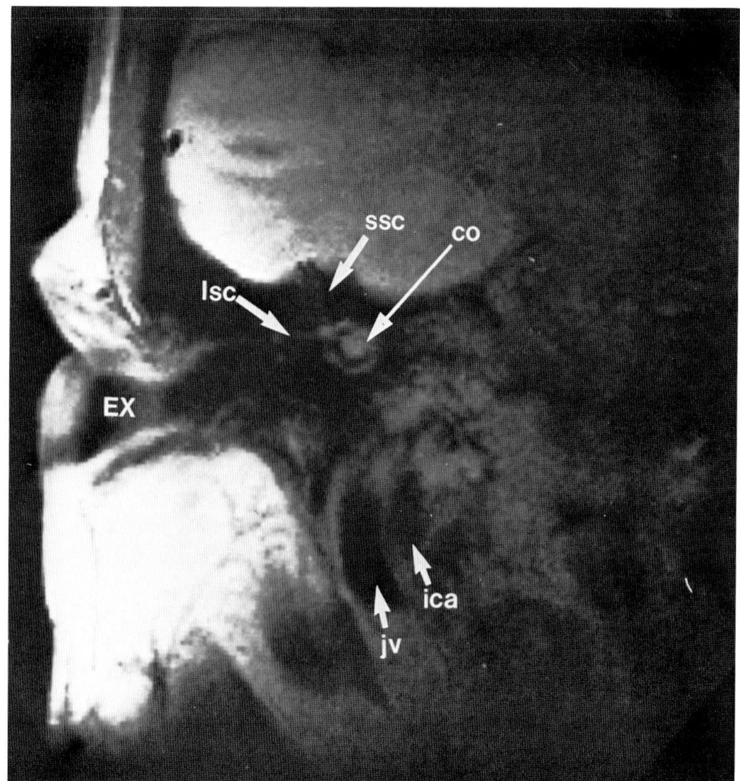
ssc
co
lsc
EX
ica
jv
B

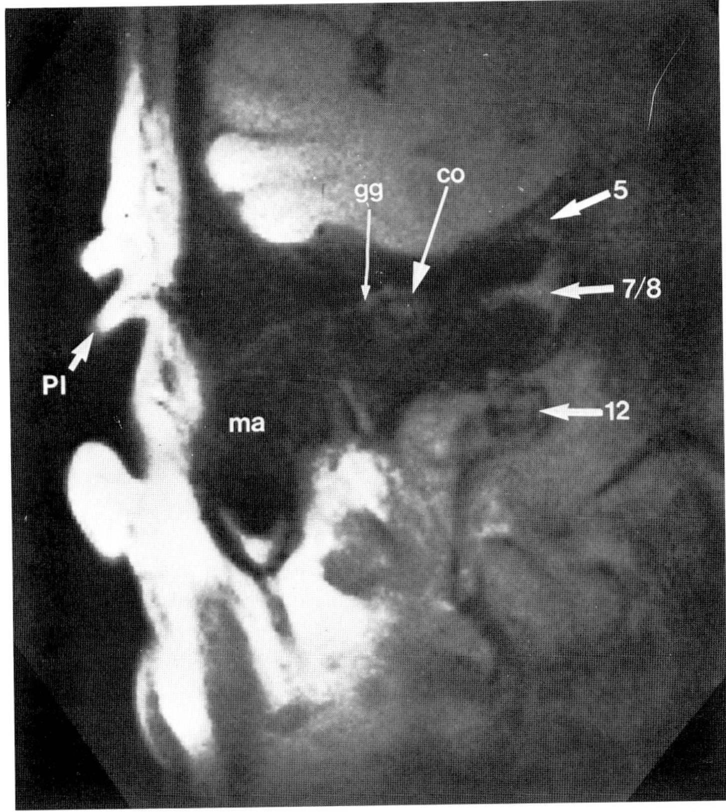

Fig. 11-33. (A–D) Coronal SE 500/30 surface coil images of the right temporal bone. (See legend for explanation of symbols.) (*Figure continues*).

4, fourth ventricle;
5, trigeminal nerve;
7, facial nerve;
7h, facial nerve, horizontal segment;
7d, facial nerve, descending limb;
8, vestibulocochlear nerve;
8c, cochlear division;
8i, inferior vestibular division;
8s, superior vestibular division;
9, glossopharyngeal nerve;
10, vagus nerve;
11, spinal accessory nerve;
12, hypoglossal nerve;
EX, external auditory canal;
LP, lateral pterygoid muscle;
M, medulla;
PE, petrous apex;
PI, pinna;
PO, pons;
S, sigmoid sinus;
TL, temporal lobe;
cer, cerebellum;
co, cochlea;
fl, flocculus;
gg, geniculate ganglion;
gsp, greater superficial petrosal nerve;
ica, internal carotid artery;
ima, internal maxillary artery branches;
jb, jugular bulb;
jv, jugular vein;
lsc, lateral semicircular canal;
ma, mastoid air cells;
mcp, middle cerebellar peduncle;
smf, stylomastoid foramen;
ssc, superior semicircular canal;
ves, vestibule.

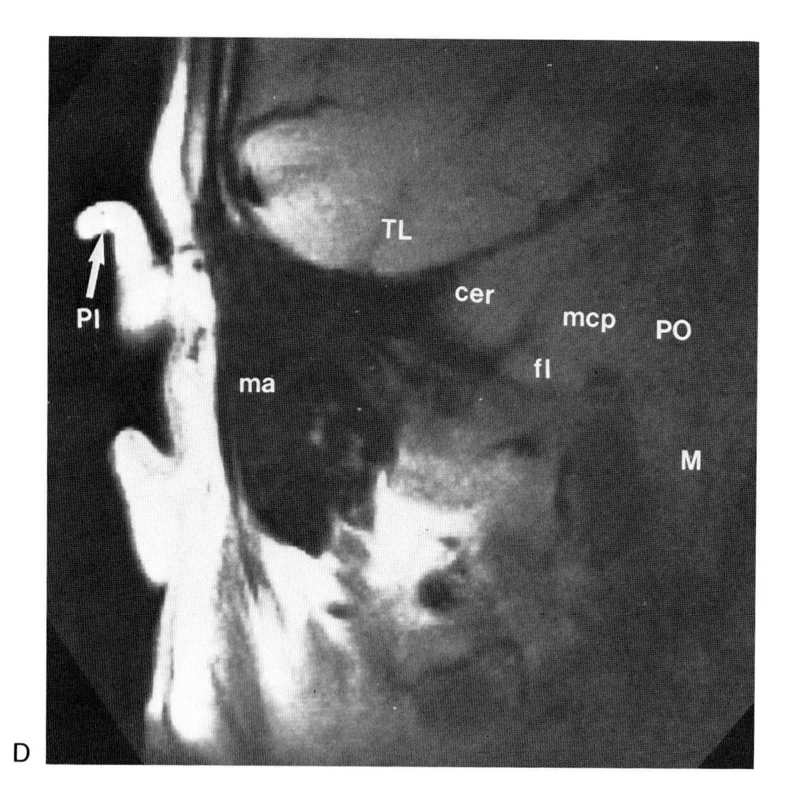

Fig. 11-33 (*continued*).

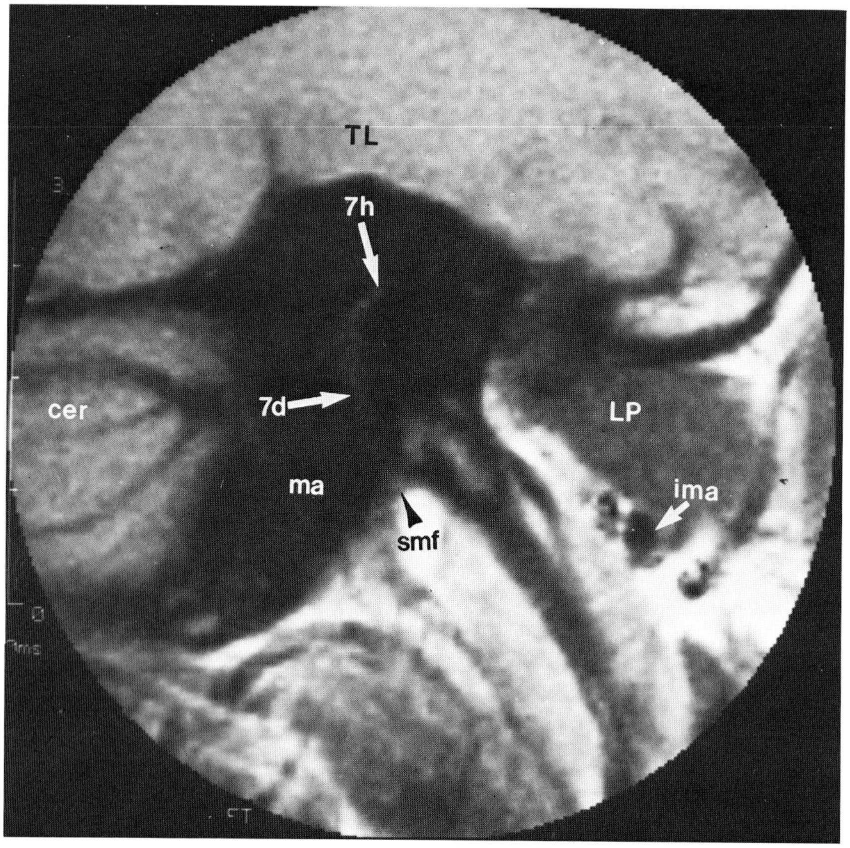

A

Fig. 11-34. (A–D) Sagittal SE 600/26 surface coil images of the right temporal bone. (See legend for explanation of symbols.) (*Figure continues*).

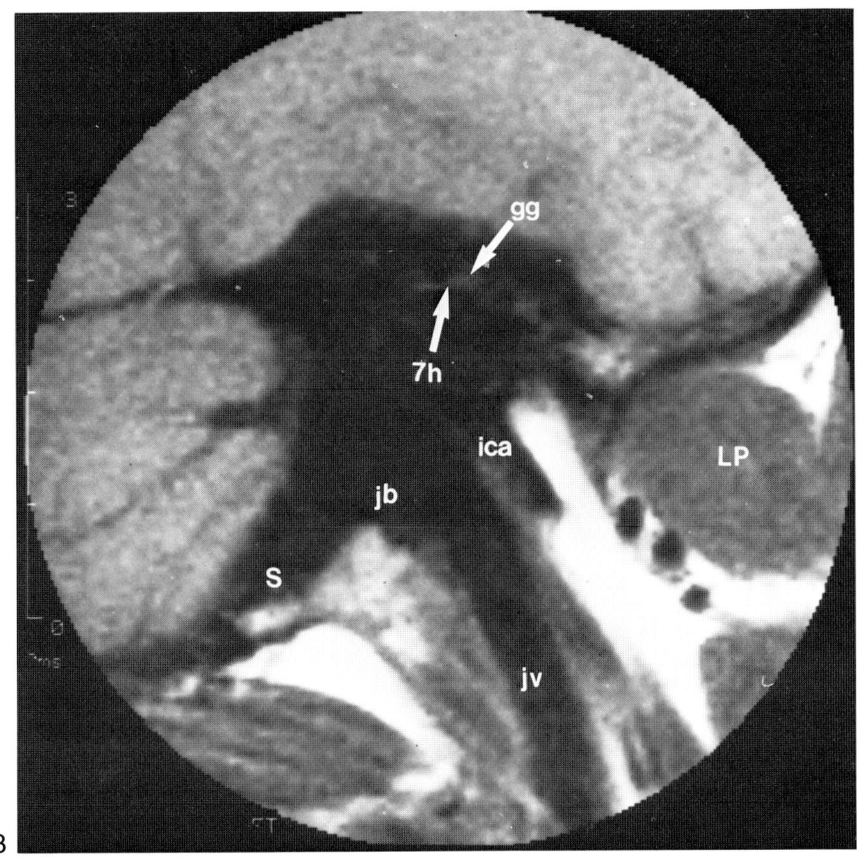

Fig. 11-34 *(continued)*. **A** and **B.**

4, fourth ventricle;
5, trigeminal nerve;
7, facial nerve;
7h, facial nerve, horizontal segment;
7d, facial nerve, descending limb;
8, vestibulocochlear nerve;
8c, cochlear division;
8i, inferior vestibular division;
8s, superior vestibular division;
9, glossopharyngeal nerve;
10, vagus nerve;
11, spinal accessory nerve;
12, hypoglossal nerve;
EX, external auditory canal;
LP, lateral pterygoid muscle;
M, medulla;
PE, petrous apex;
PI, pinna;
PO, pons;
S, sigmoid sinus;
TL, temporal lobe;
cer, cerebellum;
co, cochlea;
fl, flocculus;
gg, geniculate ganglion;
gsp, greater superficial petrosal nerve;
ica, internal carotid artery;
ima, internal maxillary artery branches;
jb, jugular bulb;
jv, jugular vein;
lsc, lateral semicircular canal;
ma, mastoid air cells;
mcp, middle cerebellar peduncle;
smf, stylomastoid foramen;
ssc, superior semicircular canal;
ves, vestibule.

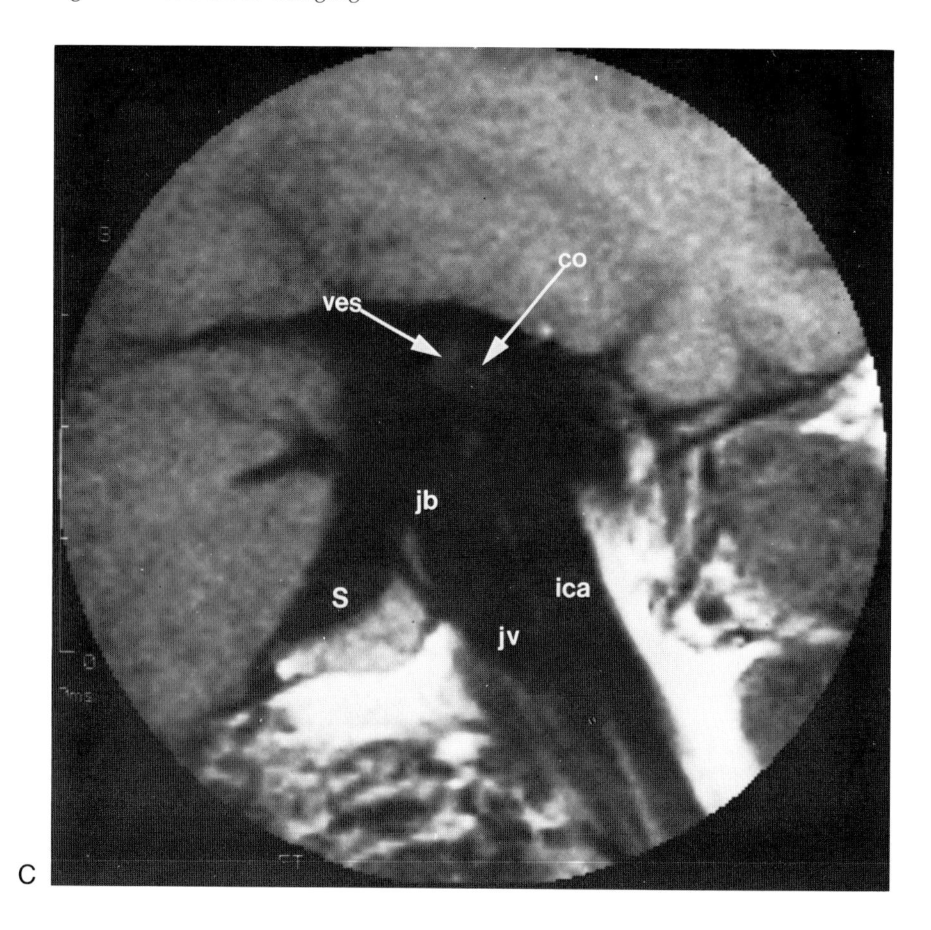

Acoustic Neurinoma

Acoustic neurinomas are slowly growing encapsulated tumors of the eighth cranial nerve that account for 5 to 10 percent of all intracranial tumors and 80 to 90 percent of those in the cerebellopontine angle.[81] Acoustic neurinomas occur most often in patients aged 30 to 60. When seen in younger patients or when bilateral, the diagnosis of neurofibromatosis should be strongly considered.[82]

Acoustic neurinomas arise from the Schwann cells of the eighth cranial nerve. The older term "neuroma" is a misnomer; these tumors are more properly called neurilemomas, schwannomas, or neurinomas. The majority of these tumors arise from the intracanalicular portion of the superior vestibular nerve. Origin from the inferior vestibular or cochlear nerves occurs in less than one-third of cases. About 5 percent of acoustic neurinomas originate medial to the porus acousticus and do not involve the intracanalicular portion of the nerve.[83] These tumors may grow to large size in the CPA before producing symptoms.

Unilateral hearing loss, tinitus, and unsteadiness are the cardinal symptoms of an acoustic neurinoma; headache and true vertigo appear late.[81,84] Hearing loss is typically high frequency, a fact attributed to the arrangement of cochlear nerve fibers subserving high frequency around the outer part of the nerve, which is more vulnerable to compression. Loss of speech discrimination often precedes pure tone loss. Despite the origin of acoustic tumors from the vestibular nerve, vestibular symptoms are often late and mild. Because these tumors grow so slowly it is thought that central vestibular mechanisms can com-

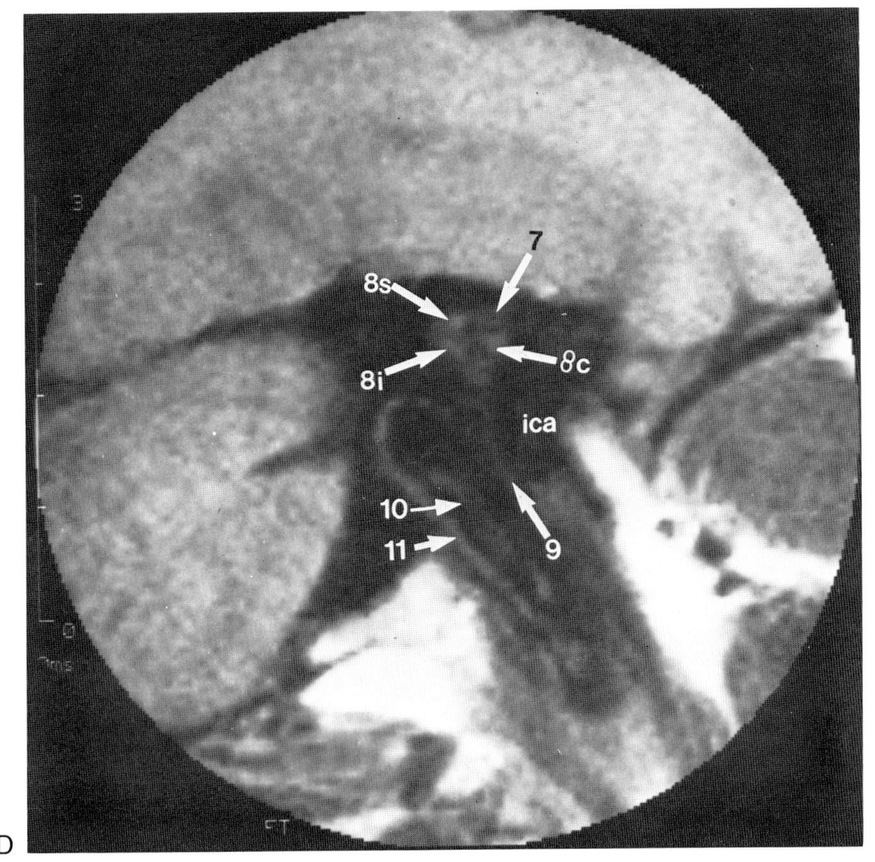

Fig. 11-34 *(continued)*. **C** and **D.**

4, fourth ventricle;
5, trigeminal nerve;
7, facial nerve;
7h, facial nerve, horizontal segment;
7d, facial nerve, descending limb;
8, vestibulocochlear nerve;
8c, cochlear division;
8i, inferior vestibular division;
8s, superior vestibular division;
9, glossopharyngeal nerve;
10, vagus nerve;
11, spinal accessory nerve;
12, hypoglossal nerve;
EX, external auditory canal;
LP, lateral pterygoid muscle;
M, medulla;
PE, petrous apex;
PI, pinna;
PO, pons;
S, sigmoid sinus;
TL, temporal lobe;
cer, cerebellum;
co, cochlea;
fl, flocculus;
gg, geniculate ganglion;
gsp, greater superficial petrosal nerve;
ica, internal carotid artery;
ima, internal maxillary artery branches;
jb, jugular bulb;
jv, jugular vein;
lsc, lateral semicircular canal;
ma, mastoid air cells;
mcp, middle cerebellar peduncle;
smf, stylomastoid foramen;
ssc, superior semicircular canal;
ves, vestibule.

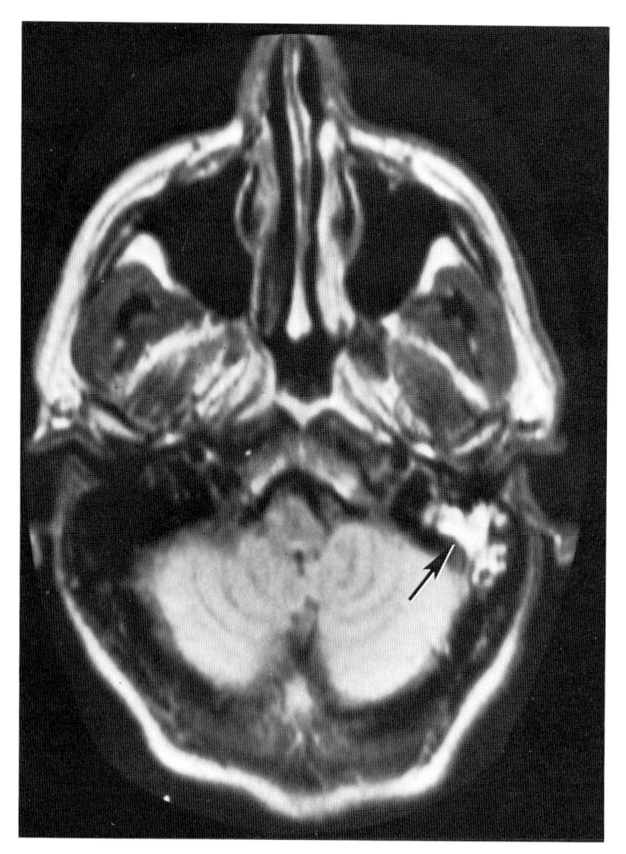

Fig. 11-35. The presence of fluid in the mastoid air cells may produce high signals that obscure temporal bone pathology.

pensate for progressive peripheral vestibular imbalance.

Neurologic signs, other than vestibulocochlear dysfunction, are correlated with tumor size and occur in less than 50 percent of patients with small tumors in recently reported series.[81] Early findings include nystagmus, impaired corneal reflex, and ataxia. Despite the close relationship of the facial nerve to these tumors, clinical facial weakness is a late and subtle finding, seldom more than flattening of the nasolabial fold. Large tumors at the CP angle, are more likely to have associated cranial nerve dysfunction and papilledema.[81]

A number of paraclinical and electrophysiologic tests may aid in the diagnosis of acoustic neurinoma.[81] Vestibular testing (calorics and electronystagmography) are quite sensitive for detecting larger tumors; however, they are insensitive for small tumors and nonspecific for the population of acoustic tumor suspects (including those with Meniere disease, otosclerosis, and vestibular neuronitis). Audiometry typically shows a high frequency, retrocochlear pattern of hearing loss. Stapedius reflex testing is also quite sensitive. Brain stem auditory evoked response testing is the most sensitive single electrophysiologic test, having a sensitivity of over 95 percent and a false positive rate of less than 10 percent.

Gas-CT cisternography has in recent years become the gold standard for diagnosis of small acoustic neurinomas. The presence or absence of tumor can be confidently decided in over 98 percent of studies.[86] Larger tumors can frequently be diagnosed using only intravenous contrast enhanced CT.

Several comparison studies have demonstrated that MRI is superior to contrast CT and equivalent to gas-CT cisternography in the diagnosis of acoustic neurinomas.[87,88] Occasionally the extent of tumor is better delineated by MRI than CT. These results should make MRI the modality of choice in evaluating acoustic tumors. The future use of paramagnetic MR constrast agents promises even greater sensitivity and specificity.[89]

Examples of acoustic neurinomas seen on MRI are presented in Figures 11-36 through 11-38. Using T1-weighted pulse sequences the tumor has similar signal to normal nerve but is distinguished by virtue of its size and mass effect. The small intracanalicular tumor (Fig. 11-38) was noted to have high signal on SE 1500/56 image aiding in its detection. This sequence is not T2-weighted enough to turn CSF white and obscure the nerve; rather it seems ideal for highlighting tumor within the canal.

Although most acoustic neurinomas behave as the examples presented here, a few are at variance. Some tumors with cystic components may have particularly long T1 and T2 values. Other histologically "hard" tumors have been found to have particularly short T1 and T2 values.[87] The recognition of such signal changes is of possible therapeutic significance. A large, but cystic tumor can sometimes be removed via a translabyrinthine approach with aspiration rather than a posterior fossa approach.

MRI may aid in the differential diagnosis of CPA tumors having similar appearance on CT.[90,91] In Figure 11-39 the diagnosis of CPA meningioma was sug-

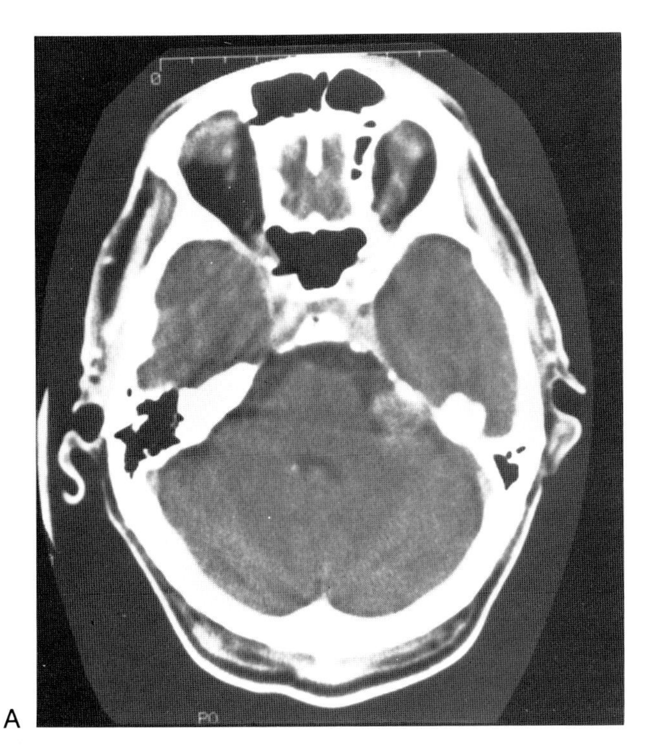

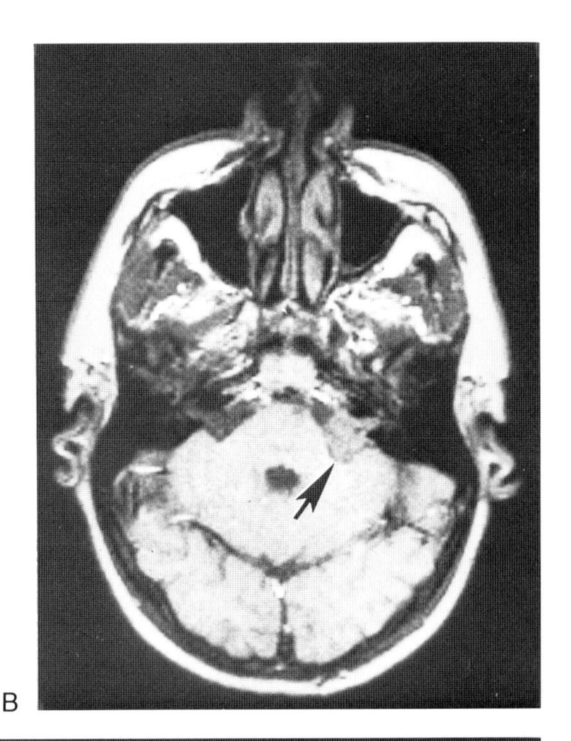

Fig. 11-36. Acoustic neurinoma. (A) CT scan shows large CPA mass. (B) Axial T1-weighted MR scan clearly shows tumor in the internal auditory canal and CPA cistern. (C) Sagittal image shows enlarged nerve (arrow) and CPA mass (arrowhead). (Courtesy Picker International.)

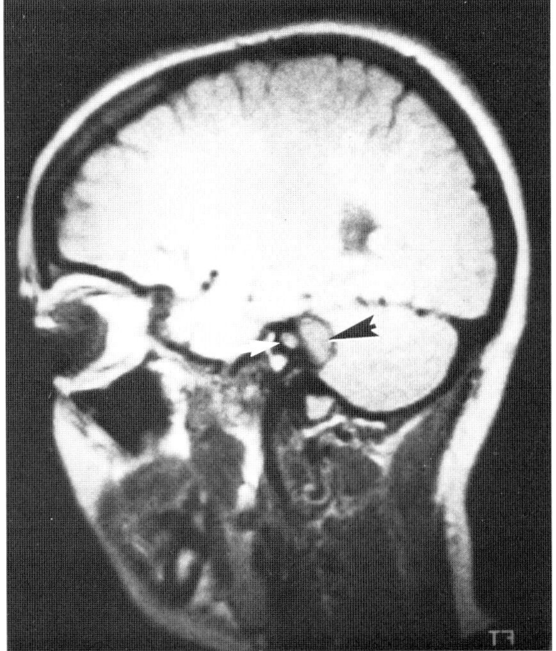

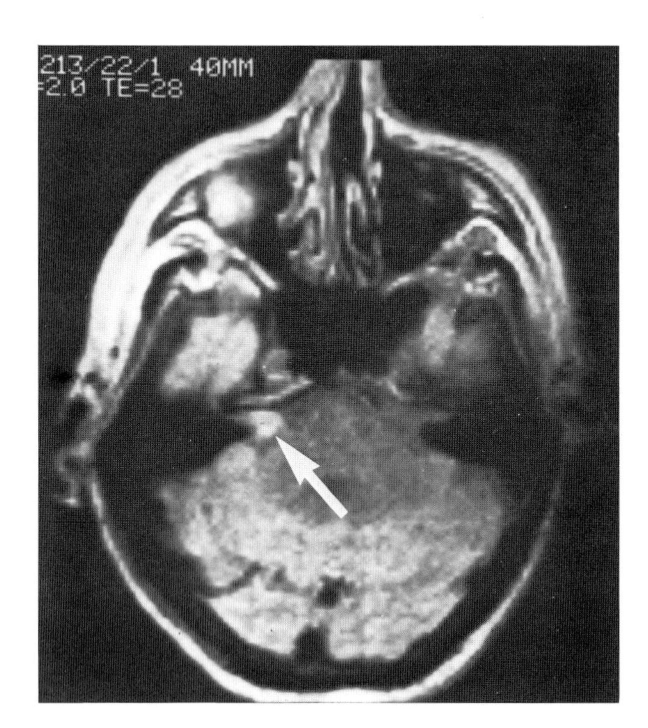

Fig. 11-37. High signal from a right acoustic neurinoma on this SE 2000/28 axial image.

A

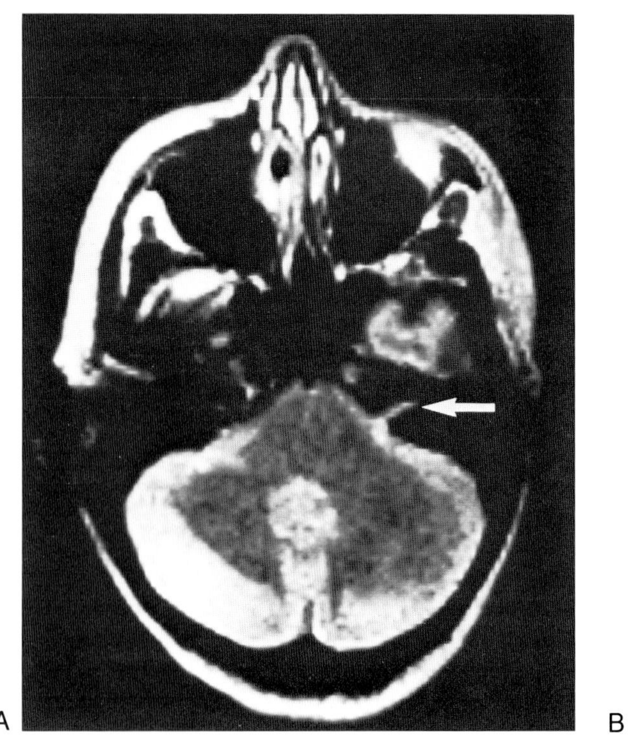

B

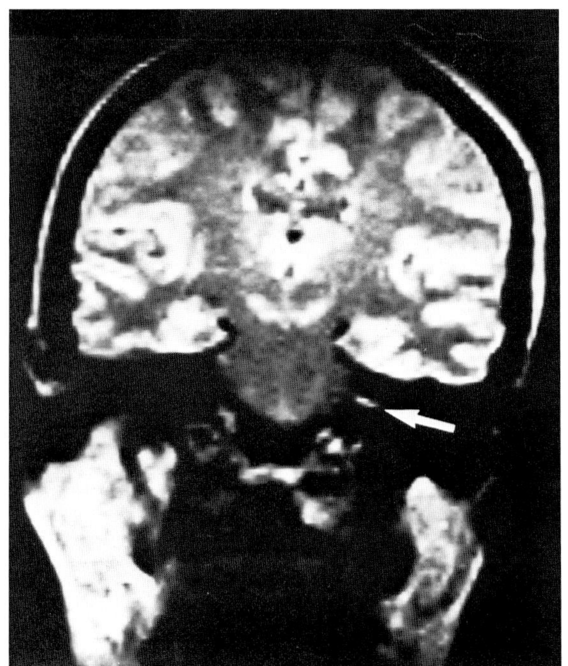

Fig. 11-38. (**A** and **B**) Small (4 mm) largely intracanalicular acoustic neurinoma detected by its high signal on SE 1500/56 and size asymmetry.

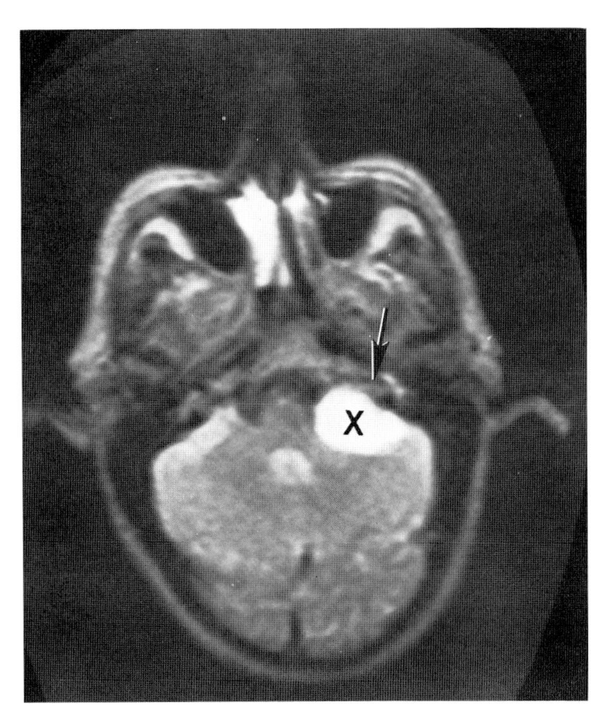

Fig. 11-39. CPA meningioma. MRI shows the tumor (*X*) is separate from the left neurovascular bundle (arrow), which is normal in size and signal.

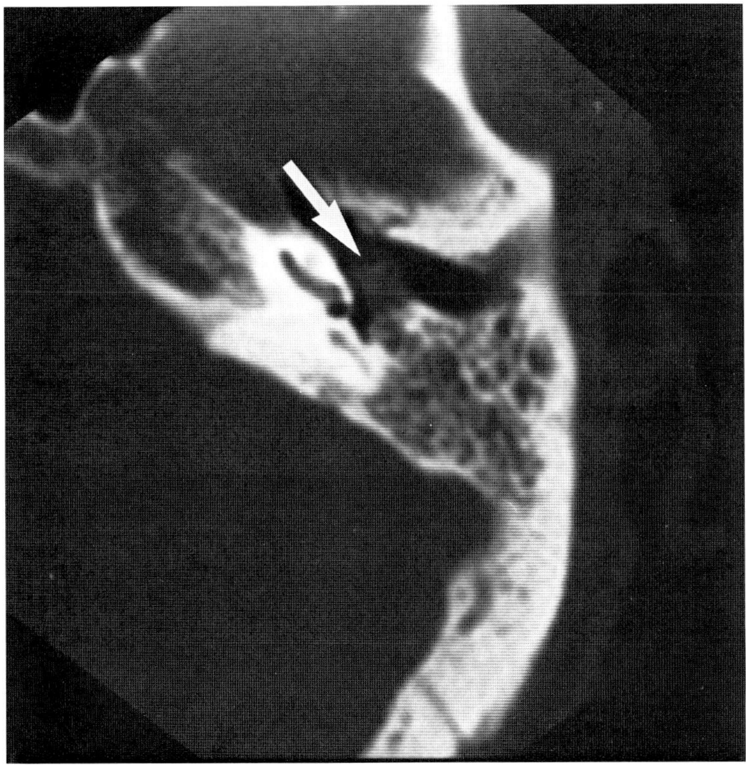

Fig. 11-40. Glomus tympanicum. **(A)** CT shows middle ear mass (arrow). The mastoid air cells are opacified with fluid. (*Figure continues*).

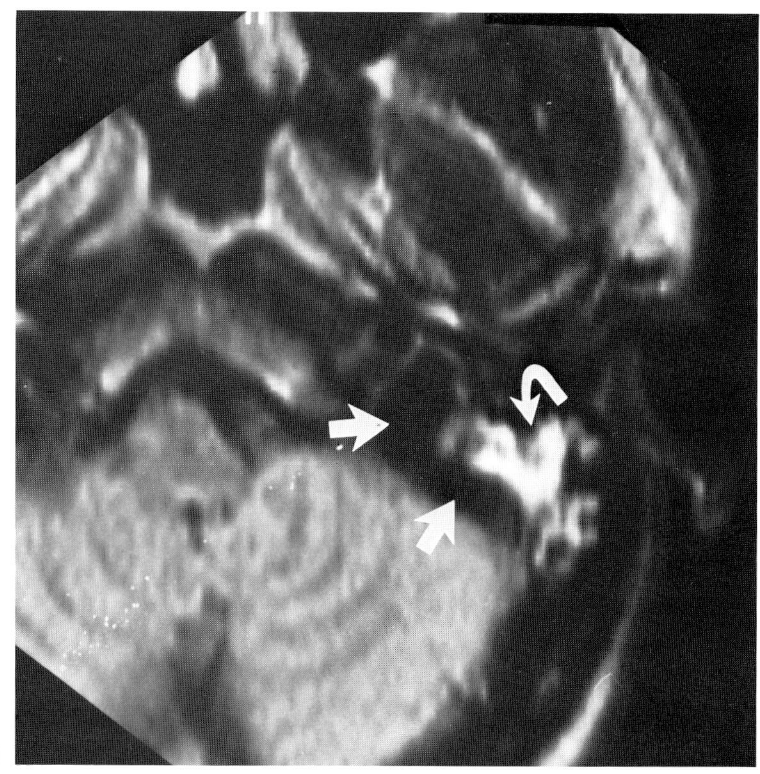

Fig. 11-40 (*continued*). **(B)** The small tumor (curved arrow) is hard to appreciate on MR images due to adjacent mastoid disease. However, patency of the sigmoid sinus (straight arrow) is well established, excluding jugular invasion or origin of this middle ear tumor.

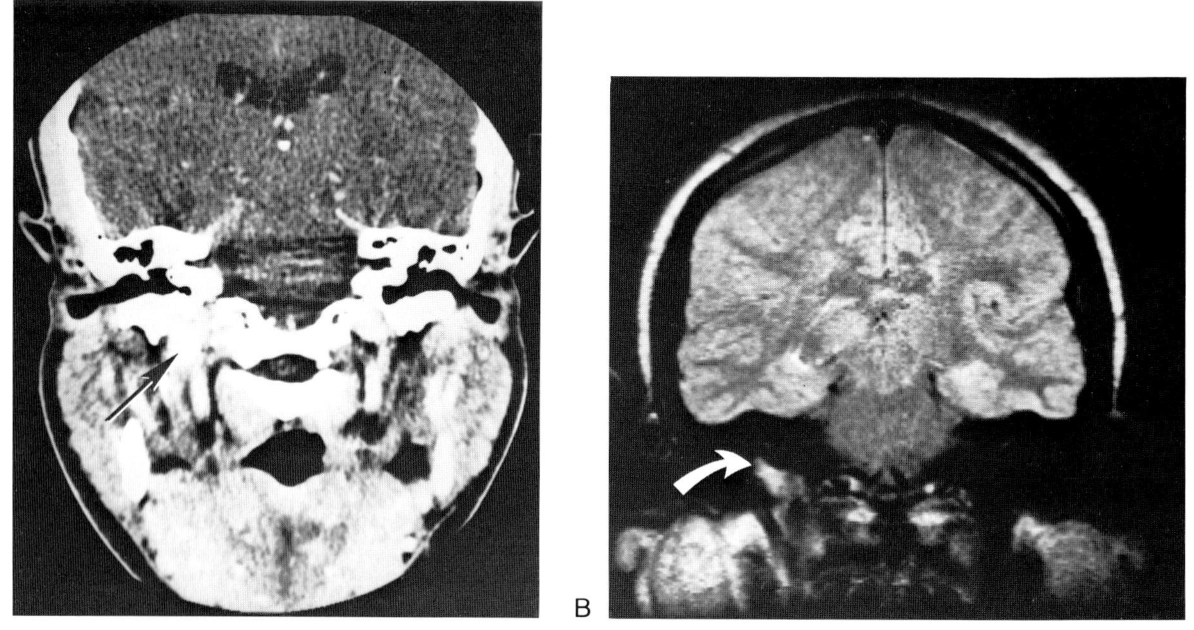

Fig. 11-41. Glomus jugulare tumor. **(A)** Coronal CT shows abnormal enhancing tissue in right jugular bulb (arrow). **(B)** Corresponding MR image. (*Figure continues*).

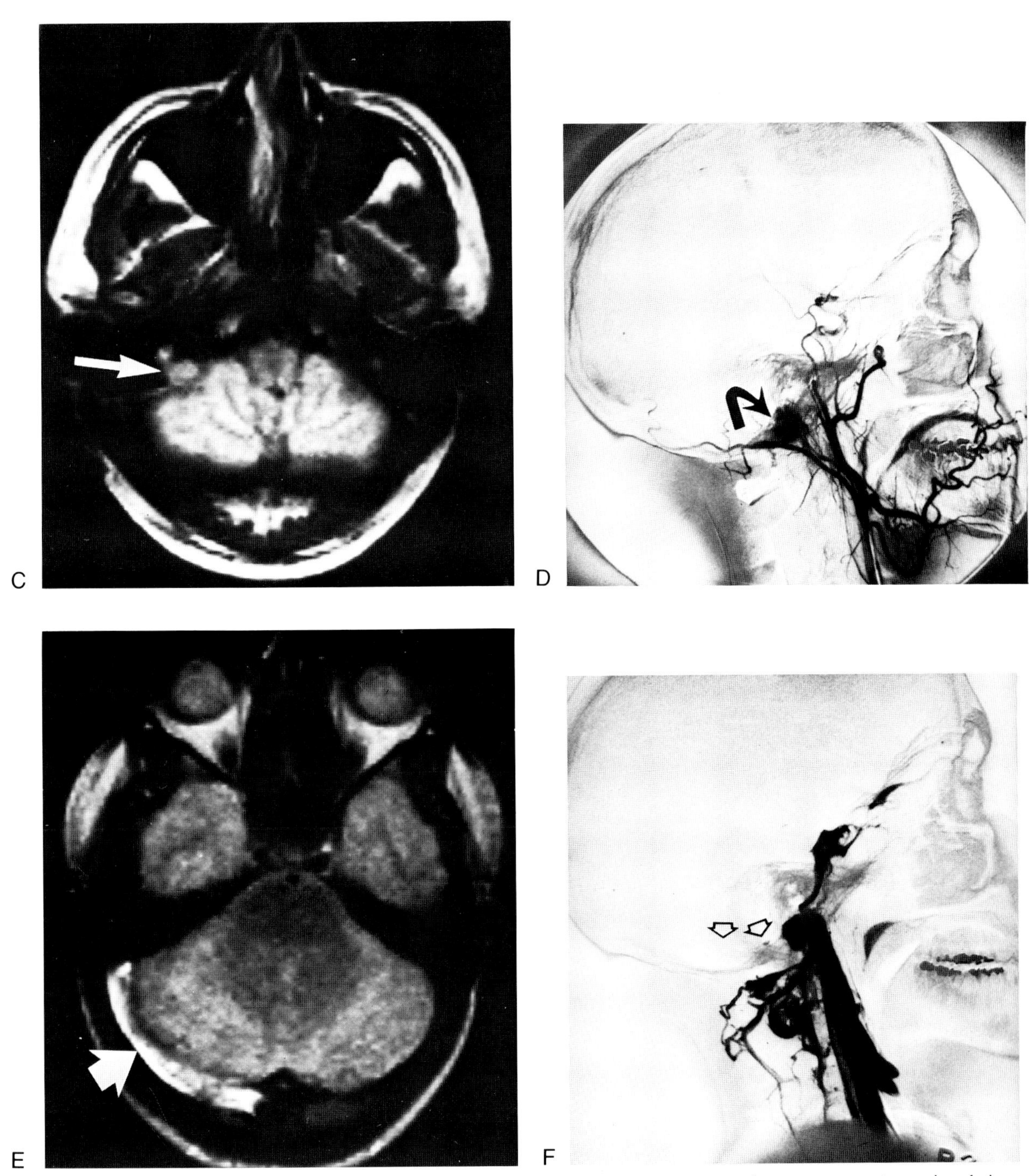

Fig. 11-41 *(continued)*. **(C)** The tumor on axial SE 2000/28 MR. **(D)** Hypervascular mass at angiography, fed by ascending pharyngeal artery. **(E)** Thrombosis of the right transverse and sigmoid sinus was produced by the tumor as evidenced by abnormal high signal (arrow). **(F)** Venography confirms signoid sinus occlusion.

gested instead of acoustic tumor because the eighth nerve could be well visualized separate from the tumor. Similarly, CPA epidermoids may be distinguished from acoustic tumors because of their different signal characteristics and anatomic separability.

Glomus Tumors

Glomus tumors of the temporal bone are neoplasms of the chemoreceptor system, more properly named paragangliomas. Traditionally these lesions have been divided into those confined to the middle ear cavity (*glomus tympanicum*) and those arising in the jugular bulb (*glomus jugulare*).[92] Extracranial paragangliomas are well known and may arise along the course of the vagus nerve, in the chest, abdomen, or retroperitoneum.

Glomus tumors are the most common neoplasms of the middle ear and are second only to acoustic neurinomas as the most common tumor of the temporal bone.[93] They occur predominantly in middle aged women. Up to 10 percent of patients have multiple tumors or tumors in other organ systems (pheochromocytoma, thyroid cancer).

Symptoms of glomus tumors may be otologic or neurologic.[94] Otologic symptoms include conductive hearing loss and pulsatile tinnitus. These are more common with glomus tympanicum tumors. Neurologic symptoms include impairment of cranial nerves V through XII, and are more frequently seen with the larger glomus jugulare tumors.

Glomus tumors grow slowly and invade adjacent bone by expansion. Metastases rarely occur. The tumors tend to grow along the paths of least resistance: along fissures, air cell tracts, vascular channels, and foramina. They may descend along and occlude the jugular vein. CNS invasion occurs late and may be the ultimate cause of death.

High resolution CT is well suited for evaluating glomus tumors, helping to differentiate them from vascular anomalies and other benign and malignant lesions.[93,95] Angiography characteristically shows a hypervascular mass, frequently supplied by the ascending pharyngeal artery. The role for MRI is uncertain although both glomus tympanicum and jugulare tumors may be recognized (Figs. 11-40 and 11-41). MR may be useful to exclude or confirm jugular vein invasion, which significantly alters the surgical approach.

REFERENCES

1. Mark L, Pech P, Daniela D et al: The pituitary fossa: a correlative anatomic and MR study. Radiology 153:453, 1984
2. Nishimura K, Fujisawa I, Togashi K et al: Posterior lobe of the pituitary: identification by lack of chemical shift artifact in MR imaging. J Comput Assist Tomogr 10:899, 1986
3. Wiener SN, Rzeszotarski MS, Droege RT et al: Measurement of pituitary gland with MR imaging. AJNR 6:717, 1985
4. Swartz JD, Russell KB, Basile BA et al: High-resolution computed tomographic appearance of the intrasellar contents in women of childbearing age. Radiology 147:115, 1983
5. Davis PC, Hoffman JC Jr, Tindall GT et al: Prolactin-secreting pituitary microadenomas: inaccuracy of high-resolution CT imaging. AJR 144:151, 1985
6. Peyster RG, Hoover ED: CT of the abnormal pituitary stalk. AJNR 5:49, 1984
7. Siedel FG, Towbin R, Kaufman RA: Normal pituitary stalk size in children: CT study. AJNR 6:733, 1985
8. Chambers EF, Turski PA, LaMasters D et al: Regions of low density in the contrast-enhanced pituitary gland: normal and pathologic processes. Radiology 144:109, 1982
9. Burrow GN, Wortzman G, Rewcastle NB et al: Microadenomas of the pituitary and abnormal sellar tomograms in an unselected autopsy series. N Engl J Med 304:156, 1981
10. Costello RT: Subclinical adenoma of the pituitary gland. Am J Pathol 12:205, 1936
11. Parent AD, Bebin J, Smith RR: Incidental pituitary adenomas. J Neurosurg 54:228, 1981
12. Shanklin WM: The incidence and distribution of cilia in the human pituitary with a description of micro-follicular cysts derived from Rathke's cleft. Acta Anat 11:361, 1951
13. Shuangshoti S, Netsky MG, Nashold BS Jr: Epithelial cysts related to sella turcica: proposed origin from neuroepithelium. Arch Pathol 90:444, 1970
14. Kovacs K: Adenohypophysial necrosis in routine autopsies. Endokrinologie 60:309, 1972
15. Roessmann U, Kaufman B, Friede RL: Metastatic lesions in the sella turcica and pituitary gland. Cancer 25:478, 1970
16. Gurling KJ, Scott GBD, Baron DN: Metastases in pitu-

itary tissue removed at hypophysectomy in women with mammary carcinoma. Br J Cancer 11:519, 1957
17. Landolt AM: Pituitary adenoma. Clinical-morphologic correlations. J Histochem and Cytochem 27:1395, 1979
18. Davis PC, Hoffman JC Jr, Tindall GT et al: CT-surgical correlation in pituitary adenomas: evaluation in 113 patients. AJNR 6:711, 1985
19. Turski PA, Damm M: The role of computed tomography in the evaluation of pituitary disease. Semin Ultras CT MR 6:276, 1985
20. Randall RV, Scheithauer BW, Laws ER Jr et al: Pseudoprolactinomas. Trans Am Clin Climatol Assoc 94:114, 1982
21. Donovan Post MJ, David NJ, Glaser JS et al: Pituitary apoplexy: diagnosis by computed tomography. Radiology 134:665, 1980
22. Reid RL, Quigley ME, Yen SSC: Pituitary apoplexy. A review. Arch Neurol 42:712, 1985
23. Davis PC, Hoffmann JC Jr, Spencer T et al: MR imaging of pituitary adenoma: CT, clinical and surgical correlation. Radiology 157(p):286, 1985
24. Lee BCP, Deck MDF: Sellar and juxtasellar lesion detection with MR. Radiology 157:143, 1985
25. Pojunas KW, Daniels DL, Williams AL et al: MR imaging of prolactin-secreting microadenomas. AJNR 7:209, 1986
26. Weissman JD, Sebok D, Weinstein M et al: High resolution MR imaging of the sella: first application at 1.5 T of three dimensional limited-volume techniques. Radiology 157(p):304, 1985
26a. Weissbuch SS: Explanation and implication of MR signal changes within pituitary adenomas after bromocriptine therapy. AJNR 7:214, 1986
27. Bilaniuk LT, Zimmerman RA, Wehrli FW et al: Magnetic resonance imaging of pituitary lesions using 1.0 to 1.5 T field strength. Radiology 153:415, 1984
28. Vignaud J, Aubin ML, Jardin C et al: MR imaging of the cavernous sinus: report of 40 pathologic cases. Radiology 157(p):286, 1985
29. Wilson CB: Neurosurgical management of large and invasive pituitary tumors. p.335. In Tindall GT, Collins WF (eds): Clinical Management of Pituitary Disorders. Raven, New York, 1979
30. D'Abrera VStE, Burke WJ, Bleasel KF et al: Carcinoma of the pituitary gland. J Pathol 109:335, 1979
31. Rubinstein LJ: Atlas of Tumor Pathology. Armed Forces Institute of Pathology, Washington D.C., 1972
32. Ahmadi J, North CM, Segall HD et al: Cavernous sinus invasion by pituitary adenomas. AJNR 6:893, 1985
33. Taveras JM, Wood EH: Diagnostic Neuroradiology. 2nd Ed. Williams & Wilkins, Baltimore, 1976
34. Greenhouse AH: Pituitary sarcoma, a possible consequence of radiation. JAMA 190:269, 1964
35. Gurley KJ, Scott GBD, Baron DN: Metastases in pituitary tissue removed at hypophysectomy in women with mammary carcinoma. Br J Cancer 11:519, 1957
36. Sheehan HL: Postpartum necrosis of the anterior lobe of the pituitary. Lancet 2:321, 1940
37. Sheehan HL, Summers SVK: The syndrome of hypopituitarism. Q J Med 18:319, 1949
38. Kaufman B, Chamberlin WB Jr: The ubiquitous "empty" sella turcica. Acta Radiol (Diagn) (Stockh) 13:413, 1972
39. Neelon FA, Goree JA, Lebovitz HE: The primary empty sella. Clinical and radiographic characteristics and endocrine function. Medicine 52:73, 1973
40. Caplan RH, Dobben GD: Endocrine studies in patients with the "empty sella syndrome." Arch Intern Med 123:611, 1969
41. Lee KF, Schatz NJ, Savino PJ: Ischemic chiasmal syndrome. Neuroophthalmology 8:115, 1975
42. Hall K, McAllister VL: Metrizamide cisternography in pituitary and juxtapituitary lesions. Radiology 134:101, 1980
43. Rudwan MA: Pituitary abscess. Neuroradiology 12:243, 1977
44. Enzmann DR, Sieling RJ: CT of pituitary abscess. AJNR 4:79, 1983
45. Sholkoff S, Kerber C, Cramm R et al: Parasellar choristoma. AJR 128:1051, 1977
46. Garduer D, Nachanakian A, Milliard JC et al: Neuroradiological aspects and therapeutic incidence of a case of posterior pituitary tumor (choristoma). J Neuroradiol 5:321, 1978
47. Baskin DS, Wilson CB: Transsphenoidal treatment of non-neoplastic intrasellar cysts. A report of 38 cases. J Neurosurg 60:8, 1984
48. Bayoumi ML: Rathke's cleft and its cysts. Edinburgh Med J 55:745, 1948
49. Fager CA, Carter H: Intrasellar epithelial cysts. J Neurosurg 24:77, 1966
50. Okamoto S, Handa H, Yamashita J et al: Computed tomography in intra- and suprasellar epithelial cysts (symptomatic Rathke cleft cysts). AJNR 6:515, 1985
51. Kuno T, Sudo M, Momoi T et al: Pituitary hyperplasia due to hypothyroidism. Radiol Clin North Am 4:600, 1980
52. Mayfield RK, Levine JH, Gordon L et al: Lymphoid adenohypophysitis presenting as a pituitary tumor. Am J Med 69:619, 1980
53. Asa SL, Bilbao JM, Kovacs K et al: Lymphocytic hypophysitis of pregnancy resulting in hypopituitarism: a distinct clinico-pathologic entity. Ann Intern Med 95:166, 1981
54. Karnaze M, Sartor K, Winthrop JD et al: Suprasellar lesions: evaluation with MR imaging. Radiology 161:77, 1986

55. Rao KCVG, Harwood-Nash DC, Fitz CR: Neurodiagnostic studies in craniopharyngiomas in children. Rev Interam Radiol 2:149, 1977
56. Petito C, DeGirolami U, Earle K: Craniopharyngiomas: a clinical and pathological review. Cancer 37:1944, 1976
57. Banna M: Radiology, p.135. In Hankinson J, Banna M (eds): Pituitary and Parapituitary Tumors. WB Saunders, Philadelphia, 1976
58. Taveras JM, Wood EH: Diagnostic Neuroradiology. 2nd Ed. Williams & Wilkins, Baltimore, 1976
59. Currarino G, Maravilla KR, Salyer KE: Transsphenoidal canal (large craniopharyngeal canal) and its pathologic implications. AJNR 6:39, 1985
60. Pussey E, Kortman KE, Bradley WG: MR imaging of craniopharyngiomas. Radiology 157(p):344, 1985
61. Becker RD, Zimmerman RD, Deck MDF: MR scanning in craniopharyngioma. Radiology 157(p):150, 1985
62. Lin S-R, Bryson MM, Gobien R et al: Radiologic findings of hamartomas of the tuber cinereum and hypothalamus. Radiology 127:697, 1978
63. Diebler C, Ponsot G: Hamartomas of the tuber cinereum. Neuroradiology 25:93, 1983
64. Judge DM, Kulin HE, Page R et al: Hypothalamic hamartoma: souce of luteinizing-hormone-releasing factor in precocious puberty. N Engl J Med 296:7, 1977
65. Johnson MA, Pennock JM, Bydder GM et al: clinical NMR imaging of the brain in children: normal and neurologic disease. AJR 141:1005, 1983
66. Lichtenstein L, Jaffe HL: Fibrous dysplasia of bone. Arch Pathol 33:777, 1942
67. Edeiken J: Roentgen Diagnosis of Diseases of Bone. 3rd Ed. Williams & Wilkins, Baltimore, 1981
68. Mabrey RE: Chordomata: a study of 150 cases. Am J Cancer 25:501, 1935
69. Lee KF, Lin S: Neuroradiology of Sellar and Juxtasellar Lesions. Charles C Thomas, Springfield, 1979
70. Kendall B, Lee BCP: Intracranial chordomas. Br J Radiol 150:687, 1977
71. Kamrin RP, Potanos JN, Pool JL: An evaluation of the diagnosis and treatment of chordoma. J Neurol Neurosurg Psychiatry 27:157, 1964
72. Congdon CC: Benign and malignant chordomas. A clinical anatomic study of 22 cases. Am J Pathol 28:793, 1952
73. Ham JS, Huss RG, Benson JE et al: MR imaging of the skull base. J Comput Assist Tomogr 8:944, 1984
74. McGinnis BD, Brady TJ, New PFJ et al: Nuclear magnetic resonance (NMR) imaging of tumors in the posterior fossa. J Comput Assist Tomogr 7:575, 1983
75. Strang C, Rannie I: Dyschondroplasia with hemangiomata (Maffucci's syndrome) reported of a case complicated by intracranial chondrosarcoma. J Bone Joint Surg 32B:376, 1950
76. Berkmen YM, Blatt ES: Cranial and intracranial artilaginous tumors. Clin Radiol 19:327, 1968
77. Sarwar M, Swischuk LE, Schechter MM: Intracranial chondromas. Am J Roentgenol 127:973, 1976
78. Daniels JF, Schenck JF, Foster T et al: Surface-coil magnetic resonance imaging of the internal auditory canal. AJR 145:469, 1985
79. Koenig H, Lenz M, Sauter R: Temporal bone region: high resolution MR imaging using surface coils. Radiology 159:191, 1986
80. New PFJ, Bachow TB, Wismer GL et al: MR imaging of the acoustic nerves and small acoustic neuromas at 0.6 T prospective study. AJR 144:1021, 1985
81. Hart RG, Davenport J: Diagnosis of acoustic neuroma. Neurosurgery 9:450, 1981
82. Kasantikul V, Netsky MG, Glasscock ME III, Hays JW: Acoustic neurilemmoma: clinicoanatomical study of 103 patients. J Neurosurg 52:28, 1980
83. Naunton RF, Petasnick JP: Acoustic neurinomas with normal internal auditory meatus. Arch Otolaryngol 91:437, 1970
84. Ojemann RG, Montgomery WW, Weiss AD: Evaluation and surgical treatment of acoustic neuroma. N Engl J Med 287:895, 1972
85. Matthew GD, Facer GW, Suh KW et al: Symptoms, findings and methods of diagnosis in patients with acoustic neuroma. Laryngoscope 88:1893, 1978
86. Solti-Bohman LG, Magaram DL, Lo WWM et al: Gas-CT cisternography for detection of small acoustic nerve tumors. Radiology 150:403, 1984
87. Curati WL, Graif M, Kingsley DPE et al: MRI im acoustic neuroma: a review of 35 patients. Neuroradiology 28:208, 1986
88. Kingsley DPE, Brooks GB, Leung AWL, Johnson MA: Acoustic neuromas: evaluation by magnetic resonance imaging. AJNR 6:1, 1985
89. Curati WL, Graif M, Kingsley DPE et al: Acoustic neuromas: Gd-DTPA enhancement in MR imaging. Radiology 158:447, 1986
90. Mikhael MA, Ciric IS, Wolff AP: Differentiation of cerebello-pontine angle neuromas and meningiomas with MR imaging. J Comput Assist Tomogr 9:852, 1985
91. Latack JT, Kartush JM, Kemink JL et al: Epidermoidomas of the cerebellopontine angle and temporal bone: CT and MR aspects. Radiology 157:361, 1985
92. Alford BR, Guilford FR: A comprehensive study of tumors of the glomus jugulare. Laryngoscope 72:765, 1962
93. Lo WWM, Solti-Bohman LG, Lambert PR: High-resolution CT in the evaluation of glomus tumors of the temporal bone. Radiology 150:737, 1984
94. Spector GJ, Druck NS, Gado MH: Neurologic manifestations of glomus tumors in the head and neck. Arch Neurol 33:270, 1976
95. Chakeres DW, LaMasters DL: Paragangliomas of the temporal bone: high-resolution CT studies. Radiology 150:749, 1984

12

The Orbit

While CT is quite well suited for evaluating orbital pathology, a number of reports have demonstrated the utility and potential for MRI.[1–15] The use of surface coils to enhance the MR signal-to-noise ratio has recently allowed improved visualization of the orbital contents.[8,13,15] The future applications of in vivo spectroscopy hold promise in the study of certain ocular diseases, such as cataract formation, corneal edema, and vitreal liquefaction.[16,17] Routine multiplanar imaging by MR may allow improved appreciation of certain anatomic relationships between masses and normal orbital structures. MR also holds certain advantages over CT for evaluating the intracanalicular, optic nerves, the chiasm, and optic tracts.[1,12]

MR IMAGING STRATEGIES AND TECHNIQUES

Patient cooperation is the key element for successful high resolution orbital MRI. Prior to scanning, ferromagnetic dental bridgework should be removed because it may sometimes cause field distortions that affect the MR signal. Additionally, all mascara should be removed, because many brands contain colbalt or iron oxides that are ferromagnetic and may produce an MR artifact (Fig. 12-1).[18] The patient should be comfortably positioned supine on the gantry and asked to gaze at a single spot in the scanner bore. (It may be helpful to paint or mark a suitable spot on the scanner interior.) If the patient cannot maintain a fixed gaze, the eyes should be closed, allowing random movements that will "average out" to produce a relatively stable image for scanning.

Orbital surface coils are generally preferred, if available, because they allow the highest signal-to-noise ratios and greatest spatial resolution. Most surface coils are supplied as goggles or loops that attach to the patient's head (Fig. 12-2A). Some manufacturers supply orbital coils, which are mounted independent of the patient (Fig. 12-2B). This latter configuration has certain theoretical advantages in minimizing artifacts from patient motion, but the former can also provide good images and both are in wide use today.

Both axial and coronal imaging seems warranted in most situations. Slice thickness should be 3 to 5 mm and a minimum image matrix of 128 × 256 should be specified. The field of view should be chosen as small as possible, preferably less than 20 cm spherical diameter. Direct sagittal images are useful in evaluating the chiasm. Oblique imaging (along the course of the optic nerve) is possible on some scanners, and this is generally preferred to simple sagittal imaging of the orbital contents. Whole-brain images are also indicated for evaluating the spread of tumors along the optic tracts as well as identifying midbrain lesions and demyelinating diseases that may produce visual symptomatology.

T1-weighted pulse sequences such as SE 500/20 provide the best anatomic delineation of orbital con-

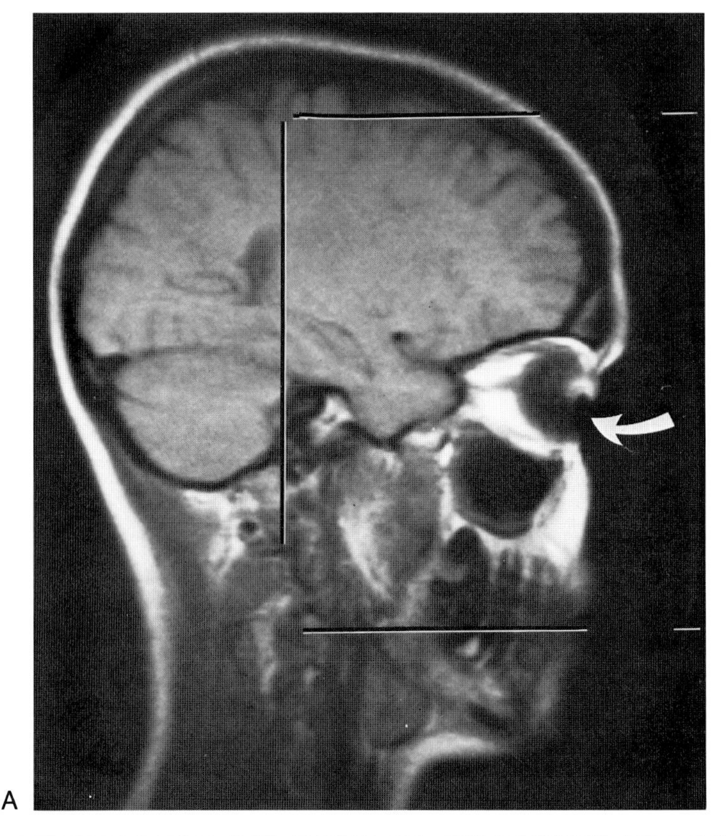

A

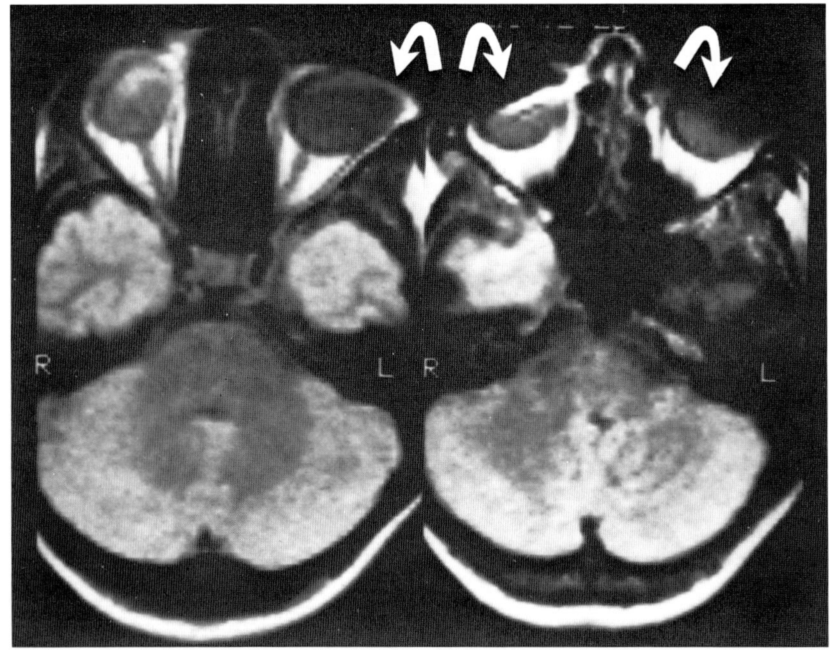

B

Fig. 12-1. (**A** and **B**) Ferromagnetic oxides in certain brands of mascara may cause local field distortions, resulting in artifacts on the MR image.

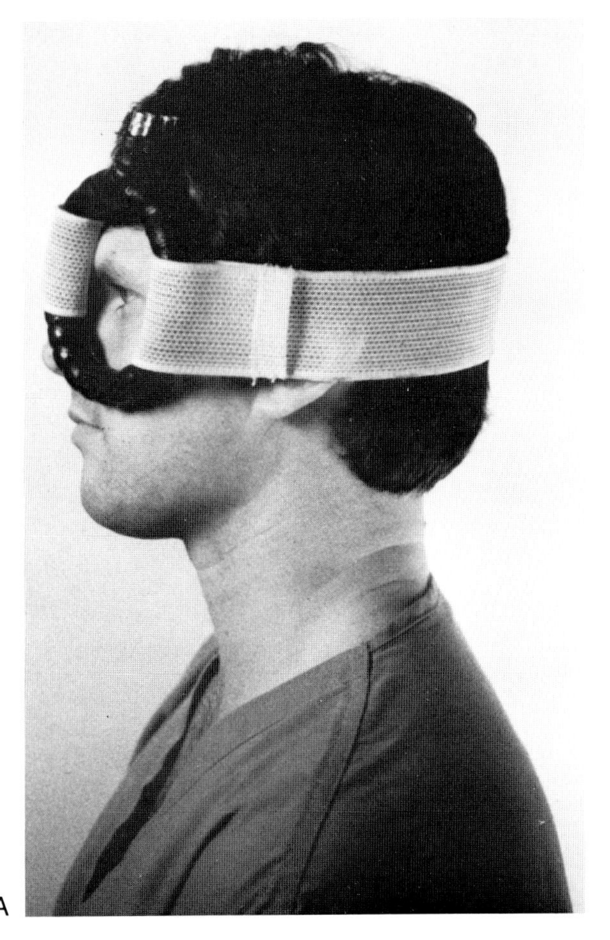

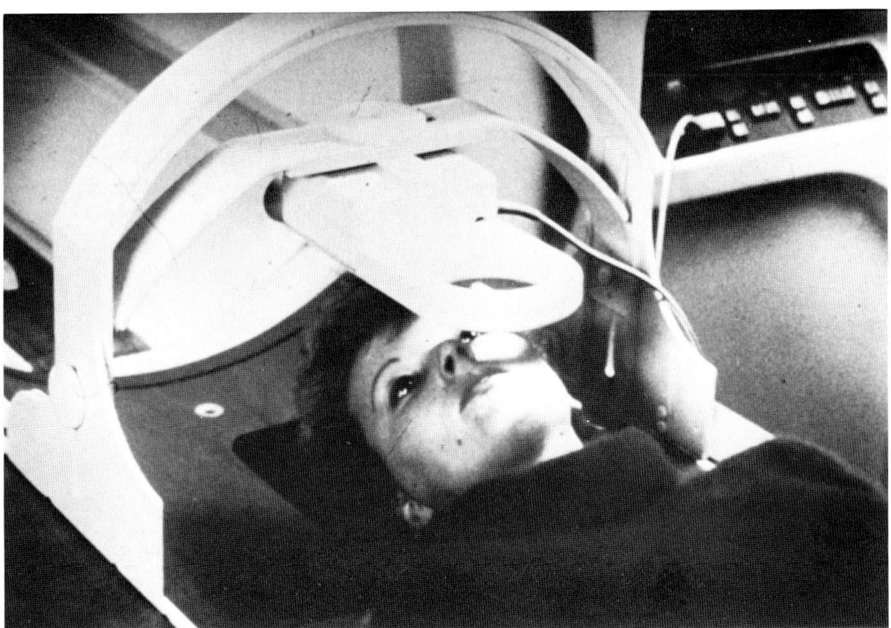

Fig. 12-2. Two methods of orbital surface coil mounting. **(A)** Attached to patient. (Courtesy of Fonar Corporation, 110 Marcus Drive, Melville, NY.) **(B)** Off the patient. (Courtesy of Siemens Medical Systems, Inc.)

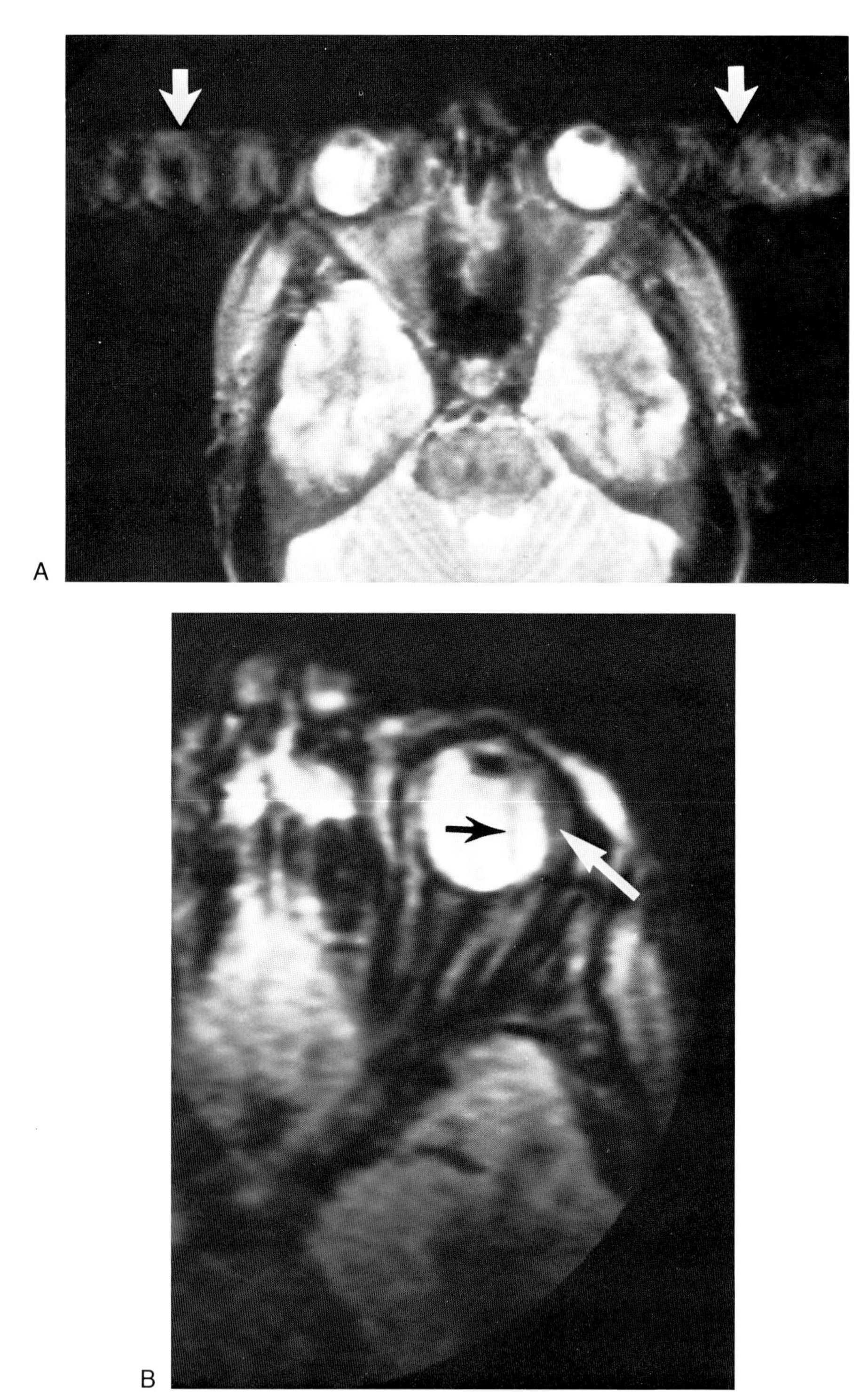

Fig. 12-3. Motion artifacts of the globe during scanning. **(A)** "Ghost" bands on either side of the orbit along the phase encoded axis. **(B)** Curvilinear bands posteriorly within the globe, not to be mistaken for retinal detachment.

tents and lesions. It is also desirable to have a T2-weighted sequence for tissue characterization as well. In the orbit this is probably best obtained by using a long TE (75 msec) but only a moderately prolonged TR (1000 msec). While this sequence retains appreciable T1-dependence it does provide improved contrast on the basis of T2 differences between tissues. While using a longer TR value would make the image more independent of T1, the prolonged imaging time incurred would likely result in more motion artifacts from the globe. Accordingly, both spatial and contrast resolution might be compromised with long TR/TE sequences. An example of such motion artifact on a long TR spin echo sequence is shown in Figure 12-3A. Motion artifacts may be minimized by using a relatively large number of excitations (4–6) for signal averaging.

Table 12-1. Average Relaxation Times of Ocular Tissues at 1.4 T

Tissue	*T1 (msec)*	*T2 (msec)*
Cornea		
Middle layer	1220	22
Internal layer	>3000	92
Sclera	838	31
Nucleus	720	21
Cortex	1229	76
Retina/choroid	1898	74
Ciliary body	1719	101
Iris	2300	162
Anterior chamber fluid	>3000	392
Vitreous	>3000	443
Optic nerve	1142	63
Fat	451	24

(Data from Gomori JM, Grossman RI, Shields JA et al: Ocular MR imaging and spectroscopy: an ex vivo study. Radiology 160:201, 1986.)

NORMAL ANATOMY

The relaxation times of various ocular tissues have been measured both by in vivo imaging and ex vivo spectroscopy[16] (Table 12-1). While the exact values are methodology dependent, the relative values of each ocular component help explain their imaging characteristics. Excellent MR contrast exists in the orbit because of the anatomic juxtaposition of tissues (fat, water, and parenchymal) that have widely varying T1 and T2 values. Subtle changes in these relaxation times may be sensitive indicators of disease, particularly with regard to the lens, cornea, or vitreous body.[14,16]

The adult globe measures approximately 24 mm in diameter.[19] Its wall consists of three neurovascular and fibrous layers (retina, choroid, and sclera, from inward out) that cannot be differentiated on CT. On MRI the retina and choroid cannot be separated; however, their signal is markedly more intense than the fibrous sclera, which appears dark on most imaging sequences (Fig. 12-4). These layers should not be confused with curvilinear motion artifacts, which are quite characteristic and seen posteriorly in the globe (Fig. 12-3B).

Three corneal layers can be distinguished by MRI performed on enucleated eyes.[16] The signal differences in these layers can be explained by their relative water and collagen contents. Early or deep corneal edema may potentially be detected by MRI. However, due to motion artifacts these layers are not usually discernable during in vivo imaging. Furthermore, because the cornea and conjunctiva are so easily visualized directly, there seems little immediate role for MRI in evaluating pathology here.

The lens consists of five concentric layers of cells, all surrounded by a thin elastic capsule. In enucleated specimens it is possible to distinguish five different MR signals, presumably corresponding to the five histologic layers.[16] In clinical practice, however, two major signals are distinguished from the lens (Fig. 12-4). The external lens layer (cortex) is hyperintense on all imaging sequences, indicating a higher water content than the deeper layers. The more central lens (nuclear portion) has shorter relation times, especially T2, and appear less intense on most sequences. Cortical cataracts have increased free water and produce prolongation of cortical T1 and T2.[17] Nuclear cataracts have increased insoluble albuminoid protein, which does not affect free-to-bound water ratios. Accordingly, no differences in relaxation times between the normal lens nucleus and nuclear cataracts have been observed.[17]

The globe's interior is divided by the lens and ciliary body into aqueous (anterior) and vitreous (posterior) compartments. The T1 and T2 values of the aque-

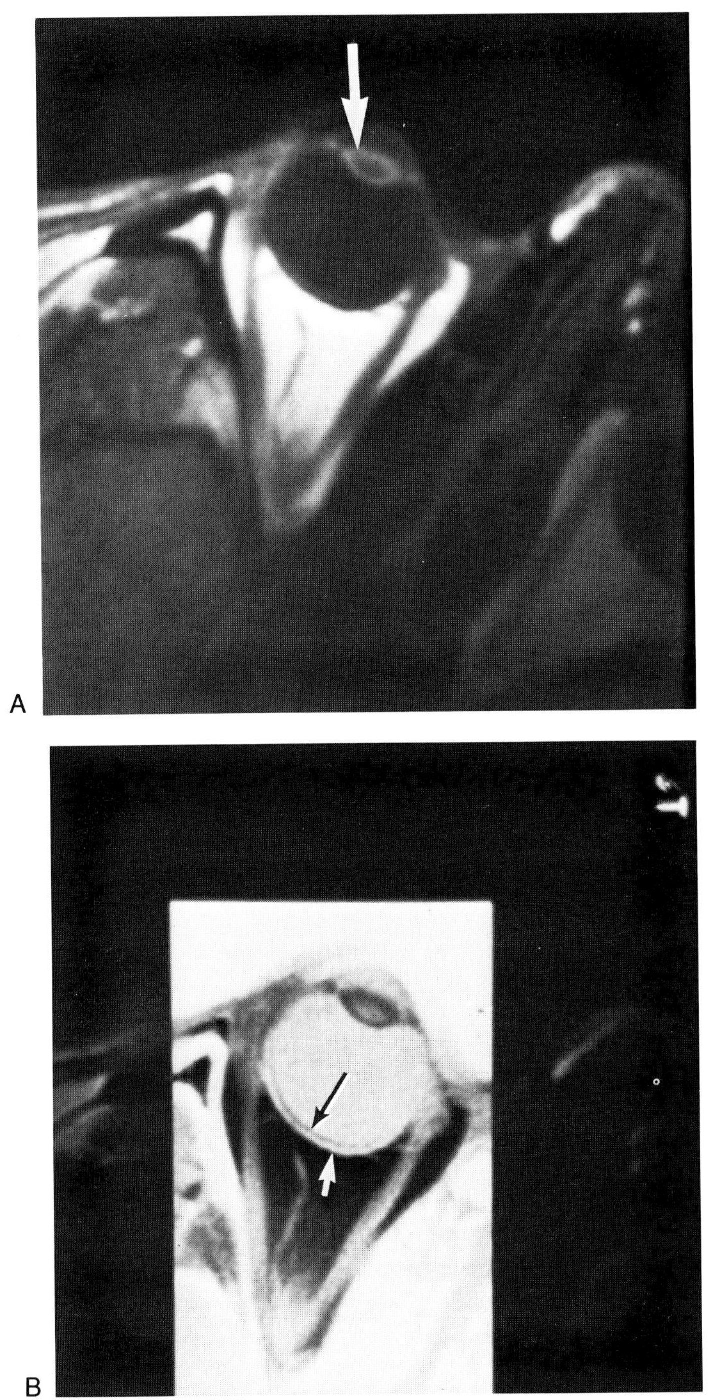

Fig. 12-4. Unique anatomy displayed by MR. **(A)** At least two layers of the lens, cortex and nucleus, are routinely visualized (arrows). SE 500/30 surface coil image. **(B)** The choroid/retina (black arrow) can be distinguished from the sclera (white arrow) SE 500/30, reverse mode of imaging in **(A).** (Courtesy Fonar Corporation, 110 Marcus Drive, Melville, NY.)

ous and vitreous humors are very long, reflecting extremely high water content (vitreous, for example, is 99 percent water, 1 percent hyaluronic acid, and small amounts of collagen). Vitreal liquefaction may result in slight T1 shortening in an ex vivo experimental model.[14]

The optic nerves are about 3 to 4 mm in diameter and 40 mm in length, extending from the globe to the optic chiasm.[19] They are divided into intraocular, intraorbital, intracanalicular, and intracranial segments. The intraorbital portion is the longest and most sinuous, allowing movement of the globe. The intraorbital portion is well seen on MRI silhouetted against adjacent fat. A frequency shift artifact may be seen at the fat-nerve interface and should be easily recognized (Fig. 12-5).[20] The ophthalmic artery crosses the nerve obliquely near its entry into the optic canal (Fig. 12-6). Partial volume effects with the low signal in this vessel can potentially hide pathology on axial MRI.[1]

More proximally the intracanalicular portion of the optic nerve is well seen by MRI. This is in distinction to CT where beam hardening artifacts from the dense bone around the optic canal obscure good visualization of the nerve. However, posterior ethmoid disease and occasionally fat in the anterior clinoid process may obscure the nerve or cause confusing signals (Fig. 12-6).

Intracranially the optic nerves may be followed into the chiasm on axial, coronal, or sagittal images (Fig. 12-6). The chiasm lies above and often slightly anterior to the sella but below the anterior cerebral arteries. Lesions of the chiasm are especially well visualized on MRI.

The seven extraocular muscles (4 recti, 2 obliques, and levator palpebrae superioris) are easily identified by MRI in various planes suitable for each. The muscles are well seen in contrast to the orbital fat with its short T1 and moderate T2 values. Frequency shift artifacts where the muscles abut fat may be noted commonly. In considering orbital masses it is convenient to divide them into those arising within the muscle cone (intraconal) and those outside the muscle cone (extraconal). MRI with its multiplanar imaging

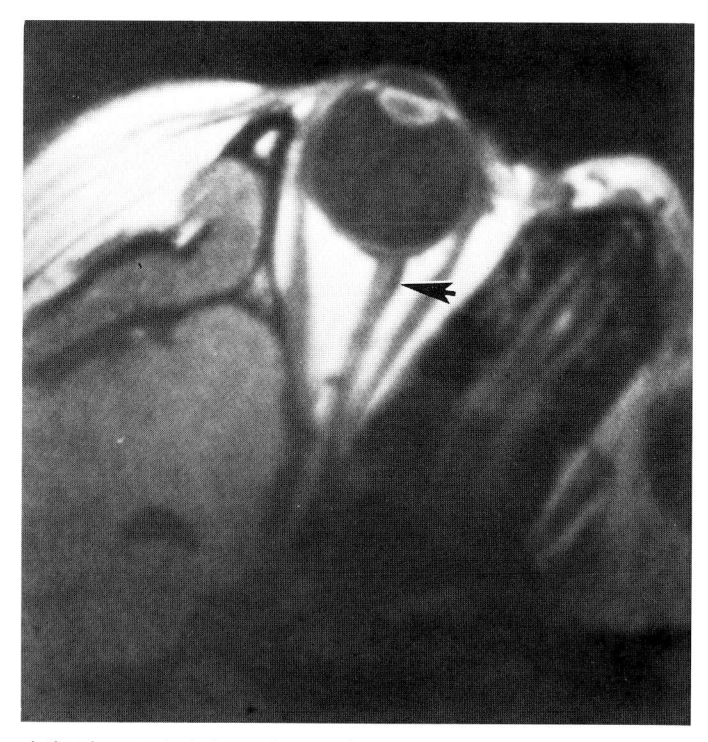

Fig. 12-5. Frequency shift (chemical shift) artifact at the junction of the optic nerve and orbital fat, displayed as a dark band.

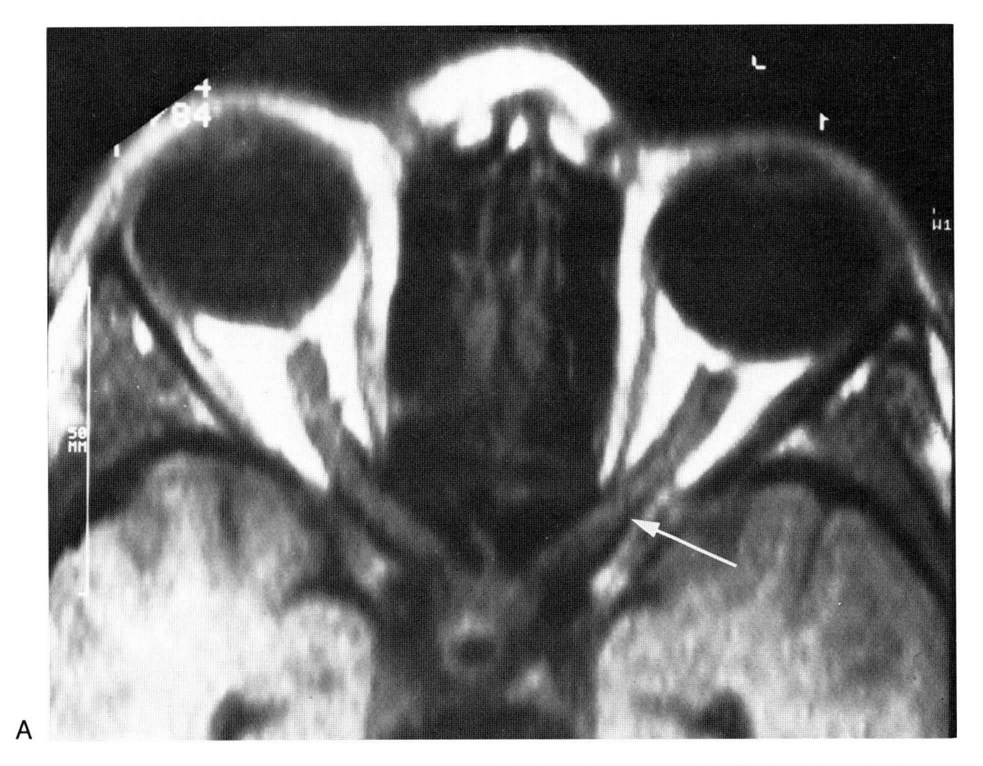

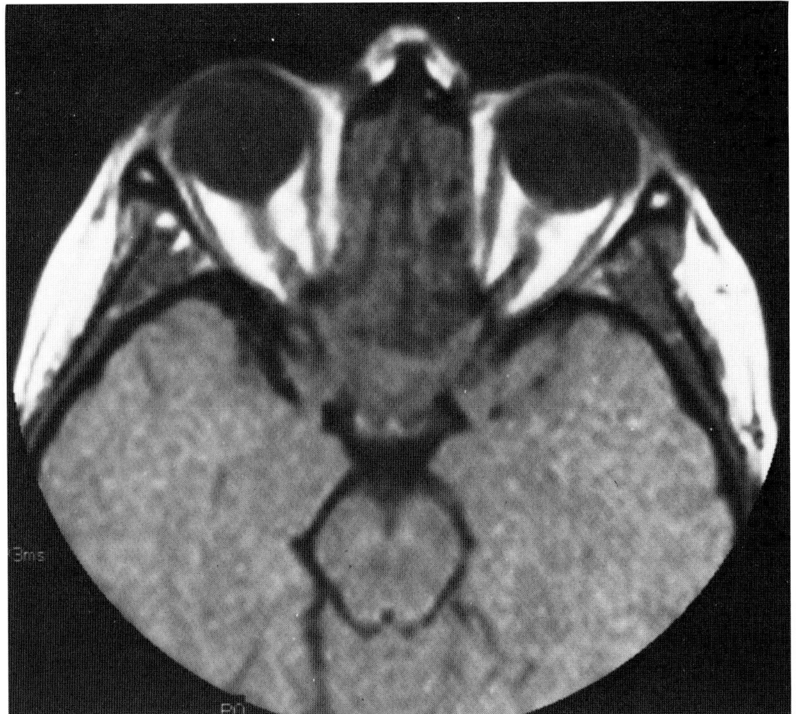

Fig. 12-6. (A) The ophthalmic artery (arrow) lies adjacent to the nerve within the optic canal. Low signal (flow phenomenon) within the artery may potentially mask high signal pathology from the nerve itself. **(B)** Sinus disease in the posterior ethmoids may also obscure visualization of the optic nerves.

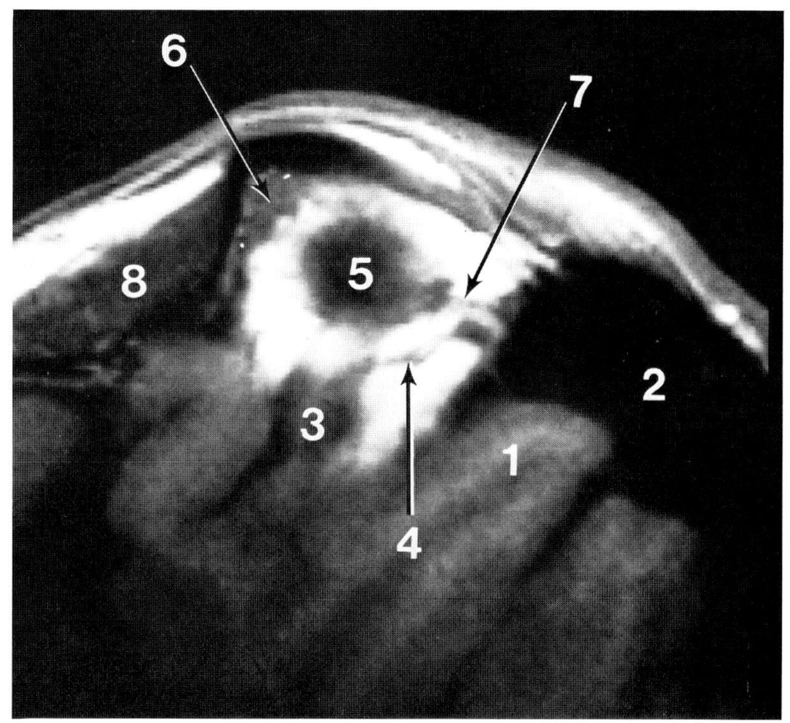

Fig. 12-7. (A–E) Axial SE 500/30 surface coil MR images of the right orbit, from superiorly to inferiorly. (*Figure continues*).

1, Superior frontal gyrus;
2, frontal sinus;
3, superior rectus/levator palpebrae superioris muscle complex;
4, superior ophthalmic vein;
5, globe (vitreous body);
6, lacrimal gland;
7, superior oblique tendon;
7′, superior oblique muscle;
8, temporalis muscle;
9, lateral rectus tendon;
9′, lateral rectus muscle;
10, orbicularis oculi muscle;
11, anterior ethmoidal artery;
12, optic nerve;
13, ophthalmic artery;
14, medial rectus muscle;
15, ethmoidal air cells;
16, fat in crista galli;
17, lateral palpebral ligament;
18, iris;
19, medial palpebral ligament;
20, lens (nuclear portion);
21, lens (cortex);
22, anterior chamber;
23, eyelid;
24, choroid/retina;
25, sclera;
26, orbital fat;
27, inferior oblique muscle;
28, inferior rectus muscle;
29, maxillary sinus;
30, tarsal plate;
31, infraorbital nerve;
32, frontal nerve;
33, sphenoidal sinus.

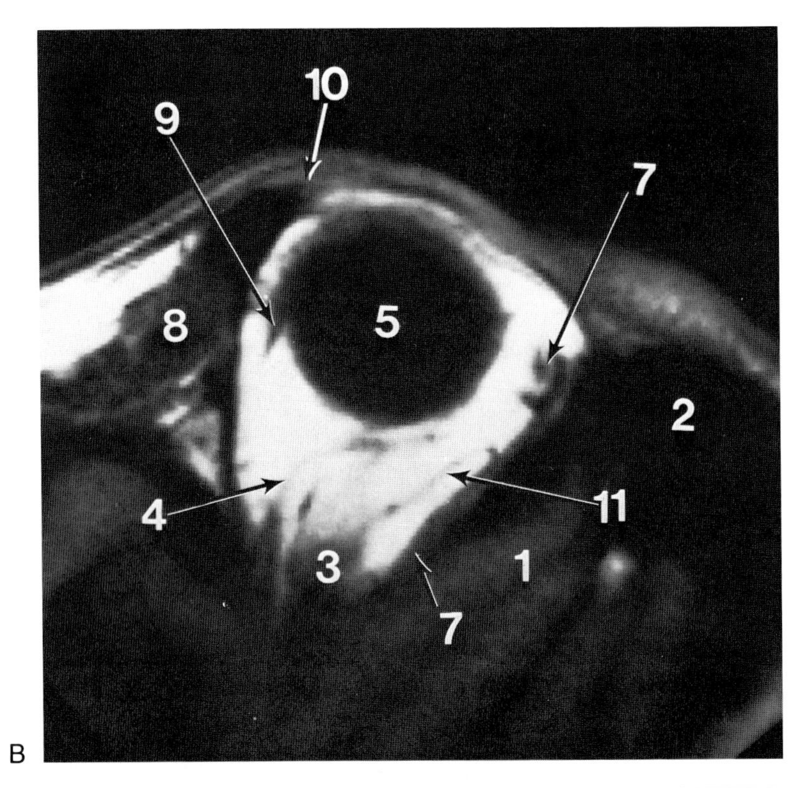

B

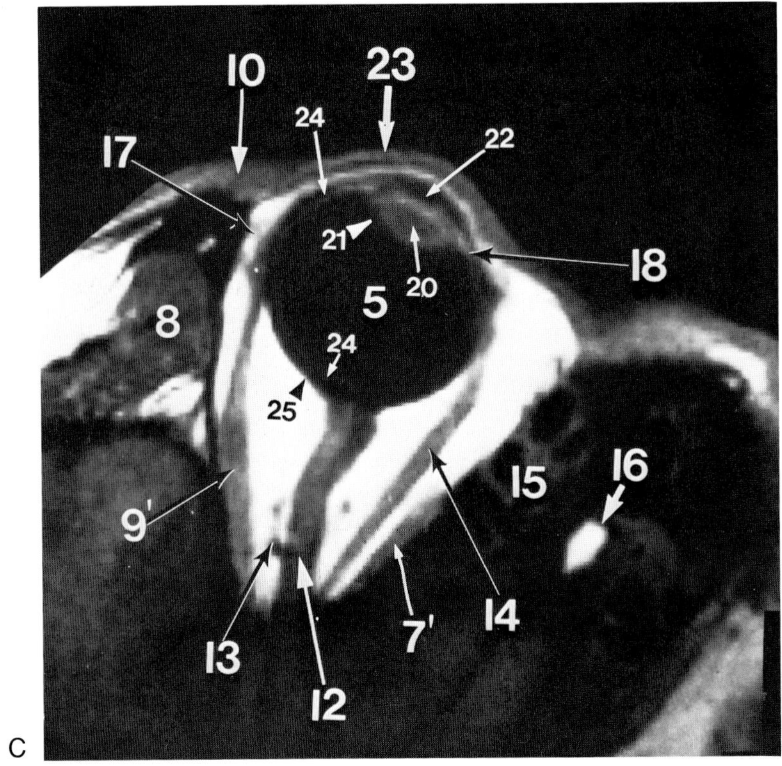

C

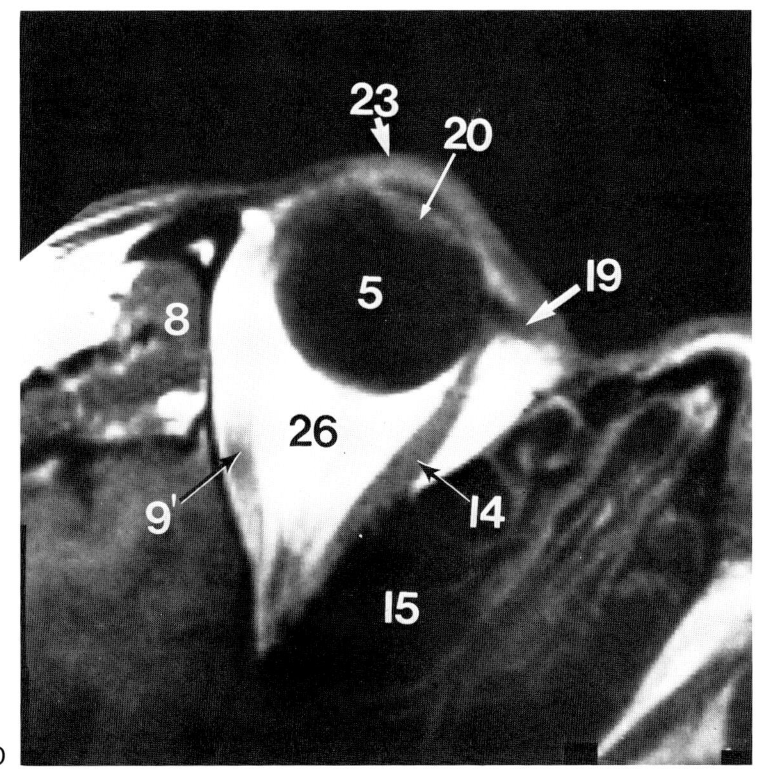

Fig. 12-7 *(continued)*. **B, C,** and **D.**

1, Superior frontal gyrus;
2, frontal sinus;
3, superior rectus/levator palpebrae superioris muscle complex;
4, superior ophthalmic vein;
5, globe (vitreous body);
6, lacrimal gland;
7, superior oblique tendon;
7′, superior oblique muscle;
8, temporalis muscle;
9, lateral rectus tendon;
9′, lateral rectus muscle;
10, orbicularis oculi muscle;
11, anterior ethmoidal artery;
12, optic nerve;
13, ophthalmic artery;
14, medial rectus muscle;
15, ethmoidal air cells;
16, fat in crista galli;
17, lateral palpebral ligament;
18, iris;
19, medial palpebral ligament;
20, lens (nuclear portion);
21, lens (cortex);
22, anterior chamber;
23, eyelid;
24, choroid/retina;
25, sclera;
26, orbital fat;
27, inferior oblique muscle;
28, inferior rectus muscle;
29, maxillary sinus;
30, tarsal plate;
31, infraorbital nerve;
32, frontal nerve;
33, sphenoidal sinus.

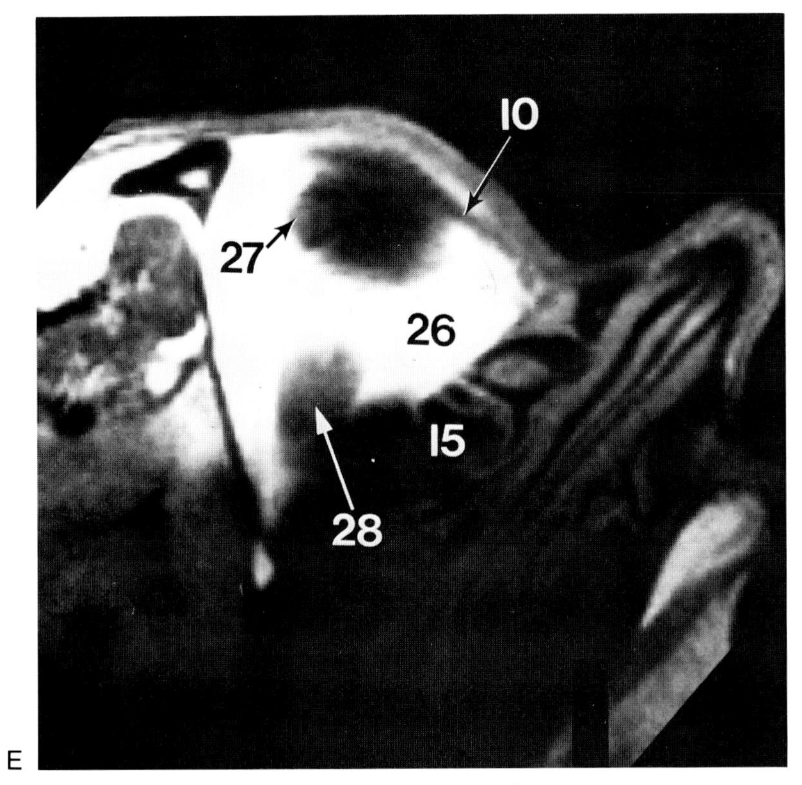

Fig. 12-7 (*continued*).

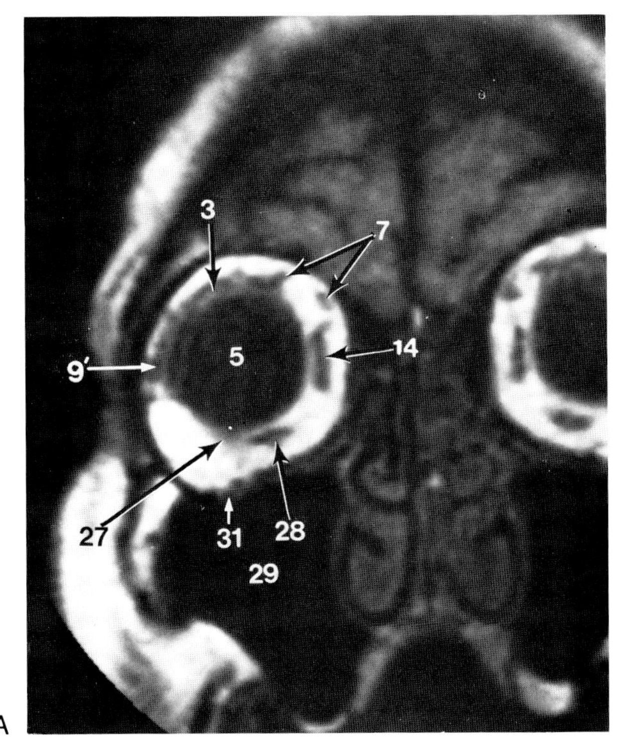

Fig. 12-8. (A–D) Coronal SE 500/30 surface coil images of the right orbit, from anteriorly to posteriorly. (*Figure continues*).

1, Superior frontal gyrus;
2, frontal sinus;
3, superior rectus/levator palpebrae superioris muscle complex;
4, superior ophthalmic vein;
5, globe (vitreous body);
6, lacrimal gland;
7, superior oblique tendon;
7′, superior oblique muscle;
8, temporalis muscle;
9, lateral rectus tendon;
9′, lateral rectus muscle;
10, orbicularis oculi muscle;
11, anterior ethmoidal artery;
12, optic nerve;
13, ophthalmic artery;
14, medial rectus muscle;
15, ethmoidal air cells;
16, fat in crista galli;
17, lateral palpebral ligament;
18, iris;
19, medial palpebral ligament;
20, lens (nuclear portion);
21, lens (cortex);
22, anterior chamber;
23, eyelid;
24, choroid/retina;
25, sclera;
26, orbital fat;
27, inferior oblique muscle;
28, inferior rectus muscle;
29, maxillary sinus;
30, tarsal plate;
31, infraorbital nerve;
32, frontal nerve;
33, sphenoidal sinus.

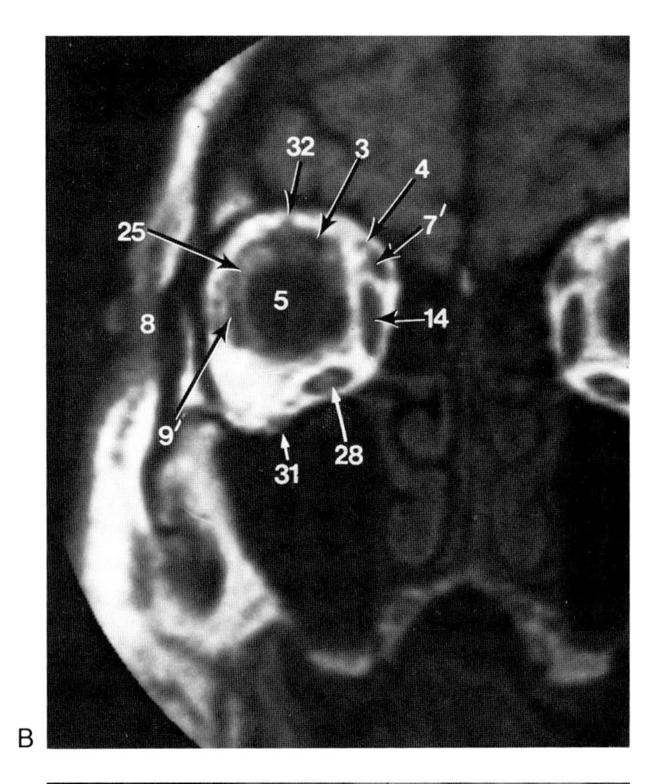

B

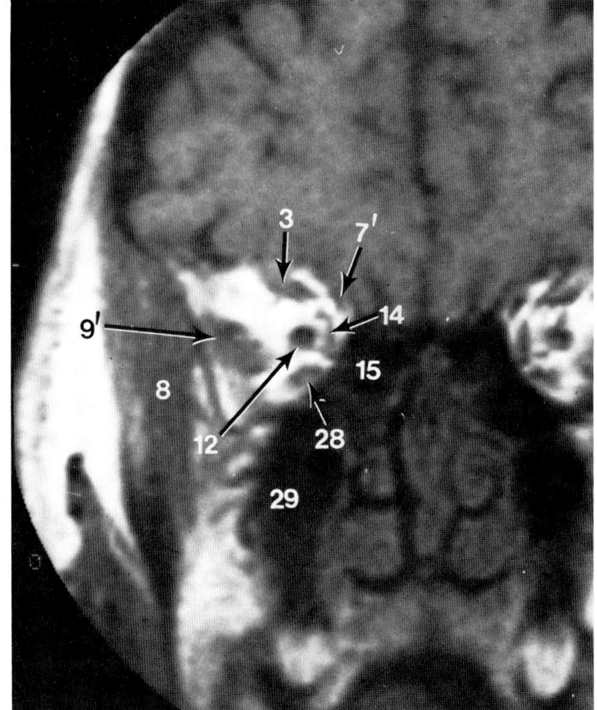

C

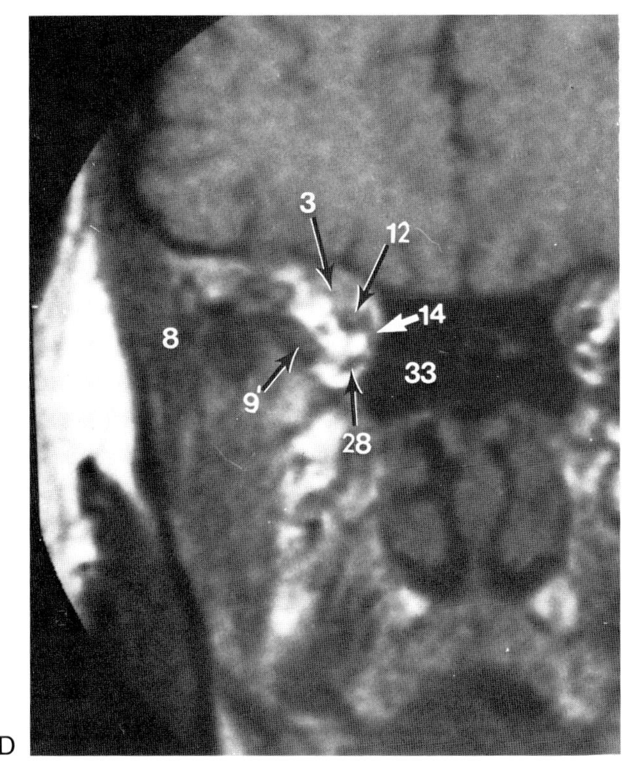

Fig. 12-8 (*continued*). **B, C,** and **D.**

1, Superior frontal gyrus;
2, frontal sinus;
3, superior rectus/levator palpebrae superioris muscle complex;
4, superior ophthalmic vein;
5, globe (vitreous body);
6, lacrimal gland;
7, superior oblique tendon;
7′, superior oblique muscle;
8, temporalis muscle;
9, lateral rectus tendon;
9′, lateral rectus muscle;
10, orbicularis oculi muscle;
11, anterior ethmoidal artery;
12, optic nerve;
13, ophthalmic artery;
14, medial rectus muscle;
15, ethmoidal air cells;
16, fat in crista galli;
17, lateral palpebral ligament;
18, iris;
19, medial palpebral ligament;
20, lens (nuclear portion);
21, lens (cortex);
22, anterior chamber;
23, eyelid;
24, choroid/retina;
25, sclera;
26, orbital fat;
27, inferior oblique muscle;
28, inferior rectus muscle;
29, maxillary sinus;
30, tarsal plate;
31, infraorbital nerve;
32, frontal nerve;
33, sphenoidal sinus.

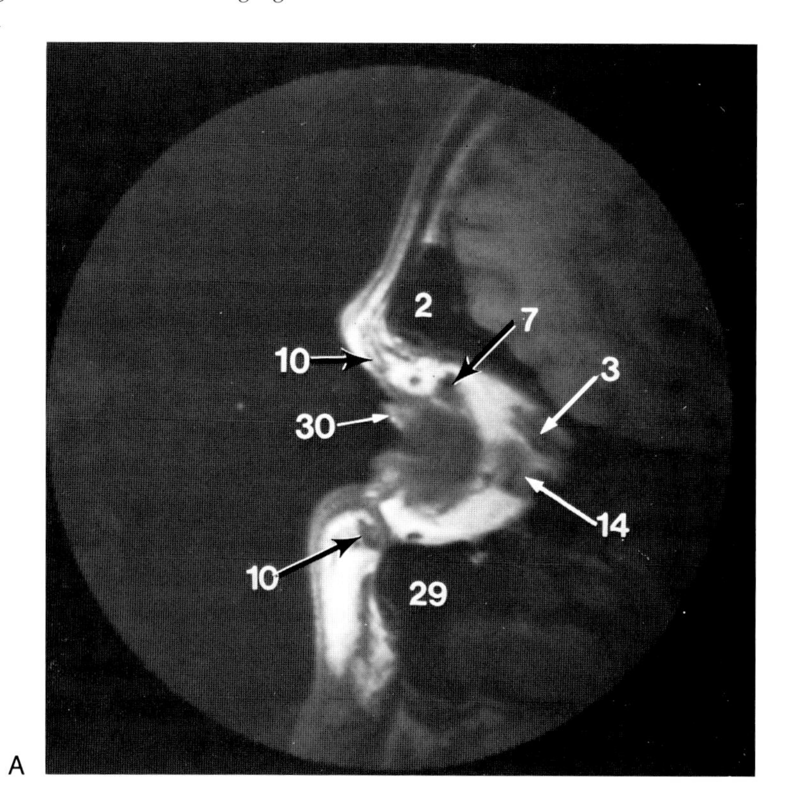

A

potential, can sometimes allow these anatomic relationships to be better appreciated than on CT.

A number of vessels may be routinely identified on orbital MRI. These include the superior ophthalmic vein and ophthalmic artery, as well as many smaller veins and arteries. The ophthalmic artery courses inferior and lateral to the optic nerve before crossing it to end as orbital and nasociliary vessels. The superior ophthalmic vein has a curvilinear course inferior and then lateral to the superior rectus muscle. It is well seen on axial or coronal images (Fig. 12-7).

The lacrimal gland is located in the superiolateral aspect of the anterior orbit. It has an elliptical shape on axial images but is perhaps better seen in the parasagittal plane (see Fig. 12-9). The medial and lateral orbital septa and lids are also easily identified on MRI separate from the orbital fat and other contents.

Multiplanar MR imaging of the normal orbit is presented in Figures 12-7 through 12-9.

OCULAR PATHOLOGY

Persistent Hyperplastic Primary Vitreous

Persistent hyperplastic primary vitreous (PHPV) is a congenital developmental ocular abnormality usually diagnosed in the perinatal period. The disorder results from persistence and hyperplasia of the embryonic hyaloid vascular system present at the 8th week of fetal life. If diagnosed in its early stages sight may be restored by microscopical excision of the persisting vessels and fibrous tissues.

Persistent hyperplastic primary vitreous is unilateral in 90 percent of cases.[21] Bilateral disease is associated with trisomy 13–15. Rarely PHPV may coexist with retinoblastoma.[22] Otherwise there are no familial tendencies or associations with maternal factors or prematurity.

The affected eye is usually small, with a shallow anterior chamber and esotropia. A retrolental fibro-

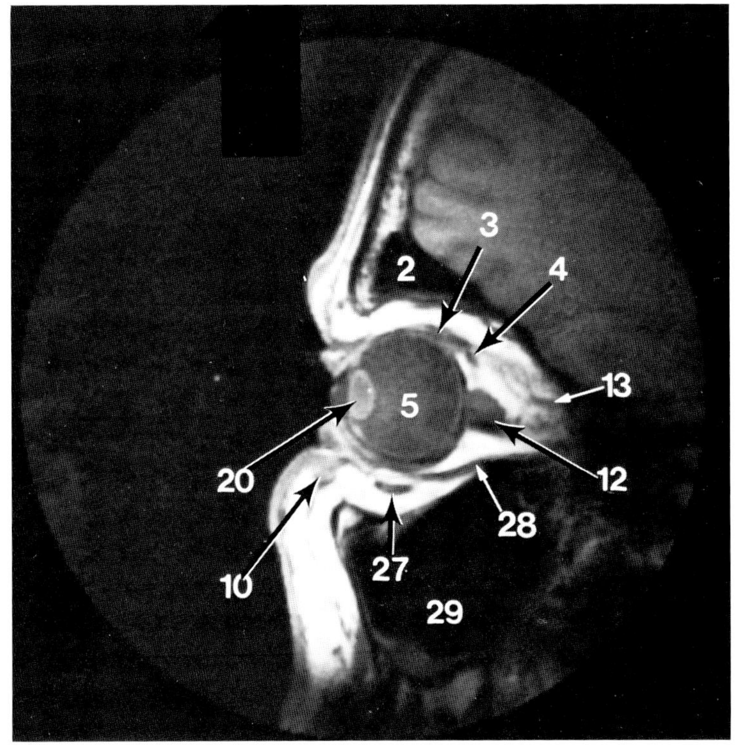

Fig. 12-9. (**A** and **B**) Sagittal SE 500/30 surface coil images of the right orbit from medially to laterally. (Courtesy of General Electric Medical Systems.) (*Figure continues*).

1, Superior frontal gyrus;
2, frontal sinus;
3, superior rectus/levator palpebrae superioris muscle complex;
4, superior ophthalmic vein;
5, globe (vitreous body);
6, lacrimal gland;
7, superior oblique tendon;
7′, superior oblique muscle;
8, temporalis muscle;
9, lateral rectus tendon;
9′, lateral rectus muscle;
10, orbicularis oculi muscle;
11, anterior ethmoidal artery;
12, optic nerve;
13, ophthalmic artery;
14, medial rectus muscle;
15, ethmoidal air cells;
16, fat in crista galli;
17, lateral palpebral ligament;
18, iris;
19, medial palpebral ligament;
20, lens (nuclear portion);
21, lens (cortex);
22, anterior chamber;
23, eyelid;
24, choroid/retina;
25, sclera;
26, orbital fat;
27, inferior oblique muscle;
28, inferior rectus muscle;
29, maxillary sinus;
30, tarsal plate;
31, infraorbital nerve;
32, frontal nerve;
33, sphenoidal sinus.

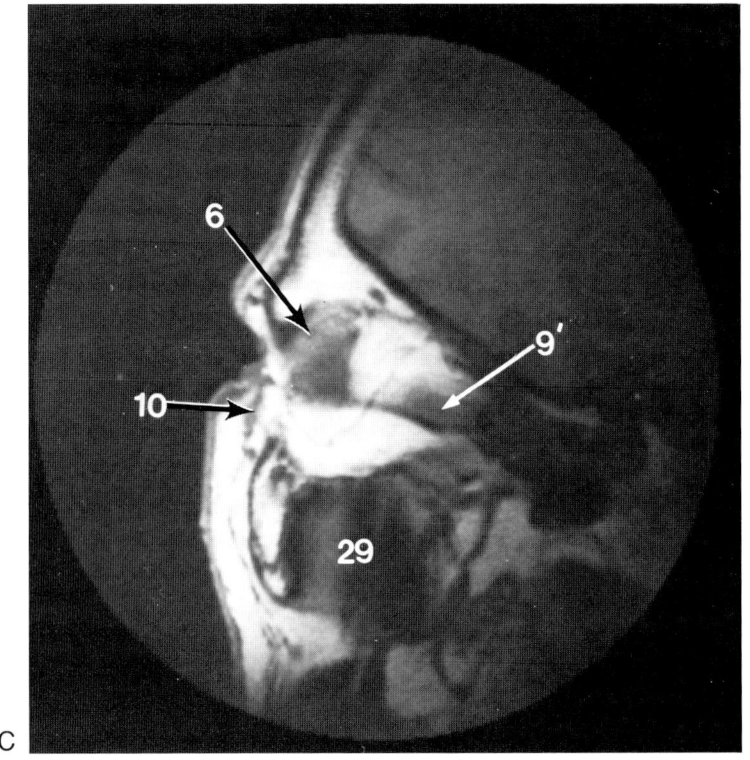

Fig. 12-9 *(continued).*

vascular mass with a central stalk may be seen. After birth the fibrous tissue progressively contracts, and may result in lens capsule rupture, glaucoma, or hemorrhage. A cataract may develop secondarily.[23]

On CT a dense but noncalcified vitreous is easily detected. A blood-vitreous level may occasionally be observed. Using narrow window settings the central fibrovascular stalk may be seen. Few cases studied by MR are available (Fig. 12-10). Sullivan and colleagues have suggested that globes with PHPV have a shorter T1 than those with other disorders such as Coate's disease or retinoblastoma.[24] It is unclear whether this relatively short T1 relates to prior hemorrhage from the friable vasculature of the PHPV or has other explanation.

Retinoblastoma

Retinoblastoma is the most common pediatric intraocular tumor, arising from primitive neural cells in the nuclear layer of the retina.[25] The tumor is usually diagnosed during the 2nd year of life when the mother or pediatrician notices a whitish mass behind the lens (leukocoria or "cat's eye"). In about one-quarter of cases the tumors are bilateral, representing an autosomal inherited disorder with variable penetrance.[26] Bilateral retinoblastomas may coexist with other cerebral neoplasms, notably pineoblastomas, a complex sometimes called trilateral retinoblastoma.[27]

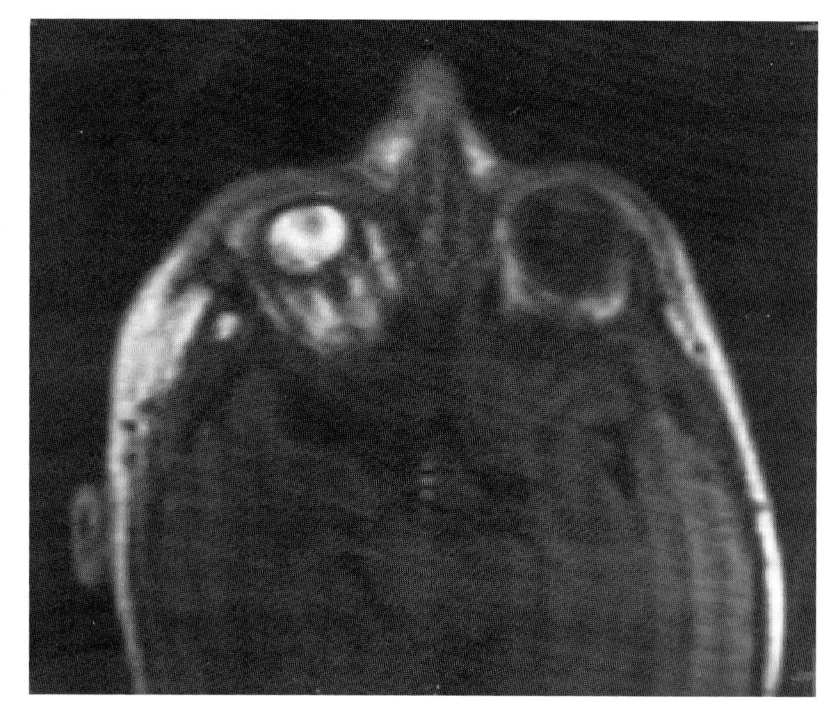

Fig. 12-10. Persistent hyperplastic primary vitreous. The right globe is small and images with high intensity (short T1) on this SE 500/30 image. Resolution on this scan did not permit visualization of the central fibrovascular bundle in the dystrophic globe. (Imaging News: Surface coils aid MRI of head, neck, and spine. Diagn Imaging 8(6):13, 1986.)

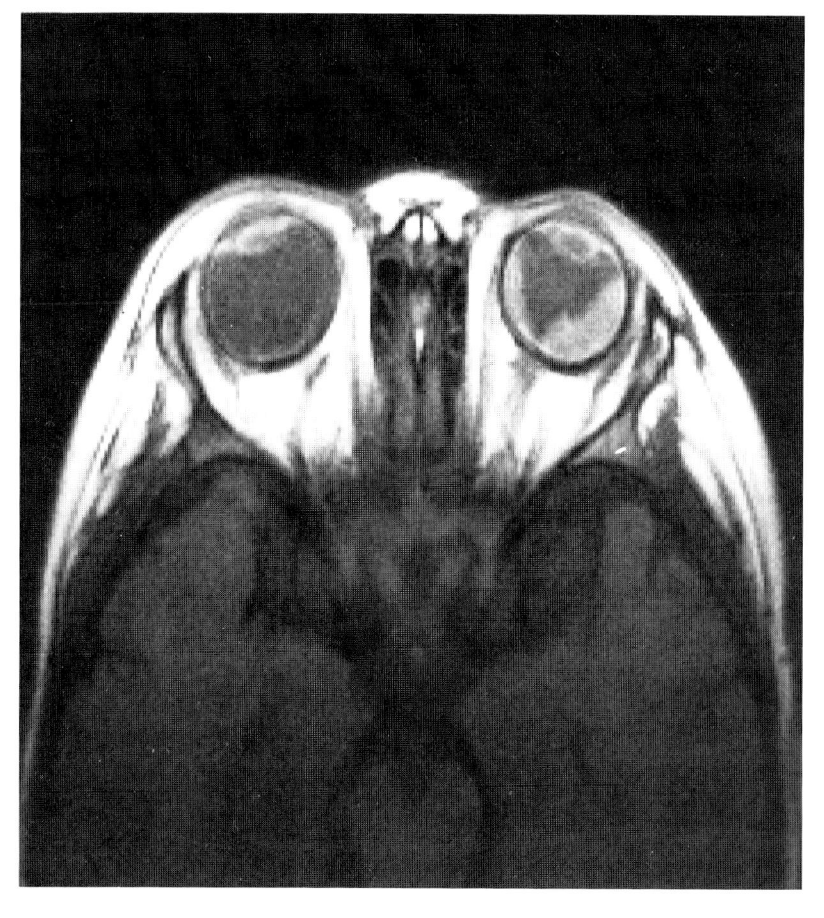

Fig. 12-11. Retinoblastoma. In this untreated case the left globe retinoblastoma is imaged as relatively high intensity compared to vitreous on this SE 500/30 image. (Imaging News: Surface coils aid MRI of head, neck, and spine. Diagn Imaging 8(6):13, 1986.)

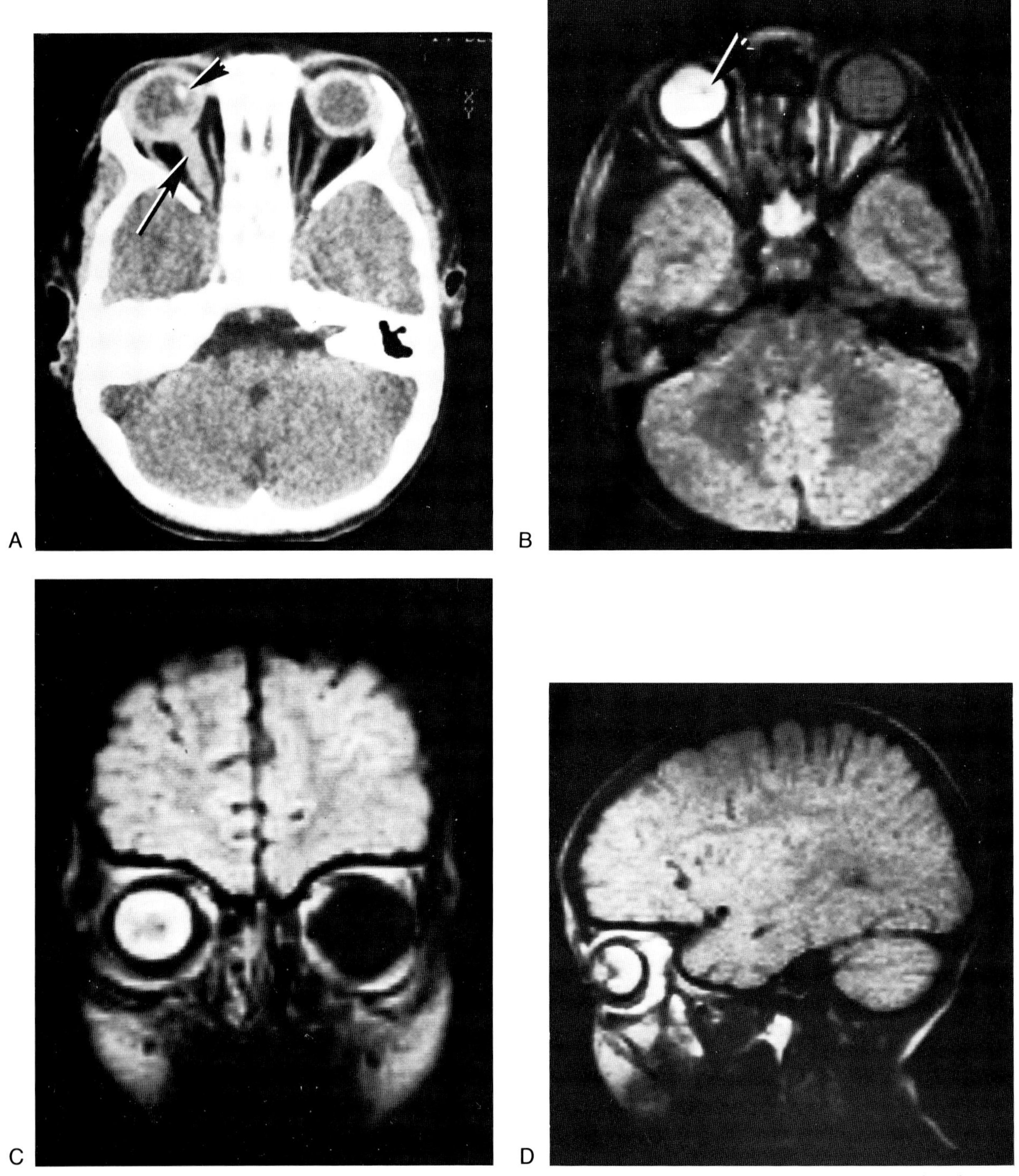

Fig. 12-12. Retinoblastoma, postradiation therapy. **(A)** CT scan shows a small focus of calcification (arrowhead). Tumor extends posteriorly along the optic nerve (arrow). **(B)** Axial SE 2000/56 MR image. The calcification is seen as a low intensity region in the globe. Postirradiation, the right globe is smaller and its vitreous has high signal, suggesting vitritis or protein leakage. (**C** and **D**) Coronal and sagittal SE 1500/28 images demonstrate the tumor centrally within the globe.

Intraocular retinoblastoma quickly outgrows its blood supply, becoming centrally necrotic and calcified. Intraocular spread of tumor is common with malignant cells implanting at multiple sites including choroid, iris, and posterior cornea. Extension along the optic nerve, posterior orbit, and even into the subarachnoid spaces is common. In about 1 percent of cases retinoblastomas undergo spontaneous regression much like those that have been treated successfully with irradiation.[28] Distant metastases (to bone, lymphatics, viscera) may also occur.

Computed tomography scanning should probably remain the first-line diagnostic modality for retinoblastoma, because it detects calcification within the tumor, a pivotal diagnostic feature. However, MRI may aid in several regards. Coexisting retinal detachment may be better appreciated by MR than CT.[29] Posterior extension to the optic chiasm may also be seen better on MR. On T1-weighted images untreated retinoblastoma is imaged as a focus on high signal intensity relative to vitreous.[10] (Fig. 12-11). In eyes that have been radiated, signal changes in the vitreous may result in the tumor being hypointense (Fig. 12-12). MR is incapable of reliably detecting intraocular calcification and small (<2 mm) retinoblastomas.

Ocular Melanoma

Malignant melanoma is the most common ocular neoplasm in adults. Thought to arise from pre-existing nevi, these tumors are found in the uveal tract (choroid, ciliary body, and iris). Choroidal melanoma, the most common type, has a higher incidence in white people over age 50 while the rarer iris melanoma occurs in younger patients.[30]

Small melanomas are often asymptomatic, while larger tumors may produce a variety of symptoms: visual loss, retinal detachment, vitreous hemorrhage, and glaucoma.[30] The tumors are highly aggressive with a propensity for extraocular extension along the optic nerves or vortex veins. Distant hematogenous metastases (especially to lung, liver, and subcutaneous tissues) are common and associated with a dismal prognosis. Iris melanomas have a less aggressive course than those from the choroid or ciliary body.[30]

Most ocular melanomas are easily diagnosed by ophthalmoscopy. Fluorescein angiography and ultrasound are also reliable. However, clinical examination is difficult or impossible in several situations: dense cataracts, opaque vitreous, intraocular blood, or proteinaceous subretinal exudate.[31] Also lesions of the posterior ciliary body may be difficult to visualize directly or by sonography. In these situations MRI may prove helpful in the recognition of an occult underlying tumor.[32] Also retro-orbital extension can be assessed on MR while it is inapparent clinically.

Table 12-2. Characteristic Relaxation Times of Melanomas and other Ocular Pathologies

Tissue	*T1*	*T2*
Melanotic melanoma	Short	Short
Fresh hematoma	No change	Short
Subacute or chronic hematoma	Short	Long
Proteinaceous effusions	No change	Long

Melanotic melanomas have characteristically short T1 and T2 values, largely unlike other ocular tumors or disease processes (Table 12-2).[29,32] This causes them to appear hyperintense on T1-weighted images and hypointense on T2-weighted images (Fig. 12-13). These signal patterns are thought to be a result of the paramagnetic properties of melanin, which causes a parallel shortening of T1 and T2 via proton relaxation enhancement. Unfortunately, amelanotic melanomas will not exhibit these signal characteristics and cannot be differentiated from other nonmelanotic tumors by MR.[32] A good review differentiating uveal melanoma from other simulating lesions has recently been published.[29]

Coat's Disease

Coat's disease is a unilateral ocular disorder primarily of childhood characterized by exudative retinal detachment and telangiectasias. The subretinal exudate is thought to arise secondarily from deposition of cholesterol esters and lipids. The clinical presentation, which includes retinal detachment, hemorrhage, and visual loss may mimic retinoblastoma or toxocariasis. In a comparison study MR demonstrated the retinal detachments, hemorrhage, and exudate more

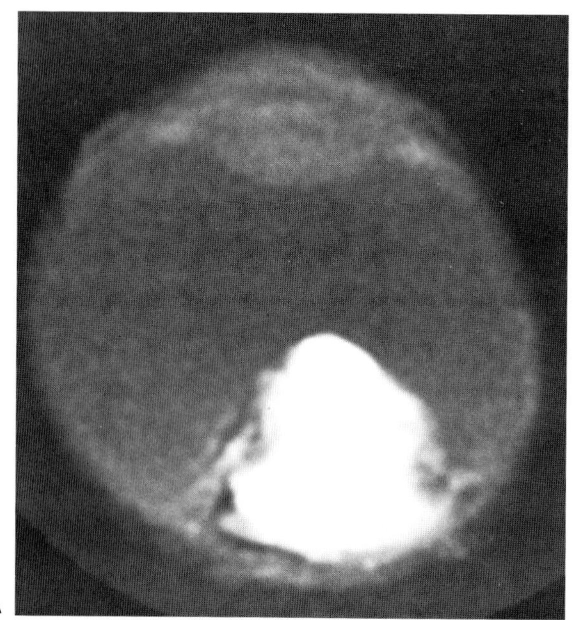

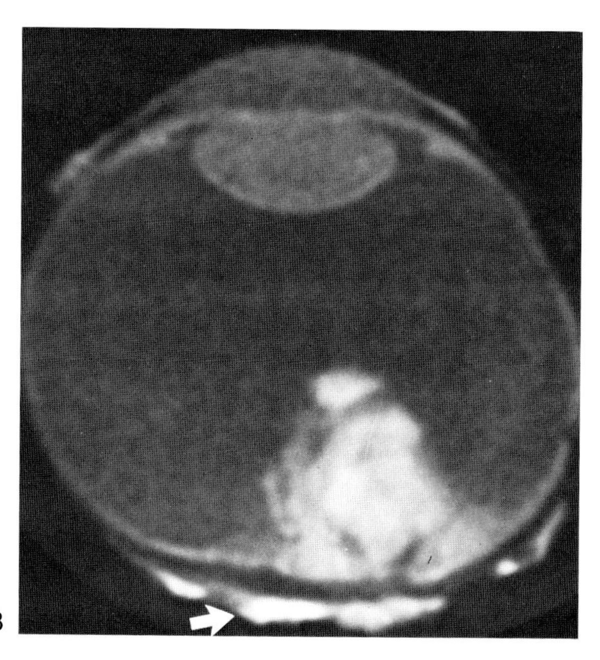

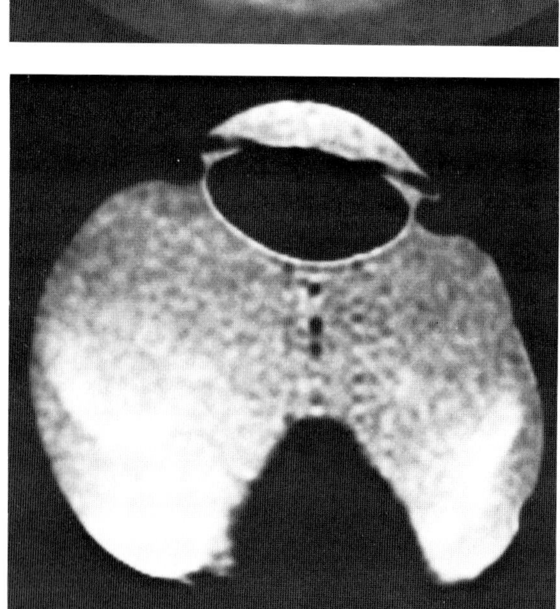

Fig. 12-13. Melanotic melanoma in an enucleated eye. **(A, B)** Contiguous T1-weighted images (SE 200/15) show melanoma to have highest intensity, except for fat (arrow). **(C)** On T2-weighted image (SE 2000/75), the melanoma is quite hypointense. These findings can be explained on the basis of very short T1 and T2 values in the tumor due to paramagnetic properties of melanin. (Gomori JM, Grossman RI, Shields JA et al: Choroidal melanomas: correlation of NMR spectroscopy and MR imaging. Radiology 158:443, 1986.)

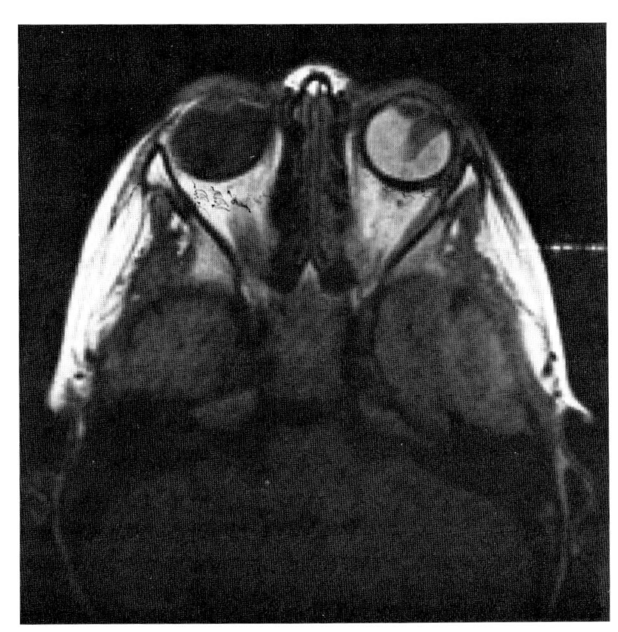

Fig. 12-14. Coat's disease. Note large subretinal exudates (arrows) in the left globe on this SE 500/30 image. (Imaging News: Surface coils aid MRI of head, neck, and spine. Diagn Imaging 8(6):13, 1986, with permission.)

clearly than CT.[32] MR was judged to be extremely valuable in assessing these patients because it helped exclude the diagnosis of underlying noncalcified retinoblastoma. An example is shown in Figure 12-14.

Miscellaneous Ocular Lesions

A wide variety of other ocular lesions are well described clinically and by CT. However, at the time of this writing, MR reports are largely anecdotal or totally lacking. Some of these diseases will be briefly considered for completeness.

Astrocytic Hamartomas

Astrocytic hamartomas are benign tumors of the retina and optic disc usually occurring with tuberous sclerosis or neurofibromatosis.[33] They are apparent as small punctate calcifications on clinical examination or CT. They are usually asymptomatic but occasionally cause retinal detachment or undergo malignant degeneration. In one case documented clinically, we were unable to observe the lesion on low field strength MR.

Drusen

Drusen are asymptomatic, calcified acellular masses of hyalinelike material found at the junction of the retina and optic nerve.[33] Found in adults exclusively, they are nearly always bilateral and easily diagnosed clinically or on CT.

Choroidal Osteomas

Choroidal osteomas are rare bony masses found in the choroid of young women. They cannot usually be distinguished radiographically from astrocytic hemartomas.

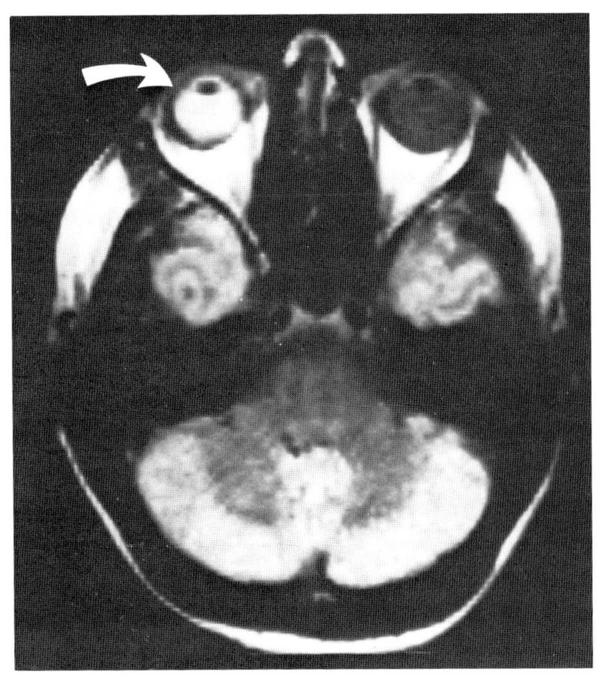

Fig. 12-15. Distant vascular occlusion to the right eye has resulted in atrophy and high signal on this SE 2000/28 image.

Phthysis Bulbi

Phthysis bulbi is a nonfunctioning, contracted, and calcified globe secondary to trauma, vascular occlusion, or infection. The globe is small and disorganized. Peripheral calcification, especially posteriorly, may be seen on plain radiographs or CT. Increased signal from within the vitreous may be noted as it undergoes liquefaction and degeneration (Fig. 12-15).[32]

Choroidal Hemangiomas

Chroidal hemangiomas are benign vascular hamartomas often associated with Sturge-Weber syndrome.[33] In a single published case the tumor was much more apparent on contrast enhanced CT than on MR.[29]

Vitreous Lymphoma

Vitreous lymphoma is a rare complication of systemic lymphoma or leukemia. Clinical and CT findings are variable. No MR reports are yet available.

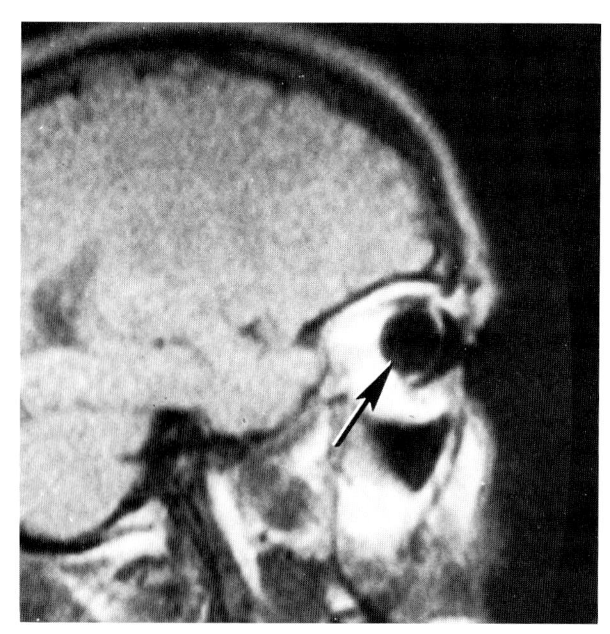

Fig. 12-16. Low intensity from an ocular prosthesis ("glass eye") seen on this SE 600/40 image.

Retinal Detachment

Retinal detachment may be post-traumatic, spontaneous, or secondary to neoplasia. MR is well suited for evaluating retinal detachment and detecting an underlying tumor.[29,32] An example of retinal detachment associated with Coat's disease is shown in Figure 12-14.

Postsurgical Changes

Postsurgical changes in the globe and orbit have been demonstrated by MR.[29] Silicone oil and bands have low signal on both T1-and T2-weighted images. A prosthetic globe also has low signal intensity on all sequences (Fig. 12-16).

TUMORS OF THE OPTIC NERVES AND PATHWAYS

Optic Gliomas

Optic gliomas are uncommon tumors of the optic nerves and pathways occurring primarily in childhood and adolescence.[34] The peak age for presentation is between 2 and 6 years, with 90 percent of cases seen before age 20. In about one-quarter of cases there is an association with neurofibromatosis. Symptoms include exophthalmos, visual loss, and strabismus, usually occurring in that order.

Optic gliomas are seen as either fusiform or irregular, nodular enlargements of the optic nerve.[35] Unsuspected involvement of the chiasm, geniculate body, or optic radiations may occur in up to one-fourth of patients. In early MR reports optic gliomas were infrequently detected.[9,10] However, these studies were performed using early generation MR scanners with relatively thick slices (7 to 10 mm). With 3 to 5 mm slices it is now felt that optic gliomas can be as reliably detected by MR as by CT.

Optic gliomas demonstrate slight prolongation of T1 and T2 values compared to the normal nerve. However, the MR diagnosis relies more upon abnormal shape, size, and contour of the nerve than from

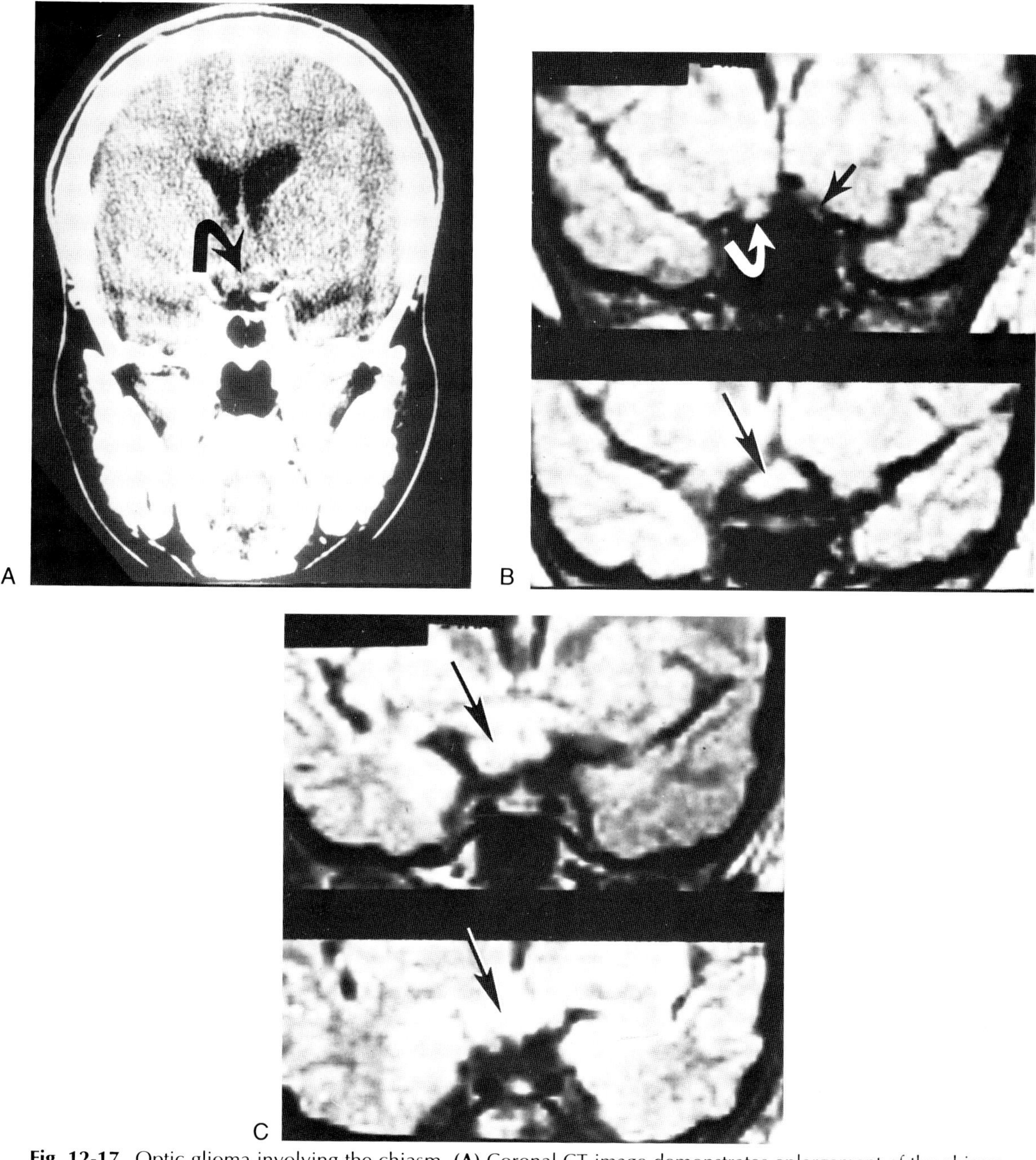

Fig. 12-17. Optic glioma involving the chiasm. **(A)** Coronal CT image demonstrates enlargement of the chiasm (arrow). **(B, C)** Successive coronal SE 1500/28 images from anteriorly to posteriorly. A glioma involves the right optic nerve (curved white arrow), but not the left (short black arrow). Extension of tumor into the chiasm and optic tract is noted (long black arrows).

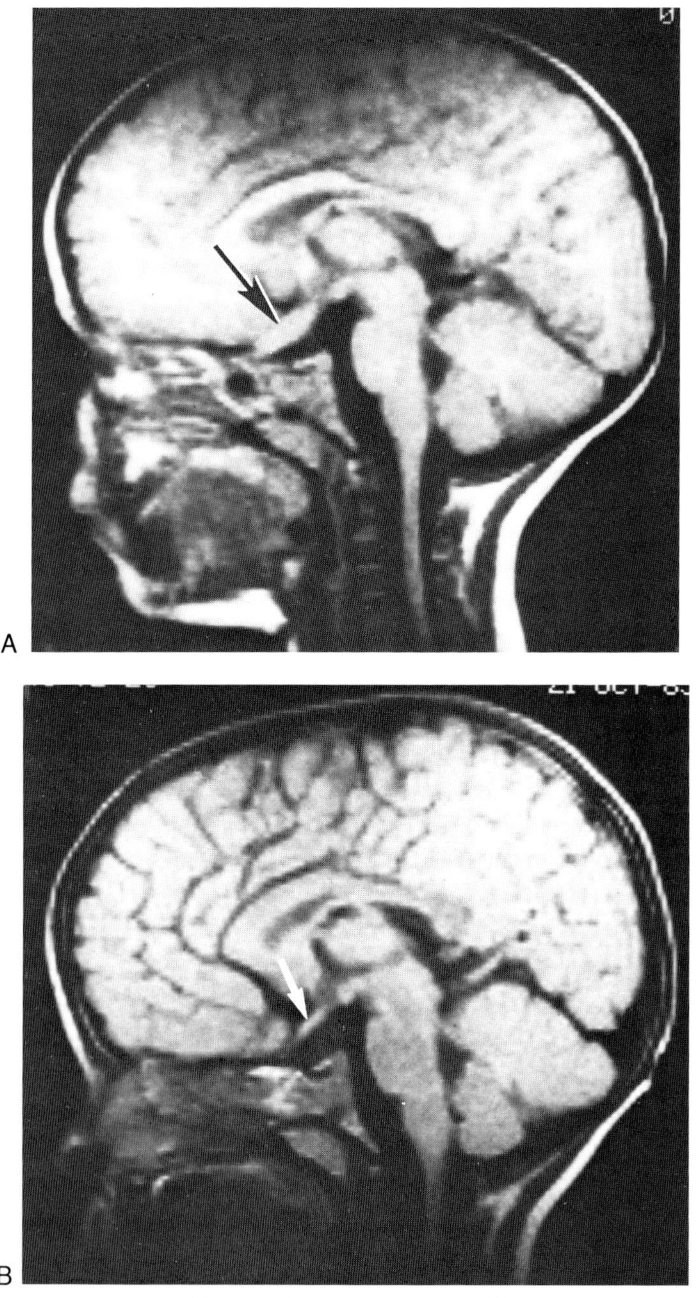

Fig. 12-18. Optic glioma in a child with neurofibromatosis. **(A)** Initial presentation. Note thickened optic nerve and chiasm (arrow) on this SE 500/28 sagittal image. **(B)** One year later, following radiotherapy. The tumor has regressed.

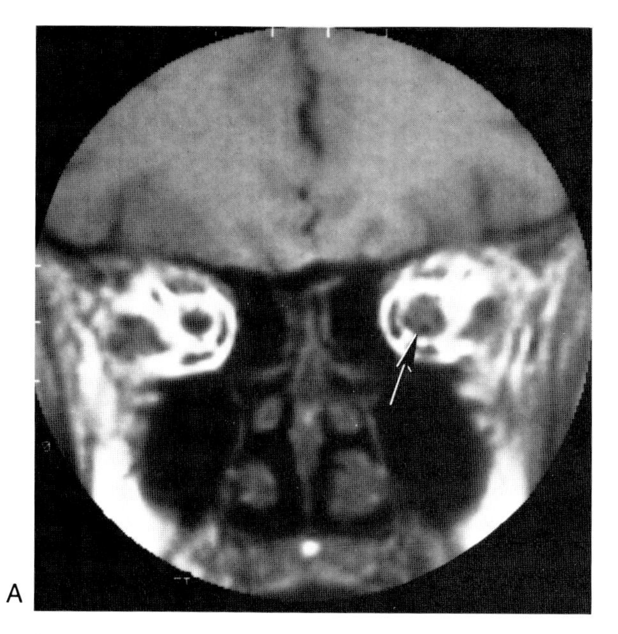

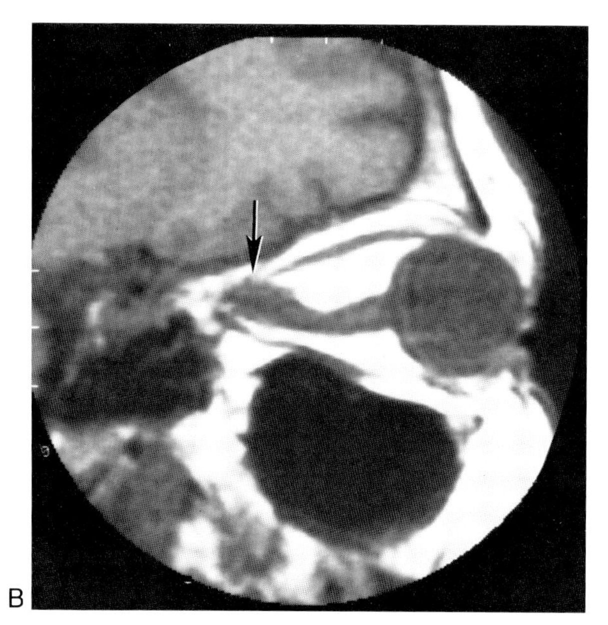

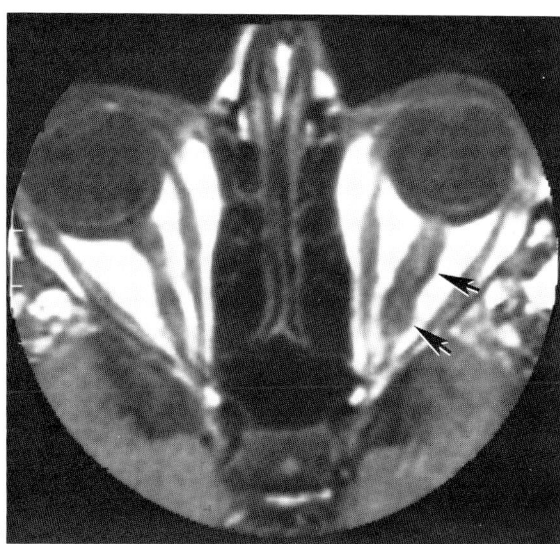

Fig. 12-19. Excrescent meningioma at the orbital apex (arrows).

recognizing abnormal signal. Extension of the tumor into the chiasm is particularly well seen on MRI (Figs. 12-17 and 12-18).

Meningiomas

Meningiomas in the orbit may originate from the optic nerve sheath, periosteum of the orbital wall, from arachnoidal rests within the orbit, or by extension from an intracranial source.[34] Meningiomas secondarily involving the orbit are more common than those originating in the orbit. Primary intraorbital meningiomas most frequently arise from the optic nerve sheath. Like meningiomas elsewhere, middle aged women are most frequently affected. Some cases are associated with neurofibromatosis.

Optic nerve meningiomas characteristically cause early visual loss, with proptosis occurring later. They are usually fusiform, but may be excrescent.[35] On CT their appearance is quite characteristic with a "tram-track" sign on longitudinal section showing encasement of the nerve by tumor.[35] Optic nerve meningiomas on MRI have not shown this sign, because their intensity is usually similar to the nerve itself.[13] Identification on MR may rest with concentric or irregular enlargement of the optic nerve contour rather than by virtue of signal changes around the nerve. Conversely we have had no difficulty visualizing non-nerve sheath meningiomas within the orbit, by virtue of their excellent natural contrast with orbital fat (Fig. 12-19).

Plexiform Neurofibromatosis of the Orbit

Neurofibromatosis (von Recklinghausen disease) has numerous manifestations in the orbit, including osseous dysplasia, neoplasms (optic gliomas, meningiomas, schwannomas), buphthalmos, and plexiform neurofibromas.[36] Plexiform neurofibromatosis (PNF) is pathognomonic of von Recklinghausen disease. Pathologically PNF consists of diffuse intertwining bundles of Schwann cells, fibroblasts, and axons. The disease is usually manifest in childhood.

Plexiform neurofibromatosis of the orbit has a characteristic appearance.[36,37] findings include enlargement of the orbit, thickening of the lids and periorbital soft tissues, buphthalmos, and proptosis. Irregular thickening of the optic nerve and enlargement of smaller intraconal and ciliary nerves may be identified. Extension into the cavernous sinus is characteristic. In our single case, MRI was no better than CT in identifying the extent of pathology (Fig. 12-20).

INFLAMMATORY DISORDERS

Thyroid Ophthalmopathy

In adults, thyroid disease is the most common cause of either unilateral or bilateral exophthalmos.[38] Ophthalmic findings, including proptosis, lid lag or stare can be seen in about one-quarter of patients with Graves' disease. However, thyroid ophthalmopathy runs an independent course. There is no correlation between clinical or laboratory findings and degree of exophthalmos.

In its classic, milder form, thyroid ophthalmopathy occurs in young females who are generally asymptomatic, being either euthyroid or slightly hyperthyroid.[39] Here the disease is bilateral, and the patient has lid lag, mild proptosis, and a prominent stare. A more severe clinical form of ophthalmopathy occurs in older patients who are thyrotoxic. These patients present with severe proptosis and ophthalmoplegia.

The pathologic basis of the exophthalmos involves two factors. First, there is inflammatory edema and lymphocytic infiltration of the extraocular muscles, which may enlarge up to eight-fold in volume. Usually the muscle involvement is bilateral, but is markedly asymmetric in 30 percent and unilateral in 5 percent.[40] The inferior and medial recti are the first and most severely involved. A second factor involved in endocrine exophthalmos is increased volume of orbital fat.[41] This may account for the proptosis of Graves' disease in the 10 percent of cases where no extraocular muscle enlargement is noted.

An example of asymmetric muscle enlargement in Graves' disease seen on MR is shown in Figure 12-21. MR apparently can offer little to the evaluation of such patients, other than its ease of obtaining direct

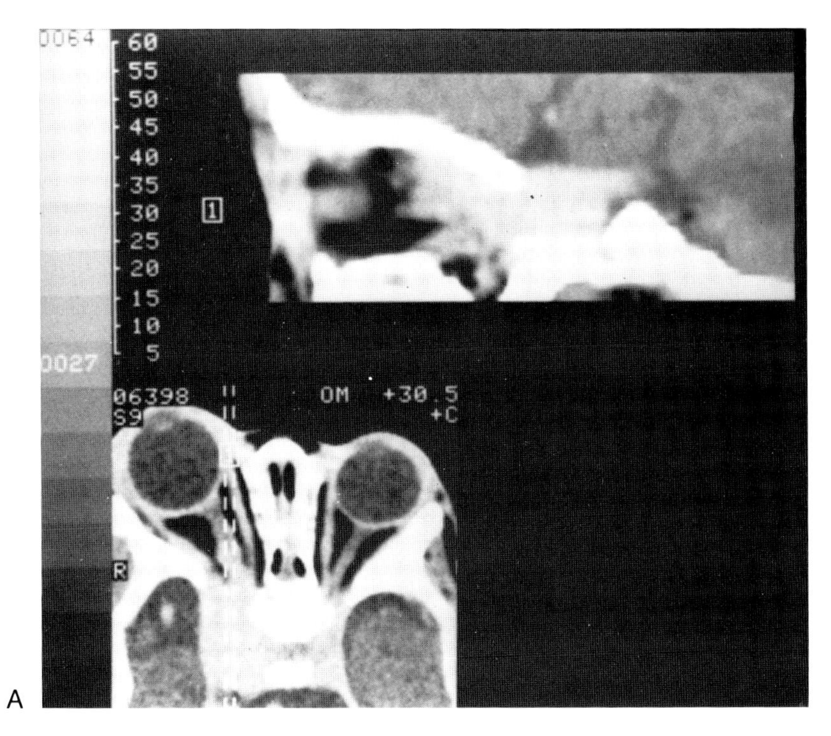

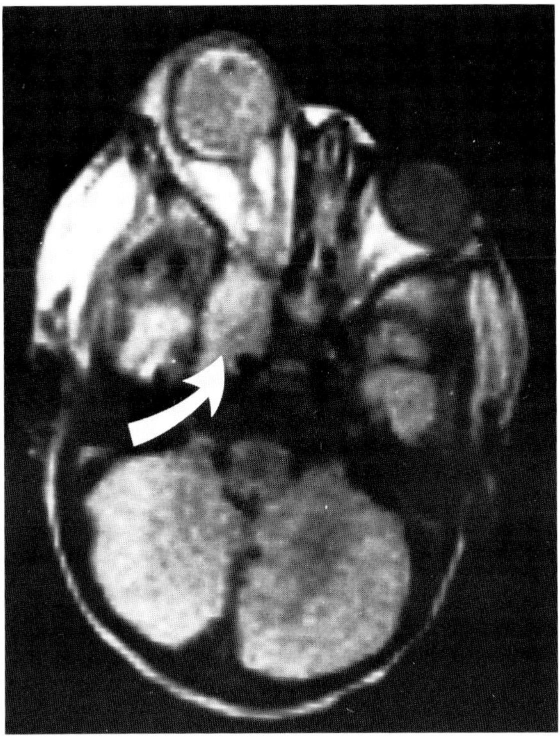

Fig. 12-20. Plexiform neurofibromatosis of the orbit seen on CT and MRI. Note excessive mass (arrow) in the cavernous sinus and middle temporal fossa as well as the orbit.

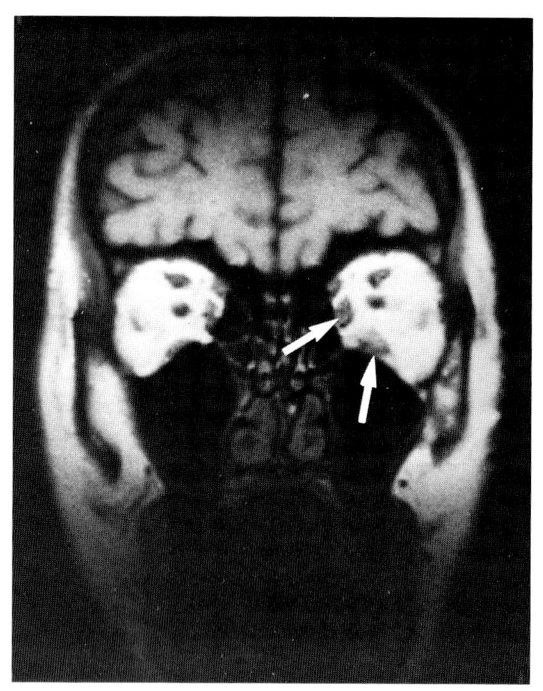

Fig. 12-21. Unilateral Graves' disease. Note enlargement of inferior and medial rectus muscles (arrows) compared to the opposite side. (Courtesy of Picker International.)

coronal images. CT should remain the modality of choice for this suspected diagnosis and in the initial evaluation of patients with proptosis.

OTHER ORBITAL MASSES

Cavernous Hemangioma

Cavernous hemangioma is the most common benign orbital tumor in adults, presenting most frequently in young adulthood and middle age.[44] Onset is usually insidious, with symptoms caused by local expansion. Proptosis, impaired ocular motility, blurred vision, and diplopia frequently occur.[43]

Pathologically, cavernous hemangiomas are composed of large endothelial-lined vascular channels enclosed in a fibrous pseudocapsule. Because arterial blood flow is sluggish, hemangiomas have a tendency for thrombosis and fibrous change. Expansion of the adjacent orbital wall is common. About 83 percent are intraconal, and 67 percent are lateral to the optic nerve.[45] However, some tumors may be both intra- and extraconal. Despite their size, characteristically a small triangle of fat at the orbital apex is spared.

Several examples of cavernous hemangioma have appeared in the MR literature.[13,15] The lesions were well defined and had long T1 and T2 values. Multiplanar imaging aided in defining the relationship of the tumor to the optic nerve. An example of a case we have seen is shown in Figure 12-22.

Cavernous hemangioma should be differentiated radiographically from the uncommon *lymphangioma*. Lymphangiomas usually present at younger ages than hemangiomas (45 vs. 22 years).[44] While hemangiomas often present insidiously by slow growth and thrombosis, lymphangiomas may produce acute symptoms from hemorrhage. Whereas most hemangiomas are intraconal, most lymphangiomas are extraconal. Finally, hemangiomas are well encapsulated and amenable to surgery; lymphangiomas are infiltrating and difficult to excise.[44]

ORBITAL PSEUDOTUMOR

Pseudotumor is a relatively common inflammatory process of the orbit that occurs mostly in teenagers and young adults.[42] The clinical and radiologic findings are highly variable since any or all of the orbital contents may be affected acutely, subacutely, or chronically. Pseudotumor is sometimes considered to be part of a spectrum of orbital lymphoid proliferations ranging from focal masses to diffuse inflammation.[43]

The etiology of orbital pseudotumor is unknown, but experimental work suggests that it is an autoimmune process, characterized by variable vasculitis and lymphocytic infiltration. Pseudotumor may be associated with a number of other diseases including Wegener's, fibrosing mediastinitis, thyroiditis, and cholangitis.[44] Pseudotumor usually shows a good clinical response to steroids.

The typical acute case is characterized by proptosis, soft tissue swelling, and impaired ocular mobility. The disease is usually unilateral but may be bilateral in up to 15 percent.[44] The sclera is thickened together

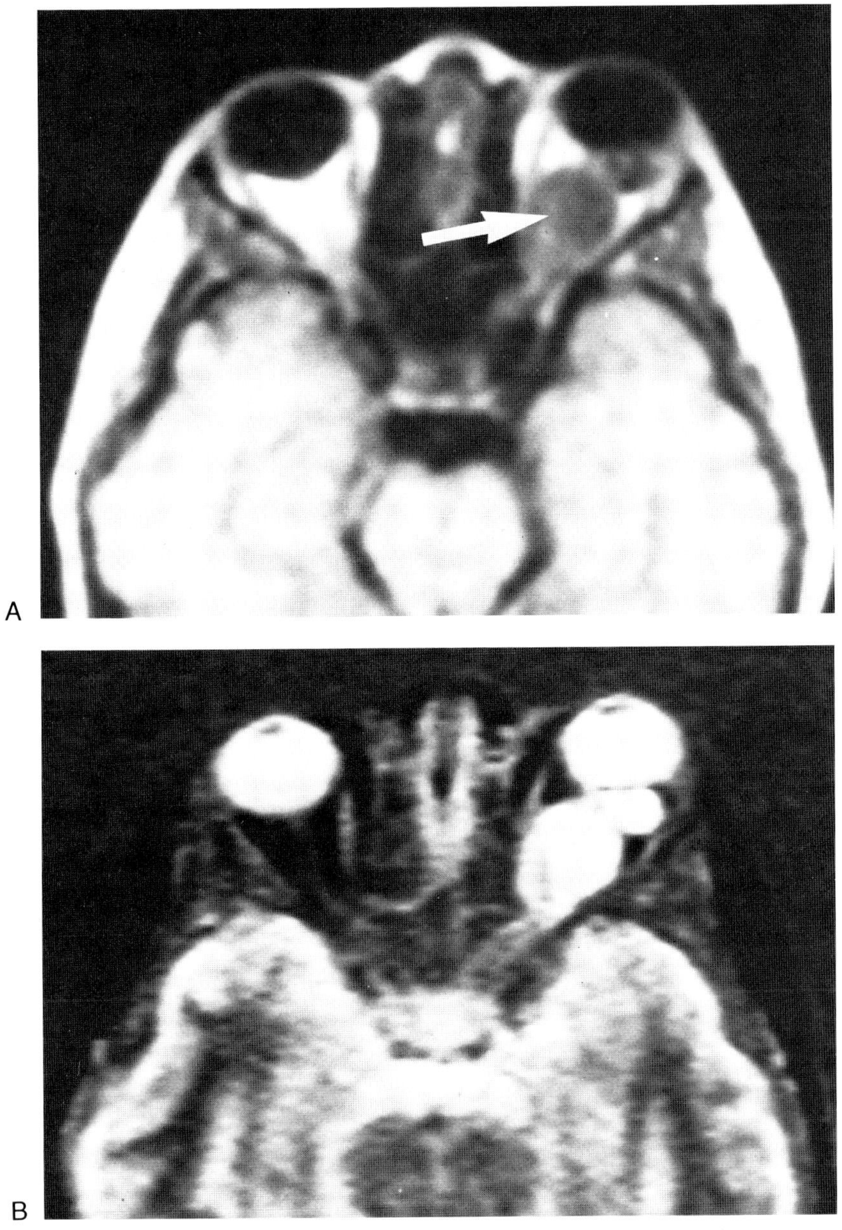

Fig. 12-22. Cavernous hemangioma. Predominantly intraconal, well circumscribed mass is seen in the left eye that focally expands the orbital wall. **(A)** SE 500/26. **(B)** IR 750/400 (Fat suppression technique).

with the extraocular muscles. The fat may also become inflamed and the orbital contents involved in a large inflammatory mass. Pseudotumor may also manifest as (1) an isolated discrete mass without scleral thickening, (2) obliteration of all retrobulbar tissue planes, (3) thickening of one or several extraocular muscles. The diagnosis is one of exclusion, biopsy, and response to steroids.

Recently two reports have suggested that MRI may add significant specificity to the CT diagnosis of

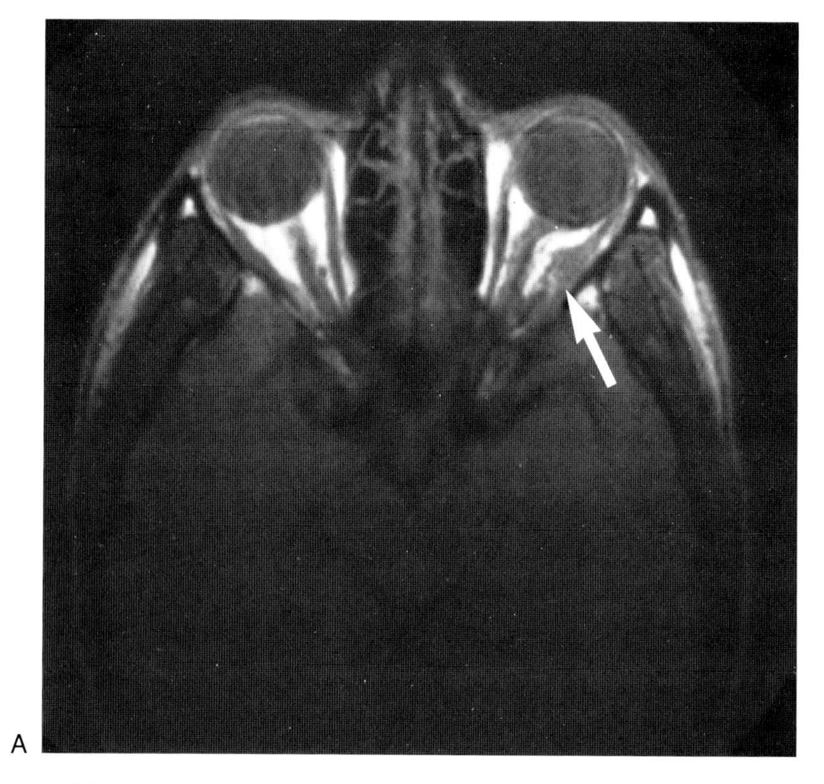

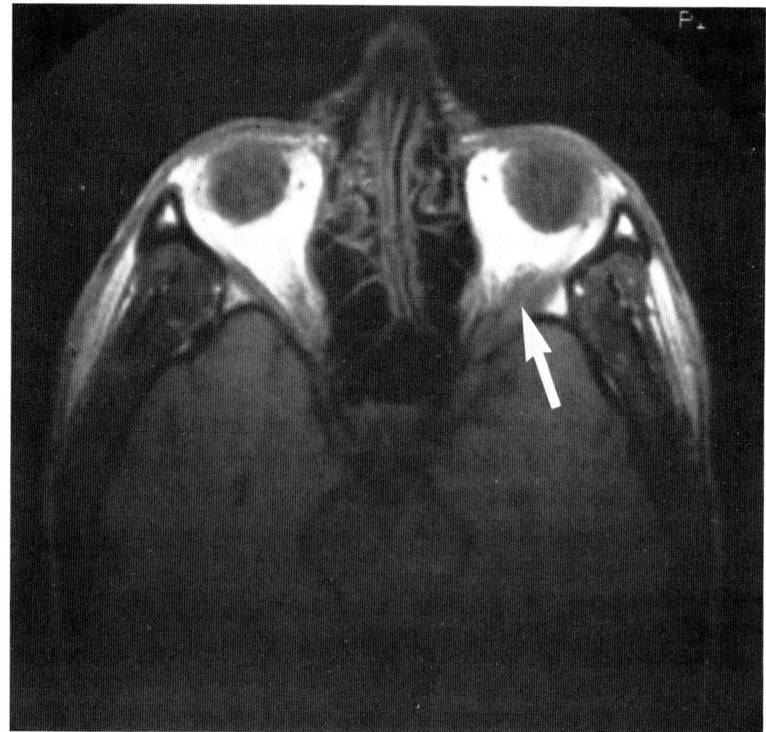

Fig. 12-23. Pseudotumor. There is inflammatory thickening of the extraocular muscles, frequently involving the sclera and orbital fat.

orbital pseudotumor.[46,47] Pseudotumor, as well as infectious myositis and sarcoid, is usually hypo- or isointense to fat and muscle on both T1- and T2-weighted pulse sequences. By contrast, orbital lymphoma and metastatic disease became hyperintense on T2-weighted sequences. An example is shown in Figure 12-23.

Extrinsic Tumors

Tumors that arise extrinsic to the orbit may involve it in two ways: by direct invasion or by metastasis. Direct invasion of the orbit is most frequently from a squamous cell carcinoma of the paranasal sinuses. Sinus carcinomas from the ethmoids or superomedial maxillary sinuses most frequently invade the orbit (Fig. 12-24). CT is more sensitive than MR at detecting early bone invasion, but MR may add anatomic information by virtue of its multiplanar capacity.

Orbital metastases from distant sites are relatively common, particularly with disseminated disease. It should be noted, however, that a significant percentage of patients present to the ophthalmologist first, before their primary tumor is discovered. Accordingly, metastatic disease should be considered in the differential diagnosis of both ocular and orbital lesions.

The patterns and types of metastases are somewhat different in children than in adults. In children, more metastases are to the orbit, while in adults more metastases are to the globe.[44] The most common childhood tumors to metastasize to the orbit are embryonal cell carcinoma, neuroblastoma, Ewings sarcoma, and leukemia. In adults the most common metastases are from carcinomas of the lung and breast.

Metastases to the globe are best seen by ophthal-

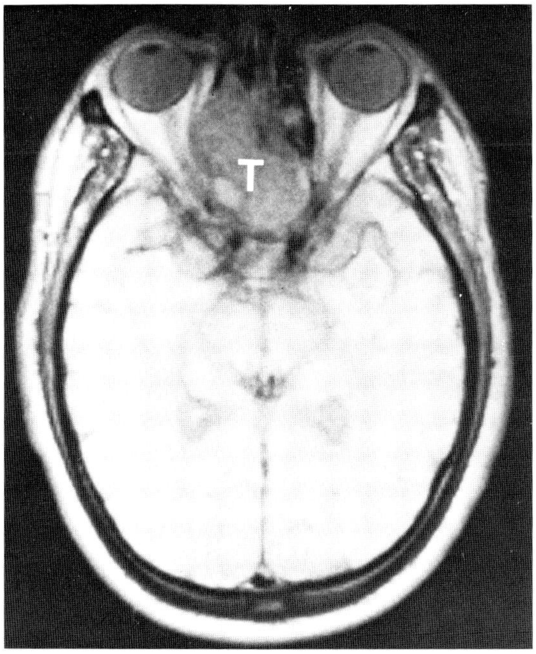

Fig. 12-24. A tumor (*T*) of the paranasal sinuses invading the orbit. (Courtesy of Picker International.)

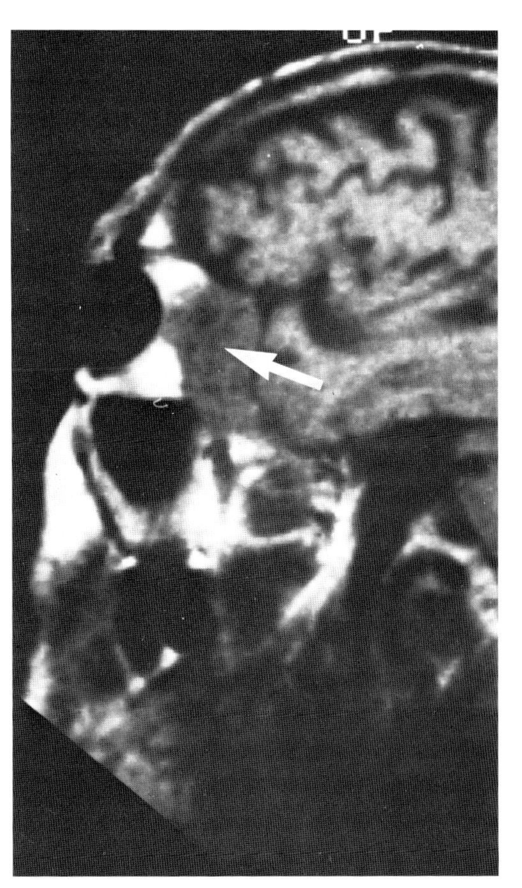

Fig. 12-25. Metastasis (arrow) to the orbit from lung carcinoma seen on this sagittal SE 500/53 image.

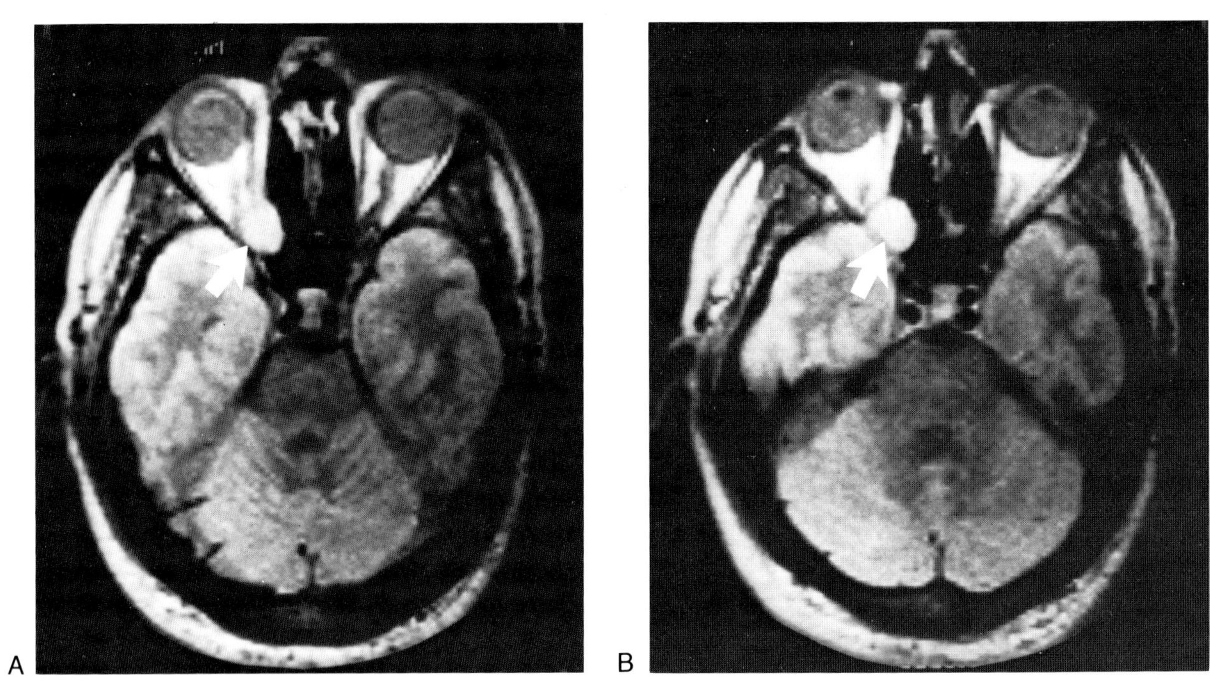

Fig. 12-26. Frontoethmoidal mucocele invading the right orbit.

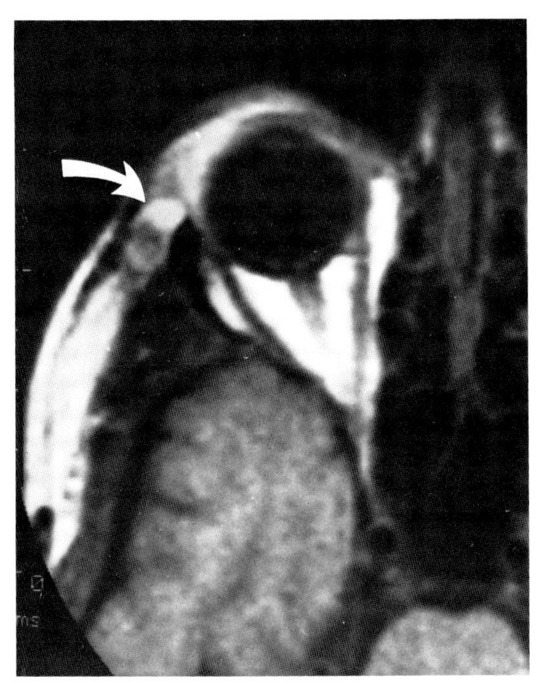

Fig. 12-27. Dermoid of the lacrimal gland.

moscopy or ultrasound. Metastases to the orbit are well evaluated by either CT or MR. Usually orbital metastases have indistinct boundaries and are diffusely infiltrating. An example shown by MRI is presented in Figure 12-25.

Mucocele

A mucocele is an expansile, epithelial-lined, fluid-filled sac arising from an occluded paranasal sinus. The mucocele begins when a sinus ostium, usually of the frontal or ethmoid, becomes occluded by inflammation, fibrosis, or mass. Continued secretion of mucus by the respiratory epithelium results in expansion of the sinus, with thinning or remodeling of the sinus wall. Frontal and ethmoid mucoceles may erode into the orbit, causing proptosis and limitation of eye movement. CT is probably better suited than MR to show this invasion and the status of the bony

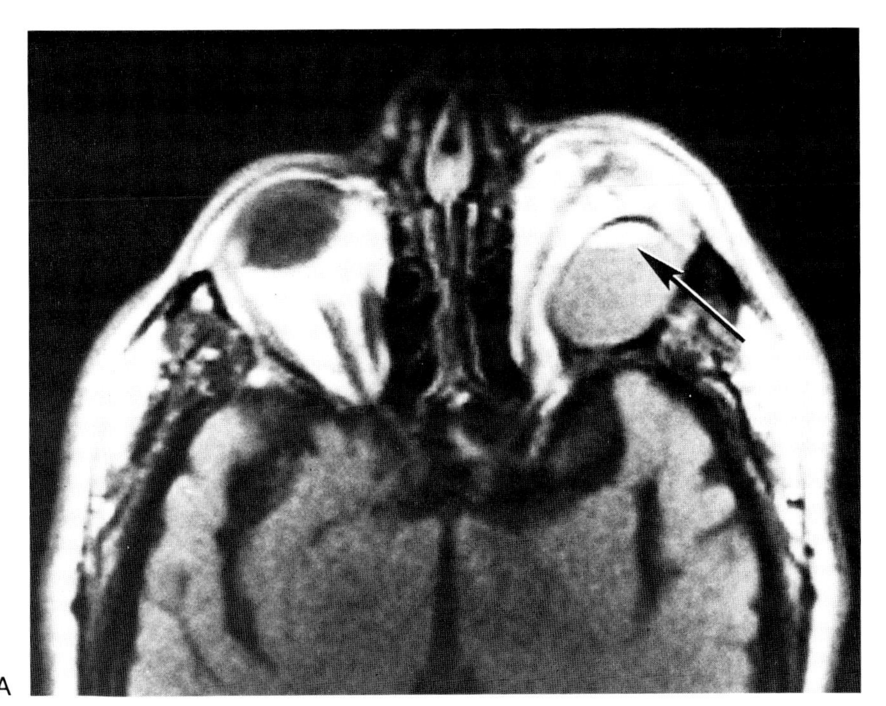

Fig. 12-28. (A and B) Large orbital dermoid. Note characteristic fat-fluid level (arrow). (Courtesy of Picker International.) (*Figure continues*).

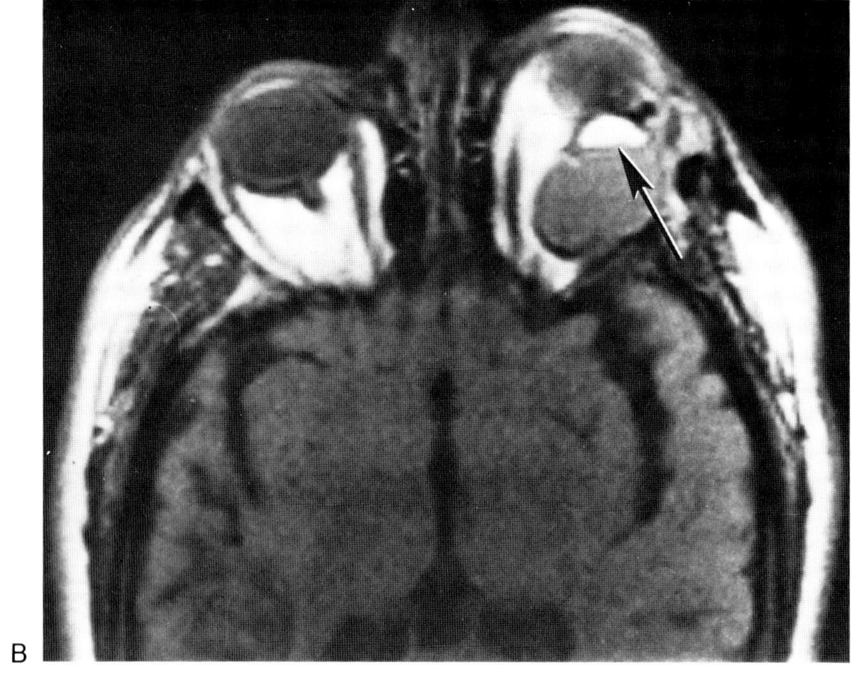

Fig. 12-28 *(continued)*.

orbital wall relative to the mucocele. Still, multiplanar imaging by MR may aid in anatomic delineation (Fig. 12-26).[13,44] The relatively short T1 value of mucocele contents may conceivably aid in establishing their diagnosis on MRI.[46]

Orbital Inclusion Cysts

During embryogenesis, ectodermal anad mesodermal elements may become abnormally sequestrated into the orbit resulting in orbital inclusion cysts. Dermoid cysts are more common than epidermoids. They frequently present during the first decade of life with swelling of the eyelid or proptosis. Many are found in or adjacent to the lacrimal gland (Fig. 12-27). They may also lie within the orbit (Fig. 12-28). Dermoids are usually well circumscribed masses with low attenuation on CT due to fat. The MR appearance is variable depending on the relative proportions of mesenchymal fat and cholesterol. A fat-fluid level may be seen in the tumor (Fig. 12-28).[15,44] Epidermoids and epithelial inclusion cysts are much less common than

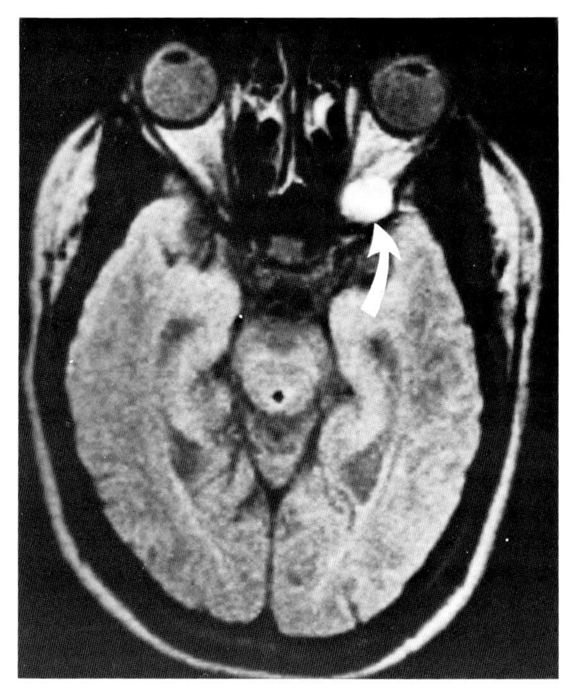

Fig. 12-29. Epithelial inclusion cyst. Sharply defined mass filled with proteinaceous fluid seen in the posterior orbit. (Courtesy of Picker International.)

dermoids. They are sharply defined masses with long T1 and T2 values due to their fluid content (Fig. 12-29).

REFERENCES

1. Daniels DL, Herfkins R, Gager WE et al: Magnetic resonance imaging of the optic nerves and chiasm. Radiology 152:79, 1984
2. Edwards JH, Hyman RA, Vacirca SJ et al: 0.6 T magnetic resonance imaging of the orbit. AJR 144:1015, 1985
3. Han JS, Benson JE, Bonstelle CT et al: Magnetic resonance imaging of the orbit: a preliminary experience. Radiology 150:755, 1984
4. Hawkes RC, Holland GN, Moore WS et al: NMR imaging in the evaluation of orbital tumors. AJNR 4:254, 1983
5. Li KC, Poon PY, Hinton P et al: MR imaging of orbital tumors with CT and ultrasound correlations. J Comput Assist Tomogr 8:1039, 1984
6. Moseley I, Brant-Zawadski M, Mills C: Nuclear magnetic resonance imaging of the orbit. Br J Ophthalmol 67:333, 1983
7. Sassani JW, Osbakken MD: Anatomic features of the eye disclosed with nuclear magnetic resonance imaging. Arch Ophthalmol 102:541, 1984
8. Schenck JF, Hart HR Jr, Foster TH et al: Improved MR imaging of the orbit at 1.5 T with surface coils. AJR 144:1033, 1985
9. Sobel DF, Kelly W, Kjos BO et al: MR imaging of orbital and ocular disease. AJNR 6:259, 1985
10. Sobel DF, Mills C, Char D et al: NMR of the normal and pathologic eye and orbit. AJNR 5:345, 1984
11. Worthington BS: NMR imaging of intracranial and orbital tumours. Br Med Bull 40:179, 1984
12. Albert A, Lee BCP, Saint-Louis L, Deck MDF: MRI of optic chiasm and optic pathways. AJNR 7:255, 1986
13. Bilaniuk LT, Schneck JF, Zimmerman RA et al: Ocular and orbital lesions: surface coil MR imaging. Radiology 156:669, 1985
14. Gonzalez RG, Cheng H, Barnett P et al: Nuclear magnetic resonance imaging of the vitreous body. Science 223:399, 1984
15. Sullivan JA, Harms SE: Surface-coil MR imaging of orbital neoplasms. AJNR 7:29, 1986
16. Gomori JM, Grossman RI, Shields JA et al: Ocular MR imaging and spectroscopy: an ex vivo study. Radiology 160:201, 1986
17. Pope JM, Chandra S, Balfe JD: Changes in the state of water in senile cataractous lenses as studied by nuclear magnetic resonance. Exp Eye Res 34:57, 1982
18. Wright RM, Swietek PA, Simmons ML: Eye artifacts from mascara in MRI. AJNR 6:652, 1985
19. Unsold R: Computed tomographic anatomy of the orbit. Int Opthalmol Clin 25:45, 1982
20. Daniels DL, Kneeland JB, Shimakawa A et al: MR imaging of the optic nerve and sheath: correcting the chemical shift misregistration effect. AJNR 7:249, 1986
21. Haddad R, Font RL, Resser F: Persistent hyperplastic primary vitreous. A clinicopathologic study of 62 cases and a review of the literature. Surv Ophthalmol 23:123, 1978
22. Morgan KS, McLean IW: Retinoblastoma and persistent hyperplastic vitreous occurring in the same patient. Ophthalmol 88:1087, 1981
23. Reese AB: Persistent hyperplastic primary vitreous. Jackson Memorial lecture. Trans Am Acad Ophthalmol Otolaryngol 59:271, 1955
24. Imaging News: Surface coils aid MRI of head, neck, and spine. Diagnostic Imaging (8(6):13, 1986
25. Reese AB: Tumors of the Eye. Harper & Row, New York, 1976
26. Zimmerman RA, Bilaniuk LT: Computed tomography in the evaluation of patients with bilateral retinoblastomas. CT: J Comput Assist Tomogr 3:251, 1979
27. Jacobiec FA, Tso MOM, Zimmerman LE et al: Retinoblastoma and intracranial malignancy. Cancer 30:2048, 1977
28. Hopper KD, Katz NNK, Dorwart RH et al: Childhood leukocoria: computed tomographic appearance and differential diagnosis with histopathologic correlation. Radiographics 5:377, 1985
29. Mafee MF, Peyman GA, Grisolano JE et al: Malignant uveal melanoma and simulating lesions: MR imaging evaluation. Radiology 160:773, 1986
30. Shields JA: Diagnosis and Management of Intraocular Tumors. CV Mosby, St. Louis, 1983
31. Shields JA, Zimmerman LE: Lesions simulating malignant melanomas of the posterior uvea. Arch Ophthalmol 89:466, 1973
32. Gomori JM, Grossman RI, Shields JA et al: Choroidal melanomas: correlation of NMR spectroscopy and MR imaging. Radiology 158:443, 1986
33. Saint-Louis LA, Haik BG, Ellsworth RM, Deck MDF: MR imaging of Coat's disease. Radiology 157(p):214, 1985

33A. Harris GJ, Williams AL, Reeser FH: Intraocular evaluation by computed tomography. Int Ophthalmol Clin 22:197, 1982

34. Peyster RG, Hoover ED, Hershey BL, Haskin ME: High resolution CT of lesions of the optic nerve. AJR 140:869, 1983
35. Rothfus WE, Curtin HD, Slamovits TL, Kennerdell

JS: Optic nerve/sheath enlargement. Radiology 150: 409, 1984

36. Zimmerman RA, Bilaniuk LT, Metzger RA et al: Computed tomography of orbital facial neurofibromatosis. Radiology 146:113, 1983
37. Reed D, Robertson WD, Rootman J, Douglas G: Plexiform neurofibromatosis of the orbit: CT evaluation. AJNR 7:259, 1986
38. Grove AS: Evaluation of exophthalmos. N Engl J Med 292:1005, 1975
39. Alper MG: Endocrine orbital disease. p. 187. In Arger PH (ed): Orbit Roengenology. Wiley, New York, 1977
40. Enzmann DR, Donaldson SS, Kriss JT: Appearance of Grave's disease on orbital computed tomography. J Comput Assist Tomogr 3:815, 1979
41. Peyster RG, Ginsberg F, Silber JH, Adler LP: Exophthalmos caused by excessive fat: CT volumetric analysis and differential diagnosis. AJNR 7:35, 1986
42. Nugent RA, Rootman J, Robertson WD et al: Acute orbital pseudotumors: classification and CT features. AJNR 2:431, 1981
43. Char DH, Norman D: The use of computed tomography and ultrasonography in the evaluation of orbital masses. Surv Ophthalmol 27:49, 1982
44. Zimmerman RA, Bilaniuk LT: The orbit. p. 71. In Lee SH, Rao CVG (eds): Cranial Computed Tomography. McGraw-Hill, New York, 1983
45. Davis KR, Hesselink JR, Dallow RL, Grove AS Jr: CT and ultrasound in the diagnosis of cavernous hemangioma and lymphangioma of the orbit CT: J Comput Tomogr 4:98, 1980
46. Sullivan JA, Harms SE: Characterization of orbital lesions by surface coil MR imaging. RadioGraphics 7:9, 1987
47. Atlas SW, Grossman RI, Savino PJ et al: Surface-coil MR of orbital pseudotumor. AJNR 8:141, 1987

Glossary

B_0 A symbol for the constant magnetic field produced by the large magnet in an MR scanner.

B_1 A symbol for the smaller rotating radio-frequency magnetic field, which in conjunction with $\mathbf{B}_0$ excites hydrogen nuclei to resonance.

Coherence Rotation or oscillation of protons harmoniously in phase with one another. The transverse component of magnetization (M_{xy}) is generated by coherent precession at resonance.

Coil Multiple loops of wire that may be used to produce a magnetic field (by passing current through it) or detect a changing magnetic field (by measuring the voltage induced in the wire).

Echo time (TE) The time in milliseconds between application of the 90° pulse and echo signal in a spin echo pulse sequence.

Fourier transform (FT) A mathematical procedure used in MR scanners to analyze and separate amplitude and frequency components of the complex detected signal. Fourier transform analysis allows spatial information to be reconstructed from the raw data. Most MR scanners in clinical use today use two-dimensional or three-dimensional Fourier transformation reconstruction techniques (2DFT or 3DFT).

Free induction decay (FID) The signal recorded in the transverse plane resulting after a radio frequency pulse has excited a system to resonance.

Frequency (f) The number of cycles or repetitions of an event per unit time, usually expressed in hertz (Hz) where 1 Hz = 1 cycle per second. When a rotation system is studied, the term angular frequency is sometimes used. Angular frequency, ω, is related to plain frequency, f, by the equation $f = 2\pi\omega$.

Gauss (G) A unit of magnetic field strength that is the approximate strength of the Earth's magnetic field on its surface. The unit gauss has been largely replaced by the larger unit tesla (T) and 1 T = 10,000 G.

Gradient magnetic field A small, supplemental magnetic field applied in addition to the large static field in an MR scanner, which causes each small volume element in a sample to experience a different magnetic field, and hence resonate at a different frequency. In this way, spatial encoding may be performed.

Gyromagnetic ratio (γ) A constant relating the nuclear MR frequency and the strength of the external magnetic field (see Larmor frequency).

Hertz (Hz) A unit of frequency equal to 1 cycle per second. The larger unit megahertz (MHz) = 1,000,000 Hz.

Inversion recovery (IR) An imaging sequence that involves successive 180° and 90° pulses, after which a heavily T1-weighted signal is obtained. The inversion recovery sequence is specified in terms of two parameters, the inversion time (TI) and the repetition time (TR), discussed in Chapter 2.

Inversion time (TI) The time between the 180° and subsequent 90° pulse in an inversion recovery sequence.

Larmor frequency (ν_0) The frequency at which nuclear magnetic resonance occurs. The Larmor frequency is related to magnetic field strength ($\mathbf{B}_0$) and a constant called the gyromagnetic ratio (γ) by an equation known as the Larmor equation.

Magnetic moment A measure of the torque exerted on a magnet or spinning proton that tends to align it with an externally applied magnetic field.

Magnetization The net effect that an externally applied magnetic field has on a material placed within it and that tends to polarize and turn that material into a magnet. Most body tissues are weakly susceptible to this process, and their protons tend to align with the field, producing a net magnetization vector. Some substances called paramagnetics have a more marked tendency to align with the field. Other substances have a tendency to align opposite to the field and are called diamagnetics.

Nuclear magnetic resonance (NMR) A phenomenon exhibited by hydrogen and certain atoms with odd numbers of protons and neutrons, such as ^{13}C, ^{19}F, ^{23}Na, and ^{31}P. When these atoms are placed in a strong external magnetic field, $\mathbf{B}_0$, and excited by a suitable induction field, $\mathbf{B}_1$, rotating at the Larmor frequency, energy absorption and emission occur, constituting NMR. NMR is often referred to simply as magnetic resonance (MR).

Paramagnetic Materials with a small but positive magnetizability. These substances usually possess an unpaired electron and include certain transition elements, rare earths, and free radicals. They may one day be used as contrast agents for MRI.

Partial saturation (PS) An NMR pulse sequence consisting of repetitive 90° pulses after which a free induction decay is recorded. This produces a strongly T1-wieghted image.

Phantom A usually oil-filled plastic cylinder of known dimensions and properties used to test and calibrate an MR scanner.

Phase The position relative to a particular part of the cycle of a periodic function, such as a sine wave.

Pixel A "picture element," the smallest discrete part of a digital image display.

Precession The wobbling of spinning protons when placed in an external magnetic field, analogous to the wobbling of a spinning top in the Earth's gravitational field.

Precessional frequency Same as the Larmor frequency. The frequency of a radio frequency pulse that induces MR as well as the frequency of the emitted NMR signal.

Pulse sequence A set of radio frequency and gradient magnetic field pulses used to induce nuclear resonance and produce an image. Examples of commonly used pulse sequences include inversion recovery and spin-echo techniques.

Radio frequency (RF) High-frequency electromagnetic waves used in MRI similar to those used in commercial radio transmission, usually in the range of megahertz (MHz).

Receiver coil A coil of wire that is placed around the patient and that detects the MR signal.

Reconstruction The computer manipulation of the many complex MR signals to decode spatially and produce an image.

Relaxation times After excitation with the applied radio frequency pulse, spinning protons tend to realign with the static field. The time it takes for this to occur is known as relaxation time. Two relaxation time constants may be defined parallel and perpendicular to the static magnetic field. These are known as the T1 and T2 relaxation times, respectively.

Repetition time (TR) The time interval between the beginning of a pulse sequence and the succeeding pulse sequence.

Resistive magnet A magnet that uses electric current in coils to generate a magnetic field and that operates at temperatures higher than superconductive range (i.e., $>-360°C$).

Resonance The rapid oscillation between high- and low-energy states of a system that occurs when the system is stimulated or excited at a specific frequency known as the resonance frequency.

RF pulse A brief burst of electromagnetic radiation at radio frequencies that excite the net magnetization vector to rotate a certain amount. Commonly used RF pulses include the 90° and 180° pulses.

Saturation A nonequilibrium state in MR in which equal numbers of spinning protons are aligned with and opposite to the static magnetic field, resulting in no net magnetization.

Saturation recovery (SR) An MR imaging pulse sequence similar to partial saturation but in which recovery from total saturation occurs before the next repetition. Like PS, SR produces a heavily T1-weighted image.

Signal-to-noise ratio (S/N or SNR) The ratio between the amplitude of the recorded signal and background noise, which distorts that signal. Signal-to-noise ratio (and hence image quality) may be improved by taking more averages of the signal, by using longer sampling times, and by sampling larger volumes.

Spin A property of nuclei with an odd number of protons or neutrons accounts for their magnetic moments and allows for MR to occur.

Spin-echo (SE) An MR pulse sequence technique in which alternating 90° and 180° pulses are applied to the sample and an echo signal recorded. The spin-echo sequence is sensitive to both T1 and T2 values.

Spin-lattice relaxation time See T1.

Spin-spin relaxation time See T2.

Superconductive magnet A magnet whose field is produced by current flowing through a substance near absolute zero.

Surface coil A specially constructed flat receiver coil which is placed directly on or near the skin surface, allowing high resolution imaging of organs and tissues that are within a few centimeters of the coil.

T1 Also called the spin-lattice or longitudinal relaxation time. This represents the time constant for spins to realign themselves after perturbation with the externally applied field. When starting from zero magnetization in a direction parallel to the field, T1 represents the time for this magnetization to grow to 63% of its final value.

T2 Also called the spin-spin or transverse relaxation time. This represents the time constant for loss of phase coherence between spins along a plane perpendicular to the static field. Starting from a magnetization in this transverse plane, T2 represents the time it takes for this magnetization to decay from 63% of its initial value.

Tesla (T) A unit of magnetic field flux density, equal to 10,000 gauss.

Transverse relaxation time See T2.

Vector A physical quantity with both magnitude and direction.

Voxel A volume element, which corresponds in three-dimensional space to the pixel.

Appendix A
Multiplanar MR Atlas of the Brain

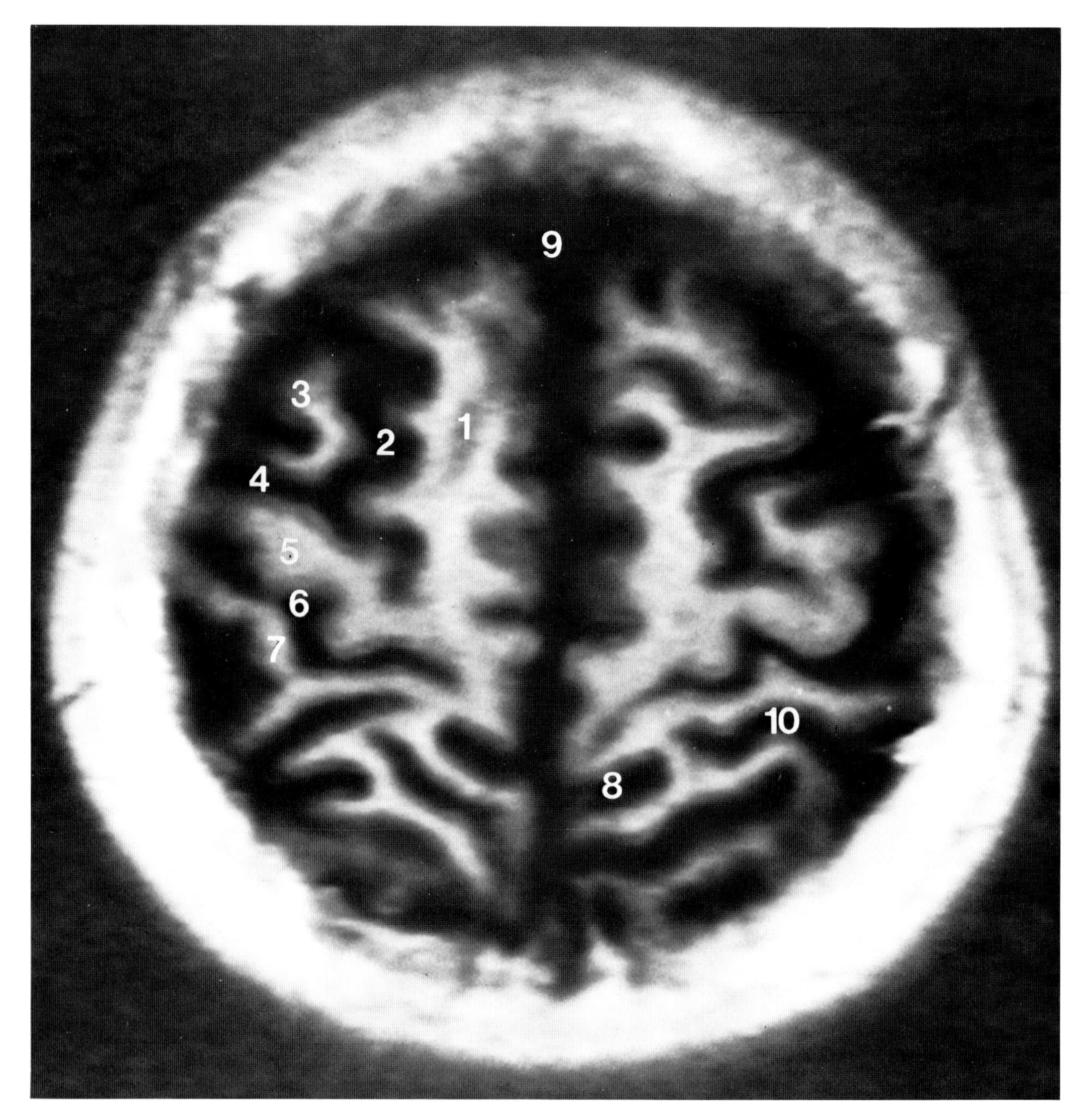

Axial 1

1. Superior frontal gyrus
2. Superior frontal sulcus
3. Middle frontal gyrus
4. Precentral sulcus
5. Precentral gyrus
6. Central sulcus
7. Postcentral gyrus
8. Pars marginalis of cingulate sulcus
9. Interhemispheric fissure
10. Postcentral sulcus

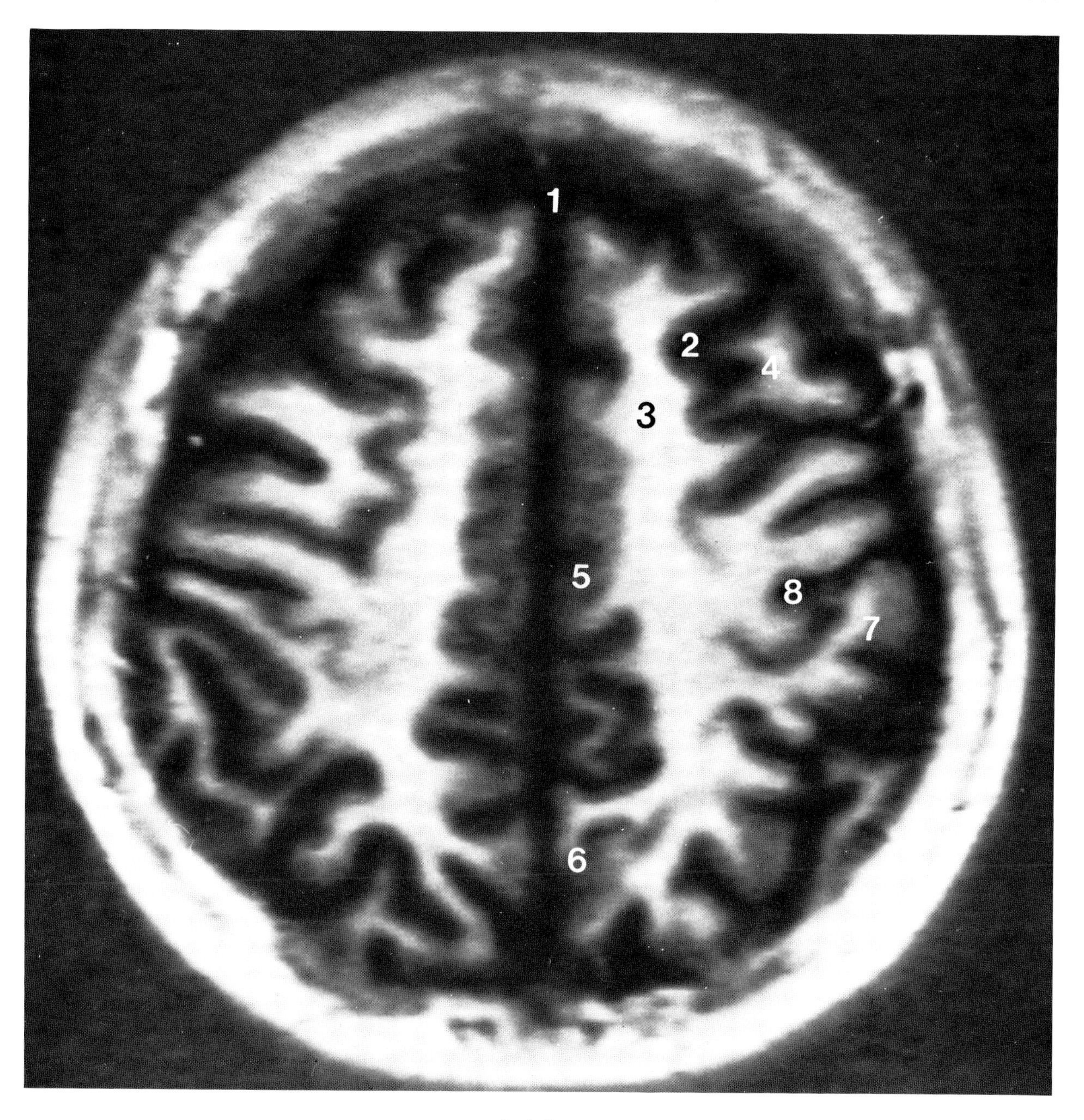

Axial 2

1. Interhemispheric fissure
2. Superior frontal sulcus
3. Superior frontal gyrus
4. Middle frontal gyrus
5. Paracentral lobule
6. Precuneus
7. Postcentral gyrus
8. Central sulcus

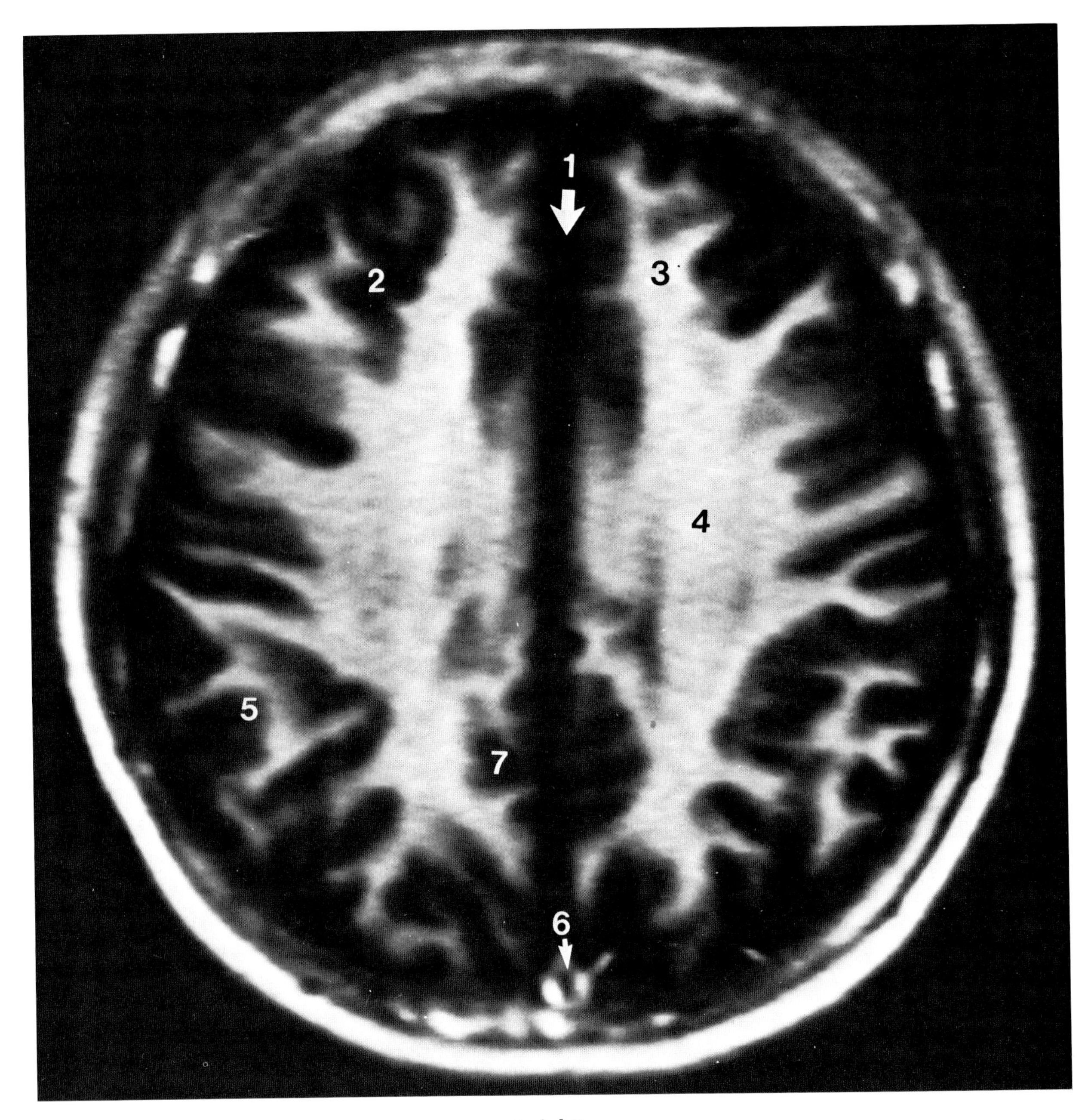

Axial 3

1. Interhemispheric fissure
2. Superior frontal sulcus
3. Superior frontal gyrus
4. Centrum semiovale
5. Superior parietal lobule
6. Superior sagittal sinus
7. Parieto-occipital sulcus

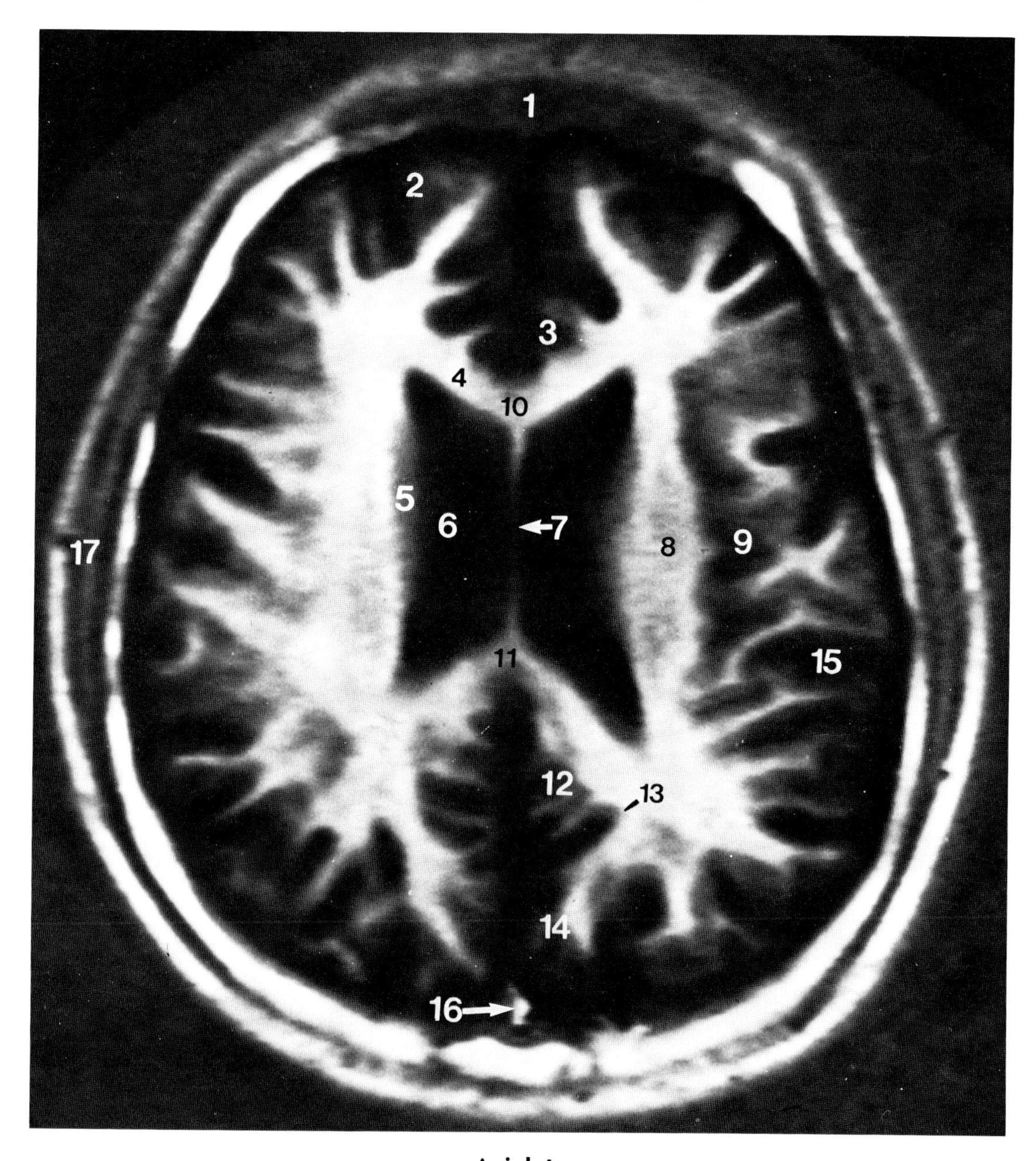

Axial 4

1. Frontal sinus
2. Frontal pole
3. Cingulate gyrus
4. Corpus callosum (forceps major)
5. Caudate (body)
6. Lateral ventricle
7. Septum pellucidum
8. Putamen
9. Insula
10. Corpus callosum (genu)
11. Corpus callosum (body)
12. Precuneus
13. Parieto-occipital sulcus
14. Cuneus
15. Sylvian fissure
16. Superior sagittal sinus
17. Temporalis muscle

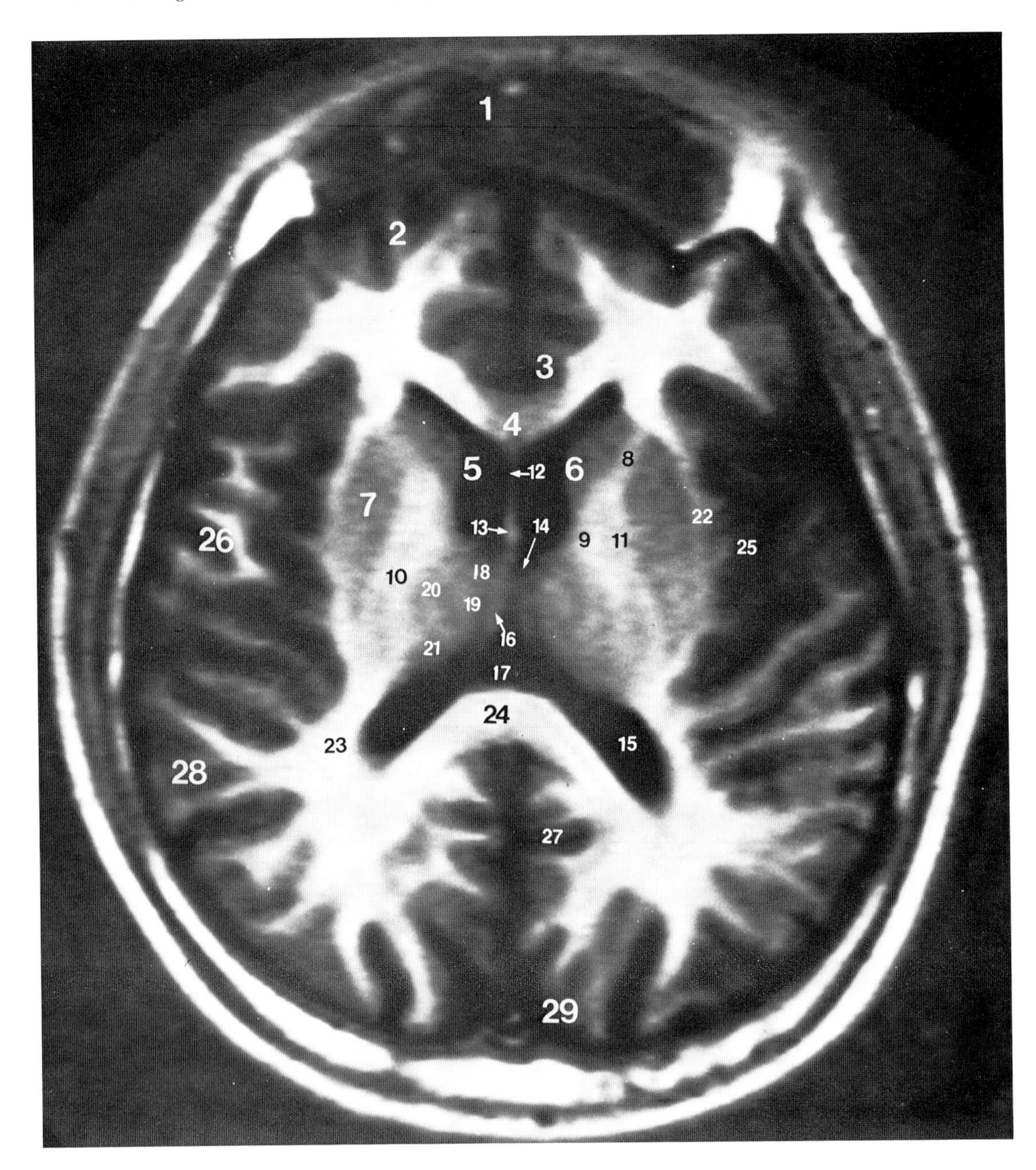

Axial 5

1. Frontal sinus
2. Frontal pole
3. Cingulate gyrus
4. Corpus callosum (genu)
5. Frontal horn
6. Caudate (head)
7. Putamen
8. Internal capsule (anterior limb)
9. Internal capsule (genu)
10. Internal capsule (posterior limb)
11. Globus pallidus
12. Septum pellucidum
13. Column of fornix
14. Foramen of Monro
15. Trigone of lateral ventricle
16. Body of fornix
17. Cistern of velum interpositum
18. Anterior thalamic nuclei
19. Dorsomedial thalamic nuclei
20. Lateral-ventral thalamic nuclei
21. Pulvinar
22. External capsule
23. Optic radiations
24. Corpus callosum (splenium)
25. Circular sulcus
26. Superior temporal gyrus
27. Parieto-occipital sulcus
28. Middle temporal gyrus
29. Cuneus

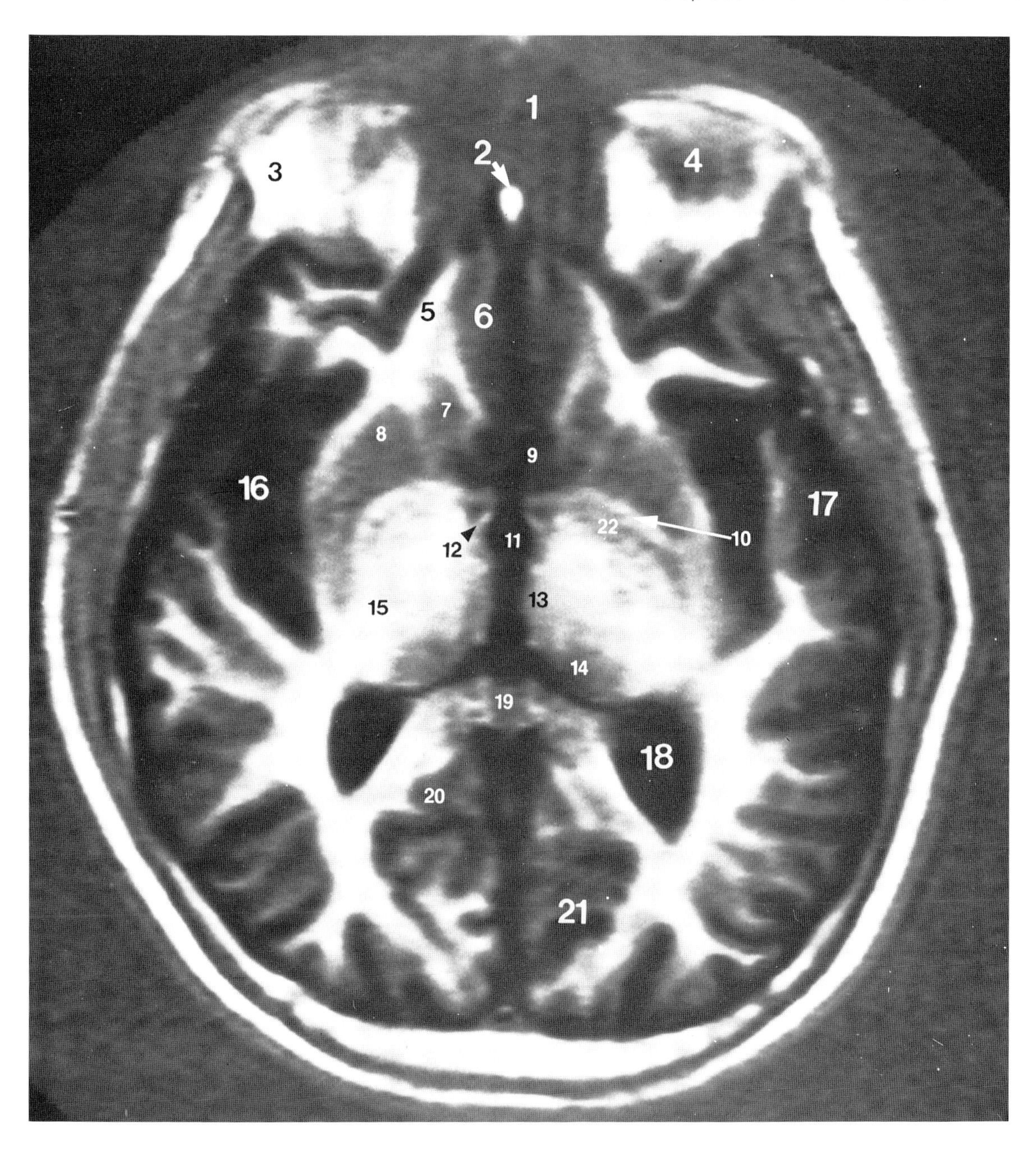

Axial 6

1. Ethmoid sinus
2. Fat in crista galli
3. Orbital fat
4. Superior rectus/levator palpebrae superioris muscle complex
5. Medial orbital gyrus
6. Gyrus rectus
7. Caudate nucleus (head)
8. Putamen
9. Septal (parolfactory) area
10. Anterior commissure
11. Third ventricle
12. Postcommisural fornix
13. Dorsomedial thalamic nucleus
14. Pulvinar
15. Posterior limb of internal capsule
16. Circular sulcus/Sylvian fissure
17. Middle temporal gyrus
18. Trigone
19. Isthmus
20. Precuneus
21. Cuneus
22. Globus pallidus

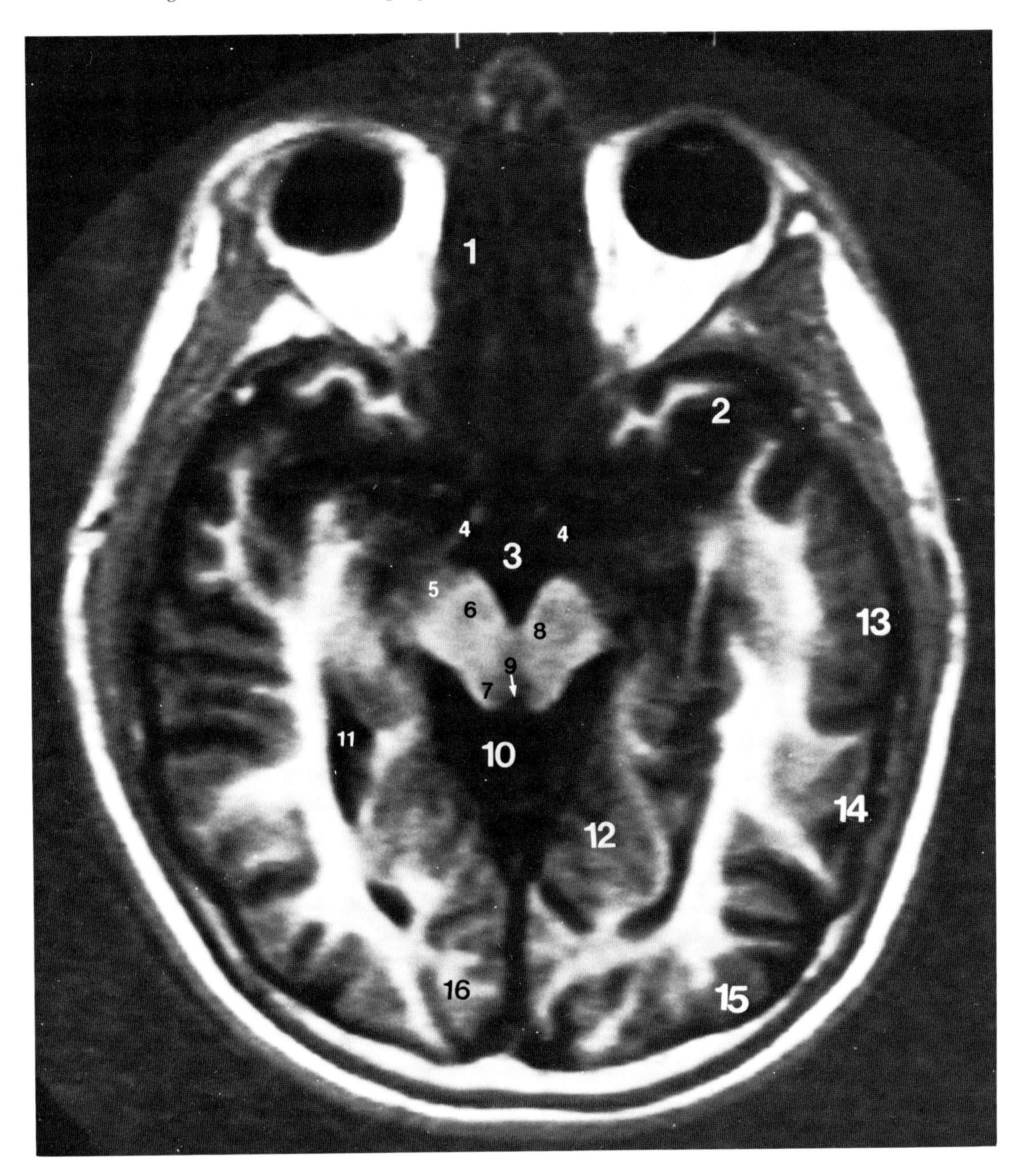

Axial 7

1. Ethmoid air cells
2. Temporal pole
3. Interpeduncular cistern
4. Optic tract
5. Cerebral peduncle
6. Substantia nigra
7. Inferior colliculus
8. Red nucleus
9. Cerebral aqueduct
10. Superior cerebellar cistern
11. Temporal horn of lateral ventricle
12. Parahippocampal gyrus
13. Inferior temporal gyrus
14. Middle temporal gyrus
15. Lateral occipital gyri
16. Cuneus

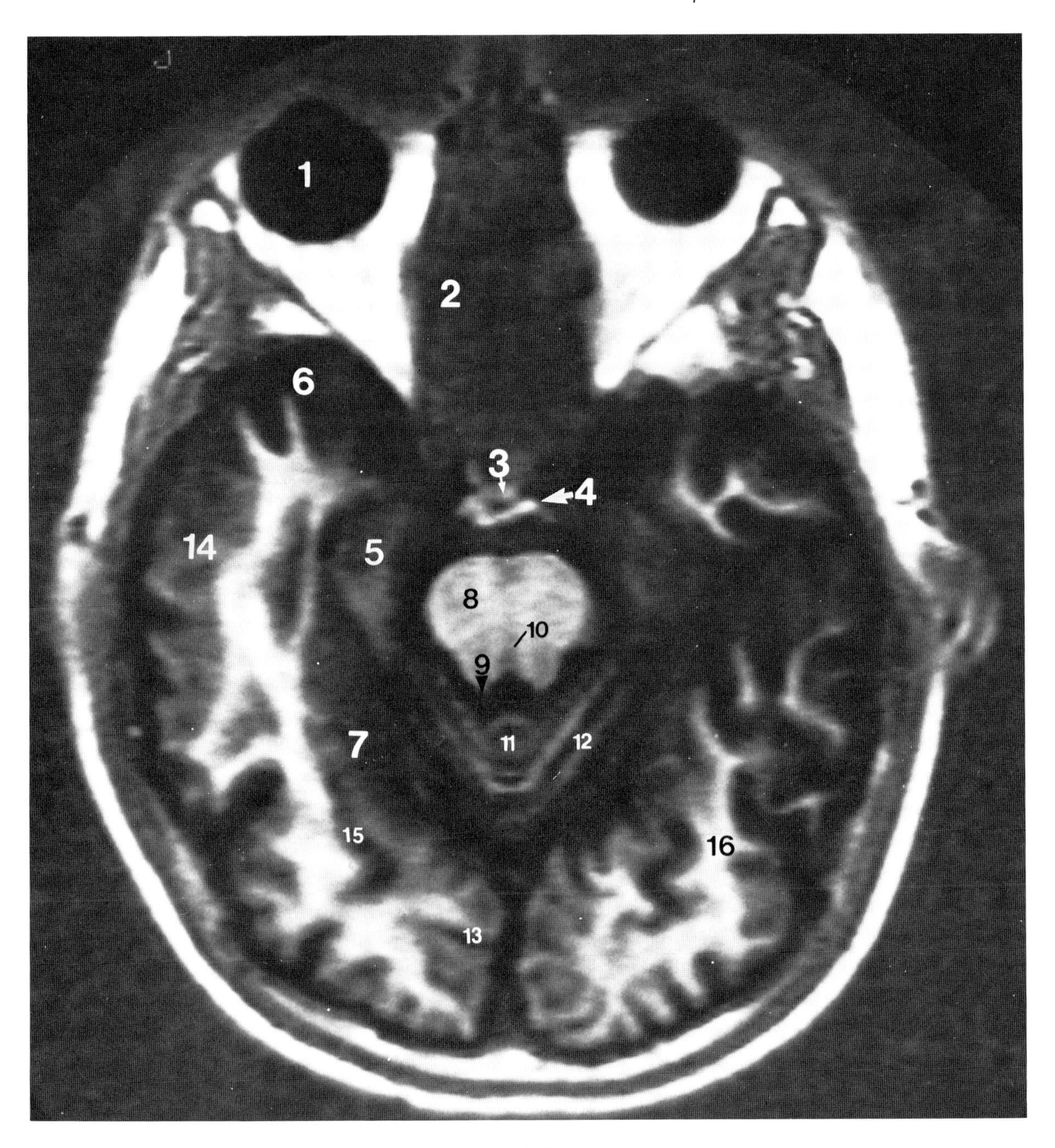

Axial 8

1. Globe
2. Ethmoid air cells
3. Pituitary
4. Dorsum sellae
5. Hippocampus
6. Temporal pole
7. Parahippocampal gyrus
8. Basis pontis
9. Superior cerebellar peduncle
10. Decussation of superior cerebellar peduncle
11. Culmen of vermis
12. Cerebellar hemisphere
13. Calcarine sulcus
14. Inferior temporal gyrus
15. Collateral sulcus
16. Occipitotemporal gyrus

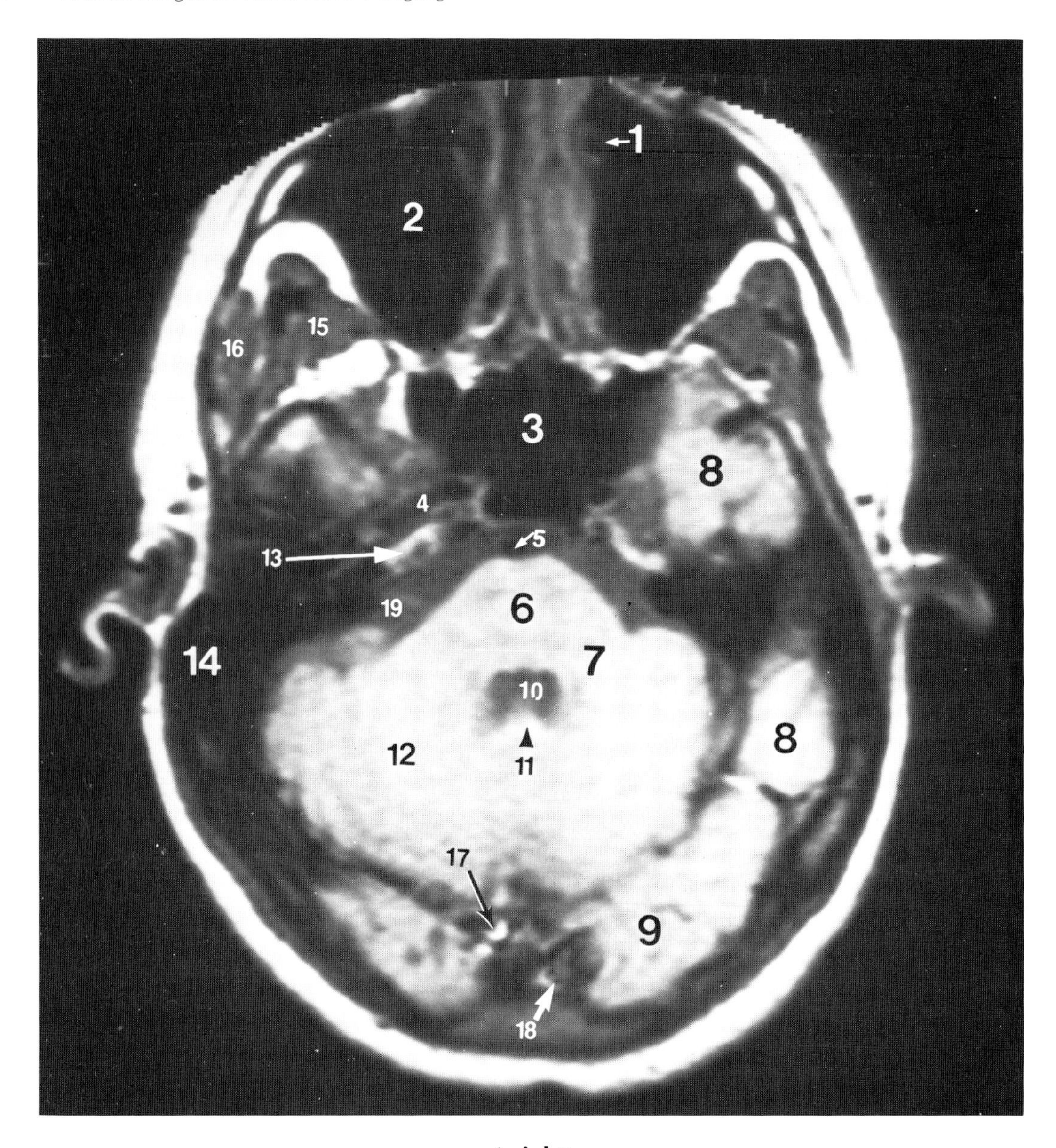

Axial 9

1. Nasolacrimal duct
2. Maxillary sinus
3. Sphenoid sinus
4. Internal carotid artery (petrous portion)
5. Basilar artery
6. Pons
7. Middle cerebellar peduncle
8. Temporal lobe
9. Occipital lobe
10. Fourth ventricle
11. Nodulus of vermis
12. Cerebellar hemisphere
13. Fat in petrous pyramid apex
14. Mastoid air cells
15. Temporalis muscle
16. Masseter muscle
17. Superior sagittal sinus
18. Transverse sinus
19. Cerebellopontine angle cistern

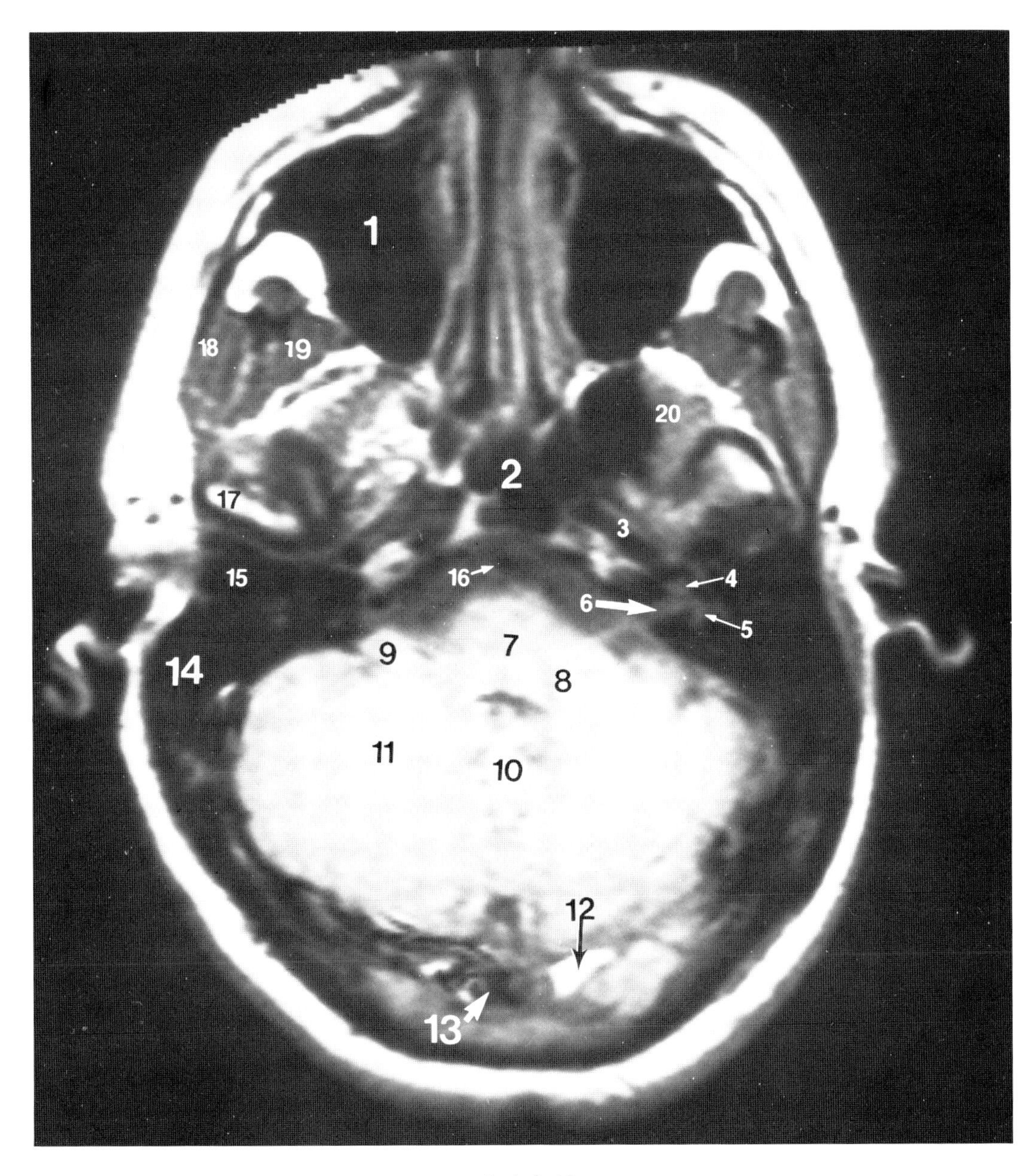

Axial 10

1. Maxillary sinus
2. Sphenoid sinus
3. Internal carotid artery (petrous portion)
4. Cochlea
5. Vestibule
6. Neurovascular bundle in internal auditory canal
7. Pons
8. Middle cerebellar peduncle
9. Flocculus
10. Vermis
11. Cerebellar hemisphere
12. Transverse sinus
13. Torcula
14. Mastoid air cells
15. External auditory canal
16. Basilar artery
17. Mandibular condyle
18. Masseter muscle
19. Temporalis muscle
20. Lateral pterygoid muscle

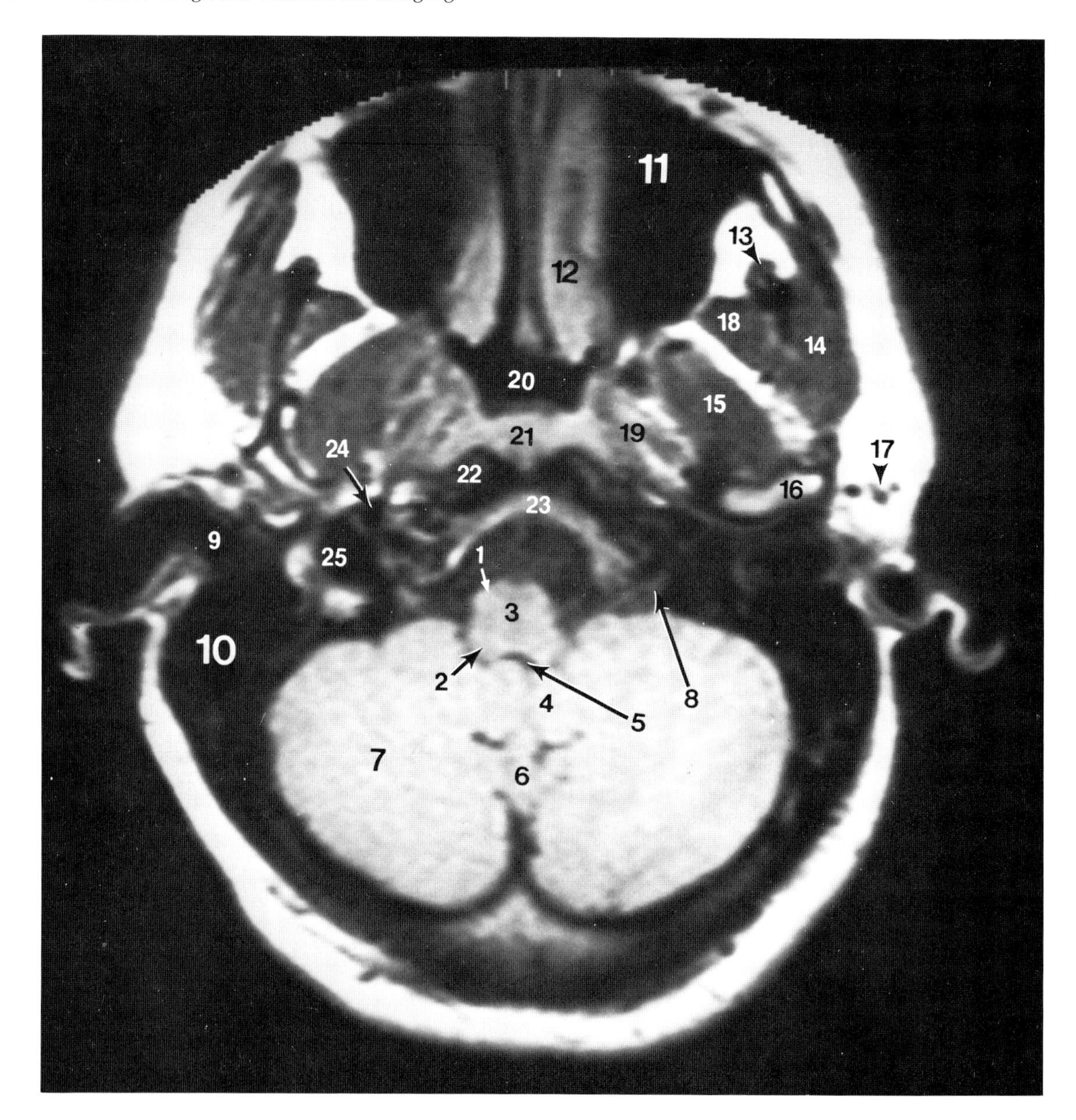

Axial 11

1. Pyramid
2. Inferior cerebellar peduncle
3. Medulla
4. Tonsil
5. Fourth ventricle
6. Inferior vermis
7. Cerebellar hemisphere
8. Hypoglossal nerve in canal
9. External auditory canal
10. Mastoid air cells
11. Maxillary sinus
12. Inferior nasal turbinate
13. Coronoid process of mandible
14. Masseter muscle
15. Lateral pterygoid muscle
16. Condyle of mandible
17. External carotid artery branches
18. Temporalis muscle
19. Medial pterygoid muscle
20. Nasopharynx
21. Adenoidal tissue
22. Longus capitis muscle
23. Clivus
24. Internal carotid artery
25. Internal jugular vein

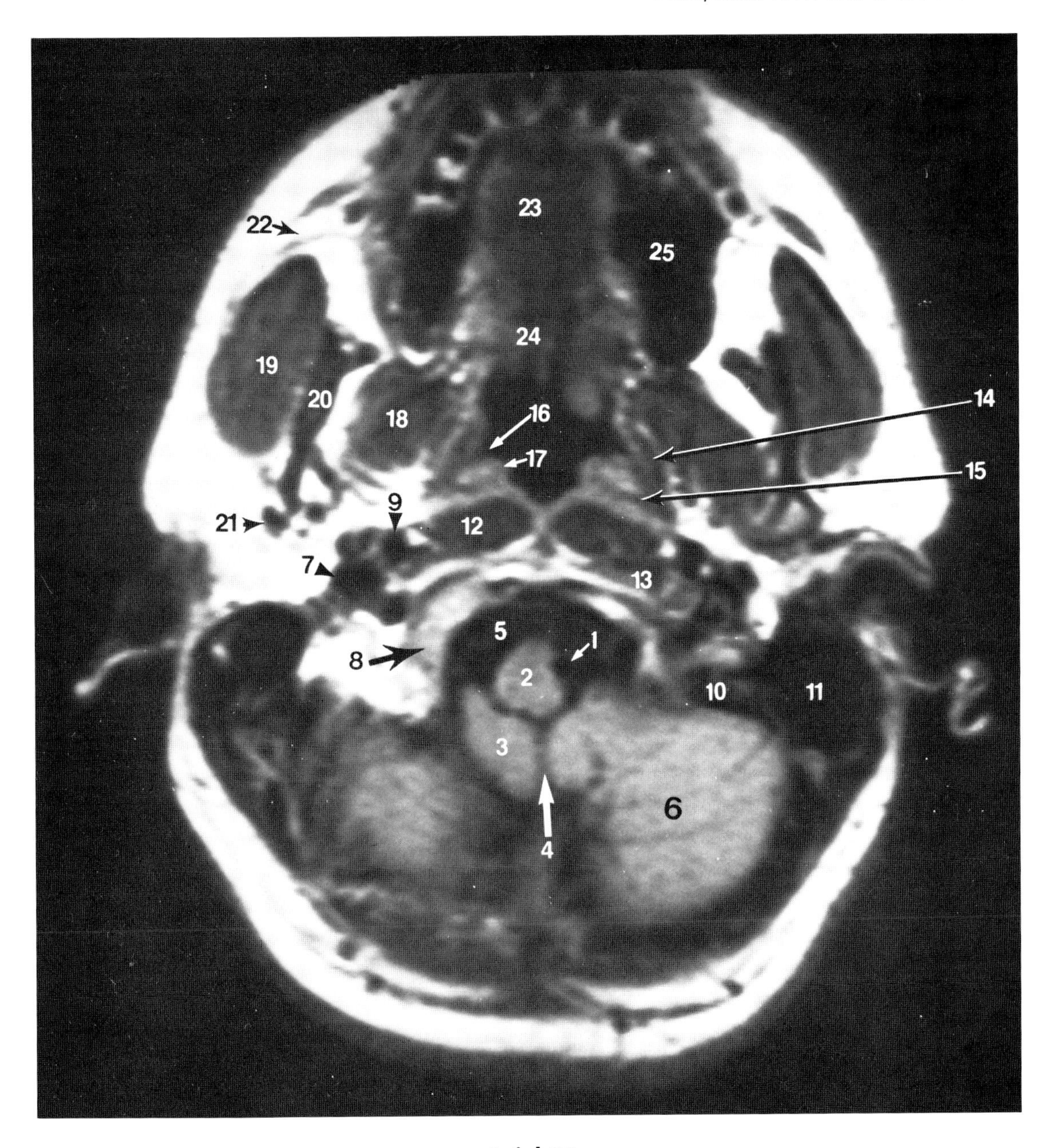

Axial 12

1. Vertebral artery
2. Medulla
3. Cerebellar tonsil
4. Vallecula
5. Foramen magnum
6. Cerebellar hemisphere
7. Internal jugular vein
8. Occipital condyle
9. Internal carotid artery
10. Sigmoid sinus
11. Mastoid air cells
12. Longus capitis muscle
13. Rectus capitis anterior muscle
14. Tensor veli palatini muscle
15. Levator veli palatini muscle
16. Eustachian tube opening
17. Torus tubarius
18. Medial pterygoid muscle
19. Masseter muscle
20. Mandibular ramus
21. Facial artery in parotid gland
22. Zygomatic bone
23. Tongue
24. Soft palate
25. Maxillary sinus

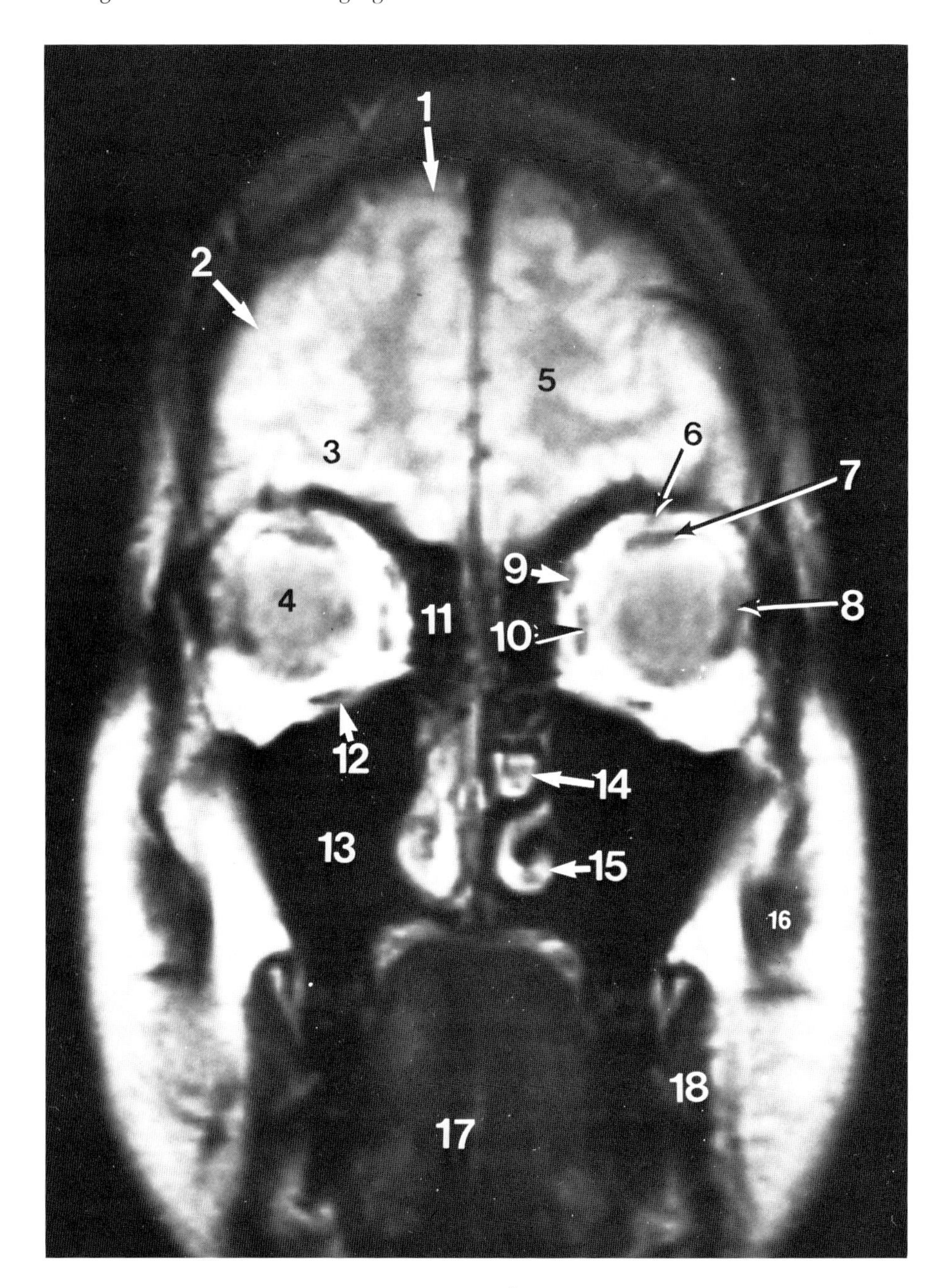

Coronal 1

1. Superior frontal gyrus
2. Middle frontal gyrus
3. Orbital gyri
4. Globe
5. Frontal pole
6. Frontal nerve
7. Superior rectus/levator palpebrae superioris muscle complex
8. Lateral rectus muscle
9. Superior oblique muscle
10. Medial rectus muscle
11. Ethmoid sinus
12. Inferior rectus muscle
13. Maxillary sinus
14. Middle turbinate
15. Inferior turbinate
16. Temporalis muscle
17. Tongue
18. Buccinator muscle

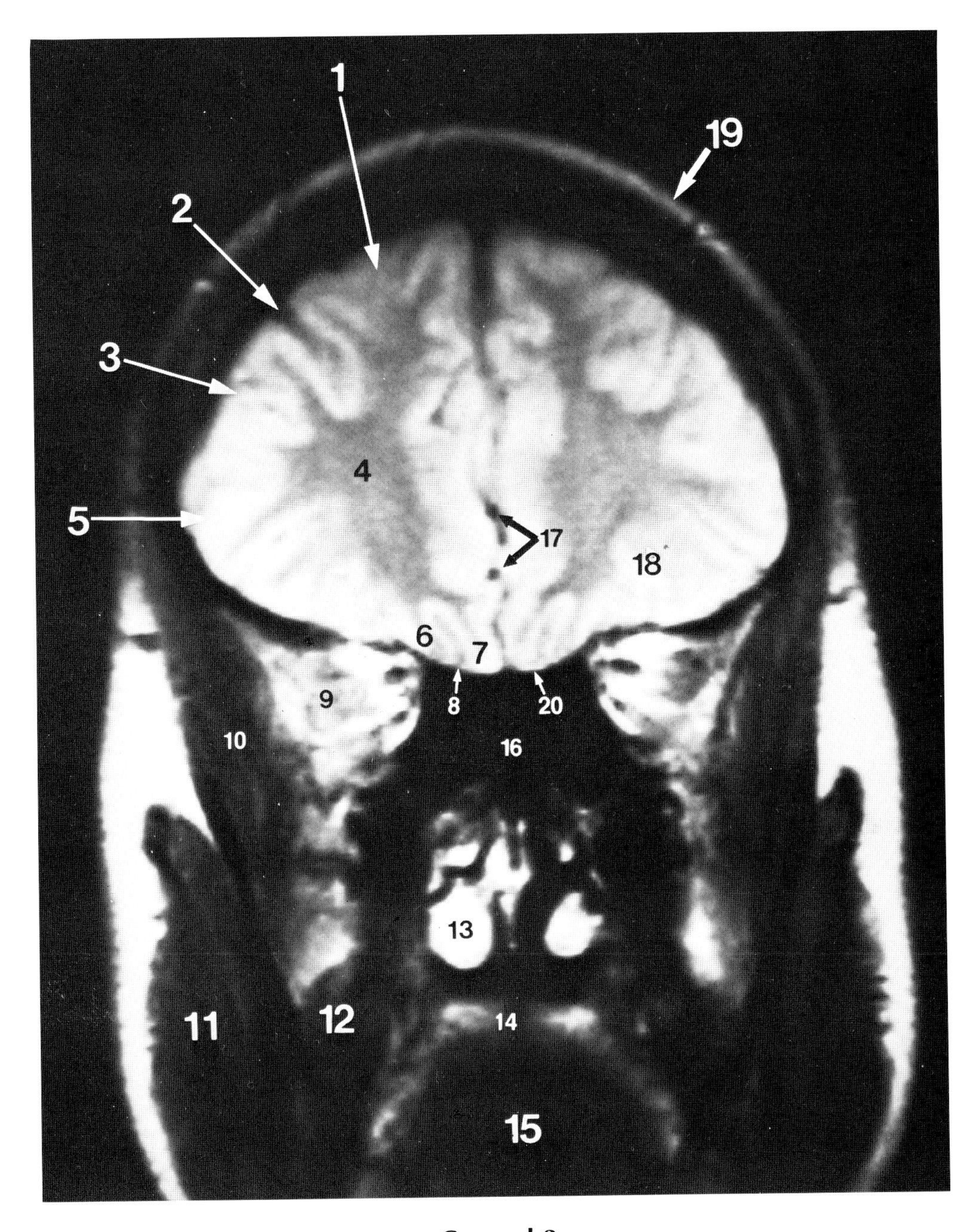

Coronal 2

1. Superior frontal gyrus
2. Superior frontal sulcus
3. Middle frontal gyrus
4. Frontal lobe white matter
5. Inferior frontal gyrus
6. Medial orbital gyrus
7. Gyrus rectus
8. Olfactory sulcus
9. Lateral rectus muscle
10. Temporalis muscle
11. Masseter muscle
12. Medial pterygoid muscle
13. Inferior turbinate
14. Soft palate
15. Tongue
16. Sphenoid sinus
17. Anterior cerebral artery branches in interhemispheric fissure
18. Lateral orbital gyri
19. Scalp fat
20. Olfactory nerve

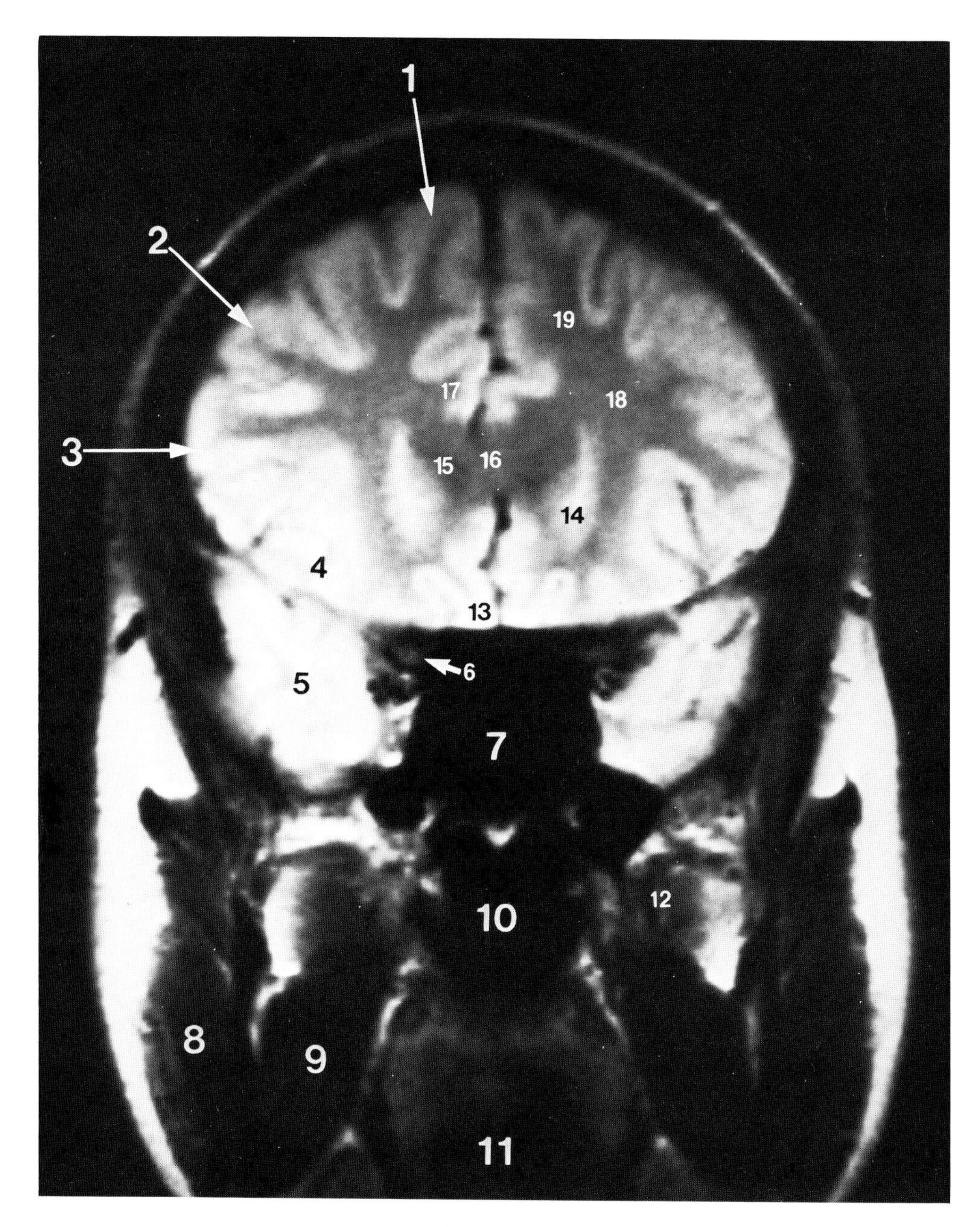

Coronal 3

1. Superior frontal gyrus
2. Middle frontal gyrus
3. Inferior frontal gyrus
4. Lateral orbital gyri
5. Temporal pole
6. Optic nerve
7. Sphenoid sinus
8. Masseter muscle
9. Medial pterygoid muscle
10. Nasopharynx
11. Tongue
12. Lateral pterygoid muscle
13. Gyrus rectus
14. Caudate nucleus
15. Frontal horn
16. Genu of corpus callosum
17. Cingulate gyrus
18. Frontal lobe white matter
19. Superior longitudinal fasciculus

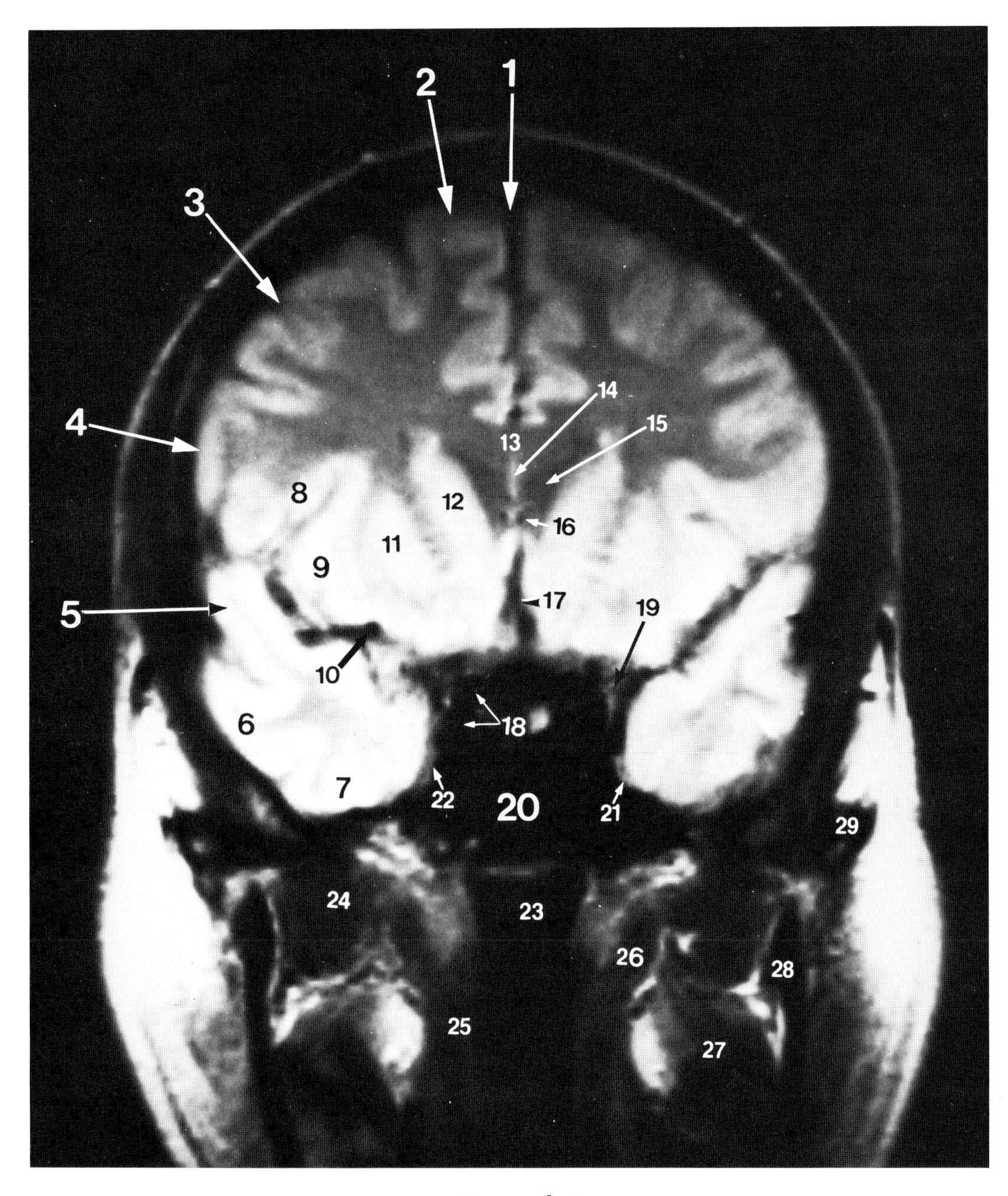

Coronal 4

1. Interhemispheric fissure
2. Superior frontal gyrus
3. Middle frontal gyrus
4. Inferior frontal gyrus
5. Superior temporal gyrus
6. Middle temporal gyrus
7. Inferior temporal gyrus
8. Opercular portion frontal lobe
9. Insula
10. Middle cerebral artery in lateral fissure
11. Putamen
12. Caudate
13. Corpus callosum
14. Septum pellucidum
15. Frontal horn of lateral ventricle
16. Septal vein
17. Interhemispheric fissure
18. Internal carotid artery
19. Cavernous sinus
20. Sphenoid sinus
21. Maxillary nerve (V_2)
22. Mandibular nerve (V_3)
23. Nasopharynx
24. Lateral pterygoid muscle
25. Pharyngeal constrictors
26. Levator veli paltini muscle
27. Medial pterygoid muscle
28. Mandible
29. Masseter muscle

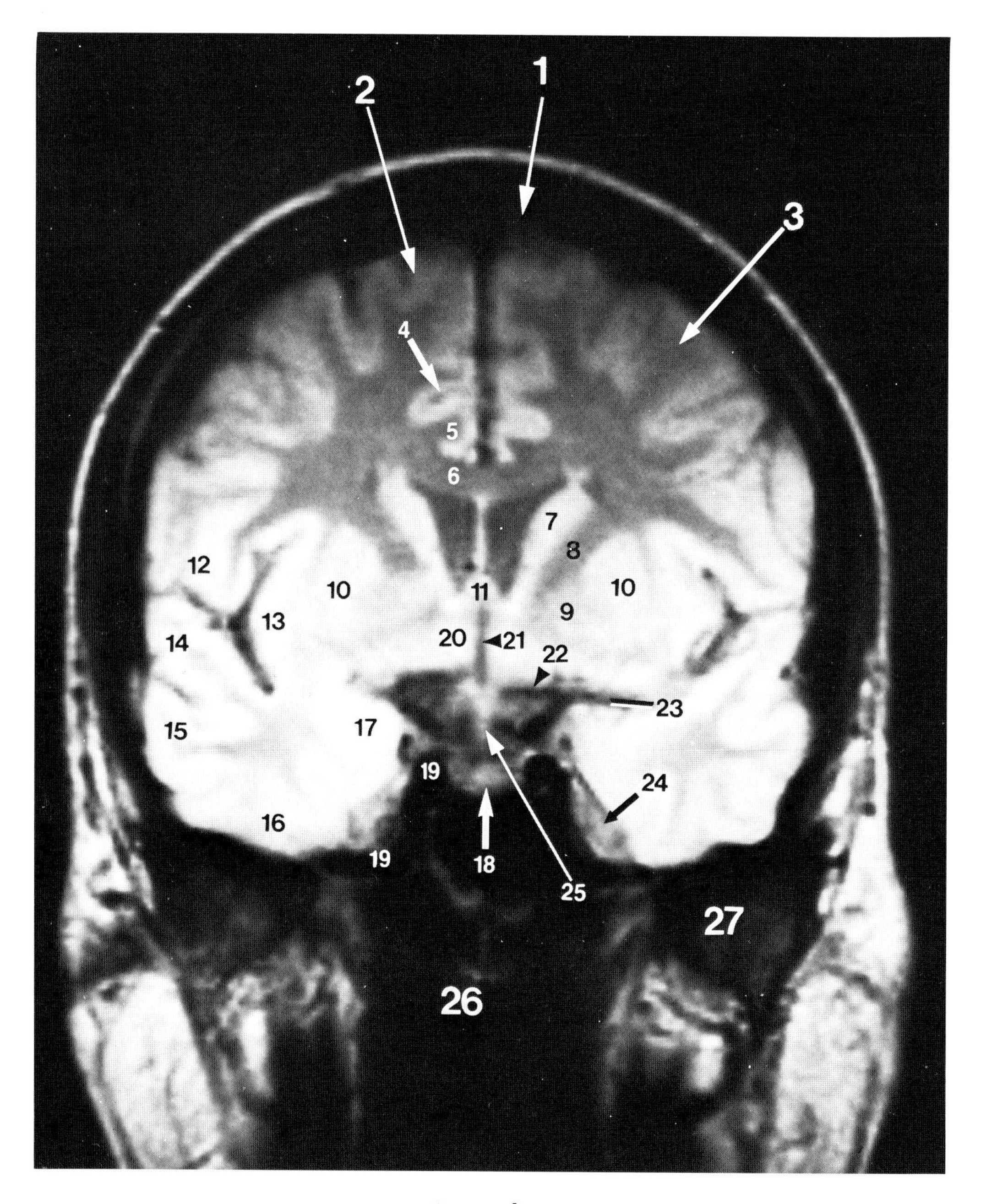

Coronal 5

1. Pacchionian granulation
2. Superior frontal gyrus
3. Middle frontal gyrus
4. Cingulate sulcus
5. Cingulate gyrus
6. Corpus callosum
7. Caudate nucleus
8. Internal capsule
9. Globus pallidus
10. Putamen
11. Fornix
12. Frontal operculum
13. Insula
14. Superior temporal gyrus
15. Middle temporal gyrus
16. Inferior temporal gyrus
17. Uncus
18. Pituitary
19. Internal carotid artery
20. Hypothalamus
21. Third ventricle
22. A1 segment anterior cerebral artery
23. M1 segment middle cerebral artery
24. Gasserian ganglion
25. Infundibulum
26. Nasopharynx
27. Lateral pterygoid muscle

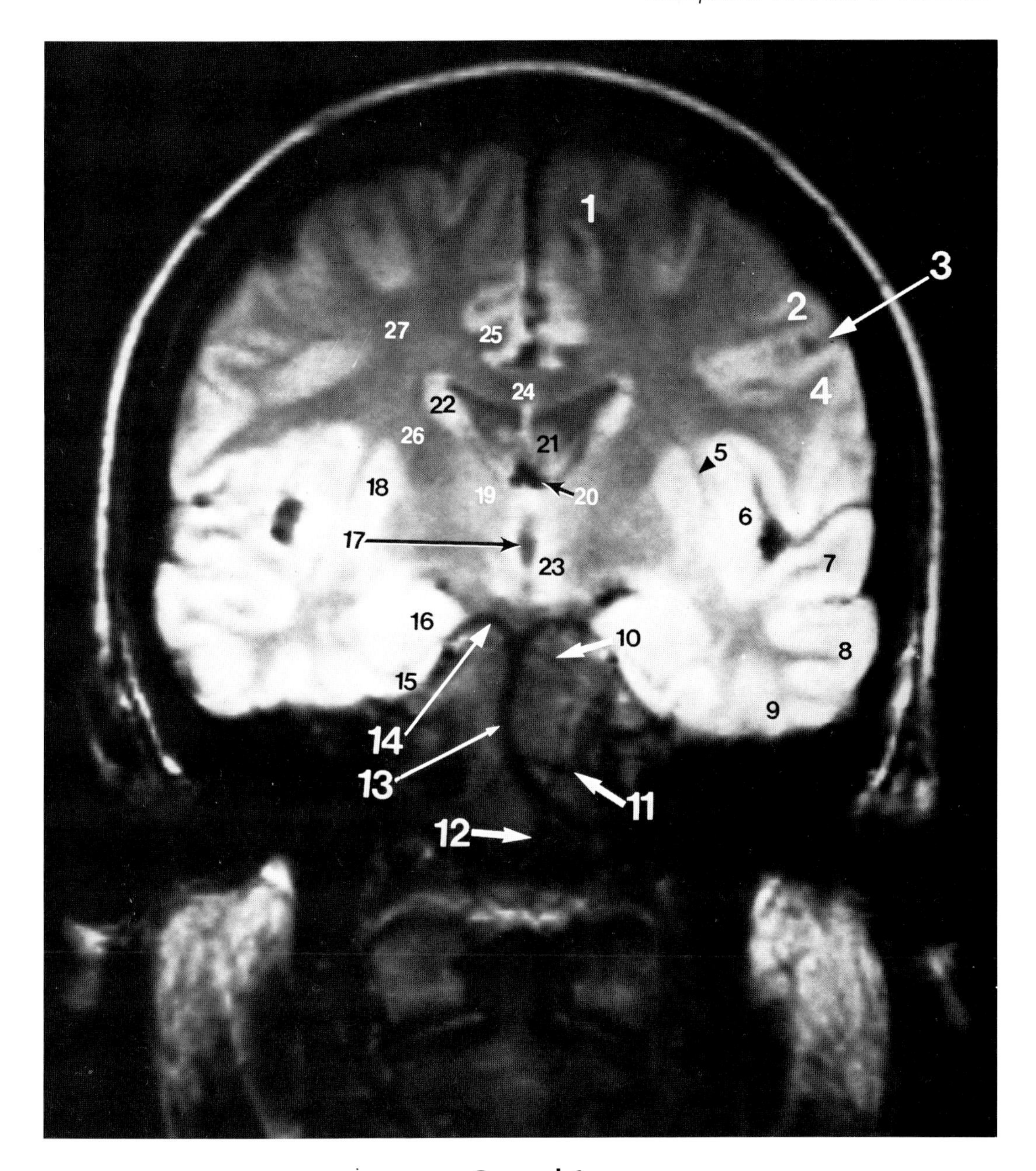

Coronal 6

1. Paracentral lobule
2. Precentral gyrus
3. Central sulcus
4. Postcentral sulcus
5. External capsule
6. Insula
7. Superior temporal gyrus
8. Middle temporal gyrus
9. Inferior temporal gyrus
10. Superior cerebellar artery
11. Anterior inferior cerebellar artery
12. Vertebral artery
13. Basilar artery
14. Posterior cerebral artery
15. Parahippocampal gyrus
16. Uncus
17. Third ventricle
18. Putamen
19. Anterior thalamic nuclei
20. Internal cerebral veins
21. Choroid plexus in lateral ventricle
22. Caudate nucleus
23. Hypothalamus
24. Corpus callosum
25. Cingulate gyrus
26. Anterior limb of internal capsule
27. Centrum semiovale

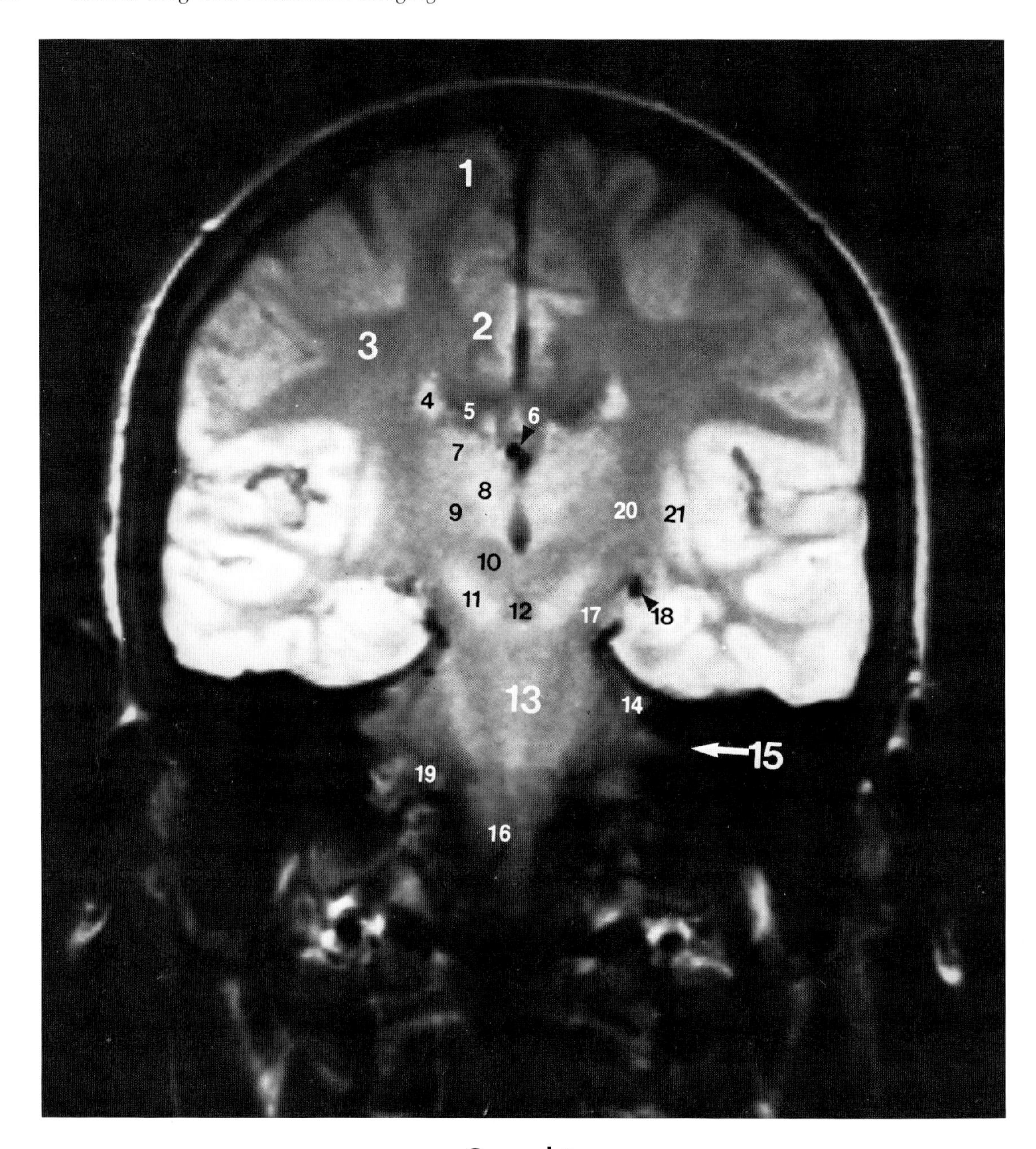

Coronal 7

1. Paracentral lobule
2. Cingulate gyrus
3. Centrum semiovale
4. Caudate nucleus
5. Lateral ventricle
6. Internal cerebral veins in roof of third ventricle
7. Lateral posterior thalamic nucleus
8. Dorsomedial thalamic nucleus
9. Ventral posterolateral thalamic nucleus
10. Red nucleus
11. Substantia nigra
12. Decussation of superior cerebellar peduncle
13. Pons
14. Trigeminal nerve
15. Vestibulocochlear and facial nerves
16. Olive
17. Cerebral peduncle
18. Posterior cerebral artery in choroidal fissure
19. Flocculus
20. Internal capsule
21. Putamen

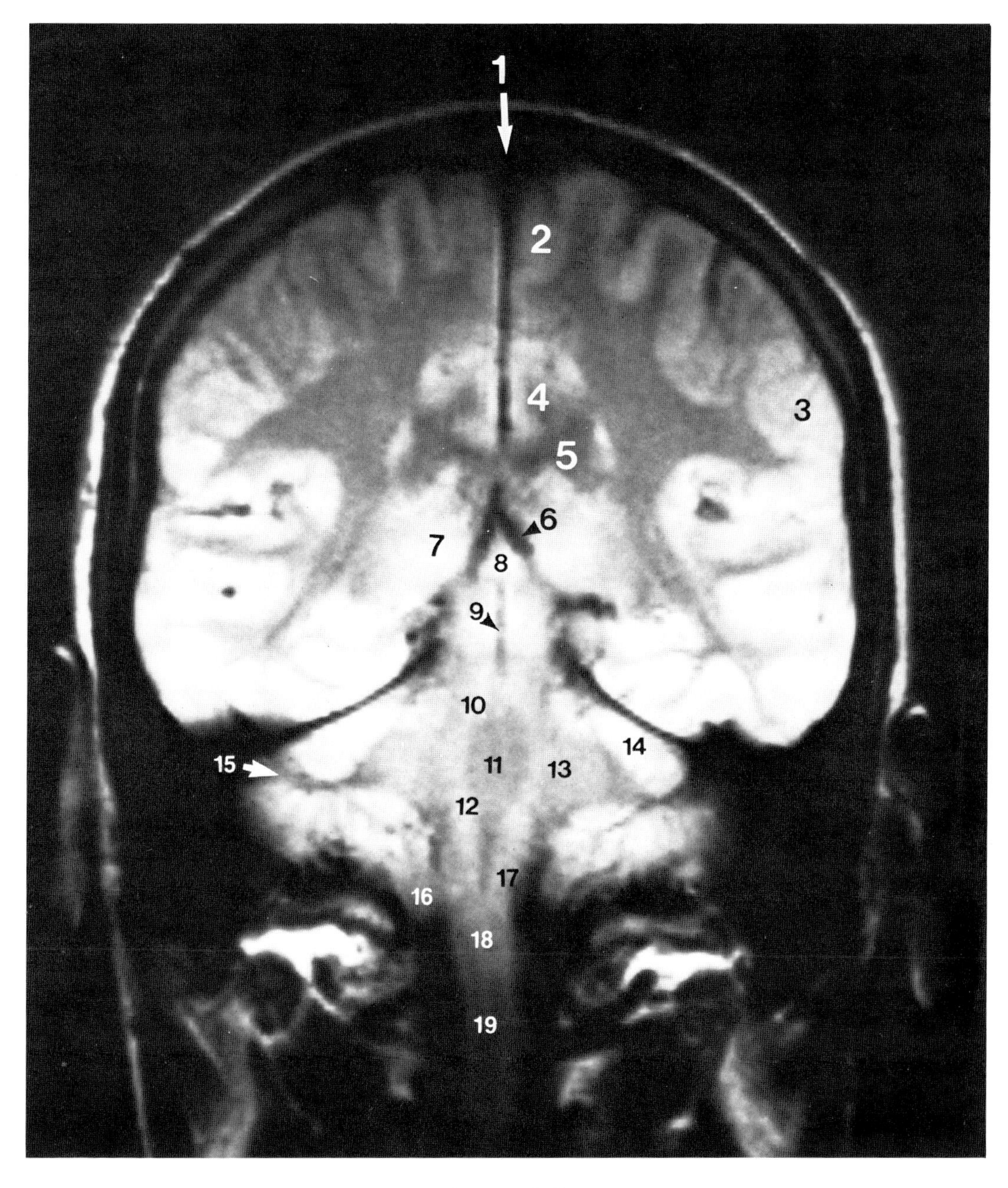

Coronal 8

1. Superior sagittal sinus
2. Paracentral lobule
3. Supramarginal gyrus
4. Cingulate gyrus
5. Lateral ventricle
6. Basal vein of Rosenthal
7. Pulvinar of thalamus
8. Pineal
9. Cerebral aqueduct
10. Superior cerebellar peduncle
11. Fourth ventricle
12. Inferior cerebellar peduncle
13. Middle cerebellar peduncle
14. Superior cerebellar hemisphere
15. Horizontal fissure of cerebellum
16. Tonsil
17. Inferior olive
18. Medulla
19. Cervical cord

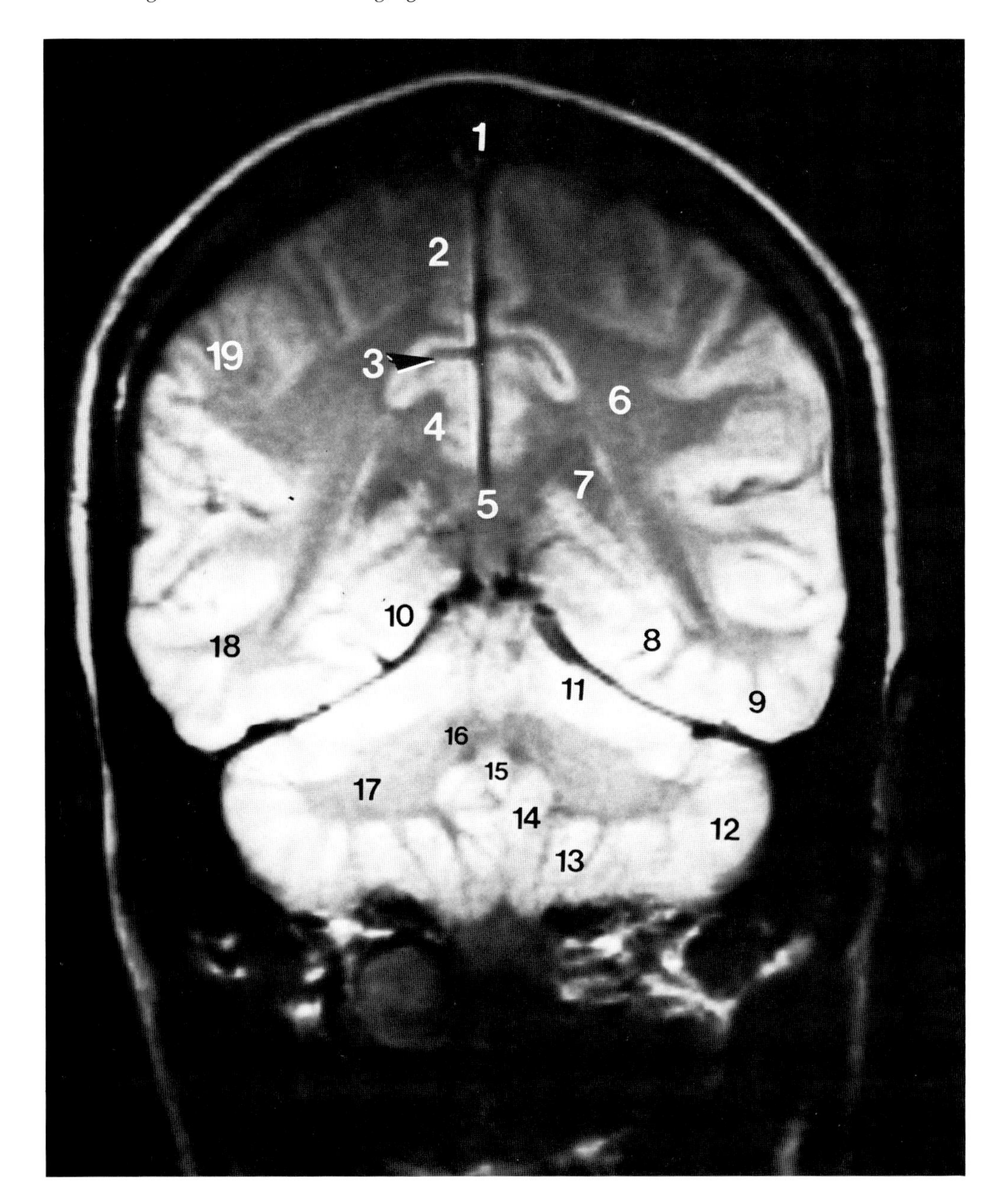

Coronal 9

1. Superior sagittal sinus
2. Precuneus
3. Cingulate sulcus
4. Cingulate gyrus
5. Splenium of corpus callosum
6. Centrum semiovale
7. Posterior horn of lateral ventricle
8. Collateral sulcus
9. Inferior temporal gyrus
10. Hippocampus
11. Superior semilunar lobule
12. Inferior semilunar lobule
13. Biventral lobule
14. Tonsil
15. Nodule of vermis
16. Dentate nucleus
17. Corpus medullare
18. Temporal lobe
19. Parietal lobe

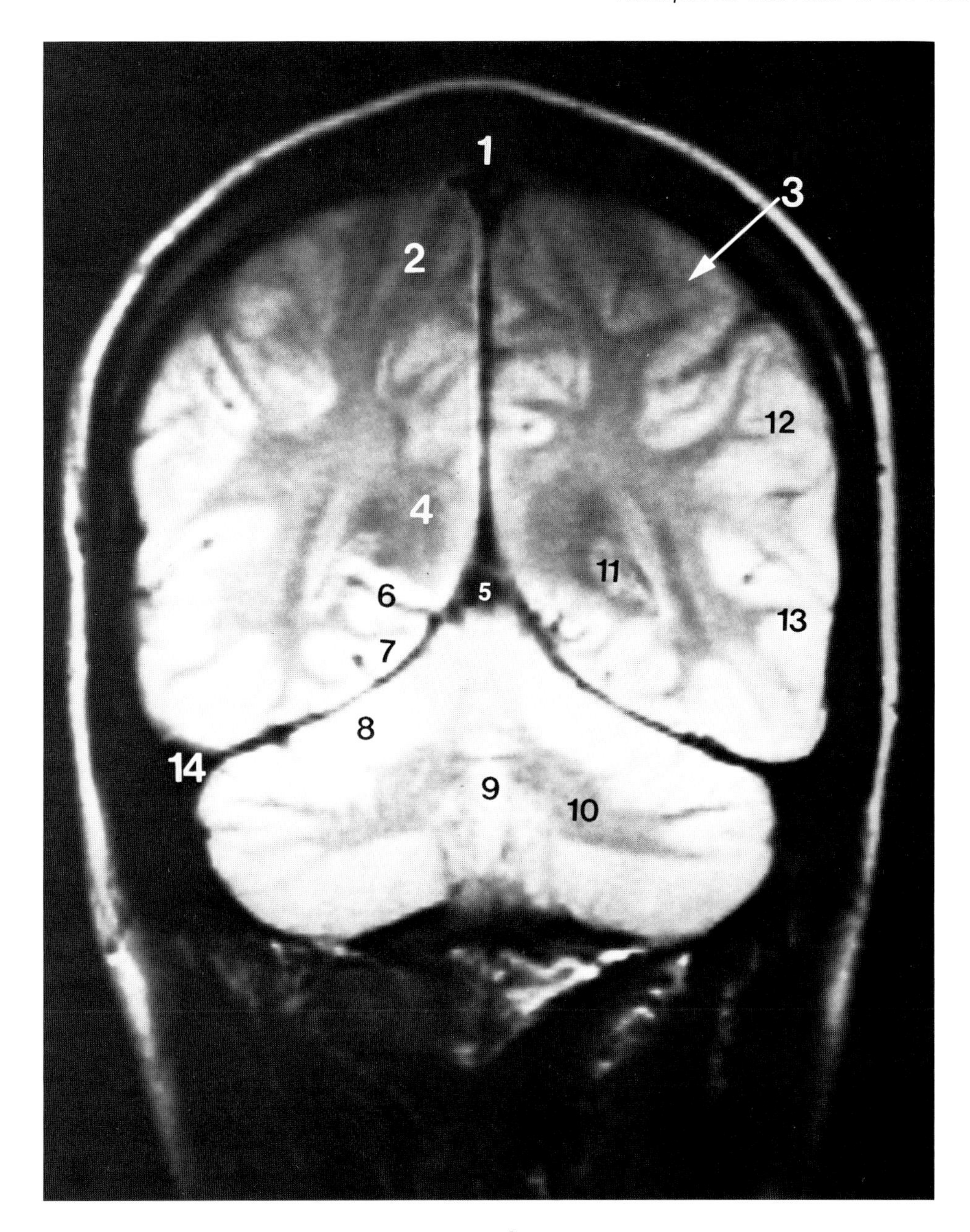

Coronal 10

1. Superior sagittal sinus
2. Precuneus
3. Superior parietal lobule
4. Forceps minor
5. Superior cerebellar veins
6. Calcarine fissure and cortex
7. Lingual gyrus
8. Superior cerebellar hemisphere
9. Vermis
10. Corpus medullare
11. Choroid plexus of occipital horn
12. Inferior parietal lobule
13. Superior temporal gyrus
14. Transverse sinus

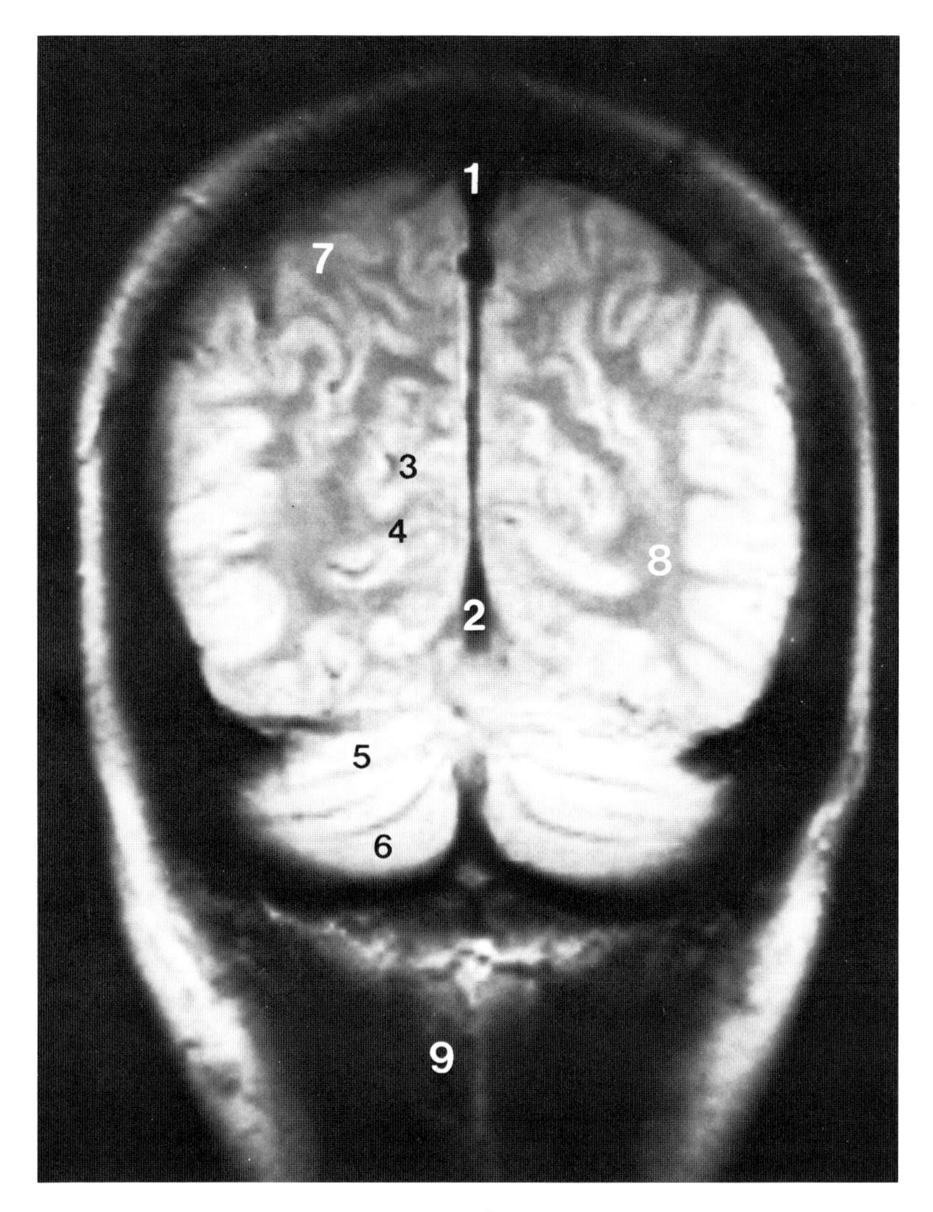

Coronal 11

1. Superior sagittal sinus
2. Straight sinus
3. Parieto-occipital sulcus
4. Calcarine sulcus and cortex
5. Superior semilunar lobule
6. Inferior semilunar lobule
7. Superior parietal lobule
8. Occipital pole
9. Semispinalis capitis muscle

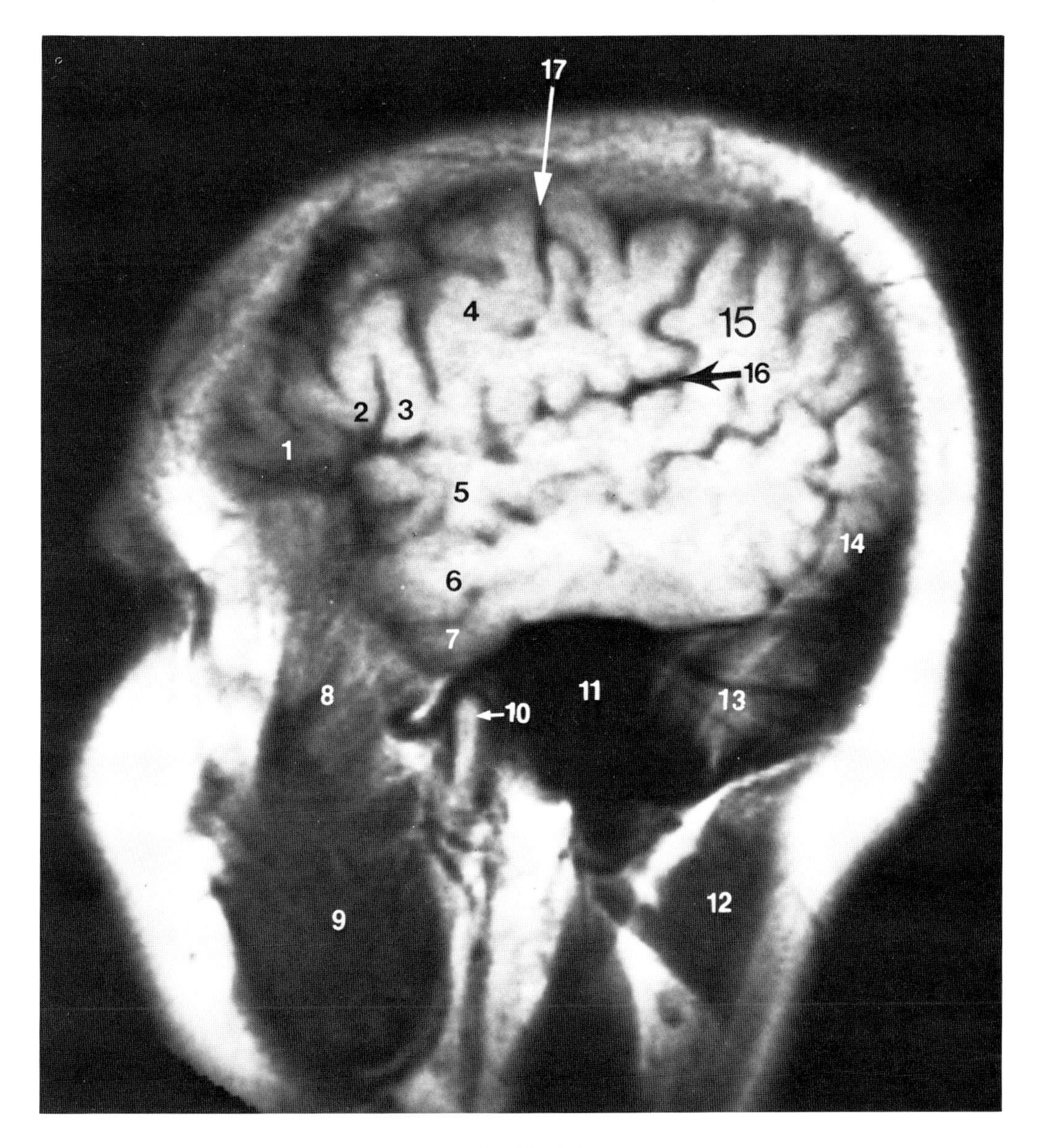

Sagittal 1

1. Inferior frontal gyrus (pars orbitalis)
2. Inferior frontal gyrus (pars triangularis)
3. Inferior frontal gyrus (pars opercularis)
4. Precentral gyrus
5. Superior temporal gyrus
6. Middle temporal gyrus
7. Inferior temporal gyrus
8. Temporalis muscle
9. Masseter muscle
10. Mandibular condyle
11. Temporal bone
12. Splenius capitis muscle
13. Cerebellum
14. Lateral occipital gyrus
15. Superior parietal lobule
16. Sylvian fissure
17. Central sulcus

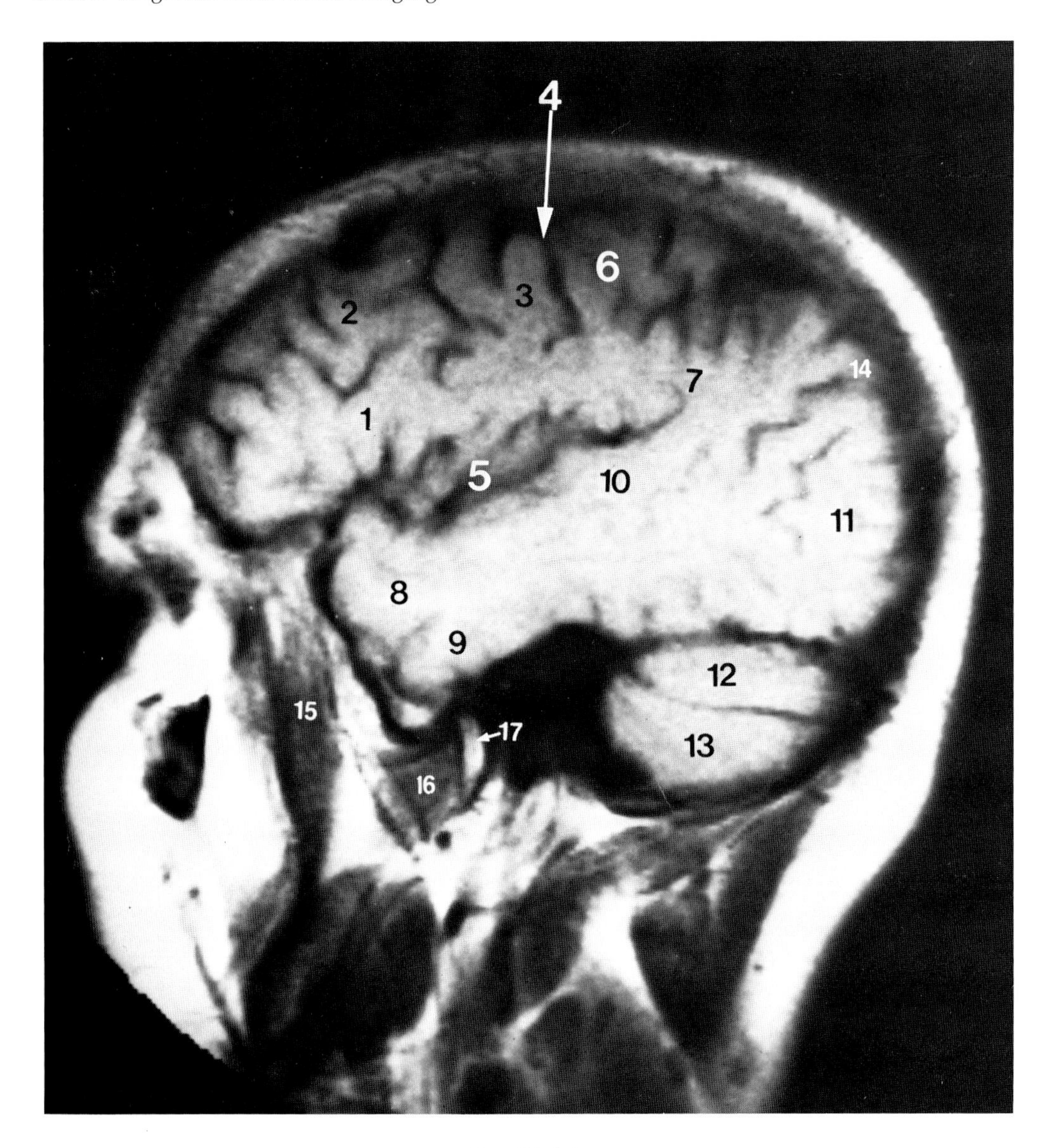

Sagittal 2

1. Inferior frontal gyrus
2. Middle frontal gyrus
3. Precentral gyrus
4. Central sulcus
5. Insula
6. Postcentral gyrus
7. Supramarginal gyrus
8. Middle temporal gyrus
9. Inferior temporal gyrus
10. Superior temporal gyrus
11. Lateral occipital gyri
12. Superior semilunar lobule
13. Inferior semilunar lobule
14. Angular gyrus
15. Temporalis muscle
16. Lateral pterygoid muscle
17. Mandible

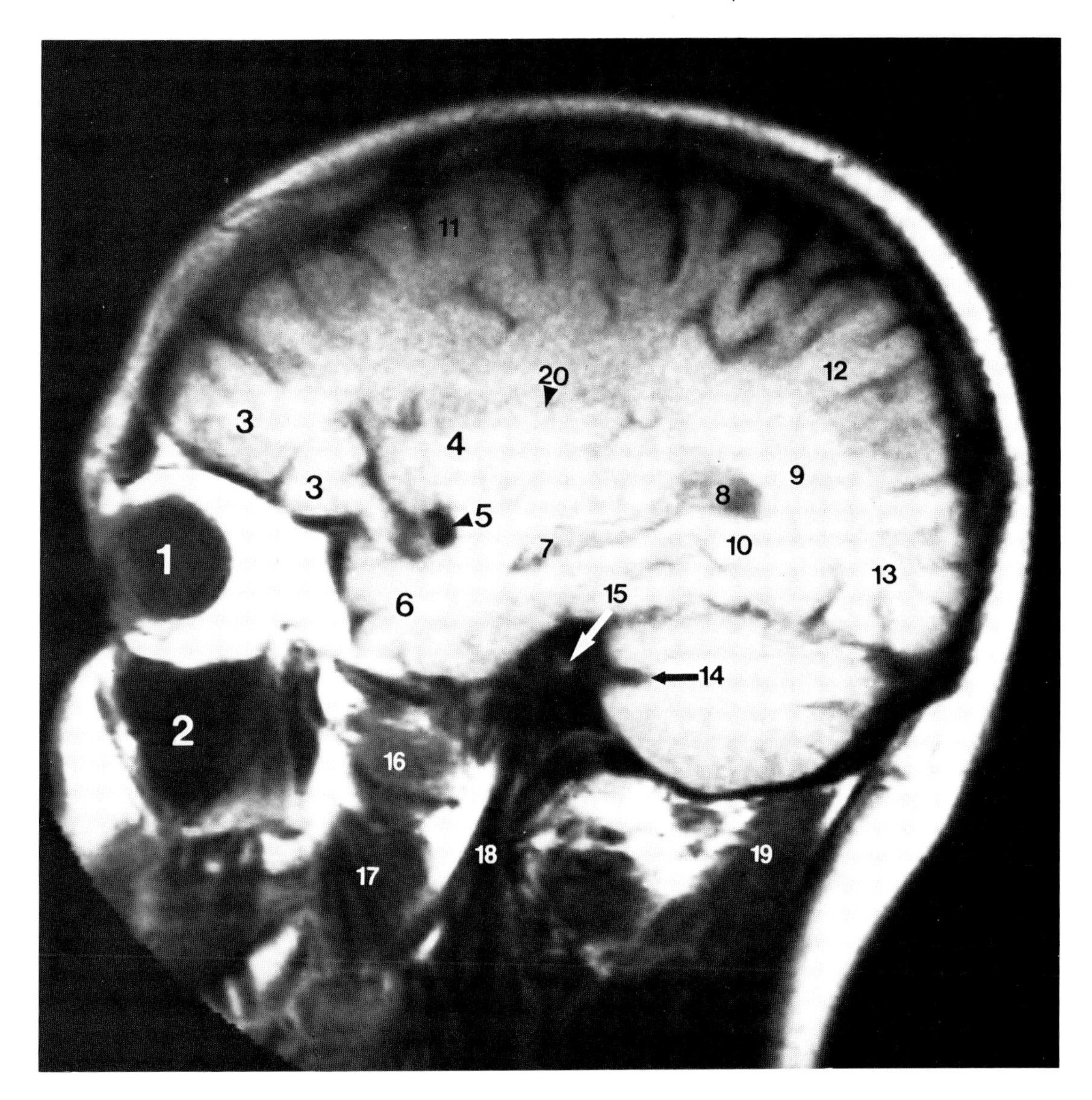

Sagittal 3

1. Globe
2. Maxillary sinus
3. Orbital gyri or frontal lobe
4. Insula
5. Middle cerebral artery
6. Temporal pole
7. Temporal horn of lateral ventricle
8. Atrium of lateral ventricle
9. Parietal visual radiations
10. Temporal visual radiations
11. Superior frontal gyrus
12. Superior parietal lobule
13. Lateral occipital gyri
14. Horizontal fissure of cerebellum
15. Neurovascular bundle in internal auditory canal
16. Lateral pterygoid muscle
17. Medial pterygoid muscle
18. Internal jugular vein
19. Semispinalis capitis muscle
20. Circular sulcus

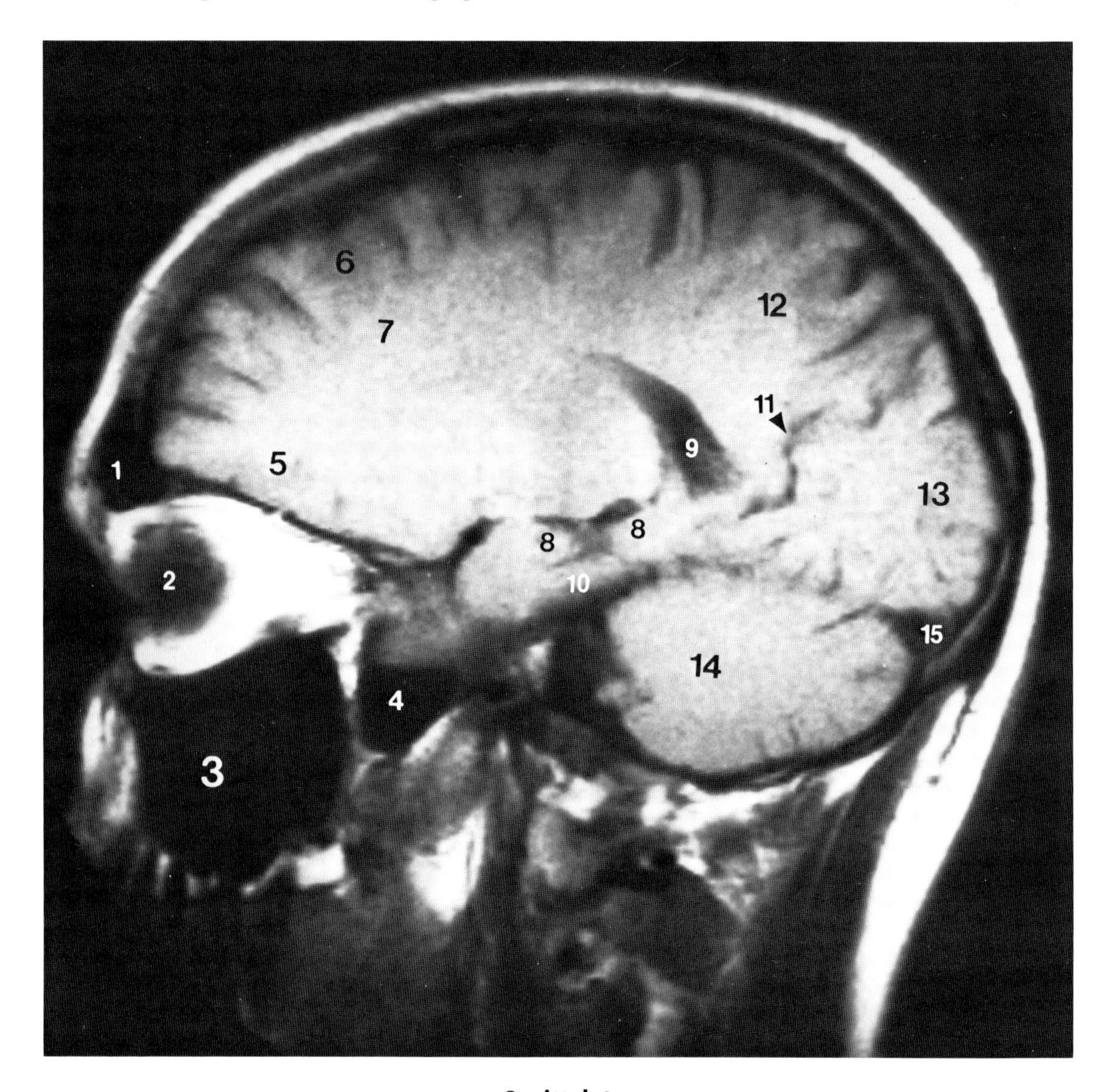

Sagittal 4

1. Frontal sinus
2. Globe
3. Maxillary sinus
4. Sphenoid sinus
5. Orbital gyri
6. Superior frontal gyrus
7. Centrum semiovale
8. Hippocampus
9. Atrium of lateral ventricle
10. Parahippocampal gyrus
11. Parieto-occipital sulcus
12. Parietal lobe
13. Occipital lobe
14. Corpus medullare cerebelli
15. Transverse sinus

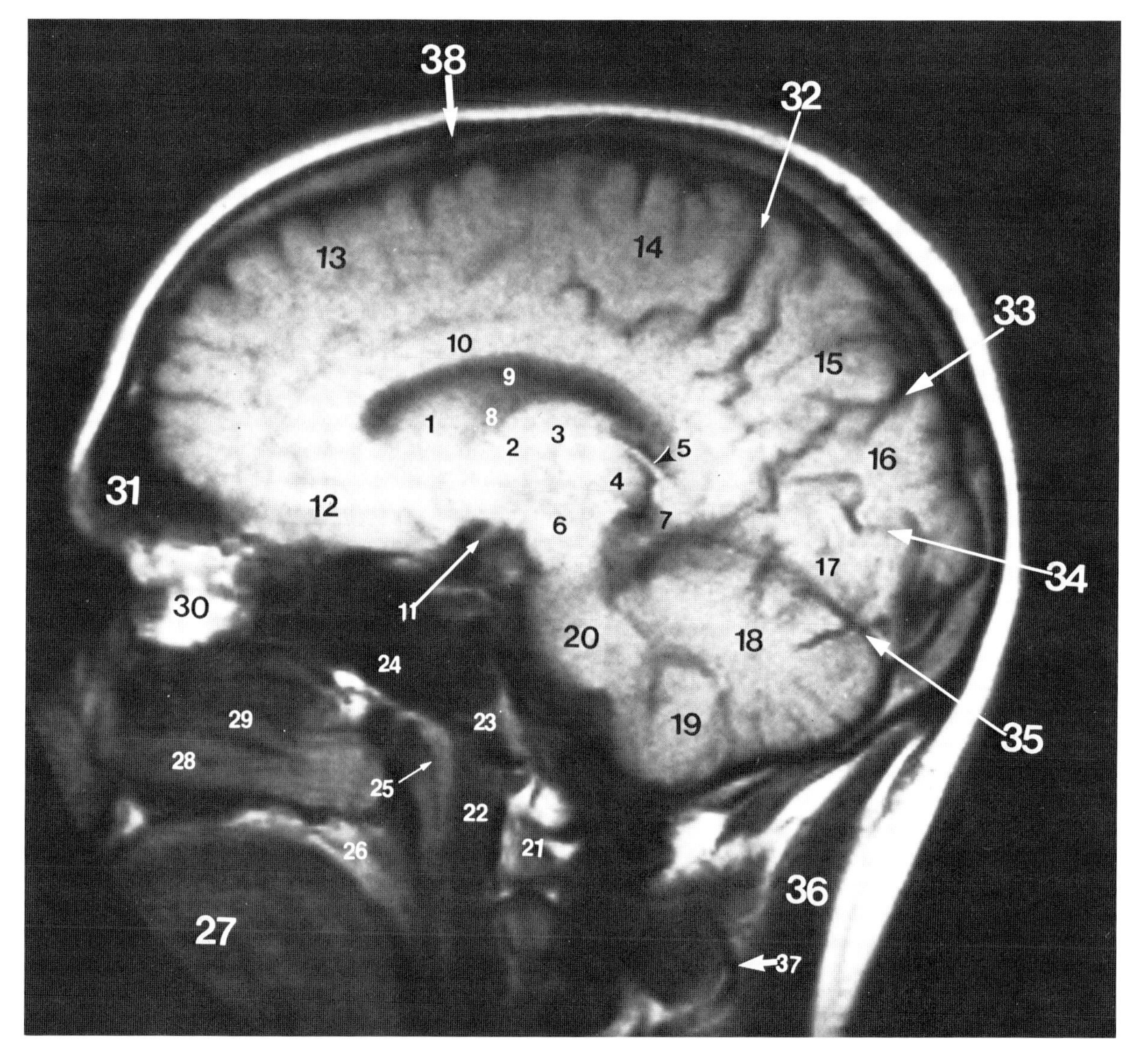

Sagittal 5

1. Caudate nucleus
2. Ventral thalamic nuclei
3. Lateral thalamic nuclei
4. Pulvinar
5. Fornix (fimbria)
6. Cerebral peduncle
7. Isthmus of cingulate gyrus
8. Thalamocaudate groove
9. Lateral ventricle
10. Corpus callosum
11. Optic tract
12. Medial orbital gyri
13. Superior frontal gyrus
14. Paracentral lobule
15. Precuneus
16. Cuneus
17. Lingual gyrus
18. Cerebellar hemisphere
19. Cerebellar tonsil
20. Pons
21. Body of C_2
22. Longus capitis muscle
23. Basiocciput (clivus)
24. Sphenoid sinus
25. Pharyngeal recess
26. Soft palate
27. Tongue
28. Inferior turbinate
29. Middle turbinate
30. Orbital fat
31. Frontal sinus
32. Marginal sulcus
33. Parieto-occipital sulcus
34. Calcarine sulcus
35. Tentorium
36. Splenius capitus muscle
37. Vertebral artery
38. Coronal suture

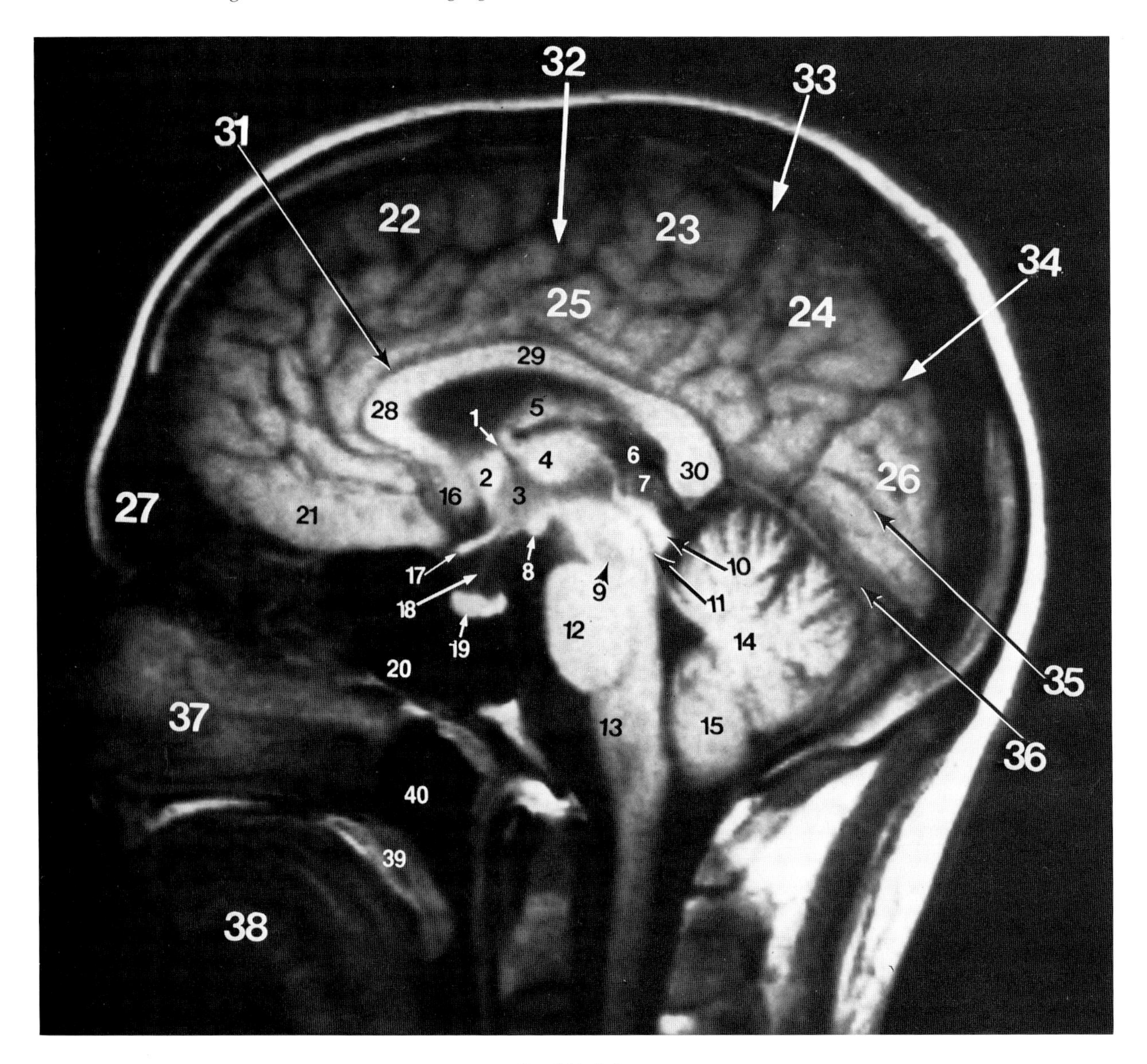

Sagittal 6

1. Foramen of Monro
2. Anterior commissure
3. Third ventricle
4. Midline nuclei of thalamus
5. Fornix
6. Cistern of velum interpositum
7. Pineal
8. Mamillary body
9. Decussation of superior cerebellar peduncle
10. Inferior colliculus
11. Cerebral aqueduct
12. Pons
13. Olive
14. Cerebellar vermis
15. Cerebellar tonsil
16. Parolfactory gyri
17. Optic chiasm
18. Infundibulum
19. Pituitary
20. Sphenoid sinus
21. Gyrus rectus
22. Superior frontal gyrus
23. Paracentral lobule
24. Precuneus
25. Cingulate gyrus
26. Cuneus
27. Frontal sinus
28. Corpus callosum (genu)
29. Corpus callosum (body)
30. Corpus callosum (splenium)
31. Callosal sulcus
32. Cingulate sulcus
33. Cingulate sulcus (pars marginalis)
34. Parieto-occipital sulcus
35. Calcarine sulcus
36. Tentorium
37. Nasal cavity
38. Tongue
39. Soft palate
40. Nasopharynx

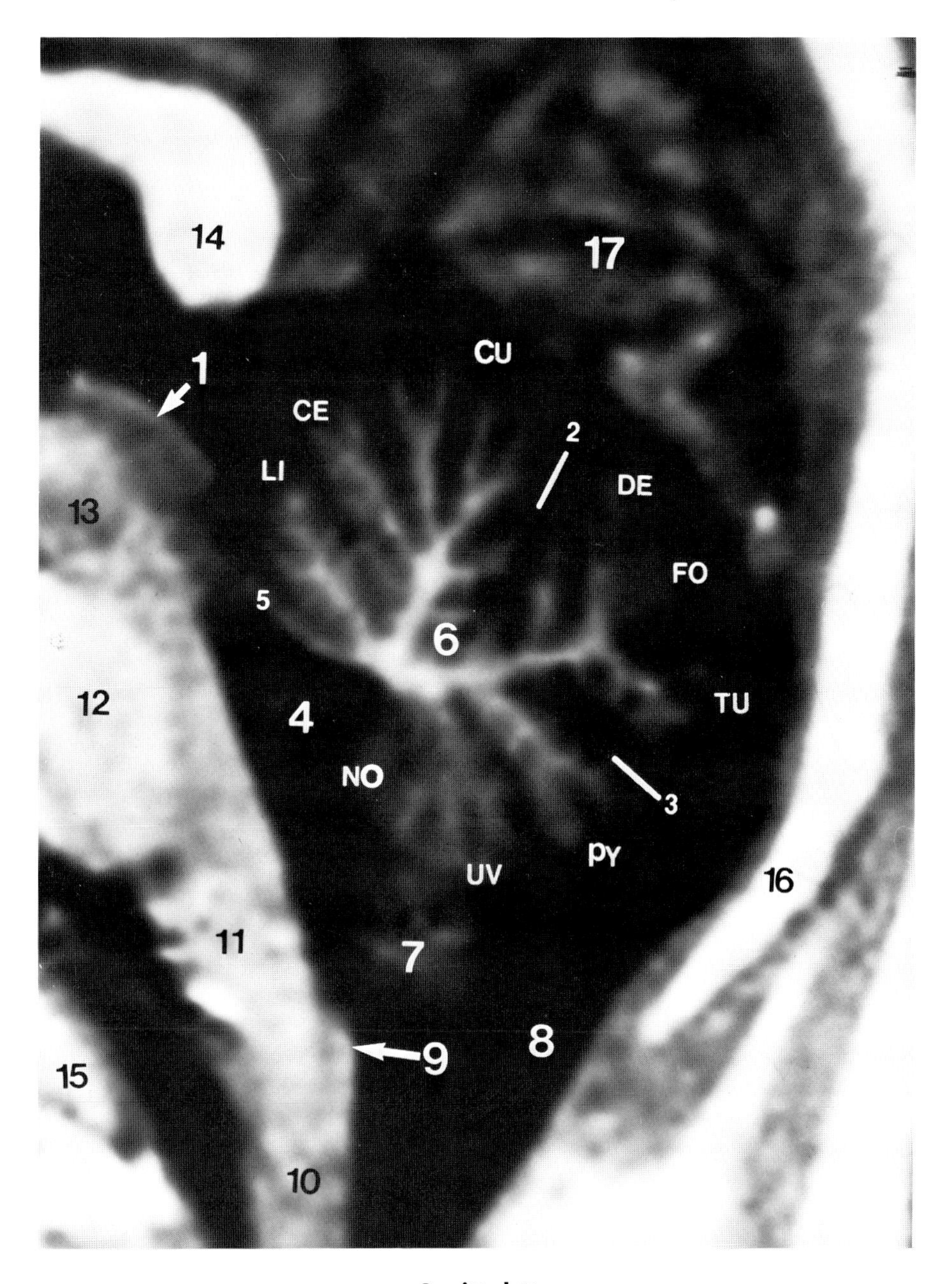

Sagittal 7

1. Quadrigeminal plate
2. Primary fissure of cerebellum
3. Prepyramidal fissure of cerebellum
4. Fourth ventricle
5. Superior medullary velum
6. Cerebellar vermis (lobes)
   LI, Lingula; CE, Central lobule; CU, Culmen; DE, Declive; FO, Folium; TU, Tuber; PY, Pyramid; UV, Uvula; NO, Nodule
7. Cerebellar tonsil
8. Foramen magnum
9. Gracile tubercle
10. Cervical cord
11. Medulla
12. Pons
13. Midbrain
14. Splenium of corpus callosum
15. Clivus
16. Occipital bone
17. Occipital lobe (cuneus)

Appendix B
Protocols for Cranial MRI

The following protocols were developed by the author and have been used successfully at the Bowman Gray School of Medicine since 1986. They are especially efficient for use on Picker Vista MR scanners operating at 0.5 T. It should be noted that these protocols may not be particularly optimal when operating at other field strengths or using another brand of scanner. Each institution should seek to develop its own protocols tailored to the needs of its patient population and the type of MR scanner. For a set of alternative protocols, the reader is referred to the excellent monograph by Haiken, Glazer, Lee et al.).

In addition to selection of imaging plane, pulse sequence, and timing parameter, several other options must be specified prior to scanning. Some of these are listed below.

1. *Matrix size:* Options available for dimensions of the image acquisition matrix on most scanners vary from 128 × 128 to 512 × 512. Larger dimension matrices produce higher spatial resolution images but require longer imaging time to achieve it. As a compromise for most cranial imaging applications a 128 × 256, 256 × 256, or 256 × 512 matrix is usually selected.
2. *Field of view (FOV)*: The field of view should be made as small as possible to encompass the anatomic region of interest. For whole brain imaging on adult patients a FOV of 20 to 25 cm (spherical) is usually ideal. When interest is only directed to a limited area, such as the brain stem or pituitary gland, the FOV should be made much smaller, say 15–20 cm. If the region of interest is larger than the selected FOV, a wrap around artifact may occur obscuring part of the image.
3. *Selection of phase and frequency gradient directions:* In general two-dimensional Fourier Transform imaging one imaging axis is encoded by phase and the other by frequency, each under operator control. Because most motion artifacts tend to occur along the phase encoded axis, the selection of this axis should be carefully chosen. For general cranial imaging the phase encoded axis should be the horizontal (x-axis). However, in specific situations (e.g., evaluating a temporal lobe lesions on coronal section, looking at the lateral wall of the orbit on axial

section) choosing the phase encoded axis to be the vertical (y-axis) will help throw motion artifacts away form the region of interest.

4. *Number of signal averages:* Increasing the number of signal averages is the single best way to improve signal-to-noise ratio in cranial MRI. However, this also results in prolongation of imaging time directly. When high field strength (1.5 T) systems are used, often one excitation is sufficient. With lower strength systems, however, at least two excitations (one average) will significantly improve image quality.
5. *Slice thickness:* For routine cranial imaging, contiguous 5 to 10 mm slices are generally preferred. For areas where anatomic detail is needed (e.g., temporal bone, sella, brain stem) thinner slices (2 to 4 mm) are recommended.

Protocols for various clinical situations follow:

Routine Brain Protocol

Uses: For patients with ill-defined symptoms and no hard neurologic signs (e.g., dizziness, headache, confusion, behavioral disorders)
Protocol: (1) Axial plane SE 2500/40,80
(2) Sagittal plane, SE 400/25

Suspected Lesion of Cerebral Hemisphere

Uses: Possible stroke
Possible tumor
Characterize hemispheric lesions seen on CT
Intracerebral hemorrhage
Seizure work-up
Protocol: (1) Axial plane SE 2500/100
(2) Coronal plane SE 1750/40,80
(3) Coronal plane SE 400/25 or IR 1500/400

Suspected Multifocal Disease

Uses: Possible multiple sclerosis, demyelinating disease
Rule out metastases
Infections/inflammatory diseases
Multi-infarct dementia
Protocol: (1) Axial plane SE 2500/100
(2) Coronal plane SE 1750/40,80
(3) Sagittal plane SE 400/25

Pituitary Lesions

Uses: Pituitary tumors < 2.5 cm
Protocol: (1) Sagittal plane SE 400/25
(2) Coronal plane SE 1500/40,80
Notes: Use thin slices (2–3 mm)
Use smallest field of view allowed, preferably <20 cm

Juxtasellar Lesions

Uses: Pituitary macroademomas > 2.5 cm
Lesions of optic nerves or chiasm
Other parasellar masses, tumors, aneurysms, granulomas
Protocol: (1) Sagittal plane SE 400/25
(2) Sagittal plane SE 1500/80
(3) Coronal plane SE 2000/40,80

Lesions of the Pineal or Midbrain

Uses: Pineal tumors
Midbrain lesions
Third ventricle
Aqueduct obstruction
Protocol: (1) Sagittal plane SE 400/25
(2) Sagittal plane SE 2000/20,100
(3) Axial plane SE 2500/20,80

Cerebellopontine Angle Mass

Uses: Rule out acoustic neuroma
Evaluate other mass lesions
Protocol: (1) Axial plane SE 1500/30,100
(2) Coronal plane SE 1500/30
(3) Axial plane SE 700/25
Notes: Occasionally one may wish to add a few T1-weighted parasagittal images to delineate brain stem lesions

Posterior Fossa Lesion

Uses: Brain stem tumor
Cerebellar tumor, stroke
Protocol: (1) Sagittal plane SE 500/25
(2) Sagittal plane SE 1500/80
(3) Axial plane SE 2000/40,80
Notes: Occasionally coronal SE 2000/40,80 is helpful

Clivus and Foramen Magnum Regions

Uses: Chiari malformation
Chordoma and extra-axial tumors
Syringomyelia, syringobulbia

Protocol: (1) Sagittal plane SE 500/25
(2) Axial plane SE 500/25
(3) Axial plane SE 2000/40,80

Base of Skull

Uses: Evaluate invasion by sinus carcinoma
Glomus tumors
Neuromas of cranial nerves
Perineural metastases
Protocol: (1) Coronal plane SE 2000/40,80
(2) Axial plane SE 1500/30,60
(3) Sagittal plane SE 500/30
Note: Coronal plane mandatory for evaluating base of skull invasion by carcinoma

Pediatric Anatomy and Myelination

Uses: Evaluate congenital craniocerebral malformations
Evaluate
Protocol: (1) Axial plane SE 400/25
(2) Sagittal plane SE 400/25
(3) Axial plane SE 3500/40,80
Notes: To better evaluate disturbances in myelination, sequence (1) should be replaced by:
IR 3000/1000 (prematures)
IR 2400/800 (0–3 months)
IR 1800/600 (3 month–2 years)
IR 1500/500 (over 2 years)

Orbit and Optic Nerves

Uses: Orbital mass
Lesion of optic nerves or chiasm
Protocol: (1) Coronal SE 500/30
(2) Axial plane SE 500/30
(3) Axial plane SE 1500/40,80
(4) Fat suppression technique, IR 1000/100 sometimes helpful
Notes: Use surface coils and thin sections (<5 mm)
Have patient remove mascara to avoid iron-cobalt artifacts.

SUGGESTED READING

Haiken JP, Glazer HS, Lee JKT et al (eds): Manual of Clinical Magnetic Resonance Imaging. Raven Press, New York, 1986

Index

Page numbers followed by f indicate figures; page numbers followed by t indicate tables.